Three-Dimensional Integrated Circuit Design

Three-Dimensional Integrated Circuit Design

Second Edition

Vasilis F. Pavlidis

Ioannis Savidis

Eby G. Friedman

Morgan Kaufmann is an imprint of Elsevier
50 Hampshire Street, 5th Floor, Cambridge, MA 02139, United States

British Library Cataloguing-in-Publication Data
A catalogue record for this book is available from the British Library

Library of Congress Cataloging-in-Publication Data
A catalog record for this book is available from the Library of Congress

ISBN: 978-0-12-410501-0

For Information on all Morgan Kaufmann publications
visit our website at https://www.elsevier.com/books-and-journals

Publishing Director: Jonathan Simpson
Acquisition Editor: Jonathan Simpson
Editorial Project Manager: Charlotte Kent
Production Project Manager: Punithavathy Govindaradjane
Cover Designer: Mark Rogers

Typeset by MPS Limited, Chennai, India

To Stella for her disarming criticism.
Vasilis

To my wife, Ana Lucia, my son, Ioannis Alexander, and to the rest of my family for their endless love and support.
Ioannis

To Laurie, my companion in life.
Eby

Contents

List of Figures

About the Authors

Vasilis F. Pavlidis received the B.Sc. and M. Eng. degrees in Electrical and Computer Engineering from the Democritus University of Thrace, Greece, in, respectively, 2000 and 2002. He received the M.Sc. and Ph.D. degrees in Electrical and Computer Engineering from the University of Rochester, Rochester, New York, in, respectively, 2003 and 2008.

He is currently an Assistant Professor in the School of Computer Science at the University of Manchester, Manchester, UK. From 2008 to 2012, he was a postdoctoral fellow with the Integrated Systems Laboratory at the Ecole Polytechnique Fédérale de Lausanne, Lausanne, Switzerland. He was with INTRACOM S.A., Athens, Greece from 2000 to 2002. He has also been a visiting researcher at Synopsys Inc., Mountain View, California with the Primetime group in 2007. His current research interests include interconnect modeling, analysis, and design, 3-D and 2.5-D integration, and other issues related to very large scale integrated (VLSI) design. He has published several conference and journal papers in these areas. He was the leading designer of the *Rochester cube* and cocreator of the *Manchester Thermal Analyzer*.

Dr. Pavlidis is on the editorial board of the *Microelectronics Journal* and *Integration, the VLSI Journal*. He also serves on the Technical Program Committees of several IEEE conferences. He is a member of the VLSI Systems and Applications Technical Committee of the Circuits and Systems Society and a member of the IEEE. He is also involved in public policy issues as a member of the ICT working group of the IEEE European Public Policy Initiative.

Ioannis Savidis received the B.S.E. degree in electrical and computer engineering and biomedical engineering from Duke University, Durham, North Carolina in 2005. He received the M.Sc. and Ph.D. degrees in electrical and computer engineering from the University of Rochester, Rochester, NY, USA, in, respectively, 2007 and 2013.

He is currently an Assistant Professor with the Department of Electrical and Computer Engineering at Drexel University, Philadelphia, Pennsylvania where he directs the Integrated Circuits and Electronics (ICE) Design and Analysis Laboratory. He has held visiting research positions with the 3-D Integration group at Freescale Semiconductor, Austin, Texas in 2007, and the System on Package and 3-D Integration group at the IBM T. J. Watson Research Center, Yorktown Heights, New York in 2008, 2009, 2010, and 2011. His current research and teaching interests include analysis, modeling, and design methodologies for high performance digital and mixed-signal integrated circuits, power management for SoC and microprocessor circuits (including on-chip dc–dc converters), emerging integrated circuit technologies, IC design for trust (hardware security), and interconnect related issues in 2-D and 3-D ICs. He has authored or coauthored over 40 technical papers published in peer reviewed journals and conferences and holds four pending patents.

Dr. Savidis is a member of the editorial boards of the *IEEE Transactions on Very Large Scale Integration (VLSI) Systems*, the *Microelectronics Journal*, and the *Journal of Circuits, Systems, and Computers*. He serves on the Organizing Committees and Technical Program Committees of many international conferences including the IEEE International Symposium on Circuits and Systems, the Great Lakes Symposium on VLSI, the ACM/IEEE System Level Interconnect Prediction Workshop, and the IEEE International Symposium on Hardware Oriented Security and Trust.

Eby G. Friedman received the B.S. degree from Lafayette College in 1979, and the M.S. and Ph.D. degrees from the University of California, Irvine, in, respectively, 1981 and 1989, all in electrical engineering.

From 1979 to 1991, he was with Hughes Aircraft Company, rising to the position of manager of the Signal Processing Design and Test Department, responsible for the design and test of high performance digital and analog integrated circuits. He has been with the Department of Electrical and Computer Engineering at the University of Rochester since 1991, where he is a Distinguished Professor, and the Director of the High Performance VLSI/IC Design and Analysis Laboratory. He is also a Visiting Professor at the Technion—Israel Institute of Technology. His current research and teaching interests are in high performance synchronous digital and mixed-signal microelectronic design and analysis with application to high speed portable processors, low power wireless communications, and power efficient server farms.

He is the author of more than 500 papers and book chapters, 13 patents, and the author or editor of 18 books in the fields of high speed and low power CMOS design techniques, 3-D integration, high speed interconnect, and the theory and application of synchronous clock and power delivery and management. Dr. Friedman is the Editor-in-Chief of the *Microelectronics Journal*, a Member of the editorial board of the *Journal of Low Power Electronics* and *Journal of Low Power Electronics and Applications*, and a Member of the technical program committee of numerous conferences. He previously was the Editor-in-Chief and Chair of the Steering Committee of the *IEEE Transactions on Very Large Scale Integration Systems*, the Regional Editor of the *Journal of Circuits, Systems, and Computers*, a member of the editorial board of the *Proceedings of the IEEE, IEEE Transactions on Circuits and Systems II: Analog and Digital Signal Processing, IEEE Journal on Emerging and Selected Topics in Circuits and Systems, Analog Integrated Circuits and Signal Processing*, and *Journal of Signal Processing Systems*, a member of the Circuits and Systems (CAS) Society Board of Governors, Program and Technical chair of several IEEE conferences, and a recipient of the IEEE Circuits and Systems 2013 Charles A. Desoer Technical Achievement Award, a University of Rochester Graduate Teaching Award, and a College of Engineering Teaching Excellence Award. Dr. Friedman is an inaugural member of the University of California, Irvine Engineering Hall of Fame, a Senior Fulbright Fellow, and an IEEE Fellow.

Preface to the Second Edition

Eight years have passed since the first edition of this book was published. During this period, considerable research has been produced in the field of three-dimensional (3-D) integration and the first commercial products of multitier DRAM memories have appeared in the market. Furthermore, the introduction of interposers forming 2.5-D systems has enabled the integration of memory and processor modules in close proximity, supporting higher bandwidth for memory-processor communication as compared to individually packaged modules.

The intention of this second edition is to cohesively integrate and present several research milestones from different, yet interdependent, aspects of 3-D integrated circuit design, reflecting recent progress in the field. The foremost goal of the book is to describe design methodologies for 3-D circuits; methodologies that exploit the innate technological diversity existing in 3-D integrated circuits. While the focal point of this book is design and analysis techniques and related methodologies, this material also provides an overview of significant manufacturing technologies for 3-D systems, including interposer-based systems, as well as demonstration of several 3-D test circuits.

3-D or vertical integration is an exciting path to boost the performance and extend the capabilities of modern integrated systems. These capabilities are inherent in 3-D integrated circuits. The former enhancement is due to the considerably shorter interconnect length in the vertical direction, and the latter improvement is due to the ability to combine dissimilar technologies within a multitier system. It is also worth noting that vertical integration, other than the stacking process, is compatible with standard two-dimensional integrated circuit design processes that have been developed over the past several decades. These distinctive characteristics make 3-D integration highly attractive as compared to other emerging technological solutions that have been proposed to resolve the increasingly difficult issue of on-chip interconnect, which limits speed and power, while also providing heterogeneous integration of diverse and complex systems.

The new fabrication steps introduced for stacking circuits require a different business model and important shifts in the supply chain as additional stakeholders are involved in manufacturing and designing 3-D systems. Unfortunately, to date, a concrete and lucrative business model is not yet in place, the primary reason for the delay in the use of vertical systems. Beyond these nontechnical issues, the challenges of providing high performance, multi-functional heterogeneous, energy efficient, and small form factor 3-D systems are due to the lack of a complete and effective design and verification flow for these types of systems. Thus, the realization of complex 3-D systems requires advanced design technologies across multiple abstraction levels. The development of these capabilities will be better supported if the physical behavior and mechanisms that govern intertier communication and manufacturing are properly understood. The focus of this book diligently serves this purpose.

Another important objective of this second edition is to integrate the several different approaches that have been developed over the past few years in the physical design process, such as power delivery, floorplanning, placement, and synchronization. The important issue of process variability in 3-D systems that surprisingly has received minimal attention is also discussed in this book.

Recognizing the importance of the vertical interconnections in 3-D circuits, the 3-D structures are central to the content of this book. Tutorial chapters are dedicated to manufacturing processes and technological challenges, complemented by an in-depth treatment of electrical models of these structures.

The vertical wires are investigated not only as a communication medium but also as heat conduits within the 3-D system. From this perspective, novel and efficient algorithms are presented for improving the signal propagation characteristics of heterogeneous 3-D systems based on the electrical behavior of the vertical interconnects. Additionally, the important role that the vertical interconnects play in global signaling and thermal amelioration is discussed throughout this book.

As thermal issues were identified at the earliest stages as an important challenge for 3-D systems, particular emphasis is placed on thermal analysis and modeling of 3-D systems in general and the vertical wires in particular. Related thermal aware physical design techniques and architectural approaches to lower the high temperatures within these systems are also reviewed.

In all of these issues, our approach has been to cover as broadly as possible the different novel techniques and ideas that have been proposed to address these issues. Thus, the material does not follow a narrow minded approach where only the results produced by the authors are presented but rather offers full coverage of the field. Additionally, as many design methods are based on existing methods for planar (2-D) circuits, several tutorial sections are included within the appropriate chapters to provide the requisite background to ensure that the reader can more easily follow the material. Finally, another useful feature of this second edition is that techniques and methodologies are often accompanied by silicon measurements from several prototype ICs designed at the University of Rochester and fabricated by MIT Lincoln Laboratories (and other facilities).

Vasilis F. Pavlidis
Ioannis Savidis
Eby G. Friedman
2017

Preface to the First Edition

The seminal source of this book is the lack of a unified treatment of the design of three-dimensional (3-D) integrated circuits despite the significant progress that has recently been achieved in this exciting new technology. Consequently, the intention of this material is to cohesively integrate and present research milestones from different, yet interdependent, aspects of 3-D integrated circuit design. The foremost goal of the book is to propose design methodologies for 3-D circuits; methodologies that will effectively exploit the flourishing manufacturing diversity existing in 3-D integration. While the focal point is design techniques and methodologies, the material also highlights significant manufacturing strides that complete the research mosaic of 3-D integration.

3-D or vertical integration is an exciting path to boost the performance and extend the capabilities of modern integrated circuits. These capabilities are inherent to 3-D integrated circuits. The former enhancement is due to the considerably shorter interconnect length in the vertical direction and the latter is due to the ability to combine dissimilar technologies within a multiplane system. It is also worth noting that vertical integration is particularly compatible with the integrated circuit design process that has been developed over the past several decades. These distinctive characteristics make 3-D integration highly attractive as compared to other radical technological solutions that have been proposed to resolve the increasingly difficult issue of on-chip interconnect.

The opportunities offered by 3-D integration are essentially limitless. The constraints stem from the lack of design and manufacturing expertise for these circuits. The realization of these complex systems requires advanced manufacturing methods and novel design technologies across several abstraction levels. The development of these capabilities will be achieved if the physical behavior and mechanisms that govern interplane communication and manufacturing are properly understood. The focus of this book diligently serves this purpose.

This book is based upon the body of research carried out by Vasilis Pavlidis from 2002 to 2008 at the University of Rochester during his doctoral study under the supervision of Prof. Eby G. Friedman. Recognizing the importance of the vertical interconnections in 3-D circuits, these structures are central to the content of this book. Tutorial chapters are dedicated to manufacturing processes, technological challenges, and electrical models of these structures. The vertical wires are investigated not only as a communication medium but also as heat conduits within the 3-D system. From this perspective, novel and efficient algorithms are presented for improving the signal propagation delay of heterogeneous 3-D systems by analyzing the electrical behavior of the vertical interconnects. Additionally, the important role that the vertical interconnects play in global signaling and thermal amelioration is described. Measurements from a case study of a 3-D circuit increase the physical understanding and intuition of this critical interconnect structure.

The short vertical wires enable several 3-D architectures for communication centric circuits. Opportunities for improved communication bandwidth, enhanced latency, and low power in 3-D systems are investigated. Analytic models and exploratory tools for these architectures are also discussed. The intention of this part of the book is to illuminate important design issues while providing guidelines for designing these evolving 3-D architectures.

The organization of the book is based on a bottom-up approach with technology and manufacturing of 3-D systems as the starting point. The first two chapters cover the wide spectrum of available 3-D technologies and related fabrication processes. These chapters demonstrate the multitechnology palette that is available in designing a 3-D circuit. Based on these technologies and *a priori* interconnect models, projections of the capabilities of 3-D circuits are provided. Physical design methodologies for 3-D circuits, which are the core of this book, are considered in the next eight chapters. The added design complexity due to the multiplane nature of a 3-D system, such as placement and routing, is reviewed. Efficient approaches to manage this complexity are presented. Extensions to multiobjective methodologies are discussed with the thermal issue as a primary reference point. As the vertical interconnects can behave either obstructively or constructively within a 3-D circuit, different design methodologies to utilize these interconnects are extensively discussed. Specific emphasis is placed on the through silicon vias, the important vertical interconnection structure for 3-D circuits. Various 3-D circuit architectures, such as a processor and memory system, field programmable gate arrays (FPGAs), and on-chip networks, are discussed. Different 3-D topologies for on-chip networks and FPGAs are explored. Novel algorithms and accurate delay and power models are reviewed. The important topic of synchronization is also investigated, targeting the challenges of distributing the clock signal throughout a multiplane circuit. Experimental results from a 3-D fabricated circuit provide intuition into this global signaling issue.

3-D integration is a seminal technology that will prolong the semiconductor roadmap for several generations. The third dimension offers nonpareil opportunities for enhanced performance and functionality; vital requirements for contemporary and future integrated systems. Considerable progress in manufacturing 3-D circuits has been achieved within the last decade. 3-D circuit design methodologies, however, considerably lag these technological advancements. This book ambitiously targets to fill this gap and strengthen the design capabilities for 3-D circuits without overlooking aspects of the fabrication process of this emerging semiconductor paradigm.

Vasilis F. Pavlidis
Eby G. Friedman
2009

Acknowledgments

The authors are thankful to several researchers who have contributed in different ways to make this second edition a much more complete manuscript. First, we would like to thank Dr. Dimitrios Velenis who contributed Chapter 8. We would also like to thank Dr. Hu Xu, Boris Vaisband, and Ioannis Papistas. Dr. Xu contributed to Chapter 17, Boris Vaisband contributed Chapter 5, and Ioannis Papistas contributed Chapter 6. Finally, the authors are also grateful to Professor Dimitrios Soudris of the National Technical University of Thrace for his important contributions to Chapter 20.

The authors would also like to thank MIT Lincoln Laboratories for providing foundry support for the 3-D test circuits and, more specifically, to Dr. Chenson Chen and Bruce Wheeler of MIT Lincoln Laboratories for their assistance during the 2nd and 3rd multiproject runs. We would also like to thank Yunliang Zhu, Lin Zhang, and Professor Hui Wu of the University of Rochester for their help in testing the 3-D circuits. Special thanks to Nopi Pavlidou for the creative design of the front cover of this second edition of the book. Finally, this book would not have been possible without the continuous support and encouragement of Charlotte Kent of Morgan Kaufmann Publishers, who enthusiastically supported the development of this second edition and thankfully prodded us throughout the writing process.

Organization of the Book

The organization of the book is based on a bottom-up approach with technology and manufacturing of three-dimensional (3-D) systems as the initial starting point. The first two chapters cover the wide spectrum of available 3-D technologies and related fabrication processes. These chapters demonstrate the multi-technology palette available in designing a 3-D circuit. Two more chapters are dedicated to the vertical interconnects including electrical models and noise characteristics. Although 3-D circuits are based on vertical interconnects, alternatives for intertier communication, for example, AC coupling, are also described. Due to the importance of cost issues for 3-D systems, the implications of cost on the more elaborate manufacturing processes for the vertical interconnects and assembly of the tiers are explored in a separate chapter.

Based on these technologies and a priori interconnect models, projections of the capabilities of 3-D circuits are provided. Physical design methodologies for 3-D circuits, the core of this book, are considered in the next 11 chapters. The added design complexity due to the multi-tier nature of a 3-D system, such as placement and routing, is reviewed. Efficient approaches to manage this complexity are presented. Appropriate models are also described in an interwoven manner with these methodologies, where the effectiveness of the design techniques and accuracy of these models are stressed. As the vertical interconnects can behave either obstructively or constructively within a 3-D circuit, different design methodologies that utilize these interconnects are extensively discussed. Extensions to multi-objective methodologies are reviewed with thermal effects as a primary issue. Thermal management techniques, including cooling, are also discussed. A test circuit to evaluate thermal coupling within and among tiers in a 3-D circuit is described. The important topics of synchronization and power delivery and distribution are also reviewed in exquisite detail. The issues of distributing the clock signal throughout a multi-tier circuit, delivering abundant current with low losses, and allocating the decoupling capacitance across a multi-tier system are addressed. Related experimental results from fabricated circuits provide physical intuition into these global issues for 3-D integrated systems.

A discussion of several 3-D circuit architectures, such as processor and memory systems, FPGAs, and on-chip networks, concludes the book. Different 3-D topologies for on-chip networks and FPGAs are explored. Novel algorithms and accurate high level delay and power models are reviewed. As the short vertical wires enable several 3-D architectures for communication centric circuits, opportunities for improved communication bandwidth, enhanced latency, and low power in 3-D systems are reviewed. Analytic models and exploratory tools for these architectures are also discussed. The intention of the last chapter of the book is to illuminate important design issues while providing guidelines for developing these evolving 3-D architectures.

3-D integration and its variants are a promising technology that will prolong the semiconductor roadmap for several generations. The third dimension offers nonpareil opportunities for enhancing the performance and diversifying the functionality by exploiting the close physical proximity of heterogeneous components, vital requirements for contemporary, and evolving integrated systems. Considerable progress in manufacturing 3-D circuits has been achieved over the last decade with several products already in the commercial marketplace. Forecasts from market analysis companies predict significant growth in these systems where more diverse technologies and functions will be vertically integrated to support a host of new and exciting applications. This second edition presents a complete treatment of design methodologies for 3-D systems and ambitiously intends to strengthen the design capabilities for 3-D circuits without overlooking the fabrication process of this seminal semiconductor paradigm. We believe this book will be a useful companion both for the newly interested audience who are earnestly excited by this fascinating technology and the experienced researchers and designers in the field as the material covers, both in depth and breadth, multiple design aspects of 3-D systems.

CHAPTER

INTRODUCTION

1

CHAPTER OUTLINE

The invention of the integrated circuit occurred in 1958 [1]. The semiconductor industry has now traversed six decades. During this time the microelectronics industry has grown tremendously, introducing electronics technology to almost all areas of human activity. The workhorse of microelectronics-based products has been the metal oxide field effect transistor (MOSFET). This invention occurred a few years after the groundbreaking demonstration of the point contact transistor by J. Bardeen, W. Brattain, and W. Shockley in 1947 [2]. The evolution of this novel device into several other types of semiconductor devices and logic circuit families is illustrated in Fig. 1.1.

Over these past six decades the MOSFET, the primary device used in modern integrated circuits, has followed an extraordinary trajectory of physical scaling [3], producing seminal improvements in speed, area, power, and reliability. Throughout this period, engineers and scientists touted the end of transistor scaling. Fortunately, industry consistently invented innovative techniques and improved manufacturing methods to overcome the potential end of the microelectronics revolution. In addition to this progress, these novel methods also lowered the cost per transistor. This situation remained unchanged until approximately the 28 nm technology node, where the cost per transistor began to rise [4].

Furthermore, the increasing difficulty in controlling the behavior of the elemental building block of integrated circuits led to new transistor structures, such as FinFETs [5], and a shift from bulk complementary metal oxide semiconductor (CMOS) to fully depleted silicon-on-insulator CMOS devices [6]. The primary reason for the introduction of this technology has been enhanced control of the electrostatic characteristics of the devices, a prerequisite for the multibillion component integration densities in modern integrated circuits.

On a different front, semiconductor processes for dynamic random access memory (DRAM) memory successfully scaled the size of the elemental memory cell that stores a single bit of information composed of one transistor and one capacitor [7,8]. The same difficulty in scaling the physical dimensions of the cells to nanometer dimensions occurred, requiring the engineering community to search for new materials and integration approaches to enhance performance and provide greater integration densities without relying on simple physical scaling of the memory cell.

Three Dimensional Integrated Circuit Design. DOI: http://dx.doi.org/10.1016/B978-0-12-410501-0.00001-0

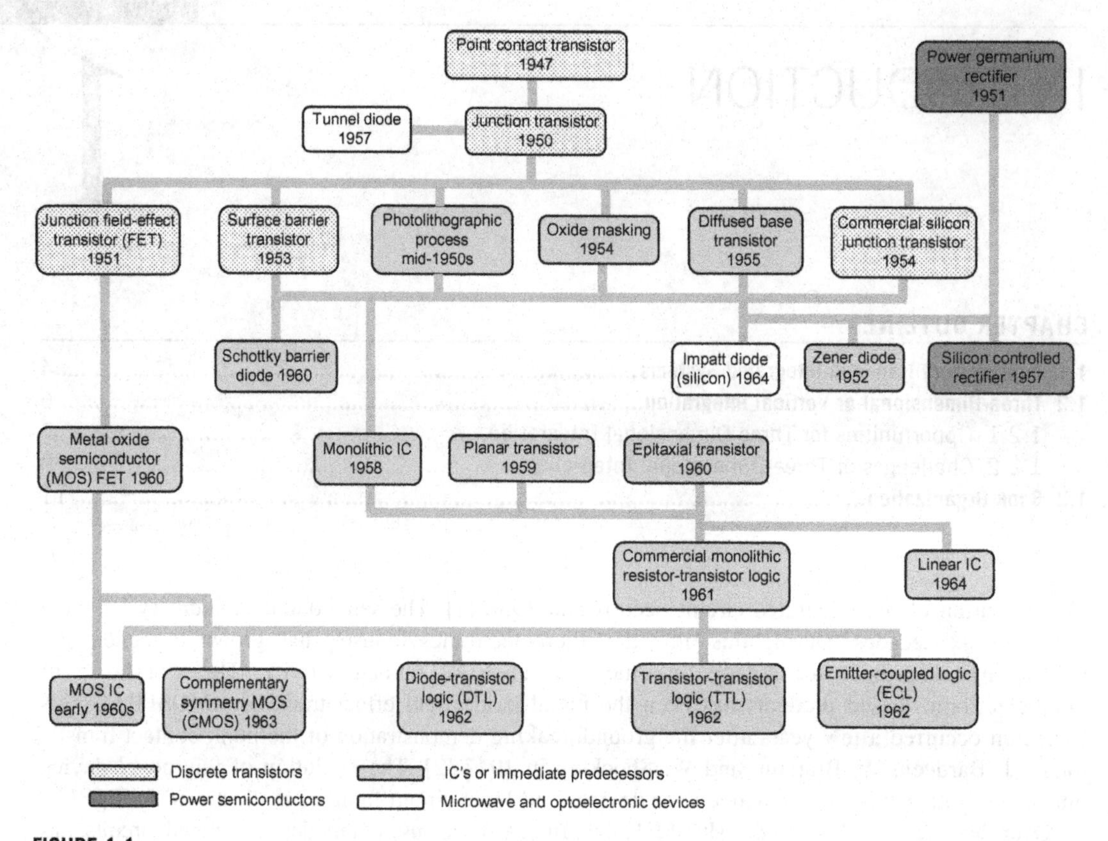

FIGURE 1.1

History of semiconductor transistors and logic styles [14].

In addition to these crucial device level issues, the advent of mobile products and the pervasiveness of handheld appliances, for example, smartphones and tablets [9], introduced new requirements and challenges for the semiconductor industry. In recent years, the outlook of the semiconductor roadmap, as described by the International Technology Roadmap for Semiconductors (ITRS) [10], predicted a crossroad for the sector, as illustrated in Fig. 1.2. These two discrete trends are driven by emerging markets and exciting new applications.

The "More Moore" path emphasizes the continuing miniaturization of transistors to deliver greater performance [10]. Important applications exploiting this approach include exascale computing [11]. Machines that can deliver this level of performance are required for scientific applications but also for novel and developing commercial applications such as Big Data and cloud computing [12].

Alternatively, the "More than Moore" path focuses on integrated systems where the predominant objective is functional diversity [10]. This diversification of functionality is the result of the growth of portable electronic products that interact with the external world in increasingly diverse ways to satisfy an immense variety of end user requirements.

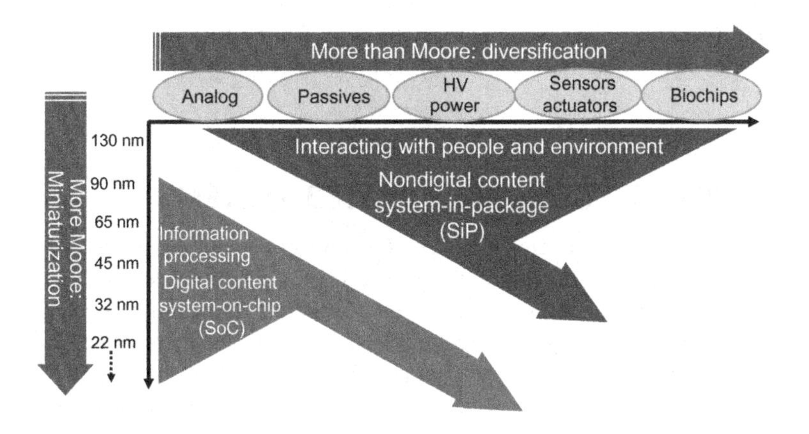

FIGURE 1.2

Evolutionary paths of the microelectronics industry [10].

The interaction of electronic entities with the environment is envisaged to grow even further with the rise of the Internet of Things (IoT), often described as the Internet of Everything. The key concept of this new era of electronic products includes miniaturized integrated systems that interact with the surrounding environment and communicate sensed data over the Internet. Consequently, a massive amount of machine to machine communication over the internet is expected.

For those companies designing and manufacturing integrated systems, these emerging markets will fundamentally affect the functional and performance objectives of these systems. For example, IoT products will be predominantly power driven to ensure the greatest autonomy while simultaneously integrating heterogeneous components capable of sensing the ambient, while processing, storing, communicating, and actuating on related data.

On the high performance end of the application spectrum, where computationally powerful machines targeting Big Data applications are required, power remains a primary objective. Issues such as reliability, packaging, and electricity cost all deeply depend on the dissipated power. Although these machines at first glance do not seem to require *technological* heterogeneity, *processing* heterogeneity is inherent to these systems to satisfy the different performance-power levels. Moreover, as these systems place massive demands on main memory and storage, low power (and/or nonvolatile) memories with sufficient bandwidth will become an important objective. Several innovative memory technologies are currently under development, which are expected to satisfy these impending requirements. These technologies exploit different transport mechanisms, unlike charge-based storage used over the past decades [13].

Memory modules will need to be located closer to the processing elements, otherwise memory-processor communication will remain the primary performance bottleneck. This situation requires innovation at several fronts including novel interconnect materials and integration strategies.

One primary path for increased integration, performance, and heterogeneity is three-dimensional (3-D) integration. The design and manufacture of these systems are the topic of this book. Interconnect related issues since the earliest days of the integrated circuit industry and the impending performance bottleneck caused by the interconnect are discussed in Section 1.1. A short historical overview of vertical 3-D integration along with some milestones in the development of 3-D systems is introduced in Section 1.2. An outline of this book is presented in Section 1.3.

1.1 INTERCONNECT ISSUES IN INTEGRATED SYSTEMS

During the infancy of the semiconductor industry, the connections among the active devices within an integrated circuit presented an important obstacle to higher performance. The significant capacitance of the interconnects required large drivers and hindered the significant increase in performance available from the transistors. The deleterious effects of the interconnects, such as greater delay and noise coupling, had already been noticed from the earliest days of integrated circuits [15,16]. The invention of the integrated circuit temporarily alleviated some of these early interconnect related problems by placing the wires on-chip. The interconnect length was significantly reduced, decreasing the propagation delay and power consumption while enhancing yield. From a performance point of view, the delay of the transistors dominated the overall delay characteristics. Over the next three decades, the on-chip interconnects were not the major focus of the IC design process, as performance improvements reaped from scaling the devices were much greater than any degradation caused by the interconnects.

With continuous technology scaling, however, the delay, noise, and power of the interconnect grew in importance [17,18]. A variety of methodologies at the architectural, circuit, and material levels have been developed to address these interconnect performance objectives. At the material level, manufacturing innovations, such as the introduction of copper interconnects and low-k dielectric materials in the mid-1990s, helped to continue the improvements in performance gained from scaling [19–23]. This improvement is due to the lower resistivity of the copper as compared to aluminum previously used in the interconnects, and the lower dielectric permittivity of the novel insulator materials as compared to silicon dioxide (SiO_2).

Multilevel interconnect architectures [24,25], shielding [26], wire sizing [27,28], and repeater insertion [29] are only a handful of the many methods employed to cope with interconnect issues at the circuit level. Multilevel interconnect architectures, for example, support multiple levels of metal layers with different cross-sections [25], as illustrated in Fig. 1.3. Each group of layers typically consists of multiple metal layers routed in orthogonal directions with the same cross-section. The key concept of this structure is to utilize wires of decreasing resistance to connect those circuits

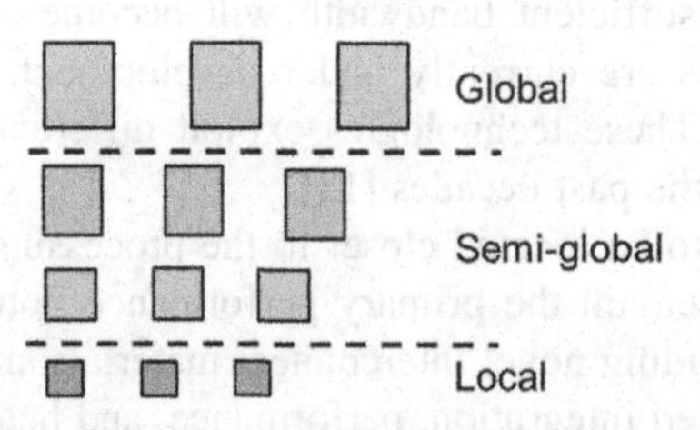

FIGURE 1.3

Interconnect system composed of groups of local, semi-global, and global layers. The metal layers in each group are typically of different thickness.

located farther away. Therefore, the farther the distance, the thicker and wider the wire. The increase in the cross-sectional area of the wires is shown in Fig. 1.3. The thickness of the levels, however, is limited by the fabrication technology and related reliability and yield concerns.

Varying the width of the wires, also known as wire sizing, is an important means to manage the interconnect impedance characteristics. Wider wires lower the interconnect resistance, decreasing the attenuative behavior of the interconnect. Although wire sizing typically has an adverse effect on the power dissipated by the interconnect, proper sizing techniques can also decrease the power consumption [28,30].

Other practices do not modify the physical characteristics of the propagation medium. Rather, by introducing additional circuitry and wire resources, the performance and noise tolerance of an interconnect system can be enhanced. For instance, in a manner similar to the use of repeaters in telephone line systems, a properly designed interconnect system with buffers (also known as repeaters) amplifies the attenuated signals, recovering the originally transmitted signal that is propagated along a line. Repeater insertion converts the quadratic dependence of the delay on the interconnect length to a linear function of length [31], as shown in Fig. 1.4.

Shielding is an effective technique to reduce crosstalk among adjacent interconnects. Single- or double-sided shielding, as depicted in Fig. 1.5, is commonly utilized to improve signal integrity. Shield lines can also improve interconnect delay and power, particularly in bus architectures, in addition to mitigating noise. Careful tuning of the relative delay of the propagated signals [32] and signal encoding schemes [33] are other strategies to enhance signal integrity. Despite the benefits of these techniques, issues such as increased power consumption, greater routing congestion, reduction in wiring resources, and increased area arise.

At higher abstraction levels, pipelining the global interconnects and employing error correction mechanisms can partially improve the performance and fault tolerance of the wires. The related

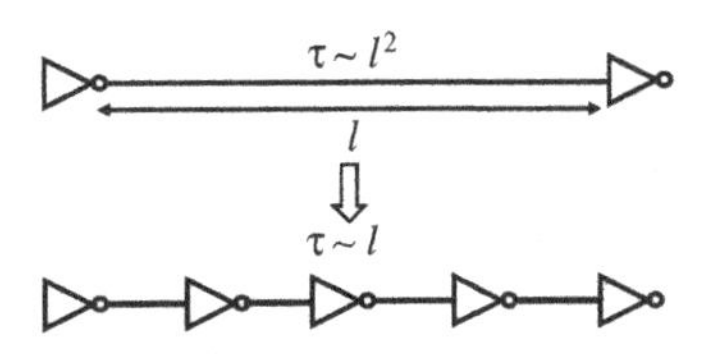

FIGURE 1.4

Repeaters are inserted at specific distances to improve the interconnect delay.

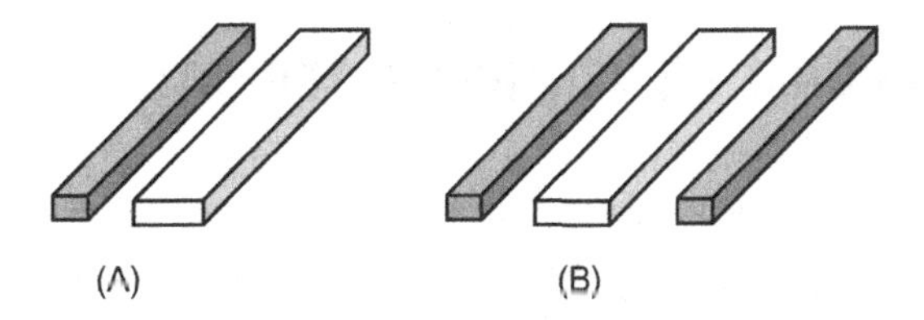

FIGURE 1.5

Interconnect shielding to improve signal integrity, (A) single-sided shielding, and (B) double-sided shielding. The shield and signal lines are, respectively, illustrated by the gray and white color.

effects of these architectural techniques in terms of area and design complexity, however, have considerably increased. Other schemes, such as current mode signaling [34], wave pipelining [35], and low swing signaling [36], have been proposed as possible solutions to the impending interconnect bottleneck. These methods, however, have limited ability to reduce the length of the wire, which is the primary cause of the deleterious behavior of the interconnect.

Novel design paradigms are therefore required that do not impede the well established and historic improvements in performance in evolving generations of integrated circuits. Canonical interconnect structures that utilize internet-like packet switching for data transfer [37], optical interconnects [38], 3-D integration, and/or combinations of these technologies are possible solutions for enhancing communication among devices or functional blocks within an integrated circuit.

On-chip networks can greatly enhance the communication bandwidth among the individual functional blocks of an integrated system, since each of these blocks simultaneously utilizes the resources of the network. In addition, noise issues are easier to manage as the layered structure of the communication protocols utilized within on-chip networks provides error correction. The speed and power consumed by these networks, however, are eventually limited by the delay of the wires connecting the network links.

Alternatively, on-chip optical interconnects can greatly improve the speed and power characteristics of the interconnects within an integrated circuit, replacing the critical electrical nets with optical links [39,40]. On-chip optical interconnects, however, remain a technologically challenging issue. Indeed, integrating a modulator and detector onto the silicon within a standard CMOS process is a difficult task [38]. In addition, the detector and modulator need to exhibit sufficiently high performance to ensure the optical links significantly outperform the electrical interconnects [40]. Furthermore, an on-chip optical link consumes larger area as compared to a single electrical interconnect line. Multiplexing the optical signals using wavelength division multiplexing (WDM) can be exploited to limit the area consumed by the optical interconnect. On-chip WDM, however, imposes significant challenges.

Volumetric integration by exploiting the third dimension greatly improves the interconnect performance characteristics of modern integrated circuits while not degrading the interconnect bandwidth. In general, 3-D integration should not be seen as a competitive but rather synergistic technology with on-chip networks, optical interconnections, and other emerging technologies and architectures. The unique opportunities that 3-D integration offers to the circuit design process and the challenges that arise from the increasing complexity of these systems are discussed in the following section and throughout this book.

1.2 THREE-DIMENSIONAL OR VERTICAL INTEGRATION

Methods to vertically interconnect circuits were first proposed in 1962 during the earliest days of integrated circuits [41,42]. Although these vertical conductors intended to connect circuits fabricated on both surfaces of a wafer, these methods demonstrated that the benefits of vertical integration were appreciated by industry as early as the 1960s. Vertical integration was proposed and supported by some of the most prominent engineers and scientists at the time, such as William Shockley [43] and Richard Feynman [44]. Early enthusiasm on 3-D integration was however not followed by development and high volume production of 3-D circuits as the evolution of planar

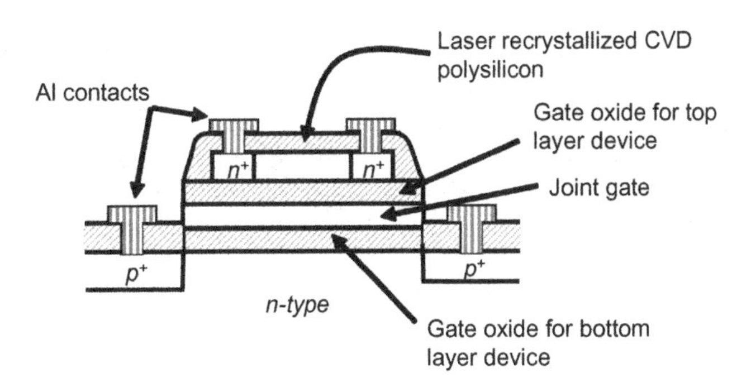

FIGURE 1.6

Cross-section of a joint MOS (JMOS) inverter [46].

processes yielded the desired improvements in transistor density, speed, and power in integrated circuits with a commensurate decrease in manufacturing cost.

Consequently, for several decades, 3-D circuits were considered a niche with limited scientific and research importance. Efforts primarily focused on monolithic fabrication of vertical circuits, several examples of which were demonstrated in the early 1980s [45]. These structures included 3-D CMOS inverters, where the p-type metal oxide semiconductor (PMOS) and n-type metal oxide semiconductor (NMOS) transistors share the same gate, greatly reducing the total area of an inverter, as illustrated in Fig. 1.6. The term joint metal oxide semiconductor (JMOS) was used for these structures to describe the joint use of a single gate for both devices [46]. Prototypes with up to three layers of active devices were demonstrated in these early vertical monolithic systems [47]. Other examples of early uses of 3-D integration include infrared detectors, where the infrared detectors, manufactured in exotic materials such as mercadmium telluride or indium phosphate, were flipped and bonded to silicon-based detector readout circuits [48].

Since the 2000s, due to the increasing importance of the interconnect and the demand for greater functionality on a single substrate, the concept of vertical integration has been revived and has become a prominent topic of research and commercial development. Over the last 10 to 15 years, 3-D integration has evolved into a design paradigm manifested at many abstraction levels, such as the package, die, and wafer. Different manufacturing processes and interconnect schemes have been proposed for each of these abstraction levels [49]. These recent approaches highlight the advantages and disadvantages of introducing 3-D integration at these levels of the design abstraction, where 3-D technology as a systems integration platform yields significant improvements in transistor density, performance, heterogeneity, form factor, and cost. The salient features and important challenges of 3-D systems are briefly summarized in the following subsections.

1.2.1 OPPORTUNITIES FOR THREE-DIMENSIONAL INTEGRATION

The quintessence of 3-D integration is the drastic decrease in the length of the longest interconnects across an integrated circuit. To illustrate this situation, consider the simple structure shown in Fig. 1.7. A common metric to characterize the longest interconnect is to assume that the length of a long wire is twice the length of the die edge. Consequently, assuming a planar integrated circuit

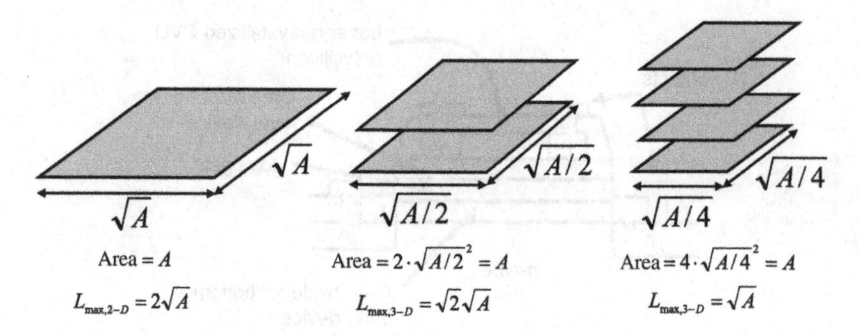

FIGURE 1.7

Reduction in wirelength where the original 2-D circuit is composed of two and four tiers.

with an area A, the longest interconnect in a planar IC has a length $L_{\max,2-D} = 2\sqrt{A}$. The same circuitry on two bonded dies requires an area $A/2$ for each tier while the total area of the system remains the same. Hence, the length of the longest interconnect for a two tier 3-D IC is $L_{\max,3-D} = 2\sqrt{A/2}$. By increasing the number of dies within a 3-D IC to four, the area of each die is further reduced to $A/4$, and the longest interconnect would have a length of $L_{\max,3-D} = 2\sqrt{A/4}$. Consequently, the wirelength exhibits a reduction proportional to $\sqrt{n}$, where n is the number of dies or physical tiers integrated within a 3-D system. Although in this simplistic example the effect of the connections among circuits located on different dies is not considered, *a priori* accurate interconnect prediction models adapted for 3-D ICs also demonstrate a similar trend due to the reduction in wirelength [50]. This considerable decrease in the length of the interconnects is a promising solution for increasing speed while reducing the power dissipated by an IC.

Another characteristic of 3-D ICs of even greater importance than the decrease in the interconnect length is the ability of these systems to include disparate heterogeneous technologies. This defining feature of 3-D integration offers unique opportunities for highly heterogeneous and diverse multifunctional systems. A real-time image processing system where the image sensor on the topmost tier captures the light, the analog circuitry on the tier below manipulates and converts the analog signal to digital data, and the remaining two tiers of digital logic process the information from the upper tiers is a powerful example of a heterogeneous 3-D system-on-chip (SoC), exhibiting considerably improved performance as compared to a planar version of the same system [51,52]. Another example, where the topmost tier can include other types of sensors, such as seismic and acoustic, and an additional tier with wireless communications circuitry are vertically integrated, is illustrated in Fig. 1.8. Several application domains can greatly benefit from vertical integration including healthcare, healthy aging, military, security, and environmental monitoring to name just a few, as the proximity of the components due to the third dimension is suitable for both high performance and low power integrated systems.

Recent products demonstrate the many merits the third dimension can bring to computing. For example, the 3-D memories recently produced by Micron [53] and SK-Hynix [54] include four DRAM memory tiers which exhibit superior performance as compared to state-of-the-art planar double data rate memories [55]. The Virtex series of Xilinx field programable gate arrays (FPGAs) also utilize advanced stacking technologies [56]. This approach alleviates the issue of large die size, improving yield and cost.

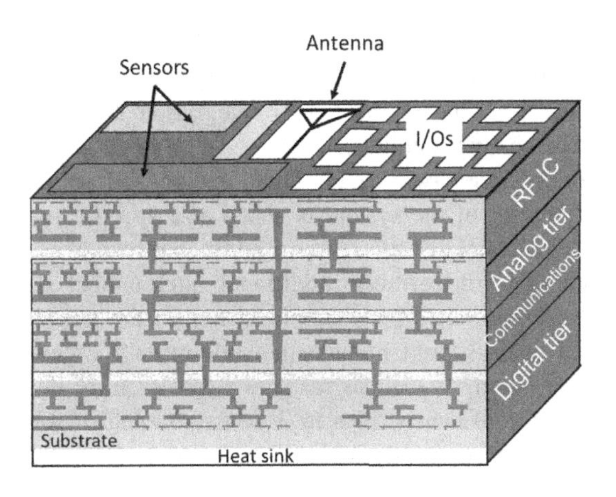

FIGURE 1.8

Heterogeneous 3-D SoC comprising sensors and processing tiers.

1.2.2 CHALLENGES OF THREE-DIMENSIONAL INTEGRATION

Developing a design flow for 3-D ICs is a complicated task with many ramifications. Despite the recent progress that has led to commercially successful 3-D systems, a number of challenges at each step of the design process have to be solved for 3-D ICs to successfully evolve into a mainstream technology. Design methodologies at the front end and mature manufacturing processes at the back end are required to provide large scale 3-D systems. Several of the primary challenges to successfully develop 3-D systems are summarized below.

1.2.2.1 Technological/manufacturing limitations

Some of the fabrication issues encountered in the development of 3-D systems concern the reliable assembly of multiple ICs, possibly from dissimilar technologies. The stacking process should not degrade the performance of the individual tiers, while guaranteeing the tiers remain reliably bonded throughout the lifetime of the 3-D system. Furthermore, novel packaging solutions that accommodate these complex 3-D structures need to be developed. In addition, the expected reductions in wirelength depend upon the vertical interconnects that propagate signals and deliver power throughout the tiers within a 3-D system.

The technology of the intertier interconnects is a primary determining factor in circuit performance. Consequently, providing high quality and highly dense vertical interconnects is of fundamental importance; otherwise, the expected speed and power improvements available from the third dimension will be diminished [57]. Alternatively, the density of the vertical interconnects dictates the granularity at which the tiers of the system can be interconnected, directly affecting the bandwidth of the intertier communication.

1.2.2.2 Testing

Manufacturing a 3-D system typically includes bonding multiple physical tiers. The stacking process can occur either in a wafer-to-wafer or die-to-wafer manner. Consequently, novel testing methodologies at the wafer and die level are required. Developing testing methodologies for wafer level integration is significantly more complicated than die level testing techniques. The considerable reduction in turnaround time due to the higher integration levels, however, may justify the additional complexity of these testing methods.

An important distinction between two- and 3-D IC testing and validation is that in the latter case only part of the functionality of the system is tested at a specific time (since only one tier is typically tested at a time). This characteristic requires additional resources, for example, scan registers embedded within each tier. Furthermore, additional interconnect resources, such as power/ground pads, are necessary. These extra pads supply power to the tier during testing. In general, testing strategies for 3-D systems should include methodologies for generating appropriate input patterns for each tier of the system, as well as managing the circuitry dedicated to efficiently test each tier within a 3-D stack. Significant strides in 3-D methodologies for test and reliability have been made and some standards are currently under consideration. Open questions however relating to these issues remain. Although the importance of test and reliability is recognized, this aspect of integrated 3-D systems has only been tangentially considered in this book. The interested reader is referred to other more appropriate sources that discuss these topics.

1.2.2.3 Global interconnect design

Design for test strategies is only a portion of the many design methodologies that require further development for 3-D ICs. The design and analysis of the global interconnect within 3-D circuits are also challenging tasks. This challenge is primarily due to the inherent heterogeneity of these systems, where different fabrication processes and disparate technologies are combined into a 3-D system. Consequently, models that consider the particular traits of 3-D technology are necessary. In these diverse systems, the global interconnect, such as the clock and power distribution networks, grow in importance. Furthermore, well developed noise mitigation techniques may not be suitable for 3-D circuits. Noise caused by capacitive and inductive coupling of the interconnections between adjacent tiers needs to be considered from a 3-D perspective [58]. For example, signal switching on the topmost metal layer of a digital tier can produce a noise spike in an adjacent face-to-face bonded analog tier. Considering the different forms of 3-D integration and the various manufacturing approaches, the development of design techniques and methodologies for the global interconnect is a primary focus in high performance 3-D systems.

1.2.2.4 Thermal issues

A fundamental concern in the design of 3-D circuits is thermal effects. Although the power consumption of these circuits is expected to decrease due to the considerably shorter interconnects, the power density will greatly increase since there is a significantly greater number of devices per unit area as compared to a planar 2-D circuit. As the power density increases, the temperature of those tiers not adjacent to the heat sink of the package will rise, resulting in degraded performance or accelerated wear out. Exploiting the performance benefits of vertical integration while mitigating thermal effects is a difficult task. In addition to design practices, advanced packaging solutions and more effective heat sinks are required to alleviate thermal effects.

1.2.2.5 CAD algorithms and tools

Other classic problems in the IC design process, such as partitioning, floorplanning, placement, and routing, will need to be revisited in an effort to develop efficient solutions that support the complexity of 3-D systems. Considerable effort has been invested in proposing novel algorithms for 3-D systems. No cohesive physical design flow, however, currently exists that seamlessly consolidates all or some of these algorithms and techniques into a complete back end design flow for 3-D integrated systems. Also, the solutions supported by commercial electronic design automation tools relate to specific tasks and require considerable manual intervention.

Furthermore, a capability for exploratory design is required to facilitate the front end design process. For example, design entry tools that provide a variety of visualization options can improve comprehension while managing the greater complexity of 3-D systems. In addition, as diverse technologies are combined into a single 3-D stack, algorithms that include behavioral models for a larger variety of disparate components are needed. Moreover, the computational power of the simulation tools will need to be significantly enhanced to ensure that the entire system can be efficiently evaluated in an integrated fashion. In this book, emerging 3-D technologies and design methodologies are discussed and solutions for certain critical problems are proposed. In the following section, an outline of the book is provided.

1.3 BOOK ORGANIZATION

A brief description of the challenges that 3-D integration faces is provided in the previous section. Several important problems are considered and a variety of techniques to address these problems are presented throughout this book. In the second chapter, different forms of vertically integrated systems are discussed. 3-D circuits at the package and die integration levels are reviewed. Some of these approaches, such as wire bonded stacked die and through silicon via (TSV)-based integration, have become commercially available. Although vertical integration of packaged or bare die offers substantial improvements as compared to planar multichip packaging solutions, the increasing number of I/Os hampers potential advancements in performance. This situation is primarily due to manufacturing limitations hindering the aggressive scaling of the off-chip interconnects to satisfy high I/O requirements.

Consequently, in the third chapter, emphasis is placed on those technologies that enable 3-D integration, where the interconnections between the noncoplanar circuits are achieved by short vertical vias. These interconnect schemes provide the greatest reduction in wirelength and therefore, the largest improvement in speed and power consumption. Specific fabrication processes that have been successfully developed for 3-D circuits are reviewed.

The predominant type of vertical interconnections is the through silicon vias. Due to the important role of this interconnect, Chapter 4, Electrical Properties of Through Silicon Vias, is dedicated to models of the electrical behavior of this structure. Models of differing complexity are discussed, and the appropriate model for the specific TSV process and 3-D system is described.

The effects of through silicon vias on the noise characteristics of the substrate of the tiers are presented in Chapter 5, Substrate Noise Coupling in Heterogeneous Three-Dimensional ICs. The noise due to different types of substrates is evaluated and appropriate noise models for each type of substrate are offered. Mitigation techniques to suppress the noise from the TSVs are also discussed.

An alternative approach to TSVs to provide intertier communication is contactless interconnects. Existing efforts to enable contactless communication through inductive links are described in Chapter 6, Three-Dimensional ICs with Inductive Links. Models of crosstalk noise from inductive links to adjacent interconnects are presented. Furthermore, measures to avoid interference for specific types of interconnects are provided. A multiobjective algorithm for the design of inductive links to support intertier communication is also described.

A theoretical analysis of interconnections in 3-D ICs is offered in Chapter 7, Interconnect Prediction Models. This investigation is based on *a priori* interconnect prediction models. These stochastic models are used to estimate the distribution of the length of the on-chip interconnects. The remaining sections of this chapter apply the interconnect distribution model to demonstrate the opportunities and performance benefits of vertical integration.

Cost issues for vertical integration are discussed in Chapter 8, Cost Considerations for Three-Dimensional Integration. The diverse processing steps introduced in each of the manufacturing processes of TSVs have different cost requirements. Models that capture these cost implications are constructed in this chapter. In addition, a comparison between the manufacturing cost of 2.5-D and 3-D systems is provided. Several important aspects, such as the area of the active die and the prebond test coverage, are included in these models.

The following three chapters focus on issues related to the physical design of 3-D ICs. The complexity of the 3-D physical design process is discussed in Chapter 9, Physical Design Techniques for Three-Dimensional ICs. Several approaches for classical physical design issues, such as floorplanning, placement, and routing, from a 3-D perspective, are extensively reviewed. Certain fundamental methods and algorithms used in the physical design of planar circuits are also discussed to enhance the understanding of these techniques targeted to 3-D systems.

Beyond the reduction in wirelength that stems from 3-D integration, the delay of those interconnects connecting circuits located on different physical tiers of a 3-D system (i.e., the intertier interconnects) can be further improved by optimally placing the TSVs. Considering the highly heterogeneous nature of 3-D ICs including the nonuniform impedance characteristics of the interconnect structures, a methodology is described in Chapter 10, Timing Optimization for Two-Terminal Interconnects, to minimize the delay of the intertier interconnects. An interconnect line that includes only one TSV is initially described. The location of the TSV that minimizes the delay of a line is analytically determined. Any degradation in delay due to the nonoptimal placement of the 3-D vias is also discussed. To incorporate the presence of physical obstacles, such as logic cells and prerouted interconnects (for example, segments of the power and clock distribution networks), the discussion in this chapter proceeds with interconnects that include more than one TSV. An effective heuristic is described for placing TSVs to minimize the overall delay of a multi-tier interconnect.

By extending the heuristic for two-terminal interconnects, a near-optimal heuristic for multiterminal nets in 3-D ICs is described in Chapter 11, Timing Optimization for Multiterminal Interconnects. Necessary conditions for locating the TSVs are described. An algorithm that exhibits low computational complexity is also presented. The improvement in delay achieved by placing the TSVs for different via placement scenarios is discussed. For the special case where the delay of only one branch of a multiterminal net is minimized, a simpler optimization procedure is described.

Based on this approach, a second algorithm is presented. Finally, the sensitivity of this methodology to the interconnect impedance characteristics is demonstrated, depicting a significant dependence of the delay on the interconnect.

In the next two chapters, techniques for 3-D ICs are extended to thermal design and management. Thermal models of different complexity and accuracy are presented in Chapter 12, Thermal Modeling and Analysis, including models for liquid cooling. Based on these thermal models and thermal analysis techniques, thermal management methodologies for 3-D systems are presented in Chapter 13, Thermal Management Strategies for Three-Dimensional ICs. Both physical and architecture level methods are discussed. These techniques either lower the overall power of the 3-D stack or carefully distribute the power densities across the tiers of a 3-D system to satisfy local temperature limitations. Moreover, design techniques that utilize additional interconnect resources to increase the thermal conductivity within a multi-tier system are also discussed.

A prototype circuit to enhance the understanding of thermal effects in 3-D structures is described in Chapter 14, Case Study: Thermal Coupling in 3-D Integrated Circuits. The test circuits are used to evaluate thermal coupling among the tiers of a three tier stack and between blocks located within the same physical tier. The effects of the TSVs on thermal coupling are also discussed.

The important issue of synchronization is the topic of Chapter 15, Synchronization in Three-Dimensional ICs. Clock tree synthesis techniques under diverse design objectives for multi-tier systems are described. As these techniques extend clock tree synthesis methods for planar circuits to 3-D structures, standard algorithms and methods used in the synthesis of planar clock trees are presented. Global clock distribution networks for 3-D circuits are also discussed in this chapter.

Following the treatment of clock tree synthesis algorithms, a prototype circuit investigating different global 3-D clock distribution networks is described in Chapter 16, Case Study: Clock Distribution Networks for Three-Dimensional ICs. A variety of clock networks, such as H-trees, rings, tree-like, and trunk-based, are explored in terms of clock skew and power consumption to determine an effective clock distribution network for 3-D ICs. A prototype test circuit composed of these networks has been designed and manufactured with the 3-D fabrication process developed at MIT Lincoln Laboratories (MITLL). The design and modeling process and related experimental results are also included in this chapter.

The effects of variations on the behavior of clock distribution networks are reviewed in Chapter 17, Variability Issues in Three-Dimensional ICs. The combined effects of die-to-die and within die variations are modeled for clock paths that span more than one tier. The distribution of skew for different clock networks is evaluated based on this model, and enhanced topologies that reduce skew variability are described. The skew model is extended to include power supply noise, and design guidelines for clock trees considering both power supply noise and process variations are presented.

The behavior of power supply noise depends strongly on the design of the power distribution network. The issues of power delivery and distribution for 3-D circuits are discussed in Chapter 18, Power Delivery for Three-Dimensional ICs. Integrating power delivery components into one tier of the 3-D stack allows smaller currents to propagate through the overall power distribution system, decreasing the power supply noise. The role of the vertical interconnects in distributing power and ground within a 3-D stack is discussed. Different approaches for distributing the decoupling capacitance throughout a multi-tier stack are treated coherently with the distribution of the TSVs to ensure that the power supply noise satisfies specified limitations. To quantify the effect of the TSVs on the switching noise, a 3-D test circuit is described that is used to evaluate the characteristics of several power distribution topologies, as described in Chapter 19, Case Study: 3-D Power Distribution Topologies and Models.

Exploiting the advantages of 3-D integration requires the development of novel circuit architectures. A 3-D version of a microprocessor-memory system is, therefore, discussed in Chapter 20, 3-D Circuit Architectures. Major improvements in throughput, power consumption, and cache miss rate are demonstrated. Communication centric architectures, such as a network-on-chip, are also discussed. On-chip networks are an important design paradigm to address the interconnect bottleneck, where information is communicated among circuits within packets in an internet-like fashion. The synergy between these two design paradigms, NoC and 3-D, can be exploited to significantly improve performance while decreasing the power consumed in communication limited systems.

Research on the design of 3-D ICs has only recently begun to produce commercially viable products. Many challenges remain unsolved and significant effort is required to provide effective solutions in the design of 3-D ICs. The major foci of this book are summarized in the last chapter, and general conclusions are offered regarding directions for research that will contribute to the maturation of this exciting solution in next generation multifunctional heterogeneous systems.

CHAPTER

MANUFACTURING OF THREE-DIMENSIONAL PACKAGED SYSTEMS

2

CHAPTER OUTLINE

With the ongoing demand for greater functionality resulting in polylithic (multiple IC) products, longer off-chip interconnects plague the performance of microelectronic systems. The advent of the system-on-chip (SoC) in the mid-1990s primarily addressed the increasing delay of the off-chip interconnects. By integrating all of the components on a monolithic substrate the overall speed of the system is enhanced while the power consumption is decreased. To assimilate disparate technologies, however, several difficulties must be overcome to achieve high yield for the entire system while mitigating the greater noise coupling among the dissimilar blocks within the system. Additional system requirements for the RF circuitry, passive elements, and discrete components, such as the decoupling capacitors, which are not easily integrated due to performance degradation or size limitations, have led to increased need for technological innovations. Three-dimensional (3-D) or vertical integration and system-on-package (SoP) have been introduced to overcome many of these inherent SoC constraints. The concepts of vertical integration and SoP are described, respectively, in Sections 2.1 and 2.2. Several technologies for 3-D packaging are reviewed in Section 2.3. The emerging packaging paradigm of 2.5-D integration is often

Three-Dimensional Integrated Circuit Design. DOI: http://dx.doi.org/10.1016/B978-0-12-410501-0.00002-2

considered a stepping stone towards stacked 3-D circuits and is described in Section 2.4. The cornerstone of 2.5-D integration is a physical component that is interposed between the die(s) and the package substrate; hence, the term interposer is broadly used to describe this component. As this technology is on a rather evolutionary path, several manufacturing processes for interposers are also reviewed in this section. The technological implications of 3-D integration at the package level are summarized in Section 2.5.

2.1 STACKING METHODS FOR TRANSISTORS, CIRCUITS, AND DIES

One of the first initiatives demonstrating 3-D circuits was reported in 1981 [45]. This work involved the vertical integration of p-type metal oxide semiconductor (PMOS) and n-type metal oxide semiconductor (NMOS) devices with a single gate to create an inverter, considerably reducing the total area and capacitance of the inverter. A cross-section of this inverter is illustrated in Fig. 2.1. Several other approaches to 3-D integration have, however, been developed. Both at the package and circuit level, bare or packaged die are vertically integrated, permitting a broad variety of interconnection strategies. Each of these vertical interconnection techniques exhibits different advantages and disadvantages. Other more esoteric technologies for 3-D circuits have also been proposed. A crucial differentiation among the various 3-D integration approaches is, therefore, maintained in this chapter to avoid confusion. Two primary categories of 3-D systems are discerned, namely system-in-package (SiP) and three-dimensional integrated circuits (3-D IC). The criterion to distinguish between SiP and 3-D IC is the interconnection technology that provides communication for the circuits located on different tiers within a 3-D system. In SiP, through silicon vias (TSVs) with high aspect ratios are typically utilized. Due to the size of these vias, high vertical interconnect density cannot be achieved. Consequently, these interconnects provide coarse grain connectivity among circuit blocks located on different tiers. 2.5-D systems also contain vertical interconnects including TSVs. The diameter of these TSVs is on the order of several tens of micrometers, supporting a higher interconnection density as compared to standard SiP. Alternatively, in 3-D ICs, fine grain interconnection among devices on different tiers is achieved by narrow and short TSVs.

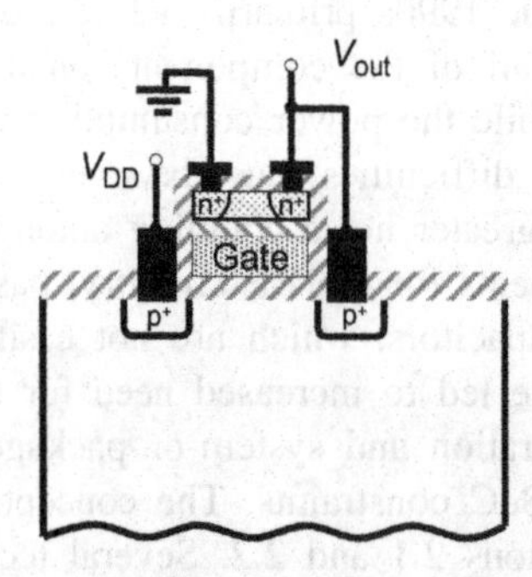

FIGURE 2.1

Three-dimensional stacked inverter [45].

2.1.1 SYSTEM-IN-PACKAGE

A SiP is described henceforth as an assemblage of either bare or packaged die along the third dimension, where the interconnections through the *z-axis* are primarily achieved through the following means:

- Wire bonding
- Vertical interconnects along the periphery of the die/package
- Long and wide, low density vertical interconnects (in an array arranged across the die/package)
- Metallization between the faces of a 3-D stack

Die or package bonding is achieved by utilizing a diverse collection of materials, such as epoxy and other polymers. Some examples of SiP structures are illustrated in Fig. 2.2, while each of these manufacturing techniques is discussed in Section 2.3.

2.1.2 TRANSISTOR AND CIRCUIT LEVEL STACKING

3-D ICs can be conceptualized either as a sequential or a parallel process. For a sequential process, the devices and metal layers of the upper tiers of the stack are grown on top of the first tier. Hence, the 3-D system can be treated as a monolithic structure. Alternatively, a 3-D IC can be created by bonding multiple wafers or bare die. The distinctive difference between an SiP and a 3-D IC is the vertical interconnect structure. Communication among different die within a 3-D IC is achieved by

- High density short and thin TSVs (placed at any point across the die not occupied by transistors);
- Capacitive coupling among parallel metal plates located on vertically adjacent die;
- Inductive coupling among inductors located on vertically adjacent die.

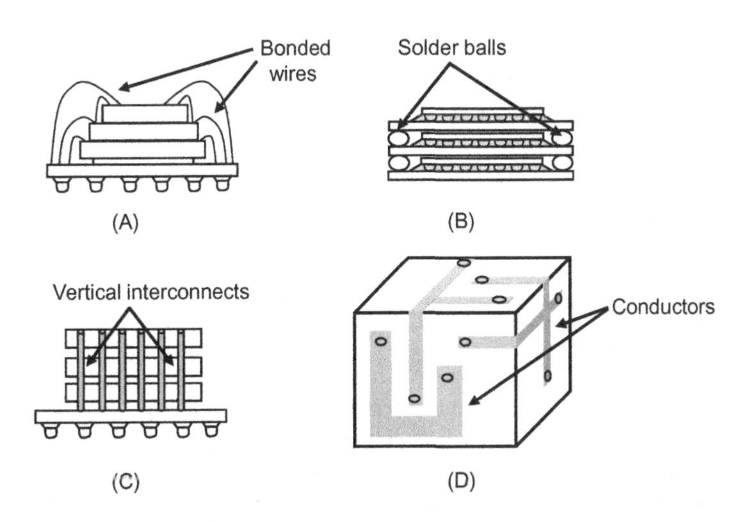

FIGURE 2.2

Examples of SiP technologies. (A) Wire-bonded SiP, (B) solder balls at the perimeter of the tiers, (C) area array vertical interconnects, and (D) interconnects on the faces of the SiP.

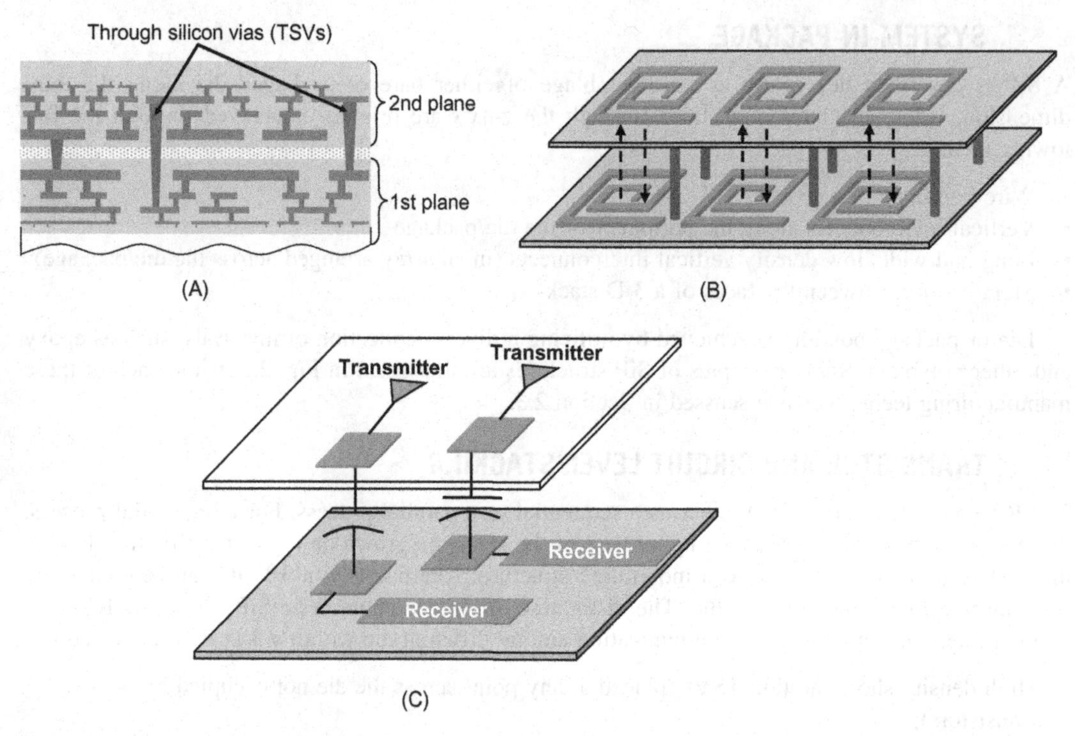

FIGURE 2.3

Different communication schemes for 3-D ICs. (A) Short TSVs, (B) inductive coupling, and (C) capacitive coupling.

Some characteristic examples of these communication mechanisms for 3-D ICs are illustrated in Fig. 2.3. Although 3-D ICs can be characterized as a subset of SiP, the advantages of 3-D vertical interconnects as compared to the interconnect mechanisms utilized in SiP are substantially greater. In addition, 3-D ICs face fewer performance limitations as compared with the SiP approach, unleashing the potential of 3-D integration. Several technologies for 3-D ICs are discussed in Chapter 3, Manufacturing Technologies for Three-Dimensional Integrated Circuits.

2.2 SYSTEM-ON-PACKAGE

As portability has become a fundamental characteristic in many consumer products, systems with a small footprint, low weight, long battery life, and equipped with a variety of features typically attract a significant market share. Communications is the most remarkable feature of these products. Wireless communication, in turn, requires RF circuitry that includes passive components, such as resistors and high Q inductors, that cannot be easily and cost effectively fabricated on the silicon. In addition, off-chip capacitors of different sizes are required to manage the voltage droop and simultaneous switching noise at the package level [59].

Integrating all of these components within a package and not as discrete components on a printed wiring board contributes significantly to system miniaturization. SoP is a system level concept that includes passive components manufactured as thin films. High performance transceivers and optical interconnects can also be embedded within a package [60]. State-of-the-art micro-vias provide communication among the embedded circuits and the mounted subsystems [61]. Micro-vias are vertical interconnects that connect the various components of an SoP through the laminated layers of the SoP substrate. Consequently, SoP supports wiring among the various subsystems as in an SiP and can also contain system components integrated within the package.

A wide spectrum of subsystems that gather, process, store, and transmit data can be included in an SoP. Sensors and electromechanical devices that record environmental variables, analog circuits that process and convert the analog signals to digital data, digital circuits such as application specific integrated circuits (ASICs), digital signal processing (DSP) blocks, and microprocessors, memory modules, and RF circuits that provide wireless communication, are a short and non-exhaustive list of components that can be mounted on an SoP. Some of these subsystems can be vertically integrated, making an SoP a hybrid technology that combines planar ICs and 3-D circuits, as illustrated in Fig. 2.4. As the dies are vertically stacked, SiP can also support diverse technologies with a smaller form factor as compared to an SoP. The interconnection density among these dies cannot, however, compensate for the increasing I/O densities due to the inherent limitations of wire bonding. Alternatively, interposers fully support heterogeneity with high interconnection density at the die level, enabling energy efficient and high performance chip-to-chip communication. The performance and power benefits at the IC level are supported by 3-D integration, where short and thin vertical wires can be manufactured.

Several manufacturing and design challenges, as illustrated in Fig. 2.5, must be addressed to ensure that the various forms of 3-D integration can be economically produced. From a technology perspective, a 3-D system should exhibit yield similar to a 2-D circuit. Since each tier within a stack can be separately prepared with a high yield fabrication process, the primary challenge is to guarantee a high yield bonding process. The bonding process should ensure reliable interconnectivity of circuits located on different tiers. In addition, fabrication of the vertical vias is a critical component contributing to the total system yield. Finally, advanced packaging technologies provide highly efficient heat transfer from the 3-D system to the ambient environment.

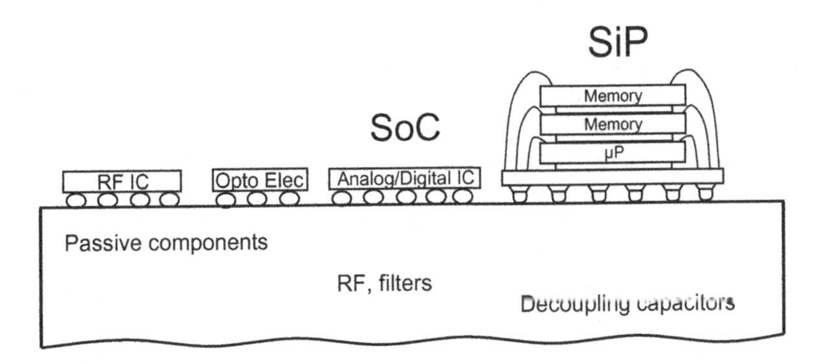

FIGURE 2.4

Typical SoP, which can include both SiP and SoCs.

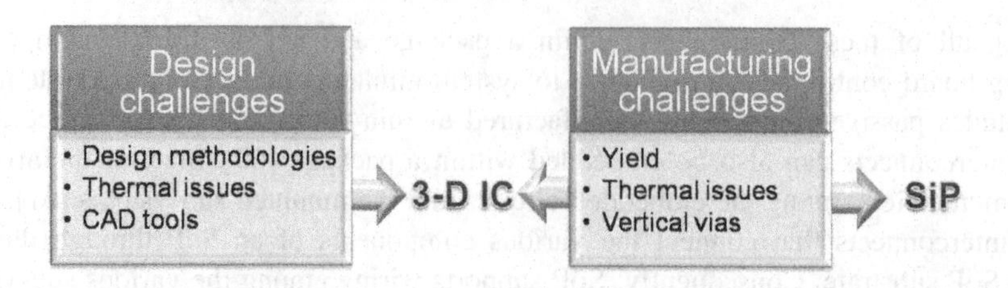

FIGURE 2.5

Manufacturing and design challenges for 3-D integration.

Overcoming these manufacturing challenges results in improved SiP systems, where reliable interconnect mechanisms that are more efficient than wire bonding can be employed, improving the performance of this form of 3-D integration. The potential of vertical integration, however, can be best exploited if several design challenges, as shown in Fig. 2.5, are also satisfied, resulting in high volume production of 3-D circuits. Design (and test) methodologies for 3-D circuits are currently in development.

Specific steps of the design flow for digital circuits support the third physical dimension, but these stages are semi-automated, requiring manual intervention with the design process, potentially leading to suboptimal systems. In addition, no heterogeneous systems are currently supported by commercial EDA tools. This situation can be attributed to the current lack of a substantial market for vertical systems, which does not justify the investment by EDA companies in sophisticated design tools for 3-D integration. Alternatively, academia has developed several open source tools for 3-D circuits, where these tools emphasize physical design tasks, such as placement and floorplanning (see Chapter 9, Physical Design Techniques for Three-Dimensional ICs), and thermal analysis and management (see Chapter 12, Thermal Modeling and Analysis and Chapter 13, Thermal Management Strategies for Three-Dimensional ICs).

Mitigating both manufacturing and design issues will greatly enhance the evolution of 3-D systems, considerably expanding the application pool for vertical integration. Homogeneous 3-D circuits, such as memories, have recently appeared in the consumer market [62,63]. These strides demonstrate that the adoption of vertical integration in the semiconductor industry progresses, albeit not at the desired rate.

Relating to 3-D packaged systems, interposer-based products have also been developed [63,64]. The evolution of interposers can help to overcome issues related to SoP, as illustrated in Fig. 2.6. For any of these packaging technologies, there are a few basic requirements.

The development of a reliable package substrate that enables the embedding of both passive and active components is a challenging task. Minimum warpage and a low coefficient of thermal expansion (CTE) are basic requirements for package substrates or interposers. Materials with a low CTE in the range of 8–12 ppm/°C have been developed [61]. Microstip lines of 50 Ω impedance are fabricated with low loss off-chip wiring in the package to enhance the on-chip global interconnects, which are responsible for the longest on-chip delays. Consequently, conductor geometries and dielectrics with low losses are necessary [61,65].

Dielectric materials also play an important role to provide both high performance and reliability. Dielectrics with high permittivity increase the charge stored within a capacitor, while low

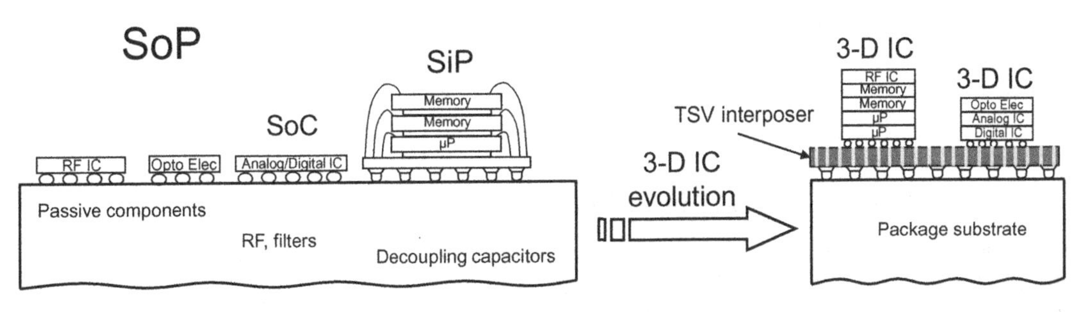

FIGURE 2.6

System miniaturization through the integration of sophisticated 3-D ICs.

Table 2.1 High Performance Dielectric Materials [61]

Dielectric Material	Dielectric Constant	Loss Tangent @ 1 GHz	Modulus (GPa)	X, Y, CTE (ppm/°C)
Polyimide	2.9–3.5	0.002	9.8	3–20
Benzocyclobutene (BCB)	2.9	<0.001	2.9	45–52
Liquid crystal polymer (LCP)	2.8	0.002	2.25	17
Polyphenylene ether (PPE)	2.9	0.005	3.4	16
Poly-norbornene	2.6	0.001	0.5–1	83
Epoxy	3.5–4.0	0.02–0.03	1–5	40–70

permittivity is required for those dielectrics that provide electrical isolation among neighboring interconnects, thereby reducing the coupling capacitance. A metric characterizing the loss of a dielectric material is the loss tangent (tan δ). This metric describes the ratio of the energy dissipated by the capacitor formed between the conductors and the energy stored in the capacitor. Low tangent loss for achieving low latency signal transmission and high modulus for minimum warpage across the package is desirable. In addition, dielectrics should exhibit excellent adhesion properties to avoid delamination and withstand, without any degradation in material properties, the highest temperature used to manufacture the package. Some dielectric materials that have been developed for SoP with the relevant material properties are listed in Table 2.1.

2.3 TECHNOLOGIES FOR SYSTEM-IN-PACKAGE

A SiP constitutes a significant and widely commercialized variant of vertical integration. The driving force towards SiP is the increase in packaging efficiency, which is characterized by the die to package area ratio, the reduced package footprint, and the decreased weight. Homogeneous systems consisting of multiple memory dies and heterogeneous stacks including combinations of memory modules and an ASIC or a microprocessor are the most common type of SiP, which has been commercialized by several companies [66,67]. The typical communication mechanism among different tiers of the stack is realized through wire bonding, as discussed in the following subsection.

2.3.1 WIRE BONDED SYSTEM-IN-PACKAGE

Wire bonding for a SiP is a commonplace technique. Due to the simplicity and low cost of the process, the required time to fabricate such an SiP is substantially shorter as compared to the turn-around time for an SiP that utilizes other interconnect techniques. Several wire bonded SiP structures are depicted in Fig. 2.7. The parameters that determine the packaging efficiency of these structures are the thickness of the bonded die and spacers, thickness of the adhesive materials, capability of providing multiple rows of wire bonding, and the size of the bumps used to mount the bottom die of the stack onto the package substrate. Alternatively, the performance of the stack is determined from the length of the bonding wires and the related parasitic impedances. Communication among the die of the SiP is primarily achieved through the package substrate and, as shown in Fig. 2.7C, through wire bonding from one IC substrate to another IC substrate.

The manufacturing process for the structures shown in Fig. 2.7 comprises the following key steps. All of the stacked dies or wafers are thinned with grinding and polished to a thickness of 50 to 75 μm [68]. Although aggressive wafer thinning is desirable, the die or wafer must be of sufficient thickness to avoid curling and cracking during the bonding process. The length of the

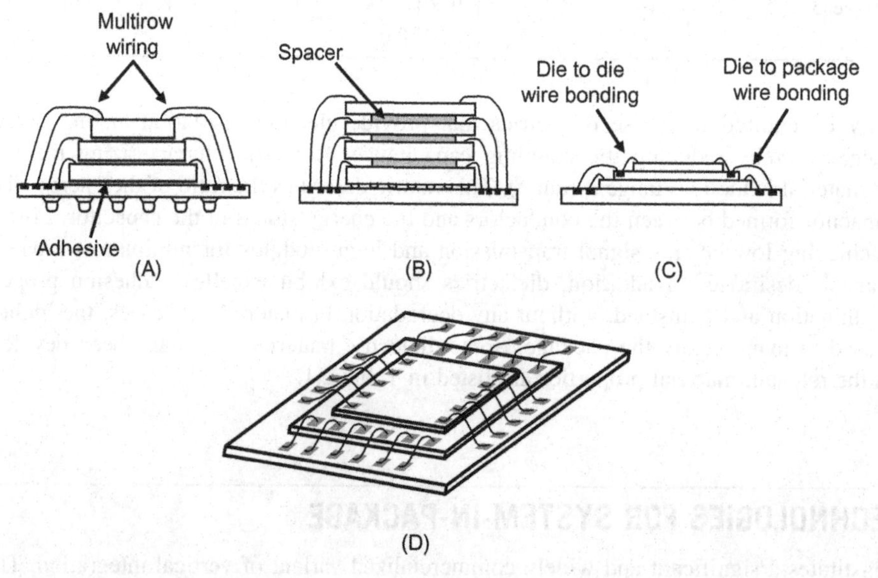

FIGURE 2.7

Wire bonded SiP. (A) Dissimilar dies with multiple row bonding, (B) wire bonded stack delimited by spacer, (C) SiP with die-to-die and die-to-package wire bonding, and (D) top view of wire bonded SiP.

Table 2.2 Loop Overhang Requirements Versus Die Thickness [68]

Die Thickness (μm)	Overhang (mm)
100	2.0
75	1.0
63	0.7

Table 2.3 Required Bonding Wire Pitch for Multiple Row Bonding [41]

Row no.	Wirebond Pitch (μm)
1	40
2	25/50
3	20/60

bonding wires or, equivalently, the loop overhang, also depends upon the thickness of the attached wafers. In Table 2.2 the overhang requirement as a function of the die thickness is reported.

The bottom die of the stack is bumped and mounted in a flip chip manner onto the package substrate. The remaining die are successively adhered to the stack through thin film epoxies. In certain cases, spacers between subsequent die are required. These spacers provide the necessary clearance for the bonding wire loop. Epoxy pastes and other thick film materials are utilized with a thickness ranging from 100 to 250 μm [68]. A uniform gap between the attached die is required across the area of the spacer and, therefore, spheres of spacer material can be utilized for this purpose. With these spacers, similar die with quite diverse areas can be stacked on an SiP.

Wire bonded SiP composed of up to four or five die have been demonstrated [68], while two die stacking is quite common [69]. Increasing the number of die on an SiP is hampered by the parasitic impedance of the bonding wires and the available interconnect resources. For example, bonding wires can exhibit an inductance of a few nH. Consequently, a low loop profile is required to control the impedance characteristics of the wire. Additionally, multiple rows for bonding can increase the interconnect bandwidth. For multiple row wiring, low height bonding wires are also necessary. The wire pitch for three bonding rows is listed in Table 2.3. Although the off-chip interconnects are decreased with wire bonding, limitations due to parasitic interconnect impedances and constraints on the number of die that can be wire bonded have led to other SiP approaches that provide greater packaging efficiency while exhibiting higher performance. These improved characteristics are enabled by replacing the bonding wires with shorter through hole vias or solder balls [69]. This SiP technique is discussed in the following subsection.

2.3.2 PERIPHERAL VERTICAL INTERCONNECTS

To overcome the constraints of wire bonding, SiPs utilize peripheral interconnects formed with solder balls and through hole vias. Several approaches based on this type of interconnect have been developed, some of which are illustrated in Fig. 2.8 [66]. A greater number of die can be integrated

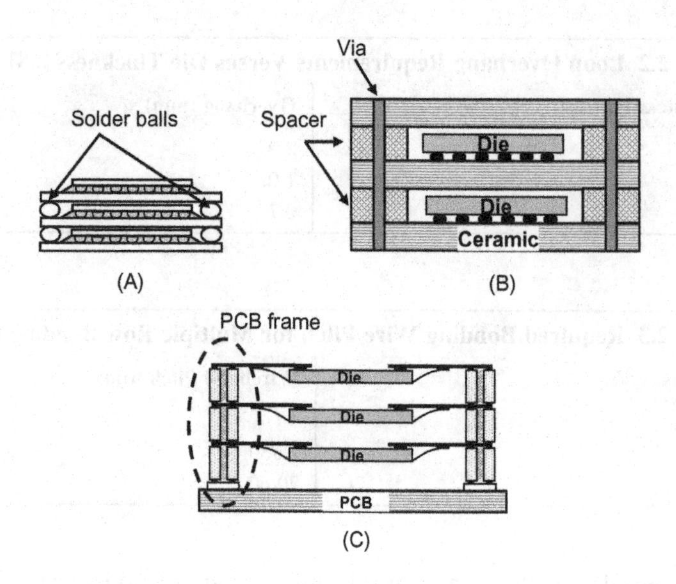

FIGURE 2.8

SiP with peripheral connections. (A) Solder balls, (B) through hole via and spacers, and (C) through hole via in a PCB frame structure.

with this approach as the constraints due to the wire parasitic impedances and the length are relaxed. An interconnect structure that provides a reduced pitch connection between the die and the package is included in each tier of an SiP, as shown in Fig. 2.8. This structure also provides the wiring between the die and the solder balls or through hole vias at the periphery of the system. Several companies have developed similar SiP technologies. Micron Technology uses the method shown in Fig. 2.8B to produce high density static random access memory (SRAM) and dynamic random access memory (DRAM) memory ICs [70]. In another technique developed by Hitachi, vertical pillars are attached onto the PCB, called "PCB frames," to accommodate the vertical interconnects [71]. The I/O signals are transferred through tape adhesive bonds to the PCB frames, as illustrated in Fig. 2.8C. High density memory systems also use this method.

The basic steps of a typical process for SiP with peripheral interconnects are depicted in Fig. 2.9 [72]. Bumps are initially deposited on the interposers, while solder plated polymer spheres are attached at the periphery of the printed wire board interposers to communicate among the die of the stack. Note that the geometry of these interposers differs substantially from modern interposer technologies, where back-end-of-the-line processes for nanometer scale silicon ICs are utilized. Alternatively, the physical size of the wires on these early interposers is on the order of tens of micrometers (e.g., 75 μm [72]).

Next the IC is attached to the interposer through a flip chip technology. Vertical stacking of the ICs through the solder balls follows. The solder balls constitute an important element of the stacking process as these joints provide both the electrical connections as well as the mechanical support for the SiP. In addition, the height of the solder ball core provides the necessary space between successive interposers where the die is embedded. Solder balls with a height greater than 200 μm can

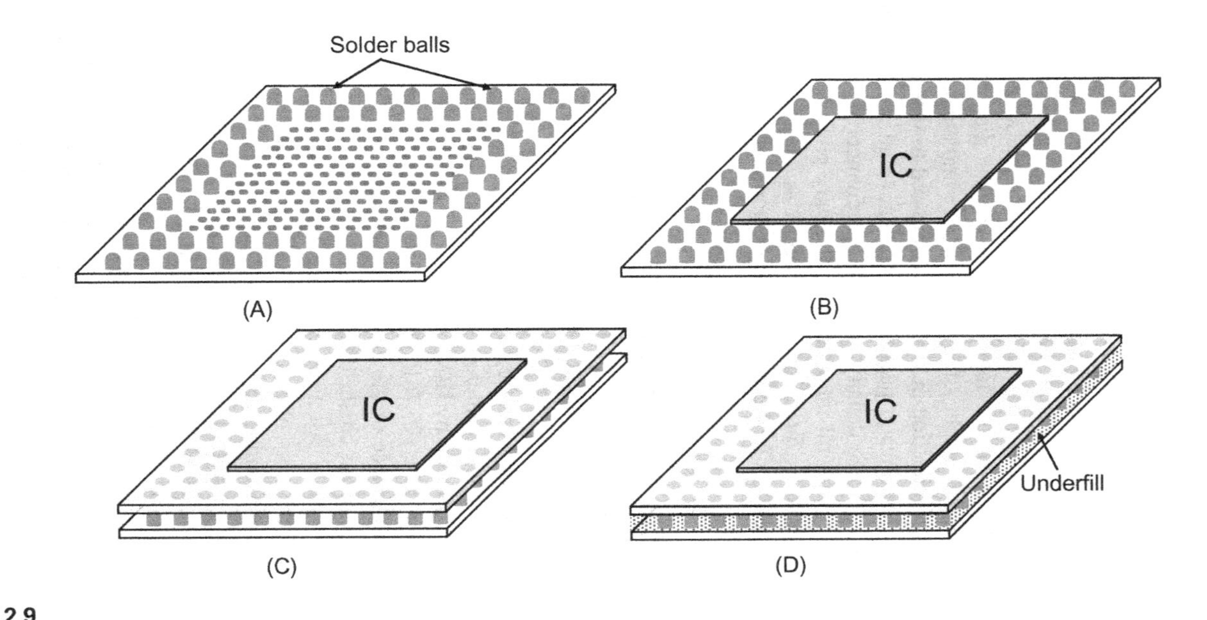

FIGURE 2.9

Basic manufacturing phases of an SiP. (A) Interposer bumping and solder ball deposition, (B) die attachment, (C) tier stacking, and (D) epoxy underfill for enhanced reliability.

therefore be utilized. If the solder balls collapse, the reliability of an SiP can be severely affected. Finally, to further reinforce the durability of this SiP structure, an epoxy underfill between adjacent tiers is applied. A similar technique has been developed by the Interuniversity MicroElectronics Center (IMEC), where the 3-D stack is achieved by directly soldering two contiguous solder balls ("ball on ball" bonding) [73].

Further enhancements of this bonding technique results in a decreased diameter of the vertical interconnects by utilizing through hole electrodes and bumps with a diameter and depth ranging, respectively, from 20 to 30 μm and 30 to 50 μm [74]. The through hole electrodes are etched and gold (Au) bumps are grown on the front and back side of the interposers to bond the IC. An important characteristic of this method is that only low temperatures are used to compress the stack. In addition, the thickness of the interposers is smaller than 50 μm, providing greater packaging efficiency. The expected height of a ten tier SiP implemented with this method is approximately 1 mm, a significant reduction in height as compared to wire bonding techniques.

Although through hole vias yield greater packaging efficiency and mitigate certain performance issues related to wire bonding, higher density vertical interconnect is required as interconnect demands increase. Indeed, with peripheral interconnects, the wire pitch decreases linearly with the number of interconnects. Alternatively, interconnects arranged as an array across the die of an SiP greatly increases wiring resources, while the interconnect pitch decreases with the square root of the number of interconnects [69].

2.3.3 AREA ARRAY VERTICAL INTERCONNECTS

An SiP with peripheral interconnects is inherently limited by the interconnect density and the number of components that can be integrated within a stack. Area array vertical interconnects improve the interconnect density among the tiers of a 3-D system, offering opportunities for greater integration. Solder bumps or through hole vias can be utilized for the area array interconnects; however, the processing steps for these two types of interconnects differ considerably.

In a manufacturing process developed by IBM and Irvine Sensors, the memory chips are bonded together to produce a 3-D silicon cube [75,76]. The wafers are processed to interconnect the I/O pads of the die to the bumps on the face of the cube. This metallurgical step is followed by wafer thinning, dicing, and die lamination to create the silicon cube. The stacking process is achieved through controlled thermocompression. An array of controlled collapse chip connect (C4) bumps provides communication among the tiers of the cube. The interconnect density achieved with this technology is 1,400 C4 bumps, consuming an area of 260 mm^2. Note that with this technique, all of the tiers within the cube share a common C4 bump array.

A higher contact density was developed by 3D-PLUS and THALES [77]. A heterogeneous wafer is formed by attaching the individual ICs onto an adhesive tape that provides mechanical support. The I/O pads from each IC are redistributed with two different techniques into coarser pads whose size depends upon the alignment accuracy prior to stacking the ICs. The first technique for pad redistribution utilizes two layers of copper embedded in BCB polymer. The average measured contact resistance is ~2 Ω. Another technique for developing the vertical interconnects between the tiers includes laminated films (as a dielectric) and copper. Despite the large number of steps, test vehicles have shown no signs of delamination among the adhered tiers within the SiP [77].

Prior to bonding, the wafer thickness is reduced to 150 μm with grinding and mechanical polishing. Once the target thickness is achieved, the wafer is diced and tested for IC failures or open contacts during connection of the die I/Os with the coarser pads of the package. Only known good dies (KGD) are therefore bonded, guaranteeing high yield. The 3-D system is connected through liquid adhesive and compression. The resulted SiP can contain up to six ICs with a height less than a millimeter.

In these two techniques, each tier of the stack is designed and fabricated without any constraints imposed by the assembly process of the SiP, as the vertical interconnects are placed on one face of the tier and not through the tier. Alternatively, the vertical interconnects use copper bumps and electrodes formed through the stacked circuits within the SiP [78]. The stacked dies are aggressively thinned to a thickness of 50 μm and copper (Cu) electrodes with a 20 μm pitch are formed. Square Cu bumps with a side length of 12 μm and a height of 5 μm are electroplated and covered with a thin layer of tin (Sn) (<1 μm) for improved bonding. The bonding is further enhanced by a nonconductive particle paste, which is also used to encapsulate the Cu bumps.

Two bonding scenarios with different temperature profiles have been applied to stack four 50 μm thick dies on a 500 μm interposer [78]. In the first case, a bonding force is applied at high temperature (245°C), while in the second scenario, the temperature is reduced to 150°C accompanied by a thermal annealing step at 300°C. Both of these scenarios demonstrate good adhesion among the die, while the backside warpage is less than 3 μm for a square die with a side length of 10 mm.

With array structured interconnects, the throughput among the die of the stack can be improved as compared to an SiP with peripheral interconnect and bonding wires. The solder ball or bump size, however, constitutes an obstacle for vertically integrated systems with ultra-high interconnect density. Another concern is the rewiring that is necessary with some of these techniques, which prevents achieving shorter intertier interconnects.

2.3.4 METALIZING THE WALLS OF AN SiP

Another SiP approach entails several ICs mounted on thin substrates with wiring or stacked printed circuit boards (PCBs) to yield a 3-D system. Interconnects among the circuits on different dies are metal traces on the individual faces of the cube. The interconnect density achieved with this technique is similar to a wire bonded SiP; however, a higher number of monolithic structures can be integrated within the cube.

One application of this type of SiP is a high density SRAM memory stack manufactured with a process invented at Irvine Sensors [79]. In this process, each module is separately tested before the stacking step. In this way, the KGD problem is significantly addressed as the faulty parts can be excluded or reworked during the initial steps. Fully operational ICs are selected and glued with a thin layer of adhesive material between each pair of ICs. The bonded stack undergoes a baking step with a specific curing profile. The stack is plasma etched to reveal the buried metal leads from each memory module. If required, more than one face of the cube can be etched. The dimensions of the leads are 1 μm thick and 125 μm wide. The etching step exposes the metal leads as well as the substrate of the stacked memory tiers. Passivation isolates the substrates from the metal traces placed on the sidewall of the stack. Several layers of polyimide are deposited, where the thickness of each layer is greater than the exposed metal leads. Lift-off photolithography and sputter deposition create the interconnects, including the busses, and the pads on the face of the cube with a titanium tungsten/gold (Ti-W/Au) alloy. The pads on the sidewall of the cube connected to the metal leads form a T-shape interconnect, which is referred to as a "T-connection." The resistance of a T-connection has been measured to be approximately 25 mΩ [79]. Finally, a soldering step attaches the stack onto a silicon substrate or PCB, permitting the stack to be connected to external circuitry.

Prototypes with as many as seven stacked ICs demonstrate a considerably higher integration as compared with a wire bonded SiP. These prototypes are diced from an initial stack consisting of 70 identical ICs. Shorter stacks from a taller cube contribute to a reduction in the overall fabrication time. In addition to memory products, an image capturing and processing system fabricated with a similar technique has been developed [80,81]. The entire fabrication process consists of seven basic steps. Each IC is attached onto laminated films or, if necessary, a PCB along with discrete passive components. The substrate contains metal tracks, such as copper lines plated with gold, and the I/Os of the IC are wire bonded to these tracks. Each mounted IC is tested before bonding. This testing step improves the total yield as bonding KGD is guaranteed. Testing includes validation of the interconnections connecting the IC to the laminated film or PCB. The test structures are placed in a plastic mold and encapsulated with an epoxy resin. After removal of the mold, a sawing step is applied to expose the metal tracks of each PCB within the stack. Alternatively, a deeper sawing can be applied to reveal the bonding wires rather than the metal tracks of the PCB, as shown in Fig. 2.10. The cubical structure is plated with conventional electroplating techniques. The plated surfaces of the cube are patterned by a laser to produce the interconnects connecting the tiers of the SiP. Typical dimensions for these sidewall interconnects are 6 to 7 μm thick and 500 μm wide. During the last step, the SiP is soldered to a PCB on one face of the cube.

Although the methods used in each step of this technique are well known and relatively low cost, there are several design and efficiency concerns regarding the reliability of the surface contacts. These issues include the reliability of the internal contacts of each IC to the external interconnections on the faces of the cube, and the thermo-mechanical performance of a dense SiP. Finally, a particularly long interconnect path may be required to connect two circuits from different tiers within an SiP. Consequently, utilizing thin and short vertical interconnects for communication among the tiers within a 3-D system can result in higher performance. Although these approaches

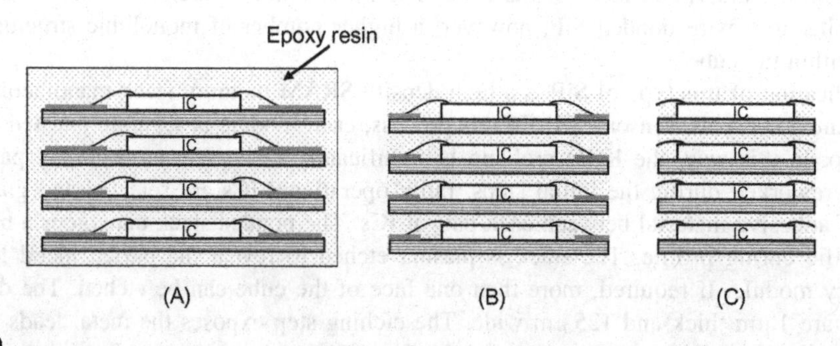

FIGURE 2.10

Cross-section of the SiP after removing the mold. (A) The SiP encapsulated in epoxy resin, (B) sawing to expose the metal traces, and (C) sawing to expose the bonding wires.

exhibit considerable potential, emphasis in the last few years has been placed on interposer technologies, the advantages being the relatively high density and low loss interconnects as compared to the other 3-D packaging approaches discussed in this section. This promising technology is discussed in the following subsection.

2.4 TECHNOLOGIES FOR 2.5-D INTEGRATION

Interposer technologies have been developed as an intermediate step towards 3-D integrated systems since there are many challenges that must be circumvented for 3-D integration to be widely adopted. These issues are not only of a technical nature and relate to, for example, the lack of a unified model and supply chain among the diverse stakeholders included in the manufacturing process of these systems. Consequently, interposers (also termed 2.5-D integration in this chapter), which originated from multichip module technologies, face fewer challenges and exhibit a lower cost than 3-D integration to become a viable solution for addressing the increasingly difficult chip-to-chip communication. Interposers support high I/O densities up to 10^6/cm^2 as compared to ceramic or organic packages, which typically provide I/O densities of 10^3/cm^2. Additionally, the interconnect pitch of the interposers is on the order of 5 μm, while for ceramic packages this pitch is an order of magnitude larger [82].

Due to the benefits of interposers, the development of fabrication processes and related materials has made significant progress, allowing interposers to evolve into a stand-alone manufacturing technology with a potentially large market rather than a short-term intermediary technology on the path to 3-D systems [44]. The principle of interposers is illustrated in Fig. 2.11, where dies are integrated onto either one or both sides of an interposer [83]. In Fig. 2.11A, the dies are interconnected through the metal layers on the interposer (usually called redistribution layers (RDL)) and connections to the package are provided by vertical wires. As shown in Fig. 2.11B, the vertical wires provide both interconnections between the package and the dies and also among the die. A double side

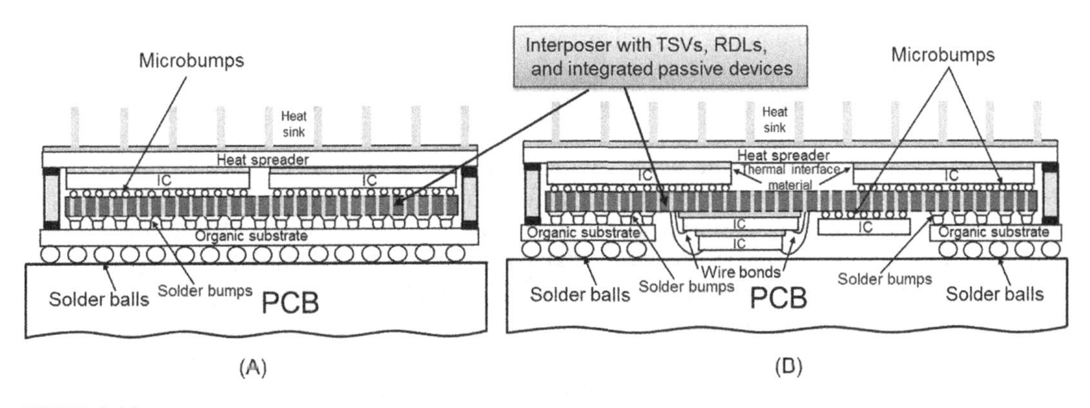

FIGURE 2.11

Interposer-based 2.5-D systems where (A) ICs are mounted on only one face of the interposer (single side), and (B) ICs are attached to both sides of the interposer (double side) [83].

interposer exhibits a smaller footprint and potentially better performance for interchip communication through the short TSVs rather than the longer RDLs [83].

Interposers are mainly divided into either active or passive, depending on whether some functionality is included in the interposer, including optical [84]. Integrating some logic within the interposer enhances the functionality supported by this interconnection medium but also increases the cost and complicates the design process of interposer-based systems [44]. Examples of active interposers are wide I/O memory/logic [85] and DRAM memory ICs [86]. Alternatively, passive interposers are encountered more often and include, in addition to multiple metal layers for signal and power and ground routing, other passive components, such as decoupling capacitors and inductors.

Interposer technology includes many approaches depending upon several factors, such as the type of interposer, raw material, vertical interconnects, and structural format. An overview of these issues is provided in the following subsections. The individual examples, however, should not be considered as standard since these processes are quickly evolving. Rather, these examples are used to describe the physical scale of these systems and the steps and issues related to the interposer manufacturing process.

2.4.1 INTERPOSER MATERIALS

The primary raw materials used for fabricating interposers are silicon and glass, although organic interposers are also being explored [87]. Early effort emphasized silicon interposers, where standard back-end-of-the-line processes and tools can be used without significant modification to provide the metallization of the interposer. Another innate advantage of a silicon interposer is matching the thermal coefficient of expansion (termed as TCE or CTE) to the silicon die(s), alleviating thermally induced stress between the attached dies and the interposer. The most common silicon interposer technology is currently based on the 65 nm process node [84].

As compared to silicon interposers, glass interposers include a great variety of glass compounds [88] tailored to exhibit specific properties based on the requirements of the target system. Glass interposers can be superior to silicon interposers. For example, the CTE can be adapted to match the carrying die, minimizing thermal stress. Furthermore, dielectric losses are much lower, making glass a better candidate for high performance applications. The mechanical reliability and chemical durability of glass interposers can also be improved by using proper material compounds [88].

Another important difference between glass and silicon interposers is formation of the glass interposers, which can be produced either in a wafer format similar to silicon interposers or as a large panel format, leading to greater manufacturing throughput and consequently lower cost [84]. Typical sizes for silicon interposers are 150 to 300 mm wafers [89], while for glass interposers in a panel format, dimensions of 730 mm × 900 mm [90] and 2,400 mm × 2,800 mm [84] for the most recent generations have been reported.

To better describe the advantages of panels rather than wafer interposers, consider that in wafers, the exposure field size (the optical field that defines a single interposer part on a wafer) does not fit when the illumination area is close to the edge of the wafer. This field partly or fully falls into the exclusion zone, thereby limiting the number of usable parts [84]. Alternatively, with rectangular panels, the number of interposer parts produced during each exposure increases,

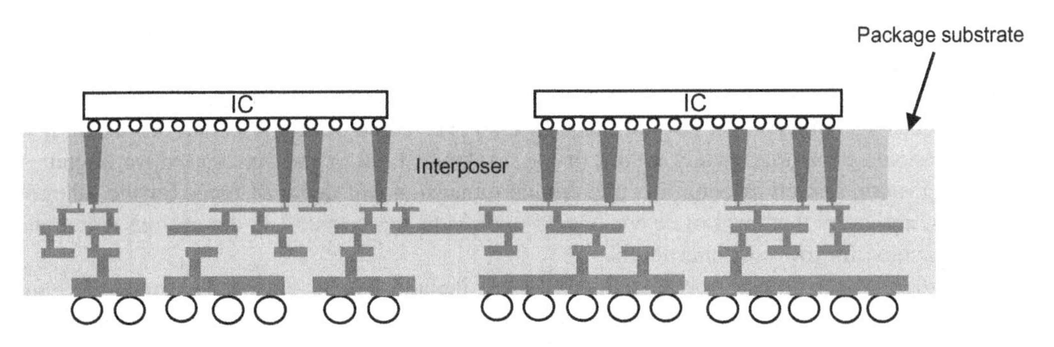

FIGURE 2.12

Embedded interposer within a package substrate enabling multi-IC integration [91].

improving the manufacturing throughput. This throughput also increases due to the considerably larger area supported by the panels, yielding a greater number of parts.

A challenge for interposers and particularly for glass panels, which are large in area, is the surface flatness of the glass and total thickness variation (TTV). Surface polishing is the primary method to reduce this variation. A process called fusion formed glass has recently been proposed [88], where the glass being formed does not come in contact with any other surface, leading to excellent surface flatness, thereby eliminating or reducing the polishing and grinding steps. In addition, this process is adaptive in the sense that glass surfaces with different CTEs can be formed to limit the thermal stress within a packaged system.

Although silicon interposers are typically soldered onto a package substrate, glass interposers can be embedded onto the package substrate with considerable improvements in the form factor of the package. This process has been demonstrated for interposers with an area of 21 mm $\times$ 14 mm and a thickness of 100 μm [91], which were embedded on an organic package substrate. In addition to removing the soldering step, this approach also does not require dispersion of underfill of the soldering (for increased reliability), which can further reduce the overall packaging cost. In this method, the interconnections between the interposer and the substrate utilize standard processes for manufacturing laminate substrates. This approach is schematically shown in Fig. 2.12. Test samples have been thermally evaluated within a temperature range of $-55°C$ to $125°C$ for 172 cycles. The resistance of several vertical wires connected in daisy chains through the interposer exhibited almost 25% variations after the thermal test [91].

2.4.2 METALLIZATION PROCESSES

Double side interposers, as depicted in Fig. 2.11B, offer high interconnect density and small form factor as compared to single side interposers. Double side metallization can, however, lead to severe warpage [84]. An important aspect of the metallization process is the dimensions of the interconnects, usually characterized as the line/space (L/S) dimensions, similar to the width and space of the on-chip metal layers. Typical L/S values are in the range of a few micrometers. Several lithographic challenges are associated with metallization. One of these challenges is the minimum resolution to produce the specified line dimensions [84]. Although an expensive dual damascene process is utilized for the on-chip back-end-of-the-line, other technologies for metallization, such as wafer level packaging, are lower cost [92].

The accuracy of the overlay of the subsequent metal layers can also incur yield loss as overlay inaccuracies can accumulate over several metal layers [93]. A rule of thumb is that the accuracy of the overlay should be approximately a third of the resolution limit of the optics used for alignment [84]. These factors, as well as control of the critical dimension and depth of focus for the lithography systems, are issues that need to be considered to provide appropriate L/S dimensions and multi-level metallization for low cost interposers.

Recent strides in interposer metallization [87–93] include five layers on the front side, with four Cu layers (of 2 μm thickness) and intermetal dielectric layers with the same thickness and one layer of the same thickness for the pads [92]. The four layers are manufactured with a plating process for Cu thick films, and the SiO_2 dielectric layers are deposited with plasma enhanced chemical vapor deposition (PECVD). A Ni/Au film is plated to form the pad layer. Backside metallization comprises two layers, where one layer is for signal routing and the other is for the pads. Metal organic chemical vapor deposition is used to construct these layers [92].

2.4.3 VERTICAL INTERCONNECTS

Another important element of interposers is the vertical interconnects, called TSVs for silicon interposers and through glass vias (TGVs) for glass interposers. The physical dimensions and electrical properties of these interconnects play a vital role in determining the electrical performance of interposer-based systems. Similar to metallization, several processes have been developed to manufacture these vertical interconnects. Due to the dissimilar materials used for the fill material of the TSVs (although unfilled vias have also been developed [92]) and for the surrounding materials, thermal stress during manufacturing can cause cracks in the liner or delamination of the metal filling. Many processes for this type of interconnects exist, implemented according to the substrate material of the interposer. Only the basic steps of the TSV manufacturing process are therefore discussed here, recognizing that many details of these steps vary among processes.

The primary steps for constructing TSVs include [83]: (1) formation of the openings either with deep reactive etching or laser drilling (typically for TGVs), (2) deposition of an adhesive layer on the sidewall of the openings based on PECVD, (3) physical vapor deposition (PVD) for the barrier and seed layers, (4) metal filling of the TSV, typically with a Cu electroplating process, (5) polishing of any excessive Cu with chemical mechanical polishing (CMP), and (6) revealing the TSVs through backgrinding, Si dry etching, low temperature passivation, and CMP.

Each of these steps affects the physical and electrical properties as well as the reliability of these interconnects. Some processes are preferred for forming the openings, e.g., laser drilling, if the interposer substrate is glass or polycrystalline silicon [89]. Each of these steps contributes to the overall cost of the TSV fabrication process, where the cost of these processing steps in descending order is PVD, PECVD, CMP, plating, and etching [44].

The process technology for TSVs has also been used to embed passive devices, more specifically inductors, which consume considerable area within the interposer substrate. This method is different from integrated passive devices within thin films, where the etching step for the TSVs is used to create deep trenches, forming the windings of the spiral inductor [94]. This technique is better suited for glass interposers, as glass supports lower losses, thereby improving the overall quality of the inductor. The process forming the trenches for the turns of the inductor includes exposure of the photosensitive glass to UV light (a wavelength range of 250 to 350 nm), an annealing process to recrystallize the glass at 580°C, and immersion in a HF solution for wet etching [94]. The etching rate is a combination of the parameters of the HF solution and the exposure time to UV light. The tested trench inductors demonstrate a Q as high as 30 at 2 GHz, achieved due to the low resistance of the thick turns formed within the trenches [94].

An alternative approach for vertically interconnecting die on a double side interposer is creation of through silicon holes (TSH) on the substrate and the growth of Cu pillars on one of the die. In other words, filling of the openings in the substrate is not performed. A conceptual illustration of this approach is shown in Fig. 2.13A, where related geometric characteristics are provided in Fig. 2.13B. The basic trait of this approach is that a metallization step is not required after forming the vertical openings, and backgrinding to reveal the TSV is avoided. A prototype technology based on this interconnection method with 256 Cu pillars has been demonstrated, where reliability tests of thermal cycling within −55 to 125°C do not exhibit any failures [83].

As the processes for all of the major stages relating to manufacturing interposers are currently in development, such as metallization and TSV or TGV formation, favoring one process over another is currently an unproductive task. Rather, an overview of some recent processes, with available information on each listed in Table 2.4, provides insight into the progress made to date. Note that this table is by no means exhaustive, but reflects the present art in fabrication processes for interposers.

Different 3-D technologies result in dissimilar performance for an integrated system, while imposing a specific cost based on the complexity of the fabrication process. A comparison of the manufacturing cost of different 3-D technologies is provided in Chapter 8, Cost Considerations for Three-Dimensional Integration.

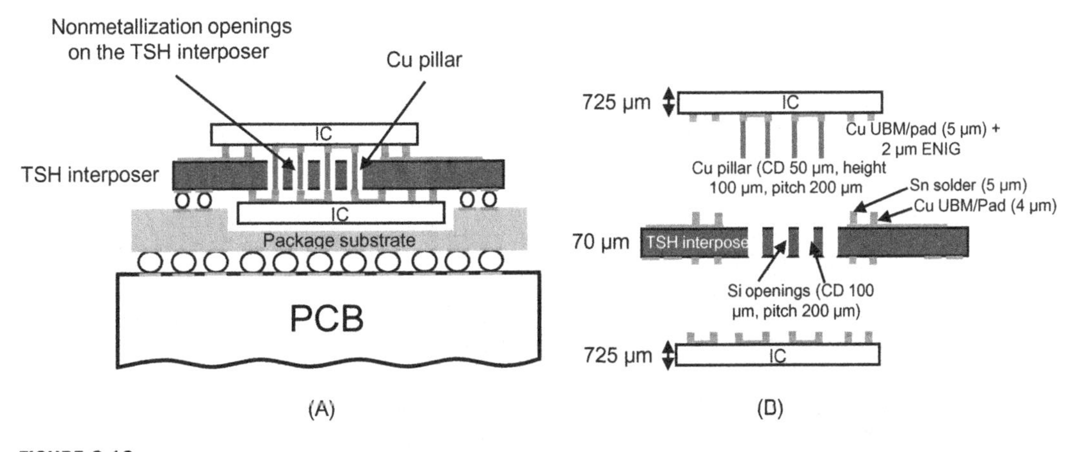

FIGURE 2.13

An SiP system comprising (A) a top IC with Cu pillars and a bottom IC with solder bumps on a TSH interposer, and (B) the dimensions of several components (not shown to scale) [83].

Table 2.4 Traits of Manufacturing Processes for Interposers

Process	Interposer Substrate					Metallization		Through Substrate Interconnects		
	Material	Format	Panel Area/ Wafer Diameter	Interposer Area	TTV	Type	L/S	Type	Diameter	Length
[95]	Glass	Wafer	370 mm × 470 mm	N/A	<4 μm	Double side	3.5 μm/ 3.5 μm	TGV	35 μm	135 μm
	Glass	Panel	150 mm	N/A	<4 μm	Double side		TGV	35 μm	~130 μm
[92]	Silicon	Wafer	150 mm	39.7 mm × 36.7 mm	N/A	Double side	2 μm (thickness)	TSV (unfilled)	80 μm	500 μm
[87]	Glass	Panel	150 mm × 150 mm	25 mm × 30 mm	3 μm	Double side	3 μm/3 μm	TGV	30 μm	N/A
[83]	Glass	N/A	N/A	10 mm × 10 mm	N/A	Double side	N/A	TSH	50 μm	100 μm
[91]	Glass	Panel	508 mm × 508 mm	21 mm × 14 mm	N/A	Double side	3 μm/3 μm (frontside) 15 μm/ 15 μm (back side)	TGV	30 μm	100 μm

2.5 SUMMARY

The technological progress and related issues regarding 3-D integration at the package level are summarized as follows:

- 3-D packaging includes bare or packaged die vertically integrated using a variety of interconnection schemes.
- SiP- and TSV-based 3-D ICs are currently the primary methods used for vertical systems. A major difference between these approaches is the type of vertical interconnects that connect the circuits located on different tiers.
- A SiP is an assemblage of either bare or packaged die connected along the third dimension, where the interconnections through the *z-axis* are primarily achieved by wire bonding, metallization on the faces of the 3-D stack, vertical interconnects along the periphery of the die/package, or low density vertical interconnects arranged as an array across the die/package.
- The aforementioned SiP interconnect structures are sorted in descending order of vertical interconnect density.
- 3-D ICs are interconnected with high density short and thin TSVs, supporting low level integration rather than die or package level integration as in an SiP.
- An SoP is a hybrid technology that combines planar ICs and 3-D circuits within a compact system. The primary difference as compared to an SiP is an SoP supports more than wiring, containing functional components within the package.
- An SiP exhibits higher packaging efficiency and a smaller footprint and weight as compared to a conventional 2-D system.
- Wire bonded SiP are limited by the impedance of the bonding wires.
- An SiP with peripheral interconnects uses solder balls and through hole vias. SiP are inherently limited by the interconnect density as the wire pitch decreases linearly with the number of interconnects.
- An SiP with array structured interconnects can improve the interconnect density as compared to an SiP with peripheral interconnects and bonding wires as the interconnect pitch decreases with the square root of the I/O (or pin) requirements.
- An interposer-based 2.5-D system offers several advantages over traditional packaging approaches such as lower cost as compared to 3-D stacked systems.
- Different materials, such as silicon and glass, in disparate formats, such as panel and wafer, are used for interposers.
- The vertical interconnects are formed by etching (TSVs) or laser drilling (TGVs) and are substantially larger than the TSVs in 3-D systems.
- Several lithography challenges relating to interposers exist, particularly due to the increasing size of the panels of the glass interposers. These challenges include line/space dimensions, alignment of metal layers, field exposure, and panel/wafer warpage.
- 3-D technologies increase the manufacturing cost of an integrated system. This additional manufacturing cost is often counterbalanced, however, by the decrease in area and increase in performance of the overall system.

CHAPTER

MANUFACTURING TECHNOLOGIES FOR THREE-DIMENSIONAL INTEGRATED CIRCUITS

3

CHAPTER OUTLINE

A system-in-package (SiP) offers a large number of advantages over traditional two-dimensional (2-D) system-on-chip (SoC), such as shorter off-chip interconnect lengths, increased packaging efficiency, and higher density. These advantages provide significant performance improvements as compared to a 2-D SoC. Manufacturing issues, however, limit the scaling of the interchip interconnects, such as the wire bonds and solder balls, within an SiP. Additionally, the inevitable increase in the delay of on-chip interconnects is not alleviated by SiP interconnect technologies. Although, an SiP employing coarse grain through silicon vias (TSVs) can improve this delay, the low density and impedance characteristics of these vertical interconnects limit the bandwidth of the intertier interconnect. The full potential of vertical integration is therefore not fully exploited. Consider, for instance, the SiPs depicted in Fig. 3.1, and assume that blocks A and B, located on different dies, are connected. The arrows, shown in Figs. 3.1A and B, represent typical interconnect paths connecting blocks A and B for each type of SiP. A 3-D IC, shown in Fig. 3.1C, provides the shortest interconnection between blocks A and B by utilizing a TSV. Consequently, 3-D ICs can achieve

Three-Dimensional Integrated Circuit Design. DOI: http://dx.doi.org/10.1016/B978-0-12-410501-0.00003-4

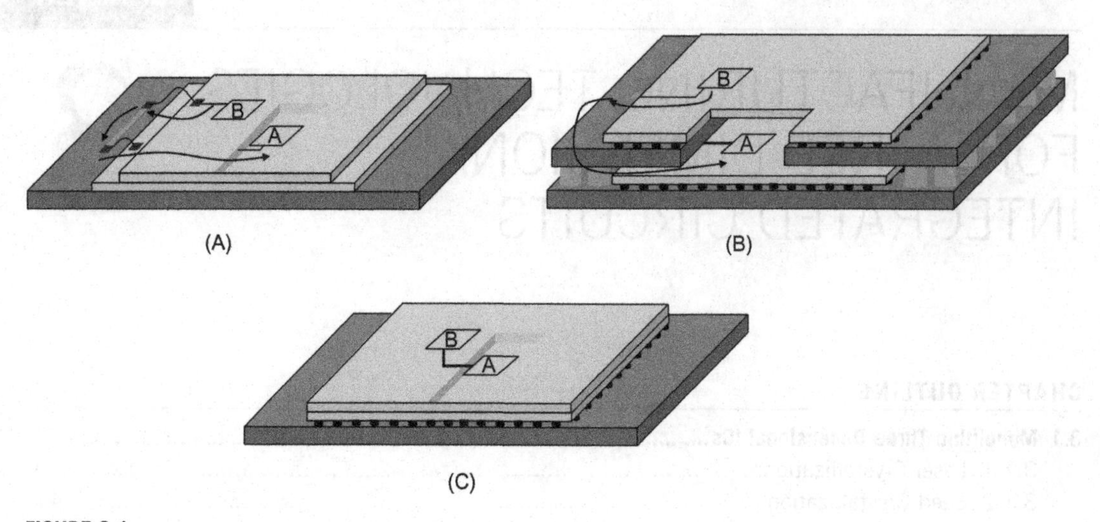

FIGURE 3.1

Typical interconnects paths for (A) wire-bonded SiP, (B) SiP with solder balls, and (C) 3-D IC with TSVs.

the greatest improvement in speed and power by decreasing the length of the long global interconnects as compared to other vertical integration technologies.

3-D ICs can be fabricated in either a batch (sequential) or parallel process. In the former case, the devices on the upper tiers of a 3-D stack are grown on top of the first tier, resulting in a purely monolithic system. Fabrication processes based on this type of 3-D circuit structure are discussed in Section 3.1. Alternatively, some ICs are prepared separately before the bonding process and bonded to form a 3-D system. Such a 3-D IC is a polylithic structure and related manufacturing processes are described in Section 3.2. Other techniques that provide contactless intertier communication have also been developed. Technologies for contactless 3-D ICs are reviewed in Section 3.3. One of the most important characteristics of polylithic 3-D ICs is the TSV, which provide an electrical connection among the circuits on different tiers, i.e., the intertier interconnects. Due to the significance of these vertical interconnects, Section 3.4 is dedicated to discussing the manufacturing process of the vertical vias and related issues. The technological implications of 3-D integration are summarized in Section 3.5.

3.1 MONOLITHIC THREE-DIMENSIONAL ICs

Monolithic or sequential 3-D ICs include stacking of circuits at the device level. Device level stacked 3-D ICs comprise layers of planar devices successively grown on a conventional CMOS or silicon-on-insulator (SOI) substrate. Several processes for monolithic devices were developed with moderate success in the 1980s; however, none of these processes reached high volume manufacturing primarily due to the decrease in manufacturing cost of transistors. However, as this cost has recently started to rise, interest in this technology has been revived. Past and more current efforts to manufacture high quality monolithic 3-D circuits are discussed in this section.

The first fabricated 3-D ICs were stacked bulk CMOS or SOI devices with simple logic circuitry. The devices comprising a logic gate can be located on different layers and, more importantly, manufactured with different technologies, such as CMOS or SOI. Independent of the technology utilized for the first device layer, these transistors are fabricated with conventional and mature processes. For the devices on the upper tiers, however, different fabrication methods are required. Several techniques, based on laser recrystallization or seed crystallization, are used to produce CMOS or SOI devices on the upper tiers and are described in Sections 3.1.1 and 3.1.2. The use of double-gate metal oxide semiconductor field effect transistors (MOSFETs) to form highly dense 3-D cells are discussed in Section 3.1.3. Recent methods on molecular bonding are presented in Section 3.1.4. Irrespective of the fabrication approach, the primary challenges for monolithic 3-D integration include fabrication of a high quality substrate for the upper layers, comparable performance of the devices in the bottommost layer with the performance of the FETs in the upper layers, and the use of low temperatures to manufacture FETs in these layers [96].

3.1.1 LASER CRYSTALLIZATION

Following the work in [97], several techniques using beam recrystallization were developed to successfully fabricate 3-D ICs. In these techniques, the first device layer is fabricated with a traditional CMOS or SOI process. Note that only a device layer is fabricated on the first tier. The interconnect layers are not fabricated at this initial stage. Fabrication of the transistors on the upper layers satisfies a twofold objective. The devices on the upper layers should exhibit satisfactory transistor electrical characteristics, such as field mobility, threshold voltages, and leakage currents (i.e., I_{on}/I_{off} ratio) and the characteristics of the transistors on the lower layer should not be degraded by the high temperatures incurred during the manufacturing process.

The first layer of devices is formed with a common MOS process. Depending upon the fabrication process, this layer can include both PMOS and NMOS devices [98] or only one of the two devices [99]. In the latter case, the complementary type of devices is fabricated exclusively on the upper layer(s). Alternatively, the upper layer(s) can include only SOI devices [100].

Prior to the development of the upper device layer, an insulating layer of SiO_2 is deposited [98]. To protect the devices of the first layer from the elevated temperatures during the recrystallization phase, a thick layer with an approximate thickness of 1 μm is deposited on top of the insulating SiO_2 layer. This layer is a standard feature of the recrystallization technique and is composed of various materials, such as polysilicon [98], phosphosilicate glass [99], and silicon nitride (Si_3N_4).

To grow the devices on the upper layers, polysilicon islands or thin polysilicon films are crystallized to single grains by an Argon laser [101]. The resulting 3-D IC is shown in Fig. 3.2, where the different layers are also indicated. The temperature during the recrystallization phase can be as high as 950°C [100], while device formation with a lower temperature of approximately 600°C has been demonstrated [98].

Since the temperature used in these manufacturing steps approaches or exceeds the melting point of the metals commonly used for interconnects, doped semiconductor materials are used to interconnect the devices located on different tiers. These materials can include n^+ doped polysilicon [98] or phosphorous-based polysilicon [100]. Alternatively, the interconnect layers can be fabricated after the upper layer devices are formed, where the contact holes are produced through reactive ion etching (RIE). The intertier interconnects are manufactured by sputtering aluminum [99]. A cross-section of a 3-D IC fabricated with aluminum interconnects is shown in Fig. 3.3.

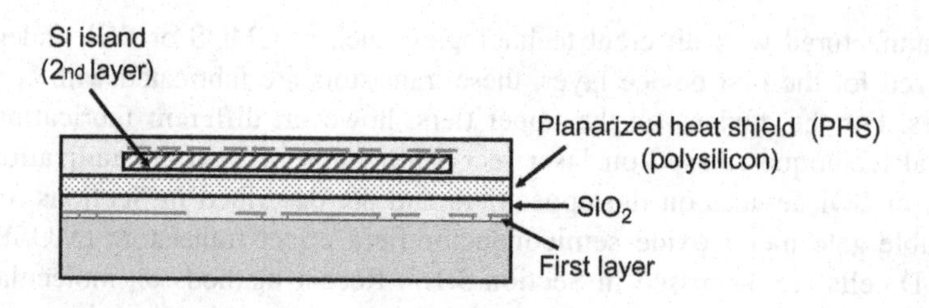

FIGURE 3.2

Cross-section of a stacked 3-D IC with a planarized heat shield to avoid degradation of the transistor characteristics on the first layer due to the temperature of the fabrication processes [98].

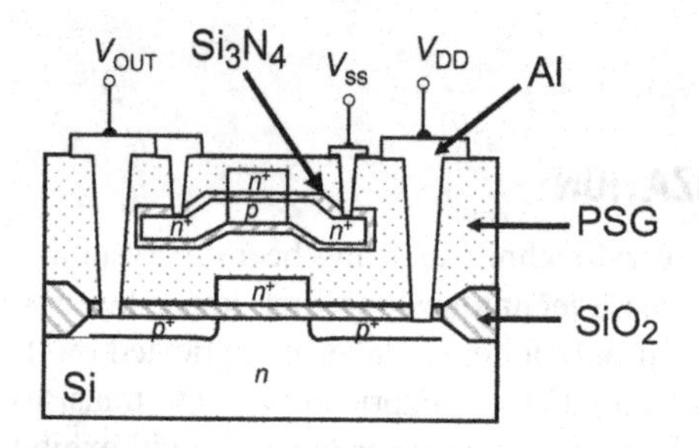

FIGURE 3.3

Cross-section of a device level stacked 3-D IC with a PMOS device on the bottom layer and an NMOS device in recrystallized silicon on the second layer.

The major drawbacks of the laser recrystallization technique are the quality of the grown devices on the upper layers and the effect of the high temperatures on the electrical characteristics of the devices on the lowest layer. Comparison of the device characteristics for the two layers shows that the threshold voltage is sufficiently controlled, while the mobility of the devices on the upper layer is slightly worse than the transistors on the bottom layer [99]. Since the mobility of the devices on the upper layers is degraded, the PMOS devices which exhibit lower mobility are fabricated on the lower layer [46]. In addition, simple shift registers [98] and ring oscillators [99] have been shown to operate correctly, demonstrating that the devices on the first layer remain stable despite the high temperature steps utilized to manufacture the devices on the upper layer.

More advanced recrystallization techniques can provide up to three device layers, where the lowest layer is a CMOS layer while the other two layers are composed of SOI devices [100]. Since there are three device layers in this structure, two isolation layers are required to protect the devices on the lower layers. Measurements from sample structures demonstrate that the threshold voltage of the devices is well controlled with low subthreshold currents for the devices on the upper layers [100]. The mobility of these transistors can, however, vary significantly.

Alternatively, E-beam has been used to recrystallize the polysilicon to form the devices on the upper layers [101]. A three-layer prototype 3-D IC performing simple image processing has been demonstrated [102]. On the topmost layer, consisting of amorphous-Si (a-Si), the light is captured and converted to a digital signal and stored on the bottom layer composed of the bulk-Si transistors.

The intermediate layer, which includes the SOI transistors, compares the digital data from one pixel to the digital data in adjacent pixels to produce an edge detection operation. Measurements show a narrow distribution of the threshold voltage of the devices on the SOI layer with satisfactory $I-V$ device characteristics. Although correct operation of the circuitry is confirmed, demonstrating the capabilities of a monolithic 3-D circuit, 34 mask steps are required, including five layers of polysilicon and two layers of aluminum interconnects. Although several recrystallization approaches have been developed, none of these techniques has demonstrated the capability of manufacturing complex, high performance 3-D circuits, primarily due to the inferior quality of the devices on the upper layers of the 3-D stack.

3.1.2 SEED CRYSTALLIZATION

Another technique to fabricate multiple device layers on bulk-Si is crystallizing a-Si into polysilicon grains. Thin film transistors (TFTs) are formed on these grains. The seed utilized for recrystallization can be a metal, such as nickel (Ni), or another semiconductor, such as germanium (Ge).

The basic processing steps of this technique are illustrated in Fig. 3.4. A film of amorphous silicon is deposited with low pressure chemical vapor deposition. A second film of low temperature oxide (LTO), SiO_2, is deposited and patterned to form windows at the drain or both drain and source terminals of the devices for Ge or Ni seeding. Consequently, two kinds of TFTs are produced. One type is seeded only at one terminal while the second type is seeded at both terminals.

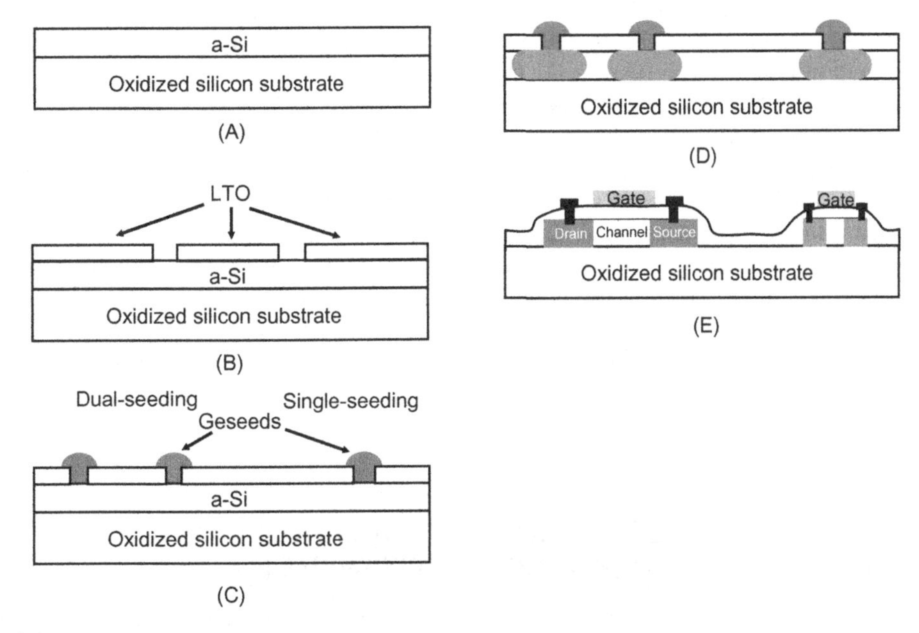

FIGURE 3.4

Processing steps for laterally crystallized TFT based on Ge-seeding. (A) Deposition of amorphous silicon, (B) creating seeding windows, (C) deposition of seeding materials, (D) producing silicon islands, and (E) processing of TFTs [103].

The deposition of the LTO SiO_2, patterning, and the deposition of the seed are additional steps as compared to a conventional process for TFTs. Thermal annealing is necessary to completely crystallize the channel films. Finally, the LTO and seeds are etched, permitting the TFTs to be fabricated with a standard process. A thermal layer of SiO_2 is used as the gate dielectric, and boron and phosphorous doping is used for the junctions. The gate electrode is formed by *in situ* doped polysilicon and the interconnections are fabricated by modified interconnect and plug technologies from conventional 2-D circuits. The peak temperature of the process is 900°C.

Certain factors can degrade the quality of the fabricated TFTs, such as the size of the grains and the presence of defects within the channel region. Controlling the size and distribution of the grains contributes significantly to the quality of the manufactured devices. Comparisons among unseeded devices, produced by crystallizing a-Si, single seeded, and dual seeded TFTs, have shown that seeded devices exhibit enhanced device characteristics, such as higher field effect mobility and lower leakage currents. The greatest improvement in these characteristics is achieved by dual seeded TFTs. However, as compared to single seeded transistors, the improvements of dual seeded transistors diminish with device size. Consequently, for small devices where the grain size of a single seeded device is approximately the entire channel region, the performance of those TFTs is close to double seeded TFTs as grain boundaries are unlikely to appear within the channel. Grain sizes over 80 μm have been fabricated [104]. High performance TFTs formed by lateral crystallization on the upper layers exhibit low leakage currents and high mobility as compared to the SOI devices on the first layer, while significantly reducing the area of the logic gates [103,104].

Selective epitaxial growth (SEG) and epitaxial lateral overgrowth (ELO) have also been combined to fabricate multiple layers of SOI devices [105,106]. The various steps of these techniques are summarized in Fig. 3.5. The first device layer is formed on a thick layer of SiO_2. The oxide is patterned with photolithography and etching procedures to define the SOI islands (Fig. 3.5A). A thinner SiO_2 layer, acting as the insulator, is deposited (Fig. 3.5B). With photolithography, a window for SEG is opened within the silicon island patterns (Fig. 3.5C). Using SEG and ELO, the patterns are filled by vertical and lateral growth of silicon within the SEG window (Fig. 3.5D). The redundant silicon is etched with chemical mechanical planarization (CMP) (Fig. 3.5E). A second layer of devices is fabricated with the same process, unlike oxide deposition, which is achieved with plasma enhanced chemical vapor deposition. The transistors are fabricated on islands using conventional SOI techniques (see Figs. 3.5F to H). The smaller island has an area of 150 nm × 150 nm, enabling a high degree of integration on a single die [105]. The photolithography and etching steps limit the dimensions of the SOI islands. Satisfactory device characteristics have been reported with stacking faults only appearing on the first device layer [105].

3.1.3 DOUBLE-GATE METAL OXIDE SEMICONDUCTOR FIELD EFFECT TRANSISTORS FOR STACKED THREE-DIMENSIONAL ICs

Similar to 2-D circuits where standard cell libraries include a compact physical layout of individual logic gates or more complex logic circuits, standard cell libraries for 3-D circuits include "volumetric" logic cells with minimum volume and a smaller load as compared to 2-D cells. Fabrication techniques that use local clusters of devices to form standard 3-D cells have been developed [107] based on double-gate MOSFETS [108,109]. The major fabrication stages for a 3-D inverter cell are

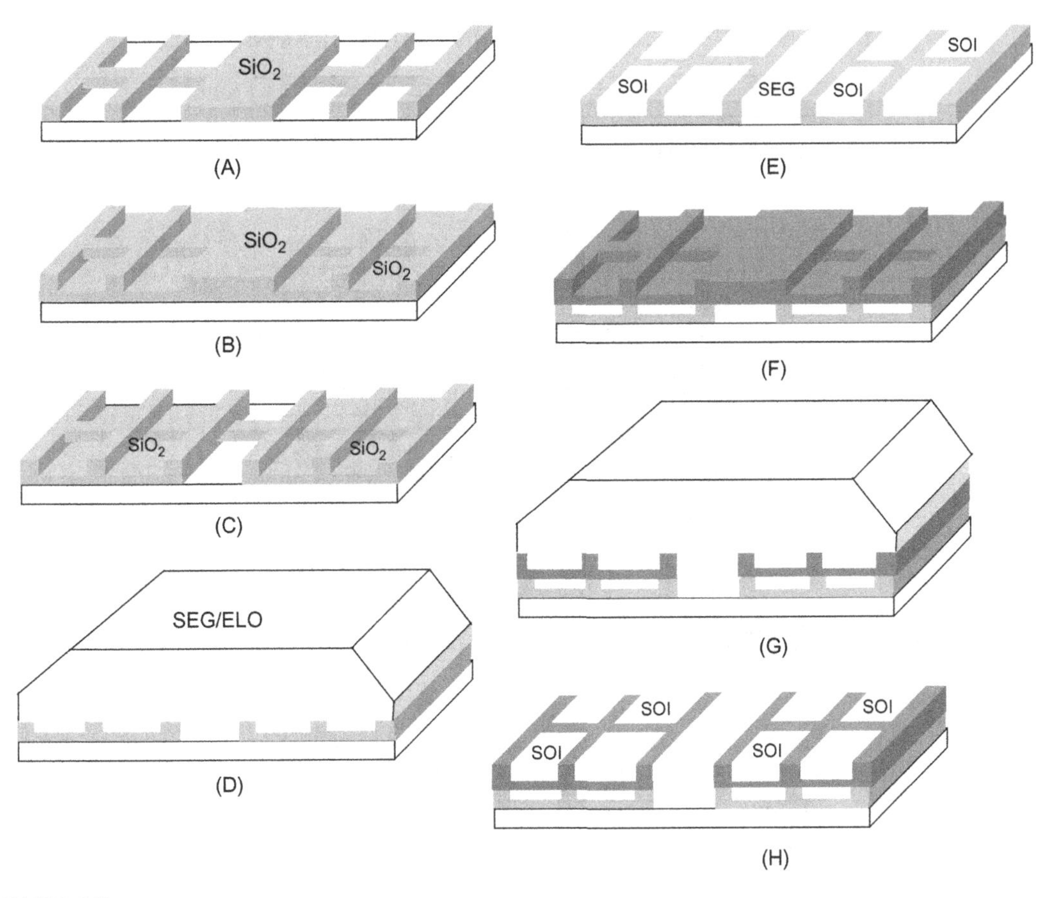

FIGURE 3.5

Processing steps for vertical and lateral growth of 3-D SOI devices [105]. (A) Definition of SOI islands, (B) silicon dioxide deposition, (C) formation of SEG window, (D) silicon growth within the SEG window, (E) etching of redundant silicon with CMP, (F) definition of upper device layer, (G) deposition of upper layer, (H) formation of SOI islands on the upper tier.

illustrated in Fig. 3.6, where a decrease of 45% in the total capacitance is achieved. Similar improvements are demonstrated for more complex circuits, such as a 128-bit adder, where a 42% area reduction is observed [107].

An SOI wafer is used as the first layer of the 3-D IC. The top silicon layer is thinned by thermal oxidation and oxide etching. A thin layer of oxide and a layer of nitride are deposited as shown in Fig. 3.6A. The wafer is patterned and shallow trench etching is used to define the active area (Fig. 3.6B). A layer of LTO is deposited to fill the trench, which is planarized with CMP, where the nitride behaves as the stop layer (Fig. 3.6C). The nitride film is removed and LTO is utilized as a dummy gate. The LTO, silicon, and oxide are patterned and etched, followed by depositing a nitride film (Fig. 3.6D). The drain and source regions for the first device layer are created with an

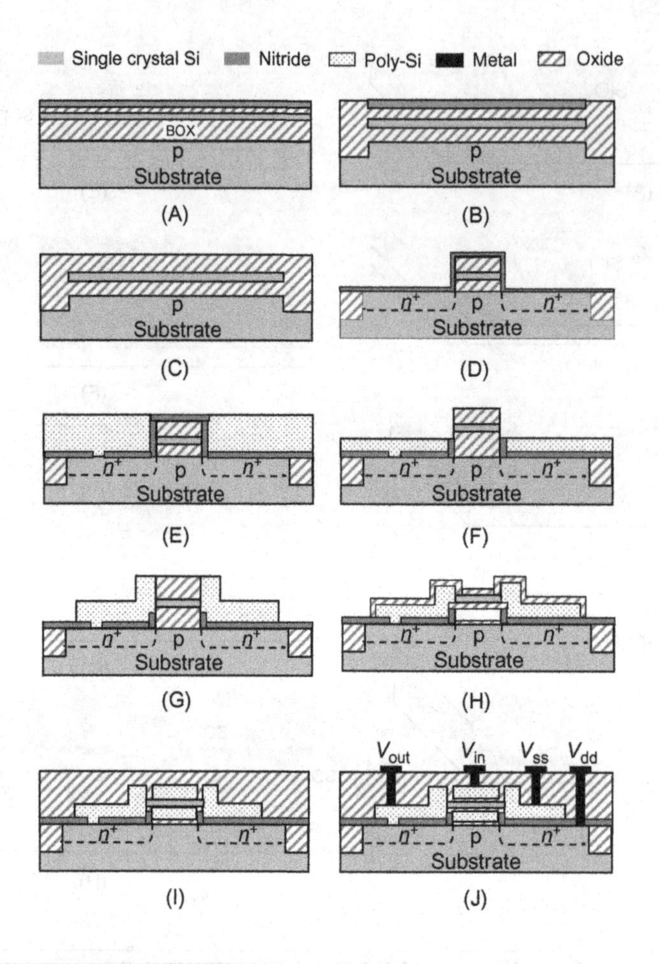

FIGURE 3.6

Basic processing steps for a 3-D inverter utilizing the local clustering approach [107]. (A) Oxide deposition, (B) wafer patterning and active area definition, (C) low temperature oxide deposition, (D) deposition of nitride film, (E) via formation at the drain side, (F) etching of the nitride film, (G) boron doping, (H) active area definition, (I) gate oxide growth by thermal oxidation, (J) deposition of doped polysilicon.

As^+ implantation. As depicted in Fig. 3.6E, a via is opened at the drain side to connect the terminals of the NMOS and PMOS transistors located on the second layer. The contact is implemented as a tunneling ohmic contact. A polysilicon layer is deposited and planarized by CMP utilizing the nitride film as the stop layer. The nitride film is etched to expose the silicon for the channel region (Fig. 3.6F). Boron doping is used to fabricate the source and drain regions of the PMOS transistor on the top layer (Fig. 3.6G). The active area for the devices is defined by removing the LTO and the buried oxide (Fig. 3.6H). Before forming the gates for the stacked transistors, the gate oxide is grown by thermal oxidation, as shown in Fig. 3.6I. Finally, *in situ* deposition of doped polysilicon forms the gate electrodes. Note that for the gate below the channel region of the PMOS transistor (see Fig. 3.6J), a slow deposition rate is required to completely fill this region.

3.1.4 MOLECULAR BONDING

Monolithic 3-D circuits utilizing molecular bonding demonstrate several advantages over the previously described approaches. An important trait of this technique is that an entire wafer can be transferred onto the bottom layer at low temperature (200°C), where the transistors for the upper layers are formed. There are several advantages associated with this approach, for example, the quality of the substrate, which is not limited by defects as in the case of seed crystallization and recrystallization-based methods. Furthermore, the integration density is the same as the bottommost layer, whereas the seed windows and the polycrystalline nature of the substrate in other approaches limit this density. Additional benefits of a high quality monocrystalline substrate are better bonding and improved control of the substrate thickness [110].

Although this technique includes the transfer of the substrate onto another wafer, resembling the case of 3-D integration with TSVs, the devices on the upper layers are not yet fabricated, which greatly differentiates the manufacturing process. The manufacturing process includes a number of sequential steps, which commences with the fabrication of the devices in the bottom layer with standard fully depleted silicon on insulator (FDSOI) processes. These steps are summarized in Fig. 3.7. A thin layer of dielectric is deposited and planarized before the transfer of the substrate. The thickness of the dielectric layer is an important factor, which can be adapted to exploit the coupling between the stacked devices [111], where thicknesses ranging from 60 nm to 110 nm

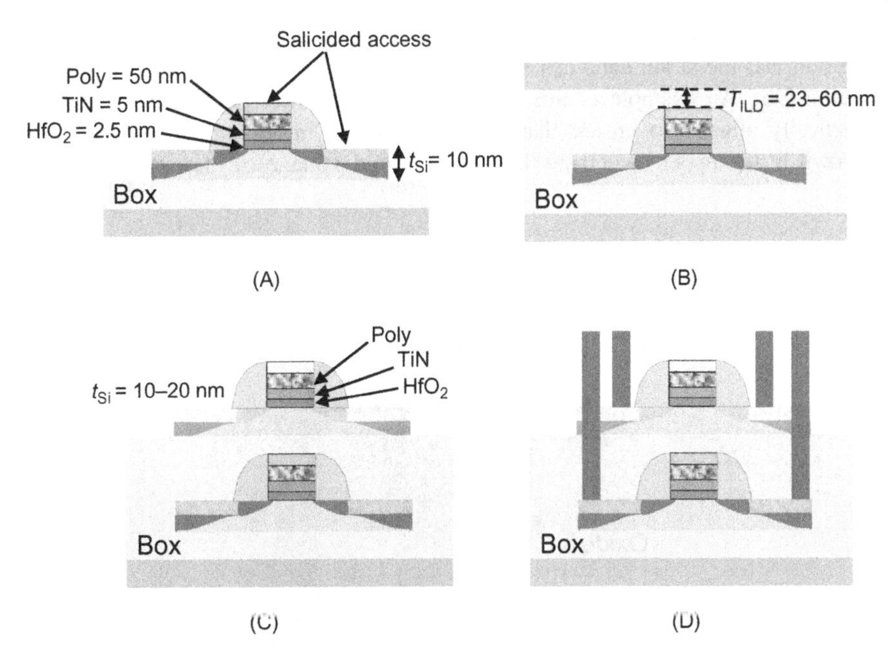

FIGURE 3.7

Sequential process for fabricating monolithic 3-D circuits, where (A) the SOI devices in the first layer are manufactured with standard SOI processes, (B) molecular bonding allows the transfer of a high quality substrate, (C) the devices in the upper layers are formed, and (D) metal contacts connect the device layers [110].

have been tested. In the next step, molecular bonding is used to transfer an SOI substrate (see Fig. 3.7B) (other substrates can also be transferred, which makes this technology appealing).

Having a high quality substrate, the devices are formed where the orientation of these devices can be controlled to achieve specific transistor characteristics [110,111]. Furthermore, non-Si devices, such as Ge p-MOSFETs [112] and devices other than bulk MOSFETs, can be grown, including FinFETs and trigate transistors [113,114] (see Fig. 3.7C). One critical step for the performance of the transistors in the upper layers is activation of the dopants, which for the bottom layer is achieved with high temperatures (e.g., >1000°C) [96]. These temperatures, however, are prohibitive for thermal activation of the dopants in the upper layer(s). Consequently, solid phase epitaxy regrowth is applied at 600°C irrespective of the type (n or p) of the transistor. The last major step of the process includes the interlayer interconnects, which are similar to standard contacts used in the BEOL of planar circuits (see Fig. 3.7D). Case studies with tungsten contact plugs have been demonstrated [96]. The similarity of an interlayer contact with a typical contact between two metal layers is illustrated in Fig. 3.8.

Since different types of devices can be manufactured, several applications can be supported by monolithic 3-D circuits. This situation is also due to monolithic 3-D technology able to support integration at both the device and cell level. Thus, the integration of different device types leads to heterogeneous applications, while integration at the cell level leads to considerable gains in performance and area for the logic and memory circuits. The disparate applications enabled by this technology are listed in Table 3.1.

Integrating monolithic circuits at the device level allows for adapting the behavior of the devices in several ways. For example, the choice of substrate, the material used for the channel, the orientation of the crystal, and the strain can each be used to produce devices that are best adapted for the target application [96]. An example of this approach is 4T SRAM (loadless) cells, where the transistors are vertically aligned to ensure that the stored bit of information back biases the access transistors (grown in the upper layer), shifting the threshold voltage of these transistors [115]. The

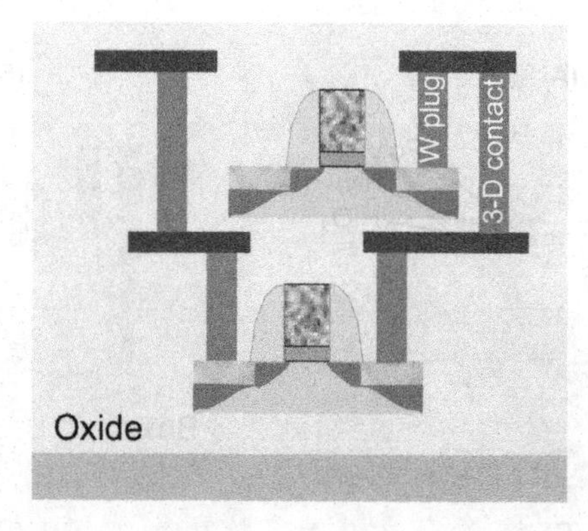

FIGURE 3.8

Monolithically stacked devices where the interlayer contacts (3-D contact) and the standard metal contacts (tungsten plug) connecting the devices are illustrated. The 3-D contact has similar traits to a standard contact connecting two metal layers [96].

Table 3.1 Potential Applications for Monolithic Integration at the Device Level [110]

Application	Layer	Example of Partitioning	Best Suited Technology
Field programable gate array (FPGA)	1	Pass gate	High performance transistors
	2	6T SRAM	Low standby power transistors
Highly miniaturized pixels	1	Photodiode and transfer gate	1 μm thick SOI with backside illumination
	2	Readout transistors	Low noise transistors with relaxed (*L*, *W*) and gate oxide
CMOS gates	1	nFET	nFET gate stack, tensile-Si or InGaAs
	2	pFET	pFET gate stack, compressive-Si or Ge or (110) Si

transistors should be carefully aligned to modulate the threshold voltage of the devices in the upper layer based on the applied gate voltage to the devices in the lower layer. Furthermore, the thickness of the interlayer dielectric (ILD) separating the two layers of transistors should be appropriately adapted. Simulations have shown that if this thickness is 300 nm, no noticeable change in the threshold voltage is observed due to the gate voltage of the transistors in the bottom layer. If this thickness decreases to 10 nm, the shift in voltage threshold can be as high as 130 mV [115].

The benefit of this approach is improved stability of the memory cell. The read noise and static noise margins are, respectively, 320 and 150 mV [115]. Additionally, the noise margins are achieved without any modification to the size of the devices. Both the transistor width and length are, therefore, the same for all four of the transistors in the SRAM cell.

Although transistor level integration is beneficial in this specific example, significant and higher gains are demonstrated when integration occurs at the cell level. In this approach, cells are placed on top of other cells, yielding more compact circuits. To compare the gains produced from each integration technique, the area savings and performance improvements are evaluated for some basic logic gates and circuits. A flip flop in these two styles exhibits a density gain of 28.5% and 38% for integrating, respectively, nFETs on top of pFETs and CMOS gate-on-gate integration. For more complex circuits, such as a 16-bit multiply accumulate unit, the savings are, respectively, 42% and 52% for the two integration styles [112]. Alternatively, if the area is maintained constant, increased performance can be achieved. Thus, for a NAND gate with a driving strength of 8×, the gain in current is 20% and 100%, respectively, when integrating at the transistor and cell level.

These examples demonstrate that integration at the cell level offers considerable improvements. Application of this integration style has also been applied to FPGA fabrics, where the fabric is partitioned to ensure that the memory cells are placed on the bottom layer whereas the logic cells are placed on the upper layer. The savings in area and energy delay product are, respectively, 55% and 47%.

Although monolithic integration exhibits appealing characteristics, integrating several layers of devices with a single stack of interconnect layers (i.e., BEOL) is a challenging task. Alternatively, fabricating a 3-D system with vertically bonded ICs or wafers that are individually processed can reduce the total manufacturing time without sacrificing the quality of the devices on the upper tiers of the stack. Consequently, parallel or polylithic integration, through where circuits from individual wafers are stacked to form multi-tier systems, is discussed in the following sections.

3.2 THREE-DIMENSIONAL ICs WITH THROUGH SILICON VIA OR INTERTIER VIA

Wafer or die level 3-D integration techniques, which utilize TSVs, are appealing candidates for 3-D circuits. Intertier vias offer the greatest possible reduction in wirelength with vertical integration. In addition, each tier of a 3-D system can be processed separately, decreasing the overall manufacturing time. As each tier of the 3-D stack is fabricated individually, a high yield process can be utilized for each tier suitably tailored to the nature of the circuit in that tier, e.g., memory or logic. A broad spectrum of fabrication techniques for 3-D ICs has been developed. These techniques support either wafer or die level 3-D stacking and are discussed, respectively, in Sections 3.2.1 and 3.2.2. The different bonding approaches and related issues are discussed in Section 3.2.3.

3.2.1 WAFER LEVEL INTEGRATION

Although none of the developed wafer level techniques has been standardized, nearly all of these methods share similar fabrication stages, which are illustrated in Fig. 3.9. The order of these stages, however, may not be the same or some stages may not be used, depending upon the TSV fabrication scheme and the bonding approach and mechanism.

Initially, CMOS or SOI wafers are separately processed to produce the physical tiers of the 3-D stack, while a certain amount of active area is reserved for the TSVs. The TSVs are etched and filled with metal, such as tungsten (W) or copper (Cu), or even low resistance polysilicon. To decrease the length of the TSVs, the wafers are attached to an auxiliary wafer, usually called a "handle" or "carrier" wafer, and thinned to a different thickness depending upon the target aspect ratio for the TSVs. After the TSVs are revealed, backside metallization provides electrical connections between the tiers. The alignment and bonding phase follows, as shown in Fig. 3.9D. Finally, the handle wafer is removed from the thinned wafer and, if required, the appropriate side of the wafer is processed and attached to another tier. A broad gamut of materials and methods exist for each of these phases, some of which are listed in Table 3.2.

Several of the proposed techniques for 3-D ICs support the integration of both CMOS and SOI circuits [120,122]. From a fabrication point of view, however, SOI facilitates the wafer thinning step as the buried oxide (BOX) serves as a natural etching stop layer. This situation is due to the high selectivity of the etching solutions. Solutions with a Si to SiO_2 selectivity of 300:1 are possible [116]. In addition, SOI technology can yield particularly thin wafers or tiers (<10 μm), resulting in short TSVs. Alternatively, SOI circuits inherently suffer from poor thermal properties due to the low thermal conductivity of the oxide [124].

The formation of the TSVs is an important issue in the design of high performance 3-D ICs. This fabrication stage includes opening deep trenches through the silicon substrate (and potentially the inter layer dielectric and metal layers), passivation of the trench sidewalls from the conductive

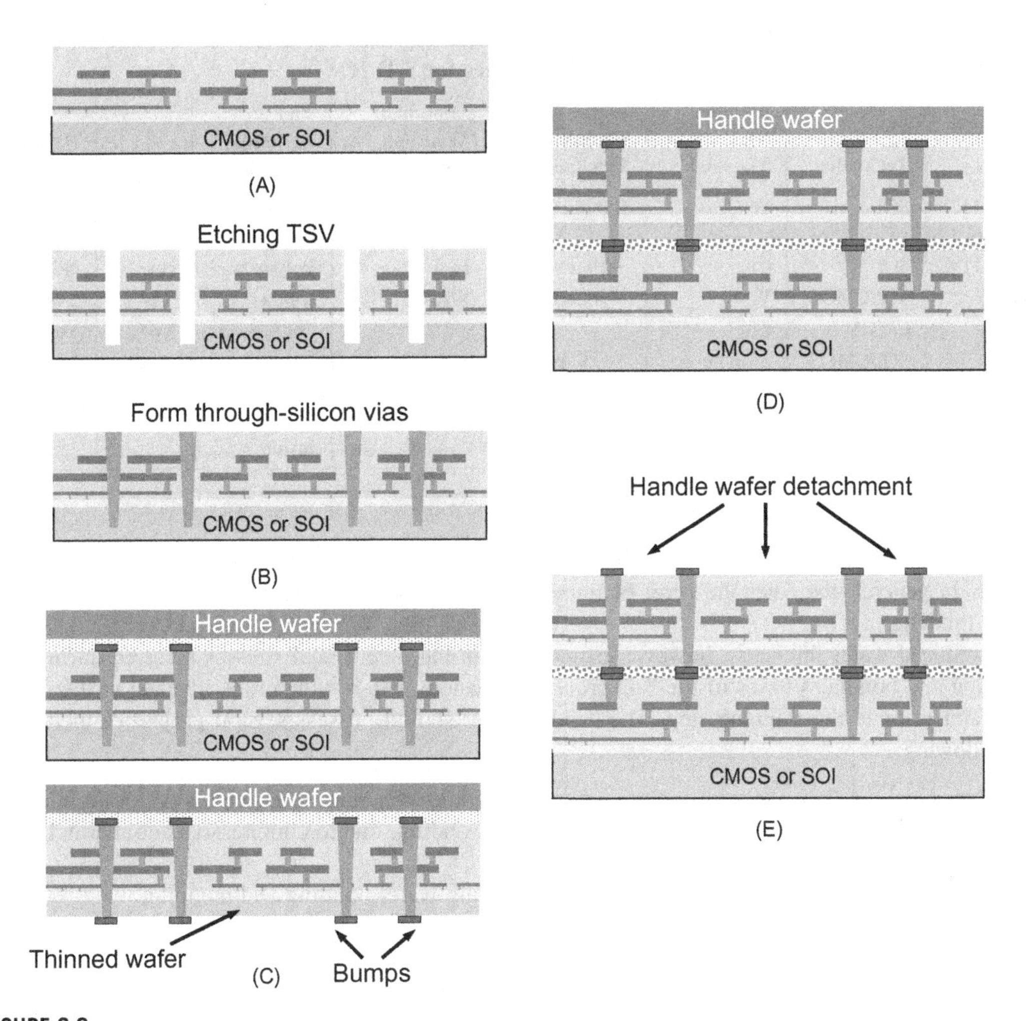

FIGURE 3.9

Typical fabrication steps for a 3-D IC process. (A) Wafer preparation, (B) TSV etching, (C) wafer thinning, bumping, and handle wafer attachment, (D) wafer bonding, and (E) handle wafer removal.

substrate (for CMOS circuits), and (partially or fully) filling the opening with a conductive material to electrically connect the tiers of the 3-D system. Due to the importance of this fabrication stage a more complete discussion of TSVs follows in Section 3.4.

Before stacking the tier, the wafers are thinned to decrease the overall height of the 3-D system and, therefore, the length of the TSVs. The diameter and length of the TSVs are interdependent and are typically characterized by the TSV aspect ratio (length/diameter). Processes that produce TSVs with aspect ratios over 20:1 [125] have been demonstrated. Typical ratios range from six to ten [126,127], where six is a common ratio. Furthermore, the minimum diameter of the TSVs depends

Table 3.2 Characteristics of Fabrication Techniques for 3-D ICs

Process From	IC Technology	Intertier Vias Material	Wafer Thinning Thickness	Alignment Accuracy (μm)	Tier Bonding Material	Bumps	Handle Wafer
[116,117]	SOI	W	500 nm	±3	Cu–Cu pads	Yes	Yes
[118,119]	SOI	Cu	~10 μm	1–2	Polymers	No	No
[120]	SOI/CMOS	W	~10 μm	N/A	Cu/Sn	Yes	Yes
[121]	CMOS	Cu		<2	Polymer	No	Yes
[122]	SOI/CMOS	n^+ poly – Si/W	7–35 μm	±1	Epoxy adhesive	Yes	Yes
[123]	SOI	Cu	~ 2 μm	<1	Oxide fusion	No	Yes

upon the preferred integration granularity in addition to the capabilities of the manufacturing processes. In general, the finer the level of integration, the more aggressive the TSV aspect ratio or wafer thinning process.

A reduced wafer thickness, however, cannot sustain the mechanical stresses incurred during the handling and bonding phases of the 3-D process. A handle wafer is therefore attached to the original wafer prior to the thinning step. The handle wafer should possess several properties including the following:

- Mechanical durability to withstand the mechanical stresses incurred during the wafer thinning process due to grinding and polishing, and during bonding, due to compressive and thermal forces.
- Thermal endurance to processing temperatures during wafer bonding and bumping.
- Chemically inert to the solutions employed for wafer thinning and polishing.
- Simple and fast removal of the thinned wafer with appropriate solvents.
- Precise wafer alignment; for example, by being transparent to light.

Thinning can proceed with a variety of methods, such as grinding and etching or both, accompanied by polishing. Various combinations can also be applied, such as a silicon wet etch followed by CMP, mechanical grinding with CMP, mechanical grinding succeeded by a spin etch, and dry chemical etching [119]. Several requirements relating to the wafer thinning step also exist, including [128]:

- Few defects and damage during thinning
- Few defects and damage during edge trimming
- Careful control of the thinning process to ensure uniform wafer thinning until the desired thickness is achieved. The total thickness variation (TTV) should remain low at the target thickness.

Accurate alignment of the thinned wafers is a challenging task. A variety of techniques are utilized to align the wafer based on the precise registration of the alignment marks. These mechanisms include infrared alignment, through wafer via holes, transparent substrates, wafer backside alignment, and inter-substrate alignment [129]. Typical alignment precisions range from 1 to 5 μm (see Table 3.2). Submicrometer accuracies have also been reported [123,129].

Once the wafers are aligned, the bonding step follows. Disparate methods to perform this step are listed in Table 3.3. If die-to-wafer or die-to-die bonding is preferred to avoid the integration of faulty dies, each wafer is diced prior to bonding and the dies are successively bonded. An increase in the turnaround time, however, is inevitable with this approach as discussed in the following subsection.

3.2.2 DIE-TO-DIE INTEGRATION

With die-to-die 3-D stacking, the wafer is diced and each die is separately tested. The known good dies are placed on supporting wafers and these carrier wafers are bonded, producing a two-tier system. This approach has a better manufacturing throughput than individual die-to-die bonding but a lower throughput than direct wafer level bonding where some of the dies can be defective.

Another approach to allow multiple die-to-die stacking rather than only two-tier systems is the cavity alignment method where multiple tiers are assembled, improving manufacturing throughput. The method utilizes a silicon substrate with metal layers, which provides mechanical support for the entire stack. A template that forms the cavity is deposited with the use of an optical microscope to obtain precise placement of the template. The template is composed of a metal plate with a

Table 3.3 Comparison of Bonding Technologies [130]

	Metal Bonding				
Bonding Type	**C4 Bonding**	**Intermetallic Compound Bonding**	**Cu–Cu Bonding**	**SiO_2 Bonding**	**Hybrid Bonding (Metal and Adhesive)**
Bonding temperature	~260°C	~260°C	~400°C	Room temp.	~300°C
Heat tolerance	<260°C	<450°C*	<1084°C	<1400°C	<400°C*
Connectivity	Mechanical and electrical	Mechanical and electrical	Mechanical and electrical	Mechanical	Mechanical and electrical
Interconnection pitch	Low	Middle	Middle	High	Middle
Chip level applicability	High	High	Medium*	NA	Medium*
Wafer level applicability	Low	Medium*	Medium*	High	Medium*
Issues	Low heat tolerance, large I/O pitch	Intermetallic compound (IMC) thickness control, reliability	Thermal and compression stress, flatness, cleaning	Flatness, voids/particles	High bonding pressure, processing, integrity, flatness

**The applicability can vary depending upon the detailed steps of the process and the requirements of the aimed application.*

thickness of 0.5 mm and exhibits a small coefficient of thermal expansion (CTE). Upon alignment, the template is bonded to the substrate with a removable adhesive [130]. The tiers of the system are placed sequentially into the cavity where bonding is achieved by compression. A final bonding step is applied to the entire system based on thermal compression, where a typical applied temperature is 250°C and the exerted force is 10 N. The process is completed by removing the template. The stages of the process are illustrated in Fig. 3.10. The alignment accuracy of this method is within ± 1 μm.

Other die-to-die integration techniques that can produce up to four-tier systems have also been demonstrated, which, although based on conventional bonding through thermal compression, utilize scrubbing and heating to enhance bonding strength [131]. In this process, the first die is of regular thickness, while the upper tiers are thinned to 50 μm. Scrubbing improves the TTV of the bump height of the TSV, and the surface flatness of the copper bumps. With the introduction of a scrubbing force the bonding temperature can be lowered to less than 200°C, while the pressure is also decreased to less than 100 MPa. Key parameters of this technique are the scrubbing time and force [131]. Tests on daisy chain connections of TSVs bonded with this approach demonstrate a resistance of approximately 170 mΩ, close to the theoretically expected resistance of 113 mΩ [131].

Alternatively, wafer scale bonding exhibits the highest manufacturing throughput. Wafer level bonding can be applied with a high initial yield for each tier within a 3-D system (the overall yield can be affected by later processing steps, e.g., bonding). Thus, wafer scale stacking results in an acceptable overall yield for the 3-D system, assuming the yield for each tier is high.

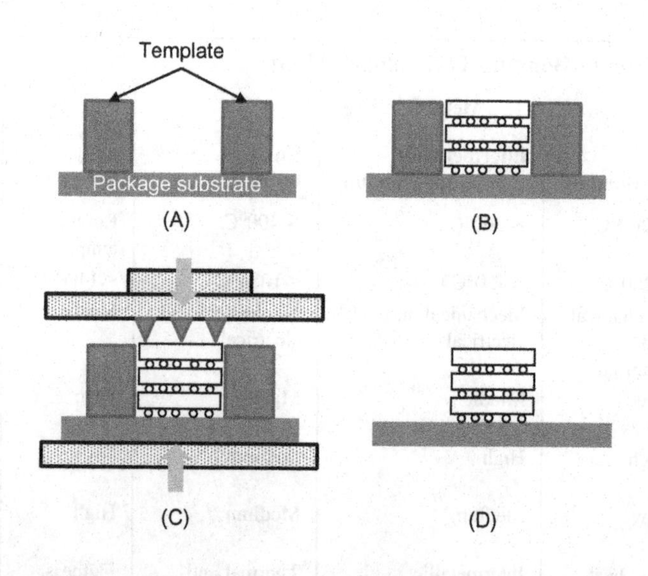

FIGURE 3.10

The cavity alignment method, (A) the cavity template is aligned and bonded to the substrate, (B) the individual tiers are placed in the cavity through compression, (C) the 3-D stack is assmbled through thermal compression, and (D) the cavity template is removed [130].

3.2.3 BONDING OF THREE-DIMENSIONAL ICs

In addition to the level of bonding, different bonding styles also exist. For example, back-to-face tier bonding is illustrated in Fig. 3.9D. In addition, face-to-face bonding can be utilized to efficiently bond two tiers. An advantage of face-to-face bonding is that if only two tiers are bonded, the TSVs can be removed, as shown by the process illustrated in Fig. 3.11. In this process, the "mother" circuit is considerably larger and provides the interconnections to the package. For this reason, two different types of bumps are utilized. Microbumps with a fine pitch (25 to 35 μm) connect the two tiers, while Cu pillars at a 185 μm pitch interconnect the two tier system to the package. Adding, however, a third tier to a two tier stack is not straightforward since TSVs are required. In this case, the third (or other) tier is added in a back-to-face manner, which requires TSVs. Back-to-back bonding is also possible, resulting, however, in longer and, therefore, lower performance TSVs.

Bonding can be achieved with adhesive materials, metal-to-metal bonds, oxide fusion, and eutectic alloys deposited among the tiers of a 3-D system [117]. A list of the traits and challenges of these disparate bonding technologies are reported in Table 3.3. Epoxies and polymers possess good adhesive properties and are widely used. Some of these polymers are listed in Table 3.4. Metal-to-metal bonding requires the growth of (square) bumps on both candidate tiers for stacking

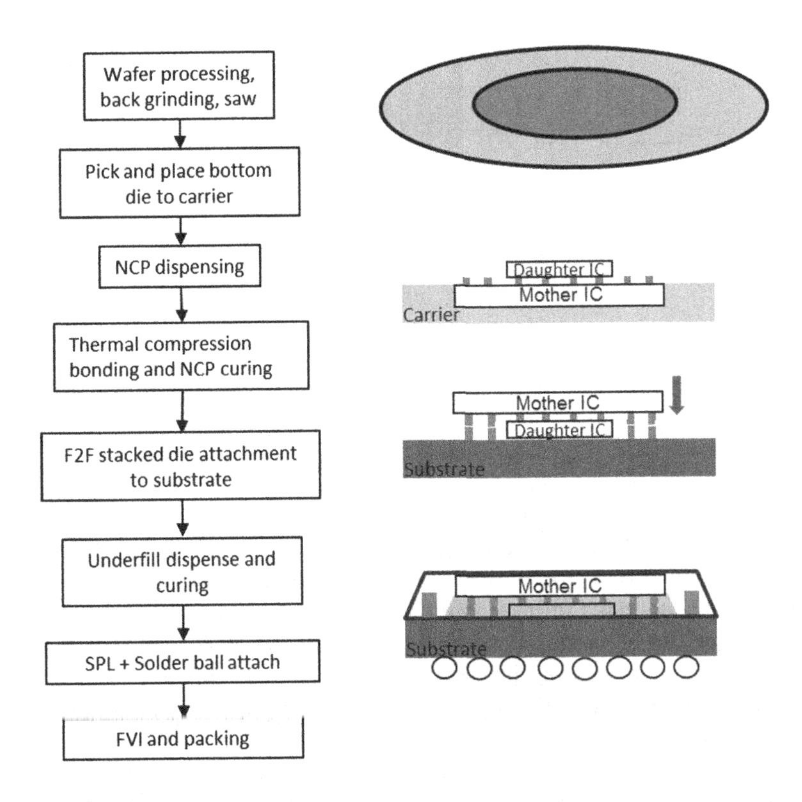

FIGURE 3.11

Process for face-to-face bonding and substrate assembly, removing the need for TSVs [132].

Table 3.4 Several Materials Used for Wafer Bonding

Polymer
Polyarylethel (FLARE) [135]
Methylsilesequionexane [136]
Benzocyclobutene (BCB) [137]
Hydrogensilsesquioxane [138]
Parylene-N [139]

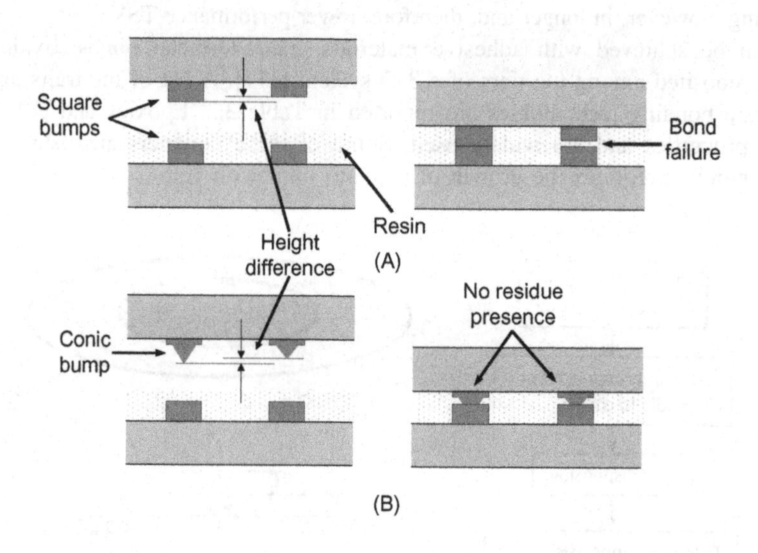

FIGURE 3.12

Metal-to-metal bonding; (A) square bumps, and (B) conic bumps for improved bonding quality [133].

and can consist of, for example, Cu, Cu/Ta, In/Au, and Cu/Sn. A portion of these bumps is deposited at the edge of the TSVs, providing enhanced electrical connection between the tiers. In the case where metals with different melting points are utilized (for example, Cu and Sn), the metal with the lower melting point is placed between the Cu pillars with the higher melting point [130]. An inter-metallic compound (IMC) is formed by applying thermal compression to ensure that the metal with the lower melting point diffuses into the metal with a higher melting point.

An epoxy filler can also be used to reinforce the bonding structure. An issue that arises in this case, however, is that the resin can result in bonding failure due to differences in the height of the metal bumps. To avoid these failures, pyramid or conic shaped bumps have been proposed [133]. The compressive bonding force causes the bumps to spread, excluding the resin from the bond region, as illustrated in Fig. 3.12. Eutectic solders are also used for tier bonding, where tin is plated on the tier surface followed by heating and compression [120]. Additionally, the temperature profile depends upon the material used for the interface and typically ranges from 200 to 400°C to not degrade the copper interconnections. Other highly refractory metals, such as tungsten or doped polysilicon, can be used if higher temperatures are necessary, increasing, however, the resistance of the TSVs.

Oxide fusion is another typical technique for bonding, where bonding films are used for this purpose. The advantages of oxide bonding include avoidance of high stresses due to the thermal compression, higher tolerance to the alignment accuracy thereby enhancing reliability, and the suitability of this approach for dense and fine pitch TSVs [128,134]. Alternatively, issues with oxide bonding primarily relate to the bonding films, where key requirements include [128,130]

- Careful control of the film composition;
- High density bonding films;
- Smooth surface of the oxide bonding films at the atomic level to strengthen the bonding density and limit the presence of nanovoids at the bonding interface;
- Clean surface of the bonding films to avoid microvoids due to the presence of particles from foreign materials and process residues.

Depending upon the bonding mechanism and material, certain requirements should be satisfied to avoid delamination and cracking of the tiers:

- Small mismatch between the CTE of the bonding material and the tiers
- Mechanical endurance in the later processing steps
- Minimum wafer warpage
- No outgassing of the adhesives from heating, which can result in void formation
- No void generation due to the presence of residues on the layer surfaces

Achieving a reliable bond throughout the lifetime of a 3-D stack is an important requirement for any bonding and/or assembly process. A host of tests based on industry standards are applied to ensure the reliability of the manufactured 3-D stack [140]. These reliability tests include [130,141]:

- Temperature cycling (e.g., −40 to 125°C)
- High temperature storage (for example, at 150°C)
- Humidity test (e.g., 85°C at 85% RH)
- Pressure cooker test (e.g., 121°C at 100% RH)
- Unbiased highly accelerated stress test (e.g., 130°C, 85% RH)
- Razor tests, where a razor blade is used to penetrate the interface between the tiers, while other tests can include bending forces applied to the bonded wafers [118].

The final processing step for a 3-D IC with TSVs is removal of the handle wafer and cleaning the tier surface if subsequent bonding is required. A variety of solutions can be used to detach the handle wafer. The time to accomplish this step also depends upon the size of the wafer [120].

3.3 CONTACTLESS THREE-DIMENSIONAL ICs

Although the majority of the processes developed for 3-D ICs utilize vertical interconnects with some conductive material, other techniques provide communication among circuits located on different tiers through coupling of electric or magnetic fields. Capacitively coupled circuits are presented in Section 3.3.1. Inductive vertical interconnects are discussed in Section 3.3.2.

3.3.1 CAPACITIVELY COUPLED THREE-DIMENSIONAL ICs

In capacitively coupled signaling, the TSVs are replaced with small on-chip parallel plate capacitors that provide intertier communication to the upper tier. A schematic of a capacitively coupled 3-D system is illustrated in Fig. 3.13 [142]. A buffer drives the capacitor. The receiver circuitry, however, is more complex, as the receiver must amplify the low voltage signal to produce a full swing output. In addition, the receiver circuit should be sufficiently sensitive and fast to detect and respond to the voltage transferred through the coupling capacitors. The maximum voltage level that can be propagated through a coupling capacitance C_c is [143]

$$\Delta V_{\max} = \frac{C_c}{C_c + C_P + \sum C_{tr}} \Delta V_{\text{in}}, \tag{3.1}$$

where C_p is the capacitance of the capacitor plate to the substrate and adjacent interconnect preceding the receiver. The denominator in (3.1) includes the capacitance of the transistors at the input of the receiver.

A 5 fF capacitance has been used in [143] to transmit a signal to several receiver architectures. This capacitance uses 20 μm × 20 μm electrodes with a separation of 2.5 μm (e.g., dielectric thickness) and a dielectric constant of 3.5. Simulations indicate correct operation for frequencies up to 500 MHz, while measurements show successful operation for signal transmission up to 25 MHz for a 0.5 μm CMOS technology. Smaller parallel plate capacitor structures and improved transceiver circuitry have also been reported, yielding a communication bandwidth of 1.23 Gb/s [144]. For a 0.13 μm CMOS technology the size of the capacitor electrodes is 8 μm × 8 μm. Comparing capacitively coupled 3-D ICs with SiP, where the intertier wiring is over the edge of the IC (the total

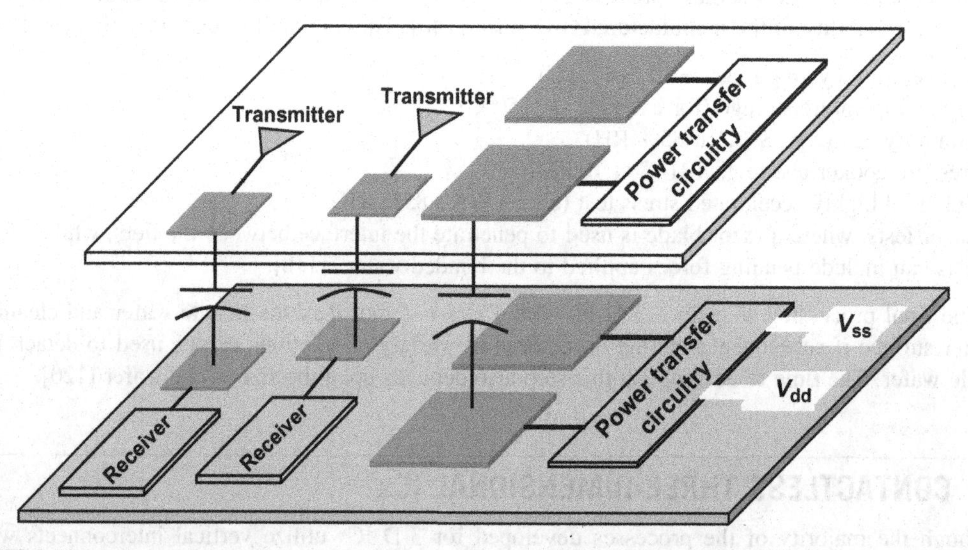

FIGURE 3.13

Capacitively coupled 3-D IC. The large plate capacitors are utilized for power transfer, while the small plate capacitors provide signal propagation.

interconnect length is on the order of a centimeter), a 30% improvement in delay is demonstrated. In addition, significant dynamic power is saved with this interconnect scheme. The static power consumed by the receiving circuitry, however, decreases the overall power savings.

The parasitic capacitance of the devices is greatly reduced with a silicon-on-sapphire technology, an early type of SOI technology [142]. This situation is particularly advantageous for power transfer circuits realized with a charge pump. The transceivers and charge pump circuits have been fabricated in a 0.5 μm CMOS technology [142], utilizing capacitors with dimensions of 90 μm × 90 μm. The distance between the two tiers is 10 μm. Decreasing the separation between the tiers can increase the coupling, requiring smaller capacitor plates to propagate a signal. The capacitors used for the power exchange should be large, though, for enhanced coupling. A prototype based on this technique provides approximately 0.1 mA current to the devices on the upper tiers. Successful operation up to 15 MHz has been demonstrated [142].

Critical factors affecting this technique are the size of the capacitors, which affects the interconnect density, the intertier distance, and the dielectric constant of the material between the tiers, which determine the amount of coupling. Face-to-face bonding is preferable as the distance between the tiers is shortened. Interconnecting a 3-D IC consisting of more than two tiers with this approach, however, is a challenging task.

3.3.2 INDUCTIVELY COUPLED THREE-DIMENSIONAL ICs

Inductive coupling can be an alternative for contactless communication in 3-D ICs. Each tier in a two-tier 3-D IC accommodates a spiral inductor, located at the same horizontal coordinates (see Fig. 3.14). The size and diameter of the inductors are based on an on-chip differential transformer structure [145]. Inductors with a diameter of 100 μm have been shown to be suitable for signal transmission, where the inductors are separated by a distance of 20 μm. In contrast to capacitively coupled 3-D ICs, signal propagation is achieved through current pulses. As compared to the capacitive coupling technique, specialized circuitry both for the transmitter and receiver is required, dissipating greater power. A current-mode driver generates a differential signal. The receiver amplifies the transmitted current or voltage pulses, producing a full swing signal. Alternatively, inductive coupling does not require as small a separation of the tiers as with the capacitive coupling technique, relaxing the demand for thin spacing between tiers. This advantage supports the integration

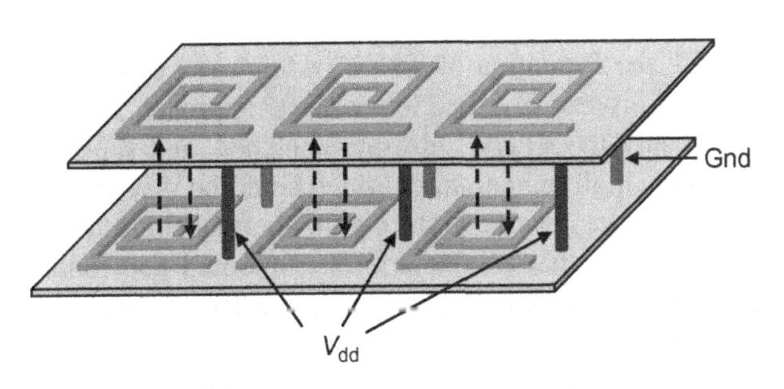

FIGURE 3.14

Inductively coupled 3-D ICs. Galvanic connections may be used for power delivery.

of more than two tiers for inductively coupled 3-D ICs. Simulations indicate 5 Gb/s throughput with 70 ps jitter at the output, consuming approximately 15 mW [145].

The main limitations of this technique include the size of the inductors, which yields a low interconnect density as compared to the size of the TSVs, increased power consumption of the transceiver circuitry, and interference among adjacent on-chip inductors. Additionally, power delivery to the upper tiers is realized with galvanic connections for high performance applications. Consequently, the fabrication process and cost for this type of 3-D IC is not greatly simplified as the TSV formation step is not eliminated for this type of applications. Alternatively, applications that lie in the low power regime can benefit more from this approach by eliminating the galvanic connections between tiers and supporting contactless power transfer in addition to wireless signaling. Due to the potential of this contactless scheme, Chapter 6, Three-Dimensional ICs with Inductive Links, is dedicated to the design and modeling of inductive links for contactless 3-D systems.

3.4 VERTICAL INTERCONNECTS FOR THREE-DIMENSIONAL ICs

Increasing the number of tiers that can be integrated into a single 3-D system is a primary objective of three-dimensional integration. A 3-D system with high density vertical interconnects is therefore indispensable. Vertical interconnects implemented as TSVs produce the highest interconnect bandwidth within a 3-D system as compared to wire bonding, peripheral vertical interconnects, and solder ball arrays. Other important criteria should also be satisfied by the fabrication process for TSVs. A fabrication process for vertical interconnects should produce reliable and inexpensive TSVs. In addition, a TSV should exhibit low impedance characteristics. A high TSV aspect ratio, the ratio of the length of the via to the diameter of the top edge, may also be required for certain types of 3-D circuits. The effect of forming the TSVs on the performance and reliability of the neighboring active devices should also be negligible.

As shown in Fig. 3.9, TSVs are formed after both the active devices (i.e., front end of line (FEOL)) and the metal layers (i.e., BEOL) on each tier of the 3-D circuit are fabricated. In this case, the TSV fabrication process is called a "via-last" approach, as also shown in Fig. 3.15. In via-last processes, the TSVs typically have large aspect ratios. TSV formation for via-last processes is a complicated step as etching is performed through several materials, such as metal and dielectric, in addition to the silicon substrate. Another approach, as depicted in Fig. 3.16, commences the fabrication of a 3-D system with the formation of TSVs. In this case, since neither the FEOL nor the BEOL are fabricated, the TSV process is called a "via-first" approach. Via-first approaches usually have faster etching rates, as only the silicon substrate is etched, and support smaller aspect ratios. However, as the FEOL requires high temperatures, care should be placed to maintain the reliability of the TSVs and control the stresses developed during the thermal cycles required by the FEOL.

A TSV manufacturing approach that encounters fewer thermal and reliability challenges is the case where TSVs are formed after the FEOL but before the BEOL. This approach is called "via-middle" and is shown in Fig. 3.17. In this case, the formed TSVs maintain a low aspect ratio as compared to via-last processes without undergoing the high temperatures due to the FEOL in via-first approaches. Some TSV processes with information about the physical and electrical traits of the TSVs are reported in Table 3.5.

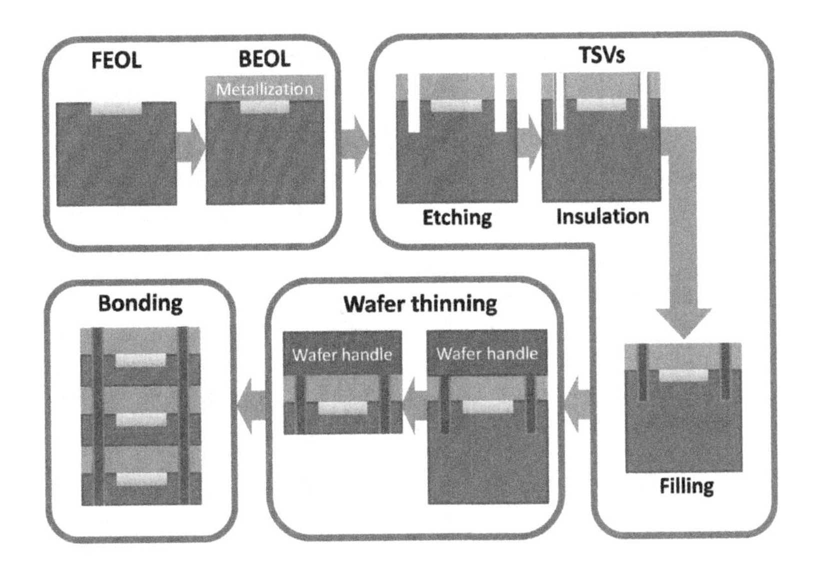

FIGURE 3.15

Basic steps of a via-last manufacturing process (not to scale).

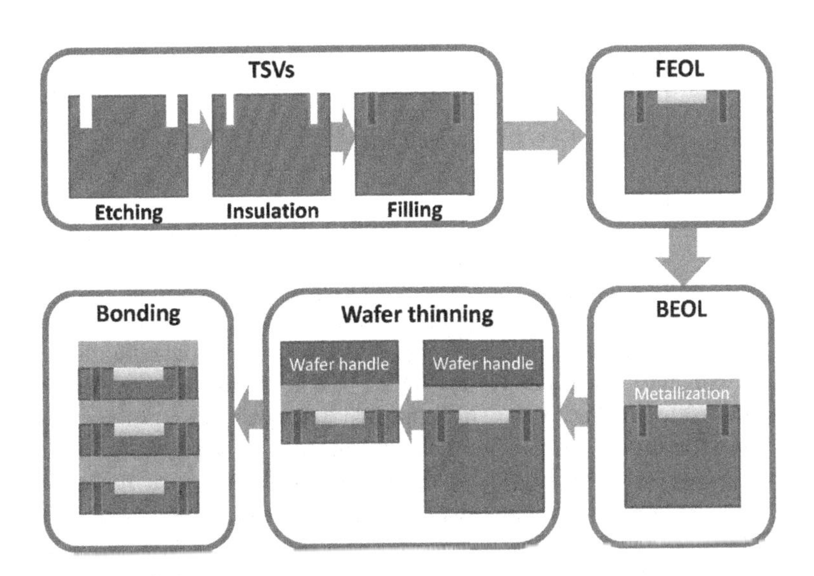

FIGURE 3.16

Basic steps of a via-first manufacturing process (not to scale).

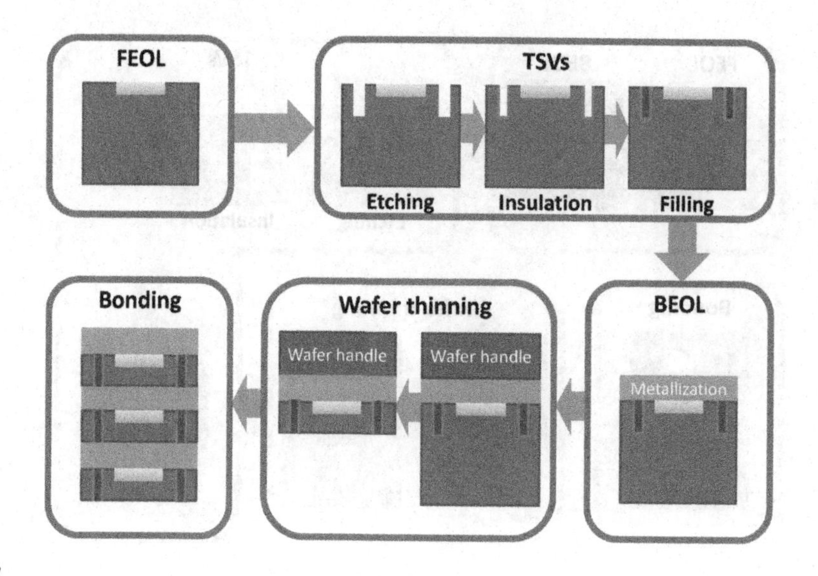

FIGURE 3.17

Basic steps of a via-middle manufacturing process (not to scale).

Table 3.5 Dimensions and Electrical Characteristics of the TSVs

Process From	Depth (μm)	Diameter (μm)	Total Resistance (mΩ)
[121]	25	4	140
[146]	30	2 × 12	230
[147]	80	5 and 15	9.4 or 2.6
[147]	150	5 and 15	2.7 or 1.9
[148]	90	75	2.4

The TSVs can be formed as blind vias exposed during the wafer thinning step. This method provides the important advantages of compatibility with existent process flows and simplicity in wafer handling [149]. A disadvantage of this approach is the effect on reliability resulting from wafer thinning and bonding. Alternatively, the TSVs can be fabricated after the wafer thinning step [150–152], as depicted in Fig. 3.18. This approach alleviates those problems related to back end processing; however, this method requires several processing steps with thin wafer handling, a potential source of manufacturing defects. Despite the disadvantages of the via-first approach, via-first, and via-middle approaches currently appear as the most promising technologies for TSVs.

The manufacturing technology for TSVs was greatly advanced by the invention of the BOSCH process in the mid-1990s, which was initially used to fabricate micro-electro-mechanical systems [153]. The BOSCH process consists of two functions, namely, etching and deposition, applied in successive time intervals of different duration (typically on the order of seconds) [149,154,155].

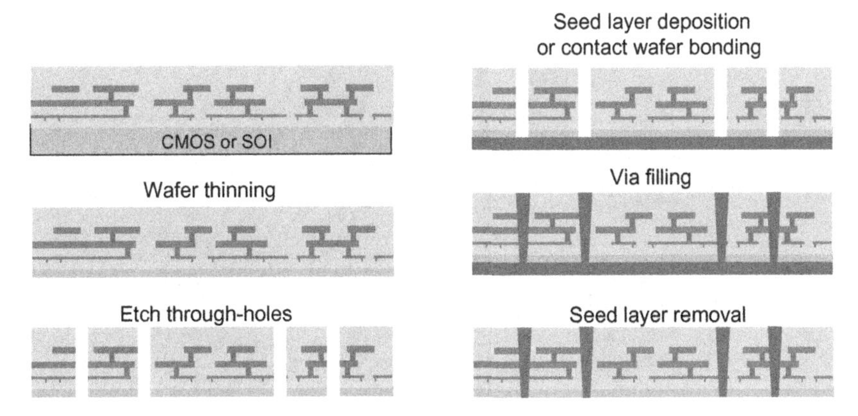

FIGURE 3.18

TSV formation and filling after FEOL (wafer thinning) and BEOL (via-last approach).

Sulphurhexafluoride (SF_6) is utilized for the etching cycles, while fluorocarbon (C_4F_8) is used to passivate the lateral walls of the TSV [153–155]. The BOSCH process is followed first by depositing a barrier, after which a seed layer is deposited. The former layer, typically consisting of TiN or TaN deposition, prevents the Cu from diffusing into the silicon as in a conventional damascene process. The latter layer is utilized for the filling step of the TSV. Copper is mostly used for via filling.

Although tungsten or low resistance poly-Si can be used for the TSV, copper has the inherent advantage of compatibility with BEOL processing and the multilayer interconnects used in modern ICs. Copper, however, may not be the appropriate material for via-first processes, as the FEOL requires high temperatures that can exceed the melting point of the copper. Although the BOSCH process is effective in etching silicon, several issues regarding the quality of the TSVs have to be considered [149], such as

- Controllability of the via shape and tapering;
- Conformal deposition and adhesion of the insulation, barrier, and seed layers;
- Void free filling with a conductive material;
- Removal of excessive metal deposition along the edges of the TSV.

TSVs have either a straight or tapered shape with different aspect ratios, as illustrated in Fig. 3.19. The aspect ratio (i.e., D/W) ultimately depends upon the thickness of the wafer that results from the wafer thinning step, where this ratio can exceed 20 [125]. Typical aspect ratios, however, are below ten. TSVs with a wide range of diameters have also been manufactured [147,149,150,153]. Tapered vias are preferred to nontapered vias, as a lateral wall with a slope and smooth surface facilitates the deposition of the barrier and seed layers and the following fill step [149]. This behavior occurs because the tapered profile decreases the effective aspect ratio of the straight segment of the TSV. Excessive tapering, however, can be problematic, leading to V-shaped vias at the bottom edge. In addition, a specific interconnect pitch is required at the bottom side of the via; consequently, tapering should be carefully controlled.

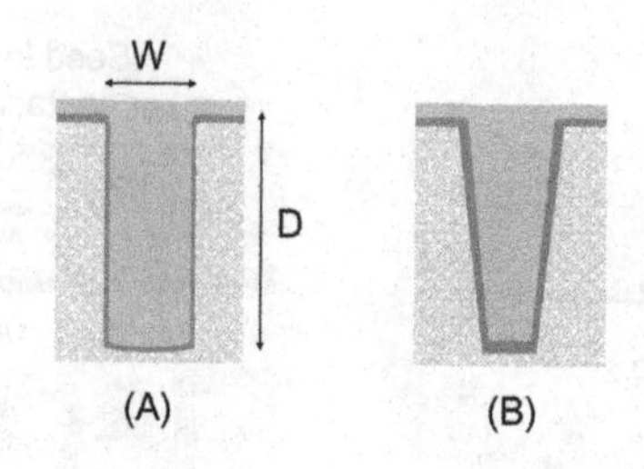

FIGURE 3.19

TSV shapes. (A) Straight and (B) tapered.

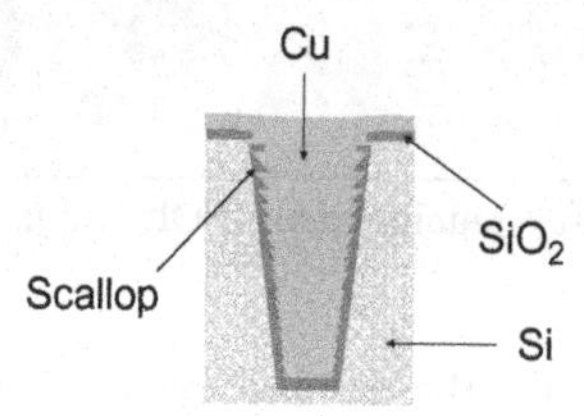

FIGURE 3.20

The scallops formed due to the time multiplexed nature of the BOSCH process.

Although the BOSCH process can etch silicon sufficiently fast, producing a smooth surface, barrier, and seed layer deposition is not a straightforward task. The BOSCH process results in a particularly rough surface, producing scallops, as illustrated in Fig. 3.20 [156]. Note that the roughness of the surface decreases with the via depth. The rough sidewall is due to the time multiplexed cycles of the etching and deposition steps of the BOSCH process. A low surface roughness is of particular importance for the steps that follow the via formation process. The scallops not only prevent a conformal barrier and seed layer deposition but also increase the diffusion of the copper into the silicon despite the presence of the barrier layer [157].

Different thermal loads can result from the following processing steps, such as wafer bonding, where thermal compression is typically utilized. Finite element analysis of TSVs with a diameter ranging from 3 to 10 μm has demonstrated that thermal loads contribute more to the induced stresses on a TSV as compared to those stresses caused by the bonding force. For specific bonding conditions, where the applied bonding force is 300 N and the temperature ranges from 300 to 400°C, the thermal loads constitute 84% of the total induced stress [158]. Simulation of the thermal loads imposed on the TSVs indicates that the thermal stress, due to the CTE mismatch between the dielectric and the metal filling, is more pronounced at the region where the scallops are sharper [156]. These stresses can crack the dielectric layer, increase the current leaking into the silicon substrate, or cause delamination of the copper within the TSV.

Another problem related to deep RIE concerns the silicon undercut below the mask at the upper and wide edge of the TSV. Uncontrolled undercut can lead to mask overhang, which can accelerate plating at the top of the TSV. Faster plating of the upper edge of the via can lead to premature closing of the via and to formation of a void inside the TSV [154].

Two key parameters can be used to adjust the surface roughness of the TSV sidewalls: the ratio of the time duration of the etching and passivation steps and the flow of the C_4F_8 [155]. In addition, variations of the BOSCH process or two step processes can be used to adjust the roughness of the TSV surface. For example, manufacturing process parameters, such as the pressure, bias power, and etching cycle time, can be varied in time rather than maintained static throughout the etching step. With this approach, the roughness of the surface has been decreased from ~ 0.05 to ~ 0.01 μm [156].

Via formation is followed by deposition of the barrier and seed layers. The primary goal of this step is to achieve a conformal profile throughout the depth of the TSV, which is increasingly difficult for TSVs with high aspect ratios. Different chemical vapor depositions are typically utilized for this step to provide good uniformity with moderate processing temperatures. Metal organic chemical vapor deposition (MOCVD) can be utilized to deposit the barrier layer. The disadvantage of this method is poor adhesion. In addition, MOCVD or atomic layer deposition used mainly for TSV with high aspect ratios are slow and expensive processes. Alternatively, physical vapor deposition (PVD) can be applied for layer deposition, which yields fair adhesion at low temperature. PVD, however, results in poor uniformity.

Via filling is achieved by electroplating, where the copper filling is grown laterally on the deposited seed layer. Alternatively, electroplating can be applied in a bottom-up manner, where a contact wafer is attached to the bottom of the device wafer, which includes the via openings. A major issue of this technique is the difficulty in removing the contact wafer, which provides the seed layer for the via filling [155]. Copper electroplating should produce uniform and void free TSVs to achieve high quality signal paths. Poor electroplating conditions that result in void formation within the TSV are illustrated in Fig. 3.21. The requirement for void free via filling can be achieved by maintaining a constant deposition rate throughout the via depth. Due to the tapered shape of the TSV, however, achieving a constant deposition rate requires the continuous adjustment of the electroplating process. Parameters that affect the TSV filling profile, which can be altered during copper deposition, include the solution composition, wafer rotation speed, applied current waveform, current pulse duration, and current density [149,154,155]. Current waveforms, such as direct current, forward pulse current, and reverse pulse current, can be applied. Since the deposition rate on the top edge of the TSV is larger than at the bottom, a low forward current density and a high reverse current density is used to maintain a constant deposition rate [149,150,155]. As filling at the bottom edge progresses, thereby closing the via opening, the current density is appropriately adjusted to maintain a fixed deposition rate.

Another alternative to avoid via filling issues is to utilize partially filled vias [159]. A partially filled via is shown in Fig. 3.22, where the different layers of the structure are also illustrated. The

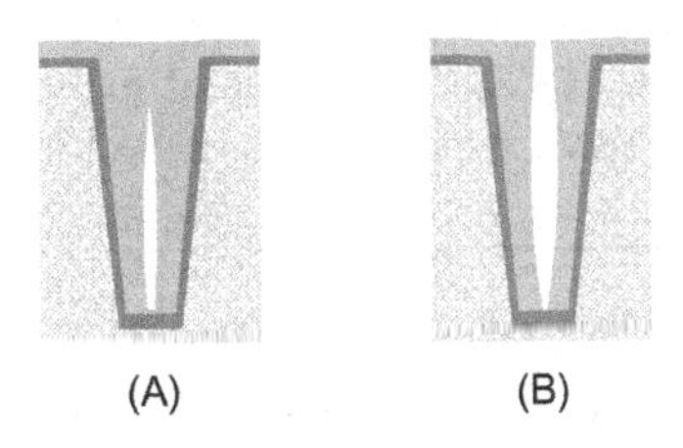

FIGURE 3.21

Poor TSV filling resulting in void formation, (A) large void at the bottom, and (B) seam void.

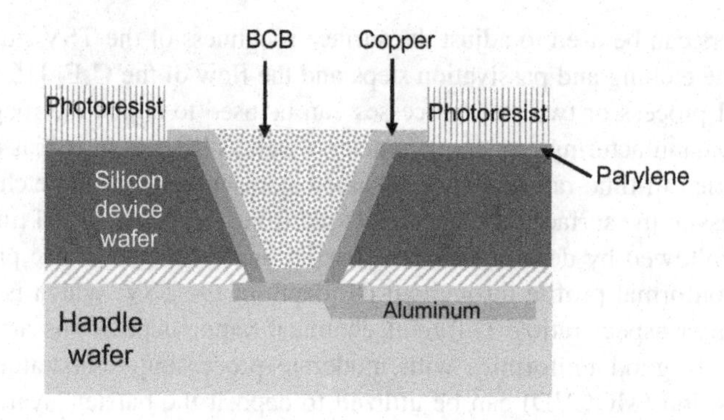

FIGURE 3.22

Structure of partial TSV and related materials [159].

Table 3.6 Resistance of Partially Filled TSV [159]

Via Bottom Diameter (μm)	Via Top Diameter (μm)	Resistance of One TSV (mΩ)
60	100	30
80	120	28
100	140	24

tapering of these TSVs ranges from 75° to 80°. The nonfilled metal volume of the etched via hole is filled with BCB, while a layer of parylene insulates the silicon substrate from the TSVs. A concern regarding these partially filled TSVs is the electrical resistance due to the reduced amount of plated metal. Experimental results listed in Table 3.6, however, demonstrate that low resistance vias can be achieved. Another important issue related to this type of TSV is that the density of metal within the volume of a 3-D circuit is reduced, while the density of the dielectric (which fills the etched via openings) is increased. This situation increases the thermal resistance in the vertical direction, which is the primary direction of the heat flow, requiring a more aggressive thermal management policy and possibly reducing the reliability of the circuits.

Any excessive metal concentrated on the top edge of the TSV should be removed prior to wafer bonding. This metal removal is typically achieved by CMP. A copper annealing step can also be present either before or after the CMP step [149]. After the thick metal residue is removed from the wafer surface, wafer thinning and bonding follows, preceded, if necessary, by a rewiring step. In addition to manufacturing reliable TSVs, TSVs should exhibit low impedance characteristics. This topic and electrical models for TSVs are discussed in the following chapter.

3.5 SUMMARY

Several manufacturing technologies and related issues for 3-D ICs are summarized as follows:

- 3-D ICs can be realized with either a sequential or parallel manufacturing process. Sequential processes produce monolithic structures, while parallel processes yield polylithic structures.

- Transistor level stacked 3-D ICs are monolithic structures that significantly reduce the total gate area and capacitance. Stacked 3-D ICs can be fabricated by laser or e-beam recrystallization and silicon growth based on semiconductor or metal seeding.
- Molecular bonding for monolithic 3-D circuits supports the integration of monocrystalline Si, which produces high quality devices for the upper layers.
- Wafer or die level 3-D integration techniques, which utilize TSVs, are an appealing candidate for 3-D circuits as these vias offer the greatest reduction in wirelength.
- In addition to the different levels of integration, several bonding styles are also possible including face-to-face, back-to-face, and back-to-back bonding.
- Fabricating 3-D ICs with TSVs typically includes the following steps: wafer preparation, TSV formation, wafer thinning, bumping, handle wafer attachment, wafer bonding, and handle wafer removal.
- TSVs can be fabricated before FEOL, after FEOL and before BEOL, and after BEOL, where the corresponding processes are called, respectively, via-first, via-middle, and via-last.
- Tier bonding can result from adhesive materials, metal-to-metal bonds, oxide fusion, and eutectic alloys.
- Coupling of electric or magnetic fields can be used to communicate among circuits located on different tiers, producing contactless 3-D ICs.
- Some limitations of contactless 3-D ICs are the size of the inductors and capacitors, small distance between tiers, and power delivery to the upper tiers.
- The formation of a TSV is largely based on the BOSCH process, which consists of etching and deposition steps, multiplexed in time. Electroplating is the most common technique for TSV filling.
- Issues with the BOSCH and electroplating techniques are rough surfaces, mask undercut, and void formation.
- To produce reliable TSVs, certain parameters of the BOSCH and electroplating techniques should be dynamically adjusted. These parameters include the solution composition, wafer rotation speed, applied current waveform, current pulse duration, and current density.

CHAPTER

ELECTRICAL PROPERTIES OF THROUGH SILICON VIAS

4

CHAPTER OUTLINE

One of the fundamental breakthroughs in the development of 3-D integrated circuits is the through silicon via (TSV). The early concept of the TSV was introduced in the late 1950s and early 1960s, first by William Shockley, followed closely by work from Merlin Smith and Emanuel Stern. These latter inventors were the first to patent a method for etching a cylindrical hole through silicon [41,42]. Sketches of these early etched holes in silicon are depicted in Fig. 4.1.

The TSV joins two separately processed wafers or dies to form a 3-D IC. Each additional device layer in a stack requires another set of TSVs to connect signals, different power domains, and various clock signals to properly operate the newly connected device tier. As the TSV is the critical component for interconnecting the separate device layers, accurate electrical characterization of the

Three-Dimensional Integrated Circuit Design. DOI: http://dx.doi.org/10.1016/B978-0-12-410501-0.00004-6

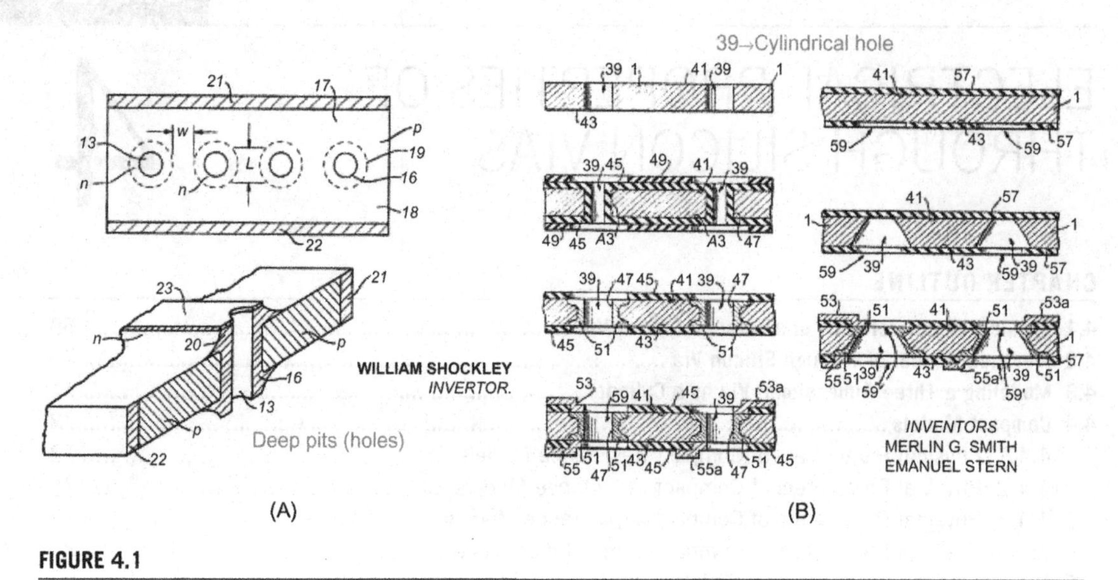

FIGURE 4.1

Early TSV from patents filed by (A) William Shockley [41], and (B) Merlin Smith and Emanuel Stern of IBM [42].

impedance of the TSV is necessary to effectively model the intertier signal delay and power characteristics, to properly interpret the impedance of the power network, and to analyze the effects of inter- and intratier clock skew and slew. The primary objective is to develop an equivalent electrical model of a TSV, as shown in Fig. 4.2. These models are often used for circuit simulation, such as the well known dynamic circuit simulator, SPICE [160].

This chapter on the electrical properties of a TSV is composed of two distinct sections. An understanding of the electrical properties of these TSVs is necessary to both model the clock distribution network described in Chapter 16, Case Study: Clock Distribution Networks for Three-Dimensional ICs, and to analyze the 3-D power delivery networks described in Chapter 18, Power Delivery for Three-Dimensional ICs. The second component of this chapter focuses on closed-form expressions of the TSV resistance, capacitance, inductance, and conductance.

Insight into the characterization and compact models of the electrical impedance of a TSV is presented in this chapter. An introduction to TSV technology is presented in Section 4.1 to better understand the physical characteristics of a TSV. An overview of the electrical modeling of a TSV is provided in Section 4.2. The technique of modeling the 3-D vias as simple cylinders without a top and bottom copper landing is discussed in Section 4.3.

Once an introduction to TSV modeling is provided, compact models of the resistance, capacitance, inductance, and conductance are presented in Section 4.4. A comparison between TSV impedance models and electromagnetic simulation is provided in Section 4.5. The process of numerically simulating a TSV is described in Section 4.6. A case study characterizing the electrical properties of the TSVs fabricated by MIT Lincoln Laboratory is described in Section 4.7. A few concluding remarks summarizing the chapter are provided in Section 4.7.

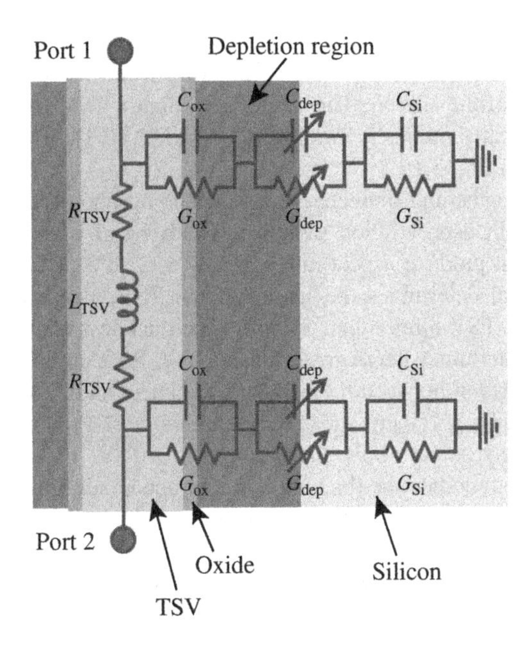

FIGURE 4.2

Equivalent π-model of a TSV.

4.1 PHYSICAL CHARACTERISTICS OF A THROUGH SILICON VIA

Beginning with the mid-1990s, significant growth has occurred in developing 3-D integrated circuits. A particular focus during the early years of 3-D integration was the development of process flows that incorporate TSV formation and wafer/die bonding into traditional fabrication processes. Some industrial leaders in this area are IBM [66,161], LETI [162], Intel [66], and Texas Instruments [66], while public funding for fabrication process flows produced important results at MIT Lincoln Laboratory (MITLL) [163] and the US National Laboratories [66]. Many more companies, universities, and national laboratories are developing TSV-based technology [66,163], enhancing the fabrication of these stacked systems.

A variety of process flows has been developed to generate stacked ICs (see Chapter 3, Manufacturing Technologies for Three-Dimensional Integrated Circuits). Process flows may vary depending upon the depth of the silicon, material of the TSV, aspect ratio of the TSV, dielectric liner thickness that insulates the TSV from the silicon, and several other physical parameters. A description of these physical parameters characterizing a TSV is provided in this section.

The TSV is typically a cylindrical structure composed of either copper, aluminum, tungsten, polysilicon (for TSVs fabricated during the front end of the line processes [164]), or solder [162,165–167]. Most current processes favor tungsten-based TSVs over other materials as the conductivity is relatively high (approximately three times the conductivity of copper), without damaging the transistors due to copper leaching into the silicon [168]. Nevertheless, a large body of work has focused on the production of copper TSVs [162,169,170]. Although the TSV is often treated as a cylinder, the shape often includes a narrowing or tapering from the surface where the transistors are patterned to the bottom of the silicon. The degree to which a TSV is tapered depends upon the fabrication process and typically ranges from close to 0 to 15° [171]. A drawing of both a straight and tapered TSV is shown in Fig. 3.19.

A wide range of TSV lengths and diameters exist depending upon the fabrication process. Most processes produce TSVs with aspect ratios (length-to-width ratio) of 10-15 to 1 [155,172–174]. Processes, however, exist that produce aspect ratios as low as 2 to 3 [172] and as high as 25 [165]. The thickness of the material determines the length of the TSV. For example, a silicon substrate thinned to 150 μm requires a TSV longer than 150 μm. The thickness of the substrate is, therefore, a primary consideration that determines the diameter of the TSV. An example of a 150 μm thick silicon substrate illustrates the correlation between the substrate thickness and TSV diameter. If a predefined aspect ratio of 10 is assumed for a 150 μm substrate, a minimum TSV diameter of 15 μm is required. Although the thickness of the substrate limits the diameter of a TSV, TSVs have been produced with different aspect ratios to accommodate the thickness of a particular substrate [162,165,172,175,176].

The dimensions of the TSV also affect the type of semiconductor fabrication process. The buried oxide of a silicon-on-insulator (SOI) process behaves like a natural etch stop layer during the wafer thinning process, which is necessary to stack the individual tiers [177]. Typical substrate thicknesses for an SOI process after wafer thinning are in the range of 10 to 20 μm [177], suggesting that TSV diameters of 1 to 3 μm can be fabricated. TSVs with diameters ranging from 1 to 60 μm and lengths of 10 to 300 μm have also been fabricated [155,169].

Additional physical parameters of the TSV worth noting are the dielectric liner and seed layers. Both the liner (typically SiON, SiN, TiN, or SiO_2 [164,178,179]) and seed layer (TiCu, TaN, or tetraethylorthosilicate (TEOS) [180]) are deposited through plasma enhanced chemical vapor deposition at temperatures of about 200°C [164], atomic layer deposition at low temperatures of 100°C [164], or electrochemical deposition [174]. The oxide liner is typically 50 to 500 nm thick and isolates the TSV from the silicon substrate [174,178,179,181]. The thickness of the oxide is increased to 1 to 3 μm for TSV diameters larger than 20 μm. The seed layer is typically 500 to 600 nm thick and is used as a bonding layer for the TSV conductor material. The material for the seed layer depends on the TSV fill material. As an example, for a copper TSV process, a titanium copper (TiCu) seed layer with a thickness ranging from 60 to 80 nm of titanium and 400 to 500 nm of copper [174] is used. A top view of the oxide layer, TiCu seed layer, and copper TSV is shown in Fig. 4.3. The combined thickness of the liner and seed layer ranges from approximately 600 nm to 1 μm [174,178].

4.2 ELECTRICAL MODEL OF THROUGH SILICON VIA

The primary technological innovation required to exploit the benefits of 3-D integration is the TSV. Much research and development have been achieved to properly characterize and model these intertier TSVs.

Accurate closed-form models of the 3-D via impedance provide an efficient method to characterize the performance of signal paths containing TSVs. These closed-formed expressions are similar in form to the models described in [182] and [183] that characterize the capacitance and

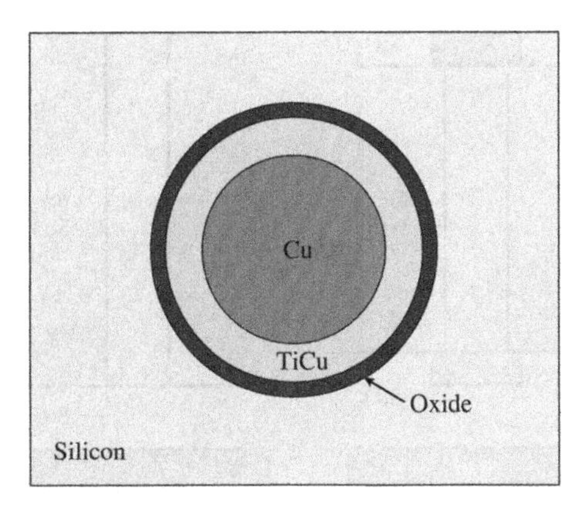

FIGURE 4.3

Top view of a TSV in silicon depicting the oxide layer, TiCu seed layer, and copper TSV.

inductance [184,185] of on-chip interconnect for very large scale integration (VLSI) circuits. Most previous work characterizing 3-D vias focused on bulk silicon and emphasized the experimental extraction of the via resistance and capacitance. Due to the large variation in the 3-D via diameter, length, dielectric thickness, and fill material, a wide range of measured resistances, capacitances, and inductances has been reported in the literature. Single 3-D via resistances can vary from 20 mΩ to as high as 350 Ω [58,186–190], while reported capacitances vary from 2 fF to over 1 pF [58,191,192]. A few measured via inductances range from as low as 4 pH to as high as 255 pH [58,169,187]. Alternatively, preliminary work on modeling 3-D vias has primarily focused on the resistance and capacitance of simple structures to verify measured *RLC* impedances while providing physical insight into 3-D via-to-via capacitive coupling [58,186,191,192]. In addition, preliminary work in modeling the electrical behavior of bundled TSVs has also been reported in [193–195]. This chapter is intended to provide intuition into both the electrical characterization of TSVs and closed-form expressions for the resistance, inductance, capacitance, and conductance of TSVs (3-D vias) based on the via length, via diameter, dielectric thickness, and fill material. The models and closed-form expressions described in this chapter also provide insight towards shielding both capacitive and inductive coupling noise among TSVs while electrically characterizing the TSV.

4.3 MODELING A THREE-DIMENSIONAL VIA AS A CYLINDER

The structural complexity of a 3-D via is treated as an equivalent cylindrical structure. A 3-D via typically includes a tapered cylindrical TSV with both a bottom and top metal landing. The electrical parameters of a 3-D via represented as a cylinder with a top and bottom copper landing (see Fig. 4.4A) is compared to an equivalent length cylindrical via without metal landings (see Fig. 4.4B). A comparison of a numerical simulation of these two TSV configurations reveals less than a 7% difference in the resistance, inductance, and capacitance. The only exceptions arise for the DC resistance and high frequency inductance when the aspect ratio is between 0.5 and 1, as

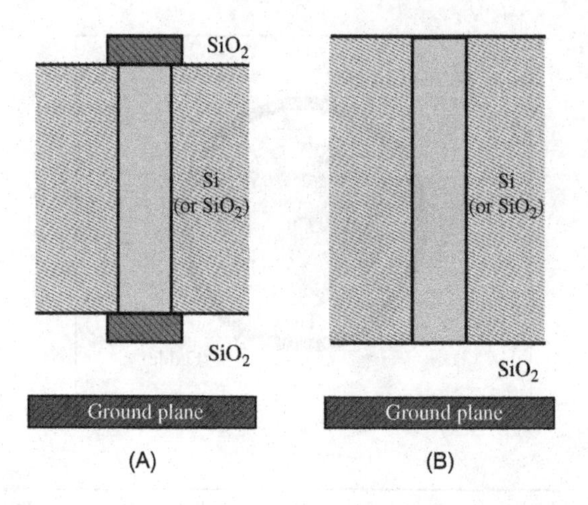

FIGURE 4.4

3-D via structure. (A) 3-D via with top and bottom copper landings, and (B) equivalent structure without metal landings.

Table 4.1 Per cent Error Between the 3-D Via Model and the Equivalent Cylindrical Model

	R		*L*		*C*
a.r.	**DC**	**1 GHz**	**DC**	f_{asym}	
0.5	41.6	3.6	0	−16.4	6.4
1	16.9	−2.0	0	−13.6	6.2
3	5.4	0.7	−0.04	−0.6	5.4
5	2.9	−0.2	−0.04	−1.4	4.9
7	2.2	−0.4	0.01	−1.4	4.0
9	1.9	0.4	0.02	−0.03	3.1

Diameter = 20 μm

listed in Table 4.1. This behavior implies that a simple cylindrical structure without metal landings is sufficient to represent a 3-D via.

The dimensions of the 3-D via are technology dependent. For an SOI process, where the buried oxide behaves as a natural stop for wafer thinning, the length $\mathcal{L}$ of the TSVs is much shorter than the bulk counterparts. The diameter D of the vias follows a similar pattern, where an SOI process utilizes a smaller diameter than in bulk technologies. These 3-D via models of an SOI process consider diameters of 1, 5, and 10 μm, while the 3-D vias in bulk silicon have a diameter ranging between 20 and 60 μm. A 20 μm diameter is chosen for this comparison between the 3-D via model and an equivalent cylindrical structure. In all cases, the via length maintains an aspect ratio $\mathcal{L}/D$ of 0.5, 1, 3, 5, 7, and 9, and the distance separating the bottom of the TSV and the ground plane is

4.5 μm. In addition, the dielectric surrounding the TSV for a bulk process is 1 μm thick, and the material filling the 3-D via in both SOI and bulk technologies is tungsten.

4.4 COMPACT MODELS

The physical characteristics contributing to the compact models of the TSV resistance, inductance, capacitance, and conductance are described in this section. The physical parameters of the compact resistance are discussed in Section 4.4.1. The physical parameters that affect the inductance are reviewed in Section 4.4.2. The physical parameters that affect the capacitance and conductance are described, respectively, in Sections 4.4.3 and 4.4.4.

4.4.1 PHYSICAL PARAMETERS OF COMPACT RESISTANCE MODELS

A brief discussion of the physical principles governing the TSV resistance is provided in Section 4.4.1.1. Frequency dependent properties that affect the resistance such as surface scattering, boundary scattering, and skin effect are described in Section 4.4.1.2. The effect of the TSV barrier thickness and temperature on the resistivity is considered, respectively, in Sections 4.4.1.3 and 4.4.1.4.

4.4.1.1 Through silicon via resistance

The resistance relates the potential difference between the two ends of a wire to the total current flowing through a wire. The electrostatic law governing the resistance is

$$R_{\text{via}} \equiv \frac{\Phi_{12}}{I} = \frac{-\int_L \vec{E}\cdot d\vec{l}}{-\int_A \sigma \vec{E}\cdot d\vec{l}}, \tag{4.1}$$

where Φ_{12} is the electrostatic potential between two points, I is the current through the wire, σ is the conductivity of the material, and E is the electric field.

The DC resistance of a TSV is determined for a nonmagnetic cylindrical wire conductor. A majority of closed-form expressions for the resistance of a TSV assumes a cylindrical wire [179,181,196–199]. The resistance of a TSV is therefore a function of the length ℓ_v, cross-sectional area πr_v^2, and conductivity of the material σ.

$$R_{\text{via}} = \frac{\ell_v}{\sigma \pi r_v^2}. \tag{4.2}$$

4.4.1.2 High frequency effects

One of the primary high frequency effects that affects the resistance of a TSV is the skin effect. At DC, the current density through a TSV is uniformly distributed across the entire cross-sectional area. As the frequency of a signal propagating through a TSV increases, the current density becomes nonuniform, and drops exponentially with distance from the surface (sidewall of the TSV). This phenomena is due to a time varying magnetic field H induced by a time varying current I, with the assumption that the current return path is sufficiently far away that any influence can be neglected [200]. The skin effect leads to current crowding at the surface, reducing the effective cross-sectional area of the TSV. The skin effect is the depth below the surface of the conductor at which the current density drops to $1/e$ (approximately 0.37) of the current density at the surface of the conductor. The skin depth is

$$\delta = \frac{1}{\sqrt{\pi f \mu_r \mu_o \sigma}}, \tag{4.3}$$

where f is the frequency, σ is the material conductivity, μ_o is the permeability of free space, and μ_r is the relative permeability of the material.

As a result of the skin effect, which reduces the effective cross-sectional area of a conductor, the resistance of a rectangular wire with width w and thickness t is

$$R \approx \frac{\ell}{\sigma 2\delta[t + (w - 2\delta)]}, \tag{4.4}$$

and the resistance of a cylindrical wire is

$$R \approx \frac{\ell}{\sigma\pi[r^2 - (r-\delta)^2]} \approx \frac{\ell}{\sigma\pi(2r - \delta)\delta}. \tag{4.5}$$

The -2δ term in (4.4) removes the overlap in area at the corners between the TSV thickness t and width w.

4.4.1.3 Effect of barrier thickness on resistivity

As previously mentioned, the TSV is isolated from the silicon by a dielectric liner that prevents direct contact to the silicon. In the case of copper-based TSVs, the dielectric liner also includes a metal barrier with a higher resistivity that prevents the diffusion of copper into the silicon. The barrier thickness is dependent on both the deposition process and the barrier material [202]. When examining a cross-sectional view of the TSV, the area occupied by the metallic current carrying portion is the conducting material deposited after the liner and barrier are formed. A thicker liner/barrier, therefore, leaves a smaller diameter to fill with the metallic conductor. Since the resistivity of the dielectric liner is much greater than the resistivity of metal, the result is an increase in resistance. Two methods exist to account for this characteristic. One method reduces the effective diameter of the TSV depending upon the thickness of the liner/barrier. For the second method, the resistivity is modified to include the effect of the liner/barrier. Banerjee in [202] applied the second technique and modified the resistivity, as described by

$$\rho_b = \frac{\rho_o}{1 - \frac{A_b}{A_{\text{metal}} \cdot \left(\frac{p_w}{2}\right)^2}}. \tag{4.6}$$

The effective resistivity due to the barrier ρ_b is inversely proportional to the area of the barrier A_b and proportional to the area of the metal conductor A_{metal}. Banerjee also considered the pitch between wires p_w, although this term can be ignored for TSVs. The bulk resistivity of the metal ρ_o is temperature dependent and is therefore adjusted based on the operating temperature.

4.4.1.4 Temperature effect

The conductivity σ of a material is dependent on the carrier concentration q and the mobility of the charge carriers μ,

$$\sigma = q\mu. \tag{4.7}$$

Charge carriers are created by ionization of the atoms comprising the lattice of a conductor. At a specific temperature, all of the atoms of a conductor are ionized, and the supply of electrons becomes constant with temperature. The increase in resistance due to increasing temperature is caused by a higher rate of collision with the lattice, impurities, and grain boundaries of the conductor material. With this increase in collision rate, the mobility of electrons is reduced.

For conductors, the loss of mobility is due to ionic scattering. The relationship between temperature and resistance includes this effect on the material resistivity. The temperature dependent resistivity is [203]

$$\rho(T) = \frac{1}{\sigma(T)} = \rho_o(T_o)[1 + \alpha(T - T_o)], \tag{4.8}$$

where $\rho(T)$ is the wire resistivity, $\rho_o(T_o)$ is the resistivity at a reference temperature T_o (often assumed to be 20°C), and α is the temperature coefficient of resistance. The temperature coefficient of resistance is $d\rho/dT$, the change in ρ with respect to the temperature. The α term for a specific material changes with the reference temperature, and therefore the relationship between resistivity and temperature only holds for a range of temperatures around this reference temperature [204].

4.4.2 PHYSICAL PARAMETERS OF COMPACT INDUCTANCE MODELS

A brief discussion of the physical principles governing the TSV inductance is provided in Section 4.4.2.1. The properties that determine the inductance such as the internal and external magnetic fields (Section 4.4.2.2), high frequency effects (Section 4.4.2.3), and the concept of magnetic loops (Section 4.4.2.4) are also described.

4.4.2.1 Through silicon via inductance

The inductance of a conductor is a measure of the magnetic field lines internal to and surrounding the wire, characterizing the ability of a conductor to couple magnetic flux or store magnetic energy. The physical layout of a conductor, current density, current direction, and signal frequency affects the inductance of a wire. The inductance of a material is defined as

$$L \equiv \frac{\oint \vec{B} \cdot d\vec{A}}{I}. \tag{4.9}$$

A critical characteristic necessary to determine the inductance of a wire is the current return path. The inductance is based on the "loop" formed as current flows into and out of an area. The current return path is not easy to identify in modern integrated circuits as this path is not a function of proximity to the source structure such as the capacitance. Current return paths in modern ICs are primarily in the power distribution network and other neighboring wires [205]. The complexity of extracting the inductance is further exacerbated since the current return paths can be tens to hundreds of micrometers from the wire under consideration. In addition, the return path does not have to be through a single wire, potentially leading to tens or even hundreds of return paths. A method proposed by Rosa in 1908 [206] and further expanded by Ruheli [184] assumes that the current return path is at infinity. This approach provides an accurate estimate of inductance as compared to numerical field solvers. Most closed-form expressions of the self- and mutual inductance of a TSV [181,191,201,207] are modifications of the inductance expressions originally developed by Rosa [206].

4.4.2.2 Internal and external magnetic field

Magnetic flux exists both internally and externally to a current carrying wire. The inductance can, therefore, be divided into internal and external components. The inductance of each component is summed to provide the total inductance of a wire. The internal component of the inductance is affected by the frequency of the signals propagating through the wire. At high frequencies, due to the skin affect (as discussed in Section 4.4.1.2), the internal component of the inductance is lower as most of the current flows at the surface of the conductor. At even higher frequencies, the internal component approaches zero as almost all of the magnetic flux is external to the conductor.

4.4.2.3 High frequency effects

As noted above, the density and cross-sectional profile of the current flowing through a wire affect the inductance of a wire. Two effects at high frequency exist (typically, frequencies exceeding 1 GHz) that change the current distribution within a conductor. The first effect is the skin effect, and the other effect, the proximity effect, is related to the skin effect.

The proximity effect is a phenomenon that occurs between two current carrying wires. When current flows in opposite directions, such as the case of a conductor and the return path, the current crowds towards the location between the two conductors, as illustrated in Fig. 4.5A. In the case where the current flows in the same direction through two separate conductors, the current moves to the opposite side of each conductor, farthest from the neighboring wire. In both cases, current flowing in the same or opposite direction, the proximity effect reduces the overall inductance of a conductor.

A third high frequency effect that changes the inductance of a wire or TSV is due to multipath current redistribution in the return path [185,205]. Many possible current return paths exist in an integrated circuit, including the power and ground networks, substrate, and nearby signal lines. Each of these returns paths exhibits a frequency dependent impedance. Current is redistributed among the different return paths to minimize the total impedance of the path. Therefore, depending upon the frequency of the signal, certain return paths are more effective in minimizing the total impedance than other paths, and the path often changes as the frequency changes from one frequency to another frequency.

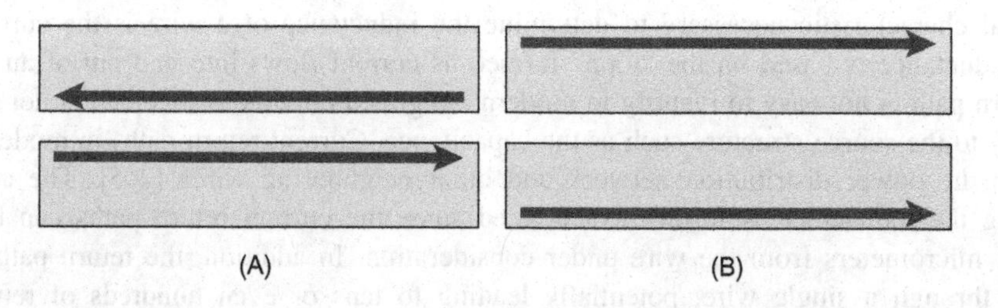

FIGURE 4.5

Current profile due to the proximity effect for (A) currents propagating in opposite directions, and (B) currents flowing in the same direction [200].

4.4.2.4 Loop inductance

The loop inductance is induced by a current flowing through a wire and a return path. In the case of multiple return paths, the total loop inductance of a wire is the sum of the partial loop inductances from each segment of the return path. When the return paths are not known, a mathematical representation of the partial loop inductance elements is determined for all wires forming a current loop. Two approaches exist to determine the effective loop inductance of a path. The first method directly calculates the partial loop inductance between two coupled segments, while the second method uses a mathematical representation of the self- and mutual inductance to all other segments before determining the loop inductance with each return path. An overview of each technique is described below, starting with the first method of directly solving for the partial loop inductance. The partial loop inductance of two segments is [208]

$$L_{ab,\text{partial}} = \frac{\mu_o}{4\pi}\frac{1}{A_aA_b}\int_{A_a}\int_{l_a}\int_{A_b}\int_{l_b}\frac{d\vec{\ell_a}\cdot d\vec{\ell_b}}{|\vec{r_a}-\vec{r_b}|}dA_adA_b, \tag{4.10}$$

where A_a and A_b are the cross-sectional area of, respectively, segments a and b, and ℓ_a and ℓ_b are the length of each respective segment.

The total loop inductance, the sum of all partial self- and mutual inductances, is therefore described by (4.11). The s_{ij} term in (4.11) is -1 when the currents flow in opposite directions and $+1$ when the currents flow in the same direction.

$$L_{\text{loop1}} = \sum_i\sum_j s_{ij}L_{ij,\text{partial}}. \tag{4.11}$$

The second method is based on the closed-form expression of the self- and mutual inductance as developed by Rosa [206]. Rosa derived the self-inductance of a cylindrical wire of a nonmagnetic material in free space. An expression for the self-inductance is

$$L_{\text{self}} = \frac{\mu_o}{2\pi}\left[\ln\left(\frac{\ell+\sqrt{\ell^2+r^2}}{r}\right)\ell + r - \sqrt{\ell^2+r^2}+\frac{\ell}{4}\right], \tag{4.12}$$

where r and ℓ are, respectively, the radius and length of the TSV, and μ_o is the permeability of free space. Most closed-form expressions of the inductance of a TSV are based on this expression by Rosa with minor modifications. One example is the expression by Pucel [181,209] in (4.13). Pucel removed the $l/4$ term, doubled the effective length of the conductor, and halved the resulting inductance to account for the image inductance caused by a vertical current proximate to a ground plane [210].

$$L_{\text{self}} = \frac{\mu_o}{4\pi}\left[\ln\left(\frac{2\ell+\sqrt{(2\ell)^2+r^2}}{r}\right)2\ell + r - \sqrt{(2\ell)^2+r^2}\right]. \tag{4.13}$$

The mutual inductance between two cylindrical wires has also been described by Rosa [206] and is applied in the closed-form expressions for a cylindrical TSV. The mutual inductance is

$$L_{\text{mutual}} = \frac{\mu_o}{2\pi}\left[\ln\left(\frac{\ell+\sqrt{\ell^2+p^2}}{p}\right)\ell + p - \sqrt{\ell^2+p^2}\right]. \tag{4.14}$$

The total loop inductance is calculated from the self- and mutual inductance of a wire and each of the return paths.

$$L_{\text{loop2}} = L_{\text{self1}} + L_{\text{self2}} - 2L_{\text{mutual21}} \tag{4.15}$$

4.4.3 PHYSICAL PARAMETERS OF COMPACT CAPACITANCE MODELS

A brief discussion of the physical principles governing the TSV capacitance is provided in Section 4.4.3.1. The effect of the silicon depletion region on the TSV capacitance is discussed in Section 4.4.3.2.

4.4.3.1 Through silicon via capacitance

Unlike the resistance and inductance (assuming that all return paths are known for each wire) of a material, where accurate closed-form expressions have been developed that relate the material properties, physical characteristics, and line dimensions, closed-form expressions of the capacitance require a model that incorporates additional effects beyond geometric and material properties. One effect is the termination of the electric field lines on neighboring metals, which plays a significant role in the total capacitance of a conductor. Two important but different considerations include: (1) field lines emanating from the conductor (TSV) and terminating at a reference ground plane on the back metal of the silicon, and (2) field lines emanating from the TSV and terminating on another TSV within the silicon. In each case, a depletion region (or accumulation region depending on the applied voltage) forms around the TSV. This effect needs to be incorporated into the closed-form expression for the capacitance of a TSV. A third effect is the fringe capacitance, as neglecting this component can greatly underestimate the total capacitance. Recent research has focused on terminating the electric field lines not only to other TSVs but also to the horizontal on-chip interconnect within a device plane [207,211], further increasing the complexity of the compact model characterizing the capacitance of a TSV.

In simple terms, capacitance is the ability of a body to store electric charge, and is given as the ratio of the charge to the applied voltage. The capacitance between two conductors is

$$C \equiv \frac{Q}{\Phi_{12}} = \frac{\oint_A \vec{D} \cdot d\vec{A}}{-\int_A \sigma \vec{E} \cdot d\vec{\ell}}, \tag{4.16}$$

where Φ_{12} is the voltage difference between the two conductors, $\vec{D}$ is the electric displacement field, $\vec{E}$ is the electric field, ℓ is any path from the negative to the positive conductor, and A is any surface enclosing the positively charged conductor. An analytic solution of Poisson's equation determines the capacitance of a material, which is dependent on the location of the ground reference. Extracting the capacitance with an electric field solver such as Ansys Q3D [212] and FastCap [213], although highly accurate, is often computationally expensive. Closed-form expressions are therefore critical to efficiently estimate the capacitance of a TSV. Including physical effects such as the depletion region, fringe capacitance, and location of the ground reference enhances the accuracy of the model.

Early research in compact models of the capacitance of a TSV assumed the silicon substrate behaved similarly to a metal with a finite conductivity. Although this assumption is somewhat correct, all field lines emanating from a TSV are assumed to terminate on the other side of the oxide barrier layer (at the interface with the silicon). This assumption produces a worst case estimate of the capacitance of a TSV. Two expressions for estimating the TSV capacitance based on the assumption that silicon behaves like a highly resistive metal are

$$C = \frac{\epsilon_{ox}}{t_{ox}} 2\pi r\ell, \tag{4.17}$$

$$C = \frac{\epsilon_{ox}}{\ln\left(\frac{r+t_{ox}}{r}\right)} 2\pi\ell, \tag{4.18}$$

where t_{ox} is the oxide thickness, r is the radius of the TSV, and ℓ is the length of the TSV. Equation (4.17) is used in [191] to compare this model with experimental data. The closed-form expression of (4.18) modifies (4.17) to more accurately represent the radius of the conductor within the dielectric without a change in the underlying assumption that the silicon behaves like a metal with low conductivity (1 to 10 S/m) [192].

From this early work, a more accurate model of the physical phenomena affecting the capacitance of a TSV has been developed. One of the most important of these phenomena is the formation of a depletion region around the TSV [181,196,207,214,215], such that the silicon can no longer be modeled as a highly resistive metal. Inclusion of the depletion region in a TSV model is further discussed in the following subsection.

4.4.3.2 Through silicon via model including the effects of the depletion region

The TSV is surrounded by silicon (see Fig. 4.6). Similar to a metal-oxide-semiconductor (MOS) capacitor [216,217], the applied voltage on a TSV either depletes the silicon of mobile charge carriers or causes accumulation [214,215]. The voltage across a TSV can also invert the surrounding silicon when a low frequency signal passes through the TSV [214]. The silicon is typically depleted as a result of a positively applied voltage on a TSV surrounded by p-type silicon. The formation of a depletion region for bulk and SOI technologies is illustrated in Fig. 4.6.

As previously mentioned, the applied voltage determines whether the TSV operates in either the accumulation region, the depletion region, or in inversion, similar to a MOS capacitor. The ratio of the TSV capacitance to the oxide capacitance as a function of the applied voltage V_g is illustrated in Fig. 4.7.

As noted in Fig. 4.7, the capacitance in accumulation is the oxide capacitance [181]

$$C_{\mathrm{TSV_{acc}}} = C_{ox} = \frac{2\pi\epsilon_{ox}\ell}{\ln\left(\frac{r_{ox}}{r_{\mathrm{metal}}}\right)}, \tag{4.19}$$

and the depletion capacitance is

$$C_{\mathrm{dep_{min}}} = \frac{2\pi\epsilon_{ox}\ell}{\ln\left(\frac{r_{\mathrm{max}}}{r_{ox}}\right)}, \tag{4.20}$$

where r_{max} is the maximum depletion radius, r_{metal} is the radius of the TSV, r_{ox} is the radius of the conductor with the oxide, and f is the length of the TSV. Since the depletion capacitance is in series with the oxide capacitance, the total TSV capacitance is reduced. The series combination of the oxide and depletion capacitance is [181,218]

$$C_{\mathrm{TSV_{min}}} = \frac{C_{ox}C_{\mathrm{dep}}}{C_{ox} + C_{\mathrm{dep}}}. \tag{4.21}$$

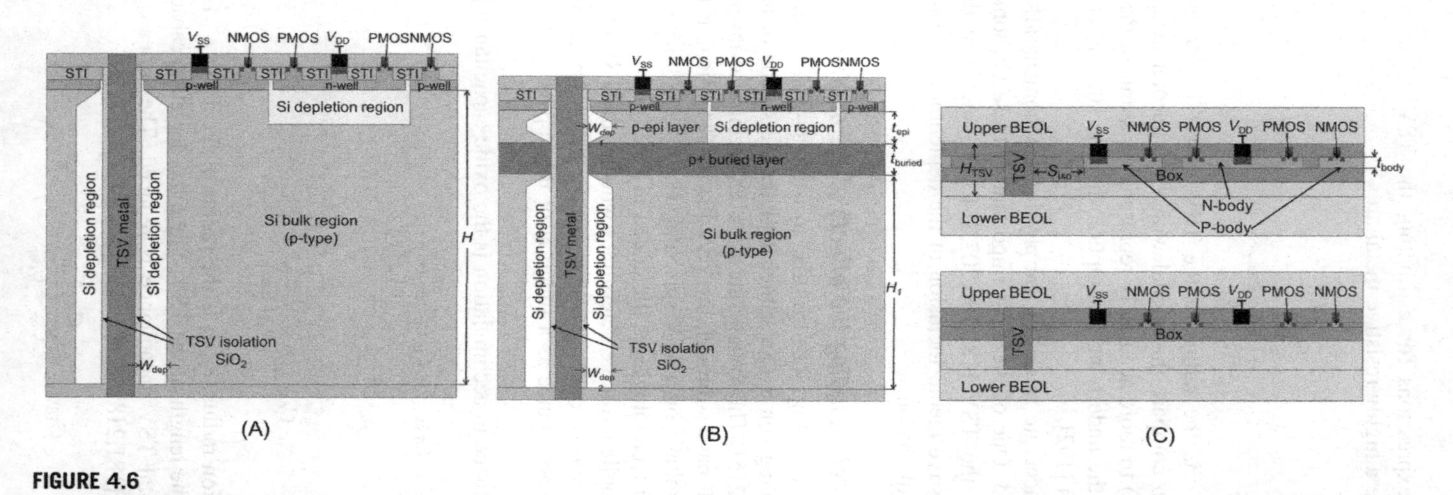

FIGURE 4.6

Cross-sectional view of different CMOS technologies with TSVs depicting the formation of a depletion region around the TSV in (A) bulk CMOS, and (B) bulk CMOS with a p + buried layer. The TSVs in either PD-SOI (shown in (C) top) or FD-SOI (shown in (C) bottom) reveal minimal formation of a depletion region [214].

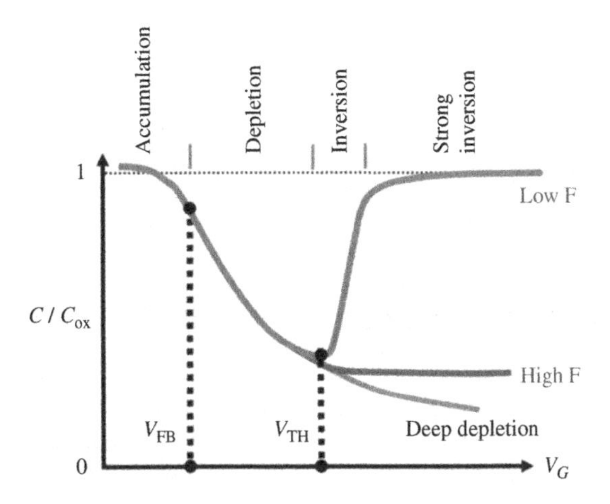

FIGURE 4.7

Ratio of the total TSV capacitance to the oxide capacitance as a function of applied voltage V_g.

4.4.4 PHYSICAL PARAMETERS OF COMPACT CONDUCTANCE MODELS

A complete model of a TSV includes the electrical conductance in a silicon substrate. The physical principles governing the conductance in a bulk silicon substrate are described in this section.

4.4.4.1 Through silicon via conductance

Two different conductive components of a TSV exist. One component is due to the dielectric liner surrounding the TSV, and the second component is due to the lossiness of the silicon substrate (with a typical conductivity of 1 to 10 S/m). An expression for the conductance between two conductors in close proximity is

$$G = \frac{\ell \pi \epsilon}{\ln\left[\frac{s}{2r} + \sqrt{\left(\frac{s}{2r}\right)^2 - 1}\right]}, \tag{4.22}$$

where, in addition to the radius r, length ℓ, and permittivity of the material ϵ, the space s between TSVs is also included [219,220]. The conductance described by (4.22) is divided in half when two equally distant TSVs are in close proximity, and by four if there are four surrounding TSVs.

A second approach to model the conductance of a TSV considers the loss tangent [221] of an inhomogeneous or lossy material. In this case, both the capacitance and conductance are strongly dependent on the frequency. Since the complex permittivity is

$$\epsilon = \epsilon' - j\epsilon'' = \epsilon' - j\left(\epsilon_b + \frac{\sigma}{\omega}\right), \tag{4.23}$$

and the loss tangent is

$$\tan \delta = \frac{\epsilon''}{\epsilon'} = \frac{\sigma}{\omega \epsilon'}, \tag{4.24}$$

the conductance of a TSV is

$$G = \omega C \tan \delta. \tag{4.25}$$

The loss tangent in (4.24) exhibits an inverse frequency dependence with ω as both ϵ and σ are also dependent on the frequency [221]. For this reason, many materials exhibit a constant dependence of $\sigma/\omega\epsilon$ on ω over a wide frequency range [221].

4.5 THROUGH SILICON VIA IMPEDANCE MODELS

A compact model of the TSV provides an efficient and accurate method to determine the TSV impedance. A large body of work exists that focuses on electrical modeling of the TSV [181,196], including bundled TSVs [196] and high frequency effects [179,207,214,222]. In addition, some research on compact TSV models has focused on including the tapered sidewall of the TSV [171,223].

General closed-form expressions for the resistance, inductance, capacitance, and conductance of the TSV structure consider the diameter, length, dielectric liner thickness, distance to ground, distance to neighboring TSVs, and tapering angle [171,214,201]. The electrical characteristics of a TSV are affected by several physical parameters including the TSV length ℓ, radius r_v, oxide thickness t_{ox}, TSV-to-TSV pitch p_v, depletion region width w_{dep}, and skin depth δ. Material parameters, which include the permittivity of the silicon (ϵ_{Si}) and oxide (ϵ_{ox}), conductivity of the silicon (σ_{Si}), resistivity of the metal ρ, and permeability of the metal μ_o are also considered. The various physical parameters and materials are illustrated in Fig. 4.8. Closed-form expressions of the TSV resistance at both DC and higher frequencies are provided in Section 4.5.1. Compact models of the inductance are described in Section 4.5.2. Closed-form expressions of the TSV capacitance and conduction are provided, respectively, in Sections 4.5.3 and 4.5.4.

4.5.1 COMPACT RESISTANCE MODEL OF A THREE-DIMENSIONAL VIA

Compact models for the resistance of a TSV are listed in Table 4.2. The relevant physical parameters that affect the resistance of the TSV are also listed. The physical parameters include the TSV length ℓ, radius r_v, and skin depth δ. Material parameters that impact resistance include the resistivity ρ and permeability μ_o (skin effect) of the metal.

A comparison of the TSV resistance models is provided in Section 4.5.1.1. A closed-form model of the resistance of a TSV is provided in Section 4.5.1.2. A comparison of the model described in Section 4.5.1.2 with numerical simulations is provided in Section 4.5.1.3.

4.5.1.1 Compact models of through silicon via resistance

Compact models of the resistance consider two types of TSV structures. The first model characterizes the TSV as a cylinder with a constant radius, and the second model represents the TSV as a tapered cylindrical TSV where the top of the TSV is wider than the bottom. The per cent error for the DC resistance of a cylindrical TSV does not exceed 1%. The per cent error increases to a maximum of 4% for frequencies up to 10 GHz when applying (4.28) with a fitting parameter (7% error without the fitting parameter). A closed-form expression for a tapered cylindrical TSV at DC (listed as (4.27) in Table 4.2) produces less than 2% error. The compact model of a tapered cylindrical

(A)

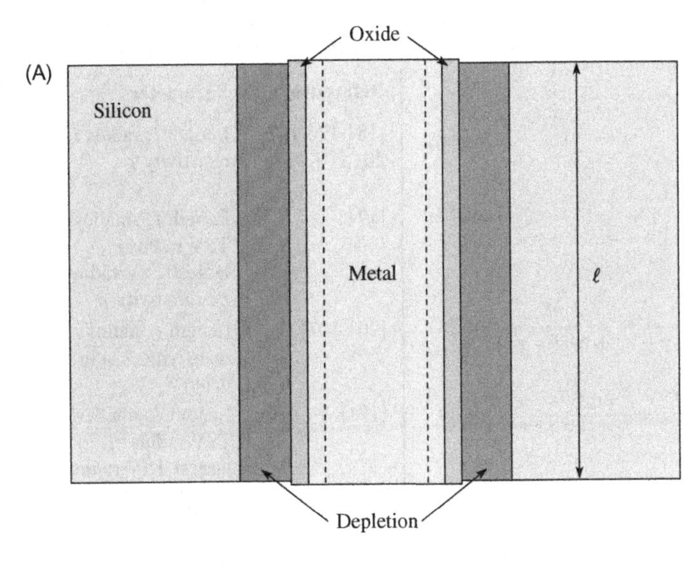

(B)

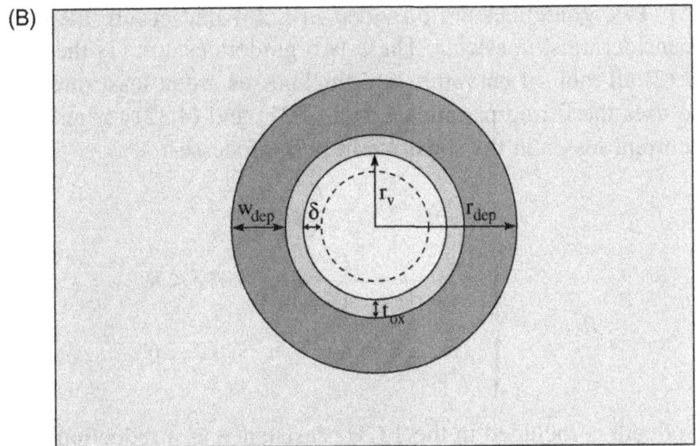

FIGURE 4.8

Physical parameters and materials used in the compact models of a TSV for a single device layer, as listed in Table 4.6. (A) Side view, and (B) top view.

TSV produces less than 6% error for frequencies up to 10 GHz when a fitting parameter is utilized in (4.29) [171].

4.5.1.2 Closed-form resistance model of a three-dimensional via

Closed-formed expressions of the 3-D via resistance at DC and 1 GHz are presented, respectively, as (4.30) and (4.31), and (4.32). The DC resistance is only dependent on the length $\mathcal{L}$, radius $\mathcal{R}$, and conductivity σ_W of tungsten. The resistance at a frequency of 1 GHz is, however, also dependent on the skin depth, the depth below the surface of a conductor where the current density drops

Table 4.2 Comparison of Closed-Form Expressions of the Resistance of a TSV

Equation (Label)	Reference	Parameters	Notes
$R_{DC} = \frac{\rho \ell}{\pi r_v^2}$ (4.26)	[181,196,197, 201,207,224]	Length ℓ, radius r_v, resistivity ρ	Constant radius
$R_{DC} = \frac{\rho \ell}{\pi r_1 (r_1 + \ell r_2)} \left[1 + \frac{1}{2} r_2^2\right]$ (4.27)	[171]	Length ℓ, smallest TSV radius r_1, largest TSV radius r_2, resistivity ρ	Tapered TSV
$R_{AC} = \frac{\rho \ell}{\pi \left(r_v^2 - (r_v - \delta)^2\right)} = \frac{\rho \ell}{\pi (2 r_v \delta - \delta^2)}$ (4.28)	[201,207]	Length ℓ, radius r_v, resistivity ρ, skin depth δ	[201] includes fitting parameter
$R_{AC} = \frac{\rho (2 + r_2^2)}{4 \pi r_2 \delta} \log\left(1 + \frac{2 r_2 \ell}{2 r_1 - \delta}\right)$ (4.29)	[171]	Length ℓ, smallest TSV radius r_1, largest TSV radius r_2, skin depth δ	Tapered TSV

by a factor of e [225]. Two guidelines are provided in [225] that ensure the high frequency resistance is valid when considering skin effects. These two guidelines are: (1) the return paths are at an infinite distance, and (2) all radii of curvature and thicknesses are at least three to four skin depths. The second guideline uses the fitting parameter α in (4.31) and (4.32) for an SOI process. In addition, α accounts for current losses in the substrate in bulk processes.

$$R_{DC} = \frac{1}{\sigma_W} \frac{\mathcal{L}}{\pi \mathcal{R}^2}, \tag{4.30}$$

$$R_{1\,\text{GHz}} = \begin{cases} \alpha \dfrac{1}{\sigma_W} \dfrac{\mathcal{L}}{\pi[\mathcal{R}^2 - (\mathcal{R} - \delta)^2]}, & \text{if } \delta < \mathcal{R} \\ \alpha \dfrac{1}{\sigma_W} \dfrac{\mathcal{L}}{\pi \mathcal{R}^2}, & \text{if } \delta \geq \mathcal{R}. \end{cases} \qquad (4.31)(4.32)$$

The effect of the skin depth is included in the 1 GHz resistance as a reduction in the cross-sectional area of the 3-D via. The skin depth δ is

$$\delta = \frac{1}{\sqrt{\pi f \mu_o \sigma_W}}, \tag{4.33}$$

where f is the frequency, and μ_o is the permeability of free space. The permeability of free space is used as neither silicon nor silicon dioxide are magnetic materials.

As noted earlier, the α term in (4.31) and (4.32) is an empirical constant to fit the resistance to the numerical simulations and is

$$\alpha = \begin{cases} 0.0472 D^{0.2831} \ln\left(\dfrac{\mathcal{L}}{D}\right) + 2.4712 D^{-0.269}, & \text{if } \delta < \mathcal{R} \\ 0.0091 D^{1.0806} \ln\left(\dfrac{\mathcal{L}}{D}\right) + 1.0518 D^{0.092}, & \text{if } \delta \geq \mathcal{R}. \end{cases} \qquad (4.34)(4.35)$$

For frequencies other than DC and 1 GHz, (4.31) and (4.32) can be adjusted to other frequencies using (4.36). The resistance of a 3-D via with a diameter of 5, 20, and 60 μm at frequencies of DC, 1 GHz, and 2 GHz is plotted in Fig. 4.9.

$$R_{f_{\text{new}}} = (R_{1\text{ GHz}} - R_{\text{DC}})\sqrt{\frac{f_{\text{new}}}{f_{1\text{ GHz}}}} + R_{\text{DC}}. \tag{4.36}$$

The proximity effect [200] is examined through simulations of two 3-D vias over a ground plane. These simulations reveal less than a 0.25% change in the resistance at all examined frequencies as compared to a single 3-D via over a ground plane. These results are therefore not included. This characteristic implies that the proximity of the second 3-D via as a return path does not significantly affect the resistance of the first via. The insignificance of the proximity effect is primarily a consequence of the relatively short length of the vias and the small space between 3-D vias, which is at least equal to the diameter of a single 3-D via.

An analysis of the per cent error between simulation and the closed-form resistance expressions as a function of frequency is provided. These results indicate less than 5.5% error between simulation and the closed-form expressions for all frequencies between DC and 10 GHz. The per cent error for two diameters, 5 and 20 μm, and three aspect ratios, 1, 5, and 9, is depicted in Fig. 4.10. The per cent variation between simulation and the model at a frequency of DC, 1 GHz, and 2 GHz is included in Section 4.5.1.3.

4.5.1.3 Per cent variation in resistance

The per cent error between the closed-form expressions described in Section 4.5.1.1 and the electromagnetic simulation for the DC, 1 GHz, and 2 GHz resistances are listed, respectively, in Tables 4.3 through 4.5. None of the errors listed in Tables 4.3 through 4.5 exceed 5.5% for all considered diameters and aspect ratios. The DC resistance based on the closed-formed expressions produces less than 2% error, while the model produces less than a 5.5% difference from numerical simulations at 1 and 2 GHz.

4.5.2 COMPACT INDUCTANCE MODEL OF A THREE-DIMENSIONAL VIA

Several compact models for the inductance of a TSV are listed in Table 4.6. The relevant physical parameters that affect the inductance of the TSV are also listed. The physical parameters include the TSV length ℓ, radius r_v, tapering angle β, and TSV-to-TSV pitch p_v. Material parameters include the permeability of the metal μ_o.

A comparison of the TSV inductance models is provided in Section 4.5.2.1. A closed-form model of the inductance of both equal and nonequal length TSVs is provided in Section 4.5.2.2. A comparison of the models described in Section 4.5.2.2 with numerical simulation is provided in Section 4.5.2.3.

4.5.2.1 Compact models of through silicon via inductance

Two basic expressions are used to model the inductance of a TSV. The expressions are listed as (4.37) and (4.38) in Table 4.6 and are derived, respectively, by Goldfarb and Pucel [210] and Rosa [206]. Goldfarb modified (4.38) to better match experimental data. Equation (4.37) was used

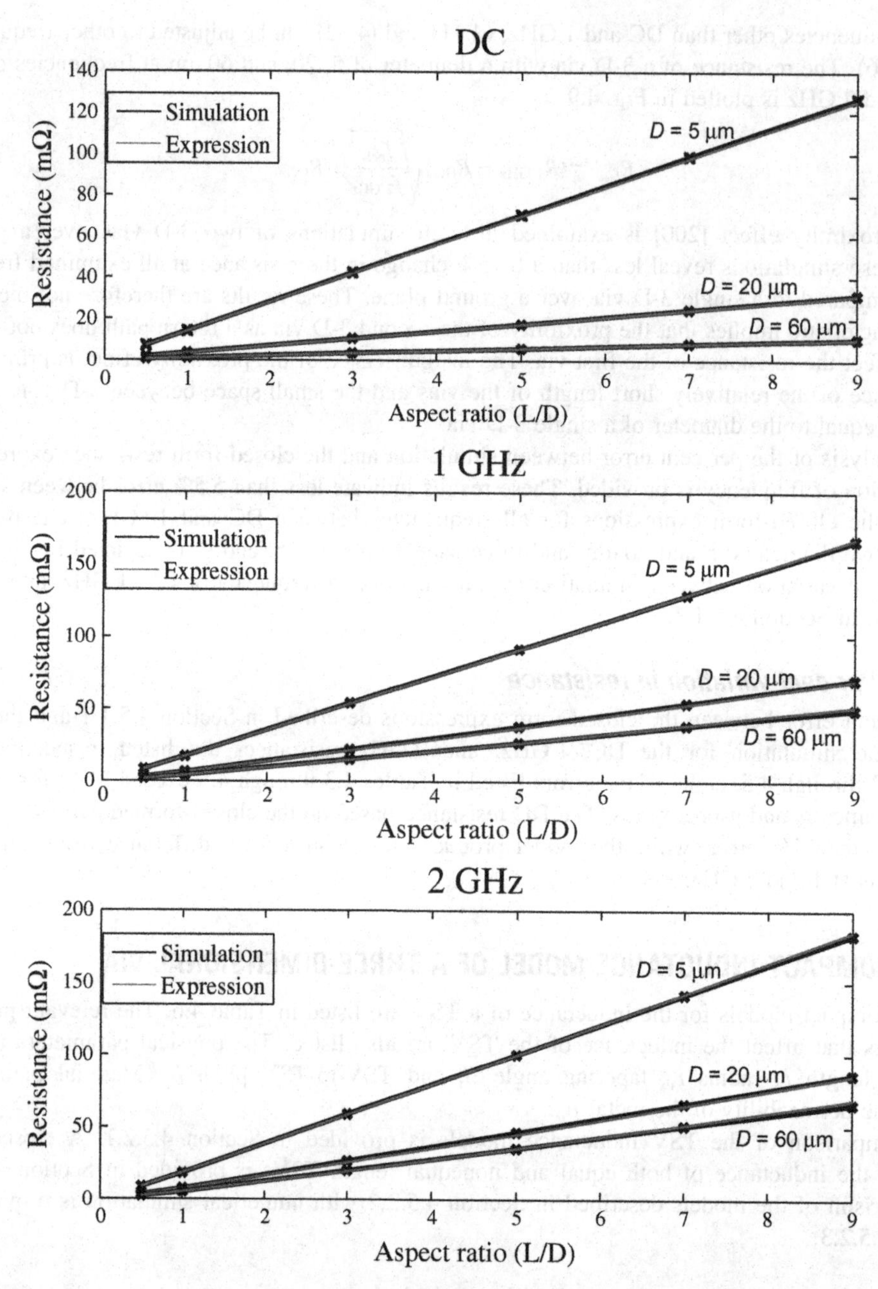

FIGURE 4.9

Resistance of a cylindrical 3-D via at DC, 1 GHz, and 2 GHz.

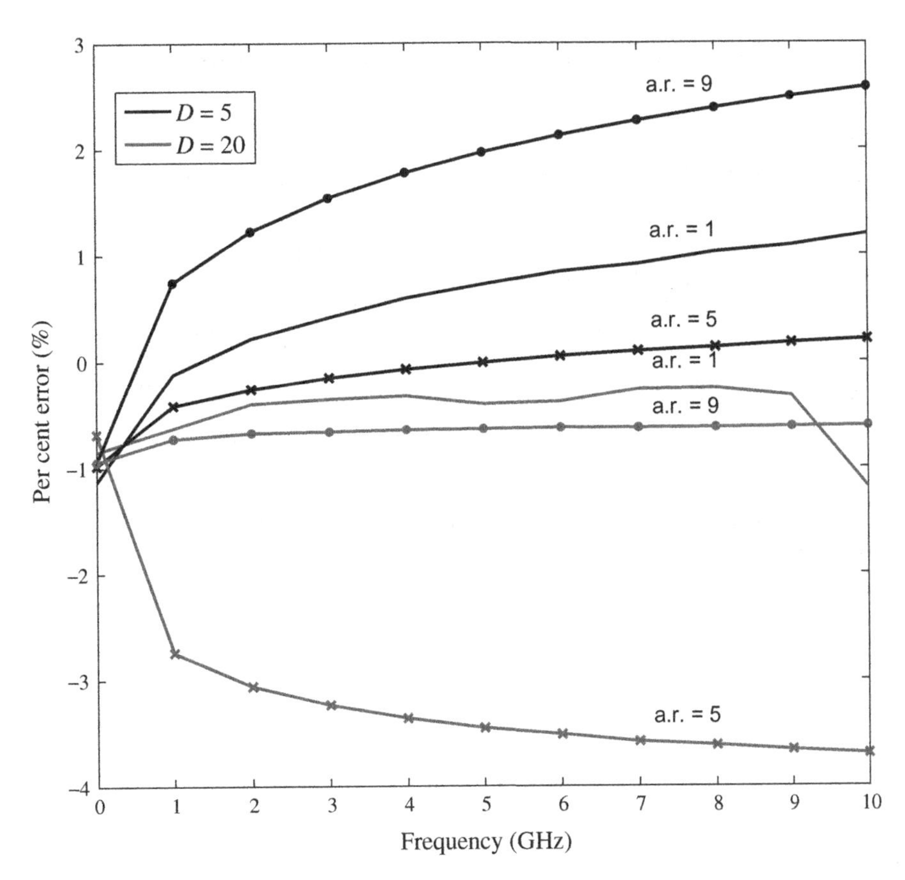

FIGURE 4.10

Per cent error as a function of frequency for the resistance of a 3-D via (a.r. = aspect ratio).

Table 4.3 Per Cent Error Between Simulations and Closed-Form Expressions of a 3-D Via, DC Resistance

	Aspect Ratio					
Diameter (μm)	**0.5**	**1**	**3**	**5**	**7**	**9**
1	−1.1	−1.1	−1.0	−1.0	−1.0	−1.0
5	−1.1	−1.1	−1.0	−1.0	−1.1	−0.9
10	−1.1	−1.0	−0.9	−0.9	−1.0	−0.9
20	−1.1	−0.9	−0.9	−0.7	−0.9	−0.9
40	−1.1	−1.1	−0.8	−0.9	−0.9	−0.9
60	−1.7	−0.9	−0.9	−0.9	−1.0	−0.9

Table 4.4 Per Cent Error Between Simulations and Closed-Form Expressions of a 3-D Via, 1 GHz Resistance

	Aspect Ratio					
Diameter (μm)	0.5	1	3	5	7	9
1	0.1	−0.1	0.04	0.1	−0.1	0.1
5	0.5	−0.1	−0.9	−0.4	0.2	0.7
10	0.9	−0.1	−0.2	0.8	0.8	1.9
20	1.4	−0.5	−3.3	−2.7	−1.7	−0.7
40	4.4	0.9	−2.7	−1.5	−0.1	1.9
60	2.3	−0.2	−2.7	0.04	1.4	4.1

Table 4.5 Per Cent Error Between Simulations and Closed-Form Expressions of a 3-D Via, 2 GHz Resistance

	Aspect Ratio					
Diameter (μm)	0.5	1	3	5	7	9
1	0.6	0.2	0.4	0.5	0.2	0.5
5	1.0	0.2	−0.8	−0.3	0.5	1.2
10	1.4	0.2	−0.1	1.2	1.3	2.6
20	1.5	−0.4	−3.7	−3.1	−1.8	−0.7
40	5.2	1.0	−2.9	−1.6	0	2.2
60	2.7	−0.4	−2.9	0.1	1.5	4.4

by Katti et al. [181] to compare his model with experimental results. The per cent error ranges from 8% to 17.3% for TSVs with a length of 400 μm and a radius of 50 to 60 μm. Rosa also developed expressions for the inductance for the case where the length ℓ significantly exceeds the radius r_v which is listed as (4.40), and for the case where the skin depth dominates, see (4.41). Both [201] and [191] use (4.41) to model the high frequency inductance of a TSV. A fitting parameter is added to both (4.38) and (4.41) in [201] to reduce the maximum error from greater than 20% to, respectively, less than 3% and 8.5%. Compact models of the DC and high frequency inductance, respectively, (4.38) and (4.41), have been verified by an electromagnetic field solver for radii of 0.5 to 30 μm and aspect ratios ($\ell/2r_v$) ranging from 0.5 to 9 [201]. Wu [191] altered (4.41) to better match experimental results with the modified expression of (4.42). Both (4.37) and (4.38) overestimate the measurements by Wu, producing errors, respectively, of 40% and 65% for an aspect ratio of 10. The errors exceed 50% and 75% for aspect ratios approaching 100. The modified expression, adding a multiplier of 0.65, reduces the maximum error to less than 25% for all aspect ratios (the radius varies from 1 to 5 μm). Equation (4.41) is modified to better match simulated results from a 3-D/2-D quasi-static electromagnetic field solver. An additional modified inductance expression is (4.43), where a fitting parameter adjusts the ratio of the length ℓ to radius r_v. The inductance from

Table 4.6 Comparison of Closed-Form Expressions of the Inductance of a TSV

Equation (Label)	Reference	Parameters	Notes
$L=\frac{\mu_o}{4\pi}\left[2\ell\ln\left(\frac{2\ell+\sqrt{r_v^2+(2\ell)^2}}{r_v}\right)+r_v-\sqrt{r_v^2+(2\ell)^2}\right]$ (4.37)	[181]	Length ℓ, radius r_v, permeability of metal μ_o	Uses Goldfarb inductance [210]
$L=\frac{\mu_o}{2\pi}\left[\ell\ln\left(\frac{\ell+\sqrt{r_v^2+\ell^2}}{r_v}\right)+r_v-\sqrt{r_v^2+\ell^2}+\frac{\ell}{4}\right]\alpha$ (4.38)	[171,191,201]	Length ℓ, radius r_v, permeability of metal μ_o, fitting parameter α	[201] added fitting parameter to Rosa inductance [206]
$L=\frac{\mu_o}{4\pi}\left[2\ell\ln\left(\frac{\ell+\sqrt{r_1^2+\ell^2}}{r_1}\right)+2r_1-2\sqrt{r_1^2+\ell^2}+\frac{\ell}{2}+\frac{4\beta\ell\left(\ell^2-(\ell+r_1)\left(\sqrt{\ell^2+r_1^2}-r_1\right)\right)}{\left(\sqrt{\ell^2+r_1^2}-\ell\right)(4r_1+\beta\ell)}\right]$ (4.39)	[171]	Length ℓ, smallest TSV radius r_1, largest TSV radius r_2, permeability of metal μ_o, tapering angle relationship $\beta(\beta=\tan\theta=\frac{r_2-r_1}{\ell})$	(4.39) goes to (4.38) as β goes to 0
$L=\frac{\mu_o}{2\pi}\ell\left[\ln\left(\frac{2\ell}{r_v}\right)-\frac{3}{4}\right]$ (4.40)	[191,206]	Length ℓ, radius r_v, permeability of metal μ_o	For $\ell\gg r_v$
$L=\frac{\mu_o}{2\pi}\ell\left[\ln\left(\frac{2\ell}{r_v}\right)-1\right]\alpha$ (4.41)	[191,201,206]	Length ℓ, radius r_v, permeability of metal μ_o, fitting parameter α	When skin depth dominates; [201] added fitting parameter
$L=0.65\frac{\mu_o}{2\pi}\ell\left[\ln\left(\frac{2\ell}{r_v}\right)-1\right]$ (4.42)	[191]	Length ℓ, radius r_v, permeability of metal μ_o	Modified [206] to match measurements

(Continued)

Table 4.6 Comparison of Closed-Form Expressions of the Inductance of a TSV *Continued*

Equation (Label)	Reference	Parameters	Notes
$L=\frac{\mu_o \ell}{2\pi}\ln\left(1+\frac{2.84\ell}{\pi r_v}\right)$ (4.43)	[196]	Length ℓ, radius r_v, permeability of metal μ_o	Modified [206]; $20\ \mu m \leq \ell \geq 140\ \mu m$, $10\ \mu m \leq r_v \geq 40\ \mu m$
$L_{21}=\frac{\mu_o}{2\pi}\left[\ell \ln\left(\frac{\ell+\sqrt{p_v^2+\ell^2}}{p_v}\right)+p_v-\sqrt{p_v^2+\ell^2}\right]\beta$ (4.44)	[171,201]	Length ℓ, TSV-to-TSV pitch p_v, permeability of metal μ_o, fitting parameter β	[201] added fitting parameter to Rosa inductance [206]
$L_{21}=\frac{\mu_o}{2\pi}\left[\ell \ln\left(\frac{\ell+\sqrt{p_v^2+\ell^2}}{p_v}\right)+p_v-\sqrt{p_v^2+\ell^2}+\frac{2\beta\ell\left(\ell^2-(\ell+p_v)\left(\sqrt{\ell^2+p_v^2}-p_v\right)\right)}{\left(\sqrt{\ell^2+p_v^2}-\ell\right)(4p_v+\beta\ell)}\right]$ (4.45)	[171]	Length ℓ, smallest TSV radius r_1, largest TSV radius r_2, TSV-to-TSV pitch p_v, permeability of metal μ_o, tapering angle relationship $\beta\left(\beta=\tan\theta=\frac{r_2-r_1}{\ell}\right)$	(4.45) approaches (4.44) as β approaches 0

(4.43) is within 3% of simulations for length ℓ ranging from 20 to 140 μm and a radius r_v ranging from 10 to 40 μm [196].

The effect of a tapered cylindrical shape on the inductance of a TSV is examined in [171]. An expression for the self-inductance is (4.39) and the mutual inductance is (4.45). Both (4.39) and (4.45) reduce to the Rosa expressions listed, respectively, as (4.38) and (4.44) where β, a measure of the tapering angle, approaches 0 [171]. The tapering angle ranges from 0 to 15 degrees, producing a maximum error of 5% at low frequencies and an error of approximately 20% at high frequencies, which decreases to 8% with a fitting parameter. Equations (4.44) and (4.45) depend upon the TSV-to-TSV pitch p_v. The p_v term is substituted for the radius of the TSV r_v in the expression for the mutual inductance.

4.5.2.2 Closed-form inductance model of a three-dimensional via

The DC and high frequency self- and mutual inductance of two equal length TSVs is included in 4.5.2.2.1. In Section 4.5.2.2.2 the DC and high frequency mutual inductance between two non-equal length TSVs is described. Non-equal length vias are used to approximate the mutual inductance between a single TSV and a stack of TSVs propagating a signal across multiple tiers.

4.5.2.2.1 Inductance of equal length three-dimensional vias

Expressions for the DC and high frequency partial self- L_{11} and mutual L_{21} inductances are provided, respectively, in (4.46) and (4.47), and (4.48) and (4.49). The DC and high frequency inductance are asymptotic, as shown in Fig. 4.16. These expressions do not consider the transition in inductance from low frequency to high frequency (in the 200 to 800 MHz range for the MITLL 180 nm CMOS technology). Note that the inductance transitions smoothly from low frequency to high frequency, indicating that the inductance of a 3-D via is bound by the DC (upper bound) and asymptotic (lower bound) inductance within this transitional range. The inductance models are based on [206] with a fitting parameter to adjust for inaccuracies in the Rosa expressions [206]. In the expressions derived by Rosa, the length $\mathcal{L}$ is assumed to be much larger than both the radius $\mathcal{R}$ (or diameter D) and the pitch P between the conductors. As $\mathcal{L}$ is not larger than $\mathcal{R}$ or P in all of the 3-D via structures considered here, α and β are used to adjust the partial inductances in [206].

$$\text{DC}: \begin{cases} L_{11} = \alpha \dfrac{\mu_o}{2\pi}\left[\ln\left(\dfrac{\mathcal{L}+\sqrt{\mathcal{L}^2+\mathcal{R}^2}}{\mathcal{R}}\right)\mathcal{L}+\mathcal{R}-\sqrt{\mathcal{L}^2+\mathcal{R}^2}+\dfrac{\mathcal{L}}{4}\right], \\ L_{21} = \beta \dfrac{\mu_o}{2\pi}\left[\ln\left(\dfrac{\mathcal{L}+\sqrt{\mathcal{L}^2+P^2}}{P}\right)\mathcal{L}+P-\sqrt{\mathcal{L}^2+P^2}\right], \end{cases} \tag{4.46)(4.47}$$

$$f_{asym}: \begin{cases} L_{11} = \alpha \dfrac{\mu_o}{2\pi}\left[\ln\left(\dfrac{2\mathcal{L}}{\mathcal{R}}\right)\mathcal{L}-1\right], \\ L_{21} = \beta \dfrac{\mu_o}{2\pi}\left[\ln\left(\dfrac{\mathcal{L}+\sqrt{\mathcal{L}^2+P^2}}{P}\right)\mathcal{L}+P-\sqrt{\mathcal{L}^2+P^2}\right], \end{cases} \tag{4.48)(4.49}$$

where

$$\alpha = \begin{cases} 1 - e^{\frac{-4.3\mathcal{L}}{D}}, & \text{if } f = DC \\ 0.94 + 0.52e^{-10\left|\frac{\mathcal{L}}{D-1}\right|}, & \text{if } f > f_{\text{asym}} \end{cases} \tag{4.50)(4.51}$$

$$\beta = \begin{cases} 1, & \text{if } f = DC \\ 0.1535 \ln\left(\frac{\mathcal{L}}{D}\right) + 0.592, & \text{if } f > f_{\text{asym}} \end{cases} \tag{4.52)(4.53}$$

The α parameter used to adjust the partial self-inductance approaches unity at DC and 0.94 at high frequencies with increasing aspect ratio $\mathcal{L}/D$. The parameter β, used to adjust the partial mutual inductance, is unity at DC and ranges between 0.49 and 0.93 at high frequencies as the aspect ratio increases, respectively, from 0.5 to 9. The Rosa expressions are most inaccurate when calculating the mutual inductance of a small aspect ratio 3-D via operating at high frequencies. A comparison of both the partial self- and mutual inductances for the adjusted Rosa expressions with numerical simulations is provided, respectively, in Figs. 4.11 and 4.12. The per cent variation between the analytic expressions and simulations for the self- and mutual inductances does not exceed, respectively, 3% and 8%.

4.5.2.2.2 Inductance of non-equal length 3-D vias

An expression for the mutual inductance at DC between two TSVs with non-equal lengths is provided in (4.54). As for the previous topologies, the DC mutual inductance is the worst case inductance, as shown in Fig. 4.13. Note also that the delay characteristics of a conductor are weakly correlated to the inductance [226]; therefore, the DC and f_{asym} inductance produces similar delay effects.

$$L_{21} = \alpha\beta\frac{\mu_o}{2\pi}\left[\ln\left(\frac{\mathcal{L}_g + \sqrt{\mathcal{L}_g^2 + P^2}}{P}\right)\mathcal{L}_g + P - \sqrt{\mathcal{L}_g^2 + P^2}\right], \tag{4.54}$$

$$\alpha = 0.8 + 0.1945\left(\frac{S}{D}\right)^{0.52}, \tag{4.55}$$

$$\beta = \beta_1 - 2e^{-0.3\frac{\mathcal{L}_y}{\mathcal{L}_g} + \beta_2}, \tag{4.56}$$

$$\beta_1 = 2.1 + 4e^{-0.375\frac{\mathcal{L}_g}{D} - 0.1}, \tag{4.57}$$

$$\beta_2 = e^{-0.21\frac{\mathcal{L}_g}{D} + 0.6} - 0.57. \tag{4.58}$$

The α parameter adjusts the partial mutual inductance between two non-equal length 3-D vias, accounting for the separation between vias. The β parameter adjusts the mutual inductance based on the ratio of the larger length 3-D via $\mathcal{L}_y$ (which represents multiple vertically stacked 3-D vias) to the shorter length via $\mathcal{L}_g$ (a single 3-D via). $\mathcal{L}_y$ is therefore always an integer multiple of $\mathcal{L}_g$. The β term is dependent on the aspect ratio $\mathcal{L}_g/D$ of a single TSV (non-stacked), as shown by (4.57) and (4.58). A comparison between the mutual inductance produced by (4.54) with numerical

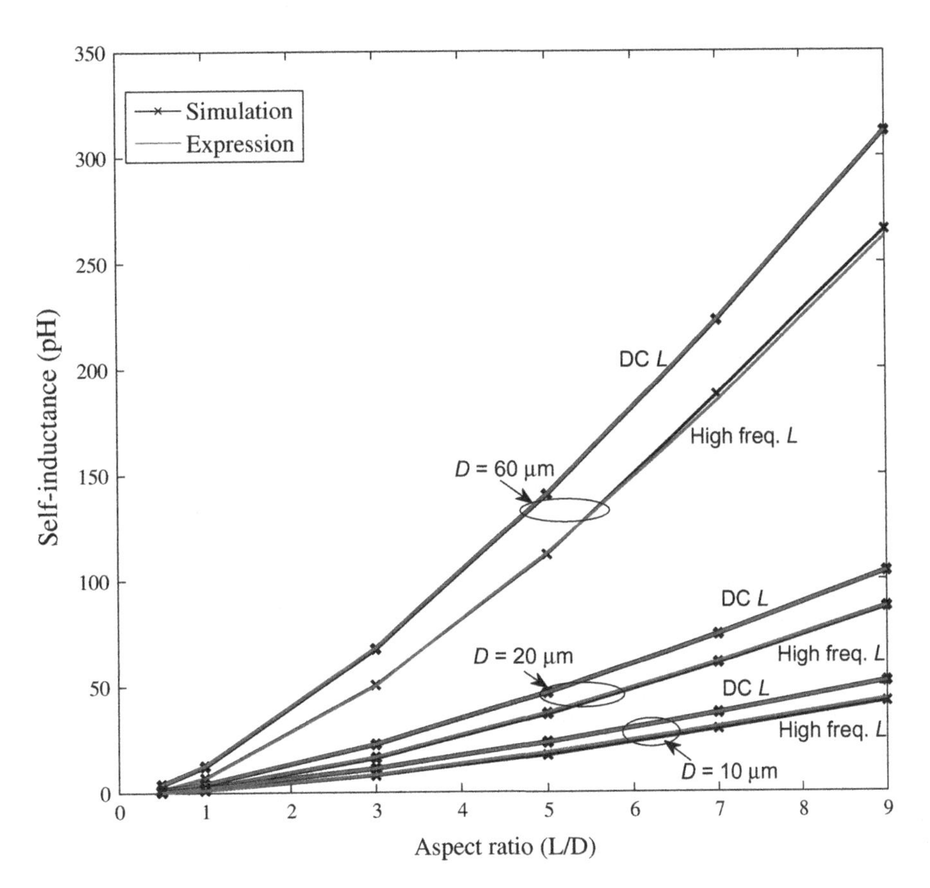

FIGURE 4.11

Self-inductance L_{11} of a cylindrical 3-D via.

simulations is provided in Fig. 4.13, and exhibits less than 8% difference between the analytic and simulated behavior.

The high frequency inductance f_{asym} is determined by multiplying the inductance produced by (4.54) with

$$\gamma = 0.955 - 1.1e^{-0.75-0.5\frac{L_g}{D}}. \quad (4.59)$$

The γ term is only dependent on the aspect ratio of a single non-stacked TSV. The self- and mutual inductance expressions produce less than an 8% variation from electromagnetic simulations.

4.5.2.3 Per cent variation in inductance

The per cent error between the closed-form expressions and electromagnetic simulation of the self- (L_{11}) and mutual (L_{21}) inductance of a 3-D via for several diameters and aspect ratios are listed, respectively, in Tables 4.7 and 4.8. The worst case DC and high frequency mutual inductance L_{21}

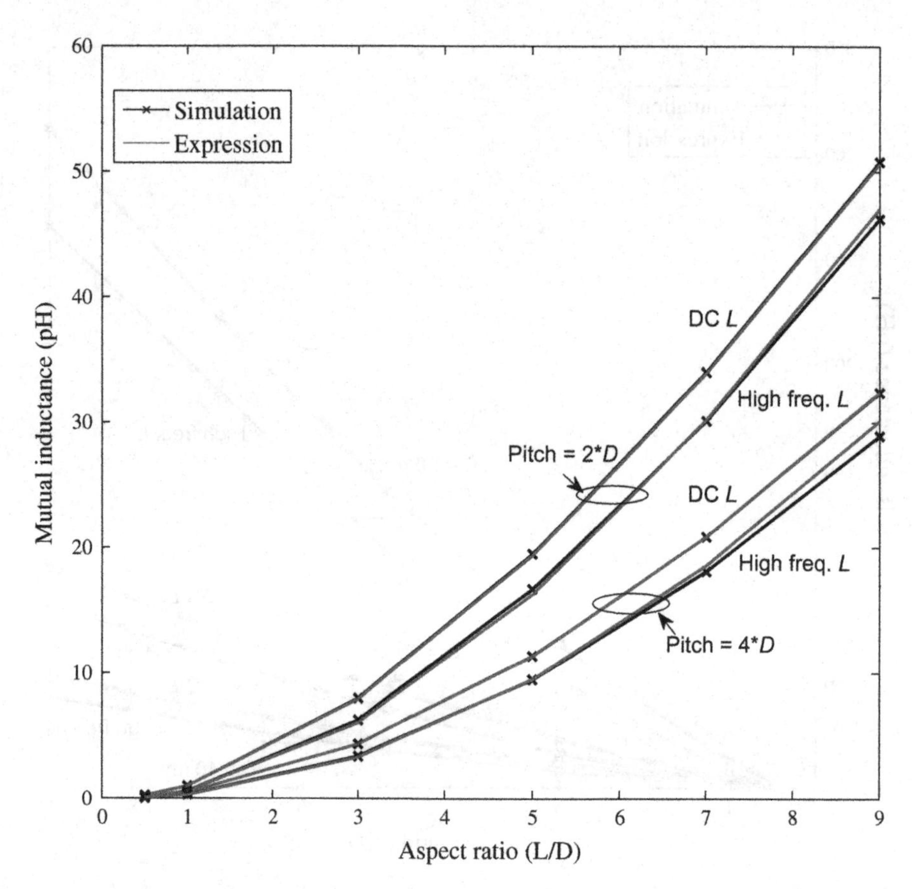

FIGURE 4.12

Mutual inductance L_{21} of a cylindrical 3-D via with a 20 μm diameter.

between two non-equal length 3-D vias are listed, respectively, in Tables 4.9 and 4.10. The error listed in these tables does not exceed 8% in all cases except for those cases where the space between the 3-D vias is at least four times the diameter and the aspect ratio is less than two. In these cases, the error does not exceed 30%. The larger error in these cases is due to the small mutual inductance (less than 0.05 pH) that exacerbates the per cent error between (4.47), (4.49), and (4.54) and the electromagnetic simulations.

4.5.3 COMPACT CAPACITANCE MODEL OF A THREE-DIMENSIONAL VIA

Several compact models for the capacitance of a TSV are listed in Table 4.11. The relevant physical parameters that affect the capacitance of the TSV are also listed. The physical parameters include the TSV length ℓ, radius r_v, oxide thickness t_{ox}, TSV-to-TSV pitch p_v, depletion region radius r_{dep} or width w_{dep}, number of body contacts n_{bc}, and distance to body contacts w_{bc}. Material

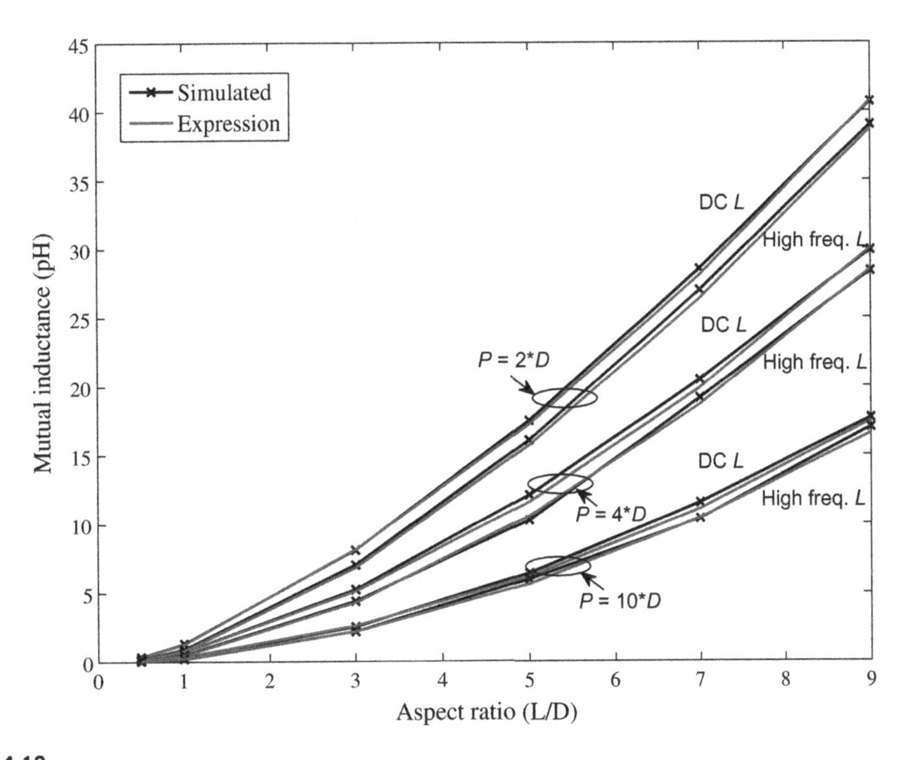

FIGURE 4.13

Mutual inductance L_{21} between two 3-D vias with different lengths ($D = 10$ μm, and $3\mathcal{L}_g = \mathcal{L}_y$).

parameters include the permittivity of the silicon (ϵ_{Si}) and oxide (ϵ_{ox}). The width of the depletion region includes the dependence on the doping concentration of the silicon. In addition, several expressions include the dependence on the applied voltage V and TSV threshold voltage V_{th}, similar to a MOS capacitor, which also exhibits a dependence on the material properties of the silicon.

A comparison of TSV capacitance models is provided in Section 4.5.3.1. A closed-form model of the TSV capacitance and capacitive coupling amongst TSVs is provided in Section 4.5.3.2. A comparison of the capacitance and capacitive coupling models described in Section 4.5.3.2 with numerical simulations is provided, respectively, in Sections 4.5.3.3 and 4.5.3.4.

4.5.3.1 Compact models of through silicon via capacitance

Several methods have been applied to the development of compact models of the TSV capacitance. These methods include (1) the worst case oxide capacitance C_{ox}, (2) a series representation of the oxide and depletion capacitance (C_{dep}), similar to an MOS capacitor, (3) a single expression encompassing both the oxide and depletion capacitance, (4) a series combination of the oxide and silicon capacitance, where a parallel plate capacitor is assumed to be formed between two TSVs in silicon [224], and (5) a series combination of the oxide, depletion, and silicon capacitance (C_{Si}) [196]. The oxide capacitance is described by (4.60) to (4.62). Most research groups use (4.61) to model the oxide capacitance of a TSV, as (4.61) more accurately models the physical dimensions

Table 4.7 Per Cent Error Between Simulations and Closed-Form Expressions of a 3-D Via, Self-Inductance (L_{11})

		Aspect Ratio					
Diameter (μm)		**0.5**	**1**	**3**	**5**	**7**	**9**
1	DC	−1.6	0	0.9	0	0	0
	High *f*	3.6	−8.3	7.7	6.8	−1.6	0.5
5	DC	0	2.9	0.5	0.2	0.1	0
	High *f*	−6.7	−8.2	−0.2	−1.1	−1.9	−2.8
10	DC	−1.6	2.4	0.7	0.2	0.08	0.04
	High *f*	3.6	0.9	6.5	8.0	4.2	3.2
20	DC	0	2.7	0.7	0.2	0.1	0.04
	High *f*	3.6	1.8	2.8	1.8	0.9	0.3
40	DC	−0.4	2.8	0.6	0.2	0.1	0.04
	High *f*	2.7	−0.9	−0.6	−0.7	−2.1	−2.1
60	DC	−0.3	2.7	0.7	0.2	0.1	0.04
	High *f*	3.0	−1.2	−0.6	0.6	−1.7	−1.2

of a TSV and reduces the error as compared to numerical simulations from approximately 10% when using (4.60) to less than 0.5% [181].

The oxide capacitance is one component of the TSV capacitance. A model that ignores the depletion region and, to a certain extent, the silicon capacitance overestimates the TSV capacitance. The depletion capacitance is determined by Salah et al. [198] and Katti et al. [181] and is listed, respectively, as (4.66) and (4.67) in Table 4.11. Katti does not provide simulation results for the depletion region capacitance in isolation since the oxide capacitance is also present, but does provide a comparison of the oxide and depletion capacitance between the analytic model and numerical simulations [181]. A maximum error of 3.5% is reported for a TSV that includes a depletion region with a maximum radius r_{dep}. Characterization of the depletion region is determined for two different TSV radii, 1 and 2.5 μm, and an oxide thickness of, respectively, 50 and 120 nm. The length is 20 μm for both radii. Katti also compares the measured TSV capacitance, but with numerically simulated geometries rather than based on a closed-form expression. However, an analysis of the oxide capacitance, based on (4.61), and the minimum TSV capacitance, using (4.67) as a function of the TSV diameter and oxide thickness is provided by Katti et al. [181]. Both the oxide and maximum depletion capacitance are considered in (4.67). The depletion capacitance developed by Salah et al. [198] is verified by electromagnetic field solvers (Q3D Extractor) by comparing the *S* parameters produced by the field solver with those parameters produced by a lumped element model. The lumped element model includes the capacitance of the oxide (4.61), depletion region (4.66), and silicon (4.68). The silicon capacitance includes a fitting parameter α that ranges from 1.5 to 3.5, and is adjusted based on the number of surrounding TSVs. The lumped model also includes a compact model of the silicon and depletion region conductance in addition to the resistance and inductance of the TSV [198].

Table 4.8 Per Cent Error Between Simulations and Closed-Form Expressions of a 3-D Via, Mutual Inductance (L_{21})

			Aspect Ratio					
Diameter (μm)	**Pitch**		**0.5**	**1**	**3**	**5**	**7**	**9**
1	$2 \cdot D$	DC	−7.7	−2.0	0	0	0	−0.4
		High *f*	0	−6.5	0	1.3	4.1	0.9
5	$2 \cdot D$	DC	−1.6	0	−0.5	−0.4	−0.2	−0.3
		High *f*	0	0	−3.8	−3.1	−1.2	0.9
10	$2 \cdot D$	DC	−1.6	2.4	0.7	0.2	0.08	0.04
		High *f*	3.6	0.9	6.5	8.0	4.2	3.2
20	$2 \cdot D$	DC	0	−1.0	−0.6	−0.5	−0.3	−0.3
		High *f*	0	0	−3.1	−2.5	0.3	1.8
	$3 \cdot D$	DC	0	−1.5	−0.4	−0.2	−0.2	−0.2
		High *f*	0	−2.5	−4.1	−2.4	0.4	4.5
	$4 \cdot D$	DC	−7.7	0	−0.2	−0.2	0	0.1
		High *f*	1.7	−3.3	−2.1	0.3	2.6	4.1
	$5 \cdot D$	DC	0	0	−0.3	−0.1	−0.1	0.1
		High *f*	19.5	0	−4.3	−1.5	0.5	2.7
40	$2 \cdot D$	DC	−2.0	−1.5	−0.7	−0.4	−0.3	−0.3
		High *f*	0	−0.9	−3.8	−1.4	−0.1	2.5
60	$2 \cdot D$	DC	−1.3	−1.3	−0.7	−0.4	−0.3	−0.2
		High *f*	0	−1.7	−4.2	−2.6	−0.1	1.4

Expressions for the silicon capacitance and conductance developed by Salah et al. also include parameters that consider the number of body contacts present in the silicon n_{bc} and the distance to the body contact w_{bc}. The per cent difference between the S_{21} parameter results produced by a lumped model and numerical simulations does not exceed 6%. The S_{21} parameters for both the model and simulation have been extracted for a frequency range from 1 to 10 GHz.

In addition to the silicon capacitance derived by Salah et al. [198], Kim et al. developed a compact model assuming that a parallel plate capacitor is formed between two TSVs in silicon [224]. The silicon capacitance model is listed as (4.69) in Table 4.11. The capacitance model of the TSV includes the oxide capacitance (4.60) and the silicon capacitance (4.69). The compact model is verified by comparing S_{21}, similar to Salah et al. [198]. The lumped model of the TSV includes the resistance, inductance, and conductance of the TSV. The per cent difference between the model and the 3-D field solver ranges from 2 to 15% for frequencies up to 20 GHz. A comparison is made for TSV radii of 5, 10, and 15 μm, distance between TSVs of 100, 120, and 140 μm, and oxide thickness of 500, 700, and 900 nm. The greatest error occurs at the lower frequency (up to 3 GHz) where the large TSV capacitance dominates, demonstrating that the reduction in TSV capacitance due to the depletion capacitance is negligible.

Table 4.9 Per Cent Error Between Simulations and Closed-Form Expressions of Two 3-D Vias with Non-equal Length, DC Mutual Inductance (L_{21})

			Aspect Ratio					
Diameter (μm)	$\frac{\mathcal{L}_{via2}}{\mathcal{L}_{via1}}$	Pitch	0.5	1	3	5	7	9
1	1	$2 \cdot D$	−7.7	−2.0	0	0	0	−0.39
	2	$2 \cdot D$	−4.0	−3.2	−2.4	−4.7	−4.0	−1.7
	3	$2 \cdot D$	−5.6	−3.1	0	−1.7	−1.4	0.3
10	1	$2 \cdot D$	−4.8	−2.0	−0.8	−0.4	−0.3	−0.2
	2	$2 \cdot D$	−6.0	−3.2	−2.6	−4.5	−4.0	−1.6
	3	$2 \cdot D$	−5.8	−3.1	0.4	−1.6	−1.6	0.2
	3	$4 \cdot D$	5.4	1.4	−2.5	−4.2	−2.9	0.4
	3	$6 \cdot D$	16.7	8.2	−1.8	−5.6	−4.1	−0.6
	3	$10 \cdot D$	28.0	20	3.2	−3.7	−3.9	−1.4
40	1	$2 \cdot D$	−2.0	−1.5	−0.7	−0.4	−0.3	−0.3
	2	$2 \cdot D$	−5.1	−3.2	−2.5	−4.5	−4.0	−1.6
	3	$2 \cdot D$	−6.3	−3.3	0.4	−1.6	−1.6	0.2

Table 4.10 Per Cent Error Between Simulations and Closed-Form Expressions of Two 3-D Vias with Non-equal Length, High-Frequency Mutual Inductance (L_{21})

			Aspect Ratio					
Diameter (μm)	$\frac{\mathcal{L}_{via2}}{\mathcal{L}_{via1}}$	Pitch	0.5	1	3	5	7	9
1	1	$2 \cdot D$	0	−6.5	0	1.3	4.1	0.9
	2	$2 \cdot D$	−7.1	−10.8	−0	−0.8	0.9	2.1
	3	$2 \cdot D$	−5.0	−15.8	0	1.3	3.1	4.6
10	1	$2 \cdot D$	0	0	0.3	3.0	4.1	5.7
	2	$2 \cdot D$	0	−1.7	−3.2	−3.9	1.3	−2.3
	3	$2 \cdot D$	−5.0	−10.1	−2.7	−2.2	−2.3	−0.8
	3	$4 \cdot D$	1.9	−7.8	−2.5	3.0	−2.5	0.4
	3	$6 \cdot D$	8.3	−2.9	−0.6	−6.7	−4.9	−2.1
	3	$10 \cdot D$	20.5	9.5	0	−5.7	0.3	−2.6
40	1	$2 \cdot D$	0	−0.9	−3.8	−1.4	−0.1	2.5
	2	$2 \cdot D$	0	−6.1	−2.9	−2.2	−2.1	1.0
	3	$2 \cdot D$	−9.8	−10.8	−1.4	−1.4	−0.2	1.3

Table 4.11 Comparison of Closed-Form Expressions of the Capacitance of a TSV

Equation (Label)	Reference	Parameters	Notes
$C_{ox} = \frac{\epsilon_o \epsilon_{ox}}{t_{ox}} 2\pi r_v \ell$ (4.60)	[191,224]	Length ℓ, radius r_v, oxide thickness t_{ox}, oxide permittivity ϵ_{ox}	
$C_{ox} = \frac{\epsilon_o \epsilon_{ox}}{\ln\left(\frac{r_v + t_{ox}}{r_v}\right)} 2\pi\ell$ (4.61)	[181,192,196, 227,228]	Length ℓ, radius r_v, oxide thickness t_{ox}, oxide permittivity ϵ_{ox}	
$C_{ox} = \frac{\epsilon_o \epsilon_{ox}(r_v + t_{ox})}{t_{ox}} 4\ell$ (4.62)	[229]	Length ℓ, radius r_v, oxide thickness t_{ox}, oxide permittivity ϵ_{ox}	
$\frac{1}{C} = \frac{1}{2\pi\ell\epsilon_{ox}} \ln\left(\frac{r_v + t_{ox}}{r_v}\right) + \frac{1}{2\pi\ell\epsilon_{ox}} \ln\left(\frac{r_v + t_{ox}}{r_v}\right)$ (4.63)	[214]	Length ℓ, radius r_v, depletion region width w_{dep}, oxide thickness t_{ox}, oxide permittivity ϵ_{ox}, silicon permittivity ϵ_{Si}	Includes both C_{ox} and C_{Si}
$C = \frac{63.34\, \ell\epsilon_o}{\ln\left(1 + 5.26 \frac{\ell}{r_v}\right)}$ (4.64)	[196]	Length ℓ, radius r_v, permittivity of free space ϵ_o	No oxide thickness parameter, no depletion region
$C = \alpha\beta \frac{\epsilon_o \epsilon_{ox}}{t_{ox} + \frac{\epsilon_{ox}}{\epsilon_{Si}} w_{dep}} 2\pi r_v \ell$ (4.65)	[201]	Length ℓ, radius r_v, depletion region width w_{dep}, oxide permittivity ϵ_{ox}, silicon permittivity ϵ_{Si}	Fitting parameters α and β adjust for, respectively, distance to ground and portion of metal contributing to C
$C_{dep} = \frac{\epsilon_o \epsilon_{Si}}{\ln\left(\frac{r_v - t_{ox} + w_{dep}}{r_v}\right)\sqrt{1 + \frac{V}{V_{th}}}} 2\pi\ell$ (4.66)	[196]	Length ℓ, radius r_v, depletion region width w_{dep}, oxide thickness t_{ox}, silicon permittivity ϵ_{Si}, applied voltage V, threshold voltage V_{th}	
$C_{dep} = \frac{\epsilon_o \epsilon_{Si}}{\ln\left(\frac{r_{dep}}{r_v + t_{ox}}\right)} 2\pi\ell$ (4.67)	[181]	Length ℓ, radius r_v, oxide thickness t_{ox}, depletion region radius r_{dep}	
$C_{Si} = \frac{0.5\, \epsilon_o \epsilon_{Si} \alpha (8n_{bc} + 1) \frac{p_v}{r_v}}{\ln\left(\frac{r_v + w_{bc}}{r_v}\right)} \pi\ell$ (4.68)	[196]	Length ℓ, radius r_v, TSV-to-TSV pitch p_v, number of body contacts n_{bc}, distance to body contact w_{bc}, silicon permittivity ϵ_{Si}	Includes fitting parameter α
$C_{Si} = \epsilon_o \epsilon_{Si} \frac{\frac{p_v}{2} + d_v + \beta}{\alpha p_v} \ell$ (4.69)	[224]	Length ℓ, diameter d_v, TSV-to-TSV pitch p_v, scaling factors α and β, silicon permittivity ϵ_{Si}	
$C_{Si} = \frac{\ell\pi\epsilon_{Si}}{\ln\left[\frac{p_v}{2r} + \sqrt{\left(\frac{p_v}{2r}\right)^2 - 1}\right]}$ (4.70)	[219]	Length ℓ, radius r_v, TSV-to-TSV pitch p_v, silicon permittivity ϵ_{Si}	

Two expressions listed in Table 4.11 produce a total TSV capacitance by either empirically fitting numerical values to match simulation or by combining the oxide and depletion capacitance into a single expression (which includes the fitting parameters). These expressions are listed, respectively, as (4.64) and (4.65). Equation (4.63), derived by Xu et al. [214], is a series combination of (4.67) and (4.61). Both (4.64) and (4.65) produce errors of less than 8% as compared to electromagnetic field simulations. The TSV capacitance of Weerasekera et al. [197] in (4.64) is compared with simulations for a range of physical parameters characterizing a TSV. The parameters include a TSV length ℓ ranging from 20 to 140 μm and radius r_v ranging from 10 to 40 μm. In comparison, Savidis and Friedman [201] examine radii of 10, 20, and 30 μm with aspect ratios ($\ell/2r_v$) ranging from 0.5 to 9. The primary difference between these two expressions is that (4.64) does not consider the depletion region, and therefore ignores the oxide thickness. The 8% error reported in [197] does, however, include inaccuracies from neglecting the oxide for thicknesses up to 1 μm.

4.5.3.2 Closed-form capacitance model of a three-dimensional via

Prior work, [191] and [192], examining the capacitance of bulk 3-D vias, neglected two important physical characteristics. The first issue is the formation of a depletion region in the bulk substrate surrounding a TSV, and the second issue is the assumption that the electrical field lines from the 3-D via terminate on a cylinder surrounding the dielectric liner of the via. Equations (4.71) and (4.72) from, respectively, [191] and [192] overestimate the 3-D via capacitance. Equation (4.73) considers both the formation of a depletion region surrounding a p-type substrate in a bulk process and the termination of the electrical field lines on a ground plane below the 3-D via. Termination of the field lines from the 3-D via to the ground plane forms a capacitance to the on-chip metal interconnect.

$$C = \frac{\epsilon_{Si}}{t_{diel}} 2\pi \mathcal{R} H, \tag{4.71}$$

$$C = \frac{\epsilon_{Si}}{\ln\left(\frac{\mathcal{R}+t_{diel}}{\mathcal{R}}\right)} 2\pi H, \tag{4.72}$$

$$C = \alpha\beta \cdot \frac{\epsilon_{Si}}{t_{diel} + \frac{\epsilon_{SiO_2}}{\epsilon_{Si}} x_{dT_p}} 2\pi \mathcal{R} H. \tag{4.73}$$

Note that (4.73) is dependent on the depletion region depth x_{dTp} in doped p-type silicon (the doped acceptor concentration N_A is, in this case, $\times 10^{15}$ cm^{-3}). The depletion region is, in turn, dependent on the p-type silicon work function ϕ_{fp}. The thermal voltage (kT/q) at $T = 300$ K is 25.9 mV, where q is the electron charge (1.6×10^{-19} C) and k is the Boltzmann constant, 1.38×10^{-23} J/K.

$$x_{dT_p} = \sqrt{\frac{4\epsilon_{Si}\phi_{f_p}}{qN_A}}, \tag{4.74}$$

$$\phi_{f_p} = V_{th} \ln\left(\frac{N_A}{n_i}\right). \tag{4.75}$$

The fitting parameters, α and β, adjust the capacitance for the two physical factors. The β parameter adjusts the capacitance of a 3-D via since a smaller component of the capacitance is contributed by the portion of the 3-D via farthest from the ground plane. A decrease in the growth of the capacitance, therefore, occurs with increasing aspect ratio. The α term adjusts the capacitance based on the distance to the ground plane S_{gnd}. As S_{gnd} increases, the capacitance of the 3-D via decreases. The α and β terms are, respectively,

$$\alpha = \left(-0.0351\frac{\mathcal{L}}{D} + 1.5701\right) S_{\text{gnd}}^{0.0111\frac{\mathcal{L}}{D}-0.1997}, \tag{4.76}$$

$$\beta = 5.8934 D^{-0.553} \left(\frac{\mathcal{L}}{D}\right)^{-(0.0031D+0.43)}. \tag{4.77}$$

The 3-D via capacitance for diameters of 20, 40, and 60 μm is shown in Fig. 4.14. The per cent variation between simulation and the model is included in Table 4.12 of Section 4.5.3.3. The error of the capacitance produced by (4.73) does not exceed 8%. In addition to a closed-form expression for the capacitance of a single 3-D via over a ground plane, an expression for the coupling capacitance between two 3-D vias over a ground plane is

$$C_c = 0.4\alpha\beta\gamma \cdot \frac{\epsilon_{Si}}{S}\pi DH. \tag{4.78}$$

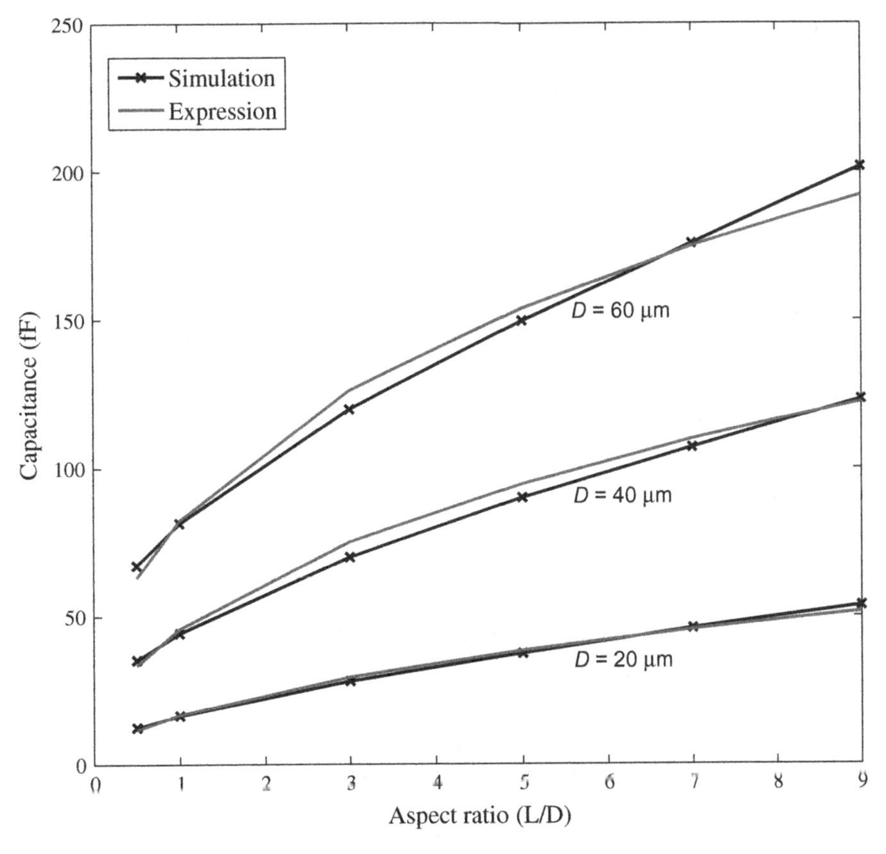

FIGURE 4.14

Capacitance of a cylindrical 3-D via over a ground plane.

Table 4.12 Per Cent Error Between Simulations and Closed-Form Expressions of a 3-D Via, Capacitance

		Aspect Ratio					
Diameter (μm)	Space to Gnd (μm)	0.5	1	3	5	7	9
20	10	−5.8	2.0	4.7	2.4	−1.0	−3.9
	30	2.4	−8.0	−5.9	−2.1	0.7	3.1
	50	6.1	−5.6	−4.4	−0.9	2.3	3.3
40	10	−4.9	3.4	7.4	5.1	2.7	−0.7
60	10	−6.2	1.2	5.3	2.8	−0.5	−4.7

The 0.4 multiplier in (4.78) adjusts the sheet capacitance between two TSVs when assuming that all electric field lines originating from half of the surface of one TSV terminate on the other TSV. Each fitting parameter (α, β, and γ) adjusts the coupling capacitance for a specific physical effect. The α term considers the nonlinearity of the coupling capacitance as a function of the aspect ratio $\mathcal{L}/D$. The effect of the separation between the TSVs to the ground plane S_{gnd} on the coupling capacitance is included in β. Note that the β term is dependent on the aspect ratio of the TSV. The γ parameter accounts for the nonlinearity of the coupling capacitance as a function of the distance S between two TSVs. The γ parameter is also dependent on the TSV aspect ratio. The pitch P, the sum of the distance between two vias and a single TSV diameter ($P = S + D$), is also included in γ. The three terms are

$$\alpha = 0.225 \ln\left(0.97\frac{\mathcal{L}}{D}\right) + 0.53, \tag{4.79}$$

$$\beta = 0.5711\frac{\mathcal{L}}{D}^{-0.988} \ln(S_{gnd}) + \left(0.85 - e^{-\frac{\mathcal{L}}{D}+1.3}\right), \tag{4.80}$$

$$\gamma = \begin{cases} 1, & \text{if } \frac{S}{D} \le 1 \\ \zeta\left[\ln\left(\frac{\mathcal{L}}{D} + 4e^{-\frac{S}{9}+51} + 2.9\right) - 10.625S^{-0.51}\right], & \text{if } \frac{S}{D} > 1 \end{cases} \tag{4.81)(4.82}$$

where ζ includes the dependence on the ratio of the pitch to the diameter and is

$$\zeta = \left(1 + e^{-(0.5+|\frac{P}{D}-4|)}\right). \tag{4.83}$$

The 3-D via coupling capacitance for a diameter of 20, 40, and 60 μm is shown in Fig. 4.15. The per cent variation between the simulations and the model is described in Section 4.5.3.4. The error of the capacitance produced by (4.78) does not exceed 15% for all aspect ratios greater than one. When the aspect ratio is between 0.5 and 1, the error between the simulations and the closed-form expression is much greater as the coupling capacitance between the 3-D vias in all cases is below 1.5 fF.

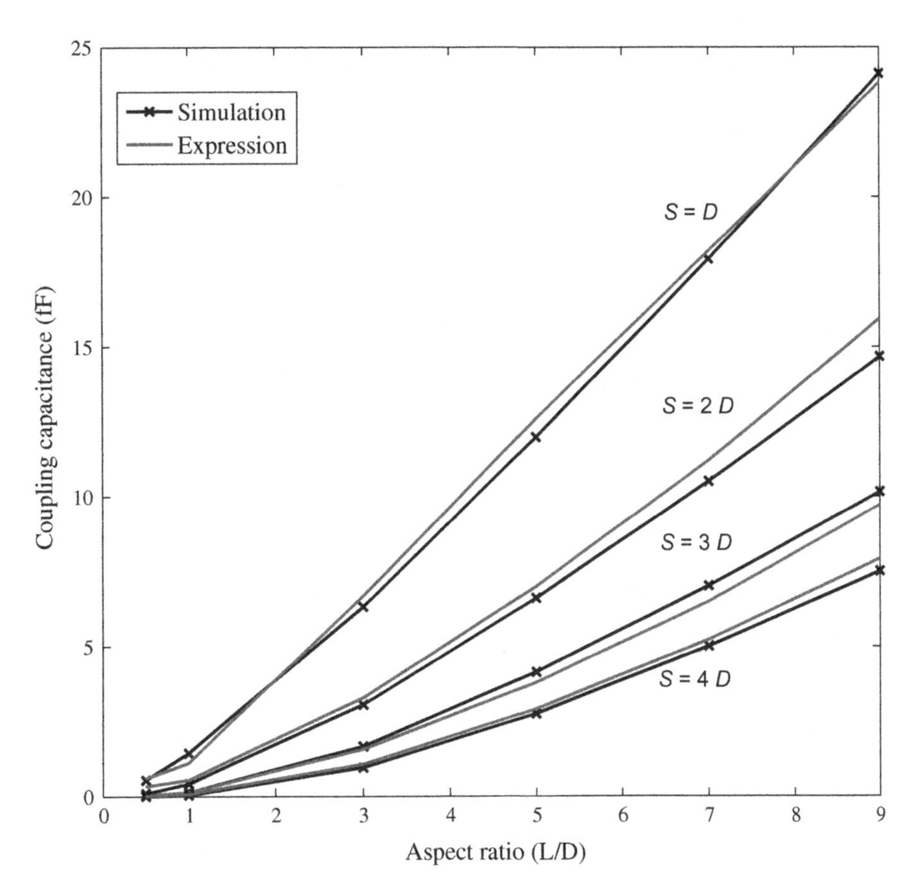

FIGURE 4.15

Coupling capacitance between two 3-D vias over a ground plane ($D = 20$ μm).

4.5.3.3 Per Cent variation in capacitance

The per cent error between the closed-form expression and the electromagnetic simulations of the capacitance of a 3-D via over a ground plane is listed in Table 4.12. All errors listed in Table 4.12 exhibit a difference of less than 8% between the expressions and simulations.

4.5.3.4 Per Cent variation in coupling capacitance

The per cent error between the closed-form expression and the electromagnetic simulations of the coupling capacitance between two 3-D vias over a ground plane is listed in Table 4.13. All errors listed in Table 4.13 exhibit a difference of less than 15% between the expressions and simulation for all aspect ratios greater than one.

Table 4.13 Per Cent Error Between Simulations and Closed-Form Expressions of a 3-D Via, Coupling Capacitance

Diameter (μm)	Space to Gnd (μm)	Space to Via (μm)	Aspect Ratio					
			0.5	1	3	5	7	9
20	10	D	9.3	−22.9	5.7	5.1	1.5	−1.2
		2·D	230	31.0	7.5	5.7	6.6	8.7
		3·D	95	0	−6.0	−8.5	−7.4	−4.2
		4·D	275	83.7	12.5	5.8	4.4	5.6
	30	D	−3.3	−16.2	4.3	4.6	1.4	−1.0
	50	D	−4.6	−14.5	3.4	3.7	0.9	−1.3
40	10	D	71	−10.1	2.3	−1.1	−3.4	−6.5
60	10	D	103	−5.9	1.4	−2.5	−5.8	−9.4

Table 4.14 Comparison of Closed-Form Expressions of the Conductance of a TSV

Equation (Label)	Reference	Parameters	Notes
$G_{Si} = \sigma_{Si} \frac{\frac{p_v}{2} + d_v + \beta}{\alpha p_v} \ell$ (4.84)	[224]	Length ℓ, diameter d_v, TSV-to-TSV pitch p_v, scaling factors α and β, silicon conductivity σ_{Si}	Fitting parameters are in μm
$G_{Si} = \frac{0.5\,\sigma_{Si}(8n_{bc}+1)}{\alpha \frac{p_v}{r_v} \ln\left(\frac{r_v + w_{bc}}{r_v}\right)} \pi \ell$ (4.85)	[198]	Length ℓ, radius r_v, TSV-to-TSV pitch p_v, number of body contacts n_{bc}, distance to body contact w_{bc}, silicon conductivity σ_{Si}	Includes fitting parameter α
$G_{Si} = \frac{\ell \pi \sigma_{Si}}{\ln\left[\frac{p_v}{2r} + \sqrt{\left(\frac{p_v}{2r}\right)^2 - 1}\right]}$ (4.86)	[219]	Length ℓ, radius r_v, TSV-to-TSV pitch p_v, silicon conductivity σ_{Si}	
$G_{dep} = \frac{\sigma_{Si}}{\ln\left(\frac{r_v + t_{ox} + w_{dep}}{r_v}\right)\sqrt{1 + \frac{V}{V_{th}}}} 2\pi\ell$ (4.87)	[198]	Length ℓ, radius r_v, depletion region width w_{dep}, oxide thickness t_{ox}, silicon conductivity σ_{Si}, applied voltage V, threshold voltage V_{th}	

4.5.4 COMPACT CONDUCTANCE MODEL OF A THREE-DIMENSIONAL VIA

A complete model of a TSV includes the electrical conductance of the silicon substrate. The physical principles governing the conductance in a bulk silicon substrate are described in this section.

Compact models for the conductance of a TSV are listed in Table 4.14. The relevant physical parameters that affect the conductance are also listed. The physical parameters include the TSV length ℓ, radius r_v or diameter d_v, oxide thickness t_{ox}, TSV-to-TSV pitch p_v, depletion region width w_{dep}, number of body contacts n_{bc}, and distance to the body contacts w_{bc}. Material parameters include the conductivity of the silicon (σ_{Si}). Similar to the compact capacitance models, several expressions include the dependence on the applied voltage V and TSV threshold voltage V_{th}.

4.5.4.1 Compact models of through silicon via conductance

The conductance of the silicon and depletion region is also listed in Table 4.14. A closed-form expression for the conductance of silicon is the same as the capacitance, but the permittivity of silicon is replaced by the conductivity. Equations (4.84) and (4.85) for the conductance of silicon are, respectively, by Kim et al. [224] and Salah et al. [198] for a lumped model representation of the TSV. The conductance model has not been verified with electromagnetic simulation, but the S_{21} parameter extracted from simulation is compared with a lumped model of a TSV. Note that the lumped model by Salah et al. also includes the depletion region conductance (as well as the capacitance), which is listed as (4.87) in Table 4.14.

The silicon capacitance and conductance are listed, respectively, as (4.70) and (4.87) [219]. These two expressions are provided for comparison with the other capacitance and conductance expressions for silicon-based TSV models.

4.6 ELECTRICAL CHARACTERIZATION THROUGH NUMERICAL SIMULATION

Characterization of the electrical properties of the TSV is often achieved through numerical simulation. The primary input of a numerical simulator is the physical structure of the item under study, including the surrounding environment and material properties. Several full wave and 2.5-D electromagnetic simulation tools are available including Quick 3D [212] and High Frequency Structural Simulator (HFSS) [230,231] from Ansys, EIP from IBM [232], FastHenry [233] and FastCap from MIT [213], EMPro 3D from Agilent [234,235], Sonnet Suites from Sonnet [236,237], and 3D EM Field Simulation from CST [238], to name just a few.

Current numerical simulators support the graphical input of shapes and structures, material properties, and shape manipulations through a graphical user interface or coded input that includes the coordinates of the shapes (rectangles, circles, spheres, prisms, etc.), shape manipulation commands, and the material properties of each shape. Several benefits and drawbacks exist for each technique. A graphical interface is easy and intuitive, allowing one to quickly form and define shapes and structures for simulation. The drawback is that any changes in the physical structure of the TSV, in particular, the length, requires a completely new set of shapes and structures to match the change in length. Alternatively, in a coded input, a steep learning curve exists to properly describe the structures, shapes, material properties, and shape manipulations for each simulation. Any change in a physical structure is, however, easily propagated to other shapes through a simple search and replace.

These field solvers produce electric field distributions, magnetic field distributions, and current and voltage distributions. In addition, field solvers compute the capacitance, conductance, inductance, and resistance matrices by solving Maxwell's equations through the method of moments (MOMs) [239,240] and finite element method (FEM) [241] techniques. Maxwell's equations are listed in Table 4.15 in both differential and integral form, as satisfying Maxwell's equations for the physical geometries and material properties is used to characterize the electrical properties of a

Table 4.15 Maxwell Equations

Law	Differential Form	Integral Form
Gauss's law	$\nabla \cdot \vec{D} = \rho$	$\oint_S \vec{D} \cdot d\vec{A} = \int_V \rho d\vec{V}$
Gauss's law (magnetism)	$\nabla \cdot \vec{B} = 0$	$\oint_S \vec{B} \cdot d\vec{A} = 0$
Faraday's law	$\nabla \times \vec{E} = \frac{\partial \vec{B}}{\partial t}$	$\oint_C \vec{E} \cdot d\vec{l} = -\frac{\partial}{\partial t} \int_S \vec{B} \cdot d\vec{A}$
Gauss's law (magnetism)	$\nabla \times \vec{H} = \vec{J} + \frac{\partial \vec{D}}{\partial t}$	$\oint_C \vec{B} \cdot d\vec{l} = \int_S \vec{J} \cdot d\vec{A} + \frac{\partial}{\partial t} \int_S \vec{D} \cdot d\vec{A}$

The relationship of the field vectors is described by $\vec{D} = \epsilon\vec{E}$ *and* $\vec{B} = \mu\vec{H}$.

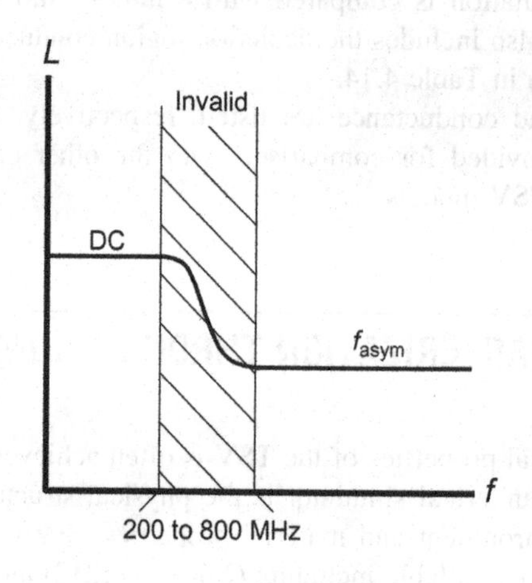

FIGURE 4.16

Frequency range applicable to Q3D models and closed-form inductance expressions.

structure. The electromagnetic field solver Ansys Quick 3-D (Q3D) is described in Section 4.6.1, while a broader analysis of the TSV impedance with numerical simulation is provided in Section 4.6.2.

4.6.1 ANSYS QUICK THREE-DIMENSIONAL ELECTROMAGNETIC FIELD SOLVER

Ansys Q3D is a software package used to examine the *RLC* impedances, coupling, and shielding behavior of TSVs [212]. Q3D is a 3-D/2-D quasi-static electromagnetic field simulator used to extract parasitic impedances and electrical parameters. The tool solves Maxwell's equations by applying the FEM [241] and the MOMs [239,240] to compute the *RLC* or *RLCG* impedance of a 3-D structure. Q3D determines the asymptotic DC and high frequency (f_{asym}) inductance, as depicted in Fig. 4.16. Q3D simulations do not provide the inductance for frequencies between DC and high frequency extrema. The closed-form expressions for the inductance, described in Section 4.5.2.1, are therefore applicable for frequencies lower and higher than the transitional frequency range (between 200 and 800 MHz).

4.6.2 NUMERICAL ANALYSIS OF THROUGH SILICON VIA IMPEDANCE

The resistance, capacitance, inductance, and conductance of a TSV varies significantly depending upon the size of the via. Numerical simulations have produced DC resistances ranging from 200 to 250 mΩ for large aspect ratio TSVs [162,191] to as small as 10 to 30 mΩ [161,181,214,242] for small TSV aspect ratios. The effect of skin depth on the AC resistance has also been explored through numerical simulations, and a near doubling of the DC resistance occurs at a frequency of 10 GHz [242].

The self-capacitance of a TSV, the capacitance assuming a ground at infinity, also varies depending upon the size of the TSV and the applied voltage. The applied voltage affects the surrounding silicon material by operating like a MOS capacitor in accumulation, depletion, or inversion. In addition, the thickness of the oxide liner also significantly affects the capacitance of the TSV [181,196,214]. The self-capacitance ranges from as low as 1 to 5 fF for small TSVs with a 1 to 2 μm diameter and a length of 10 to 15 μm to as high as 200 to 250 fF for large TSVs with a 30 to 40 μm diameter and a length of 100 to 150 μm [181,196]. The self-inductance of a TSV ranges from 2 to 5 pH for small TSVs to as large as 40 to 50 pH for larger sized TSVs (the same TSV dimensions as in the aforementioned capacitance discussion) [181,196]. The mutual inductance follows a similar trend, ranging from 1 to 2 pH for small TSVs and 10 to 25 pH for larger sized TSVs. The mutual inductance does depend upon the space between the TSVs, which decreases exponentially to a nonzero asymptotic response (see Fig. 4.16). The loop inductance can therefore range from as low as approximately 10 to 15 pH to greater than 60 pH [196].

Characterization of the TSV through numerical simulation can be computationally expensive depending upon the physical structure of the environment, as any change in the physical parameters such as the TSV diameter and length and/or the number of TSVs requires readjustment of the numerical analysis process. Numerical simulation is, therefore, often used to evaluate the accuracy of the compact models characterizing the electrical impedance of a TSV. Most of the compact models discussed in Section 4.4 are compared to numerical simulation, and a few models are also compared to experimental data. Accurate compact models, verified with numerical simulation provide a computationally efficient method to estimate the electrical impedance of a TSV.

4.7 CASE STUDY—THROUGH SILICON VIA CHARACTERIZATION OF THE MITLL TSV PROCESS

As noted in Section 4.6, a multitude of TSV lengths, diameters, and materials based on the TSV fabrication process exist. One of the more aggressive fabrication processes in terms of TSV diameter and length has been developed by the MIT Lincoln Laboratory as part of a DARPA program [163,243]. A review of the MIT Lincoln Laboratory TSV-based 3-D integration process is provided as background for the electrical characterization of a TSV, as presented in Section 4.7.1. Through numerical simulation, the resistance, capacitance, and inductance of a single via is presented in Section 4.7.2. Capacitive and inductive coupling between two 3-D vias is described in Section 4.7.3 for different separation distances and structural topologies. Intertier coupling is also discussed in Section 4.7.3. This discussion is followed in Section 4.7.4 with a review of the effects of placing a 3-D shield via between two signal vias. The effect on inductive coupling and the loop inductance as a result of the via placement is discussed in Section 4.7.5.

4.7.1 MIT LINCOLN LABORATORY THREE-DIMENSIONAL PROCESS

The MIT Lincoln Laboratory 3-D integration process is described here to provide background for a case study of via topologies. Lincoln Laboratory established a 0.18 μm low power, fully depleted silicon-on-insulator CMOS process where three independently patterned wafers are physically bonded to form a 3-D integrated structure [163,243]. Each wafer, described as either a plane or tier, includes three metallization layers for routing. The top two planes have an additional backside metal for both routing and to form the 3-D vias. The critical dimensions of a 3-D via are depicted

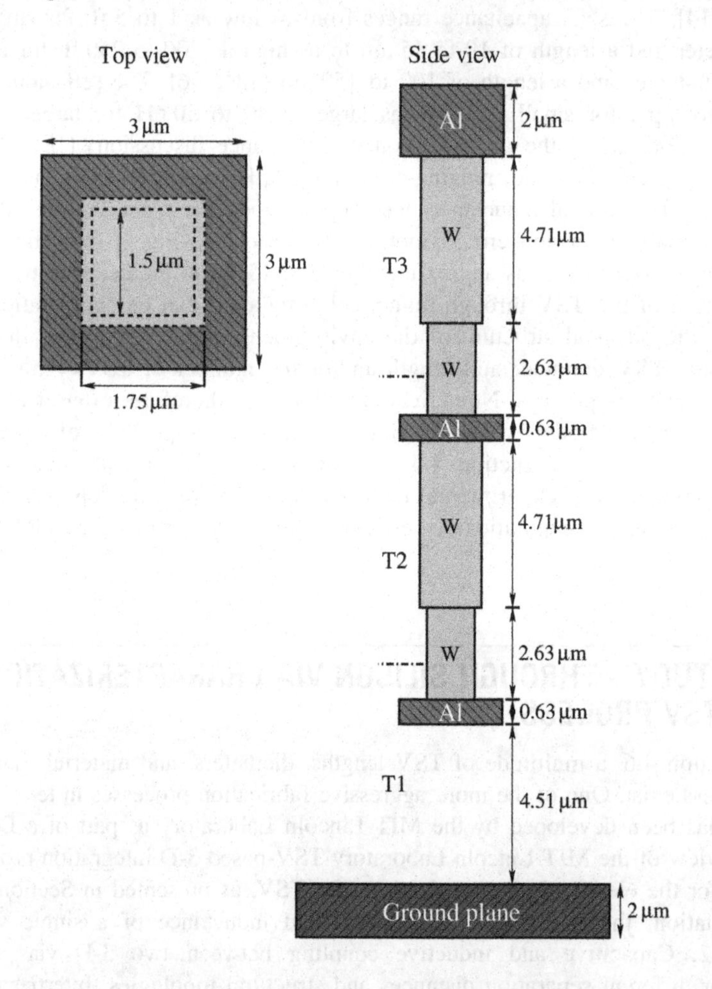

FIGURE 4.17

Critical dimensions of a 3-D via over a ground plane for the MITLL 3-D process.

in Fig. 4.17 [163]. Note that there are two intertier 3-D vias shown in the figure. One TSV passes from Tier 1 (T1) to Tier 2 (T2) while the other TSV passes from T2 to Tier 3 (T3). By overlaying the 3-D vias in this fashion, a signal can propagate from the bottom plane to the top plane. The top via (T2 to T3) or bottom via (T1 to T2) can also propagate a signal. Note that the bottom plane (T1) includes a 675 ± 25 μm thick silicon substrate, not shown in the figure, which is used for wafer handling after the top two planes are bonded to T1. As a final note, the total length of a 3-D via starting from plane T1 and terminating on plane T3 is 17.94 μm, requiring wafer thinning of planes T2 and T3 to approximately 10 μm [163]. Each 3-D via is surrounded by a combination of two dielectric layers, SiO_2 and TEOS. Both of these dielectric layers have similar relative permittivities, respectively, 3.9 and 4.0. For simulation purposes, the average of the two permittivities is assumed in the numerical simulations.

4.7.2 *RLC* EXTRACTION OF A SINGLE THREE-DIMENSIONAL VIA

Multiple 3-D via configurations with and without the ground plane for a single via are illustrated in Fig. 4.17. These configurations include T1–T2, T2–T3, and T1–T3 signal propagation paths. The 2 μm thick aluminum ground plane is located at the top of the silicon substrate. Results from these analyses are listed in Table 4.16. Characterization of the resistance, inductance, and capacitance of a single via is illustrated at DC and a frequency of 1 GHz, with and without a ground plane, and for the three TSV configurations (include T1–T2, T2–T3, and T1–T3). Based on the extracted impedances, the *RC* and *L/R* time constants are listed in Table 4.16.

These tabulated results indicate that in all cases the *L/R* time constant is approximately four orders of magnitude greater than the *RC* time constant. The ground plane does not affect either the resistance or inductance. The total via capacitance, however, is significantly affected, with the capacitance increasing by as much as 37%. The *L/R* time constant therefore remains relatively unchanged, exhibiting a maximum variation of 1.4%, whereas the *RC* time constant varies by as much as 37%.

4.7.3 *RLC* COUPLING BETWEEN TWO THREE-DIMENSIONAL VIAS

Once a single 3-D via is electrically characterized, the capacitive and inductive coupling between two 3-D vias can be determined through full wave simulation. Several configurations are considered, each with increasing separation between the two 3-D vias. A ground plane is assumed in all cases. A graphical depiction of the simulation setup and the extracted *RLC* impedances is illustrated in Fig. 4.18A. The extracted impedances for different separations between the two 3-D vias are listed in Table 4.17.

The results listed in Table 4.17 reveal several interesting characteristics in addition to the expected trend of decreasing coupling capacitance (C_{VC}) and mutual inductance (L_{VC}) with increasing via separation. As the separation between the two 3-D vias increases, the AC resistance, AC self-inductance (L_V), and capacitance of the vias approach the extracted impedances listed in Table 4.16 for a single 3-D via. The DC resistance and self-inductance (L_V) are not affected by the physical separation, and are therefore equivalent to a single 3-D via. Simulations of the intertier coupling reveal a similar characteristic where the *RLC* impedance approaches the impedance extracted for a single 3-D via with increasing spacing. Assuming an ideal oxide-to-oxide bond

Table 4.16 ***RLC*** **Impedances and Related Time Constants for a Single 3-D Via**

				DC *RL*		AC *RL* (1 GHz)		DC Time Constants		AC Time Constants	
	Configuration	C_t (fF)	C_{gc} (fF)	R (mΩ)	L (pH)	R (mΩ)	L (pH)	τ_{RC} (fs)	$\tau_{L/R}$ (ps)	τ_{RC} (fs)	$\tau_{L/R}$ (ps)
With ground plane	T1-T2	1.43	1.28	147.55	3.93	165.92	2.93	0.21	26.62	0.24	17.68
	T2-T3	1.34	1.39	152.14	4.68	171.69	3.52	0.20	30.74	0.23	20.52
	T1-T3	2.04	1.89	296.17	10.6	335.10	8.51	0.60	35.87	0.68	25.38
Without ground plane	T1-T2	1.15	–	147.62	3.92	165.85	2.92	0.17	26.56	0.19	17.58
	T2-T3	1.16	–	151.35	4.69	170.33	3.45	0.18	31.01	0.20	20.25
	T1-T3	1.49	–	296.44	10.7	335.21	8.45	0.44	35.92	0.50	25.22

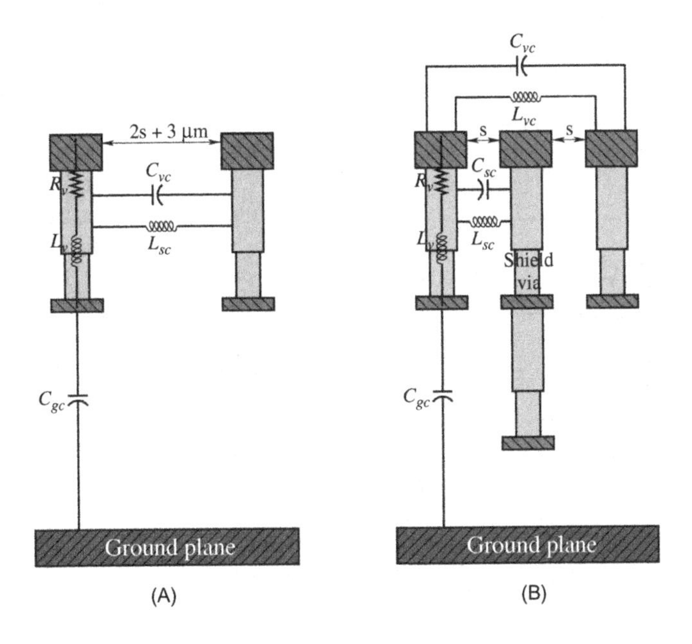

FIGURE 4.18

Circuit model for *RLC* extraction of (A) two 3-D vias, and (B) two 3-D vias with a shield via between two signal vias.

between wafers, the intertier coupling (the row labeled as T2T3–T1T2 in Table 4.17) indicates that this form of coupling cannot be neglected. Comparing the results for T2T3–T2T3 with T2T3–T1T2, the extracted electrical parameters (C and L) converge to a similar impedance with increasing spacing, which supports the need to properly model intertier coupling.

The loop inductance is also included in Table 4.17. The loop inductance is

$$L_{\text{loop}} = L_{11} + L_{22} - 2 \times L_{21}, \tag{4.88}$$

where L_{11} and L_{21} are, respectively, L_v and L_{vc} from Table 4.17. The self-inductance of the second via L_{22} is not included in either Fig. 4.18 or Table 4.17, but can be determined from (4.88). If the two 3-D vias connect the same two planes, (4.88) reduces to

$$L_{\text{loop}} = 2 \times L_{11} - 2 \times L_{21}, \tag{4.89}$$

since both vias exhibit approximately the same self-inductance.

4.7.4 EFFECTS OF THREE-DIMENSIONAL VIA PLACEMENT ON SHIELDING

A primary focus of Section 4.7.3 is to examine the extracted coupling capacitance and inductance between two 3-D vias as a function of the physical spacing. The effects of inserting a shield via between two signal vias on the coupling capacitance and inductance are discussed in this section. The simulation setup is depicted in Fig. 4.18B, and the extracted parameters are listed in Table 4.18.

Table 4.17 Extraction of *RLC* Impedances for Different Separation Between Two 3-D Vias

					DC *RL*				AC *RL* (1 GHz)			
Configuration	Separation (μm)	C_t (fF)	C_{vc} (fF)	C_{gc} (fF)	R (mΩ)	L_v (pH)	L_{vc} (pH)	L_{loop} (pH)	R (mΩ)	L_v (pH)	L_{vc} (pH)	L_{loop} (pH)
T2T3–T2T3	1	2.05	1.16	0.87	152.36	4.69	2.05	5.27	173.71	3.40	1.63	3.55
	3	1.58	0.60	0.97	152.12	4.70	1.52	6.35	172.12	3.49	1.22	4.54
	5	1.45	0.42	1.05	152.23	4.69	1.20	6.99	171.93	3.51	0.97	5.08
	7	1.40	0.31	1.04	152.03	4.69	0.98	7.41	171.88	3.52	0.80	5.44
	9	1.34	0.25	1.04	152.04	4.69	0.84	7.71	171.58	3.53	0.68	5.68
	11	1.32	0.20	1.10	152.31	4.69	0.72	7.94	171.89	3.53	0.59	5.88
	13	1.31	0.17	1.11	152.10	4.70	0.64	8.12	171.73	3.53	0.52	6.01
T1T3–T1T3	1	3.2206	1.8429	1.3836	297.92	10.662	5.3366	10.63	341.316	8.199	4.5435	7.3176
	3	2.4023	1.0010	1.4351	297.49	10.654	4.1699	12.977	237.81	8.3573	3.5556	9.6252
	5	2.2407	0.6628	1.5088	297.50	10.654	3.4101	14.489	336.938	8.4217	2.9282	10.982
	7	2.1416	0.4852	1.8334	297.25	10.660	2.8868	15.530	336.329	8.4463	2.4878	11.914
	9	2.0860	0.3792	1.6312	297.45	10.647	2.4975	16.298	336.382	8.4557	2.1554	12.594
	11	2.0650	0.3038	1.7465	297.47	10.657	2.1993	16.913	336.42	8.4648	1.8871	13.158
	13	2.056	0.2479	1.7369	297.56	10.652	1.9624	17.396	336.461	8.4714	1.6882	13.558
T2T3–T1T2	1	1.57	0.51	1.02	152.27	4.69	1.07	6.49	171.91	3.52	0.90	4.68
	3	1.43	0.33	1.06	152.31	4.69	0.91	6.80	171.88	3.52	0.76	4.95
	5	1.39	0.26	1.12	152.20	4.69	0.79	7.04	171.77	3.52	0.65	5.16
	7	1.37	0.20	1.14	152.17	4.70	0.69	7.24	171.69	3.52	0.57	5.33
	9	1.37	0.16	1.16	152.28	4.70	0.62	7.39	171.79	3.52	0.51	5.46
	11	1.28	0.14	1.10	152.14	4.70	0.55	7.52	171.65	3.52	0.45	5.57
	13	1.35	0.11	1.22	152.32	4.69	0.50	7.62	171.85	3.53	0.41	5.66

Table 4.18 *L* **and** *C* **Shielding with the Addition of a Third Shield Via**

Configuration	Separation (μm)	C_t (fF)	C_{vc} (fF)	% C_{vc} Shielded	DC L			AC L		
					L_v (pH)	L_{vc} (pH)	% L_v Reduced	L_{vc} (pH)	L_{vc} (pH)	% L_{vc} Reduced
T2T3–T1T3–T2T3	1	2.07	0.107	74.39	4.68	1.19	0.63	3.40	1.01	−4.30
	2	1.73	0.093	70.29	4.69	0.98	0.17	3.46	0.82	−2.52
	3	1.58	0.079	68.94	4.68	0.83	0.40	3.50	0.69	−1.33
	4	1.50	0.069	65.79	4.68	0.72	−0.01	3.51	0.59	−1.05
	5	1.39	0.056	66.13	4.69	0.64	−0.16	3.52	0.52	−0.39
T1T3–T1T3–T1T3	1	3.2267	0.1643	75.21	10.665	3.4051	0.147	8.1160	3.0119	−2.858
	2	2.6774	0.1393	71.29	10.652	2.8886	−0.062	8.2591	2.5264	−1.551
	3	2.4483	0.1177	68.97	10.655	2.4992	−0.068	8.3323	2.1710	−0.724
	4	2.3127	0.0993	67.31	10.645	2.2009	−0.073	8.3657	1.8853	0.095
	5	2.2023	0.08761	64.66	10.667	1.9636	−0.061	8.3679	1.6768	0.675
T2T3–T1T3–T1T3	1	2.07	0.059	76.92	4.69	0.79	0.37	3.41	0.67	−2.74
	2	1.74	0.054	73.35	4.69	0.69	0.40	3.47	0.58	−2.00
	3	1.59	0.049	69.58	4.69	0.61	0.24	3.50	0.51	−1.08
	4	1.51	0.044	67.68	4.69	0.55	0.19	3.51	0.46	−1.00
	5	1.42	0.037	66.71	4.68	0.50	0.23	3.52	0.41	−0.43

The reduction in capacitance is determined by noting the difference in coupling capacitance between two 3-D vias with and without a shield. The coupling capacitance, assuming a 1 μm space with a shield, is compared to a 5 μm space without a shield (see Table 4.17). The capacitive coupling (C_{vc}) between two signal vias is reduced by as much as 77% by including a third shield via. In addition, the effectiveness of the shield diminishes with increasing space, as additional electrical field lines terminate on the second signal line. Alternatively, the inductive coupling (L_{vc}) between the two signal lines increases due to the high mutual inductance between the shield and signal vias. The negative per cent reduction in inductance listed in Table 4.18 indicates this increase in inductive coupling. Note that the increase in inductance when an inserted shield via is present only applies to the AC inductance. The maximum increase in inductive coupling is about 4.3%. As previously mentioned, an increase in the mutual inductance produces a decrease in the loop inductance. The effect of the return path on both the DC and AC inductance is described in the following section.

4.7.5 EFFECT OF THE RETURN PATH ON THREE-DIMENSIONAL VIA INDUCTANCE

The return path can significantly affect the induced loop inductance. In addition, the distance between the signal via and the return path also contributes to the loop inductance. As the space between the signal and return path increases, the mutual inductance (L_{21}) decreases. From (4.88), a decrease in the

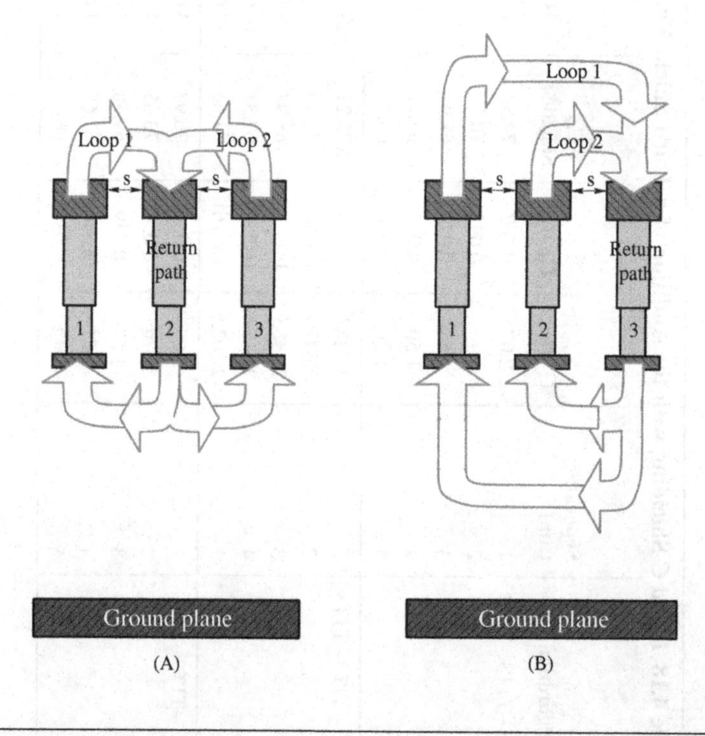

FIGURE 4.19

Effect of a return path on the loop inductance. (A) Return path placed on 3-D via 2, and (B) return path placed on 3-D via 3.

Table 4.19 Inductive Coupling and Loop Inductance for Three T2T3 3-D Vias

		DC L			AC L (1 GHz)		
	Separation (μm)	L_{11} (pH)	L_{21} (pH)	L_{31} (pH)	L_{11} (pH)	L_{21} (pH)	L_{31} (pH)
Partial inductances	1	4.69	2.05	1.20	3.36	1.57	0.99
	2	4.69	1.75	0.98	3.44	1.36	0.81
	3	4.69	1.51	0.84	3.47	1.19	0.68
	4	4.69	1.34	0.72	3.49	1.06	0.59
	5	4.69	1.19	0.65	3.50	0.95	0.52
		DC L			AC L		
	Separation (μm)	L_{loop1} (pH)	L_{loop2} (pH)	L_{loop21} (pH)	L_{loop1} (pH)	L_{loop2} (pH)	L_{loop21} (pH)
Return path conductor 2	1	5.28	5.28	1.79	3.51	3.51	1.13
	2	5.90	5.89	2.19	4.11	4.10	1.48
	3	6.36	6.36	2.50	4.52	4.52	1.74
	4	6.71	6.71	2.75	4.82	4.84	1.93
	5	6.99	6.99	2.94	5.07	5.07	2.09
Return path conductor 3	1	6.70	5.28	3.50	4.75	3.51	2.37
	2	7.41	5.89	3.70	5.25	4.10	2.62
	3	7.71	6.36	3.86	5.57	4.53	2.79
	4	7.93	6.71	3.97	5.80	4.84	2.91
	5	8.11	6.99	4.05	5.97	5.07	2.98

mutual inductance increases the loop inductance. An analysis of a three T2T3 3-D via structure illustrates the effect of the return path: the return path through the middle via (see Fig. 4.19A) and the return path through a side via (see Fig. 4.19B). The inductance listed in Table 4.19 only includes a single row of the 3×3 matrix formed with these three conductors, corresponding to the self- (L_{11}) and mutual (L_{12} and L_{13}) inductance of conductor 1 (TSV 1), as shown in Fig. 4.19. The remaining two rows are the self- and mutual inductance of the other two conductors.

These results are listed in Table 4.19 and are briefly addressed here. Placing the return path between the other two 3-D vias induces two symmetric inductive loops, as indicated by the similar loop inductance in the row labeled, "return path conductor 2" in Table 4.19. Alternatively, placing the return path on either side TSV of the configuration with three TSVs increases the loop inductance between the two side vias by 32%. The mutual loop inductance between the two loops (L_{loop21}) increases by as much as 96% when the return path is placed on one of the side vias rather than the central via. Based on these results, proper placement of the return paths is critical in minimizing the loop inductance of a 3-D via.

4.8 SUMMARY

Electrical characterization of the TSV impedance is necessary to properly model the delay and power consumption of intertier signals. The key contributions of this chapter are summarized as follows:

- Closed-form expressions of the resistance, capacitance, inductance, and conductance for intertier 3-D vias are presented.
- Compact models are often verified through field solvers and experimental data, and once verified, can be used to replace full wave simulations to estimate the impedance of a TSV.
- The complexity of compact TSV models is dependent upon effects such as frequency dependencies (both proximity and skin effects), the depletion region, and temperature.
- The primary objective of closed-form expressions for characterizing the TSV impedance is a *RLCG* electrical representation of a TSV for circuit simulation and analysis, where the electrical parameters of a TSV can be reduced to a lump *RLCG* model.
- The closed-form expressions consider the 3-D via length, diameter, dielectric thickness, pitch between vias, and spacing to ground as well as material parameters including the silicon and oxide permittivities, permeability and conductivity of metal, and silicon conductance.
- Expressions for the resistance consider a cylindrical TSV as well as a tapered cylindrical via, where the diameter of the TSV is farther into the silicon.
- Models of the resistance are described for both DC and high frequencies, where the skin effect increases the resistance of a TSV.
- Resistance models of the TSV produce less than 8% variation as compared to numerical simulations.
- Closed-form expressions for the self- and mutual inductance at DC and high frequencies are described for both a cylindrical and tapered TSV.
- Most expressions characterizing the TSV inductance are based on work originally completed by Rosa in 1908.

- Closed-form expressions of the TSV inductance are accurate within 20% of numerical simulation and the accuracy can be further improved with fitting parameters.
- Many capacitance models of a TSV have been developed as compared to only a few models of the TSV resistance and inductance.
- Closed-form expressions of the capacitance of a TSV are of varying accuracy based on the number and type of physical effects that are considered.
- Models of the TSV capacitance are within 10 to 15% of numerical simulation, whereas closed-form expressions of the coupling capacitance result in errors as high as 25%.
- Both capacitive and inductive coupling between multiple 3-D vias is dependent on the separation distance and location of the tiers within a 3-D stack.
- Closed-form expressions of the finite conductance of the silicon surrounding a TSV have been developed.
- The electrical characterization of a wire or TSV requires a field solver to accurately determine the effects of different parameters including the physical structure and dimensions, frequency, and surrounding environment.
- 3-D numerical simulations quantify the electromagnetic behavior of the resistance, capacitance, inductance, and conductance of a TSV and can be used for comparison with closed-form expressions.
- Field solvers require significant computational time; compact models are therefore often used to provide an efficient and fairly accurate method for characterizing the electrical properties of a TSV.
- The TSV-to-TSV coupling capacitance is reduced by as much as 80% by including a third shield via between two signal vias.
- Based on the extracted TSV resistance, capacitance, and inductance, the L/R time constant is much larger than the RC time constant.
- The location of the current return path within a group of TSVs determines the preferable placement of a 3-D via to reduce the overall loop inductance.

CHAPTER

SUBSTRATE NOISE COUPLING IN HETEROGENEOUS THREE-DIMENSIONAL ICs*

5

CHAPTER OUTLINE

Noise coupling is of increasing importance in the integrated circuit design process [244]. This topic is also a fundamental issue in three-dimensional (3-D) circuits where signals are distributed among multiple different tiers using through silicon vias (TSVs), creating an *electronic storm* within a 3-D system. Different types of signals (power, clock, and data) can propagate within these vertical interconnects. Several TSV processes are used in 3-D integration including via-first, via-middle, and via-last [245] (see Chapter 3, Manufacturing Technologies for Three-Dimensional Integrated Circuits). In each of these processes, the TSVs are within the substrate of a tier and connect to metal interconnects on that tier. These TSVs greatly alleviate global signaling issues. The TSVs, however, pose novel obstacles; specifically, noise coupling through the TSVs into the substrate of each layer. This noise propagates through the substrate and affects the victim circuits surrounding a TSV. The purpose of this chapter is to describe coupling noise in heterogeneous 3-D systems composed of different substrate materials. The mechanism of noise coupling from the TSVs into heterogeneous substrates, and the properties and applications of the different substrates are summarized in Section 5.1. The coupling noise and related transfer functions characterizing a noise coupling system in the frequency domain are discussed in Section 5.2. Some techniques to enhance noise isolation are offered in Section 5.3. Finally, a summary is provided in Section 5.4.

*Mr. Boris Vaisband has contributed to this chapter.

Three-Dimensional Integrated Circuit Design. DOI: http://dx.doi.org/10.1016/B978-0-12-410501-0.00005-8

5.1 HETEROGENEOUS SUBSTRATE COUPLING

A 3-D structure is an effective platform for integrating heterogeneous circuits within a single system, as illustrated in Fig. 5.1. Each tier of a 3-D IC is typically independently developed using different substrate materials for a variety of applications. Some existing work has addressed noise coupling from TSVs into the substrate in homogeneous circuits (processor/memory stacks), typically on a silicon substrate [214,246]. In this section, *heterogeneous* substrate materials are discussed, and compatible noise coupling models are proposed.

5.1.1 COMMON CIRCUITS AND COMPATIBLE SUBSTRATE TYPES

Some commonly used materials in modern integrated circuits are silicon (Si), gallium arsenide (GaAs), germanium (Ge), and mercury cadmium telluride (HgCdTe) [247–249]. Each of these substrate materials is most useful for a certain type of circuit function. Si is typically more mature and lower cost as compared to the other aforementioned materials and is, therefore, commonly used in mainstream processor and memory applications. The superior electron mobility of GaAs makes GaAs attractive for high performance analog (and some digital) applications and the direct bandgap supports optical systems. Ge is a favorable substrate material for photovoltaic and photodetector applications due to a high absorption coefficient. Special military and space applications that require high quality infrared detectors commonly use HgCdTe [250], which has a tunable bandgap ranging from 0.1 to 1 eV. This property of HgCdTe is used to detect long wavelengths of light.

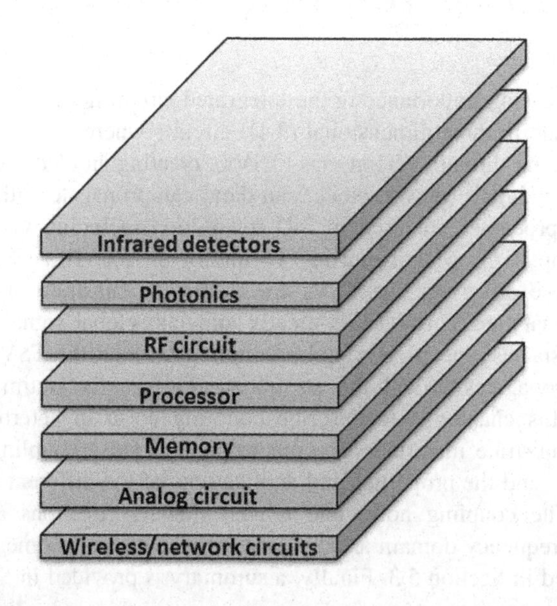

FIGURE 5.1

Heterogeneous 3-D integrated circuit.

5.1.2 RESISTIVE PROPERTIES OF DIFFERENT SUBSTRATE MATERIALS

Common circuits and compatible substrate materials are listed in Table 5.1. The electrical resistivity, which greatly affects the noise coupling process, is also listed for each substrate material. Due to the wide range of resistivities listed in Table 5.1, the individual noise coupling characteristics of each of the substrate materials are described in Section 5.2.

5.1.3 NOISE MODEL REDUCTION FOR DIFFERENT SUBSTRATE MATERIALS

Individual substrate noise models are described for each of the aforementioned substrate materials. The general noise coupling model presented in [246] is provided for convenience in Fig. 5.2A.

Table 5.1 Common Circuits and Compatible Substrate Types

Circuits	Substrate Material	Electrical Resistivity (Ωcm)
Processor/memory	Silicon (Si)	1–10
RF/analog	Gallium arsenide (GaAs)	4×10^7
Photonics	Germanium (Ge)	1.3×10^{-3}
Space applications/detectors	Mercury cadmium telluride (HgCdTe)	2

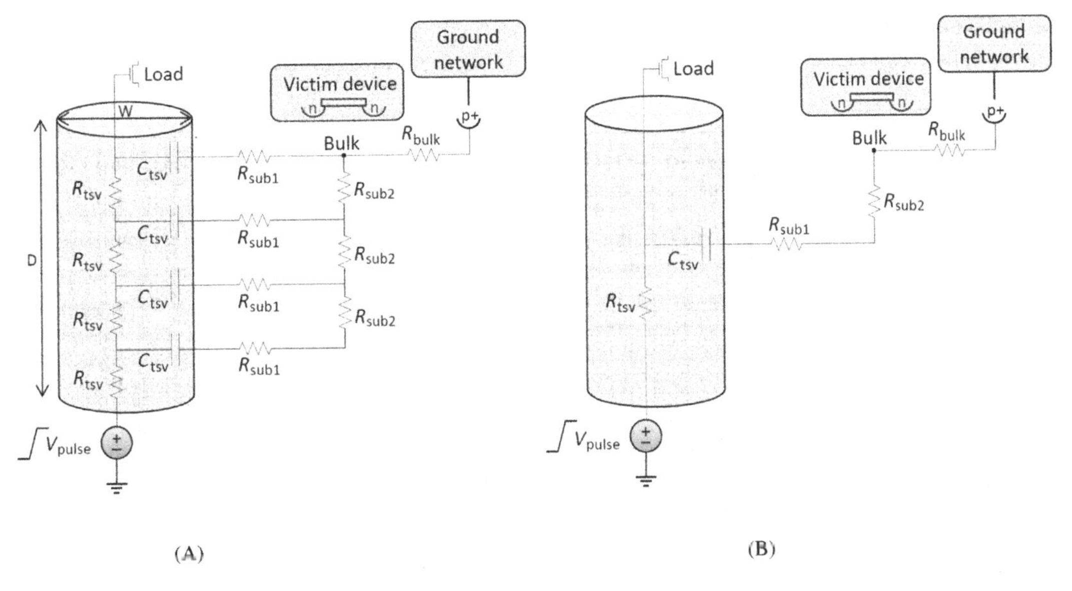

FIGURE 5.2

Model of noise coupling from TSV to a victim device through a silicon substrate. (A) General model, and (B) reduced model.

A distributed *RC* model composed of four sections is used to characterize the TSV impedance and coupling capacitance within a silicon substrate. The substrate is modeled using distributed lateral and vertical resistors. The ground network is modeled as a resistive-inductive (*RL*) impedance [205]. The model shown in Fig. 5.2A suggests the use of a distributed model for the *RC* impedances. The resistance of a TSV is [246]

$$R_{\text{tsv}} = \frac{1}{N_{\text{tsv}}} \cdot \frac{\rho_c D}{\pi(W/2)^2}. \tag{5.1}$$

The number of distributed sections of a TSV is N_{tsv}, resistivity of the conductive material within the TSV is ρ_c, and depth (length) and diameter of the TSV are, respectively, D and W. With a resistivity for copper of 2.8 μΩ-cm [251], depth of 20 μm, and diameter of 2 μm [201], the resistance of a TSV is 0.18 Ω for $1/N_{\text{tsv}} = 1$. This resistance is relatively small as compared to the output resistance of a typical digital buffer [252]. A lumped *RC* model is therefore used for a TSV [244,253], as shown in Fig. 5.2B. Another important issue is the model of the ground network. The victim device is commonly connected to the ground network through a bulk contact. The inductive behavior of this network, therefore, also has to be considered.

A comparison of a lumped model as compared to a distributed model with three sections is listed in Table 5.2 for Si, GaAs, and Ge. For Ge, a third "short-circuit" model (shown in Fig. 5.3A) is also compared. This model completely omits the resistors of the substrate since the resistance of the substrate is negligible, and therefore, the model only exhibits a coupling capacitance from a TSV to the substrate [252]. The models have been evaluated using Simulation program with integrated circuit emphasis (SPICE). A 10 ps input ramp from 0 to 1 V (V_{pulse} in Fig. 5.2) is applied to simulate switching of aggressive digital circuits. The voltage is evaluated at the victim node. Both the peak noise voltage and settling time (2% of the final value) are recorded for three different ground network inductances. Unlike coupling between adjacent interconnects [254], in this chapter,

Table 5.2 Comparison of lumped, distributed, and short circuit models for Si, GaAs, and Ge substrates, for different values of inductance of the ground network

		Si		GaAs		Ge	
Model	**Ground Inductance (nH)**	**Peak Noise (mV)**	**Settling Time (ns)**	**Peak Noise (mV)**	**Settling Time (ns)**	**Peak Noise (mV)**	**Settling Time (ns)**
Short-circuit	0.1	–	–	–	–	11.1	0
	1	–	–	–	–	645.5	1.5
	10	–	–	–	–	954.4	8
Lumped	0.1	159.8	1.57	3.8×10^{-8}	0	8.5	0
	1	162.4		3.8×10^{-8}		638.5	1
	10	186.3		3.8×10^{-8}		950.8	6
Distributed (three sections)	0.1	161.8	1.55	3.9×10^{-8}	0	8.7	0
	1	164.5		3.9×10^{-8}		637.5	1
	10	188.6		3.9×10^{-8}		950.1	6

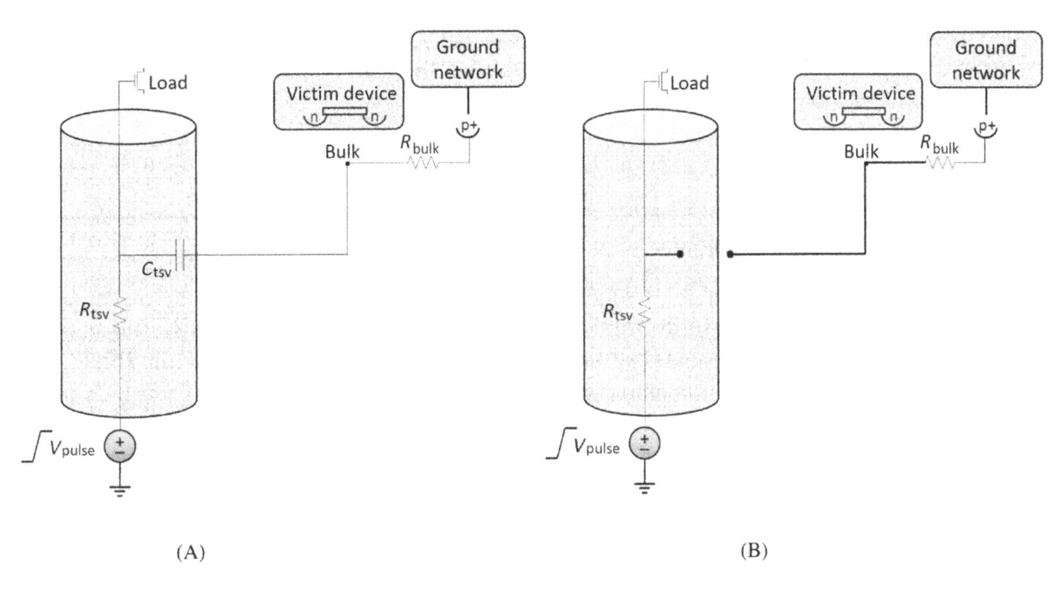

FIGURE 5.3

Noise coupling from a TSV to a victim device. (A) Short-circuit Ge substrate model, and (B) open circuit GaAs substrate model.

coupling from a signal propagating within an aggressor TSV to the substrate is described. The peak noise and settling time are sufficient metrics for evaluating transient coupling noise.

The error of a lumped model as compared to a distributed model for Si is 1.2%. A lumped model can therefore accurately characterize a silicon substrate. As observed from the results listed in Table 5.2, the inductance of the ground network significantly affects the peak noise voltage. Over a range of inductance (from 0.1 to 10 nH), a difference of 26.5 mV (14.2%) is noted in the peak noise voltage. The electrical resistivity of mercury cadmium telluride is similar to silicon. The same model, as shown in Fig. 5.2B, can therefore be used in the noise coupling analysis process of HgCdTe.

The peak noise voltage for both lumped and distributed models of GaAs is in the range of picovolts and is, therefore, negligible in most applications. The proposed model in this case is an "open circuit" model that ignores the coupling capacitance, as illustrated in Fig. 5.3B. Note from Table 5.2 that the inductance of the ground network has no effect on the peak noise voltage. This behavior is due to the resistivity of the substrate, which is sufficiently large to shunt the inductance of the ground network.

Ge is highly dependent on the inductance of the ground network. Comparing the lumped and distributed models, a distributed model provides negligible improvement in accuracy as compared to a lumped model. The worst case difference in the peak noise voltage is 0.2 mV (2.3%), while the settling time is similar. A lumped model, which incorporates fewer nodes, is therefore preferable. A short-circuit model deviates from a lumped model by 2.6 mV (23.4%) and 2 ns (25%) for, respectively, the peak noise voltage and settling time. A lumped model, similar to the model for silicon (shown in Fig. 5.2B), should therefore be used. If the circuit specifications are not particularly strict (a higher peak noise voltage and longer settling time are allowed), a short-circuit model can be used to reduce the computational effort.

5.2 FREQUENCY RESPONSE

A technology specific analysis of the frequency response of the lumped noise coupling model is offered in this section. The analysis is limited to frequencies ranging up to 100 GHz. Noise isolation improvement techniques are suggested in Section 5.2.1. The model is evaluated in SPICE, and the transfer function of the system is extracted based on the characteristics of each substrate material as discussed in Section 5.2.2. In Section 5.3, the extracted transfer functions are evaluated in MATLAB and compared to SPICE. Due to similar electrical properties of HgCdTe and Si, only Si, GaAs, and Ge as substrate materials are considered.

5.2.1 ISOLATION EFFICIENCY OF NOISE COUPLED SYSTEM

Isolation efficiency is the magnitude of the signal observed at the victim for a 1 V aggressor signal (in dB). The isolation efficiency of a noise coupled system for different substrate materials and ground network inductances is shown in Fig. 5.4, which is obtained from SPICE simulations. The isolation efficiency of Ge is strongly dependent on the frequency followed by Si, while GaAs exhibits almost no dependence on frequency due to the high resistivity of the substrate. Although Ge is strongly dependent on frequency for a wide range of frequencies (up to approximately 10 GHz), the isolation efficiency of Ge is higher than GaAs. The frequency dependent components of a Ge system lower the coupled noise at the victim. As shown in Fig. 5.4C, GaAs is independent of the inductance of the ground network. The effect of the inductance of the ground network on Si and Ge is discussed later in this section.

For Ge circuits, the resonant frequency is within a practical range of frequencies. Specific techniques to improve noise isolation should be used to avoid high coupling noise in these circuits. For Si circuits, the isolation techniques are highly dependent on the operating frequency of the circuit and the noise specifications. For a typical maximum frequency range of signal transitions in digital CMOS circuits (under 10 GHz), the isolation efficiency is high. For those circuits that require fast transitions with strict noise specifications, isolation enhancement methods should be considered. For GaAs, the isolation efficiency is −15.9 dB. Isolation techniques that operate independent of frequency should be applied to further improve the noise isolation characteristics.

5.2.2 TRANSFER FUNCTION OF NOISE COUPLED SYSTEM

To better evaluate the noise coupling mechanism, a heterogeneous system is represented as a transfer function. This system consists of an input (an aggressor signal) and output (a signal at a victim module). The isolation efficiency of the system [246,255] is described, and noise mitigation

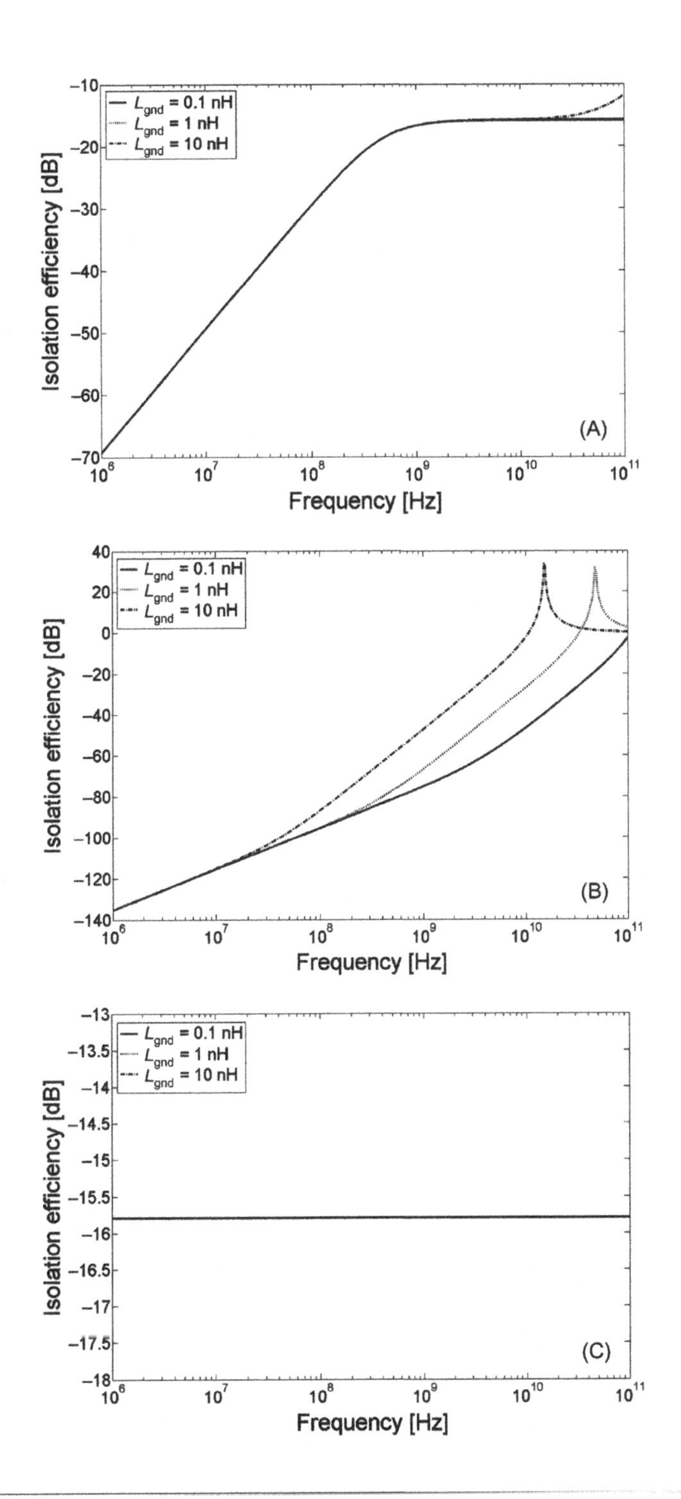

FIGURE 5.4

Isolation efficiency of a noise coupled system for different substrate materials. (A) Silicon, (B) germanium, and (C) gallium arsenide.

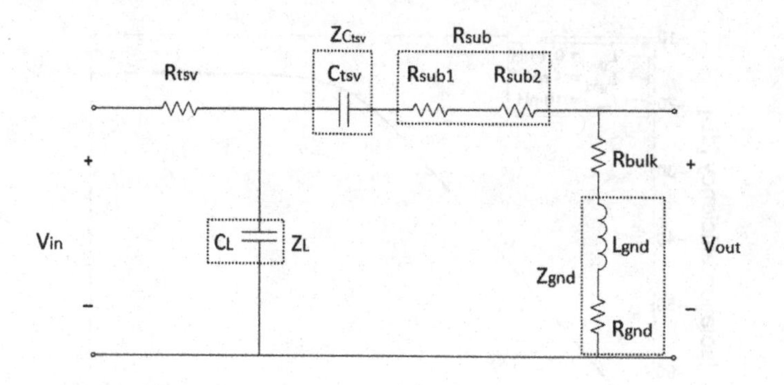

FIGURE 5.5

Equivalent small-signal model of a noise coupled system.

techniques are offered. The small-signal equivalent circuit of a noise coupled system is shown in Fig. 5.5. The following parameters are considered:

- Substrate impedance: $R_{sub} \equiv R_{sub1} + R_{sub2}$
- TSV coupling reactance: $X_{C_{tsv}} \equiv \frac{1}{\omega C_{tsv}}$
- TSV coupling impedance: $Z_{C_{tsv}} \equiv -j \cdot X_{C_{tsv}}$
- Ground network reactance: $X_{gnd} \equiv \omega L_{gnd}$
- Ground network impedance: $Z_{gnd} \equiv R_{gnd} + j \cdot X_{gnd}$
- Load reactance: $X_L \equiv \frac{1}{\omega C_L}$
- Load impedance: $Z_L \equiv -j \cdot X_L$

The transfer function is described in this chapter for a heterogeneous system based on the substrate materials discussed in Section 5.1. The transfer function of a lumped model is

$$H(\omega) = \frac{V_{out}}{V_{in}} = \frac{(R_{bulk} + Z_{gnd})Z_L}{(R_{tsv} + Z_L)(R_{sub} + R_{bulk} + Z_{C_{tsv}} + Z_{gnd}) + R_{tsv} \cdot Z_L}. \tag{5.2}$$

Reducing the transfer function produces a simpler model which requires less computational effort. The simulated load capacitance (100 fF) is relatively small. The model can therefore be treated as an open circuit assuming a small-signal model (Fig. 5.5) operating within a practical range of frequencies (1 MHz to 100 GHz). The reduced transfer function $H(\omega)$ is

$$H(\omega) = \frac{R_{bulk} + Z_{gnd}}{R_{sub} + R_{bulk} + Z_{C_{tsv}} + Z_{gnd} + R_{tsv}}. \tag{5.3}$$

Further reductions of (5.3) depend upon the substrate material for a specific layer. The substrate and bulk resistances in Si and HgCdTe are three to five orders of magnitude larger than the TSV and ground network resistances ($R_{sub}, R_{bulk} \gg R_{tsv}, R_{gnd}$) for distances as short as 10 μm between the aggressor and victim. Expression (5.3) therefore reduces to

$$H(\omega) = \frac{R_{bulk} + j \cdot X_{gnd}}{R_{sub} + R_{bulk} + j(X_{gnd} - X_{C_{tsv}})}. \tag{5.4}$$

For Ge, the substrate and bulk impedances are of the same relative magnitude as the other components of the transfer function, therefore, (5.2) cannot be further reduced. The transfer function for Ge is

$$H(\omega) = \frac{R_{\text{bulk}} + Z_{\text{gnd}}}{R_{\text{sub}} + R_{\text{bulk}} + Z_{C_{\text{tsv}}} + Z_{\text{gnd}} + R_{\text{tsv}}}. \tag{5.5}$$

The substrate and bulk resistances in GaAs are significantly larger (approximately six orders of magnitude) than all of the other components within the noise coupled system. The transfer function therefore reduces to

$$H(\omega) = \frac{R_{\text{bulk}}}{R_{\text{sub}} + R_{\text{bulk}}}. \tag{5.6}$$

Substituting the substrate and bulk parameters and worst case distance from the aggressor TSV to the victim (10 μm) leads to $H(\omega) \approx 0.16$. In units of dB, the isolation efficiency for GaAs is $20\log H(\omega) \approx -15.9$ dB, as shown in Fig. 5.4C.

5.3 TECHNIQUES TO IMPROVE NOISE ISOLATION

After obtaining the reduced transfer function of the system for each substrate type, some design considerations for decreasing the coupling noise are offered in this section. The objective is to minimize $|H(\omega)|$ by adjusting different manufacturing and design parameters; hence, to lower the noise coupled from an aggressor to a victim. Several techniques are offered here to improve noise isolation in heterogeneous 3-D circuits. The effect of the ground network inductance on the noise isolation is discussed in Section 5.3.1. The impact of the distance between an aggressor TSV and a victim on the noise isolation characteristics is described in Section 5.3.2.

5.3.1 GROUND NETWORK INDUCTANCE

The tradeoff between thinner and more resistive, and thicker and more inductive metal interconnect should be considered in the design of power distribution networks in integrated circuits (see Chapter 18, Power Delivery for Three-Dimensional ICs). In 3-D ICs, identifying the inductive return paths is more complicated as compared to 2-D circuits, since these paths can span the entire 3-D structure. Special emphasis should therefore be placed on low inductance ground lines. As shown in Fig. 5.4, low inductance ground networks directly improve the isolation efficiency of the coupled noise system for both Si and Ge. For Ge, low inductive ground networks are particularly important. The worst case difference in isolation efficiency for an inductive ground network is 73.5 dB. For a ground network with an inductance of 10 nH, the resonance frequency is 15.1 GHz, while for an inductance of 0.1 nH, the resonance frequency is above the practical range of frequencies (>100 GHz). The resonance frequency is $f_{\text{res}} = 1/2\pi\sqrt{LC}$, where the capacitance of the system is C and the inductance of the ground network is L. As shown in Fig. 5.4B, a lower ground network inductance can shift the resonant frequency out of the practical range of frequencies.

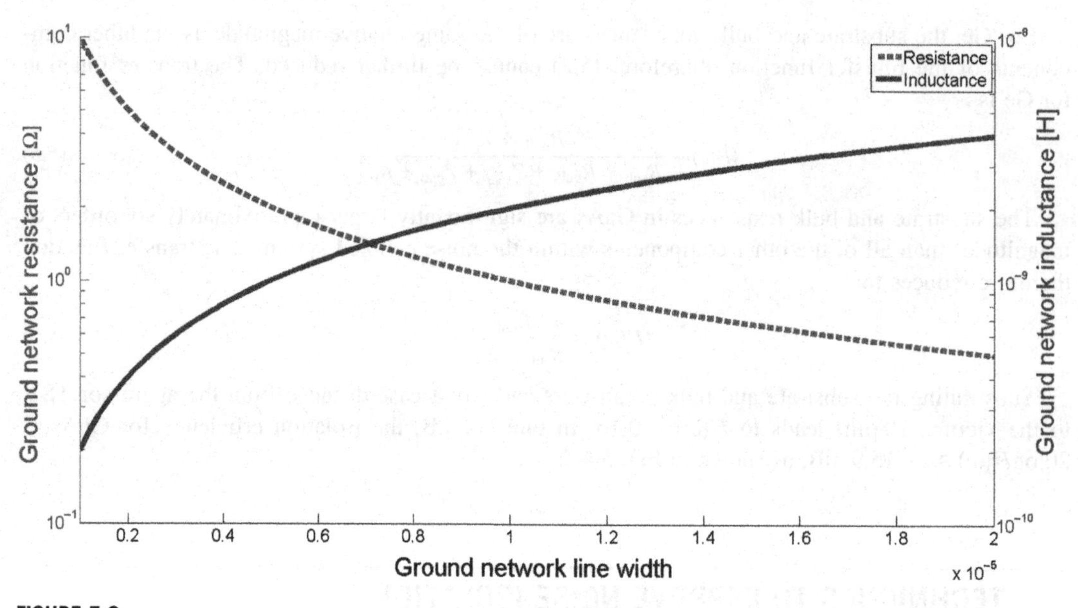

FIGURE 5.6

Resistance and inductance versus line width of the ground network. The ground network is composed of copper-based interconnects.

To further validate this technique, a tradeoff between the inductance and resistance is considered for each substrate material. The resistance and inductance as a function of the line width of the ground network [185] are shown in Fig. 5.6.

Simulation program with integrated circuit emphasis (SPICE) simulations of the isolation efficiency of each of the substrate materials are shown in Fig. 5.7. For a Si substrate, the practical range of frequencies is below 100 GHz, and the line width has no effect on the ground network inductance. A minimum line width should therefore be used. For Ge, a tradeoff exists between the resistance and inductance of the ground network. For wide lines, the peak isolation efficiency is lower than for narrow lines. The worst case difference between a line width of 2 and 20 μm is 8.2 dB. For frequencies below 56 GHz, the isolation efficiency of a narrow line (2 μm) is better than a wide line (20 μm). The line width of the ground network in Ge should therefore be chosen according to the signal transition frequency. For GaAs, the isolation efficiency is independent of the line width. The smallest allowable width should therefore be used.

5.3.2 DISTANCE BETWEEN AGGRESSOR AND VICTIM

The distance from an aggressor module "A" on tier *m* to a victim module "V" on tier *n* is shown in Fig. 5.8. The depth (length) of a single TSV and horizontal distance (on tier *n*) from the TSV to the victim are, respectively, *D* and *l*. The distance between modules "A" and "V" is therefore

$$d_{AV} = \sqrt{D \cdot |m-n|^2 + l^2}. \quad (5.7)$$

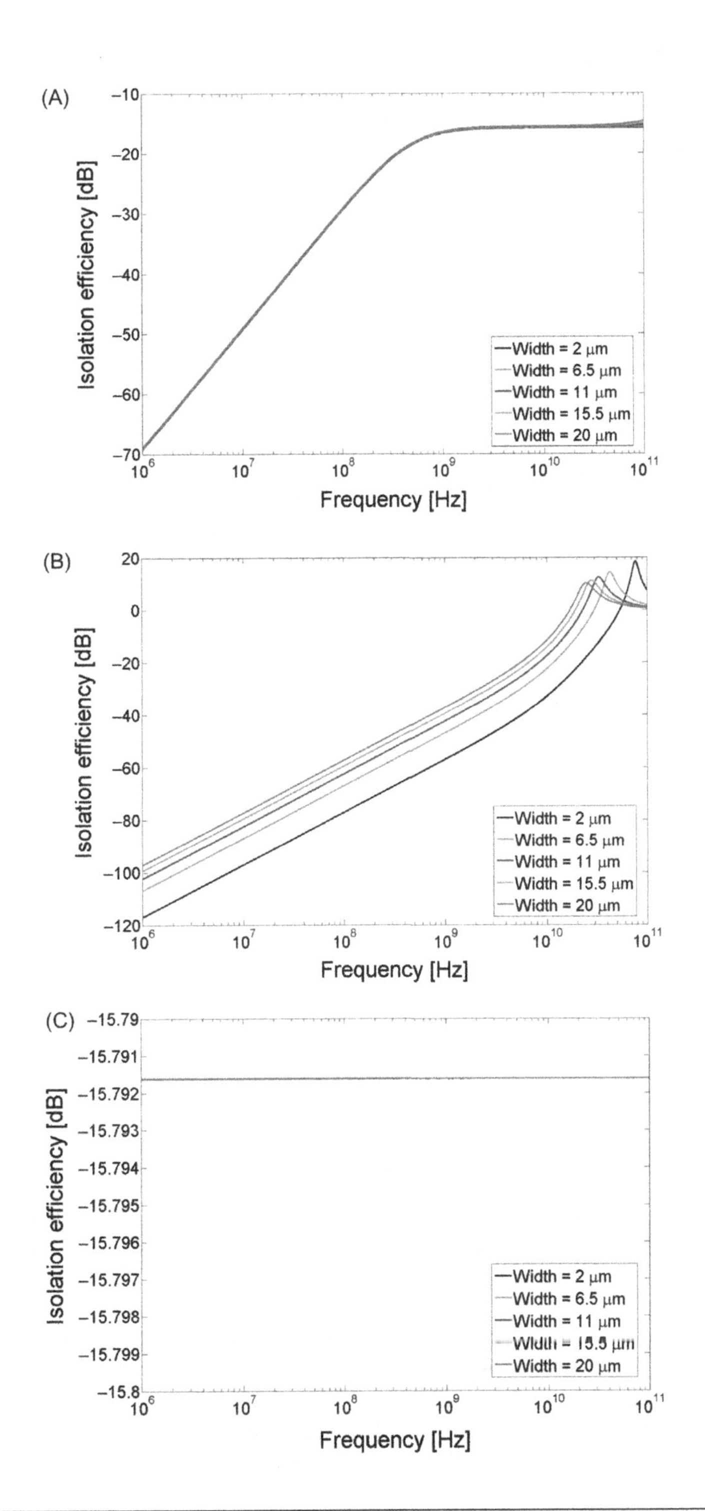

FIGURE 5.7

Isolation efficiency of a noise coupled system as a function of the line width of the ground network for different substrate materials. (A) Silicon, (B) germanium, and (C) gallium arsenide.

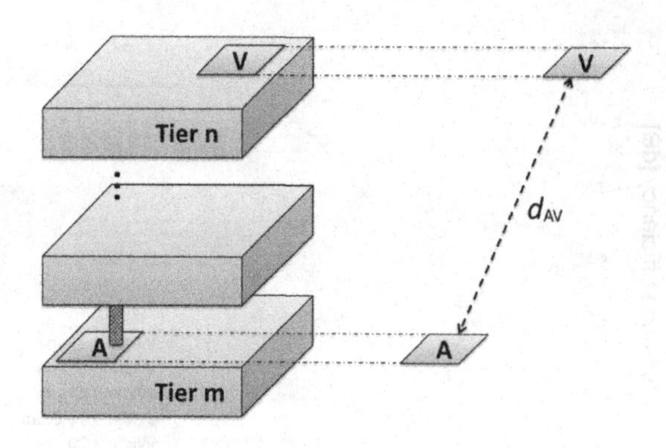

FIGURE 5.8

Distance from aggressor module "A" on tier *m* to victim module "V" on tier *n*.

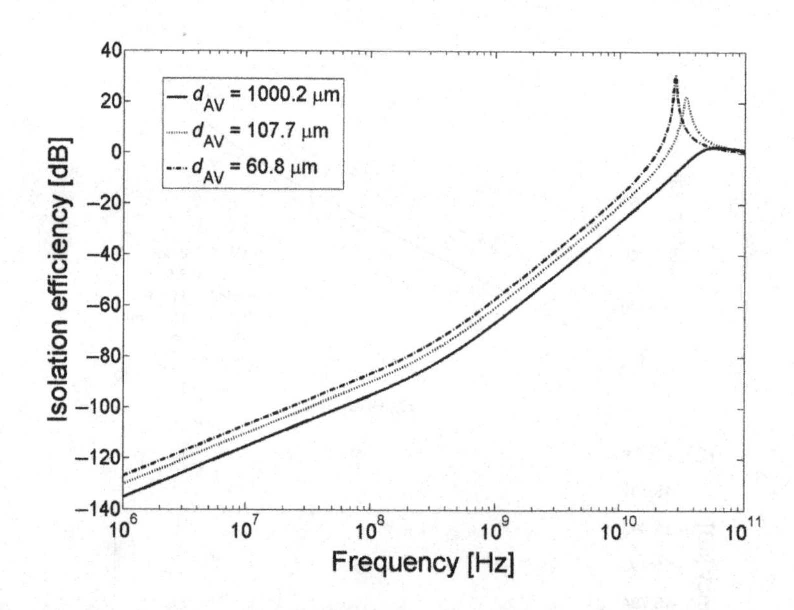

FIGURE 5.9

Effect of distance between an aggressor and victim on the isolation efficiency for a Ge substrate. The resonant frequency is observed at the peak isolation efficiency due to the increasing reactance of the ground network.

The effect of d_{AV} on the isolation efficiency of Ge is shown in Fig. 5.9. The substrate thickness is dependent on the length of the TSV D, ranging from 20 to 60 μm, and the different processes for manufacturing heterogeneous substrate materials. Similarly, the lateral distance l is evaluated over a range from 10 to 1000 μm. An improvement of 38.5 dB in isolation efficiency

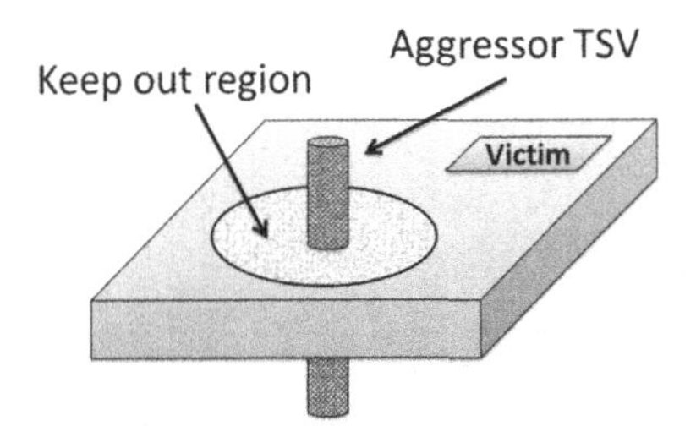

FIGURE 5.10

Keep out region around an aggressor TSV. The victim modules (Victim) should be placed outside this region.

is demonstrated for a d_{AV} of 1000.2 μm as compared to a d_{AV} of 60.8 μm. Placing the victim circuits farther from those TSVs carrying the aggressor signals therefore significantly improves the noise isolation characteristics. Alternatively, a thicker substrate or a larger number of tiers between the aggressor and victim modules only slightly improves the isolation efficiency due to the low impedance of the TSVs.

A keep out region (shown in Fig. 5.10) is a circular area surrounding an aggressor TSV in which a victim should not be placed to ensure the coupled noise is lower than N_{max} (maximum allowable noise coupling, in dB). The radius of the keep out region is l, such that $20\log|H(\omega, l)| < N_{max}$. The magnitude of the transfer function in (5.4)–(5.6) for Si, Ge, and GaAs is, respectively, (5.8), (5.9), and (5.10).

$$|H(\omega, l)| = \left[\left(\frac{R_{bulk}(R_{sub}(l)+R_{bulk})+X_{gnd}(X_{gnd}-X_{C_{tsv}})}{(R_{sub}(l)+R_{bulk})^2+(X_{gnd}-X_{C_{tsv}})^2} \right)^2 + \left(\frac{X_{gnd}(R_{sub}(l)+R_{bulk})-R_{bulk}(X_{gnd}-X_{C_{tsv}})}{(R_{sub}(l)+R_{bulk})^2+(X_{gnd}-X_{C_{tsv}})^2} \right)^2 \right]^{1/2}, \tag{5.8}$$

$$|H(\omega, l)| = \left[\left(\frac{(R_{bulk}+R_{gnd})(R_{sub}(l)+R_{bulk}+R_{tsv}+R_{gnd})+X_{gnd}(X_{gnd}-X_{C_{tsv}})}{(R_{sub}(l)+R_{bulk}+R_{tsv}+R_{gnd})^2+(X_{gnd}-X_{C_{tsv}})^2} \right)^2 + \left(\frac{X_{gnd}(R_{sub}(l)+R_{bulk}+R_{tsv}+R_{gnd})-(R_{bulk}+R_{gnd})(X_{gnd}-X_{C_{tsv}})}{(R_{sub}(l)+R_{bulk}+R_{tsv}+R_{gnd})^2+(X_{gnd}-X_{C_{tsv}})^2} \right)^2 \right]^{1/2}, \tag{5.9}$$

$$|H(\omega, l)| = \frac{R_{bulk}}{R_{sub}+R_{bulk}}. \tag{5.10}$$

Although (5.8) to (5.10) are dependent of l, it is difficult to provide a closed-form expression in l. A design space for each of the substrate materials is therefore generated according to the relevant expression, as shown in Fig. 5.11. Both frequency and l are based on the maximum

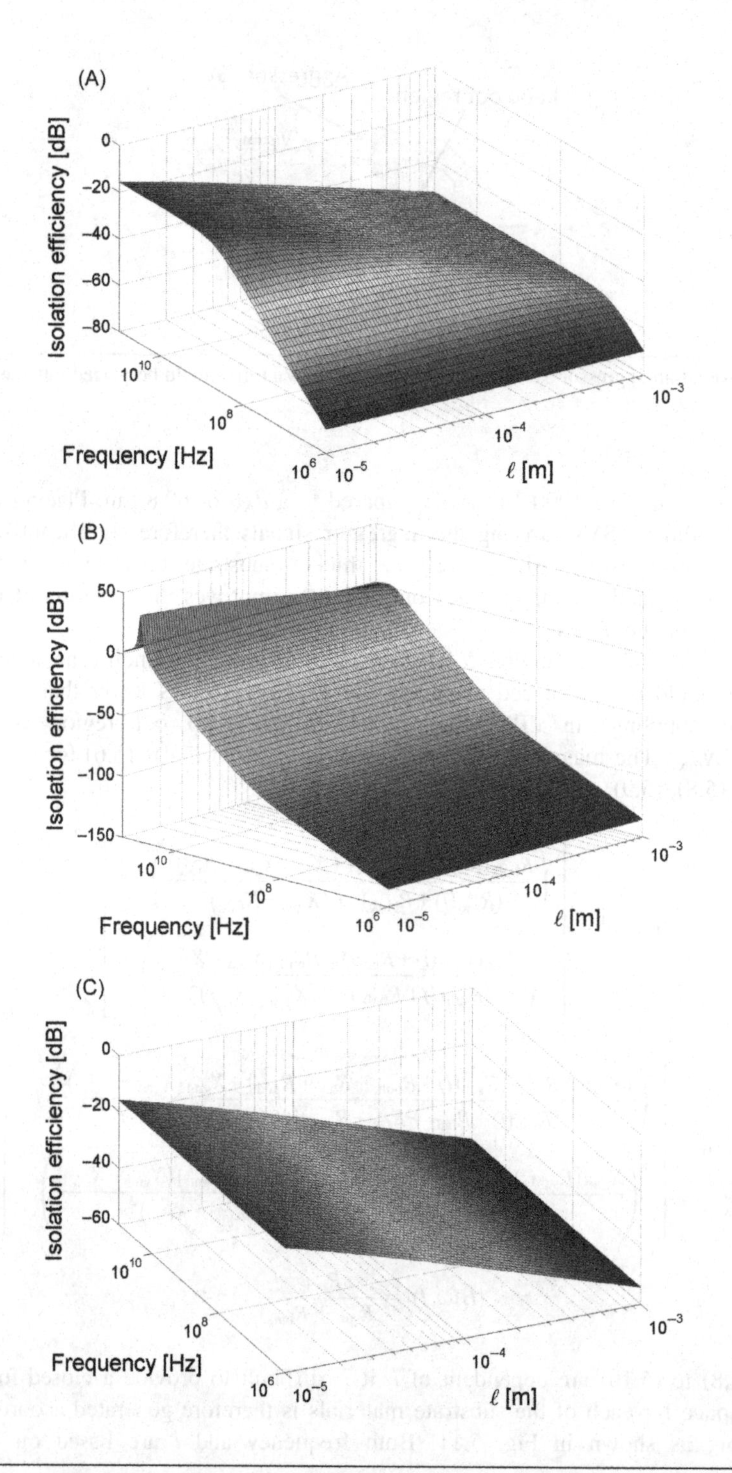

FIGURE 5.11

Isolation efficiency versus frequency and radius of keep out region for different substrate materials. (A) Si, (B) Ge, and (C) GaAs.

coupling noise (N_{max}). The design space for Si, Ge, and GaAs generated from (5.8) to (5.10) is shown in Fig. 5.11. Each plot describes the isolation efficiency of a coupled noise system in terms of frequency and l.

As shown in Fig. 5.11, the noise at the victim is less at low frequencies and increasing l. An increase in l rapidly lowers the coupled noise in both Si and GaAs. Alternatively, in Ge, the dependence of the isolation efficiency on l is weak. This behavior is due to the negligible substrate resistivity, leading to a stronger dependence on the frequency of the noise coupled system. The resonant frequency for Ge is illustrated in Fig. 5.11B. The design space around the resonant frequency should therefore be avoided.

To quantify the keep out region within the design space, a horizontal surface, described here as the "base surface," can be added to N_{max}. An example of a keep out region in a Si substrate is illustrated in Fig. 5.12. In this case, $N_{max} = -40$ dB and the keep out region is above the horizontal surface. This surface describes the minimum distance between an aggressor and victim to maintain the isolation efficiency below N_{max} for any frequency within the relevant range. Similar design spaces can be generated based on the transfer function for other design parameters (e.g., TSV diameter, TSV filling material, impedance of the ground network, and size of the victim).

A comparison between the transfer function and SPICE simulations for Si, Ge, and GaAs is provided in Fig. 5.13. This comparison is obtained from Fig. 5.11 at $l = 10$ μm and a ground network inductance of 1 nH. Discrepancies smaller than 1 dB for all substrate materials are noted.

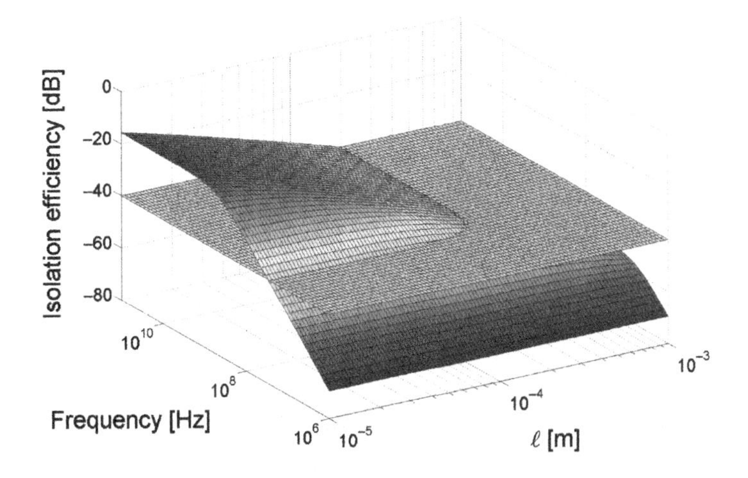

FIGURE 5.12

Keep out region around aggressor TSV for $N_{max} = -40$ dB. The victim modules should be placed on the isolation efficiency surface below the base surface.

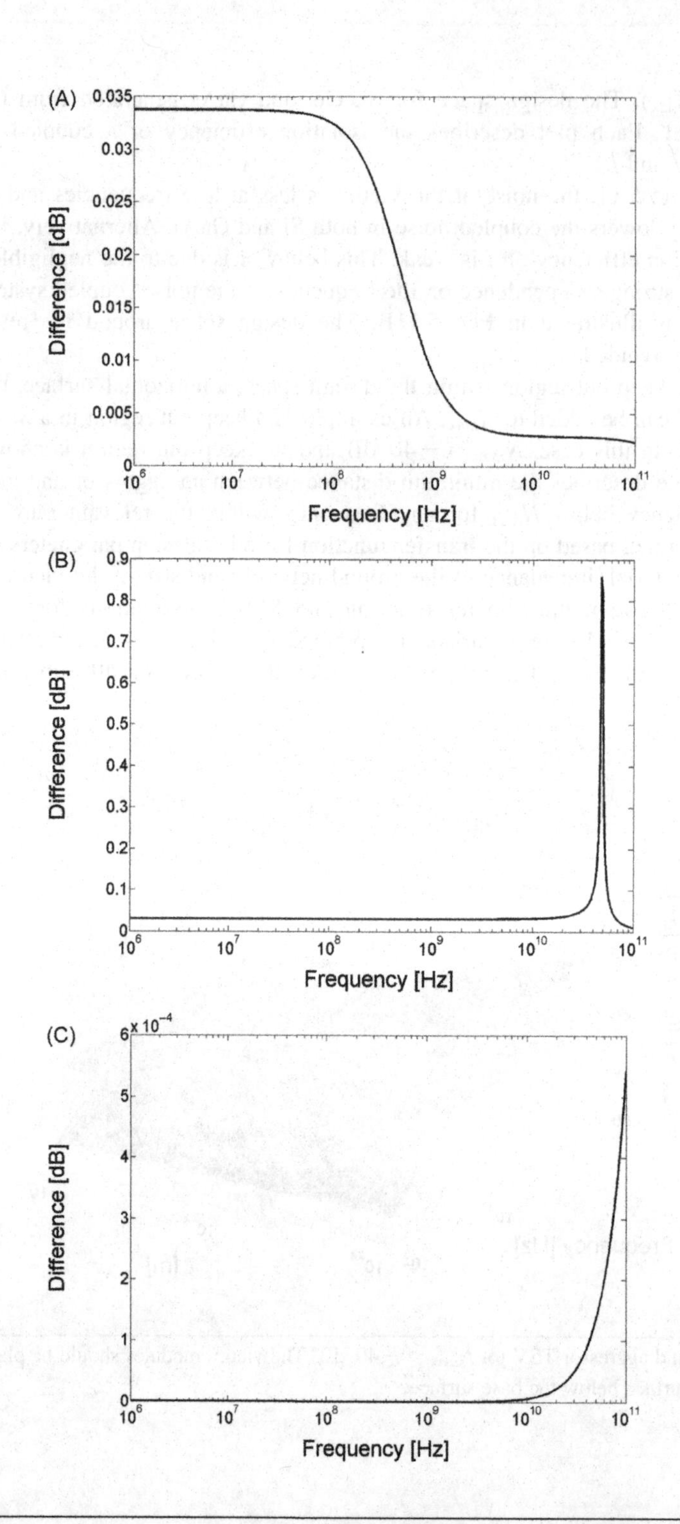

FIGURE 5.13

Comparison between SPICE model and extracted transfer function for different substrate materials. (A) Si, (B) Ge, and (C) GaAs.

5.4 SUMMARY

A complex electronic storm exists within heterogeneous 3-D systems. This multiconductor system can greatly degrade the behavior of a 3-D circuit. Models of noise coupling in heterogeneous 3-D integrated circuits are presented in this chapter. These models consider different substrate materials within a heterogeneous 3-D system.

- A lumped model is sufficient for Si and Ge substrates, producing a peak noise voltage error of, respectively, 26.5 and 0.2 mV as compared to a distributed model.
- For Ge, a short-circuit model can be used to satisfy less stringent noise constraints.
- The electrical properties of HgCdTe are similar to silicon; the model used for silicon is therefore appropriate for this type of substrate.
- GaAs substrates are highly resistive, efficiently isolating the victim from the aggressor. An open circuit model is therefore used for GaAs substrates.
- A noise coupled system can be represented as a transfer function to evaluate the isolation efficiency characteristics. Minimizing the magnitude of the transfer function, and therefore, lowering the coupled noise, is the objective.
- The transfer function can be reduced based on substrate specific parameters. Each reduced transfer function can be utilized to generate a design space for different material and design parameters.
- The reduced transfer functions are compared to SPICE simulations and good agreement is observed within a practical range of frequencies (up to 100 GHz).
- Variation of the ground network inductance, an isolation improvement technique, can shift the resonant frequency outside of the range of frequencies of interest, leading to lower coupling noise.
- A keep out region, increasing the horizontal distance between an aggressor TSV and a victim is also appropriate as an isolation improvement technique. A design space is described in terms of the keep out region and frequency for each substrate material.

CHAPTER

THREE-DIMENSIONAL ICs WITH INDUCTIVE LINKS*

6

CHAPTER OUTLINE

Wireless intertier communication enables new opportunities in three-dimensional (3-D) integrated circuits. Contactless interfaces based on inductive or capacitive coupling have been proposed as substitutes for through silicon vias (TSVs) and microbumps, as discussed in Chapter 3, Manufacturing Technologies for Three-Dimensional Integrated Circuits. Using a wireless intertier communication scheme, the development of multifunctional, heterogeneous 3-D systems is enhanced. Disparate technologies can be seamlessly combined within the same package without requiring elaborate level shifting circuitry and electrostatic discharge protection required by galvanic interconnects.

Moreover, wireless interchip communication employs passive elements within each tier of a 3-D stack, which in turn can be manufactured using conventional two-dimensional (2-D) processes. Consequently, there is no need for additional masks that increase fabrication costs as compared with TSV manufacturing where additional processing steps, such as TSV etching and copper filling, are

*Mr. Ioannis Papistas has contributed this chapter.

Three-Dimensional Integrated Circuit Design. DOI: http://dx.doi.org/10.1016/B978-0-12-410501-0.00006-X

required [256]. As a result, wireless schemes provide inexpensive communication and possibly exhibit higher yield as compared to wired schemes. The basic features of wireless intertier communication schemes are presented in Section 6.1. The design of the on-chip inductors for intertier communication is discussed in Section 6.2, while the supporting circuitry for signal transmission and recovery are presented in Section 6.3. Challenges related to the design of the inductive links are considered in Section 6.4. Issues related to wireless power transfer across the tiers of a 3-D integrated system are briefly reviewed in Section 6.5. A short summary is provided in Section 6.6.

6.1 WIRELESS ON-CHIP COMMUNICATION INTERFACES

Wireless intertier communication, also termed AC coupling, is based on either capacitive or inductive links. Capacitive coupling uses the electric field formed between the two electrodes of a capacitor to communicate, where each electrode lies on a separate tier within a 3-D system. Alternatively, inductive links manipulate the magnetic flux between concentrically placed inductors on different tiers. Due to the superior performance of inductive links as compared to capacitive communication, only inductive coupling is discussed in this chapter. The advantages and limitations of wireless communication along with an overview of the inductive coupling operation are described in the following subsection.

6.1.1 INDUCTIVE LINKS

The most versatile of the wireless on-chip communication schemes are inductively driven interfaces. Inductive links modulate the magnetic flux formed between concentrically placed inductors in different tiers within a 3-D system. Current mode signaling is used; therefore, greater communication distances can be achieved than with capacitive coupling, which is comparable to the length of a TSV (i.e., several tens of micrometers) [257]. Moreover, with inductive coupling, 3-D systems are no longer limited to face-to-face integration. Face-up, face-down, and back-to-back bonding is also supported due to low eddy current losses within the silicon substrate [258].

A transmitter and receiver circuit along with the on-chip inductors form a communication link, as depicted in Fig. 6.1. The transmitter, usually an H-bridge circuit, drives the inductor supplying a transmission current I_T, while the receiver senses, amplifies, and rectifies the induced voltage V_R into a digital pulse. A first order model of an inductive link is depicted in Fig. 6.2, illustrating the coupled on-chip inductors L_T and L_R. The parasitic resistance and capacitance of the on-chip spiral inductors are also depicted in this figure [259].

The inductively coupled link ideally behaves as a first order differentiator [260] $j\omega M$, where M is the mutual inductance coupling between the transmitter and receiver inductors. However, due to the parasitic resistance and capacitance of the inductors, the link behaves as a band pass filter, with a peak at the resonance frequency f_{SR}, as determined by the self-inductance and parasitic capacitance of the inductors [261],

$$f_{SR} = \frac{1}{2\pi\sqrt{LC}}. \tag{6.1}$$

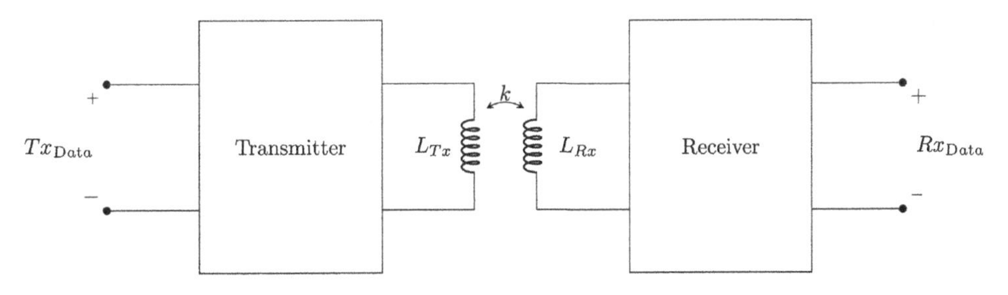

FIGURE 6.1

An inductive link between the transmitter and receiver circuits including the coupled on-chip inductors.

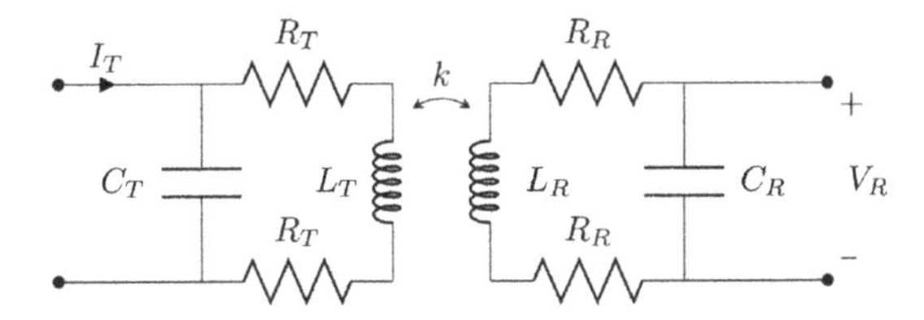

FIGURE 6.2

Equivalent circuit of an inductive link including the parasitic resistance and capacitance of the on-chip inductors.

The operation of the inductive link can be further explained by considering the link transfer function,

$$V_R = \frac{1}{(1 - \omega^2 L_R C_R) + j\omega C_R R_R} \cdot j\omega M \cdot \frac{1}{(1 - \omega^2 L_T C_T) + j\omega C_T R_T} \cdot I_T, \tag{6.2}$$

where the term $1/((1 - \omega^2 L_i C_i) + j\omega C_i R_i)$(for $i = R, T$) behaves as a second order low pass filter and constrains the bandwidth due to the losses of each inductor.

Due to the short communication distance between the two inductors, pulse modulation is chosen as the communication scheme over carrier modulation. Pulse modulation is a simpler, low power communication scheme as opposed to carrier modulation, where complex analog and RF circuits are required. In the case of inductive links, the current I_T of the transmitter produced by the digital pulse Tx_{data} is modeled as a Gaussian pulse [261], as illustrated in Fig. 6.3A,

$$I_T = I_P \exp\left(-\frac{4t^2}{\tau^2}\right), \tag{6.3}$$

where I_P and τ are, respectively, the magnitude and width of the pulse. Assuming an ideal inductive link, the voltage induced on the receiver V_R, is the derivative of the transmitted pulse,

$$V_R = -MI_P \frac{8t}{\tau^2} \exp\left(-\frac{4t^2}{\tau^2}\right), \tag{6.4}$$

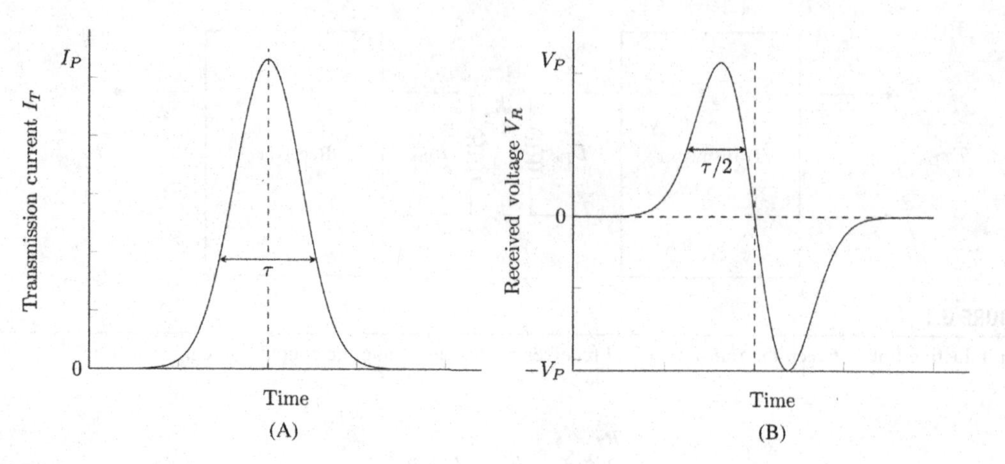

FIGURE 6.3

Model for pulse modulation. (A) Current of the transmitter modeled as a Gaussian pulse and (B) voltage induced on the receiver [260,261].

as depicted in Fig. 6.3B. This voltage is described by a Gaussian double pulse given by [262a]

$$w(t) = \frac{2\sqrt{e}At}{\tau} exp\left(-\frac{2t^2}{\tau^2}\right), \tag{6.5}$$

with an amplitude,

$$A = V_P = 2\sqrt{\frac{2}{e}}\frac{MI_P}{\tau}. \tag{6.5a}$$

The width τ of the current pulse is one of the primary parameters characterizing an inductive link. The sensitivity margin of the receiver is directly related to the width of the pulse, which is equal to $\tau/2$. Consequently, the pulse width also determines the operational bandwidth of the link. From the function describing the dependence on frequency of the induced voltage V_R, a peak frequency f_p is observed. To avoid aliasing (or intersymbol interference), the operating frequency of the link should be greater than $2f_p$. Therefore,

$$f_{SR} > 2f_p = \frac{2\sqrt{2}}{\pi\tau} \approx \frac{0.9}{\tau}, \tag{6.6}$$

provides a lower bound on the operating frequency of the communication scheme [261,262].

6.2 ON-CHIP INDUCTORS FOR INTERTIER LINKS

The objective of an inductive link is to provide a high mutual inductance M while maintaining the self-resonance frequency above the aliasing boundary, $f_{SR} > 2f_p$. The design of the coil should therefore satisfy these two guidelines. The objective of a high mutual inductance or, in other words, intertier coupling, is explored in Section 6.2.1. Coil optimization with respect to link bandwidth specifications is discussed in Section 6.2.2.

6.2.1 INTERTIER COUPLING EFFICIENCY

Intertier coupling efficiency is intrinsically associated with the speed, power, and area of the inductive link. The overall quality of the link therefore greatly depends upon the coupling efficiency. A first order analysis of the coupling efficiency of inductive links is presented in this subsection.

Assuming an inductor with an outer diameter d_{out} and a separation distance X, the coupling efficiency between the inductors is inversely proportional to X/d_{out} [261]. Since the communication distance is usually limited by technology, the outer diameter of the coil is the primary parameter for adjusting the coupling efficiency. For a specific communication distance X, the coupling efficiency for decreasing the outer diameter of the coil is illustrated in Fig. 6.4.

According to Fig. 6.4 the coupling efficiency can be divided into four discrete regions, the cubic, square, linear, and saturation regions. The difference among the regions depends upon the ratio X/d_{out}. For $X > d_{out}$, k is less than 0.1 and the coupling efficiency lies within the cubic region, which implies that the coupling efficiency is a cubic function of X. In this region, the magnetic flux between the two inductors is severely attenuated. Communication can be achieved by utilizing multiple amplification stages at the receiver tier. Roughly the minimum k for supporting low power intertier communication is 0.1 and is achieved at $d_{out} = X$, where the square region is approached.

The square region lies between $d_{out} = X$ to $d_{out} = 3X$, or $k \approx 0.65$. Synchronous receivers, in general, are used for communication when operating within the square region [261]. Synchronous circuits are preferred as these circuits exhibit lower bit error rates for low coupling as compared to asynchronous transceivers. Increasing the outer diameter allows the inductive link to enter the linear region, where asynchronous sensing schemes can be utilized. For diameter d_{out} greater than $5X$,

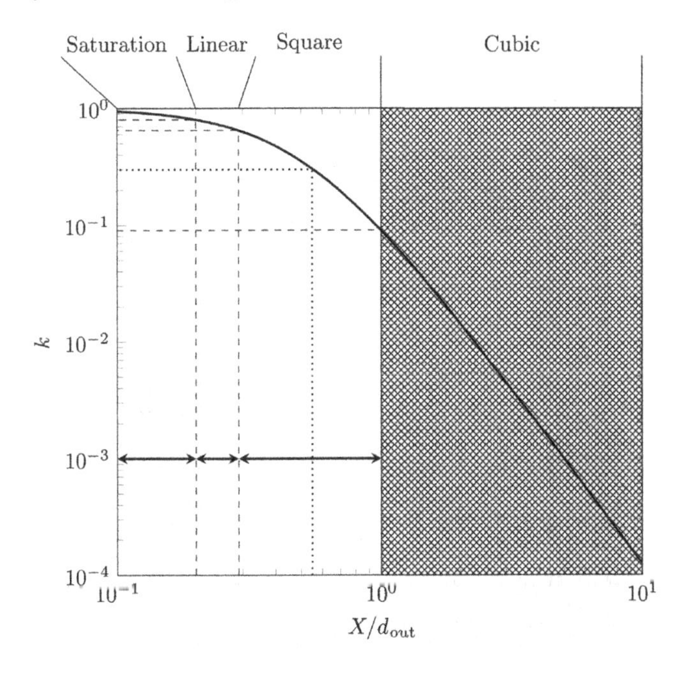

FIGURE 6.4

Coupling efficiency for distance X and decreasing outer diameter d_{out} [261].

the coupling efficiency saturates. The greater intertier coupling in the saturation region is usually exploited in wireless power transfer schemes, where increased efficiency is crucial. For data transmission schemes, the suggested coupling efficiency is $k \approx 0.3$, or $d_{out} = 2X$. In Section 6.3, both synchronous and asynchronous inductive link receivers are considered where communication occurs within the different coupling regions.

6.2.2 GEOMETRY AND ELECTRICAL CHARACTERISTICS OF INDUCTOR

The bandwidth of the link can be enhanced by proper design of the inductive link to achieve a specific coupling coefficient k. Optimization of the coil is achieved by altering the geometry of the inductor. The electrical characteristics of the inductive link are also determined by the geometry of the inductor.

The geometry of the on-chip spiral inductor depends upon four parameters, the outer diameter d_{out}, the number of turns n, the line width w, and the line spacing s, as illustrated in Fig. 6.5. As mentioned in the previous subsection, the minimum outer diameter is limited by the communication distance to achieve a target coupling efficiency k. Consequently, the number of turns is chosen to manipulate the characteristics of the inductive link [259–262]. The line width and spacing are constrained by the fabrication process. Minimum line dimensions are usually employed, although a rule of thumb for the line width is 1% to 2% of d_{out} to reduce the parasitic resistance of the thin wires [261].

To quantify the effects of the inductor windings on the link characteristics, the complex relationship between the inductor geometry and the corresponding electrical parameters is explored here. The self-inductance of the spiral inductor structure is quadratically related to the number of turns,

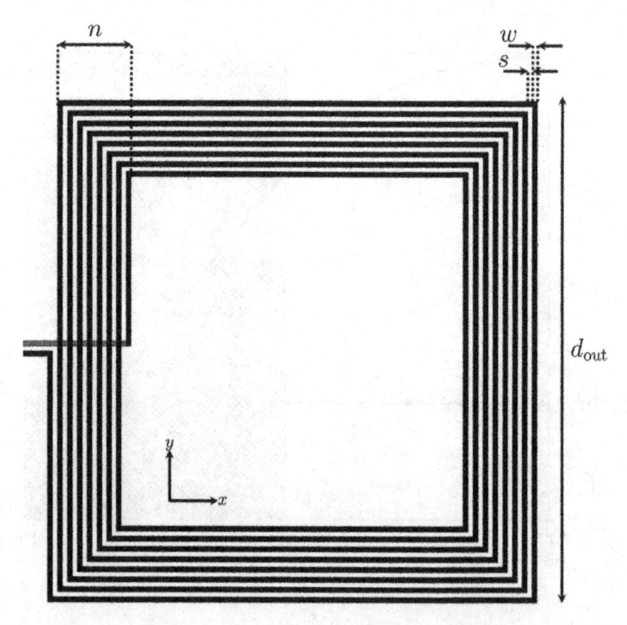

FIGURE 6.5

Square spiral on-chip inductor with n=7 turns, illustrating the geometric parameters.

$$L \propto d_{\text{out}} n^2. \tag{6.7}$$

Alternatively, the parasitic capacitance is linearly proportional to the number of turns

$$C \propto d_{\text{out}} n. \tag{6.8}$$

As a result, the self-resonance frequency is a function of $1/\sqrt{n^3}$,

$$f_{\text{SR}} \propto \frac{1}{2\pi d_{\text{out}} \sqrt{n^3}}. \tag{6.9}$$

Consequently, changing the number of turns n can adjust the link performance to satisfy the specifications of the communication scheme [261]. Based on the discussion in Section 6.2.1 and this subsection, the different aspects of the design process of the inductor are combined into a design flow that produces inductive links that satisfy specific performance, area, and power constraints.

6.2.3 DESIGN FLOW FOR INDUCTIVE LINK COILS

The design process of an inductive link scheme can require numerous iterations and be a time consuming process. A flow for designing the coils for an inductive link is illustrated in Fig. 6.6 and includes possible design iterations to satisfy power, performance, and/or area constraints.

To communicate across a distance X the outer diameter d_{out} is chosen to satisfy a minimum coupling coefficient k_{min}. The outer diameter of the inductor is increased until the coupling coefficient k satisfies the minimum coupling constraint k_{min}. If this constraint is satisfied, the link bandwidth is improved by adjusting the inductor geometry; specifically, the number of spiral turns and trace width.

The bandwidth of the inductive link is dependent on the communication scheme. Consequently, the bandwidth should satisfy the performance constraint of the link. Based on the coupling coefficient k, minimum performance requirements, and area requirements, a synchronous or asynchronous scheme is chosen for the intertier signaling.

The transceiver circuit requires a specific output data rate, which is associated with the power consumption of the link. The link bandwidth can be refined to improve the link characteristics. The power consumption is the final design criterion to be assessed. Refinements to the design of the transceiver and the outer diameter of the inductor improve the power utilization of the link.

6.3 TRANSMITTER AND RECEIVER CIRCUITS

Once the design of the coil is determined, the transmitter and receiver circuits that underpin the intertier communication are considered. Depending upon the coupling efficiency between the coils, a synchronous or asynchronous transceiver scheme can be utilized. The design of the synchronous inductive links is discussed in Section 6.3.1, while asynchronous sensing circuits are described in Section 6.3.2. High performance communication schemes with burst data transmission are reviewed in Section 6.3.3.

6.3.1 DESIGN OF SYNCHRONOUS INDUCTIVE LINK TRANSCEIVERS

For low to moderate coupling efficiencies ($0.1 < k < 0.6$), synchronous transceiver schemes are preferable. Synchronous receivers exhibit the advantage of noise immunity when sampling is off.

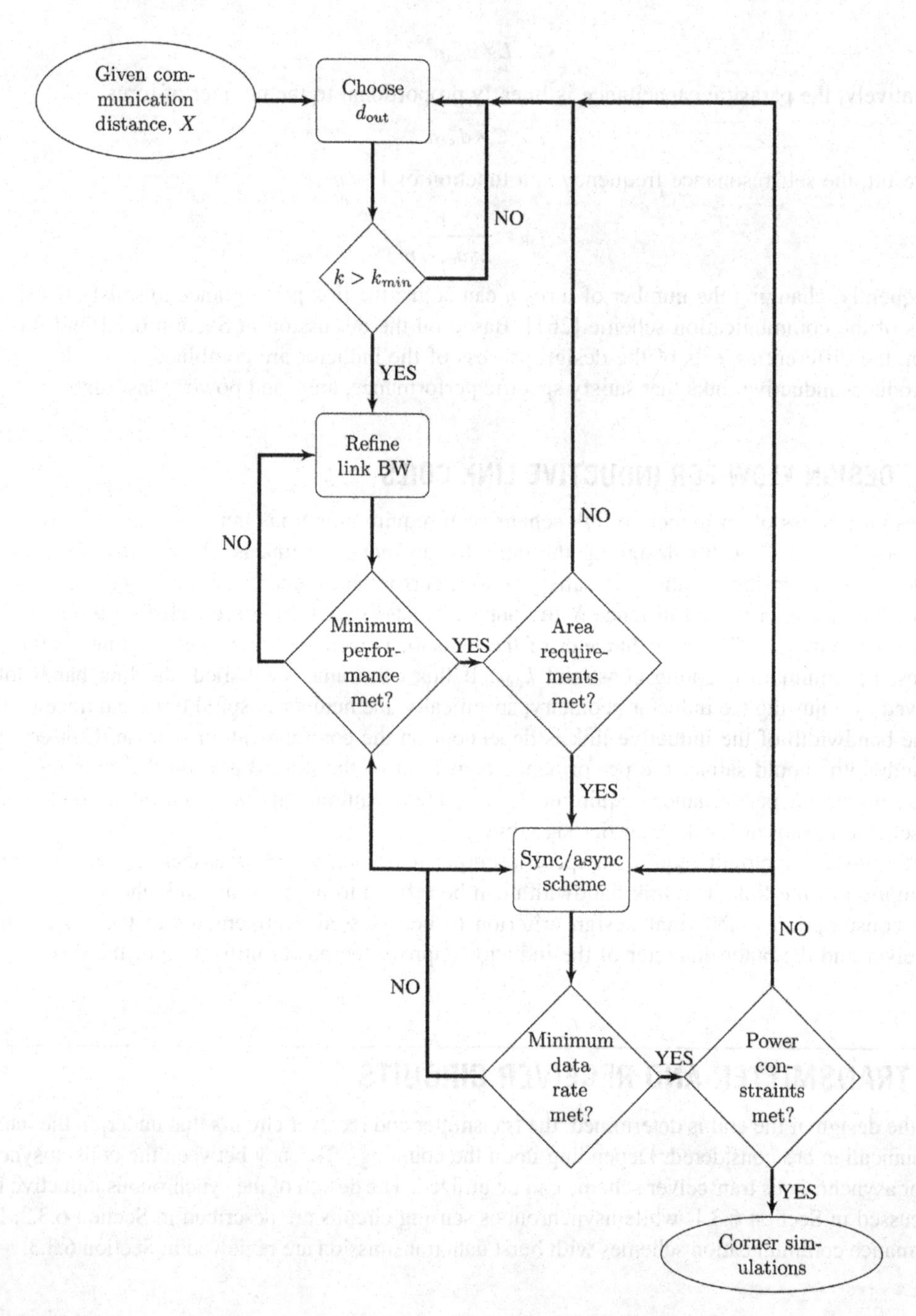

FIGURE 6.6

Flow diagram for the design of the coils in an inductive link under power, performance, and area constraints.

As a result, noise is only induced when the received data are sampled. Consequently, the signal-to-noise ratio is improved and reliable communication is ensured.

A synchronous transceiver is schematically illustrated in Fig. 6.7 [261,262]. The transmitter is composed of an H-bridge circuit driving an inductor, augmented by a digital circuit for timing and signal shaping control. Alternatively, the receiver circuit comprises a sense amplifier for signal sensing and an SR latch to capture the received signal.

For the transmitter of the synchronous inductive link, a delay generator is used along with a clock signal to shape the pulse width, denoted by τ. The pulse width is the primary factor affecting the sensitivity of the receiver. For a pulse with width τ, a positive edge in the transmitted data is sensed by transistors P1 and N2 (see Fig. 6.7), driven, respectively, by the inverter of block G1 and the OR gate of block G2, producing a positive pulse. The amplitude of the generated pulse depends upon the size of the NMOS transistor N2. Likewise, a negative pulse is generated for a falling edge through transistors P2 and N1.

Magnetic flux carries the generated pulses from the transmitter to the inductor at the receiver. The current generated at the receiver inductor is biased and amplified through the biasing resistance R_b, with a resistance in the $k\Omega$ range. The differential output of the resistance is passed to the sense amplifier through a differential amplifier stage. Two distinct operating stages of the sense amplifier exist, the precharge and evaluation stages. The amplifier alternates between those two stages depending upon the clock signal Rx_{clk}.

If the clock signal Rx_{clk} is low, the receiver circuit operates in the precharge phase, where transistors P4 and P5 are switched on. The SR latch stores the data. Alternatively, if Rx_{clk} is high, the PMOS transistors are off. Depending upon the polarity of N4 and N5, respectively, the voltage at either node V_{LP} or node V_{LN} is pulled to logic high, driving the other input to a logic low. Thus, the polarity of the transmitted data Tx_{data} is replicated. In the following precharge phase, the SR latch maintains the polarity of the pulse and generates the digital signal Rx_{data}.

6.3.2 ASYNCHRONOUS DATA TRANSMISSION AND RECOVERY

When operating in the linear region, asynchronous data transceivers for communication are typically used. The size of the transceiver circuit is much smaller since the timing control blocks are less complex. The effect is increased communication speed of the inductive link since the data rate of the channel is not limited by the timing circuit. Alternatively, the interface bottleneck is either the switching frequency of the transistors or the bandwidth supported by the pair of inductors (whichever is lower).

The asynchronous transceiver, shown in Fig. 6.8 [264], features an H-bridge circuit driving the inductor, similar to the equivalent synchronous circuit. Nevertheless, the timing circuit is removed, reducing overall area and yielding higher performance in terms of both communication speed and power. The timing circuit is also removed from the receiver circuit, where a hysteresis comparator is used rather than a sense amplifier.

In the receiving tier, a hysteresis comparator senses the induced data. The hysteresis comparator includes a gain stage (devices P3–N4 and P6–N5) and a latch circuit (see Fig. 6.8). The induced voltage V_R is biased at voltage $V_b = V_{\mathrm{DD}}/2$ and is inputted differentially to the hysteresis comparator. Depending upon the polarity of the induced signal, the hysteresis comparator is driven to either

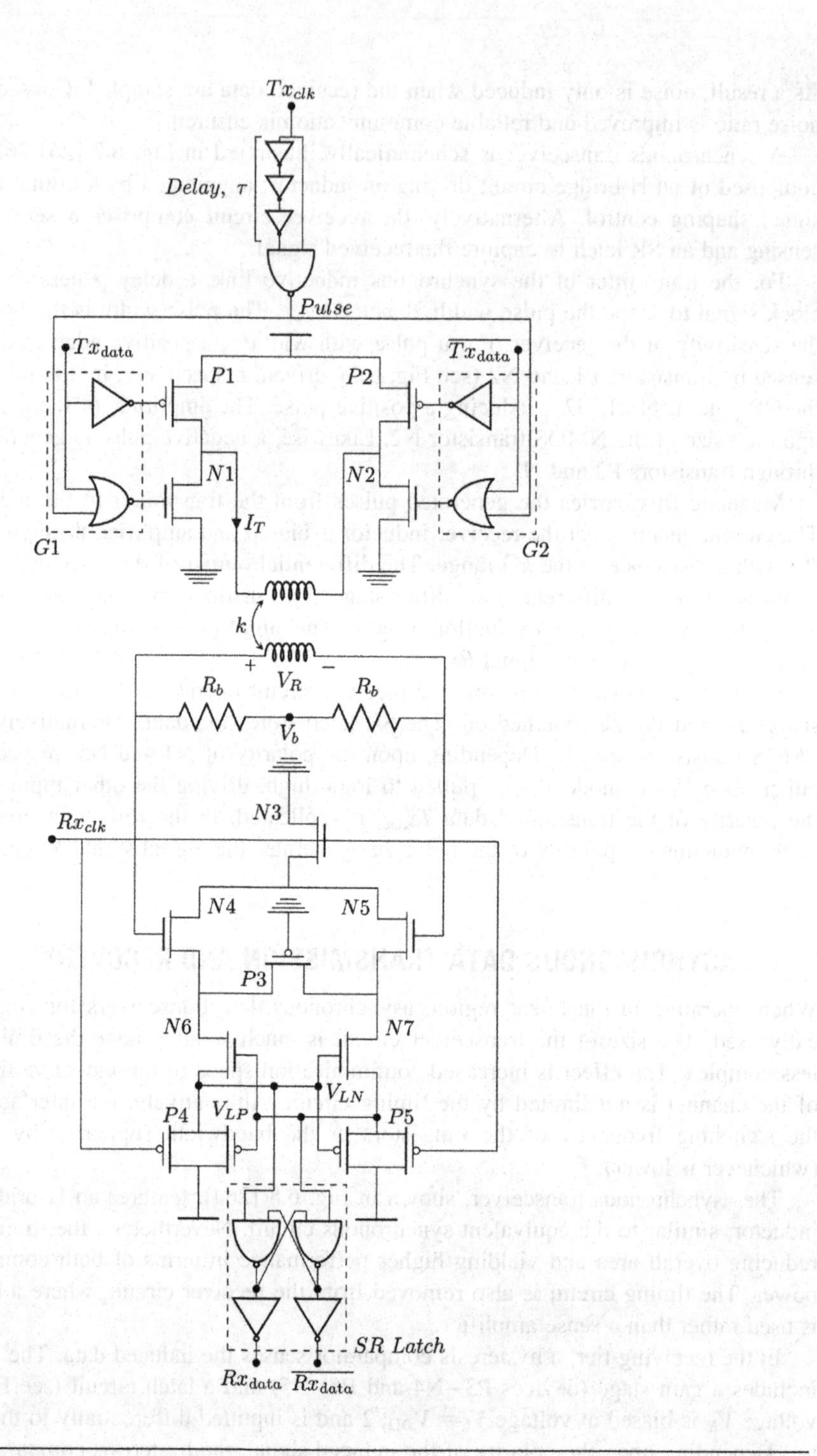

FIGURE 6.7

Transceiver circuit of a synchronous inductive link [259,263].

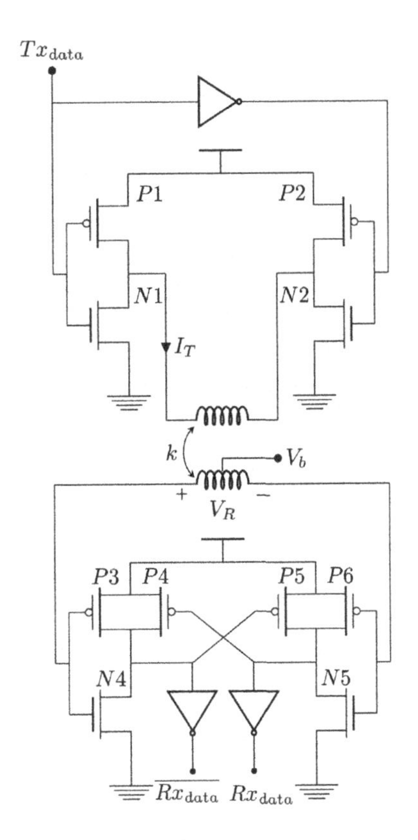

FIGURE 6.8

Transceiver circuit of an asynchronous inductive link [264].

an on or off state, reproducing the transmitted data signal without requiring a clock signal. The sensitivity of the receiver circuit depends upon the change in hysteresis voltage ΔV of the comparator. The change in the hysteresis voltage has the opposite polarity as compared to the output signal Rx_{data}, and is modulated by a latch circuit. For a positively induced pulse, the signal Rx_{data} transitions high and the latch circuit reduces the voltage threshold of the comparator by the change in the hysteresis voltage, $V_{\text{TH,comp}} = V_{\text{TH,inv}} - |\Delta V|$, where $V_{\text{TH,comp}}$ is the threshold voltage of the comparator and $V_{\text{TH,inv}} = V_{\text{dd}}/2$, the threshold voltage of the gain devices. The threshold voltage of the comparator facilitates the transition for the following negative pulse. The output signal is therefore held high until a negative edge is received. Similarly, an equivalent procedure is followed for a negative pulse.

As mentioned previously, an asynchronous communication scheme requires increased coupling efficiency to cope with signal reliability. Increased coupling efficiency can be achieved with larger inductors, which, however, lead to increased area for the inductive link. To mitigate the greater area, multiplexing techniques are applied, such as burst data transmission [264].

6.3.3 BURST DATA TRANSMISSION

Due to the size of the inductors in the inductive link, only a few links can fit within a circuit without requiring excessive area. Increasing the number of on-chip inductors can lead to area consuming or, in many cases, overly congested circuits, since the inductors are placed within the global interconnect layers. An efficient method is needed to exploit the available bandwidth of the inductive link without further increasing area. Consequently, data multiplexing schemes, such as the approach illustrated in Fig. 6.9, are utilized to reduce the number of inductive links.

Data multiplexing provides several advantages when applied to inductive links. The number of individual inductive links decreases, reducing proportionally the area requirements. Moreover, fewer transmitter circuits dissipate less power. Depending upon the circuit, an inductive link transmitter can demand high power. An array of transmitters can consume significant power as

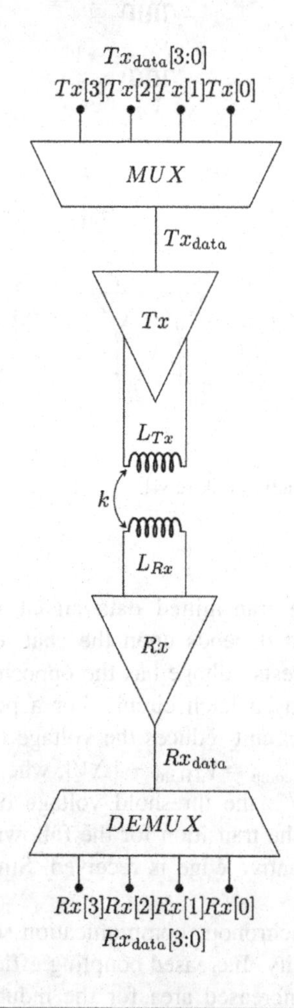

FIGURE 6.9

Block diagram of an inductive coupling scheme with burst transmission [264].

compared to a single inductive link transmitter combined with a multiplexing scheme. Thus, less power is consumed if burst transmission is employed.

The performance of the burst transmission scheme depends upon the clock frequency of the multiplexer and demultiplexer as the clock signal is the primary bandwidth bottleneck of the circuit. To achieve a high frequency, a low power clock generator, such as a ring oscillator circuit with a counter, is often preferred to control the length of the transmission. An inductive link with burst transmission for mobile devices is described in [264], where a 400 MHz clock is assumed for the processor. As a result, a 3.2 GHz clock is generated by a local ring oscillator for multiplexing and demultiplexing eight signals. To avoid the use of a second ring oscillator on the receiving tier, the clock signal is transmitted over an inductive link.

6.4 CHALLENGES FOR WIRELESS ON-CHIP COMMUNICATION

Wireless intertier communication provides plenty of advantages for 3-D integration, particularly for heterogeneous circuits. Nevertheless, challenges arise that can hinder the use of wireless interfaces for on-chip communication. Wireless systems are based on passive inductors and capacitors, which consume significant area as compared to active devices. Area optimization is therefore an important issue in the design of inductive links. Moreover, wireless interfaces suffer from coupling noise that should be considered during the design process. Coupling noise is crosstalk between inductive links due to the close proximity or noise induced by the inductive link on other vital and noise sensitive on-chip components, such as the power delivery network. Moreover, the lack of a wired interconnect interface introduces challenges to delivering power to all of the tiers except for the tier connected to the package substrate. A performance and area analysis of the inductive links is described in Section 6.4.1. Crosstalk issues between inductive links are explored in Section 6.4.2, while crosstalk noise effects caused by the inductive link on the power delivery system are discussed in Section 6.4.3.

6.4.1 PERFORMANCE AND AREA ANALYSIS

Recent works on TSV signaling and bandwidth optimization indicate that TSVs can achieve high data rates [265,266]. Inductive links are also capable of achieving performance levels comparable to TSVs [257]. The area requirements between the two communication schemes, nevertheless, are considerably different. TSVs occupy both silicon and wiring area, while inductive links primarily require interconnect resources. To model this situation, a performance-to-area metric is introduced to provide a fair comparison [257],

$$\text{eff} = \frac{\text{BW}}{A}, \tag{6.10}$$

where BW is the communication bandwidth achieved by each interface and A is the area.

To evaluate performance, an *RC* model of the TSV is utilized which includes parasitic capacitive coupling with the substrate and other neighboring TSVs. Different state-of-the-art TSV pitches are considered to model the effect of coupling on TSV performance. Moreover, the redistribution layer of the TSV is also considered in the model, since this layer adds parasitic resistance and capacitance and dominates the delay of the intertier interfaces. These models are frequency dependent, since communication in a high performance system is assumed. The behavior of the model changes significantly at high frequencies.

An asynchronous inductive link transceiver is evaluated based on a state-of-the-art on-chip inductor [264]. The performance of the inductive link is evaluated with SPICE, achieving a maximum data rate of 20 Gbps. The inductors considered in this analysis are chosen to ensure that the communication distance supported by the inductive link is comparable to the length of the TSVs. Both inductive links and TSVs are placed in an $N \times M$ array to form a high performance communication interface.

The bandwidth-area efficiency for each interface is listed in Table 6.1. To determine the interconnect efficiency, a six metal layer interconnect technology is assumed. The silicon area efficiency $\text{eff}_{s,x}$, interconnect area efficiency $\text{eff}_{i,x}$, and total efficiency eff_x are reported for each type of interface x (i.e., TSV or IL). As the TSV pitch increases, the overall efficiency drops by 80%. The TSV link efficiency, however, remains almost five times greater than the efficiency of an inductive link. The efficiency of the interconnect layers follows the overall efficiency, with the TSV interface presenting a significantly higher performance-to-area ratio.

The silicon area efficiency, however, leads to a different result. Since the transceiver used by the inductive link is a small circuit, the silicon area efficiency of the inductive link is an order of magnitude greater than the efficiency of a TSV. The silicon area underneath the on-chip inductor can be used to either reduce the area overhead of the interface, or improve the performance-to-area efficiency of the inductive interface.

A multiplexing scheme can improve the performance of the inductive interface by reducing the number of on-chip inductors. The same multiplexing scheme is compared with the TSV interface. For simplicity, a 1 Gbps data rate is assumed for a communication channel. Multiplexing decreases the area of a TSV link, although multiplexing limits the data rate due to the latency imposed by the multiplexing–demultiplexing circuitry. Depending upon the diameter of the TSV and the size of the multiplexer, the area efficiency of the TSV interface departs from the aforementioned bounds. For an inductive link, however, considerable silicon area is available. This area can be reserved for multiplexing without incurring any additional area as compared to a TSV link. A tradeoff between the multiplexing ratio and the performance-to-area efficiency of the interfaces exists to determine the highest area efficiency to achieve a specific performance.

A multiplexing ratio of up to 12:1 is used by the multiplexer described in [267]. The interface efficiency is depicted in Fig. 6.10 for each scheme. For the TSV interface, two trends can be observed. For a TSV pitch smaller than 20 μm, the interface performs better without multiplexing.

Table 6.1 Performance-to-Area Efficiency for Single Interface Links

Type of Interface	$\text{eff}_{s,x}$ (Mbps/μm^2)	$\text{eff}_{i,x}$ (Mbps/μm^2)	eff_x (Mbps/μm^2)
TSV, 20 μm	71.2	11.8	71.2
TSV, 30 μm	21.7	3.6	21.7
TSV, 40 μm	13.8	2.3	13.8
IL, 84 μm	2.7×103	1.4	2.8

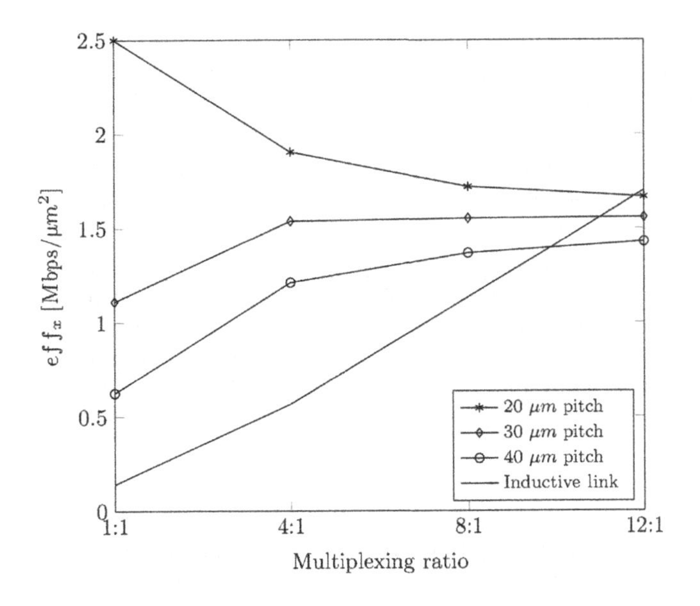

FIGURE 6.10

Efficiency of TSV and inductive interfaces with higher multiplexing density.

As the multiplexing ratio increases, the efficiency drops. However, for a TSV with a pitch greater than 30 µm, multiplexing increases system efficiency. As shown in Fig. 6.10, eff_{TSV} asymptotically approaches 1.5 Mbps/µm^2.

This behavior occurs since the multiplexing circuits require greater area as compared to the TSVs. For a data rate of 1 Gbps and no multiplexing, the size of the interface circuit is similar to the area occupied by the TSVs. A decreasing TSV diameter leads to a quadratic increase in the area efficiency of the interface. However, if signal multiplexing is included, this behavior changes. As the multiplexing ratio increases, the efficiency of the interface saturates to a level depending upon the ratio of the pitch of the TSV to the area of the multiplexer.

The performance-to-area density for each communication scheme exhibits comparable performance, while the area significantly differs. Combined with a high density multiplexing scheme, an inductive link can exhibit comparable or even improved performance density as compared to a state-of-the-art TSV interface. Nevertheless, TSVs with a sub-20 µm pitch provide the best efficiency without signal multiplexing.

6.4.2 CROSSTALK BETWEEN INDUCTIVE LINKS

Inductive links control magnetic flux and employ pulse modulation for near communication between different tiers within a 3-D system. Unavoidably, crosstalk noise exists between different inductive links when placed in high density arrays in high performance interconnect interfaces. A method based on time division multiplexing can be used to mitigate crosstalk between inductive links, as discussed in this subsection.

Electromagnetic coupling between inductive links occurs due to the high density required for increased performance and area efficiency. A straightforward method to reduce interlink noise is to increase the distance between neighboring inductors. For a space of $s_{IL} = 2d_{out}$, where d_{out} is the outer diameter of the inductor, the induced noise is negligible and the nearby channels do not affect the inductive link [268]. However, doubling the distance between links effectively quadruples the area required by the communication interface. An area efficient solution is, therefore, needed to improve the performance of a link while reducing crosstalk noise.

An effective way to reduce crosstalk between inductive links is to avoid simultaneous operation of adjacent links. An example of this approach is time division interleaving, as described in [268] for an 1 Tbps inductive coupling transceiver. A four-phase clock is generated by a phase interpolator, dividing the operation of a block of transceivers into several time slots. To achieve a data rate of 1 Gbps/link, a 1 GHz clock is applied and each inductive link occupies an operating time slot of 250 ps per clock cycle. 1,024 transceivers operate in parallel to achieve the 1 Tbps performance objective.

These results indicate that without the time interleaving technique, the crosstalk noise induced in adjacent links is similar to the voltage amplitude of the received signal (50 mV). The bit error rate of the channel in this case is significantly higher. However, for a two-phase interleaving scheme, the induced crosstalk noise is reduced to half (25 mV), improving the performance of the communication channel. A four-phase time division solution further improves performance, where an amplitude of 10 mV is observed for the crosstalk noise.

6.4.3 CROSSTALK NOISE ON ADJACENT ON-CHIP COMPONENTS

A byproduct of contactless on-chip communication is interference with other components within a system due to the wireless nature of the scheme. For example, adjacent interconnect layers, analog circuits, or sensitive digital circuits may be affected by electromagnetic waves emitted from an inductive link. Alternatively, interconnects could also affect the behavior of the inductor. Interactions between the inductor and interconnect layers, as well as noise induced by the inductive link and mitigation techniques to dampen this noise, are described in this subsection. The crosstalk produced by a single inductor link is discussed in Section 6.4.3.1. The more complex case of crosstalk due to an array of inductive links is described in Section 6.4.3.3. The resulting noise due to the crosstalk on diverse power distribution topologies is evaluated in Section 6.4.3.4.

6.4.3.1 Crosstalk effects due to an inductive link

The effect on the performance of an inductive link due to the interconnect wires in the vicinity of the spiral inductors is explored in [269]. Different types of power lines have been evaluated to model the effect of inductive links. Experimental results demonstrate that interconnects that form a closed loop are susceptible to eddy currents, reducing the performance of inductive links. Denser interconnect topologies, such as a mesh grid, exhibit higher transimpedance losses within an inductor, requiring greater transmission power to achieve the same performance.

Conversely, the effect of an inductive link on sensitive digital circuits has also been considered. An SRAM module is assumed with a sensing voltage of 50 mV. Placing an inductive link in the vicinity of an SRAM module induces a coupling noise pulse of 1 mV. The effect of the inductive link is therefore negligible as compared to the sensing voltage of the SRAM, and no read errors are induced due to the inductive link. Consequently, inductive links can be used in combination with SRAM in high density applications, such as a memory interface for a processor.

Since the on-chip inductors are placed on the topmost interconnect layers, crosstalk can, however, occur between the inductive link and a nearby power delivery network [270]. The induced noise from the inductive link adds to the other noise components of the power delivery network such as *IR* drops due to the wire resistance and simultaneous switching noise (SSN) Ldi/dt caused by device switching.

Each closed path formed within a power delivery network is susceptible to eddy currents and, consequently, voltage fluctuations induced by the inductor. Closed paths are formed between two or more power or ground wires. The amplitude of the induced voltage on the power delivery network depends upon the geometric and electrical characteristics of the return path, altering the coupling between the inductor and the power delivery network. Furthermore, the induced voltage depends upon the magnetic flux density that changes according to the spatial position of the power delivery network with respect to the inductor.

Standard design methods and CAD tools for power distribution networks consider *IR* drop and Ldi/dt noise [271]. Traditional design processes for power delivery networks, however, do not consider the additional noise originating from the interchip inductors in contactless 3-D systems. To address the combined effects of noise for power delivery networks, including SSN and resistive *IR* drops, the spatial alignment of the power and ground (P/G) loops with respect to the inductive link should be considered during the design process. Crosstalk coupling between the inductors and the P/G wires can lead to high levels of power noise. Measures are necessary to remedy this situation.

6.4.3.2 Case study

A structure consisting of an inductive link and a return path through a power delivery network has been evaluated with Ansys high frequency structural simulator (HFSS) to quantify the noise induced by an inductive link [272]. The simulated structure is illustrated in Fig. 6.11. A flip chip face-to-back integration approach is assumed. The length of the interconnect is denoted as l_{PDN}. Distance δ_c denotes the spatial separation between the geometric center of the inductor C_1 and the geometric center of the return path through the interconnect loop. A power loop is assumed; however, the same investigation also applies to the ground loop.

The noise amplitude varies considerably with the relative location of the interconnect with respect to the center of the inductors due to the different level of magnetic flux flowing through the power loop. To model this behavior, parameter δ_c is used to notate the relative location between the power loop and the inductors, where δ_c is swept within the range $[-d_{out}, d_{out}]$. Three positions with respect to the inductor, C_1, C_2, and C_2', are shown in Fig. 6.11, where, respectively, $\delta_{c,C_1} = 0$ μm, $\delta_{c,C_2} = -d_{out}/2$, and $\delta_{c,C_2'} = d_{out}/2$. For $\delta_c \geq |d_{out}|$, the amplitude of the noise is negligible, as shown in Fig. 6.12 by the crosshatched area. A maximum induced noise $V_{noise,max}$ of 39.5 mV is observed, where the power loop is close to the windings of the inductor. Alternatively, a considerable dip in the induced noise is observed for $\delta_c \pm d_{out}/2$, decreasing to $V_{noise,min}$ of 4.6 mV.

This outcome is due to the opposite direction of the magnetic flux between the two wires of the power loop, where the power loop is placed at either C_2 or C_2', as shown in Fig. 6.11. This behavior indicates that there is no need to place the power loop far from the inductive link to avoid crosstalk.

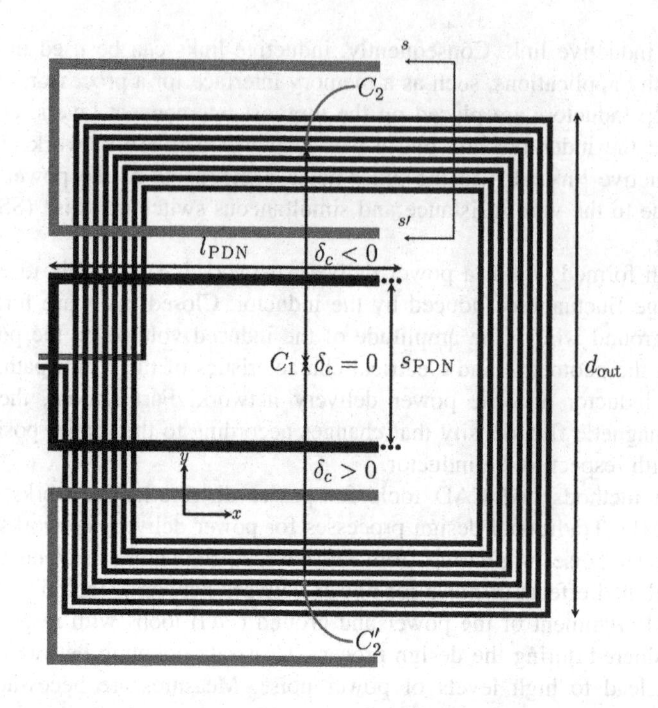

FIGURE 6.11

Top view of a structure comprising an inductive link and the return path through a power delivery network placed in different locations.

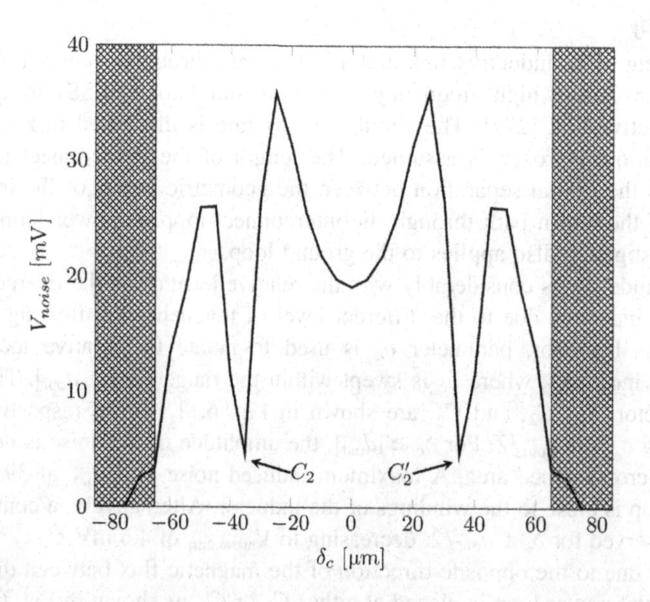

FIGURE 6.12

Noise induced by an inductive pair for varying δ_c.

Instead, due to the opposite direction of coupling between the wires of the loop and the turns of the inductor, the crosstalk is low without greatly altering the power delivery network.

6.4.3.3 Crosstalk noise effects caused by inductive link arrays

Inductive links are typically placed in a high density array to increase the communication performance of the interface. To estimate the noise in this case, an array of 4×4 inductive links is assumed with a spacing of 30 μm between each link to reduce crosstalk during simultaneous transmission. Each inductive link consists of transceiver and multiplexing circuits, consuming a current of $I = 7.1$ mA, modeled as uniformly distributed current sources across each inductor. The array is supplied by a power grid that utilizes the global and intermediate metal layers of a 65 nm process node. The grid produces an IR drop of 13.8 mV, which satisfies typical power supply noise constraints (<5% to 10% of V_{dd}). The power loops on the topmost global layer are illustrated in Fig. 6.13 as solid lines. The power loops span the entire array connecting to C4 bumps placed symmetrically at the periphery of the array. A spacing of $s_{C4} = 150$ μm and a diameter of $d_{C4} = 75$ μm are assumed for the bumps, satisfying the minimum area requirements for C4 pads [273]. The surrounding C4 bumps are assumed to supply the total current drawn by the array of inductors at a nominal $V_{dd} = 1.1$ V of 1.1 V. The bumps are connected to the power and ground lines through, respectively, the resistances $R_{\text{dist},P}$ and $R_{\text{dist},G}$.

The accumulated noise from a row within the inductive link array is illustrated in Fig. 6.14. The dashed line denotes the noise of an inductor at the edge of the array while the dash dotted line is the noise of an inductor at the middle of the line. The IR drop is superimposed on both curves. The solid line is the total noise for a row within the inductor array, based on

$$V_{\text{noise,acc}} = 2V_{\text{noise},l} + (N-2)V_{\text{noise},m}, \tag{6.11}$$

where $V_{\text{noise},l}$ is the noise generated from the inductors placed at the edge of the loop. $V_{\text{noise},m}$ is the noise produced by the remaining $N-2$ inductors coupled to the power delivery network loop, where $N = 4$ for the example illustrated in Fig. 6.13. Spatial alignment of the power grid over certain positions, such as C_2 and C_2' depicted in Fig. 6.11, results in minimum noise caused by the inductor. The corresponding positions to produce minimum noise for the structure shown in Fig. 6.13 are also shown in Fig. 6.14 for different δ_c.

Considering the first row of the inductive links, as shown in Fig. 6.13, the power loop is placed at position C_2, yielding a total noise $V_{\text{noise},C_2} = 35.4$ mV of 35.4 mV. Moreover, the ground loop can be placed at C_2', where the accumulated noise $V_{\text{noise},C_2'}$ is 59.1 mV. The noise is slightly higher as compared to the power loop due to the increased IR drop. The overall power noise is less than 8.2%. The initial placement of the power grid exhibits low IR drop. The induced noise can however violate the allowed power noise constraint, reaching $V_{\text{noise,max}}$ of 320 mV if the worst case position for both the power and ground loops is assumed. Proper placement of the power grid is therefore required to satisfy a low noise constraint despite interference from the inductive links.

6.4.3.4 Noise sensitivity of power network topologies

A similar analysis can be applied to determine the sensitivity to noise for other power delivery topologies, which depends upon the geometric characteristics of these networks. The sensitivity to induced noise is described in [274] for different power delivery network topologies,

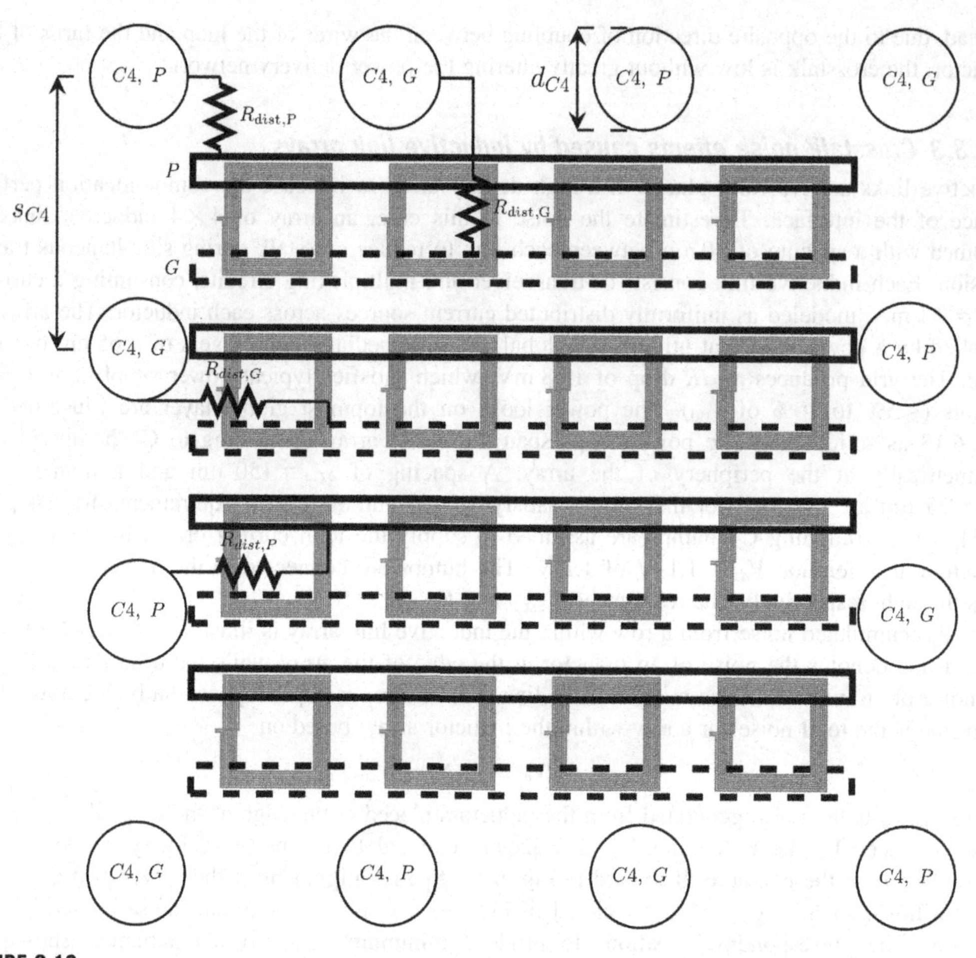

FIGURE 6.13

Array of inductive links and P/G loops connected to C4 supply pads. The power and ground lines are depicted, respectively, by solid and dashed lines.

assuming an interdigitated power delivery network with either a paired type-I or a paired type-II topology [185,275,276]. Several power delivery network topologies are depicted in Fig. 6.15.

The center and pitch of an elemental power delivery network segment, denoted, respectively, as c_{PDN} and $pitch_{PDN}$, are also shown in Fig. 6.15. As the width of the loop w_{loop} between the power or ground wires is different for each topology, $pitch_{PDN}$ is used for all of the topologies to provide a fair comparison. The center of the power delivery network segment c_{PDN} describes the relative spatial position of the power delivery network with respect to an inductive link. The same inductive link structure is considered, as shown in Fig. 6.13, to evaluate the accumulated noise produced by an inductive link.

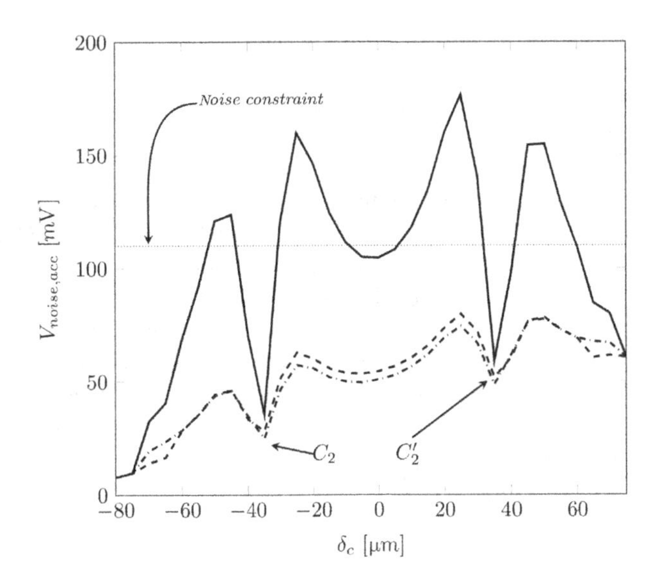

FIGURE 6.14

Parasitic noise induced on a power wire depending upon the distance of the wire from the inductor.

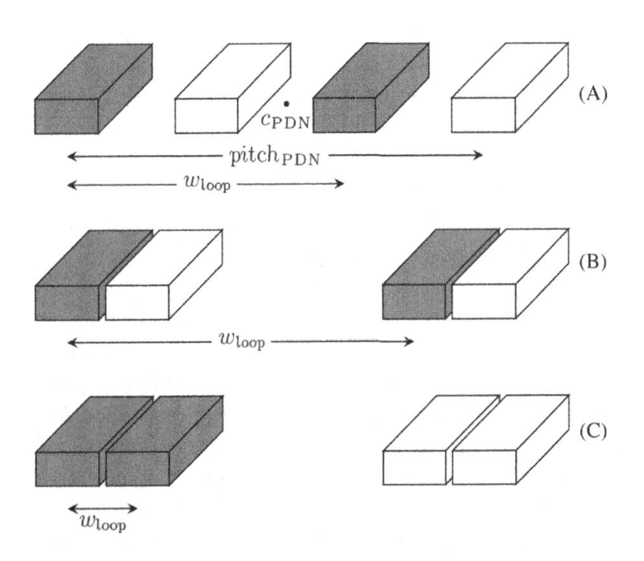

FIGURE 6.15

Power delivery network topologies. (A) Interdigitated P/G–P/G topology, (B) paired type-I P/G–P/G topology, and (C) paired type-II P/P–G/G topology [275].

An analysis of a 4×4 array of inductive links is reported in [274]. Several cases are listed in Table 6.2 for each power delivery network topology for pairs of ($\text{pitch}_{\text{PDN}}, \delta_c$) that satisfy a 10% noise constraint. The induced noise $V_{\text{noise,acc}}$ described by (6–11) is also provided, which considers the induced noise on both the power and ground loops.

The interdigitated topology exhibits a higher sensitivity to induced noise as compared to the paired topologies. Even in the location where the induced noise is minimum, the added *IR* drop for the power and ground loops increases the total noise to prohibitive levels. Overall, a minimum noise of 13.9% V_{dd} is noted, surpassing the 10% noise constraint.

Alternatively, both paired topologies include a ($\text{pitch}_{\text{PDN}}, \delta_c$) configuration where the overall noise is below the target limit. For the type-I topology, a noise level of 7.1% is achieved, well below the target limit. The overall noise primarily occurs due to the induced noise, while the added *IR* drop is relatively small for valid spatial positions. For the type-II topology, the ground loop *IR* drop significantly increases the overall noise due to the greater distance from the pads. Nevertheless, positions exist where the overall noise level is tolerable, such as the positions shown in Fig. 6.13, exhibiting an overall noise of 7.7% (listed in Table 6.2).

6.5 INTERTIER POWER TRANSFER

Inductive links enable intertier communication without the need for expensive TSVs or microbumps. Moreover, through silicon communication can be supported within multiple tiers due to the low attenuation of the magnetic flux passing through the resistive substrate, in comparison, for example, to the attenuation of the electric field in capacitive links [258]. Consequently, face-up and back-to-back bonding styles can also be exploited as well as face-to-face integration. The challenge of delivering power to the upper tiers without a wired solution potentially increases cost and manufacturing complexity. Wireless power transfer is discussed in this section as a means to overcome this problem.

Similar to inductive data links, a transmitter inductor generates magnetic flux that couples to the powered tiers through the inductor at the receiver. The received magnetic flux induces a current in the receiver inductor that is rectified and regulated to behave as a DC power source, as illustrated in Fig. 6.16. On-chip wireless power transfer is achieved in several ways. Standard inductive coupling and resonant inductive coupling are examples of wireless power delivery schemes [277].

With standard inductive coupling, the parasitic impedance of the inductors, primarily the parasitic impedance of the transmitter inductor, affects the performance of the interface circuitry. The quality factor $Q = \omega L / R_s$ of the inductor must be increased to enhance power transmission, where R_s is the parasitic resistance of the inductors. A low parasitic resistance R_s is therefore a primary issue in wireless power delivery. With a lower resistance, higher currents can flow through the high Q inductor, providing greater power efficiency or support longer signal transmission distances. The operating frequency of the inductive interface is determined by the characteristics of the integrated inductors. The interface operates as a band pass filter and, therefore, appropriate selection of the operating frequency will enhance the power efficiency. Note that the transmitter and receiver inductors can exhibit different physical and electrical characteristics.

A wireless power transmission interface for 3-D integrated systems is described in [278]. A power transmission of 2.5 mW is achieved for a face-to-face 3-D configuration with a

Table 6.2 Accumulated Noise for Different Power Delivery Network Geometries

Topology	Geometry		$V_{noise,acc}$ mV	IR−DropmV		Total (mV)	Overall Noise
	$pitch_{PDN}$ (μm)	δ_c (μm)		Power Loop	Ground Loop		
Interdigitated	45	−45	108	5.25	18.1	131.3	13.9% (fail)
	45	70	92	52.8	61.3	153.8	18.1% (fail)
Paired Type-I	45	−40	56.1	11.1	14.1	81.3	7.3% (pass)
	60	−40	59.2	14.2	6.1	79.5	7.2% (pass)
	105	−50	69.2	35.2	27.7	132.1	12% (fail)
	105	50	69.2	56.2	63.7	189.1	17.1% (fail)
Paired Type-II	45	−35	58.1	9.7	27.2	95	8.6% (pass)
	75	−45	65.3	16.7	26.3	108.3	9.8% (pass)
	90	0	28.5	10.8	45.5	84.8	7.7% (pass)

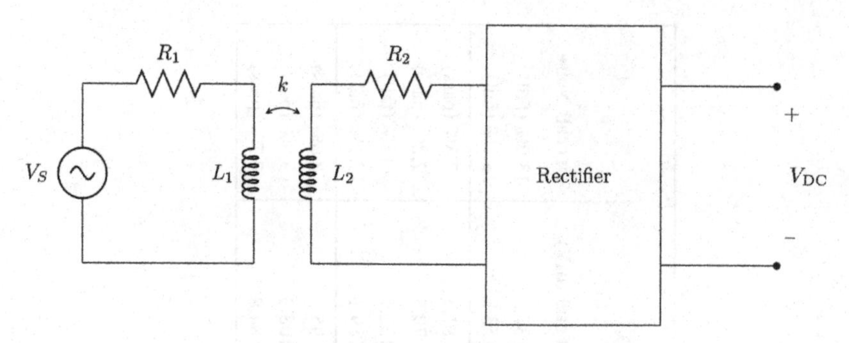

FIGURE 6.16

Wireless power transmission for standard inductive coupling.

communication distance below 20 μm. A ring oscillator drives the transmitter inductor, while cross-coupled PMOS diodes rectify the voltage at the receiver side. A resistive load of 100 Ω is connected to the rectifier, matching the source resistance, thereby maximizing the power transfer efficiency.

Another contactless power delivery system with improved efficiency is reported in [279]. An H-bridge circuit drives the transmitter inductor on the top tier. The substrate of the top tier is thinned to maintain operation of the link under different communication distances, specifically 15, 50, and 250 μm. Alternatively, the receiver consists of a sensing inductor and four cross-coupled diodes which function as a rectifier and a resistive load. The transceiver operates as a band pass filter with a center frequency of 140 MHz.

The power delivery efficiency depends upon the distance between the coupled inductors and the operating frequency of the transceiver. The efficiency of the power delivery scheme is dependent on the parasitic resistance and capacitance of the inductors. Since the transceiver behaves as a band pass filter, below the cutoff frequency, a higher frequency increases the reactance of the inductor, thereby more efficiently transferring power. Once the operating frequency passes beyond the center frequency, the parasitic capacitance of the inductor becomes the dominant loss factor of the filter, lowering the power transfer efficiency.

Alternatively, the communication distance affects the maximum output power delivered through a link. A longer communication distance decreases the coupling efficiency of the two inductors, reducing the available output power at the receiving tier. At a distance of 15 μm between the transmitter and receiver inductor, the transferred output power is 35 mW at the cutoff frequency [279]. As the distance increases to 50 μm, however, the output power is reduced to about 20 mW and, at 250 μm to 2 mW. Inductors of 100 μm × 100 μm achieve a maximum efficiency of about 10%.

Power transmission via standard inductive coupling, therefore, suffers from low efficiency (less than 30%), large coils, and limited transmission length, as reported in [277,280]. Alternatively, resonant inductive coupling can enhance standard inductive coupling. Wireless power transfer using strong magnetic resonance was first proposed in [281] for meter scale applications. Nevertheless, the fundamental concept behind strong magnetic resonance coupling is applicable to integrated systems with on-chip components.

In resonant inductive coupling, the circuit shown in Fig. 6.16 is altered with a capacitor in series or in parallel with an on-chip inductor. An *LC* resonant tank transfers power due to the resonance of the inductor and the capacitor. This circuit supports longer coupling distances, yielding a higher power efficiency. Up to nine times increased performance of resonance inductive coupling as compared to standard inductive coupling has been demonstrated [282].

Challenges remain for the resonant inductive coupling scheme; specifically, the inductor. The characteristics of the inductor should be chosen to achieve maximum power transfer efficiency while satisfying the constraints posed by a specific system. Furthermore, the choice of operating frequency plays an important role in maximizing the transferred power. For maximum power transfer, load matching is also necessary. Load impedance matching does not waste power, ensuring higher efficiency at the interface.

6.6 SUMMARY

Wireless on-chip communication with inductive links is described in this chapter. Several opportunities for contactless 3-D circuits are available with inductive links; however, specific design challenges also arise. The primary concepts discussed in the chapter are summarized as follows:

- Inductive links can be fabricated by conventional CMOS manufacturing processes without the need for additional masks, resulting in reduced manufacturing cost and fast prototyping of 3-D systems.
- Heterogeneous 3-D systems benefit from inductive interfaces as no requirement exists for intertier level shifters. Seamless integration of different technology nodes and disparate integration technologies can therefore be supported.
- Several transceivers for inductive links have been explored. Depending upon the intertier coupling mechanism, synchronous or asynchronous schemes can be used to communicate.
- Burst data transmission and data multiplexing can increase the efficiency of the inductive link. An inductive link requires interconnect resources, allowing the silicon underneath to be available for improving the performance of the intertier interface.
- Crosstalk between inductive links can be lowered by using time division multiple access schemes. A phase interpolator permits inductive links to operate within specific time slots without reducing the performance of the interface circuitry.
- The effect of the noise generated by the inductive link on an adjacent power delivery network can reach harmful levels if not properly considered. Appropriate design and placement of the power delivery network with respect to the inductive link are necessary for robust power delivery.
- Alternative methods for power delivery should be considered for fully contactless 3-D integrated systems. Wireless power transfer using resonant inductive links can deliver power efficiently to several tiers within a 3-D system.

CHAPTER

INTERCONNECT PREDICTION MODELS

7

CHAPTER OUTLINE

In nanoscale technologies, the interconnect resistance has increased significantly, profoundly affecting the signal propagation characteristics across an IC. On-chip interconnects have, therefore, become the primary focus in modern circuit design. Considering the broad gamut of interconnect issues, such as speed, power consumption, and signal integrity, beginning the interconnect design process early in the overall design cycle can prevent expensive iterations at later design stages, which increase both the design turnaround time and the cost of developing a circuit. Early and accurate estimates of interconnect related parameters, such as the required metal resources, maximum and average wirelength, and wirelength distribution are, therefore, indispensable for deciding upon the nature of the interconnect architecture within a circuit.

Interconnect prediction models aim to provide a distribution of the length of the nets within a circuit without any prior knowledge of the physical design of the circuit. This *a priori* distribution can be utilized in a general interconnect prediction framework, producing estimates for various design objectives. Interconnect prediction models have existed for a long time. Recently, these models have evolved to consider interconnect issues, such as placement and routing for interconnect dominated circuits, interconnect delay, and crosstalk noise. The majority of these models are based on either empirical heuristics or rules [283,284]. A well known rule, Rent's rule, forms the basis for the majority of interconnect prediction models. This rule and a related interconnect prediction model for two-dimensional (2-D) circuits are discussed in the following section. Several interconnect prediction models adapted for three-dimensional (3-D) circuits are presented in Section 7.2. Projections for 3-D circuits obtained from these models that demonstrate the opportunities of vertical integration are discussed in Section 7.3. The key points of this chapter are summarized in Section 7.4.

7.1 INTERCONNECT PREDICTION MODELS FOR TWO-DIMENSIONAL CIRCUITS

The cornerstone for the vast majority of wirelength prediction models is an empirical expression known as Rent's rule developed in the early 1960s [283]. Although the derivation of this rule was

Three-Dimensional Integrated Circuit Design. DOI: http://dx.doi.org/10.1016/B978-0-12-410501-0.00007-1

based on partitions of circuits consisting of only tens of thousands of gates, the applicability of this rule has been demonstrated for numerous and more complex circuits containing millions of gates [285–287]. Rent's rule is described by a simple expression,

$$T = kN^p, \tag{7.1}$$

correlating the number of I/O terminals of a circuit block with the number of circuit elements that are contained in this block. T and N are, respectively, the number of terminals and elements of a circuit block. k is the average number of terminals per circuit element, and p is an empirical exponent. A circuit element is an abstraction and does not refer to a specific circuit; consequently, a circuit element can vary from a simple logic gate to a highly complex circuit. In addition, each circuit block has a limited number of terminals and a specific circuit element capacity. The type of circuit and partition algorithm determines the specific value of k and p [283]. Although the parameter p has initially been determined to be in the range of $0.57 \leq p \leq 0.75$, greater values for p have also been observed. In general, the parameter p increases as the level of parallelism of a system increases [284,288].

Determining the interconnect length distribution is equivalent to enumerating the number of connections $i_{2\text{-D}}(l)$ for each possible interconnect length l within a circuit. Such an enumeration, however, is a formidable task. A less computationally expensive approach can be utilized to circumvent this difficulty [289,290]. Beginning with an infinite homogeneous sea of gates as illustrated in Fig. 7.1, the number of interconnects within an IC with length l is

$$i_{2\text{-D}}(l) = M_{2\text{-D}}(l) I_{gp}(l), \tag{7.2}$$

where $I_{gp}(l)$ is the expected number of connected gate pairs at a distance l in gate pitches and $M_{2\text{-D}}(l)$ is the number of gates resulting in such pairs. To obtain the number of gate pairs connected with an interconnect of length l, Rent's rule is employed, where a circuit element is considered to

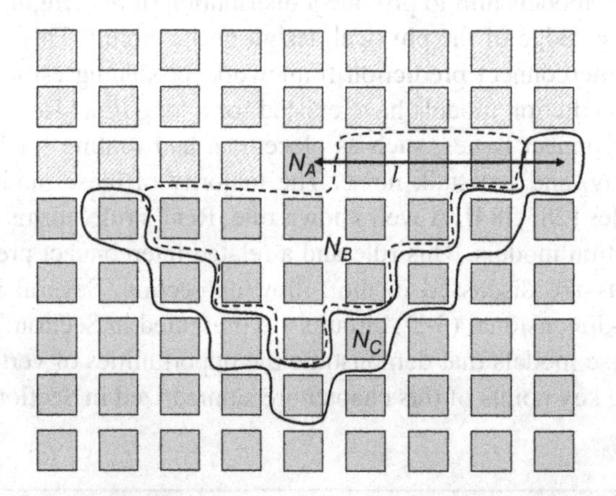

FIGURE 7.1

An example of the method used to determine the distribution of the interconnect length. Group N_A includes one gate, group N_B includes the gates located at a distance smaller than l (encircled by the dashed curve), and N_C is the group of gates at distance l from group N_A (encircled by the solid curve). In this example, $l = 4$ (the distance is measured in gate pitches).

be equivalent to a simple gate. Additionally, the total number of terminals, including both the I/O terminals and the terminals connecting different circuit blocks for any partition of the original circuit, remains the same. Furthermore, Rent's rule can be applied to any partitioning algorithm, since a different partitioning technique alters the empirical exponents included in Rent's rule, rather than the expression describing this rule. Consequently, the expected number of interconnects between two groups of gates N_A and N_C located at a distance l is

$$I_{gp}(l) = \frac{ak}{N_A N_C}\left((N_A+N_B)^p + (N_B+N_C)^p - (N_B)^p - (N_A+N_B+N_C)^p\right), \tag{7.3}$$

where N_B is a group of gates within a distance l from group N_A. To approximate the number of interconnects between two groups of gates, N_A and N_C, separated by l gate pitches, as shown in Fig. 7.1, the following observation is used.

For a sufficiently large number of gates N, those gates that comprise group N_C form a partial Manhattan circle (as the Manhattan distance is the metric used to evaluate the distance among the gates or, equivalently, the interconnect length among the gates). This situation applies to most of the gates within a circuit except for those gates close to the periphery of the IC. Consequently, from Fig. 7.1,

$$N_A = 1, \tag{7.4}$$

$$N_B(l) = \sum_{k=1}^{l-1} 2k = l(l-1), \tag{7.5}$$

$$N_C(l) = 2l, \tag{7.6}$$

where N_B corresponds to the number of gates encircled within the partial Manhattan circle formed by the gates in group N_C. The gates included in N_C shape a partial but not full Manhattan circle as the crosshatched gates are not considered as part of N_C. This elimination of gates during the enumeration process ensures that a gate is not counted more than one time. With these assumptions and by employing a binomial expansion, an approximate expression for $I_{gp}(l)$ is

$$I_{gp}(l) \cong ak\frac{p}{2}(2-2p)l^{2(p-2)}. \tag{7.7}$$

From (7.7), the number of gate pairs connected by an interconnection of length l is a positive and decreasing function of length for $p<1$. The factor a in (7.7) considers the number of terminals acting as sink terminals and is

$$a = \frac{f.o.}{1+f.o.}, \tag{7.8}$$

where $f.o.$ is the average fanout of a gate. By incorporating the number of gate pairs $M_{2\text{-D}}(l)$, which can be determined by simple enumeration, and using the binomial approximation for $I_{gp}(l)$, the interconnect length distribution $i_{2\text{-D}}(l)$ is

$$i_{2-\text{D}}(l) = \frac{ak}{2}\Gamma\left(\frac{l^3}{3} - 2\sqrt{N}l^2 + 2Nl\right)l^{2(p-2)} \text{ for } 1 \le l < \sqrt{N}, \tag{7.9}$$

$$i_{2-\text{D}}(l) = \frac{ak}{6}\Gamma\left(2\sqrt{N}-l\right)^3 l^{2(p-2)} \text{ for } \sqrt{N} \le l < 2\sqrt{N}, \tag{7.10}$$

where Γ is a normalization factor. The cumulative interconnect density function (c.i.d.f.) that provides the number of interconnects with a length smaller or equal to l is

$$I_{2\text{-D}}(l) = \int_{1}^{l} i_{2\text{-D}}(z)dz. \tag{7.11}$$

Expressions (7.9)–(7.11) provide the number and length distribution of those interconnects within a homogeneous (i.e., circuit blocks with the same capacity and number of pins) 2-D circuit consisting of N gates. By considering more than one tier, where each tier is populated with identical circuit blocks, the wirelength distribution for a homogeneous 3-D circuit can be generated. Various approaches have been followed to determine the wirelength distribution of a 3-D circuit, as described in the following section.

7.2 INTERCONNECT PREDICTION MODELS FOR THREE-DIMENSIONAL ICs

Several interconnect prediction models for 3-D ICs have been developed, providing early estimates of the wirelength distribution in a 3-D system. Although these kinds of models lack any knowledge of the physical design characteristics of a circuit, such models can be a useful tool in the early stages of the design flow to roughly predict various characteristics of a system, such as the maximum clock frequency, IC area, required number of metal layers, and power consumption. Specifically, for 3-D ICs, such models are of considerable importance as a mature commercial process technology for 3-D ICs does not yet exist. Interconnect prediction models are, therefore, an effective means to estimate the opportunities of this innovative and developing technology.

A wirelength prediction model for 3-D circuits should consider not only the interconnections among gates located within the same tier but also the interconnections among gates on different tiers. Consequently, in 3-D circuits, the partial Manhattan circles formed by the gates in N_C, as illustrated in Fig. 7.1, become partial Manhattan spheres as gate pairs exist not only within the same tier (intratier gate pairs) but also among gates located on different tiers (intertier gate pairs). These spheres are formed by partial Manhattan circles of reduced radius, which enclose the intertier gate pairs. A decreasing radius is utilized for the intertier gate pairs to consider the vertical distance among the physical tiers. This decreasing radius produces a spherical region rather than a cylindrical region for enumerating the intertier gate pairs. Such a hemisphere is illustrated in Fig. 7.2, where the vertical distance between adjacent tiers is assumed to be equal to one gate pitch. Considering different lengths for the vertical pitch results in hemispheres with different radii.

The partial Manhattan spheres are employed to determine the number of gates within groups N_A, N_B, and N_C [291–293]. To consider the particular conditions that apply to those gates located close to the periphery of the circuit, which do not result in partial Manhattan spheres, the gates are separated into two categories. The first category is the "starting gates," which denotes those gates that can form a hemisphere with radius l. Typically, these gates are located close to the center of the circuit, and the "non-starting gates," which denotes those gates that cannot form partial spheres of radius l, are roughly located close to the periphery of the circuit [294]. An example of starting and nonstarting gates for a single tier is shown in Fig. 7.3. Gate P can therefore be a starting gate forming a partial circle with radius $l = 2$, depicted by the solid line. In addition, gate Q cannot be a

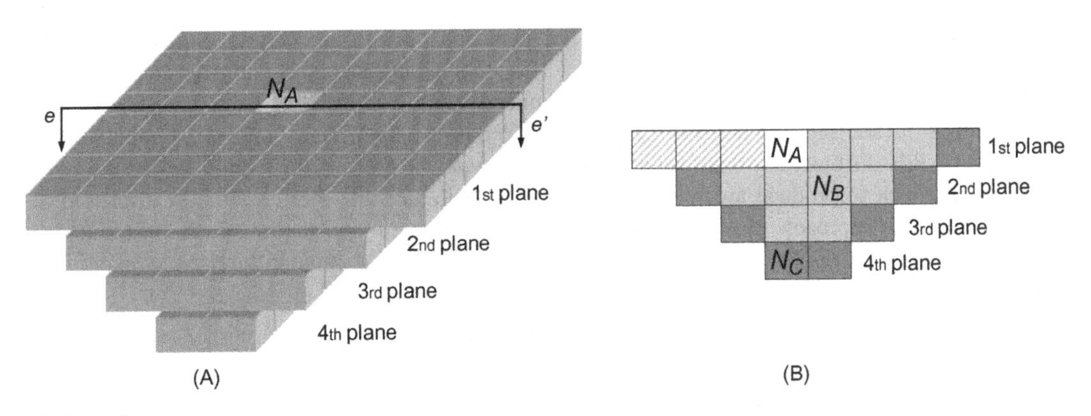

FIGURE 7.2

An example of the method used to determine the interconnect length distribution in 3-D circuits. (A) Partial Manhattan hemisphere, and (B) cross-section of the partial Manhattan hemisphere along e-e'. The gates in N_B and N_C are shown, respectively, with light and dark gray tones.

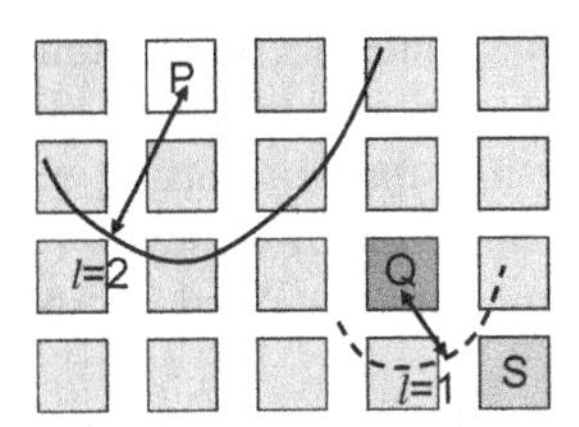

FIGURE 7.3

Example of starting and nonstarting gates. Gates P and Q can be starting gates while S is a nonstarting gate.

starting gate for gate pairs at a distance $l = 2$. Alternatively, gate Q can be a starting gate for gate pairs at a distance $l = 1$, as indicated by the dashed curve. Gate S is a nonstarting gate as a circle cannot be formed with a gate located below this gate for any distance l. The same definition applies the case of multiple physical tiers where the circles are substituted by spheres.

As in a 2-D wirelength model, a circuit element consists of only one gate. The gates included in groups N_A, N_B, and N_C, for a 3-D circuit, are [291]

$$N_A = 1, \tag{7.12}$$

$$N_B(l) = \sum_{i-1}^{l-1} N_C(i), \tag{7.13}$$

$$N_C(l) = \frac{M_{3-\mathrm{D}}(l)}{N_{\mathrm{start}}(l)}, \tag{7.14}$$

where $M_{3\text{-D}}(l)$ is the number of gate pairs at a distance l in a 3-D circuit,

$$M_{3-\mathrm{D}}(l) = M_{2-\mathrm{D}}(l) + 2\sum_{i=1}^{n-1} M_{2-\mathrm{D}}(l - id_v). \tag{7.15}$$

The first term in (7.15) considers the intratier gate pairs, while the second term includes the intertier gate pairs formed in the remaining $n-1$ tiers. The necessary expressions to determine the number of starting gates $N_{\text{start}}(l)$ are presented in Appendix A.

Evaluating the starting and nonstarting gate pairs requires a significant number of summations. Alternatively, this differentiation between the different types of gates can be removed, resulting in a simpler approach to produce an interconnect length distribution for 3-D circuits, as discussed in [295–297]. In this approach, an infinite sea of gates is considered, and the physical constraint that gates located at the periphery of the circuit cannot form a partial Manhattan circle is relaxed. The average number of gates that belong to groups N_A, N_B, and N_C for a 3-D circuit is

$$N_A = 1, \tag{7.16}$$

$$\begin{aligned} N_B \approx\; & (l(l-1) + 2(l-d_v)(l-d_v-1) + l(l-1) \\ & + 2(l-2d_v)(l-2d_v-1) + \cdots + l(l-1) + \cdots \\ & + 2(l-(n-1)d_v)(l-(n-1)d_v-1) + \cdots + l(l-1))/n, \end{aligned} \tag{7.17}$$

$$\begin{aligned} N_C(l) = & (2l + 4(l-d_v) + 2l + 4(l-2d_v) + \cdots + 2l + \cdots \\ & + 4(l-(n-1)d_v) + \cdots + 2l)/n, \end{aligned} \tag{7.18}$$

where n is the number of tiers within the 3-D system and d_v is the intertier distance between two adjacent tiers. The number of gate pairs that are l gate pitches apart, $M_{\text{3-D}}(l)$, is the sum of the gate pairs within a tier $M_{\text{intra}}(l)$ and the gate pairs in different physical tiers $M_{\text{inter}}(l)$. By removing the distinction between starting and nonstarting gates, the boundary conditions of the problem are altered. Consequently, a normalization factor Γ' is required to ensure that the total number of interconnects of a circuit implemented in both two and three dimensions is maintained constant.

A similar distribution for interconnections in 3-D circuits can be determined by changing the number of gates that correspond to a circuit element, simplifying the enumeration process. Thus, for a 3-D circuit consisting of n tiers, the expected interconnections between cell pairs (and not gate pairs) are enumerated [298,299]. Due to this modification, the length distribution for the horizontal and vertical wires can be separately determined from the prediction model for 2-D circuits, as described in Section 7.1. Horizontal wires are considered only as intratier interconnects in each tier of a 3-D circuit. Alternatively, the vertical wires include one or more intertier vias and can also include horizontal interconnect segments.

Since a cell is composed of n gates, the expected interconnections between cell pairs and not gate pairs are enumerated. In other words, the groups N_A, N_B, and N_C become

$$N_A = n, \tag{7.19}$$

$$N_B(l) = \sum_{k=1}^{l-1} 2nk = nl(l-1), \tag{7.20}$$

$$N_C(l) = 2nl. \tag{7.21}$$

To determine the distribution of the vertical wires, all possible connections between two cells are considered. The possible interconnects between a pair of cells are shown in Fig. 7.4. Using (7.19)–(7.21) and by slightly modifying the expressions for the interconnect length distribution of the model presented in Section 7.1, the distribution of horizontal and vertical wires in 3-D circuits is obtained.

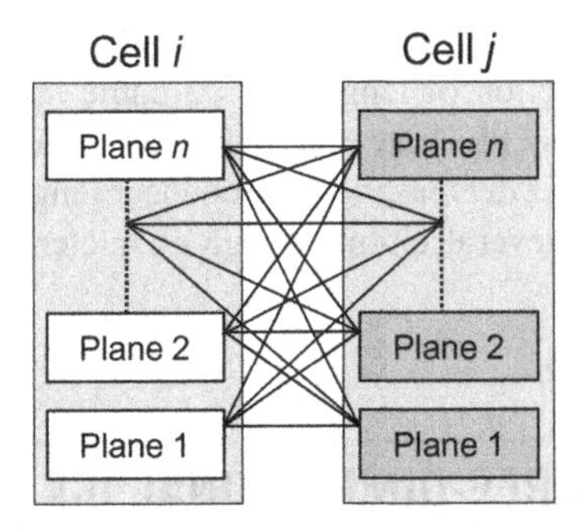

FIGURE 7.4

Possible vertical interconnections for two cells with each cell containing n gates.

In all of the interconnect length distribution models for 3-D circuits presented in this section, an enumeration of the gate pairs is required. This process typically involves several finite series summations. Alternatively, a hierarchical approach can be adopted to generate the wirelength distribution within a 3-D circuit [300]. With such an approach, the 3-D system is partitioned into K levels of hierarchy in descending order. The entire system is therefore partitioned into eight subcircuits at level $K-1$, while the lowest level (level 0) consists of single gates. The interconnect distribution of all of the interconnect lengths smaller than the maximum length (i.e., $l \leq l_{max}$) is determined by the following relationship,

$$I(l) = S(l)q(l), \tag{7.22}$$

where $S(l)$ is the structural distribution of the number of all possible gate pairs. The structural distribution is physically equivalent to the number of gate pairs $M_{2\text{-D}}(l)$, as applied to the model presented in Section 7.1. The process of calculating $S(l)$ proceeds, however, with the method of generating polynomials [288]. A generating polynomial of a length distribution is equivalent to the moment generating polynomial function of that distribution and can be described by

$$g(x) = \sum_{l=0}^{\infty} S(l)x^l, \tag{7.23}$$

where each term provides the number of gate pairs connected with an interconnect of length l. Generating polynomials can provide a more efficient representation of the length distribution, as more complex interconnections among circuit elements can be generated by combining simple interconnections and exploiting symmetries that can exist within different circuit interconnection topologies. Consequently, the polynomials evaluated at the kth level of the hierarchy can be used to construct the polynomials of the $(k+1)^{th}$ level. An analysis of the technique of generating polynomials is found in [301,302]. The occupation probability $q(l)$, which is the probability of a gate pair being connected with an interconnect of length l, is based upon Rent's rule [283] and Donath's approach [303] and, for a 3-D circuit, is

$$q(l) = Cl^{3(p-2)}, \tag{7.24}$$

where C is a normalization constant in proportion to that constant used in the 2-D prediction model, as discussed in the previous section.

Comparing (7.24) with (7.7), a considerable decrease is noted in the number of interconnects, as this number decays faster as a function of length (i.e., a cubic relationship rather than a quadratic function of length). This faster decay illustrates the reduction in the number of long global interconnects in 3-D circuits. Exploiting this decrease in interconnect length, 3-D integration is expected to offer significant improvements in several circuit design characteristics, as discussed in the following section.

7.3 PROJECTIONS FOR THREE-DIMENSIONAL ICs

A priori interconnect prediction models described in the previous section provide an important tool to estimate the behavior of 3-D systems and, consequently, assess the potential of this novel technology. The introduction of the third dimension considerably alters the interconnect distribution in ICs. The wirelength distribution of a four million gate circuit for different numbers of tiers is illustrated in Fig. 7.5 [291]. By increasing the number of tiers, the length of the global interconnects decreases. In addition, the number of these global interconnects decreases, while the number of short local interconnects increases such that the total number of interconnects in the IC is essentially conserved.

The third dimension can be used to improve several characteristics of an IC [304,202]. For example, tradeoffs among the maximum clock frequency, the required number of metal layers, and the area of an IC can be explored to achieve different design objectives. Consider a multi-tier

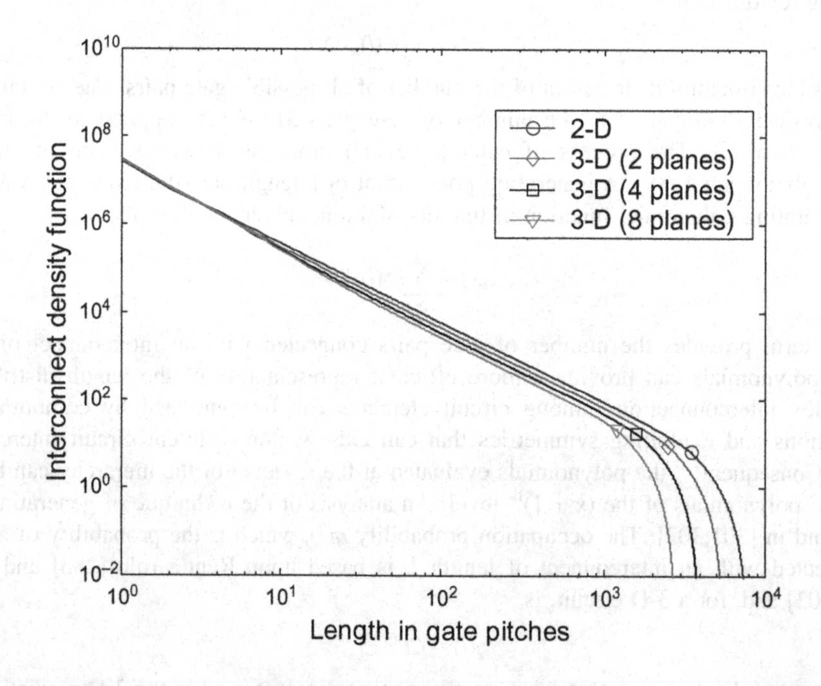

FIGURE 7.5

Interconnect length distribution for a 2-D and 3-D IC [291].

interconnect architecture where the metal lines in each tier have a different aspect ratio and pitch. In addition, each tier consists of orthogonally routed multiple metal layers.

As a smaller area per tier is feasible in a 3-D IC, the length of the corner-to-corner interconnects significantly decreases. Consequently, for a constant clock frequency, several global interconnects on the upper tiers can be transferred to the local tier with a smaller aspect ratio. The number of metal layers can therefore be reduced in 3-D circuits. Alternatively, if the number of metal layers is not reduced by transferring the global interconnects to the local tier interconnects, the clock frequency naturally increases as the longest distance for the clock signal to traverse is drastically reduced. For heterogeneous 3-D ICs, the clock frequency is shown to increase as $n^{1.5}$, where n is the number of tiers within a 3-D IC [305]. Furthermore, the third dimension can be used to reduce the total area of an IC. A 2-D circuit in multiple tiers decreases the length of the interconnects, creating a timing slack. If the clock frequency is maintained the same as that of a 2-D IC, the wiring pitch can be reduced, decreasing the total required interconnect area. The decrease in the interconnect pitch results in an increase in the capacitive characteristics of the wires, which can consume the added timing slack provided by the introduction of the third dimension. Consider a 100 nm ASIC consisting of 16 M gates, a 3-D IC with two tiers produces a 3.9× improvement in clock frequency, an 84% reduction in wire limited area, and a 25% decrease in the number of metal layers for each tier within a 3-D circuit [292].

Since the introduction of the third dimension results, in general, in shorter interconnects, 3-D ICs consume less power as compared to 2-D ICs. Using (7.12)–(7.15) and the expressions in [306] to generate the interconnect distribution for a 3-D system, a circuit composed of 16 M gates is evaluated. In Fig. 7.6, the variation of the gate pitch, total interconnect length, and power consumption are plotted versus the number of tiers for two different values of the Rent's exponent. A value of $p = 0.5$

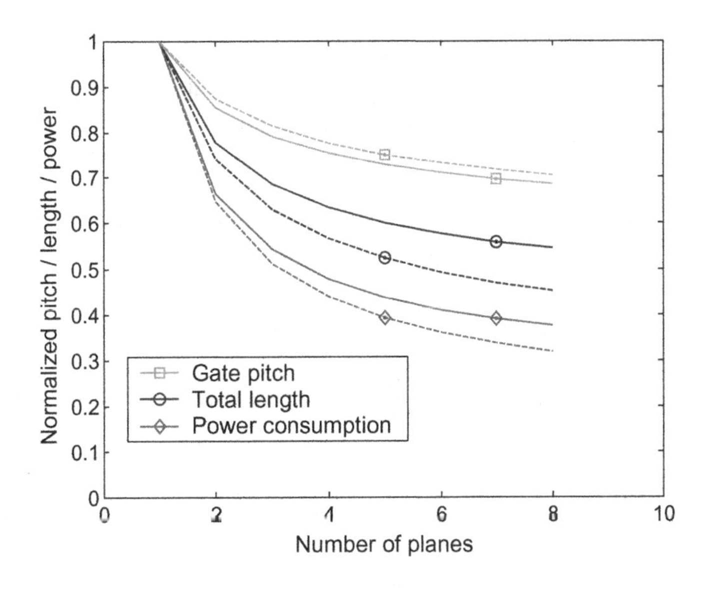

FIGURE 7.6

Variation of gate pitch, total interconnect length, and interconnect power consumption with the number of tiers.

and $p = 0.6$ corresponds, respectively, to a family of solid and dashed curves. Since higher values of p are related to systems with higher parallelism, a greater reduction in interconnect length can be achieved for these systems by increasing the number of tiers, as indicated by the curves notated by a circle. Alternatively, highly serial systems, or equivalently, small values of the Rent's exponent permit higher device densities, since these circuits require fewer routing resources and, therefore, the gate pitch is significantly reduced with the number of tiers. For greater values of p, the decrease in gate pitch is smaller, as illustrated by the dashed and solid curves notated by the square.

The interconnect power consumption depends both on the total interconnect length and the gate pitch. As both of these parameters are reduced with the number of tiers, significant savings in power can be achieved with 3-D ICs. The sensitivity of the power consumption with respect to the Rent's exponent depends upon the sensitivity of the gate pitch and the total interconnect length. Since these two parameters change in opposite directions with respect to the Rent's exponent, the sensitivity of the power consumption to this parameter is less significant (see the curves marked by the diamond in Fig. 7.6). Overall, an improvement in power consumption of 33%, 52%, and 62% is achieved, respectively, for two, four, and eight tiers [306].

This analysis only considers wire limited circuits and ignores the power dissipated by the devices. A recent analysis for 3-D circuits designed with the MITLL technology [307] also demonstrates the advantages of the third dimension. In this investigation of 3-D circuits [308], a fast Fourier transform (FFT) and an open reduced instruction set computer (RISC) platform system-on-chip (ORPSOC) [309] representing, respectively, a low power and a high performance circuit example are considered. The circuits are designed in a 180 nm silicon-on-insulator technology node. Predictions for other technology nodes are also extrapolated. The signal delay, power consumption, and silicon area for 2-D and 3-D versions of these circuits are listed, respectively, in Tables 7.1 and 7.2. The clock skew and power consumption are reported in Table 7.3.

In addition to ASICs, general purpose systems, such as microprocessors, are also expected to benefit considerably from vertical integration. As the speed and number of cores within a microprocessor system increase, the demand for larger and faster cache memories also grows. A larger cache memory inherently increases the latency of transferring data to the cores; the memory bandwidth, however, increases. Additionally, cache misses become highly problematic. As the total area of a microprocessor system increases, the required time for transferring data from the processor to the main memory, due to a cache miss, also increases. 3-D integration can significantly reduce the burden of a cache miss [310]. Realizing a microprocessor system in multiple physical tiers decreases the length of the

Table 7.1 Characteristics of 2-D Circuits [308]

Technology Node	Circuit	Delay (ns)	Power (mW)	Area (mm^2)
180 nm	FFT	25.81	1050	11.61
	ORPSOC	17.80	3298	18.66
90 nm	FFT	21.14	439	4.18
	ORPSOC	14.82	1944	6.76
45 nm	FFT	14.16	334	2.32
	ORPSOC	9.16	1215	3.26

Table 7.2 Characteristics of 3-D Circuits [308]

Technology Node	Circuit	Delay (ns)	Power (mW)	Area (mm^2)
180 nm	FFT	21.49	952	4.01
	ORPSOC	14.78	2933	6.72
90 nm	FFT	19.54	394	1.48
	ORPSOC	10.02	1551	2.42
45 nm	FFT	13.01	256	0.82
	ORPSOC	8.33	1029	1.29

Table 7.3 Clock Skew and Power Consumption [308]

		Skew (ps)		Power (mW)	
Technology Node	Circuit	2-D	3-D	2-D	3-D
180 nm	FFT	264.3	213.4	240.5	201.9
	ORPSOC	469.6	260.6	838.9	482.9
90 nm	FFT	167.0	88.8	61.0	44.6
	ORPSOC	317.6	213.0	694.2	419.1
45 nm	FFT	106.3	68.1	66.4	48.3
	ORPSOC	194.9	115.1	798.1	487.9

data and address busses used for transferring data between the memory and the logic. In addition, by placing the memory on the upper tiers, which consumes significantly lower power as compared to the power consumed by the cores, the power density can be confined within acceptable levels that will not exacerbate the overall cost of the system. Considerable improvements in speed and the power consumed during the data transfer process in a RISC processor with several memory tiers in a 3-D system have been demonstrated [311]. As compared to printed circuit board (PCB) and multi-chip module (MCM) implementations of the same system, a RISC processor and memory system integrated on a three-tier 3-D circuit exhibits an improved performance of, respectively, 61% and 49%.

7.4 SUMMARY

Interconnect length distribution models for 3-D circuits are analyzed in this chapter. Based on these models, a variety of opportunities that vertical integration offers to the IC design process is discussed. The major points of this chapter are summarized as follows:

- Several *a priori* interconnect prediction models based on Rent's rule have been developed for 3-D ICs, unanimously demonstrating a significant decrease in the interconnect length for these circuits.

- Rent's rule correlates the number of terminals of a circuit with the number of modules that constitute the circuit and the average number of terminals in each module. The granularity of the module can vary from a single gate to an entire subcircuit.
- The interconnect distribution within a 3-D circuit can be determined by several methods such as exhaustive enumeration and the method of generating polynomials.
- Vertical integration drastically alters the distribution of the interconnect lengths in a circuit, ensuring that the number and length of the global interconnects decreases while the number of local interconnects increases.
- By introducing the third dimension, a variety of circuit characteristics can be improved. For example, the performance can be increased assuming a constant total area and number of metal layers. The number of required metal layers can be decreased for a fixed clock frequency and total area. Furthermore, the total area can be reduced while maintaining a fixed clock frequency.
- For heterogeneous 3-D ICs, the clock frequency is shown to increase as $n^{1.5}$, where n is the number of tiers within a 3-D IC.
- A microprocessor system in multiple physical tiers can reduce unacceptably long data transfer times and alleviate the effect of cache misses.

CHAPTER

COST CONSIDERATIONS FOR THREE-DIMENSIONAL INTEGRATION*

8

CHAPTER OUTLINE

Adopting a new technology, in addition to the advantages of system performance, power, and reliability, should also provide benefits in terms of the cost to process each unit. This improvement has occurred with semiconductor manufacturing and scaling of device feature size with each new technology generation. The manufacturing cost per transistor has dropped with device scaling while the transistor operates at higher speeds and consumes less power.

For three-dimensional (3-D) integration technology [49,69] to be widely adopted, the processing cost of the 3-D features and the effect of vertical die stacking on the overall financial cost of a system should be considered in comparison to the benefits in system functionality. These costs are discussed in this chapter. The cost for processing different TSV structures is discussed in Section 8.1. The build-up options for a passive Si interposer substrate, the main component for 2.5-D integration, are considered in Section 8.2. A comparison between the processing costs of 2.5-D and 3-D integration is discussed in Section 8.3. The primary topics of this chapter are summarized in Section 8.4.

*Dr. Dimitrios Velenis contributed this chapter.

Three-Dimensional Integrated Circuit Design. DOI: http://dx.doi.org/10.1016/B978-0-12-410501-0.00008-3

8.1 THROUGH SILICON VIA PROCESSING OPTIONS

The through silicon via (TSV) [312] is the fundamental technology that enables vertical stacking of functional dice. Depending upon the processing stage of the TSVs within the overall process flow, this technology can be characterized as either *TSV middle* or *TSV last*, as discussed in Chapter 3, Manufacturing Technologies for Three-Dimensional Integrated Circuits. TSV middle is processed in a wafer after the active devices have been manufactured and prior to fabrication of the on-chip interconnect metal layers. The connection to the bottom side of the TSV is achieved by thinning the wafer and revealing the TSV from the backside. Alternatively, TSV last is processed after completing the full wafer fabrication flow and backside thinning. In this case, a TSV structure contacts the interconnect layers of a die from the backside of the thinned wafer. A cross-sectional representation of both TSV middle [313] and TSV last [314] is shown in Fig. 8.1.

In this section, different TSV options are compared in terms of processing complexity and cost. A 3-D cost model developed at IMEC is applied to evaluate the wafer-level processing cost of different TSV integration approaches [315]. The cost of the tools, infrastructure, personnel, equipment maintenance, and materials is considered in the analysis of each process flow. Furthermore, different TSV geometries are considered to evaluate trends in TSV processing cost with decreasing TSV dimensions.

An overview of the processing steps in the TSV middle and TSV last flows, together with the different TSV geometries, is presented in Section 8.1.1. The cost of each processing step in the different TSV flows is evaluated in Section 8.1.2 for TSVs of different geometries and aspect ratios. The overall cost of the different TSV processing flows is compared in Section 8.1.3, and scaling trends of TSV dimensions are also presented.

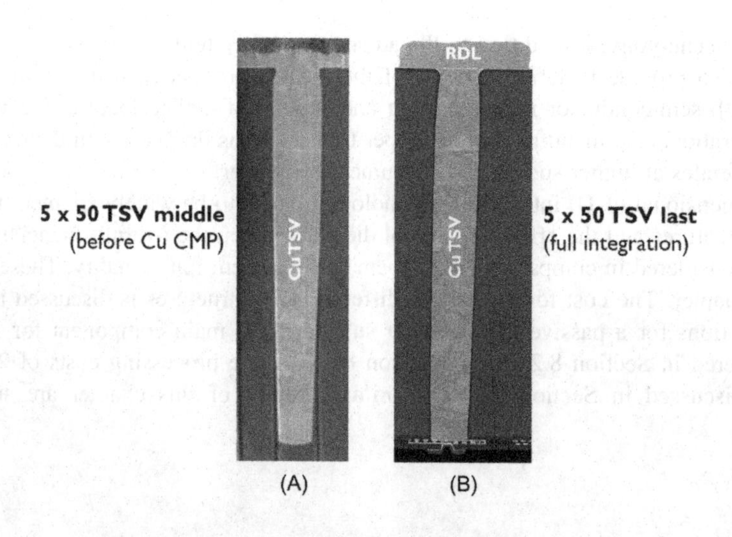

FIGURE 8.1

Cross-sectional view. (A) TSV middle, and (B) TSV last.

8.1.1 TSV FLOWS AND GEOMETRIES

The processing flows considered for TSV middle [313] and TSV last [314] are illustrated in Fig. 8.2. A number of similar processing steps are indicated in each flow: lithography, TSV etch, oxide liner deposition, barrier and seed deposition, copper (Cu) plating, and chemical mechanical planarization (CMP). Thinning the wafer from the backside is also included in both flows, although in the TSV last flow thinning is completed *before* processing the TSV, while in the TSV middle flow, thinning of the wafer is processed *after* TSV processing. Differences between the flows are: (1) opening of the in-via liner after deposition (liner etch) in the TSV last flow, and (2) opening of the TSV liner after thinning the wafer in the TSV middle flow. These different processing steps are illustrated in the rows shown in Fig. 8.2.

Different TSV geometries are evaluated for each TSV flow. The size of the TSV is given by the TSV diameter and depth using the notation *TSV diameter* × *TSV depth*, with each dimension in micrometers. In the case of TSV middle, sizes of 10 × 100, 5 × 50, 3 × 50, and 2 × 40 μm^2 TSVs are considered. In the case of TSV last, dimensions of 10 × 100, 10 × 50, 5 × 50, and 2 × 20 μm^2 TSVs are evaluated, as summarized in Table 8.1. Processing of a TSV middle 5 μm × 50 μm is considered as the reference flow (process-of-reference—POR) for all of the TSV geometries. The processing costs are therefore normalized to the cost of the 5 × 50 TSV middle flow [313].

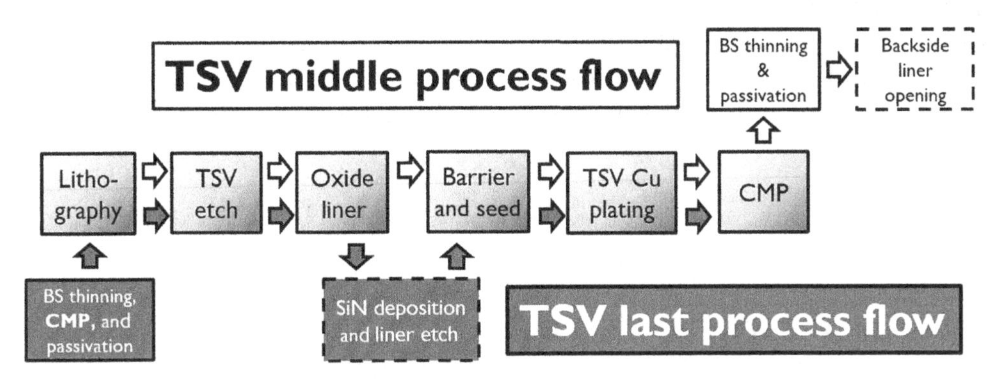

FIGURE 8.2

Processing flow for TSV middle and TSV last considered in terms of cost and complexity.

Table 8.1 TSV Geometries Considered in the Cost Comparison. The 5 × 50 TSV Middle is the POR

TSV Size: Diameter (μm) × Depth (μm)				
TSV middle	10 × 100	5 × 50	3 × 50	2 × 40
TSV last	10 × 100	10 × 50	5 × 50	2 × 20

8.1.2 COST COMPARISON OF THROUGH SILICON VIA PROCESSING STEPS

In this section, the effects of TSV geometry and process flow on the cost and complexity of each process step are explored. The requirement for different processing conditions depending upon the TSV dimensions is indicated for each step, and the effect on subsequent steps (such as CMP) is also reviewed.

8.1.2.1 Through silicon via lithography

A lithography step is executed to develop the TSV structure. The primary difference between the TSV middle and TSV last flow is that the lithography for the TSV middle is processed at a front-end-of-line fabrication compatible tool. In the case of TSV last, an outsourced semiconductor assembly and test (OSAT) fabrication compatible tool is used which is considered to be low cost. This difference results in a TSV lithographic cost which is approximately 30% more expensive for the TSV middle flow as compared to the TSV last flow. The same tool throughput and material costs for photoresist are considered for both flows. Furthermore, no difference in lithography cost is observed due to the different TSV dimensions. A comparison of the lithography processing cost for the different TSV flows and dimensions is shown in Fig. 8.3. The processing cost is normalized to the cost of the POR flow, the 5 × 50 TSV middle.

8.1.2.2 Through silicon via silicon etch

A deep silicon etch process is considered as the first step after lithography in producing TSV structures. The processing time for TSV etch depends upon the TSV depth and aspect ratio. TSVs that are deeper and narrower require additional etch cycles. The effect of the TSV geometry on the cost

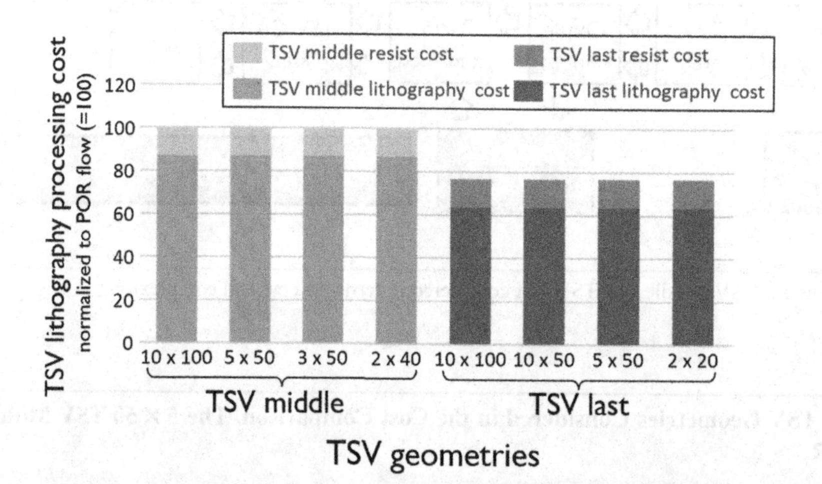

FIGURE 8.3

Comparison of TSV lithography cost for different TSV process flows and geometries. The difference in cost between TSV middle and TSV last is due to different process equipment.

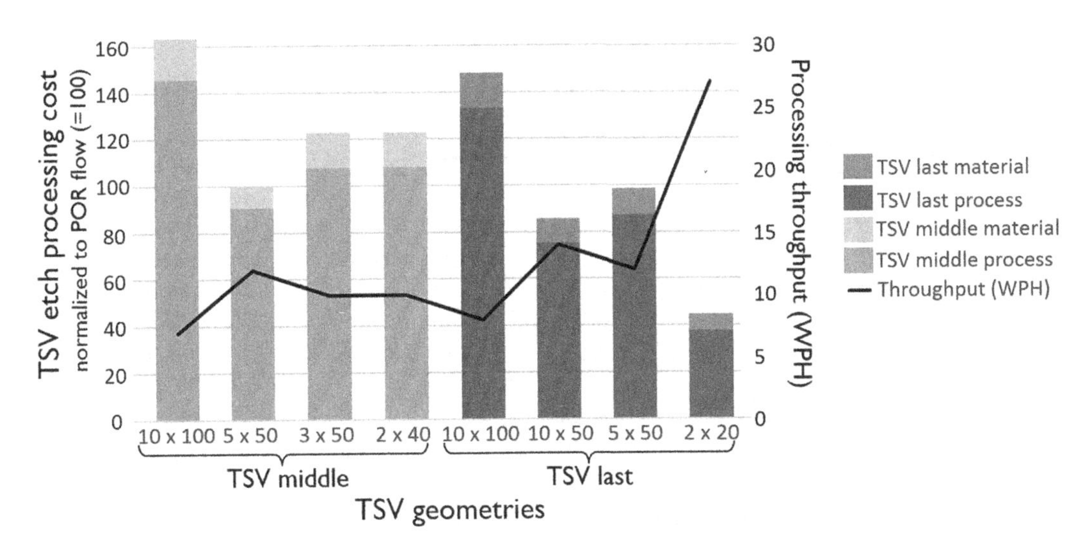

FIGURE 8.4

Comparison of processing cost of the TSV etching step for different TSV geometries. The processing cost is normalized to the cost of etching a 5 × 50 TSV middle structure.

of a TSV etch step is illustrated in Fig. 8.4. In addition, the effect of the etching time on cost is illustrated in terms of the processing throughput, expressed as wafers per hour (WPH). As shown in Fig. 8.4, shorter TSVs exhibit a faster etching time (for example, the 5 × 50 vs. the 10 × 100 TSV process). In addition, those TSVs with a more aggressive aspect ratio require more time to etch (the 3 × 50 TSV middle as compared to 5 × 50 TSV middle, or the 5 × 50 TSV last as compared to 10 × 50 TSV last). The cost of the various gases used during the etching process is treated as a material cost. Furthermore, all of the etch processing costs are normalized to the cost of etching a POR structure, a 5 × 50 μm^2 TSV middle structure.

8.1.2.3 Through silicon via liner processing

Following the opening of the TSV hole, an insulating oxide liner layer is deposited. In the case of TSV middle, different liner deposition approaches are applied, depending upon the aspect ratio of the TSV. In the case of aspect ratios up to 1:10, a tetraethylorthosilicate (TEOS) oxide is deposited with different thicknesses (800 nm for a 10 × 100 TSV and 400 nm for a 5 × 50 TSV). However, as the TSV aspect ratio increases (for 3 × 50 and 2 × 40 TSVs), a more conformal liner layer is required with higher uniformity in the liner thickness along the TSV [317]. A plasma enhanced atomic layer deposition (PEALD) oxide deposition approach is therefore preferred to achieve a liner thickness of 100 nm, capped with a 30 nm silicon nitride (SiN) layer deposited at the wafer field. The various liner processing options for the different TSV geometries are illustrated in Fig. 8.5.

Additionally, in the case of the TSV middle flow, the deposited oxide liner layer on the field is eventually removed by CMP. The effect of CMP liner polishing on the overall processing cost is illustrated in Fig. 8.5, where the cost of the liner CMP process and the CMP slurry required to

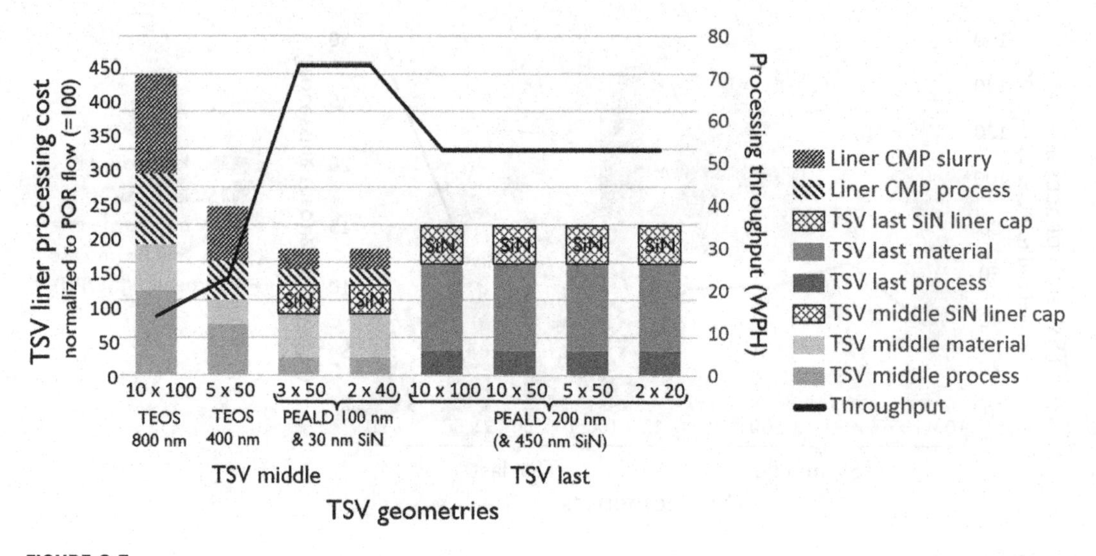

FIGURE 8.5

Comparison of processing cost to deposit the TSV oxide liner for different TSV geometries. The processing cost is normalized to the cost of the liner deposition for a 5 × 50 TSV middle structure. In the case of TSV middle, the oxide liner at the field of the wafer is removed by CMP. For TSV last, no CMP polishing of the liner is necessary.

polish the liner is considered for the TSV middle flow. The cost of the CMP depends upon the thickness of the deposited liner.

In the case of a TSV last flow, the liner needs to be opened (etched) at the bottom of the via prior to further processing; therefore, a highly conformal layer is required. For this reason, PEALD oxide deposition is used for all TSV geometries processed with the TSV last approach. Furthermore, the liner at the top of the TSV should be protected when opening the TSV liner at the bottom of the via. Liner protection is achieved by depositing a capping layer of 450 nm SiN on top of the deposited liner. Finally, in the case of TSV last, it is not necessary to polish the liner. No additional CMP cost is therefore incurred.

8.1.2.4 Through silicon via liner opening for through silicon via last flow

In the case of the TSV last process flow, the oxide liner layer at the bottom of the TSV is opened after deposition to provide a conductive path to the metal layers of the bottom device plane. This step is one of the differentiating steps between the TSV middle and TSV last flows.

A comparison of the processing cost of the liner etch step for different TSV last geometries is illustrated in Fig. 8.6. Note that processing times vary depending on the size of the TSV, resulting in variations in process throughput. Longer processing times (resulting in smaller tool throughput) are required for the narrower TSV diameters. In this case, a 5 × 50 TSV last process is used as a reference. All of the processing cost is normalized to the cost of the liner opening of a 5 × 50 TSV last structure.

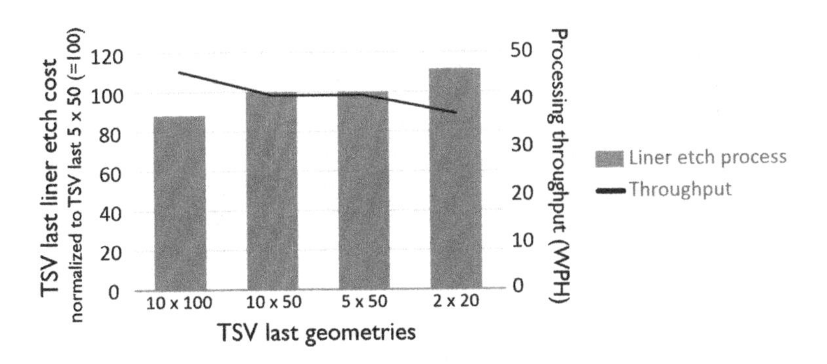

FIGURE 8.6

Comparison of in-via oxide liner etch processing cost for different TSV last geometries. The TSVs with a smaller diameter require longer liner etch processing time. The processing cost is normalized to the process cost of a 5×50 TSV last flow.

8.1.2.5 Through silicon via barrier and Cu seed processing

The first steps towards metallization of a TSV structure are the deposition of a barrier layer to prevent Cu diffusion, followed by a deposition of a Cu seed layer to allow the Cu filling of the TSV through electroplating. Two main requirements for these layers are continuity and uniformity throughout the TSV structure. With highly scaled TSV geometries, achieving these requirements is a challenge; different deposition options are therefore considered for different TSV sizes [316,317].

In the case of a TSV middle flow with an aspect ratio of 1:10 (i.e., 10×100 and 5×50), a physical vapor deposition (PVD) tantalum (Ta) barrier layer followed by a PVD Cu seed layer is applied [313]. Different layer thicknesses are required to achieve continuity for the different TSV sizes: For a 10×100 TSV, a Ta layer of 180 nm followed by a 1,500 nm Cu seed is used. In the case of a 5×50 TSV, layers of 130 nm Ta and 800 nm Cu are used, as illustrated in Fig. 8.7.

With further scaling of the TSV diameter and increasing TSV aspect ratios, however, the PVD deposition process reaches a limit and alternative deposition technologies such as atomic layer deposition (ALD) processing are considered. As shown in Fig. 8.7, ALD deposition of 12 nm titanium nitride (TiN) has been demonstrated as a successful barrier layer for a 3×50 TSV middle structure [316]. PVD Cu deposition with a thickness of 1,500 nm is an option for the seed layer to achieve continuity through the entire TSV, albeit with variations in thickness. Another option for a conformal seed is the combination of 5 nm ALD ruthenium (Ru) with 30 nm electro-less (ELD) Cu deposition [320]. Finally, a third metallization option combines the deposition of a 17 nm tungsten nitride (WN) barrier layer followed by ELD of nickel boron (NiB) that acts as the seed for the Cu plating [317]. As shown in Fig. 8.7, these last two options are scalable and can also be applied to 2×40 TSV middle structures.

The barrier and seed layers are removed by CMP after the TSV process flow. The cost of the CMP process depends upon the thickness of the deposited layers. Therefore, as shown in Fig. 8.7, the thin layers deposited by the ALD and ELD methods offer an advantage from a cost perspective when also considering the processing cost of the CMP.

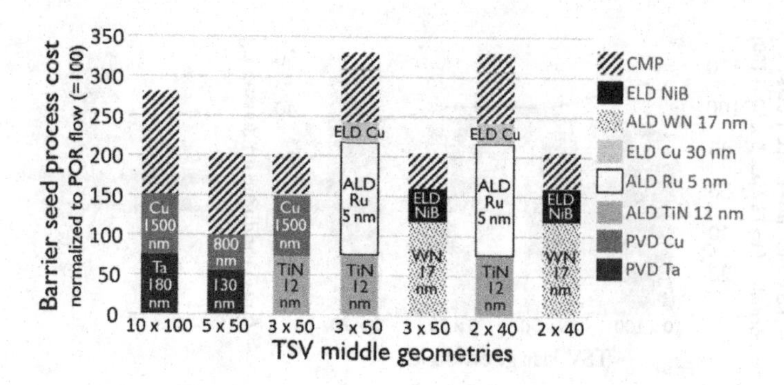

FIGURE 8.7

Cost comparison of barrier seed process for TSV middle geometries. Non-PVD deposition approaches can be applied to TSV sizes of 3×50 and 2×40. The processing cost is normalized to the process cost of a 5×50 TSV middle flow.

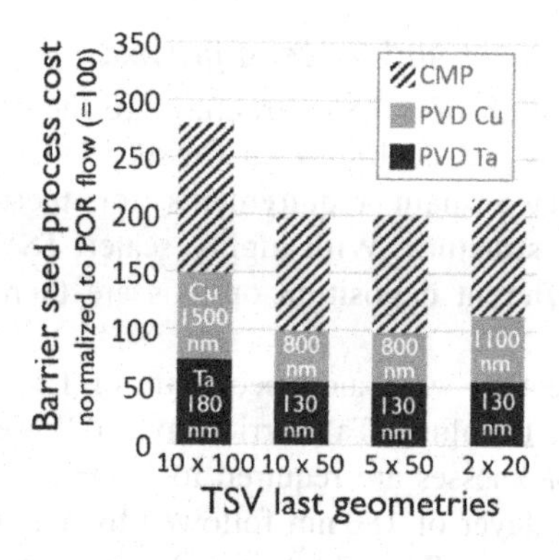

FIGURE 8.8

Cost comparison of barrier seed process for TSV last geometries. PVD processing is applied for all TSV last sizes. The processing cost is normalized to the process cost of 5×50 TSV middle flow.

In the case of a TSV last flow, the TSV geometries considered here allow the use of PVD processes for barrier and seed deposition. Depending upon the TSV geometry, different deposition thicknesses are required to achieve continuous coverage within the TSV, as shown in Fig. 8.8. For smaller diameter TSVs (i.e., 2×20), a thicker seed layer is deposited. The cost of the CMP process to remove the barrier seed layers also depends upon the thickness of the deposited material, as shown in Fig. 8.8.

8.1.2.6 Through silicon via Cu plating and effect on chemical mechanical planarization

Filling the TSV structure with Cu is achieved with a bottom-up electroplating approach that produces void free TSV plating [318]. As shown in Fig. 8.9, the plating time and material cost for TSV filling depends upon the depth and diameter of the TSV structure.

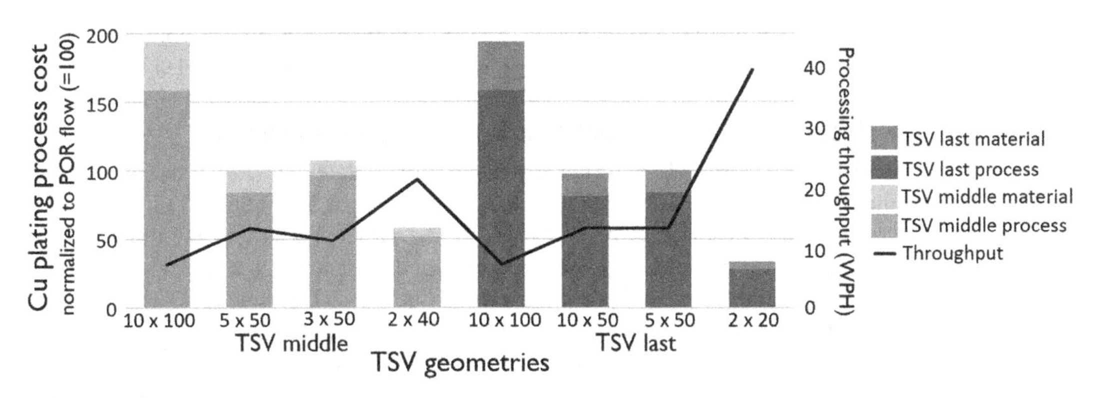

FIGURE 8.9

Cost comparison of TSV Cu plating process for both TSV middle and TSV last flows for different TSV geometries. The cost comparison considers processing and material costs and is normalized to the process cost of the 5 × 50 TSV middle (POR) flow.

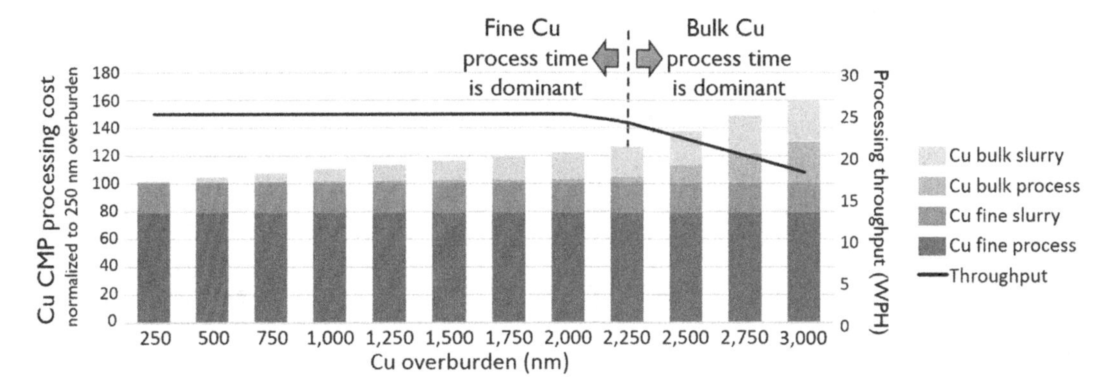

FIGURE 8.10

Cost of Cu CMP for different Cu overburden thicknesses. The fine Cu polish step is the dominant cost component for Cu thicknesses up to 2,000 nm.

TSV plating overburdens the Cu plated on top of the entire wafer area which is removed by CMP. The Cu CMP process has two stages: (1) a bulk Cu polish that is relatively fast and removes most of the plated Cu, and (2) a fine Cu polish that is slower and clears the remaining Cu traces from the wafer surface (including the Cu seed layer). The fine Cu polish step removes the final layer of Cu (approximately 250 nm) from the top of the wafer.

To estimate the effect of the Cu overburden on CMP cost, the combined effect of the two polishing steps is considered. Different initial Cu overburden thicknesses are considered, from 250 nm up to 3,000 nm, as shown in Fig. 8.10. Note in Fig. 8.10 that for a Cu overburden thickness up to 2,000 nm, the fine Cu polish step dominates the processing time, material cost, and throughput of the CMP process. The CMP cost therefore exhibits low sensitivity to the thickness of the Cu overburden. When the total Cu overburden thickness exceeds 2,000 nm, the time to polish the bulk Cu is dominant and the throughput of the CMP depends upon the overall Cu overburden thickness.

8.1.2.7 Through silicon via chemical mechanical planarization processing

Polishing the deposited layers during TSV formation electrically isolates the TSVs and allows the processing of additional layers, such as the BEOL stack (in the case of TSV middle) or redistribution layers and/or microbumps (in the case of TSV last). In the previous sections, the relative cost to remove each deposited layer by CMP has been described. It is shown that the cost of CMP depends upon the material being removed and the material thickness. Different CMP slurries are required depending on the material used, which can greatly affect the overall cost.

To compare the polishing cost of different materials used in TSV processing, the same thickness of 100 nm is assumed for all materials. In this case, one of the parameters that determines the CMP cost is the polishing time, which depends solely on the material. The polishing time determines the throughput of the CMP equipment for each material. Another parameter is the cost of the slurry, since the selection of slurry depends on the material being polished. The cost of the slurry also depends on the polishing time; however, for materials that require the same polishing time, the type of slurry is the cost differentiator. In Fig. 8.11, the materials used for the different TSV processing

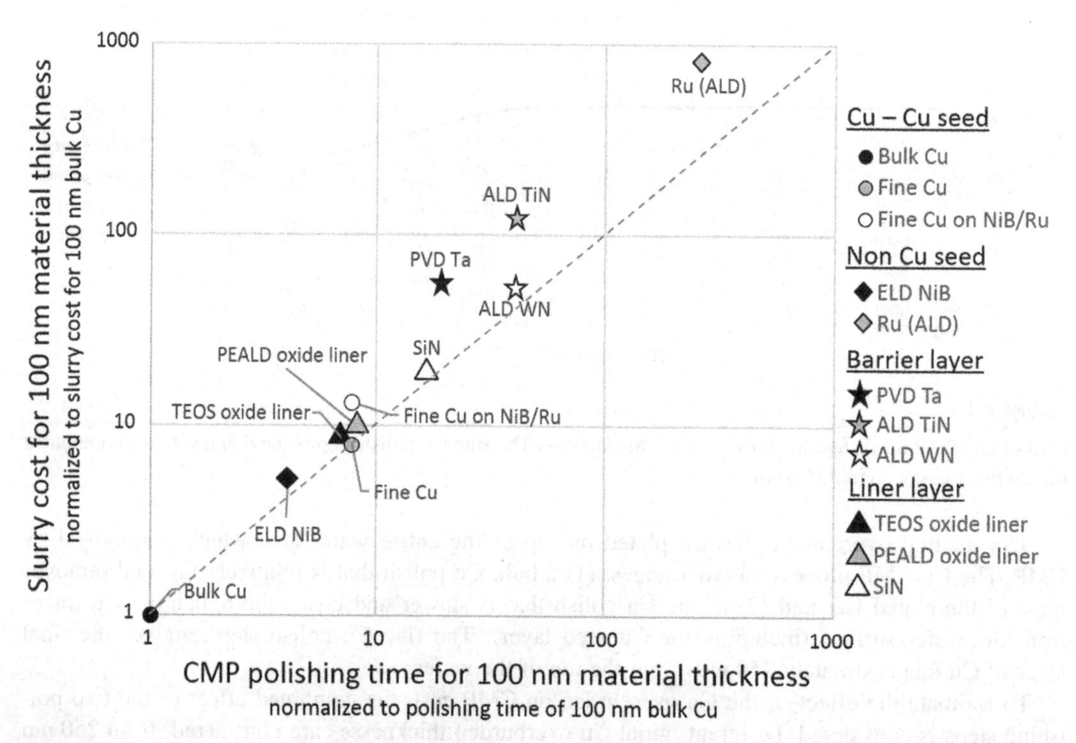

FIGURE 8.11

CMP benchmark of deposition materials used in TSV processing. For each material, a thickness of 100 nm is considered to estimate polishing time and slurry consumption.

options are depicted with respect to CMP time and slurry cost (assuming 100 nm of material thickness). For each material, both the polishing time and the slurry cost are normalized to polish a bulk 100 nm thick Cu layer.

The different materials are grouped in Fig. 8.11 based on the function within the TSV structure. For the plated Cu and Cu seed (illustrated in Fig. 8.11 with dots), polishing the fine Cu layer requires more time than bulk Cu (for the same 100 nm thickness) and the slurry cost scales accordingly. In the case where the fine Cu polishing is above a NiB or Ru seed layer, a different, more expensive slurry is used, requiring the same approximate processing time.

In the case of non-Cu seed layers such as NiB or Ru (illustrated as diamonds in Fig. 8.11), the cost for Ru polishing can be quite high. The actual Ru thickness is, however, only 5 nm (rather than 100 nm), allowing the use of Ru as a viable option for TSV processing.

For barrier materials such as PVD Ta, ALD TiN, and ALD WN (illustrated in Fig. 8.11 as stars), the polishing rate for the Ta layer is faster than for the ALD layers. However, in an actual TSV structure, the thickness of the ALD layers is an order of magnitude thinner than the thickness of the Ta layer (as previously shown in Fig. 8.7), which can significantly affect the cost of the barrier CMP.

In the case of TSV oxide liner materials such as TEOS oxide, PEALD oxide, and SiN capping (illustrated as triangles in Fig. 8.11), the polishing cost of the TEOS and PEALD oxides is comparable. In a TSV structure, however, the thickness of the PEALD layer is approximately half of the thickness of the TEOS layer, which greatly lowers the liner polishing cost, as shown in Fig. 8.5. Polishing the SiN capping layer is more expensive than the oxide layers. However, in the TSV last flow, polishing the SiN capping layer is an option, eliminating the CMP step.

8.1.2.8 Backside processing

Processing a TSV structure requires the wafer to be thinned to a thickness comparable to the TSV depth. Thinning of the wafer is accomplished from the backside by a grind and recess etch step [319]. Depending upon the type of TSV process (TSV middle or TSV last), different processing steps are performed at the backside, as illustrated in Fig. 8.2. These steps and a comparison of the corresponding processing costs are presented in this subsection.

In both the TSV middle and TSV last flows, the wafers are initially ground from the backside followed by cleaning steps. In the case of the TSV middle, recess etching of the Si is completed following a cleaning step to reveal the TSV, with the oxide liner layer fully enclosing the Cu TSV core. As a next step, backside passivation is deposited, followed by a layer of photoresist to protect the passivation layer. During the next step, the protective resist and oxide liner are etched to reveal the Cu core within the TSV. Stripping the remaining photoresist from the backside of the wafer is the final etching step.

In the case of TSV last processes, polishing the Si is completed following the backside grinding and cleaning steps. The polishing is accomplished via CMP to remove any damage caused by the grinding process, and to provide a pristine surface for the subsequent TSV last processing steps. For the TSV last process, a 7 μm thick Si layer is removed by CMP. After the CMP step, a passivation layer is applied at the backside and the TSV last flow is initiated, as illustrated in Fig. 8.2.

The backside processing for the TSV middle and TSV last flows are illustrated in Fig. 8.12. No practical difference exists in the common backside processing steps required for different TSV geometries with either the TSV middle or TSV last flows. Comparing the overall cost of the

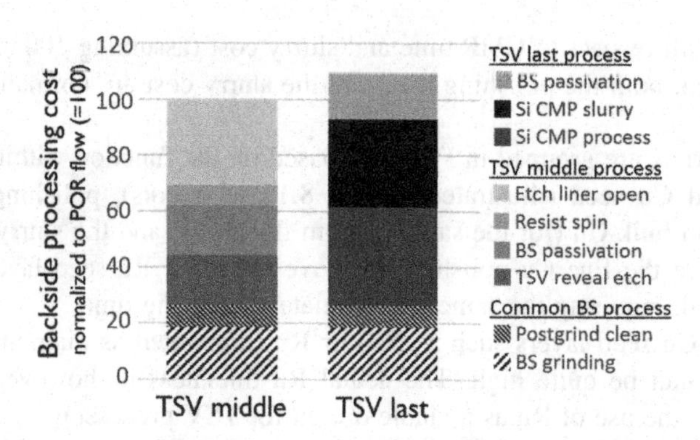

FIGURE 8.12

Cost benchmark of backside processing steps for TSV middle and TSV last flows.

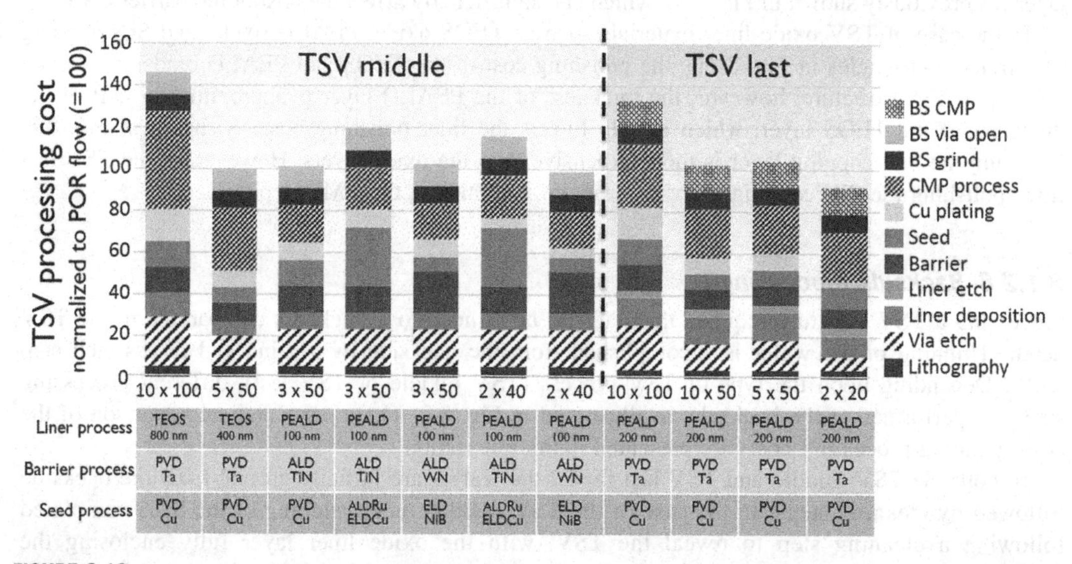

FIGURE 8.13

Benchmark of overall processing costs for different TSV geometries for both TSV middle and TSV last flows.

processing steps for the TSV middle flow with the corresponding cost in the TSV last flow, the TSV last backside processing costs are approximately 10% higher due to the Si CMP.

8.1.3 COMPARISON OF THROUGH SILICON VIA PROCESSING COST

Following the cost benchmarking of the different processing steps in the TSV middle and TSV last integration flows, the cost of the overall fabrication process is compared for different TSV geometries. This comparison is illustrated in Fig. 8.13.

Following the cost benchmark depicted in Fig. 8.13, fabrication of particular TSV geometries is described to evaluate the effect of the individual processing steps. The process flows for these TSV geometries are compared in the following subsections.

8.1.3.1 Processing of the 5× 50 through silicon via geometry

The overall processing cost for the 5 μm × 50 μm and 10 m × 100 μm TSV flows is shown in Fig. 8.14 to emphasize the difference between the TSV middle, and TSV last flows for a particular TSV dimension. In the case of the 5 × 50 TSV middle flow, the oxide liner is removed by CMP, which increases the cost of CMP processing. However, in the case of a TSV last process, the additional cost for the PEALD liner, etching the TSV liner within the via, and the backside CMP results in higher processing cost by up to 10% for the 5 μm × 50 μm TSV flow.

8.1.3.2 Processing of the 10× 100 through silicon via geometry

In the case of the 10 × 100 TSV last geometry, skipping the CMP of the oxide liner reduces the overall processing cost about 9% as compared to the 10 × 100 TSV middle flow, as shown in Fig. 8.14. Removal of the thick (800 nm) TEOS oxide liner layer is the reason for the lower CMP cost between the two TSV flows.

8.1.3.3 Scaling through silicon via geometries

For the TSV middle flows shown in Fig. 8.15, scaling the TSV size is possible while maintaining reasonable processing costs. This situation is achieved by replacing the thicker PVD deposited layers used in the larger TSV geometries with thinner ALD layers for the scaled TSVs. Although

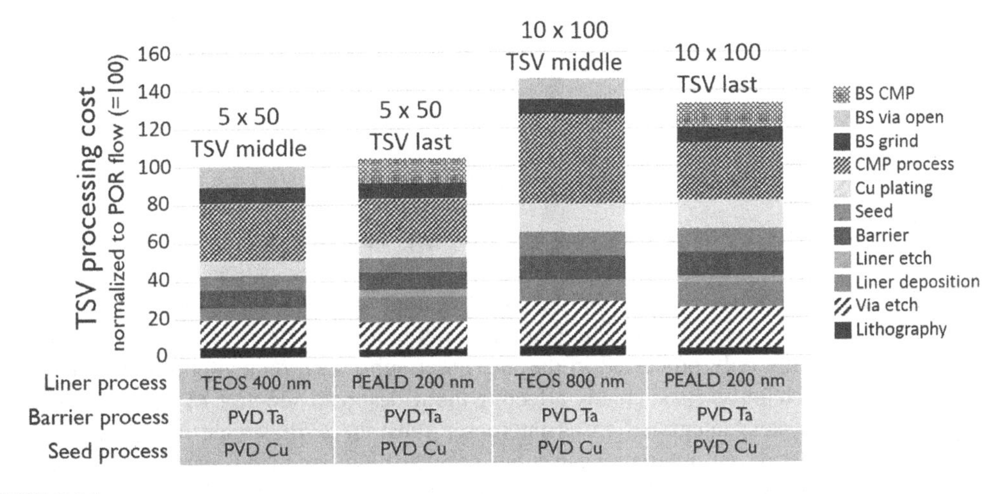

FIGURE 8.14

Cost benchmark of the 5 × 50 and 10 × 100 TSV flows. For the 5 × 50 TSV middle, polishing the oxide liner increases the cost of the CMP step. For the 5 × 50 TSV last process, the liner deposition, liner etch steps, and backside CMP are the primary cost differentiators. For the 10 × 100 TSV middle process, polishing the oxide liner increases the overall processing cost by up to 9% as compared to 10 × 100 TSV last flow.

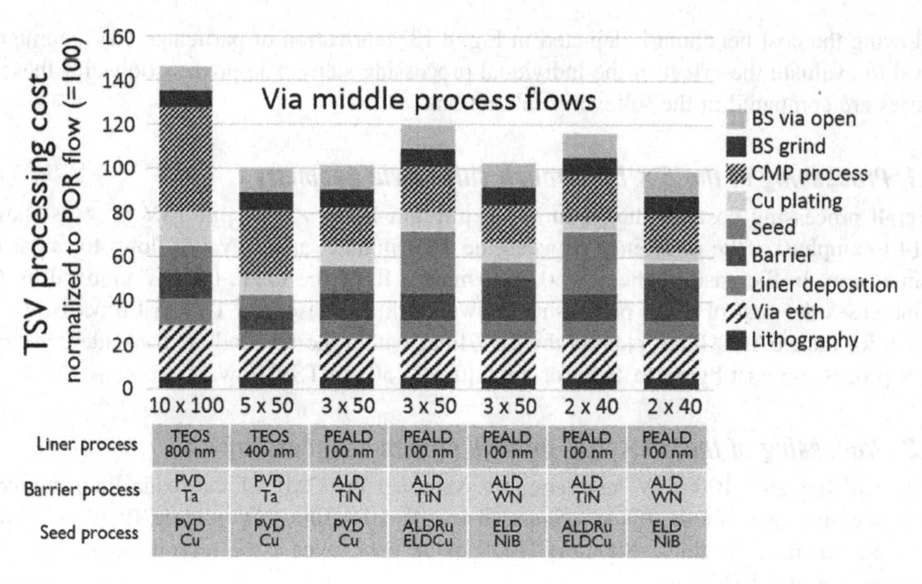

FIGURE 8.15

Cost benchmark of TSV middle processing flows for different TSV geometries. Note that the TSV size and pitch are scaled while maintaining the TSV processing cost.

the processing cost of the ALD layers is higher, the cost for CMP polishing of these thinner layers is lower. This trade-off controls the processing cost as the size of a TSV middle is scaled from 5 × 50 to 3 × 50 and 2 × 40 TSV sizes. Scaling the TSV diameter enables scaling of the TSV pitch as well, therefore higher TSV density can be achieved without increase in cost.

8.2 INTERPOSER-BASED SYSTEMS INTEGRATION

An alternative to multiple die vertical stacking is 2.5-D integration of dice on an interposer substrate [321]. In this way, fine pitch interconnections are provided between active dice. This approach addresses the manufacturability and thermal behavior of multi-die 3-D stacks [322]. An additional benefit of 2.5-D integration is that this technology does not require TSVs to be processed within the active dice (the TSVs are only processed in the interposer substrate), removing any uncertainty in the component supply chain, an issue greatly affecting the widespread adoption of 3-D technology.

An example of a system integrated on an interposer substrate is illustrated in Fig. 8.16. High-density interconnections on the interposer substrate provide high bandwidth data transfer between memory and the central processing unit. The interposer contains TSVs that provide connectivity between the system components and the outside world. Additional components provide system level I/O and sensor based functionality integrated onto the interposer substrate and in close proximity to the processing unit. Placing all of the components in close proximity benefits system performance and functionality while reducing power dissipation and overall size.

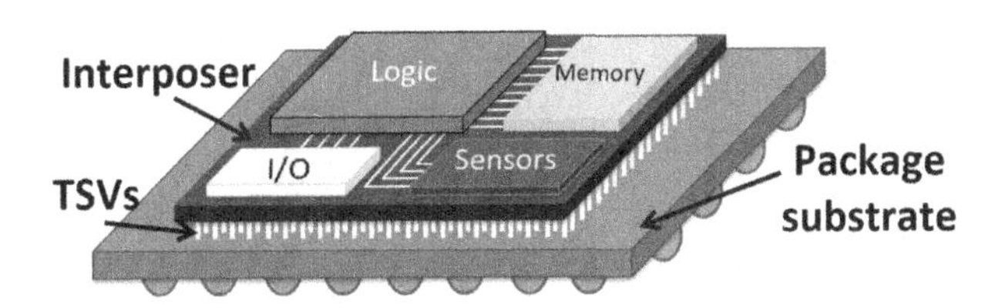

FIGURE 8.16

System integrated on top of an interposer substrate. The TSVs connect the interposer to the package substrate.

An interposer for interdie interconnections offers the advantage that no TSVs are required within the functional dice, as compared to vertical stacking of dice (i.e., 3-D integration). Only minor modifications to existing components are therefore necessary [323]. Furthermore, flexible system reconfiguration is enabled since a new component can be added, updated, or replaced by only redesigning the interposer substrate. Partitioning a system into different integrated circuits and selectively scaling different components to optimize the overall system performance and cost is also possible. In addition, the use of an interposer thermally isolates the high power system components.

All of these advantages of an interposer for system integration require manufacturing an additional substrate layer; the additional cost of the interposer on the overall system cost should therefore be considered. Different options for manufacturing an interposer substrate are described in this section and compared in terms of processing cost. Minimum requirements for an interposer substrate include interconnections among the stacked component dice and connections to the outside world through TSVs. In addition, system level power delivery planes are possible with or without incorporating decoupling capacitors within the power distribution network.

The effect of different interposer manufacturing options on system cost is evaluated using the 3-D cost model developed at IMEC [315]. In Section 8.2.1, the processing cost of the various features of an interposer substrate is described. The cost of different interposer configurations is discussed in Section 8.2.2.

8.2.1 COST OF INTERPOSER MANUFACTURING FEATURES

In this section, the cost of processing different features on an interposer substrate is presented. These features include TSVs, die-to-die interconnect, power and ground planes, metal–insulator–metal (MIM) capacitors, microbumps for functional die stacking, and backside Cu pillars to connect the interposer to the package substrate. Combinations of these structures provide options for an interposer substrate to satisfy different application requirements. A comparison of the relative processing costs per wafer of these components is illustrated in Fig. 8.17.

8.2.1.1 Cost of through silicon via processing

TSVs are fabricated within the interposer substrate to provide connectivity between the stacked functional dice and the package substrate. For the TSVs within an interposer substrate, no strict area constraints exist, and therefore, extreme TSV scaling is not necessary. The interposer thickness, however, is important. To enhance mechanical support, reducing die warpage and improving the ability to handle thin dice necessitate the silicon substrate to not be extremely thin. An

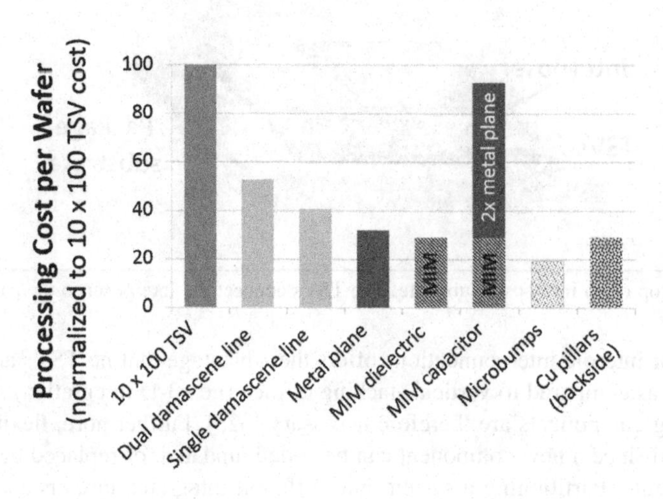

FIGURE 8.17

Comparison of processing costs per wafer for different features of an interposer substrate. All of the processing costs are normalized to the wafer cost of processing a 10 μm × 100 μm TSV middle flow.

interposer composed of TSVs with 100 μm depth and 10 μm diameter (10 × 100 TSV middle) is therefore considered. The processing cost of the 10 × 100 TSV middle is the reference to compare the cost of the other interposer components, as shown in Fig. 8.17. A benchmark of the TSV processing cost for different TSV geometries and flows is also illustrated in Fig. 8.13.

8.2.1.2 Cost of die-to-die interconnect processing

The interconnect metal layers providing connectivity among the stacked dice are processed on the interposer substrate. To support high bandwidth communication among the functional dice, low resistance interconnect is required. Thick copper interconnect lines and dielectric thicknesses of up to 2 μm are therefore considered [326]. These lines are manufactured by a dual damascene process to achieve higher line density and enhanced reliability as compared to alternative interconnect processes using semi-additive electrochemical plating [315].

The processing cost per wafer of the dual damascene interconnect lines is illustrated in Fig. 8.17, normalized to the cost of processing a 10 × 100 TSV middle flow. Note that the dual damascene process forms both the metal lines and the metal vias to the underlying metal interconnect layer. For the lowest metal layer on top of the TSVs, no vias (or substrate contacts) are necessary. This metal layer is in direct contact with the TSVs and can be processed with a single damascene process. The cost of a single damascene line is also shown in Fig. 8.17.

8.2.1.3 Cost of processing metal planes and metal–insulator–metal capacitors

In addition to the metal lines connecting the functional stacked dice, the metal planes that distribute system level power and ground can also be integrated onto an interposer substrate. These power planes use a single Cu damascene process with a metal thickness of 1 μm [324]. Furthermore, the coexistence of both power and ground planes on the same interposer substrate enables the fabrication of a MIM capacitor that can function as a decoupling capacitor within the power distribution network [325]. The MIM capacitor is formed by deposition of a stack of TaN/insulator/TaN layers on top of a planarized metal layer. A second metal plane on top of the stack is the top capacitor terminal, as discussed in [321].

Processing thick metal planes can induce significant stress on the interposer substrate, introducing the risk of die warpage. Leveraging the interaction among the MIM capacitors, thick interconnect lines, and dielectric layers can mitigate the induced stresses during manufacturing of the interposer while restraining die warpage [324].

The cost of processing a single damascene metal plane and the MIM layer stack is illustrated in Fig. 8.17. The cost of a MIM capacitor, comprised of two metal planes with a dielectric insulator in between, is also shown in Fig. 8.17.

8.2.1.4 Cost of processing microbumps and Cu pillars

Communicating signals and power among the functional stacked dice and interposer substrate is achieved by vertical microbumps. High density interconnects are required for wide I/O applications; therefore, small pitch microbump structures are necessary. Cu–Sn–Cu microbump structures with a 20 μm pitch are described in [326]. Furthermore, the interposer connects the backside to the package substrate where a more relaxed interconnect pitch is possible. These connections use Cu pillars with a 170 μm pitch and up to 50 μm thickness.

The processing cost for the microbump structures and Cu pillar structures is illustrated in Fig. 8.17. The larger dimensions and increased metal volume of the Cu pillars require additional processing time, which increases the process cost per wafer as compared to manufacturing microbumps.

8.2.2 INTERPOSER BUILD-UP CONFIGURATIONS

Exploiting the interposer features described in Section 8.2.1, different interposer configurations can serve different applications. In these systems, the interposer substrate includes microbumps and thick interconnect lines for signaling among the stacked functional dice. Additionally, TSVs and Cu pillars for signaling and power can be accessed at the backside of the interposer and package substrate.

Different interposer systems exhibit different features. Multiple layers of metal lines in the interposers increase routing complexity. Entire metal planes can be used to distribute system level power or ground. Two power metal planes separated by a dielectric can provide a low impedance power distribution network with decoupling capacitance to reduce power supply noise.

Three examples of different interposer configurations are shown in Fig. 8.18. A substrate with a single thick metal layer on top of a reference power plane is shown in Fig. 8.18A. This interposer can be used for applications with a limited number of lines connecting the functional dice. If additional routing is required, a second thick metal layer can be utilized in place of the metal plane, as illustrated in Fig. 8.18B. Finally, an interposer substrate with a MIM capacitor between the power and ground planes and two layers of thick metal lines is depicted in Fig. 8.18C.

A breakdown of the wafer processing cost for each interposer substrate is shown in Fig. 8.19. The cost of each component is normalized to the wafer processing cost of 10 μm × 100 μm TSV middle flow. Note that replacing a metal plane with a routing metal layer has almost no effect on the processing cost. Adding the functionality of a MIM capacitor, however, increases the processing cost of an interposer with two routing layers by up to 40%, as illustrated in Fig. 8.19.

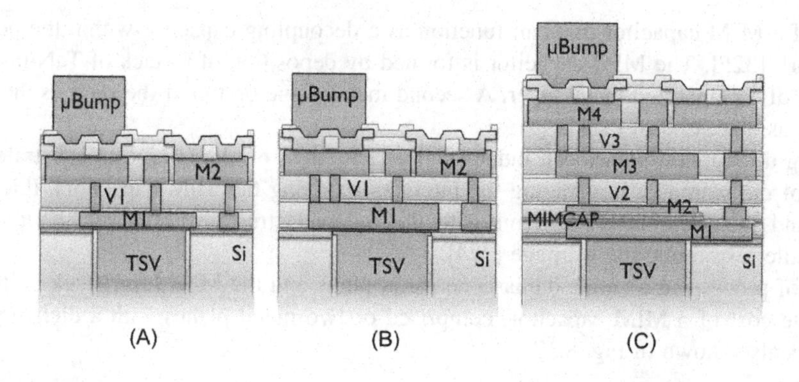

FIGURE 8.18

Different interposer configurations: (A) single metal layer over a power metal plane, (B) two thick metal layers, and (C) MIM capacitor between power and ground metal planes with two thick metal layers.

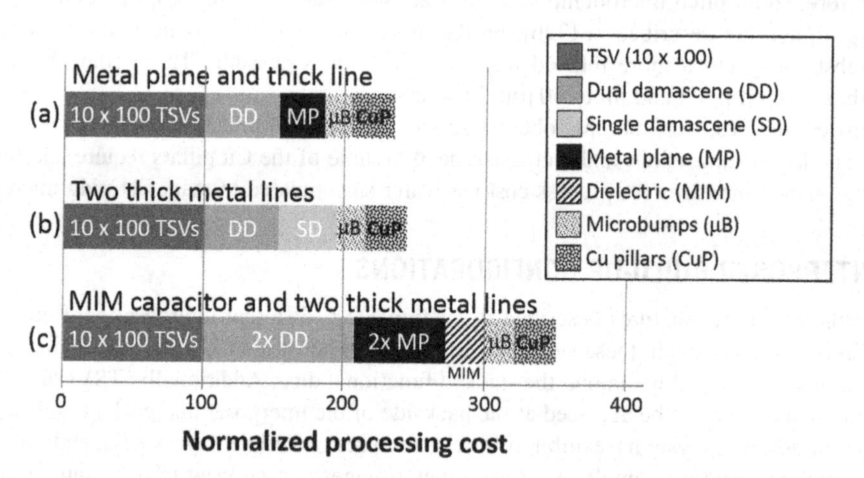

FIGURE 8.19

Comparison of wafer-level processing cost for different interposer structures. The cost of each component is normalized to the cost of the 10 μm × 100 μm TSV.

8.3 COMPARISON OF PROCESSING COST FOR 2.5-D AND THREE-DIMENSIONAL INTEGRATION

In Sections 8.1 and 8.2, the processing cost for different TSV structures and features of a Si interposer substrate are described. These integration schemes can be used to analyze and compare the cost of different 3-D stacking approaches: (1) die-to-wafer (D2W) stacking, (2) wafer-to-wafer (W2W) stacking, and (3) 2.5-D interposer-based stacking. These stacking approaches are illustrated in Fig. 8.20.

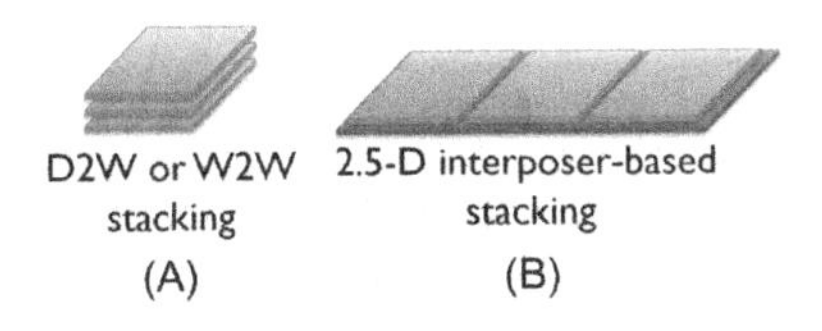

FIGURE 8.20

3-D stacking approaches: vertical stack of three active dice, (A) D2W or W2W stacking, and (B) 2.5-D interposer-based stacking.

For each 3-D stacking approach, the required processing components and complexity of the 3-D processing flows are reviewed. In addition, process yield and prestack testing of the stacked components are evaluated, and the effect on processing cost is described. Furthermore, the effect of die size on system yield and 3-D integration cost is discussed.

This section is organized as follows: The different stacking schemes and components that enable vertical integration are presented in Section 8.3.1. A comparison of the processing cost of the different components enabling 3-D integration is discussed in Section 8.3.2. The complete 3-D integration flow, together with different options for prestack testing, are compared in Section 8.3.3. The effect of the size of the active dice on the integration cost is reviewed in Section 8.3.4. Further exploration of the impact of prestack testing of the interposer in relation to processing yield and the size of the active dice is discussed in Section 8.3.5.

8.3.1 COMPONENTS OF A THREE-DIMENSIONAL STACKED SYSTEM

A 3-D stacked system with three active dice is considered here. All of the dice are assumed to be processed in the same CMOS technology (28 nm bulk CMOS is assumed) and are the same size to enable W2W stacking. The die area, however, can vary to evaluate the effects of die size and number of die per wafer on the processing yield and system cost.

The processing yield of the active dice is determined from the Poisson yield model, $Y = e^{(-AD)}$ [327], where A is the die area and D is the defect density per unit area. The defect density is assumed to be 5% for the particular process technology, while the die area can vary.

It is further assumed that the active dice are tested prior to stacking with a total fault coverage of 95%. Consequently, testing can detect 95% of the faults on a die, where the faulty dice are scrapped; however, there remains a possibility of 5% faulty dice. In the cases of D2W stacking and stacking on an interposer substrate, only those active dice that pass prestack test are selected for stacking. In the case of W2W stacking, the location of each die on a wafer is set; therefore, no prestack test of the active dice is performed. The cost for testing the active dice is considered to be proportional to the die area.

In the case of D2W and W2W stacking, the TSVs are placed between the active dice to enable the vertical interconnections. The stacking interface among the dice uses microbumps. An illustration of the resulting stack of three dice is shown in Fig. 8.21A. The bottom die in the stack uses Cu pillars of 50 μm height to connect to a laminate package substrate (the cost of a package substrate is not considered in this cost comparison). The top die in the stack does not contain TSVs; the top die is stacked face-to-face to the die below it.

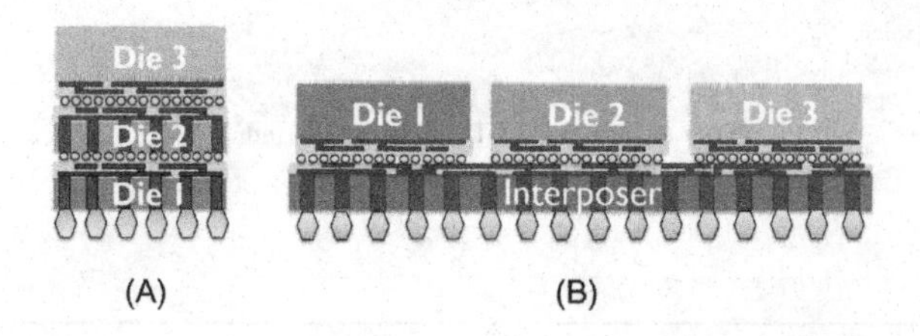

FIGURE 8.21

3-D integration technologies. (A) Three die stack. The stacking interface between the dice is microbumps. TSVs are fabricated on die 1 and die 2 to enable vertical signal propagation. (B) Three active dice on an interposer substrate. The active dice are stacked using microbumps. The TSVs are fabricated within the interposer die to provide access to the backside. The interposer is connected to the package substrate (not shown) by Cu pillars.

In interposer-based systems, the active dice are stacked face-to-face on a silicon interposer substrate using microbumps between layers. The interposer provides the multiple interconnections among the active dice. Furthermore, the interposer provides backside access to the package substrate using TSVs for the power supply and I/O signals. No TSVs are fabricated within the active dice. An interposer substrate with three stacked dice is shown in Fig. 8.21B.

The interposer substrate die should be sufficiently large to accommodate the three dice stacked on top. It is assumed that the interposer die extends 100 μm around the active die area. As the size of the active dice can vary, the interposer die is resized to fit the stacked dice.

8.3.2 COST OF THREE-DIMENSIONAL INTEGRATION COMPONENTS

To evaluate the cost of a 3-D system, the processing cost for the components that enable 3-D stacking should be considered in addition to the processing cost per wafer of the active dice. In the case of a 3-D system where the active dice are stacked on top of each other, as illustrated in Fig. 8.21A, the additional cost includes:

1. For die 1: Microbumps on the front side, TSVs (with backside processing and thinning), and Cu pillars at the backside.
2. For die 2: Microbumps on the front and backside and TSVs (with backside processing and thinning).
3. For die 3: Microbumps on the front side.

The TSVs processed within the active dice are considered to be 5×50 TSV middle structures [313], as described in Section 8.1.1.

In the case of an interposer-based system, the microbumps are processed on the face of the active die to interface with the interposer substrate. Furthermore, an additional component is the interposer die itself. Processing all of the features on the interposer dice, as shown in Fig. 8.21B, should therefore be considered:

1. Microbumps on top of the interposer.
2. Interconnect to provide connections among the stacked active die.

3. TSV processing (including temporary wafer bonding and wafer thinning).
4. Cu pillar processing at the backside to connect to the package substrate.

An interposer substrate with two thick metal lines, as illustrated in Fig. 8.18B, is considered in this comparison. The TSVs processed within the interposer die have a diameter of 10 μm and and a height of 100 μm (10 × 100 TSV middle process), as discussed in Section 8.1.1.

In both vertically stacked and interposer-based systems, the microbumps are assumed to have a 40 μm pitch and a bump diameter of 25 μm [326]. The Cu pillars have a minimum pitch of 170 μm and a diameter of 50 μm.

A comparison of the processing cost per wafer for the components that enable 3-D stacking is shown in Fig. 8.22. The processing steps for temporary wafer bonding, wafer thinning, and die stacking are also included. Notice in Fig. 8.22 that the "die pick and place" step is a die level process, and the wafer-level cost depends upon the number of dice per wafer. In this case, active dice with size of 10 mm × 10 mm are assumed, resulting in 568 dice on a 300 mm wafer. Only those dice that pass a prestack test, however, are chosen for stacking. Assuming a 5% per cm^2 defect density and prestack testing of the active dice with 95% fault coverage, 541 dice will pass testing. The cost for the "die pick and place" operation therefore assumes 541 die per wafer.

8.3.3 COMPARISON OF THREE-DIMENSIONAL SYSTEM COST

A comparison of the cost of vertically stacked systems with three different 3-D integration approaches is presented in this section. The approaches are: D2W stacking, W2W stacking, and a

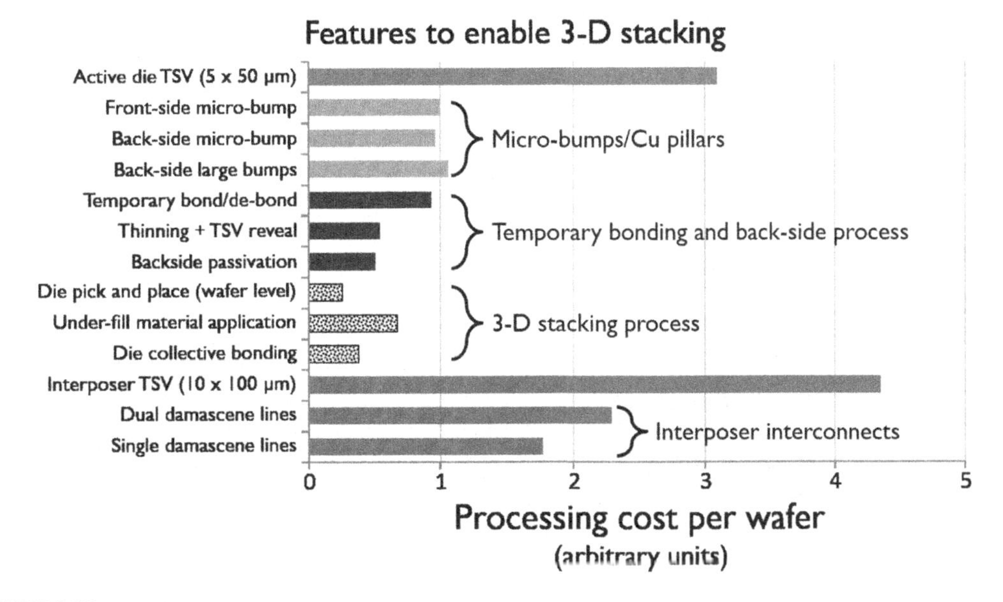

FIGURE 8.22

Comparison of processing cost per wafer to enable 3-D stacking. The features are processed either on the active dice and/or on the interposer substrate. For the die pick and place step, processing of 541 die/wafer is assumed.

2.5-D interposer-based system. The cost comparison considers a system with three active dice, 10 mm × 10 mm in size.

For each approach, the cost of the 3-D system is determined by evaluating the cost of the individual processing components. The cost of each active die includes CMOS processing, additional processing for those features that enable 3-D stacking (TSVs, wafer thinning, microbumps), and the cost of prestack testing. Furthermore, yield losses due to imperfect 3-D stacking (the stacking yield is assumed to be 98%) and test fault coverage (assumed at 95%) are also considered. The cost of each stacked component and the cost of the total 3-D system for each stacking approach is depicted in Fig. 8.23.

As shown in Fig. 8.23, in the case of D2W and W2W stacking, the 3-D enabling cost for die 1 and die 2 is greater than the corresponding cost for die 3 since the TSVs are processed within dices 1 and 2. Furthermore, the cost of prestack testing of the active dice is included in the D2W approach. The cost of the D2W stack is the sum of the processing costs of the individual active dice, 1, 2, and 3, and the cost due to compound yield losses. Compound yield losses are due to imperfect yield of the 3-D stacking process (shown in Fig. 8.23 as stacking yield cost) and the non-perfect testing of the active dice, assuming a fault coverage of 95% (shown as die yield cost). These yield losses are due to a small portion of faulty active dice that are misplaced during test and contribute to the cost due to die yield loss.

In the case of W2W stacking, no testing of the active dice is required since the position of the stacked dice is fixed. The processing cost of the individual active dice is therefore lower than the D2W approach. There is, however, an increased cost due to yield loss due to stacking of faulty active dice, as illustrated in Fig. 8.23.

In the case of an interposer-based system, an additional system component needs to be manufactured: the interposer die. The processing yield of the interposer die is assumed to be 99% and the interposer dice are tested with 100% fault coverage. Processing of the interposer die is an additional

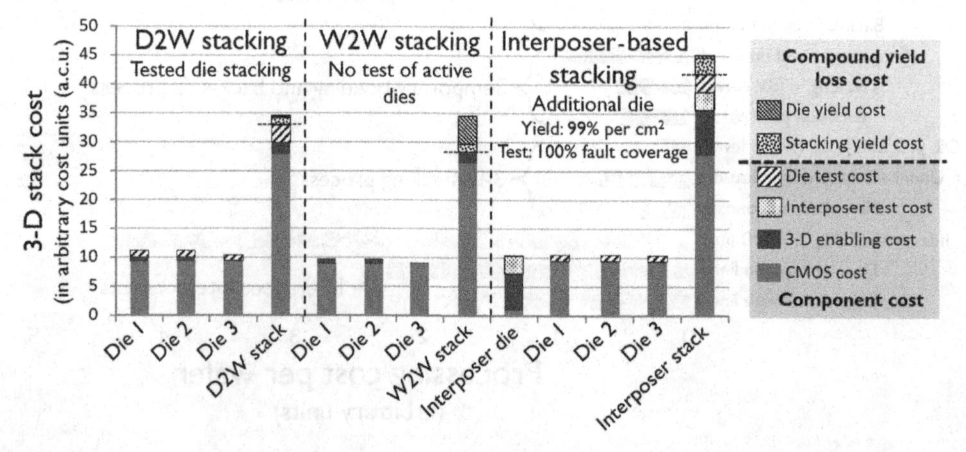

FIGURE 8.23

Cost comparison of different stacking approaches and components. The cost of the compound yield losses is illustrated for each stacking approach. An area of 10 mm × 10 mm is considered for the active dice.

3-D enabling cost, as illustrated in Fig. 8.23. Note in Fig. 8.23 that the cost due to stacking yield loss is higher for an interposer-based system. This characteristic is due to the additional stacking operation required to place the interposer die on the package substrate, affecting the overall stacking yield.

8.3.4 DEPENDENCE OF THREE-DIMENSIONAL SYSTEM COST ON ACTIVE DIE SIZE

In this section, the effect of the active die size on the cost of a 3-D system is evaluated. Different active die sizes, ranging from 3 mm × 3 mm to 15 mm × 15 mm, are considered. The same assumptions as in the previous sections are valid regarding component processing, assembly yield, and prestack test fault coverage. Considering a variable active die size, the process cost per die area is determined, including the cost of CMOS processing of the active dice, additional 3-D enabling cost (including the interposer processing cost for the 2.5-D interposer system), and the cost for testing the active dice and interposer. In addition, the cost of the compound yield loss due to 3-D stacking is reported. Summing the cost due to process with the cost due to compound yield loss produces the total cost per die area. The process cost per die area and the total cost per die area are shown in Fig. 8.24 for the three different stacking approaches.

As shown in Fig. 8.24, the stack cost per unit area for all three integration approaches increases with the size of the active dice. This increase in cost is due to the smaller number of dice per wafer and the lower utilization of the total wafer area. Furthermore, with increasing die size, the processing yield of the active dice is less; thereby increasing the process cost of each system.

Considering the D2W stacking approach, the total cost of the 3-D stack per unit area is almost a constant offset higher than the process cost over the entire range of active die area. This constant offset in cost is the effect of prestack testing of the active dice that greatly reduces yield loss due to faulty active dice. The yield losses in this case are due to the nonperfect yield of the 3-D stacking process (98%) and the small fraction of active dice that are misplaced during prestack testing.

In the case of the W2W stacking approach, as shown in Fig. 8.24, the process cost per unit area is lowest. This low cost is due to no prestack test of the active dice. However, when the total cost

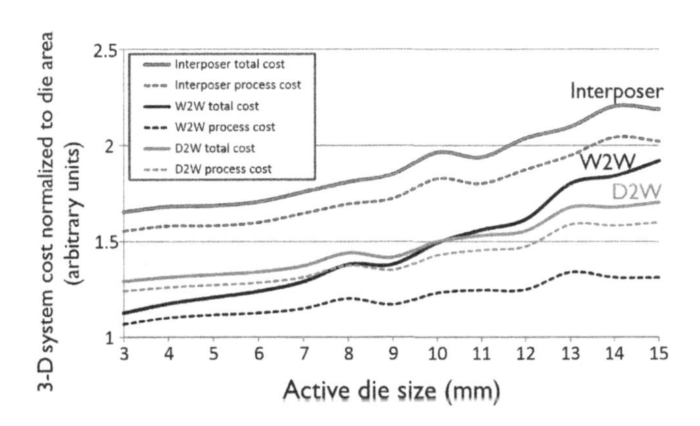

FIGURE 8.24

Process cost and total cost of a 3-D system per die area as a function of active die size. Three different 3-D integration approaches are considered: D2W, W2W, and 2.5-D interposer.

of the 3-D stack is considered, the effect of the compound yield losses is quite significant. Particularly for large active dice, the cost of the compound yield losses becomes comparable to the process cost of the stacked dice. Alternatively, as shown in Fig. 8.24, for small active die, the overall 3-D stack cost per unit area for the W2W approach is lower than the cost of the same system with D2W stacking.

For a 2.5-D interposer-based system, processing the additional component (interposer die) affects the process cost of the system. In the particular case that the interposer process yield is high (99%) and the interposer die is tested prior to stacking (with 100% fault coverage), as shown in Fig. 8.24, the total cost of the 2.5-D interposer system is an almost constant offset higher than the process cost.

8.3.5 VARIATION OF INTERPOSER PROCESS YIELD AND PRESTACK FAULT COVERAGE

Further discussion is presented in this section on the cost of an interposer-based system. The processing yield of an interposer die is varied from 99% per cm^2 to 90% and 80% per cm^2. Furthermore, the fault coverage of interposer die testing prior to stacking is varied from 100% to 50% and 0% (no testing). Active dice with a size of 10 mm × 10 mm are assumed to be stacked on the interposer. The effect of the variable process and test parameters on the cost of a 3-D interposer-based system is illustrated in Fig. 8.25.

As shown in Fig. 8.25, when the processing yield of an interposer is high (99%), the cost of testing the interposer die incrementally adds to the total system cost. It is therefore preferable to not test the interposer and tolerate the small cost due to interposer yield losses. In this case, the

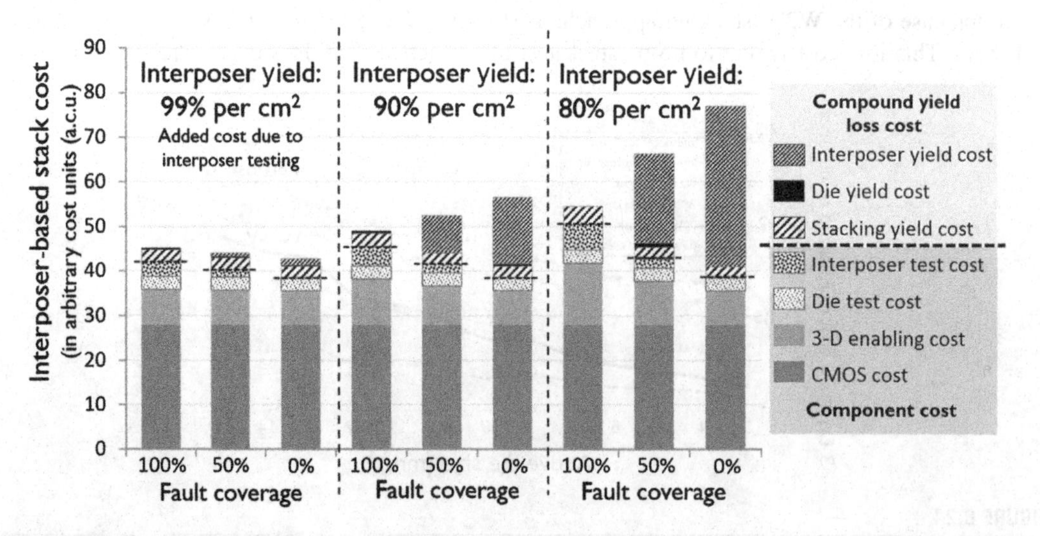

FIGURE 8.25

Effect of interposer processing yield and test fault coverage on the cost of an interposer-based 2.5-D system.

loss of the good active dice stacked on the interposer is lower than the cost to test an interposer die with a processing yield of 99% per cm^2.

As the interposer processing yield drops to 90% and 80% per cm^2, prestack test of the interposer die becomes advantageous. As shown in Fig. 8.25, the cost for testing an interposer is lower than the cost due to losses from nonperfect interposer processing. In these low interposer yield cases, the higher the fault coverage for interposer testing, the lower the cost due to yield loss.

The cost of a 3-D interposer system across a wide parameter space can be described by these experiments. In addition to variations in processing yield and fault coverage, differences in the active die size between 3 mm × 3 mm and 15 mm × 15 mm are considered. The cost of a 2.5-D interposer-based system per unit area is evaluated for each parameter, as illustrated in Fig. 8.26.

The interposer processing yield per cm^2 is varied along different *rows* in Fig. 8.26 with the corresponding yield noted to the left of each row (respectively, 99%, 90%, and 80% yield per cm^2). Similarly, the fault coverage for the interposer dice is varied in different *columns*. The fault coverage for interposer testing is reported at the top of the corresponding column in Fig. 8.26 (respectively, 100%, 50%, and 0% fault coverage).

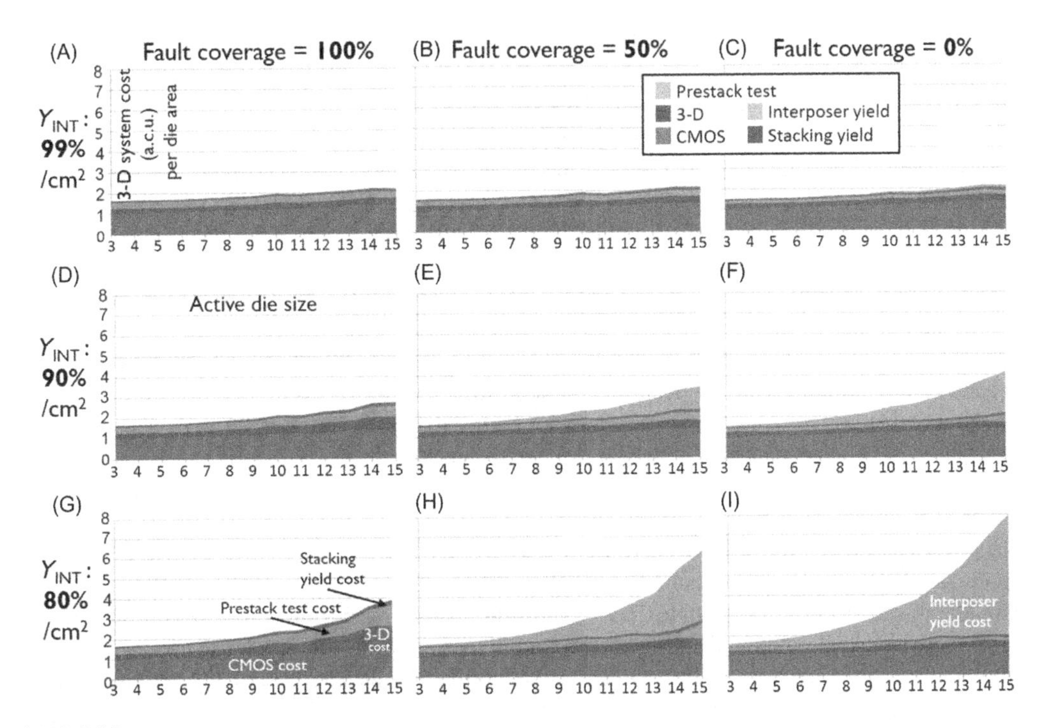

FIGURE 8.26

Cost of 2.5-D interposer-based system in terms of the size of the stacked active die, interposer die processing yield (Y_{INT}), and fault coverage (FC) of interposer prestack testing. (A) Y_{INT} = 99%, FC = 100%, (B) Y_{INT} = 99%, FC = 50%, (C) Y_{INT} = 99%, FC = 0%, (D) Y_{INT} = 90%, FC = 100%, (E) Y_{INT} = 90%, FC = 50%, (F) Y_{INT} = 90%, FC = 0%, (G) Y_{INT} = 80%, FC = 100%, (H) Y_{INT} = 80%, FC = 50%, and (I) Y_{INT} = 80%, FC = 0%.

As shown in Fig. 8.26, as the processing yield of the interposer decreases, the system cost increases. This increase can vary dramatically if no testing of the interposer die occurs, as shown in Fig. 8.26I. Furthermore, the system cost also increases with increasing size of the active die. The active dice become more expensive, requiring large interposers for system integration. Therefore, particularly in the case of large active dice, stacking these dice on an untested interposer substrate with low processing yield can cause a significant increase in the cost of a 3-D system.

8.4 SUMMARY

- The primary cost contributors in TSV processing are CMP, deep Si etch, and barrier and seed deposition. These processing cost components are valid for different TSV geometries, considering both TSV middle and TSV last process flows.
- The TSV middle flow provides a simpler process without requiring protection and in-via etching of the TSV oxide liner.
- For TSV middle flows, the deposited oxide liner should be removed by CMP. This step can significantly increase the cost of the CMP step, particularly for thick oxide liner layers.
- In the case of TSV last processes, an option exists to not remove the oxide liner deposited on the surface of the wafer, simplifying the TSV CMP step.
- Certain processing steps, such as lithography, can be performed less expensively using OSAT grade tools. However, Si CMP at the wafer backside is required before processing the TSVs.
- The oxide liner needs to be opened within the via. These steps increase the processing complexity of a TSV last process.
- Different TSV geometries, such as the TSV depth and aspect ratio, affect the processing cost for similar fabrication flows.
- The TSV size and pitch can be scaled while maintaining reasonable process cost. This situation is achieved by utilizing ALD deposition and depositing thinner layers.
- The processing cost of ALD deposition may be higher than alternative PVD approaches. Deposition of thinner materials, however, lowers the cost of the CMP.
- Reducing the CMP cost for thinner ALD materials introduces a tradeoff that can be exploited to maintain the processing cost of scaled TSVs.
- Stacking active dice on a silicon interposer produces high-density interdie connections without processing TSVs between the dice.
- The interposer features can include TSVs, die-to-die interconnect, power and ground planes, MIM capacitors, microbumps, and backside Cu pillars.
- Combinations of these structures provide varying options for an interposer substrate that satisfy various requirements at different processing costs.
- Three different stacking approaches are considered to compare processing costs at the system level: D2W stacking, W2W stacking, and 2.5-D integration on Si interposers.
- In the case of a D2W process, prestack testing of the stacked components is critical to eliminate bad dice from stacking, limiting yield losses.

- Prestack test is more important when stacking large dice which are more expensive and exhibit lower process yield. In this case, the D2W approach provides the lowest cost for integrating a multidie 3-D system.
- For W2W stacking of active dice, prestack test of the dice is not necessary, saving testing cost.
- W2W stacking produces the lowest *processing* cost per stacked system among the three stacking approaches considered in the comparison.
- The effect of compound yield losses on cost is significant in the case of W2W stacking of large stacked dice, where the processing yield is lower, increasing the likelihood of stacking a bad die above a good die.
- Large active dice are more expensive and the cost of the scrapped W2W stacks is also higher.
- W2W stacking can be beneficial for 3-D systems with relatively small dice. In this case, many stacks can be produced with a W2W process.
- Stacking of small dice increases the probability of good stacks, and the total stack cost (including compound yield losses) is lower.
- In the case of an interposer-based 2.5-D system, the interposer substrate adds to the overall system cost.
- If the yield for interposer processing is high, the additional processing cost of the interposer makes a 2.5-D system more expensive than 3-D stacks of active dice.
- In the case of large interposer substrates, high processing yield and/or good fault coverage of the interposer make 2.5-D integration a viable solution.
- If the interposer process yield is high, prestack testing of the interposer die may increase overall system cost.
- This cost comparison recognizes the multiple benefits provided by interposer-based 2.5-D system integration such as improved thermal characteristics, limited processing of the active dice (no TSVs), flexibility in system components, and simplifying the supply chain.

CHAPTER

PHYSICAL DESIGN TECHNIQUES FOR THREE-DIMENSIONAL ICs

CHAPTER OUTLINE

A variety of recently developed and emerging fabrication processes for three-dimensional (3-D) systems are reviewed in Chapter 2, Manufacturing of Three-Dimensional Packaged Systems, and Chapter 3, Manufacturing Technologies for Three-Dimensional Integrated Circuits. A complete and effective design flow, however, for 3-D ICs has yet to be demonstrated. This predicament is due to the additional complexity and issues that emerge from introducing the third dimension to the integrated circuit design process. Existing techniques for 3-D circuits primarily focus on the back end of the design process. Floorplanning, placement, and routing techniques for 3-D circuits are discussed, respectively, in Sections 9.2, 9.4, and 9.5. Many of these techniques are based on physical design methods for planar circuits. Some of these methods for floorplanning and placement of planar circuits are presented, respectively, in Sections 9.1 and 9.3.

The primary difference in the physical design of 3-D circuits is the management (or planning) of the intertier interconnects (i.e., the through silicon vias (TSVs)), which differ from the intratier interconnects that occupy silicon area. Treating the TSVs the same as the horizontal interconnects results in underperforming systems (both lower speed and higher power). During the past few years, physical design techniques for 3-D circuits have evolved to consider these issues, producing

Three-Dimensional Integrated Circuit Design. DOI: http://dx.doi.org/10.1016/B978-0-12-410501-0.00009-5

increasingly more compact circuits, which effectively translate to shorter wirelength and less area. Furthermore, many of these techniques consider thermal objectives or are primarily focused on mitigating thermal hot spots in 3-D ICs. Although all of these physical design techniques share several characteristics, due to the importance of thermal issues in 3-D integration, a discussion of thermal management techniques is deferred to Chapter 13, Thermal Management Strategies for Three-Dimensional ICs. In addition to physical design techniques, a brief discussion on layout tools for 3-D circuits is presented in Section 9.6. A short summary relating to physical design methodologies for 3-D circuits is included in the last section of this chapter.

9.1 FLOORPLANNING TECHNIQUES

The predominant design objective for floorplanning a two-dimensional (2-D) circuit has traditionally been to achieve the minimum area while interconnecting the circuit blocks with minimum length wires and with no overlap among the circuit blocks [328]. These two objectives are employed in most floorplanning algorithms, which can be classified as either slicing [329] or nonslicing [330,331]. For any circuit that comprises N blocks (where the size of these blocks has evolved over time from a handful of logic gates to thousands of logic gates), each with width w_i and height h_i, area driven techniques target minimum area floorplans that fit the total area A of the blocks, described by

$$A = \sum_{i=1}^{N} w_i h_i. \tag{9.1}$$

Both fixed outline [332–334] and free outline techniques have been developed. For fixed outline methods, the outline can be set with an increase in A by a multiplication factor $(1 + \beta)A$, where β is the percent area of the floorplan not occupied by blocks and is typically called "whitespace."[1]

Although a fixed outline floorplan implies a fixed area, this assumption is typically not the case for the aspect ratio of the outline which can vary within specified limits. Fixed outline algorithms are more suitable for hierarchical floorplanning, where a variable outline is employed for the cells within the individual blocks and a fixed outline is used at the highest level, producing a compact floorplan [332]. Consequently, depending upon the floorplanning technique, the blocks can be either variable or fixed area and can also either have a constant (hard block) or variable (soft block) aspect ratio. The primary difference between a fixed and variable aspect ratio is that soft blocks produce more compact floorplans [330,335,336], but the solution space explodes as each block can assume any aspect ratio [332].

Another important issue in floorplanning is the representation of the blocks. Many representations have been proposed such as the transitive closure graph [335], *O*-trees [336], corner block list [337], and sequence pair (SP) [338]. For the SP technique, at least one floorplan with minimum area exists. Several floorplanning methods for 3-D circuits utilize the SP representation technique [339]. The SP is therefore discussed in greater depth to assist the reader in understanding the main features of this popular technique.

[1]This term is used throughout the chapter to indicate the area not used by circuit cells except for buffers, shielding, decoupling capacitors, and TSVs.

9.1.1 SEQUENCE PAIR TECHNIQUE

The salient feature of the SP technique is the formation of a pair of orderings of blocks that is floorplanned (or packed) within some area outline, as illustrated in Fig. 9.1. To generate a pair of sequences, several lines are formed according to three rules. The lines should not cross (1) the boundary of other modules other than the module for which the line is drawn, (2) previously drawn lines, and (3) the boundary of the outline. Two types of lines are drawn for every block, which are the "positive step line" and the "negative step line." The positive step lines are formed for each block as follows. A line segment begins from the upper right corner of a block (say block b) and moves upwards and to the right in a stepwise manner towards the upper right corner of the circuit boundary, obeying the three rules. Additionally, another line segment begins from the lower left corner of the block and moves downwards and to the left towards the lower left corner of the circuit boundary. These two line segments with the diagonal line through b form the positive step line for block b. In a similar manner the negative step lines are formed, where the line segments for each block begin, respectively, from the upper left and lower right corners of the block. These segments are drawn by moving, respectively, up and left and down and right.

An example of these lines for the circuit blocks shown in Fig. 9.2A is illustrated in Figs. 9.2B and C, where the relevant lines for block d are depicted in this figure with dashed lines. The lines can be linearly ordered, leading to the two orderings for the blocks notated as Γ_+ and Γ_- corresponding, respectively, to the order of the positive and negative step lines. Based on Fig. 9.2, the SP is $\Gamma_+ = \{\text{fegbacd}\}$ and $\Gamma_- = \{\text{gaedbcf}\}$, which also provides useful information for ordering the blocks. Note that each block appears only once in each sequence and relates in four unique ways with the other blocks. For example, assuming two blocks s and s', s can only appear after/before s' in Γ_+/Γ_-. To determine the relative position among the blocks, the four disjoint sets can be described where [338]

$$\mathcal{M}^{aa}(s) = \{s'|s' \text{ is after } s \text{ in both } \Gamma_+ \text{ and } \Gamma_-\}, \tag{9.2}$$

$$\mathcal{M}^{bb}(s) = \{s'|s' \text{ is before } s \text{ in both } \Gamma_+ \text{ and } \Gamma_-\}, \tag{9.3}$$

$$\mathcal{M}^{ab}(s) = \{s'|s' \text{ is after } s \text{ in } \Gamma_+ \text{ and before } s \text{ in } \Gamma_-\}, \tag{9.4}$$

$$\mathcal{M}^{ba}(s) = \{s'|s' \text{ is before } s \text{ in } \Gamma_+ \text{ and after } s \text{ in } \Gamma_-\}. \tag{9.5}$$

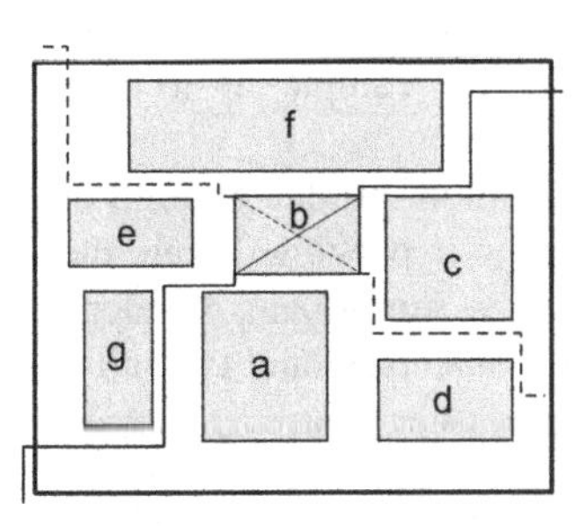

FIGURE 9.1

Example of positive (shown with solid lines) and negative (shown with dashed lines) step lines for block b.

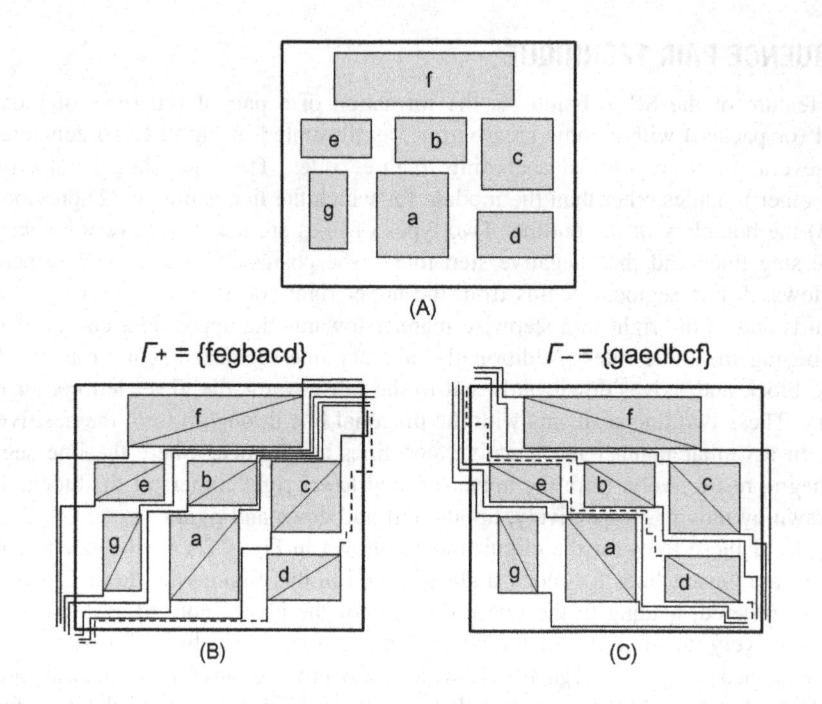

FIGURE 9.2

Example of SP representation, where (A) is a group of blocks comprising a floorplan, (B) positive step lines for these blocks, and (C) negative step lines for the blocks [338].

Based on these definitions, the following theorem has been proven [338]: assuming the SP (Γ_+, Γ_-) produced with the use of positive/negative step lines, the relative order of the blocks can be determined. Consequently, if $s \in \mathcal{M}^{bb}(s')$, s is left of s', if $s \in \mathcal{M}^{aa}(s')$, s is right of s', if $s \in \mathcal{M}^{ba}(s')$, s is above s', and if $s \in \mathcal{M}^{ab}(s')$, s is below s'.

Based on this theorem, SP (Γ_+, Γ_-), and assuming that the circuit area is divided into an $m \times m$ grid, the optimal packing can be reached in $O(m^2)$ by utilizing the longest path algorithm [340]. In this process, horizontal and vertical constraint graphs are employed. Construction of these graphs is commonly used in physical design techniques [328]. More recent methods based on the SP representation have further reduced the complexity to produce floorplans in $O(m \log(m))$ [340] and $O(m \log(\log(m)))$ [341].

In addition to an efficient representation, producing high quality floorplans requires appropriate objective functions, where if the problem is packing driven, the objective is to minimize area, while connectivity driven functions emphasize minimizing wirelength. Many floorplan methods utilize cost functions that target both of these objectives and typically are of the form,

$$\text{cost} = c_1 \times \text{area} + c_2 \times \text{wirelength}, \tag{9.6}$$

where the coefficients c_1 and c_2 indicate the importance of each of the two objectives. Area can be described, for instance, by $(1 + \beta)A$. Irrespective of the formulation of the floorplanning problem, determining the area of the blocks within a floorplan is a straightforward process. Alternatively, at this early stage, wirelength cannot be accurately known. Among different models for calculating the wirelength, the half perimeter wirelength (HPWL) model is often utilized, where the efficiency of this metric for 2-D circuits has been evaluated by many techniques [342]. An example of HPWL is illustrated in Fig. 9.3.

The two objectives in (9.6) indicate a multiobjective optimization process even for the simplest floorplanning cost functions, requiring efficient solvers. Simulated annealing (SA) [343] is a popular solving technique that has been extensively applied to floorplanning and placement problems. The annealing process progresses towards an efficient floorplan by applying different block movements, such as swapping two blocks or rotating a block. Issues with this rather classic approach are that the wirelength is determined after each block movement and the solver only considers the size of the blocks, not the available whitespace. A more compact floorplan may be possible by moving a block to a different location within an outline of the floorplan (assuming this outline is fixed).

To address these issues of increasing computational time, approaches to handle incremental moves of the blocks have recently been explored which consider the whitespace. The notion of slack is borrowed from static timing analysis [344] and a spatial slack is assigned to a block sequence where the SP representation is employed. Similar to critical timing paths, a critical path for a floorplan is a sequence of blocks in the x- or y-direction adjacent and constrained with respect to each other to ensure that changing the site of a block causes neither an overlap or an increase in floorplan area. No whitespace in a specific path direction exists when a block is moved, similar to zero slack for the delay of a critical path. The slack analogy from timing analysis requires that all of the slacks (i.e., whitespace) of paths (i.e., sequences) of blocks must be determined for each direction, which is straightforward if the SP representation is followed. An example of slack computation for two different floorplans is depicted in Fig. 9.4. Subtracting the x coordinate of a block from the results when the blocks are packed, respectively, from left-to-right and from right-to-left, yields the x-slack of the block. Subtracting the y coordinate of this block from results when the blocks are packed, respectively, from bottom-to-top and top-to-bottom, yields the y-slack of the block.

The spatial slack information can lead to more efficient block moves to reduce the floorplan area. Block paths (or sequences) with zero slack are appropriate candidates for moves as these zero slack block paths determine the span of the floorplan along a physical direction. Moving a block

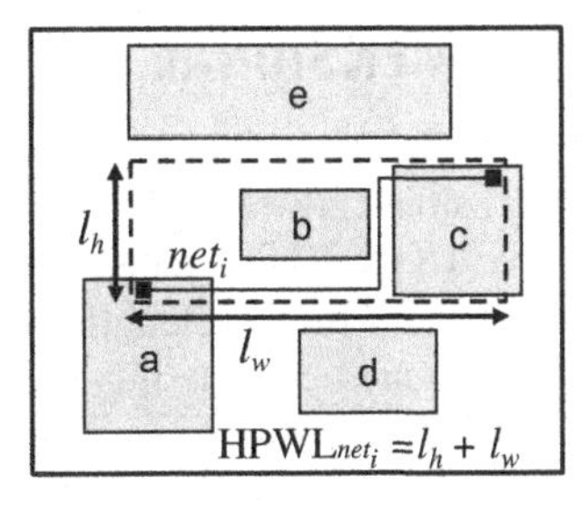

FIGURE 9.3

Example of a net bounding box connecting pins from blocks a and c. The HPWL metric is the half length of the perimeter of the net bounding box. The solid line shows a possible net route to connect pins of blocks a and c marked by the solid squares.

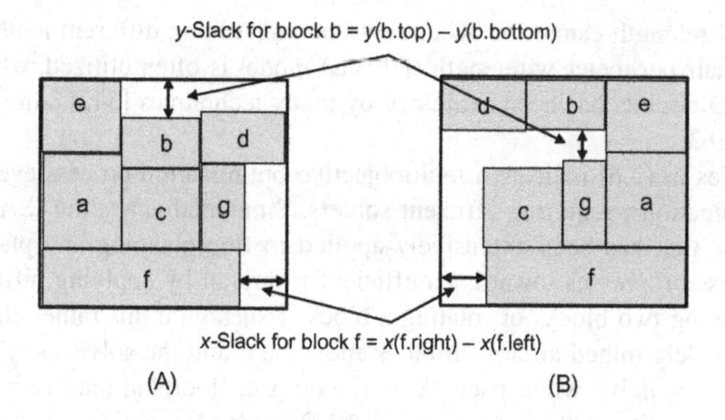

FIGURE 9.4

Example of computing slack where (A) the blocks are floorplanned in left-to-right and top-to-bottom manner, and (B) the blocks are floorplanned in right-to-left and bottom-to-top mode [332].

from a path with high slack along a direction only mildly affects the length of the floorplan along this direction. Therefore, blocks within paths with zero slack along one direction but a large slack in the other direction are good candidates for single block moves. For example, a block with close to zero slack in both directions is a good candidate to be moved to a path with a large slack in both directions [332]. Application of these slack-based criteria for guiding block moves have led to improved results in producing valid floorplans, where a maximum 15% whitespace of the total block area is allowed and the outline is fixed. Considering whitespace in the floorplaning process becomes more important in 3-D circuits as significant area is consumed by the TSVs. Moreover, determining the wirelength for intertier connections where the TSVs are placed within whitespace regions can render the traditional HPWL metric a low accuracy estimate of the actual wirelength. These challenges require a different approach when floorplanning 3-D circuits.

9.2 FLOORPLANNING THREE-DIMENSIONAL ICs

An efficient floorplanning technique for 3-D circuits should adequately handle three important issues: representation of the third dimension, any related increase in the solution space, and the allocation (number and position) of the TSVs. These requirements affect specific aspects of the floorplanning process for 3-D circuits, such as the wirelength metric and the block representation. Techniques that investigate the first two issues considering the increased solution space are discussed in Section 9.2.1. The TSVs in these techniques are either assumed to be integrated within the 3-D circuit blocks (or cuboids) during floorplanning or to be placed anywhere across the entire tier as long as no overlap exists. In Section 9.2.2, floorplanning methods that bound the allocation of TSVs within a 3-D circuit are reviewed. These techniques emphasize the effect of inserting TSVs on several design objectives. These results demonstrate that a planning step for TSV allocation is indispensable when floorplanning 3-D circuits.

9.2.1 FLOORPLANNING THREE-DIMENSIONAL CIRCUITS WITHOUT THROUGH SILICON VIA PLANNING

Certain algorithms incorporate the 3-D nature of the circuits, such as a 3-D transition closure graph (TCG) [345], sequence triple [346], and a 3-D slicing tree [347], where the circuit blocks are notated by a set of 3-D modules that determine the volume of a 3-D system. Utilizing these notations for the circuit blocks, an upper bound for 3-D slicing floorplans is determined [348]. In 3-D slicing floorplans, a tier successively bisects the volume of a 3-D system. The upper bound for 2- and 3-D slicing floorplans is illustrated in Fig. 9.5. The coefficient r shown in Fig. 9.5 is the shape flexibility ratio and denotes the maximum ratio of the dimensions of the modules (i.e., max (width/height, depth/height, width/depth)). This ratio is assumed to be greater than two. The use of this ratio implies that all modules are assumed to be "soft blocks." Moreover, as observed in Fig. 9.5, V_{total} is the sum of the volume of the blocks that comprise the target system, and V_{max} denotes the maximum volume among the blocks. In general, 3-D floorplans result in larger unused space as compared to 2-D slicing floorplans primarily due to the highly uneven volume of the 3-D modules. For high flexibility ratios, however, this gap is considerably reduced and, in certain cases, the upper bound for 3-D floorplans is smaller than in 2-D floorplans.

Although a high flexibility ratio offers more compact floorplans with less unused area, this assumption may not be practical as changing the length of a circuit block along the z-direction (i.e., changing the number of tiers spanned by a block) can greatly affect the number of intertier connections required by this block. This number can affect, in turn, the number of blocks violating the

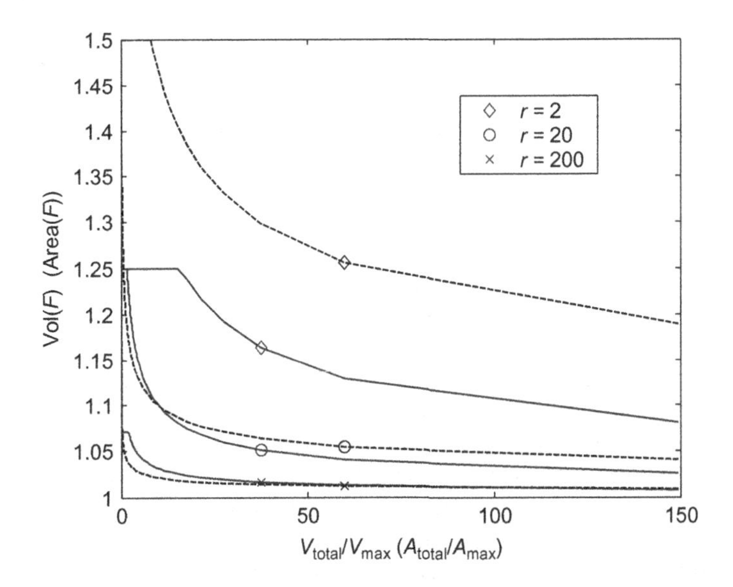

FIGURE 9.5

Upper bound of area and volume for two- and three-dimensional slicing floorplans (F) depicted, respectively, by the solid and dashed curve for different shape aspect ratios. V_{total} (A_{total}) and V_{max}(A_{max}) are, respectively, the total and maximum volume (area) of a 3-D (2-D) system [348].

assumption that the volume of the block remains constant for all values of r. Furthermore, treating blocks as soft (variable r) is not an option for IP blocks developed for 2-D systems. These legacy circuits can often be hard macros where the size of these blocks is unlikely or not permitted to change if, for example, some circuits originate from a third party source. This situation can lead to less compact floorplans with significant whitespace. Although this restriction results in area overhead, the whitespace can, alternatively, be used to insert TSVs for signaling, power and ground distribution, and thermal management, thereby avoiding additional area overhead.

Another computational issue relates to the notation of the dimensions of the blocks as continuous variables. The use of continuous variables is convenient as analytic methods can be used to accurately determine bounds or optimally solve similar problems. The third dimension, however, should be treated as a discrete variable, which can produce suboptimal solutions if continuous analytic methods are applied, since the blocks must be placed into a small number of physical tiers. Consequently, 3-D systems are more efficiently described as an array of two-dimensional planes where the circuit blocks are treated as rectangles rather than cuboids placed on any of the planes constituting the 3-D circuit [349–351]. This approach reduces complexity; despite the number of blocks increasing with the number of stacked tiers. The combinations rise drastically, exacerbating the solution space for floorplanning a 3-D system. This second challenge for 3-D floorplanning, which is to effectively explore the solution space within reasonable time, has led to multistep approaches that often are more efficient for floorplanning 3-D circuits than a single step approach.

In 3-D circuits, a multistep floorplanning technique commences by partitioning the blocks among the tiers of a stack [352]. An illustration of single and multistep approaches is shown in Fig. 9.6. In single step floorplanning algorithms, the floorplanning process proceeds by assigning the blocks to the tiers of a stack followed by simultaneous intratier and intertier block swapping, as depicted in Fig. 9.6. Alternatively, a multistep approach does not allow intertier moves after the partitioning step. The reason for this constraint is that intertier moves among blocks result in a formidable increase in the solution space, greatly affecting the computational time of a single step floorplanning algorithm. Indeed, assuming N blocks witihin a 3-D system consisting of n tiers, a flat floorplanning approach increases the number of candidate solutions by $N^{n-1}/(n-1)!$ times as compared to a 2-D circuit consisting of the same number of blocks. The solution space for floorplanning 2-D circuits based on the TCG technique [335], and 3-D circuits with a single and multistep approach are listed in Table 9.1. Consequently, a multiphase approach can be used to significantly reduce the number of candidate solutions.

The partitioning scheme adopted in the multistep approach plays a crucial role in determining the compactness of a particular floorplan, as intertier moves are not allowed when floorplanning the tiers. Different partitions correspond to different subsets of the solution space which may exclude the optimal solution(s). The objective function for partitioning should therefore be carefully selected. Due to the silicon area of the TSVs, reducing the number of intertier connections is a reasonable objective. This objective typically minimizes the number of TSVs across the entire stack. Thus min-cut algorithms, such as the hMETIS algorithm [353], can be employed to produce 3-D partitions. However, other alternatives exist where the number of TSVs is not an objective for partitioning but rather a constraint. The partitioning can, for instance, be based on minimizing the estimated total wirelength of the system [352]. For example, a partitioning problem based on minimizing the estimated total wirelength can be described as

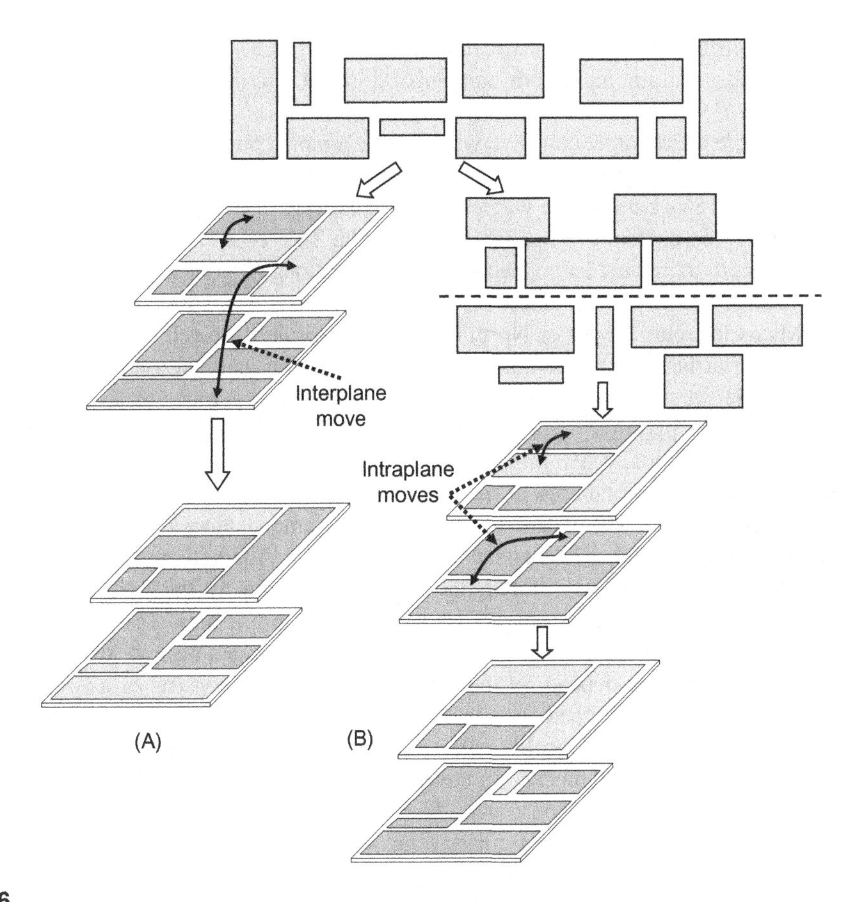

FIGURE 9.6

Floorplanning strategies for 3-D ICs. (A) Single step approach, and (B) multistep approach [354].

Table 9.1 Solution Space for 2-D and 3-D IC Floorplanning [352]

	2-D IC	*n*-Tier 3-D IC	
Characteristic	**TCG**	**TCG 2-D Array**	**Multistep**
Solution space	$(N!)^2$	$N^{n-1}(N!)^2/(n-1)!$	$((N/n)!)^{2k}$
Ratio	1	$N^{n-1}/(n-1)!$	$1/(C_n^{n/k} \cdots C_{n/k}^{n/k})^2$

$$\text{minimize} \sum_{\text{net}} EL_{\text{net}}, \tag{9.7}$$

$$\text{subject to } TN_{\text{via}} \leq TV_{\text{max}}, \tag{9.8}$$

where EL_{net} is the estimated interconnect length connecting two blocks of a 3-D circuit, which can contain both horizontal and vertical interconnect segments. This estimate is not based on the traditional HPWL metric but is probabilistically determined [352]. The total number of intertier vias is denoted by TN_{via}, and the maximum number of allowed intertier vias within a 3-D system is notated by TV_{max}.

The issue is therefore which target objective will produce a more compact floorplan with shorter wirelength. If the cut size is proportionally related to the total wirelength, minimizing the cut size decreases not only the area of the TSVs but also the wirelength. Alternatively, if the dependence of wirelength on the number of TSVs is loose, from (9.7) and (9.8), where the TSV count approaches the upper bound may be a better choice, as more intertier connections will reduce long horizontal wires.

A 3-D floorplanner for two tiers based on [350] has been applied to the benchmark circuits included in the Microelectronic Center of North Carolina/Gigascale Research Center (MCNC/GSRC) benchmark suite [355] to better evaluate this tradeoff. The partition is based on [356] where a fixed cut size is utilized. The cut size is neither the minimum nor the maximum, as these extrema cut sizes limit the flexibility of the partition algorithm [354]. Rather, a fixed cut size lying between these bounds is employed. In addition, a 5% imbalance in area between the two tiers is allowed, and the whitespace is set to 20% of the total area of the blocks. Results of this evaluation indicate that partitioning is not important for circuits where the interconnects among blocks exhibit a narrow wirelength distribution (e.g., the *n*100, *n*200, and *n*300 benchmarks). Alternatively, circuits that include interconnects with a wide distribution of lengths are affected more by the partitioning step (e.g., ami33 and ami49). In addition, results characterizing the relationship between the total wirelength and the number of vertical interconnects across the tiers of the stack demonstrate that the total interconnect length does not strongly depend on the cut size if the circuit consists of a small number of highly unevenly sized blocks. This behavior can be attributed to the significant computational effort required to optimize the area of the floorplan rather than the interconnect length. Alternatively, in circuits composed of uniformly sized blocks, an inverse relationship between the number of vertical vias and the interconnect length is demonstrated. These results are useful indicators when floorplanning a multitier circuit, although the fixed cut size has only been applied to two tier circuits. It is unclear whether the same behavior occurs in stacks with more than two physical tiers.

The partitions are an input to another phase of the multistep process where the floorplan of each tier of a 3-D circuit is generated. Note that the floorplan of each of the tiers is simultaneously produced. In [352], the circuit blocks are represented in three dimensions by the corner block list method [337], while in [354], the SP [338] is used to represent the floorplans. In both of these techniques, SA is employed to produce the floorplans. For single step approaches, where partitions are not available, the starting point for the SA engine is generated by randomly assigning the blocks to the tiers of the system to balance the area of the individual tiers. The SA process progresses by swapping blocks between two tiers (for single step approaches) or changing the location of the blocks within one tier. A candidate solution is therefore perturbed by selecting a tier within the 3-D stack and applying one of the moves described in [337]. The expected wirelength and number of vertical vias are reevaluated after each modification of the partition (which can be a computationally expensive step, as mentioned in the previous section), where the algorithm progresses until a floorplan is obtained at the target low temperature of the SA algorithm.

Application of the technique in [352] (with bounded but not fixed TSVs) to the MCNC and GSRC benchmark suites [355] with a comparison to the TCG-based 2-D array and the combined bucket and 2-D array techniques (CBA) [351] is provided in Table 9.2. A small reduction, on the order of 3%, in the number of vertical vias and a significant reduction of approximately 14% in

Table 9.2 Multistep Floorplanning Results [352]

Benchmark	TCG-Based 2-D Array			CBA			Multistep Floorplanning		
	Area	Wirelength	Vias	Area	Wirelength	Vias	Area	Wirelength	Vias
ami33	3.52E + 05	23,139	106	3.44E + 05	23,475	111	4.16E + 05	21,580	108
ami49	1.49E + 07	453,083	191	1.27E + 07	465,053	203	1.42E + 07	420,636	198
*n*100	53,295	97,066	704	51,736	90,143	752	54,648	74,176	733
*n*200	51,714	198,885	1487	50,055	175,866	1361	55,944	142,196	1358
*n*300	74,712	232,074	1613	75,294	230,175	1568	79,278	213,538	1534
Avg.	1.00	1.17	1.03	0.96	1.14	1.02	1.00	1.00	1.00

wirelength is exhibited, while the total area increases by almost 4% for certain benchmark circuits.

Although the techniques discussed in this section utilize a different metric for wirelength, both the HPWL and the probabilistic wirelength estimate in [352] neglect the increase in wirelength due to the placement of the TSVs and the wires. Including the effect of the TSVs in the floorplanning process requires different metrics and approaches, as discussed in the following subsection.

9.2.2 FLOORPLANNING TECHNIQUES FOR THREE-DIMENSIONAL ICs WITH THROUGH SILICON VIA PLANNING

The complexity of 3-D integration requires different wirelength metrics and advanced cost functions that include several objectives beyond area and wirelength (A/W) to produce efficient floorplans for 3-D circuits. These objectives can consider, for example, the communication throughput among the circuit blocks and/or the number of intertier vias. The effect of utilizing more accurate metrics than the traditional model of the half perimeter of the bounding box of the nets to estimate the length of the intertier nets is discussed in Section 9.2.2.1. Approaches to integrate TSV planning with the floorplanning process are presented in Section 9.2.2.2, while techniques treating the TSV planning as a post-floorplanning step are reviewed in Section 9.2.2.3. Practical issues in inserting TSVs within whitespace also used by other resources are considered in Section 9.2.2.4, where some nonconventional approaches to floorplanning are briefly mentioned.

9.2.2.1 Enhanced wirelength metrics for intertier interconnects

In addition to the cut size or, equivalently, the number of connections between tiers, the processing technique to bond the tiers of a 3-D circuit also affects the partition step in a multistep floorplanning methodology. The various bonding mechanisms employed in a 3-D system contribute in different ways to the final floorplan as these bonding styles support different densities of vertical wires (i.e., TSVs) with dissimilar electrical characteristics.

For example, front-to-front bonding produces a large number of short intertier vias, improving the performance of those modules with a high switching activity. Furthermore, a block with a large area can be divided into two smaller blocks assigned to adjacent tiers and employ front-to-front bonding to minimize the effect of the physical separation on the performance and power consumption. Alternatively, intertier vias utilized in front-to-back bonding can adversely affect the performance of a 3-D system if not used with caution due to the overhead in active area of these interconnections. In addition, TSVs require a different approach for determining the wirelength of a 3-D system. This situation occurs if a TSV is placed outside the bounding box of the net containing the TSV.

To better explain this situation, an example is depicted in Fig. 9.7 where blocks within a three-tier system are illustrated. A net connects the pins $p_{1,1}$, $p_{1,2}$, and $p_{1,3}$, (depicted with solid squares), respectively, of blocks b_1, b_2, and b_3, employing TSVs $v_{1,12}$ and $v_{1,23}$ (depicted with solid circles). In Fig. 9.7A, all of the pins of this net across the stack are projected onto tier 1. The projected pins from tiers 2 and 3 are shown with empty squares. If the classic HPWL is employed to determine the length of this net, this length is $L_{HPWL} = w + h$, as determined by the rectangle drawn with a dashed-dotted line. HPWL, however, does not include the segments of the nets to the TSVs,

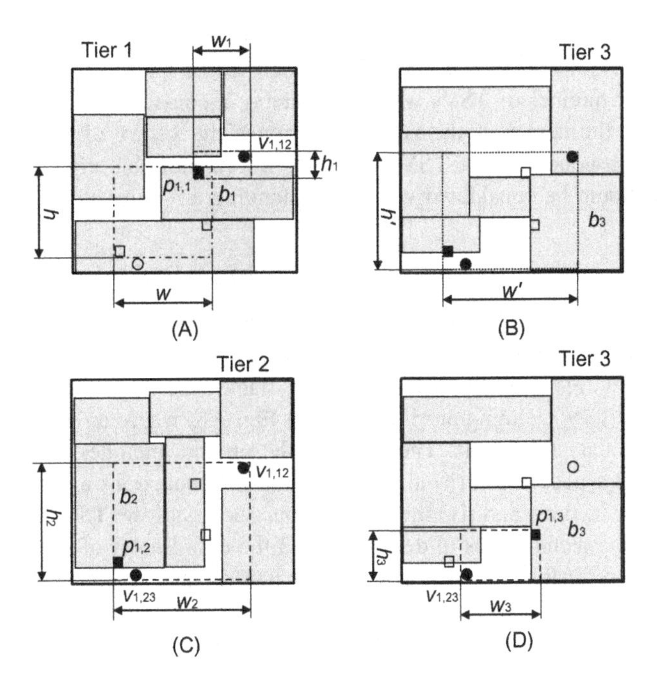

FIGURE 9.7

Different metrics to determine the length of a 3-D net, (A) the classic HPWL metric including only the pins of the net in all tiers, (B) an extended bounding box including the TSV locations, (C) the bounding box of the segments of the net within tier 2, and (D) the bounding box of the segment of the net belonging to tier 3.

thereby, for this specific example, underestimating the length. To address this issue, the bounding box of the net assumed in Fig. 9.7A is extended to include the TSV locations [357]. The new bounding box plotted in Fig. 9.7B with the dotted line results in a new length equal to $L_{\text{HPWL-TSV}} = w' + h'$ which is greater than L_{HPWL}. To determine $L_{\text{HPWL-TSV}}$, all of the pins and TSVs are projected onto a single tier, tier 3 in this specific example. If the location of the TSVs is contained within the dashed-dotted bounding box of the net shown in Fig. 9.7A, $L_{\text{HPWL}} = L_{\text{HPWL-TSV}}$. This situation is however unlikely to occur, in particular if floorplanning is at the block level, where longer interconnects typically occur. Although including the TSV locations when determining the length of the nets produces a better estimate as compared to HPWL, greater accuracy is achieved if the individual segments of the net in each tier are considered. In this approach, the bounding box—typically between a pin and a TSV—is determined for each tier. Consequently, the total length of the net is $L_{\text{HPWL-NET-SEG}} = (w_1 + h_1) + (w_2 + h_2) + (w_3 + h_3)$, where the length of each segment is assumed to be the HPWL of the bounding box of this segment depicted by the dashed lines. Note that, for this example, $L_{\text{HPWL-TSV}} = (w_2 + h_2) < L_{\text{HPWL-NET-SEG}}$.

All of these metrics have been used to estimate wirelength during floorplanning of 3-D circuits. Intuitively, these metrics produce similar results if the floorplanning process generates whitespace close to the pins connected to the intertier wires. The TSVs in this case can be placed within the bounding box determined solely by the pins projected onto a single tier. Furthermore, the whitespace does not include a single TSV but rather several TSVs; therefore, the notion of a TSV island is introduced [357]. Each TSV island is associated with a capacity, which can be adapted to accommodate a greater number of TSVs without incurring a considerable length overhead for certain nets. In this case, the center of the island determines the vertex of the bounding box rather than a single TSV. Consequently, the TSVs can have a nonnegligible effect on the length of the intertier nets, which should be considered when floorplanning a 3-D circuit, as discussed in the following subsection.

9.2.2.2 Simultaneous floorplanning and through silicon via planning

These advanced wirelength metrics are useful to perform accurate TSV planning for 3-D floorplans while producing short wirelengths to achieve a target stack. The steps of a general floorplanning methodology including TSV planning are illustrated in Fig. 9.8, where an SA engine and SP representation produce candidate floorplans. The input of the process includes: (i) a set of blocks with fixed dimensions (fixed aspect ratio), (ii) a netlist indicating connections among the blocks, (iii) the dimensions and number of tiers, and (iv) the physical parameters of the TSVs for the target technology. The objective of the technique is to determine: (1) the coordinates of each block including the tier number, (2) the coordinates and size of the TSV islands, and (3) for every TSV within each tier, the TSV island in which a TSV is allocated. Related constraints are: (1) no block coordinate can exceed the dimensions of the tier (assuming a fixed outline), (2) no block overlap is allowed, and (3) the total area of the TSVs assigned to a TSV island does not exceed the area of this island.

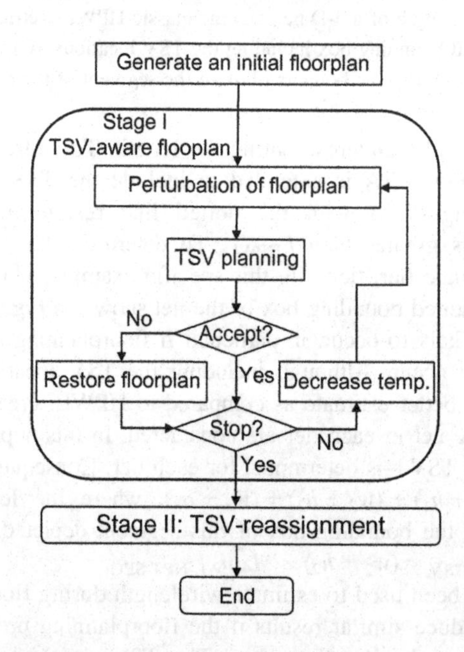

FIGURE 9.8

Flow of two stage floorplanning methods considering the TSV locations [357].

The process, shown in Fig. 9.8, proceeds as follows. A floorplan is randomly produced and successive perturbations are performed to generate low cost floorplans, where the cost function is [357]

$$\text{cost} = \text{area} + c_3 \times \text{wirelength} + c_4 \times AR\_penalty. \quad (9.9)$$

The last term penalizes any change in the aspect ratio of the floorplan. Note that this expression does not differ considerably from (9.6) used in 2-D circuits. The area now includes, however, the area of the TSVs, and the wirelength metric $L_{\text{HPWL-TSV}}$ is considered rather than L_{HPWL}.

The perturbations of a solution are based on the notion of slacks, as described in Section 9.1, where the horizontal and vertical slack for each block is determined. The perturbations include: (1) intertier swapping of a block or a pair of randomly chosen blocks, and (2) moves of blocks between tiers to balance the area among the tiers. The blocks in congested tiers are given a higher probability to be relocated. (3) The spatial slack information is used to perform a block move, where blocks of opposite slack are placed close to each other.

Based on the floorplan produced after applying these perturbations, the cost function shown in (9.9) determines whether the most recent iteration of the floorplan is accepted. If accepted, the TSV planning step commences. This step progresses by assigning in a greedy way a TSV into a TSV island. As the annealing temperature cools, a more detailed TSV assignment follows. During the greedy assignment process for each intertier net, the bounding box (see Fig. 9.7B) can contain several whitespaces, which are used as TSV islands. If several of these TSV islands for a tier exist, a TSV is placed within any of these islands with equal probability, yielding no increase in wirelength. This practice may, however, lead to some islands with more TSVs than is physically possible. The area of these islands is expanded to fit the excessive number of TSVs. The resulting overlap between the TSV islands and surrounding blocks is resolved during the finer assignment step. The reason for allowing the overlap is that no increase in wirelength occurs if as many TSVs as possible are kept within those islands contained within the bounding box of a net. The increase in area, however, should not outweigh the savings in wirelength.

Consequently, the stage of TSV refinement uses two pieces of information to decide how best to allocate the TSVs into an island. The TSVs are sorted according to the total area of the candidate TSV island(s) where a TSV is placed. Candidate TSV islands are contained within the bounding box of a net, as illustrated in Fig. 9.9, where some noncandidate TSV islands for a specific net are also shown. The smaller the area of the island, the higher the priority for a TSV. As this island can be more quickly filled, the TSV may be placed within another TSV island, potentially beyond the bounding box. To avoid these expensive placements, if the candidate island for a TSV has been filled, allocation of this TSV is deferred to a later step in the process.

At the end of this stage, some TSVs may not have been assigned to any of the candidate TSV islands. Two options can be followed, either expanding the area of a candidate TSV island or assigning one TSV to a noncandidate TSV island (i.e., a TSV island not inscribed within the bounding box of the net containing the TSV). In both cases, the wirelength relates to the nets crossing the expanded TSV island or the additional wire segment needed to reach the noncandidate island. A linear cost function to consider the increase in wirelength caused by either option is used in [357]. The option with the lower cost is selected, leading to an iteration to produce a floorplan if an expansion is chosen. If the additional wire connecting a pin to a TSV island outside the bounding box of the net is longer than the increase in wirelength due to the expansion of a candidate

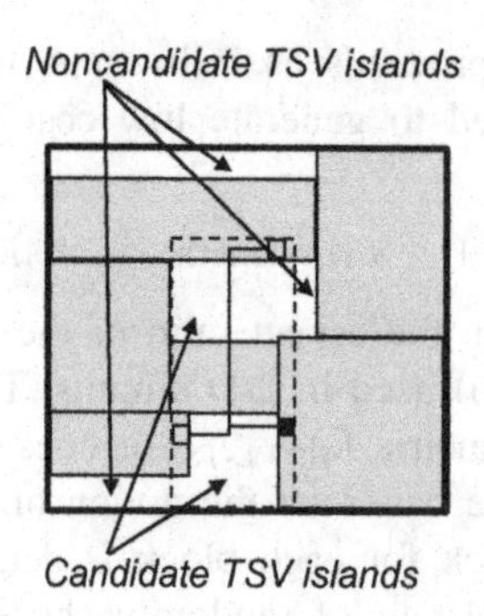

FIGURE 9.9

Whitespace within the bounding box of the intertier net can be used for placing a TSV without increasing the wirelength. This whitespace defines the *candidate TSV islands*. The whitespace outside the bounding box describes *noncandidate TSV islands*, as placing a TSV into these regions increases the wirelength.

Table 9.3 Flooplan With and Without TSV Planning [357]

# of Tiers	Circuit	TSV Aware				TSV Unaware			
		Success Rate	Avg WL	TSVs	Time (s)	Success Rate	Avg WL	TSVs	Time (s)
3	*n*100	100%	160,825	833.2	1195.91	80%	157,480	888.8	22.68
3	*n*200	100%	310,924	1509.1	7720.45	0%	339,768	1689.5	87.38
3	*n*300	100%	424,585	1899.7	21155.10	0%	440,954	2019.3	159.02
4	*n*100	100%	148,748	1171.4	1306.39	90%	165,940	1290.5	23.38
4	*n*200	100%	291,091	2179.0	8237.10	0%	367,602	2431.5	94.45
4	*n*300	100%	391,694	2730.6	21450.50	10%	448,905	2865.0	234.12

TSV island, the latter option is chosen. The computational cost required to generate a new floorplan should however be considered. In addition, this choice implicitly favors wirelength reduction over area minimization, which may increase the difficulty of the SA algorithm to produce a high quality floorplan, particularly if greater emphasis is placed on the area or aspect ratio term in (9.9).

A limitation of this flow is that the TSVs are treated individually, potentially leading to suboptimal results. To improve the quality of the solution, another stage of TSV reassignment, notated as stage II in Fig. 9.8, is added to the flow, where the main task is to reassign TSVs among the TSV islands. The area and number are fixed to further reduce the wirelength. The task is formulated as a minimum cost maximum flow problem which can be optimally solved in polynomial time [358].

The technique described in [357] has been applied to benchmark circuits used in physical design problems, such as the *ami*, *n*100, *n*200, and *n*300 benchmark circuits, and is compared with a TSV unaware floorplanning method where the length of the nets is L_{HPWL}. In addition, the location of the TSVs is considered after a floorplan is generated and not during the floorplanning process. These results, listed in Table 9.3, demonstrate improved wirelength while satisfying a fixed outline constraint with an increase in computational time (attributed to the TSV planning steps within the floorplanning process).

9.2.2.3 Through silicon via planning as a post-floorplanning step

A different path to plan and insert TSVs can also be followed where the size of a TSV island is insufficient to fit the assigned TSVs. In the previous subsection, this situation is handled by increasing the size of the whitespace or allocating a TSV to an island located farther away. The increase in size of the TSV islands is considered in the next floorplanning iteration. Alternatively, this increase can be achieved by avoiding a floorplanning iteration if area is borrowed from other whitespace regions. Although this borrowing shifts blocks, this process can be performed by shifting only specific blocks in the vicinity of this island rather than floorplanning the entire circuit. In this way, any fixed outline constraint is easily satisfied. To perform these block shifts, however, information describing the slack of all of the blocks of the circuit should be available. The technique, discussed in [339], utilizes this information to treat TSV planning as a post-floorplanning step.

A fixed outline 3-D floorplan with no overlaps among blocks is the primary input of this technique, where the coordinates of each block are x_b and y_b, and n_b is the physical tier in which the block is placed. Other inputs include the number of TSV islands, where each island has a specific capacity and dimensions ($h_{\text{TSV\_island}}$, $w_{\text{TSV\_island}}$), and a netlist describing the connections among the blocks. As $L_{\text{HPWL-NET-SEG}}$ provides an improved estimate of the wirelength, this metric is employed to determine the total wiring of the circuit. Although $L_{\text{HPWL-NET-SEG}}$ is a better estimate, $L_{\text{HPWL-TSV}}$ is faster to determine and is, therefore, used if the granularity of the floorplanning/placement is at the gate level rather than at the block level. Employing TSV islands at the gate level results in unacceptably long wires since the TSV should be placed adjacent to the connecting cell. Additionally, at the block level, providing capacity for each TSV island is not a straightforward task and can affect the quality of the overall floorplan. The decision for the TSV island capacity can be based either on user experience or on a probabilistic allocation of capacities similar to the assignment process of inserting TSVs into islands [357]; however, no evidence exists as to whether these input parameters affect the resulting floorplan.

The approach of [339] clusters intertier connections to TSV islands, and assigns each of these clusters to a whitespace. The clustering step employs the notion of a "virtual die," which is depicted in Fig. 9.10. The bounding box of the nets is projected onto the virtual die, and the intersection of several boxes forms a cluster with the respective nets. The capacity of the available TSV islands determines the number of clustered nets or, alternatively, the number of bounding boxes projected onto the virtual die of a TSV island.

An *intersection* graph is defined to determine the intersection of the overlapping bounding boxes, where the vertices correspond to bounding boxes while overlaps among blocks are expressed by edges. To determine the common region, a number of cliques are determined where the size of each clique satisfies the capacity of the target TSV islands. This step is NP-complete [359].

With the area of the whitespace across each tier known, the TSV islands are assigned to these areas where some whitespace may not be used for a TSV island or can be shared by more than one TSV island. Additionally, a net can include more than one TSV contained within several TSV islands. As previously discussed, the TSV assignment process can be achieved probabilistically or

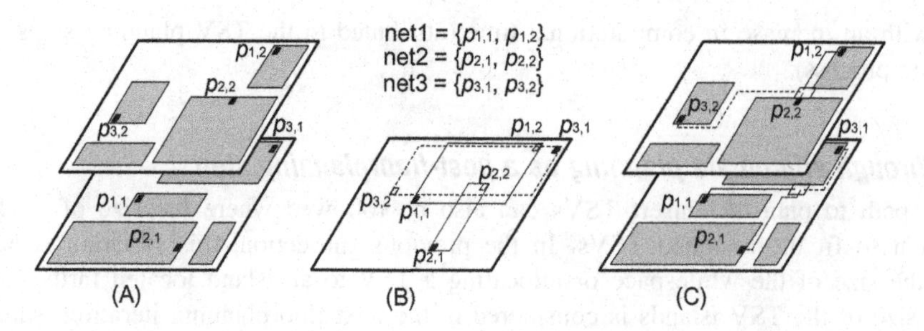

FIGURE 9.10

A two tier floorplan with three intertier connected nets, (A) the blocks and pins, (B) the virtual die with the projection of the bounding box of each net, and (C) the routed nets and corresponding TSV island are shown. The notation $p_{i,j}$ is the pin of net i in tier j. The pins connected by each net are also indicated in the figure.

with some dynamic metric which considers the decreasing available whitespace as the TSV assignment process proceeds. An example of this metric is

$$D(c) = c.whitespace \div |c.assigned\_nets|, \tag{9.10}$$

which considers the available whitespace of a cluster as compared to the number of assigned nets within a cluster. Although clusters with high scores are prioritized to reduce the number of unassigned nets, by the end of the process, nets can exist which have not been assigned to a cluster. Failing to insert a TSV island means that greater whitespace is necessary, which can be achieved through two different schemes [339]. Additional whitespace can be made available by adding channels of whitespace between blocks to accommodate the TSVs as well as buffers and local logic. This straightforward method increases the overall area of the floorplan. Alternatively, the available whitespace can be redistributed to provide the required area within each tier.

Greater whitespace can be achieved by shifting the blocks, which eliminates unnecessary use of whitespace (for example, outside the common region of the projected bounding boxes of the nets, see Fig. 9.10B), thereby increasing the available area for other islands. Block shifting is performed in two ways, producing successful TSV assignments, where (1) the blocks are shifted at the beginning of the TSV planning process, and (2) the blocks are successively shifted during the TSV island insertion step. To determine the available shifting opportunities, the notion of spatial slack is used as previously discussed. The use of slack allows the fragmented and otherwise unsuitable whitespace to be consolidated among blocks in the x and y directions without increasing the area, leading to a more compact floorplan.

Experiments on standard benchmark circuits have shown that initial shifting performs better than iterative shifting in terms of the total wirelength, although the latter shifting is performed dynamically. A rationale for this preference is that during initial shifting none of the TSV islands has been assigned and full slacks can be exploited, albeit only once. During iterative shifting, the gradual insertion of TSV islands quickly dissipates the available slack, decreasing the likelihood of redistributing whitespace to provide additional space for the remaining TSV islands. Alternatively, channel insertion is achieved by first proportionally inflating the dimensions of the blocks and then contracting the blocks to the original dimensions after a floorplan is produced. The increased whitespace can, in this case, efficiently fit the TSV islands; yet an approximately 10% increase in the total area of the circuit is incurred. Moreover, the total wirelength is shorter due to iterative block shifting but longer than from the initial block shifting.

Note that all of the techniques discussed in this section assign a group of TSVs to some empty space but do not assign individual TSVs within this space (this problem is considered in Chapter 10, Timing Optimization for Two-Terminal Interconnects, and Chapter 11, Timing Optimization for Multiterminal Interconnects, where early timing driven techniques for TSV placement are presented). The coordinates at the center of the TSV islands are used to provide wirelength estimates, resulting in pessimistic estimates and potentially producing a suboptimal placement of TSVs.

Although enhanced wirelength metrics and TSV islands improve the quality of the floorplan, inserting TSV islands can still fail despite increasing the floorplan. This limitation can be alleviated if an alternative approach is followed, where the pins of the block, the bounding box of the nets, are moved along the boundary of the blocks to provide greater flexibility in allocating TSVs within each tier. Alternatively, pin assignment for each block is another degree of freedom to reduce wirelength. Moving pins leads to disparate bounding boxes that contain more or larger capacity TSV islands. This approach is also applied as a post-floorplanning technique to minimize wirelength, where the solution to the optimization problem places both the TSVs of the intertier net as well as the pins of each block [360]. The technique adapts the problem to ensure that a minimum cost maximum flow algorithm is applied. A solver with a time complexity of $O(|V|^2|E| \log(C|V|))$, where V and E denote the vertices and edges of the related graph constructed for the target problem and C is the maximum cost of the arcs within the graph [358]. The technique optimally solves the case where a block is connected to several blocks through single and/or multiple fan-out nets. The algorithm determines the position of the pins and TSVs to minimize the wirelength of the nets. Alternatively, for the algorithm to be optimally solved, the multipin nets can be decomposed into several two-pin nets. This task requires the use of additional source terminals. The technique proceeds by replicating pins for the specific net at the source block. Once the minimum cost maximum flow algorithm terminates, the replicated pins are mapped back into the original pin, which can produce a nonoptimal floorplan. Furthermore, in practical systems, blocks are connected in many different ways. Many blocks behave both as a source and destination for different nets. A straightforward procedure to cope with this issue is to successively apply the single block source—multiple block destinations problem to all blocks within a system where the blocks are randomly chosen. The optimality of this process is however not guaranteed. In addition, HPWL is used as the wirelength metric which performs poorly in 3-D circuits. Consequently, although the technique reduces the wirelength by simultaneously manipulating pins and TSVs, additional space for decreasing the wirelength is possible through enhanced TSV planning.

9.2.2.4 Practical considerations for floorplanning with through silicon via planning

The use of whitespace has to date been limited to the important resource of TSVs. Unoccupied regions in 2-D circuits have, however, traditionally been used for repeaters and decoupling capacitors, particularly in the case where the circuit is comprised of hard macros. Consequently, in 3-D circuits, repeaters and TSVs share the same whitespace. Furthermore, TSV planning techniques which only focus on assigning TSVs to the available whitespace do not consider the effect that this assignment can have on the non-optimal placement of repeaters for intertier nets. Since these nets tend to be the longest interconnect in a multitier system, this situation can degrade system performance. To address this problem, repeaters and TSVs should be simultaneously placed to better utilize the available whitespace.

Employing the concept of (independent) feasible regions [361,362], the largest polygon should be placed to satisfy local timing constraints. Simultaneous buffer and TSV insertion improve timing closure for these intertier nets [363]. If this feasible region overlaps the whitespace, the intersection between these two areas represents a valid location for these buffers. The intertier nets, however, require a different treatment as the feasible regions should be determined for each tier before placing buffers in those tiers. Consequently, a two-step process is followed where the feasible regions for each tier spanned by the target net are determined, while simultaneously, TSVs are inserted within these regions.

The second step is allocating buffers and TSVs to each tier. To better describe this process, an example of an intertier net is illustrated in Fig. 9.11, where the pins of the net are in tiers one and three, and k buffers are assumed to be needed. The feasible region for the second tier is shown in Fig. 9.11. The feasible regions within each tier are similarly determined. Since all of these regions are candidates for buffers driving the same net, a buffer connection graph is employed to link these regions together. An example of this graph is shown in Fig. 9.11B for the net depicted in Fig. 9.11A. The rows in this graph represent the tiers spanned by the net, and the columns represent the number of inserted buffers ($k = 3$ in this example). Thus, the vertex indicated by (row_i, $column_j$) means that the jth buffer can be placed in tier i. A vertex does not exist in (row_1, $column_3$), which means that the third buffer cannot be placed in the feasible region of that tier. The edges of the graph indicate whether a TSV can be placed within the feasible regions of the tiers connected by this TSV. The projections of the two feasible regions from two adjacent tiers intersect, resulting in a nonempty (available) area for placing a TSV. In this example, edge e means that a TSV can be

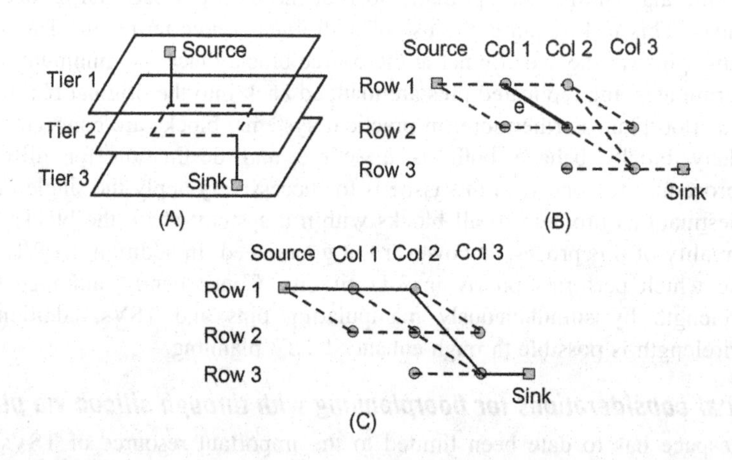

FIGURE 9.11

A three tier circuit, (A) the independent feasible region for a two pin net starting from tier 1 and terminating in tier 3 is shown by the dashed rectangle, (B) the allowed row (intertier) and column (intratier) connections are depicted with dashed lines, and (C) a potential route for this net is shown by the solid line. The dots illustrate available locations for buffers in each row (tier) [363].

placed between tiers 1 and 2 connecting the buffers in the corresponding feasible regions while satisfying existing timing constraints. The edges among vertices in the same row simply indicate that routes between buffers of the same tier are possible.

With this graph describing the connections of the buffers and TSVs, the location of these items need to be determined, preferably in a form of a path connecting the pins of a net. An example path is illustrated in Fig. 9.11C, where other paths are also shown. The choice of the path is supplemented by other objectives in addition to timing and wirelength, such as congestion, by adapting the weight of the edges. The path with the lowest cost is chosen for each net.

The complexity of this technique is $O(mn^2M_{\text{buf}})$, where m is the number of nets, n is the number of tiers, and M_{buf} is the greatest number of buffers required for a net among the P interconnections. Considering that the complexity of the method is linear with the number of nets and that the number of buffers and tiers are low, buffer and TSV planning can be integrated with floorplanning. Similar to other techniques, whitespace redistribution offers a means to further improve the efficiency of this technique. A heuristic is employed [357] based on the constraint graphs along the two physical directions x and y, which provide the spatial slack of each block as in [339]. This heuristic, however, does not shift any of the blocks during the TSV allocation process but rather assigns a TSV to a region, expanding the area of the whitespace during subsequent floorplanning iterations.

The blocks can be shifted only after all of the TSVs and buffers have been allocated. Breadth first traversals along the horizontal and vertical directions within each tier are applied to expand the feasible regions based on spatial slack information. If the greatest possible shift of blocks does not enable the assignment of all TSVs, the floorplan is perturbed, initiating a new iteration. SA is utilized for floorplanning but the typical linear cost function is extended and annealing proceeds in three different stages, where a different cost function is utilized to emphasize a different objective. Thus, the first stage emphasizes area as the modules are initially far apart (high temperatures of the annealing process). As the compactness of the floorplan increases, timing is also added to the cost function, as described by

$$\text{cost} = \text{area} + c_5 \times \text{wirelength} + c_6 \times \text{timingviolation}, \tag{9.11}$$

where the timing violations are included in the cost function. If during this phase, the temperature slowly decreases, another term is added to the cost function (which constitutes the third stage) which satisfies the timing of those intertier wires with unassigned buffers and TSVs. Note that the same perturbations as in [352] are applied, where intertier moves for blocks are disallowed.

The performance of the simultaneous treatment of buffers and TSVs has been explored on two MCNC and other synthetic benchmark circuits. Some of these results are reported in Table 9.4, where these benchmark circuits are floorplanned in a single tier without (F2D/NWR) and with whitespace redistribution (F2D/WR) using the SA approach from [356]. These results are compared to the 3-D floorplanner without (F3D/NWR) and with (F3D/WR) whitespace redistribution where a four tier system is assumed. A comparison of the A/W for these scenarios shows that F3D/WR performs best where redistributing the whitespace slightly improves the floorplans (4.8% and 2.4% for, respectively, wirelength and area). The disadvantage of whitespace redistribution is, however, a higher number of buffers and TSVs, as listed in, respectively, columns 9 and 10 to improve the timing of the circuit. Therefore, additional nets can satisfy the timing requirements, as listed in columns 11 and 12 of Table 9.4.

Table 9.4 Comparison of 2-D and 3-D Floorplans With and Without Whitespace Redistribution for Simultaneous Buffer and TSV Planning [363]

Circuit	Area (mm^2)	Wirelength (mm)	B/#B	Net Timing Met	Net Timing Failed	Area (mm^2)	Wirelength (mm)	TSV/ #TSV	B/#B	Net Timing Met	Net Timing Failed
	F2D/NWR					F3D/NWR					
ami33	1359.16	5922.63	52.30%	241	70	1317.72	3937.08	76.29%	58.48%	279	84
ami49	1373.10	8921.08	79.73%	353	156	1095.68	5781.72	79.37%	68.16%	381	164
	F2D/WR					F3D/WR					
ami33	1251.02	5745.40	33.05%	213	43	1350.52	3519.18	92.17%	86.04%	317	46
ami49	1067.57	9663.90	68.35%	330	188	1068.64	5185.88	95.85%	94.33%	432	113

9.2.2.5 Microarchitecture aware three-dimensional floorpanning

Before closing this discussion on the various aspects of 3-D floorplanning, it is worth noting that disruptive approaches for floorplanning 3-D circuits have been developed, where the length of the nets is weighted differently. Thus, a communication-based objective can utilize information from the microarchitectural level, resulting in floorplans with a higher number of instructions per cycle (IPC) [364]. In a 3-D system, blocks that communicate frequently can be assigned to adjacent tiers, decreasing the interconnect length of the interblock connections. The communication throughput is also increased while reducing the power consumed by the system. Alternatively, blocks with high switching activities should not overlap in the vertical direction to ensure that the temperature profile of the system remains within specified limits. Consequently, the communication throughput is carefully balanced with operating temperature.

A multi-objective floorplanning approach targeting microprocessor architectures is illustrated in Fig. 9.12, where a variety of tools characterize different parameters of the functional blocks within a processor. The CACTI [365] and GENESYS [366] tools provide an estimate of the speed, power, and area of the processor. The SimpleScalar simulator [367] combined with the Watch [368] framework records the information exchanged across the system to predict the power consumption of each benchmark circuit. A hierarchical approach is utilized where the SA engine is replaced by a slicing algorithm based on recursive bipartitioning [369]. This algorithm distributes the functional blocks of the processor onto the tiers of the 3-D stack to decrease computational time.

The additional objectives include area and wirelength (A/W), area and performance (A/P), area and temperature (A/T), and area, performance, and temperature (A/P/T). Based on evaluating

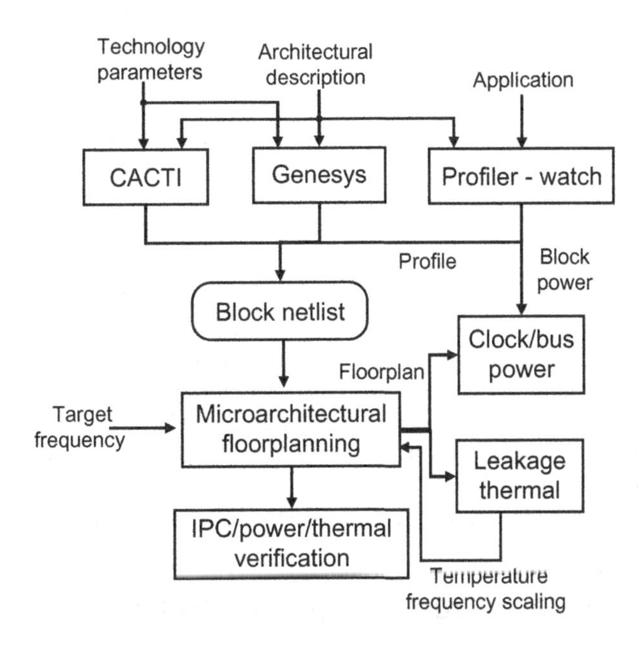

FIGURE 9.12

Design flow of microarchitectural floorplanning process for 3-D microprocessors [364].

MCNC/GSRC benchmark circuits, A/W achieves the minimum area as compared to the other objectives, decreasing by almost 40% the interconnect length as compared to a 2-D floorplan of the same microarchitecture. A/P increases the IPC by 18% over A/W, while simultaneously increasing the temperature by 19%. The more complex objective A/P/T generates a temperature close to A/W, while the IPC increases by 14%. In general, the performance generated by the A/P/T objectives is bounded by the performance provided by the A/T and A/P objectives. In addition, A/P/T achieves higher performance as compared to A/W with a similar temperature [364].

9.3 PLACEMENT TECHNIQUES

Placement algorithms traditionally target minimizing the overall area of a circuit and the interconnect length among the cells, while reserving space for routing the interconnect. A brief discussion of the different approaches used for placing 2-D circuits is offered in this section, where specific techniques are described in greater detail to provide the necessary background for applying 2-D placement techniques to 3-D circuits.

As with floorplanning, SA is also applied to placement problems [370], but the large number of cells in the placement step as compared to the fewer number of blocks during floorplanning can result in excessive computational time [371]. Other placement methods are based on partitioning, where the netlist and area of a circuit are successively partitioned, minimizing the number of connections between juxtaposed partitions during each partitioning step. Some placers rely on partitioning methods including, for example, Capo [372] and Fengshui [373].

In addition to stochastic and min-cut placement methods, another category includes analytic-based placers where an appropriate cost function is optimized for a specific single objective or multiple objectives through a combination of diverse optimization methods. The cost functions can be, in general, nonlinear and quadratic (as compared to typical linear cost functions assumed in SA-based techniques). Examples of nonlinear placers include NTUplace [374] and mPL [375].

In quadratic placement, the wirelength is described as a quadratic cost function which can be optimized effectively through a system of linear equations. Quadratic placement is based on several methods, such as partitioning [376], force directed [377,378], and warping [379]. In partition-based methods, the cost function is optimized at each level of the partition to place the blocks and/or cells with minimum wirelength [376]. Alternatively, in warping, the outline of the circuit is changed to indirectly move the circuit blocks during placement [379].

Alternatively, the force directed method applies diverse forces to bring the modules closer or to spread them apart, depending upon the intended objective [380]. Several placers based on the force directed method have been developed for 2-D circuits and exhibit useful results. This method has been employed for placement within multitier systems, where objectives unique to these systems are considered. Before discussing these extensions to 3-D circuits, the basic characteristics of the force directed method are reviewed in the following subsection, providing background for this method.

9.3.1 PLACEMENT USING THE FORCE DIRECTED METHOD

The force directed method is utilized in several placement algorithms due to the low computational time to produce a legal placement. Attractive or repelling forces are applied to each of the components (cells or blocks comprised of many cells) being placed. These components are successively moved within a specified area until the forces cancel each other.

Using an analogy from physics, the components can be thought of as being connected through elastic springs exerting forces on these components. A placement is produced when this system of elastic springs reaches a state of minimum energy. Since the derivative of energy is force, this state of minimum energy is achieved when the position of the components ensures that the sum of forces among the components is zero.

To determine the position(s) of equilibrium (or minimum energy) for the components comprising a circuit, a system of linear equations is solved. A large variety of analytic and numerical methods exists to solve this system, each with different computational efficiencies. The quality of the placements, however, depends primarily on the accuracy of the model of the forces characterizing the different properties or, equivalently, placement objectives of the circuit.

The primary circuit parameter modeled as a force is the interconnect length between components, where a force corresponds to a two pin connection between a pair of components. A wirelength model is used, as with floorplanning, since the length is not known at the time of placement. The choice of model is crucial, affecting both the quality of the placement and speed of convergence. Based on these observations, the main steps of force directed placement methods are described in this section, starting from a basic (and early) formulation towards more complex expressions that incorporate more forces in addition to a wirelength driven force.

Application of the force directed method to the placement problem begins by considering a number of components N comprising a circuit. This set can be further distinguished into movable N_m and fixed N_f location components. Fixed components can, for example, refer to hard macros, where the positon on the $x-y$ plane is constrained due to timing or I/O requirements. The position of each component i is described by the coordinates at the center of this component (x_i, y_i) where another subscript (m or f) is used whenever necessary to indicate whether the component i is movable or not. With this notation, the position of all of the components and placement is described by a $2N$-dimensional vector,

$$\vec{p} = (x_1, \ldots, x_i, \ldots, x_N, y_1, \ldots y_i, \ldots, y_N)^T. \tag{9.12}$$

A quadratic cost function to minimize wirelength is a primary objective for placement. The Euclidean distance among pairs of blocks is a suitable metric to describe the length of the connections among the blocks. Note that this wirelength model is quite different from the Manhattan distance used to describe routes in integrated circuits and the net models discussed in Section 9.1 used in floorplanning. Based on the model of a Euclidean distance, the overall cost of a placement in matrix notation is [371]

$$B = \frac{1}{2}\vec{\mathbf{p}}^{\mathbf{T}}\mathbf{C}\vec{\mathbf{p}} + \vec{\mathbf{d}}^{\mathbf{T}}\vec{\mathbf{p}} + const, \tag{9.13}$$

where $\mathbf{C}$ is a $2N \times 2N$ symmetric matrix and $\vec{d}$ is a $2N$-dimensional vector. The derivative of (9.13) is

$$\nabla B = \mathbf{C}\vec{\mathbf{p}} + \dot{\mathbf{d}}, \tag{9.14}$$

yielding a system of linear equations. Setting (9.14) equal to zero and solving for $\vec{p}$ provides the position of the components that minimizes (9.13). This system of equations is only true if B is convex, which is ensured if $\mathbf{C}$ is positive definite or semidefinite. This property of $\mathbf{C}$ applies to systems that include both only movable [381] and a mixture of movable and fixed location components [382]. In addition, the cost function of (9.13) is separable into x- and y-directions, permitting each direction to be treated independently.

The elements of matrix $\mathbf{C}$ and vector $\vec{d}$ are formed by considering the Euclidean distance (or any other appropriate (quadratic) wirelength model) between pairs of movable components and pairs of one movable and one fixed component. These distances are described by $(x_i-x_j)^2+(y_i-y_j)^2$ for any pair of movable components i and j. In the case of fixed components, this expression is adapted to reflect this situation, $(x_i-x_{jf})^2+(y_i-y_{jf})^2$, where the subscript f denotes that the j component cannot be displaced. As function B is separable, each direction can be independently solved.

In the following discussion, the x-direction is analyzed by considering the matrix $\mathbf{C_x}$ and vector $\vec{d}_x$. A similar treatment applies to the y-direction. Consequently, expanding the squared terms for the x coordinate $(x_i-x_j)^2=x_i^2-2x_ix_j+x_j^2$, the first and third terms contribute to the diagonal elements of matrix $\mathbf{C_x}$ at, respectively, rows i and j. The second term results in negative entries in matrix $\mathbf{C_x}$ at, respectively, rows and columns i and j, and j and i. Vector $\vec{d}_x$ is formed by expanding the square of the differences between a movable and a fixed component (x_{if}, y_{if}), while the term x_{jf}^2 contributes to the constant term in (9.13). The x-direction of (9.13) is $B_x=(1/2)\mathbf{x^TCx}+\mathbf{x^Td_x}+const$. Similarly, (9.14) is

$$\mathbf{F_x^n}=\nabla B_x=\mathbf{C_xx}+\mathbf{d_x}, \tag{9.15}$$

where the notation $\mathbf{F_x^n}$ indicates the force along the x-direction due to connections among the components. If the only applied force is due to the interconnections of components, the resulting placement contains significant overlap among the components. This behavior can be understood by observing that the Euclidean distance decreases by bringing the connected components closer to each other. To avoid illegal placements, other forces can also be included within the cost function. Consequently, (9.14) is recasted as

$$\nabla B=\mathbf{C}\vec{\mathbf{p}}+\vec{\mathbf{d}}+\vec{\mathbf{e}}, \tag{9.16}$$

where vector $\vec{e}$ describes these additional forces. These forces greatly affect the quality of the placement and should therefore be carefully chosen. For example, the new forces can remove or decrease the overlaps among the components caused by the force $\mathbf{F^n}$ due to the interconnections. To achieve this objective, a *spread* force $\mathbf{F^{move}}$ gradually removes the overlap among components due to the wirelength force acting upon these components. A *hold* force $\mathbf{F^{hold}}$, opposite to the wirelength force $\mathbf{F^n}$, allows the spread force to iteratively move the components to those locations that nullify forces, thereby minimizing the wirelength. Consequently, the total force applied to the components is

$$\mathbf{F^n}+\mathbf{F^{hold}}+\mathbf{F^{move}}. \tag{9.17}$$

The spread force is also modeled as the force of an elastic spring connected between the present location of a movable component and some other target location, which for the x-direction is [383]

$$\mathbf{F_x^{move}}=\mathbf{w_x}(\mathbf{x}-\mathbf{x^t}), \tag{9.18}$$

where the vector $\mathbf{w_x}$ is the spring constant. The spring constant affects the convergence of the placement, and $\mathbf{x^t}$ are the target locations of the components during a placement iteration. As $\mathbf{F^{move}}$

reduces the overlap among components, relating the amplitude of this force with the density of the components across the circuit area is a useful approach. Consequently, the target locations relate to the density of components across the circuit area. Thus, this force can be described by a general supply and demand system [371,383], which includes the density of the components as the demand and the available area for placing these components as the supply. A balanced demand and supply system imposes the constraint,

$$\int_{-\infty}^{\infty}\int_{-\infty}^{\infty} D_{\text{comp}}^{\text{dem}}(x,y) = \int_{-\infty}^{\infty}\int_{-\infty}^{\infty} D_{\text{comp}}^{\text{sup}}(x,y). \tag{9.19}$$

To determine the demand of the components at each point (x, y), a rectangle function R is used, leading to

$$R(x,y;\, x_{ll}, y_{ll}, w, h) = \begin{cases} 1, & \text{if } 0 \le x - x_{ll} \le w, \\ & \text{and } 0 \le y - y_{ll} \le h, \\ 0, & \text{otherwise,} \end{cases} \tag{9.20a–c}$$

where x_{ll} *and* y_{ll} are the coordinates of the lower left corner of a component with width w and height h. From R the demand for component i at point (x, y) is

$$D_{\text{comp},i}^{\text{dem}}(x,y) = d_{\text{comp},i} \cdot R\left(x, y;\, x_i' - \frac{w_i}{2}, y_i' - \frac{h_i}{2}, w_i, h_i\right), \tag{9.21}$$

where the component i is located at (x_i', y_i') (the center of the component is described by this point) with dimensions w_i and h_i. The coefficient $d_{\text{comp},i}$ captures the density of each component and is set to one in [383], where this coefficient is also used to remove/add some whitespace around each component. Based on these definitions, the total demand at (x, y) is equal to the number of components placed at that point, assuming that $d_{\text{comp},i} = 1$,

$$D_{\text{comp}}^{\text{dem}}(x,y) = \sum_{i=1}^{N_m+N_f} D_{\text{comp},i}^{\text{dem}}(x,y), \tag{9.22}$$

including both the movable and fixed location components. Similarly, the supply at each point across the placement area is

$$D_{\text{comp}}^{\text{sup}}(x,y) = d_{\text{sup}} \cdot R(x, y;\, x_{\text{chip}}, y_{\text{chip}}, w_{\text{chip}}, h_{\text{chip}}), \tag{9.23}$$

where the lower left corner of the circuit area is at $(x_{\text{chip}}, y_{\text{chip}})$ and the dimensions of the circuit are $(w_{\text{chip}}, h_{\text{chip}})$. The coefficient d_{sup} is determined by considering the ratio between the area of the blocks over the overall available area for the circuit,

$$d_{\text{sup}} = \sum_{i=1}^{N_m+N_f} (d_{\text{comp},i}, A_{\text{comp},i})/A_{\text{chip}}, \tag{9.24}$$

where $A_{\text{chip}} = w_{\text{chip}}\, h_{\text{chip}}$. Based on these expressions, the overall supply and demand system is

$$D(x,y) = D_{\text{comp}}^{\text{dem}}(x,y) - D_{\text{comp}}^{\text{sup}}(x,y), \tag{9.25}$$

which is treated as charge distribution with a nonzero value within A_{chip}. This "charge" produces some electrostatic potential across the circuit area, which can be determined by solving Poisson's equation. With the appropriate boundary conditions (typically Dirichlet boundary conditions), a unique solution exists for the electrostatic potential through

$$\left(\frac{\partial^2}{\partial x^2}+\frac{\partial^2}{\partial y^2}\right)\Phi(x,y)=-D(x,y). \tag{9.26}$$

Based on (9.19) to (9.28), the target locations $\mathbf{x}^{\mathbf{t}}$ can be determined. Solving (9.26) however increases the computational time. If the gradient of the potential determines where to move the block, this increase in time can be avoided. The target location for a component i is therefore

$$x_i^t = x - \frac{\partial}{\partial x}\Phi(x,y)\big|_{(x_i,y_i)}. \tag{9.27}$$

The hold force $\mathbf{F}_{\mathbf{x}}^{\mathbf{hold}}$ is opposite to the wirelength force,

$$\mathbf{F}_{\mathbf{x}}^{\mathbf{hold}} = -(\mathbf{C}_{\mathbf{x}}\mathbf{x}+\mathbf{d}_{\mathbf{x}}). \tag{9.28}$$

Having defined all constituent forces, (9.17) is equated to zero and the placement of the components is determined. An iterative process is required as the target points depend upon the existing location and density of each of the components, as described by (9.19) to (9.26). Consequently, an initial placement is obtained by solving (9.14) when only the wirelength force is present. As only this force is applied, the component placement exhibits significant overlap. This initial placement is later refined until the overlap of the components is reduced to a user defined level (e.g., 20% in [383]). During this iterative refinement process, the potential Φ is used to compute the target locations ($\mathbf{x}^{\mathbf{t}}$, $\mathbf{y}^{\mathbf{t}}$). Expression (9.17) is set equal to zero and solved, producing a new location for the components, thereby modifying the density D and, in turn, potential Φ. The process is repeated until the overlap constraint is satisfied. A final detailed placement can be employed to remove any remaining overlaps.

The quality of the solution as well as the speed of convergence for the force directed method is a function of specific parameters integrated within the expressions characterizing the different forces. For example, the spring constant in (9.18) moves components at a greater or shorter distance. A definition of this coefficient is

$$w_i \equiv \frac{A_{\text{comp},i}}{A_{\text{avg}}}\cdot\frac{1}{N_m}, \tag{9.29}$$

where A_{avg} is the average area of the components. This definition encourages moving higher area components to greater distances as compared to moving smaller components. Furthermore, the coefficient can remain constant during the entire placement procedure or adjusted during each iteration, offering placements of higher quality. The disadvantage of this dynamic adjustment process is increased computational time. Another important parameter that affects the quality of these results is the choice of wirelength model. An elaborate model is used in [383]. Net models for 3-D circuits are revised to include TSVs, as discussed in the following section.

9.4 PLACEMENT IN THREE-DIMENSIONAL ICs

In vertical integration, a "placement dilemma" arises in deciding whether two circuit cells sharing a large number of interconnects can be more closely placed within the same tier or placed on an adjacent physical tier, decreasing the interconnection length. Placing the circuit blocks on an adjacent tier can often produce the shortest wire connecting these blocks. An exception is the case of small blocks within an SiP or system-on-package (SOP) where the length of the intertier vias is greater than 100 μm. Since intertier vias consume silicon area, possibly increasing the length of some interconnects, an upper bound for this type of interconnect resource is necessary. Alternatively, sparse utilization of the intertier interconnects can result in insignificant savings in wirelength.

Several approaches have been adopted for placing circuit cells within a volume, including SA as the core solving engine, force directed placers, and analytic placement [384–387]. Some of these techniques also consider the TSV placement process simultaneously with placing the circuit cells. In the following subsections, placement tools based on these methods are discussed.

9.4.1 FORCE DIRECTED PLACEMENT OF THREE-DIMENSIONAL ICs

The force directed placement method presented in Section 9.3.1 has been extended to perform placement for multitier circuits, where the placement occurs at the cell level. As the third physical dimension is introduced in 3-D circuits, the exerted forces need to be properly adjusted. Additionally, similar to floorplanning techniques for 3-D circuits, the TSVs require different approaches, simultaneously placing the TSVs with the circuit cells or following the cell placement step. Other issues specific to TSVs, such as crosstalk between TSVs, can also be considered by modifying or adding new forces to the classic formulation of the force directed method. These topics are the foci of this subsection.

Since the location of circuit cells in a 3-D system is described by three coordinates (x, y, z), the inclusion of the z-direction in (9.13) is a reasonable yet not straightforward extension. The issue stemming from including a wirelength force $\mathbf{F}_{\mathbf{z}}^{\mathbf{n}}$ is that this force may collapse the majority of the cells into a tier containing the I/O terminals (as these terminals are typically located in only one tier). Mitigating this behavior may require a significant change in the algorithm. Alternatively, a partition step can be employed to place the cells among the tiers. The partition step, which can reduce or maximize (depending upon the partition objective) the number of TSVs, eliminates the need for $\mathbf{F}_{\mathbf{z}}^{\mathbf{n}}$, as the tier assignment of cells is not allowed to change. Alternatively, intertier moves are not permitted, decreasing the design space, trading off computational time with the quality of the placement.

With a partitioned structure, the force directed method can be applied on a per tier basis, where the technique is adapted to consider the several tiers and TSVs. As the placement of each tier occurs separately, each plane has a different density of components and therefore a different electrostatic potential Φ. The related expressions are adapted to reflect this situation, leading to the following expressions for the density $D_{\text{tier},d}$, potential $\Phi_{\text{tier},d}$, and target location of each cell $x_{i,z=d}^{t}$ [388], respectively,

$$D_{\text{tier},d}(x,y) = D_{\text{cell},z=d}^{\text{dem}}(x,y) - D_{\text{tier},d}^{sup}(x,y), \tag{9.30}$$

$$\left(\frac{\partial^2}{\partial x^2} + \frac{\partial^2}{\partial y^2}\right)\Phi_{\text{tier},d}(x,y) = -D_{\text{tier},d}(x,y), \text{ and} \tag{9.31}$$

$$x_{i,z=d}^{t} = x_{i,z=d} - \frac{\partial}{\partial x}\Phi_{\text{tier},d}(x,y)\big|_{(x_i,y_i)}. \tag{9.32}$$

The remaining issue is controlling the allocation of the TSVs. Two possible TSV placement flows are illustrated in Fig. 9.13 [388]. Both placements begin with a netlist of cells partitioned within the available number of tiers, where the required number of TSVs for each tier is determined. These TSVs are inserted as additional cells into each tier during the second step. As shown in Fig. 9.13A, both the TSVs and circuits are simultaneously placed, following the force directed method. Alternatively, as shown in Fig. 9.13B, the TSVs are initially placed uniformly across the area of each tier followed by the placement of the cells. In this case, an additional step, where the intertier nets are assigned to the TSVs, is required. Both of these processes conclude with routing each tier based on standard commercial 2-D tools.

The 3-D netlist produced after partitioning does not include the physical information related to the TSVs. Thus, simultaneously placing the TSVs and cells requires an update of the tier density in (9.30) for both processes. In the first case, where the TSVs are treated as cells, the cell density rises (the first term in (9.30)). In the second case, where the TSVs are placed prior to the cells, the supply decreases (the second term in (9.30)) since the TSVs are effectively placement obstacles for the cells. Consequently, in either case, the presence of TSVs increases the spatial density, resulting in a larger move force close to the location of the TSVs (particularly for preplacing the TSVs where the location is known at the beginning of the process). Another important modification is that this 3-D placement technique requires a suitable wirelength model, as this feature greatly affects the quality of the placement [383]. The notion of net splitting is used [388], providing a more precise wirelength estimation (i.e., $L_{HPWL\text{-}NET\text{-}SEG}$) as compared to $L_{HPWL\text{-}TSV}$.

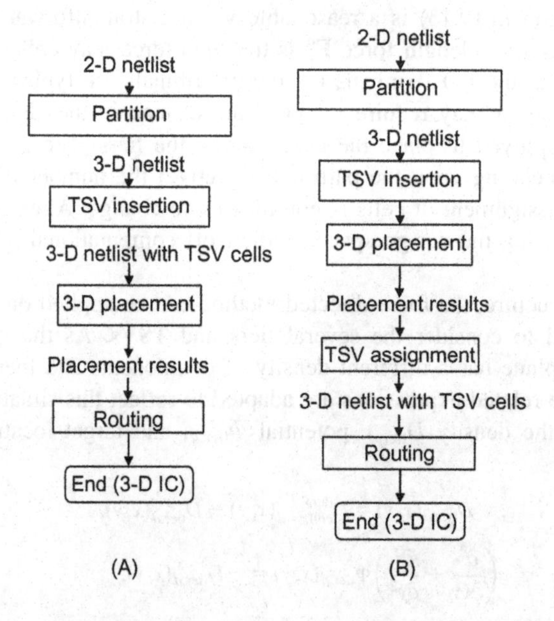

FIGURE 9.13

Two force directed placement processes, (A) the TSVs and circuit cells are placed simultaneously, and (B) the TSVs are placed prior to the circuit cells and behave as placement obstacles [388].

The cell placement shown in Fig. 9.13B avoids the preplaced TSV locations for the cells. Note that this process is orthogonal to the process used in floorplanning techniques where the cells are placed and the remaining whitespace is available for the TSVs. In this process, the TSV occupy some area and constitute placement obstacles for the cells. Once the placement of the cells is complete, the intertier nets are assigned to the TSVs. This assignment process assumes that only one TSV exists between two consecutive tiers along a given path [383]. Although the assignment problem can be addressed with standard optimization methods, such as integer linear programming, a heuristic is often employed [388]. The reason is that the search space for an exact method becomes prohibitively high. For example, for an intertier net that spans four tiers and an assignment among 20 TSVs per tier, this search space demands evaluating 8,000 different combinations [388].

The heuristic is based on constructing a minimum-spanning tree (MST) for every intertier interconnect. Starting from the shortest edge of the tree, the nearest TSV to this edge is assigned. If a need for more TSVs exists, the second TSV to the shortest edge is selected and the process is repeated. An example where the shortest edge spanning all three tiers is assigned to the closest (available) TSV is depicted in Fig. 9.14.

Both of these 3-D placement variants are applied to the IWLS 2005 benchmark circuits [389] and some industrial benchmark circuits. Some of the related results along with the characteristics of the benchmark circuits are reported in Tables 9.5 and 9.6. The comparison indicates that the simultaneous placement of TSVs and cells performs, respectively, 8% and 15% better in terms of

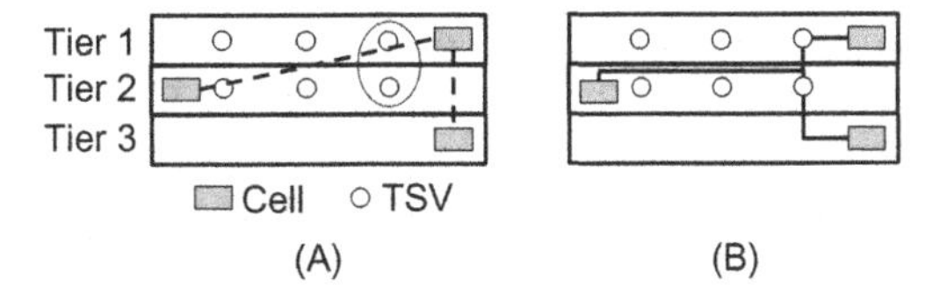

FIGURE 9.14

TSV assignment based on the MST of a net, where the closest TSV to the shortest edge of the net is inscribed by the dotted eclipse [388].

Table 9.5 Comparison of Various Metrics Between 2-D and 3-D (Four Tier) Benchmark Circuits Placed With the Two Techniques Shown in Figs. 9.13A and 9.13B

Circuit	WL (μm)		Metal Layers	Run Time (s)	Area (μm^2)	# of TSVs
	TSV Co-placement	TSV Pre-placement				
Ind2	284,340 (0.85)	310,677 (0.93)	4	53 (0.73)	58,564 (1.30)	1302
Ethernet	1,401,059 (0.91)	1,513,381 (0.98)	4	1287 (1.00)	341,056 (1.16)	3866
des_perf	1,911,731 (0.78)	2,197,209 (0.90)	4	950 (0.69)	386,884 (1.18)	3856

Cell occupancy is 80% and the number of intertier wires is set during partition between 3% and 5% of the total number of nets. The numbers in parentheses indicate ratios as compared to a 2-D placement [388].

Table 9.6 Characteristics of the Benchmark Circuits Listed in Table 9.5 [388]

Circuit	# of Gates	# TRs	# of Nets	Profile
Ind2	15K	106K	15K	Inverse discrete cosine transform (DCT)
Ethernet	77K	729K	77K	Ethernet IP core
des_perf	109K	823K	109K	DES (data encryption standard)

wirelength, while the additional computational time for the TSV assignment step is only a few seconds. Although these results favor simultaneous placement, TSV preplacement may serve other design objectives more important than wirelength.

9.4.2 OTHER OBJECTIVES IN PLACEMENT PROCESS

The introduction of other objectives into the placement process, where the force directed method is employed, adds new constituent forces that characterize these objectives. Examples of these objectives include temperature, reliability, and crosstalk noise. The last objective is integrated into the placement process to reduce coupling between adjacent TSVs. This issue is particularly important in TSV islands (discussed previously) as several signal TSVs are contained within a small area.

The standard method to reduce interconnect coupling includes either increasing the space between wires or inserting ground lines to shield the interconnect. Combining these objectives mitigates TSV coupling better than individually applying any one technique [390]. As TSVs are also modeled as cells, the force directed method is employed, where new forces are added to (9.17) to spread the TSVs and the related shield lines. Shield insertion, however, is not implemented by the force directed method but as a separate step.

This TSV coupling aware method consists of several steps: (1) an initial placement, based on the standard force directed method (essentially a form of (9.17), is set to zero and the resulting system of equations is solved), (2) TSV spreading where the TSVs are placed farther from each other to reduce coupling, (3) ground TSVs are inserted as shields, and (4) any overlap among TSVs and circuit cells is removed to produce an overlap free placement. With these steps, both methods to reduce TSV coupling (increase in space and insertion of shields) are applied to 3-D circuits. Although these steps can be reduced—also decreasing the computational time—by assuming fixed positions for the cells and only moving the TSVs to other permissible, more distant locations (i.e., where whitespace exists), the decrease in coupling is not as effective [390].

Consequently, considering these previous steps, an initial placement can be produced by applying (9.17) to reduce wirelength, where the TSVs are modeled as another component to be placed. To increase the distance among the TSVs, the move force (9.18), which applies to all of the cells, is replaced with a spread force applied only to pairs of TSVs. This spread force along the x-direction is

$$\mathbf{F}_x^{\text{spread}} = \mathbf{w}_{\text{spread},x}(\mathbf{x} - \mathbf{x}^{t}), \tag{9.33}$$

where the matrix $\mathbf{w_{spread}}$ is not based on (9.29) but is user defined. In addition, the target locations during spreading of the TSVs is not based on the placement density across the circuit but rather on another force related to the keep out zone (KOZ) of the TSVs. An expression for this force is

$$\mathbf{F_x^{KOZ}} = e^{-(d/\mathrm{KOZ})^p}, \tag{9.34}$$

where d is the distance between a pair of TSVs (or any other component) and p is a smoothing factor to avoid the discontinuous behavior of the response to a step function. Note that the concept of KOZ is that if the distance d between two TSVs is greater than KOZ, no force should be exerted on the TSV. However, application of a step function may lead to convergence issues which can be avoided by setting the smoothing parameter to $p = 10$ [390]. Applying this force to pairs of TSVs produces the target locations $\mathbf{x^t}$ for the next iteration, where, for this step, the distance d is set to twice the KOZ radius. This choice inserts a TSV as a shield at a later step if necessary. Based on these modifications, the adapted system of forces more effectively spreads the TSVs as compared to (9.17) [390],

$$\mathbf{F^n} + \mathbf{F^{hold}} + \mathbf{F^{spread}}. \tag{9.35}$$

Iteratively solving (9.35) until the TSVs reach the target locations and no forces are exerted on the TSVs completes this step.

Inserting the TSV shields in the next steps casts the problem as a minimum cost maximum flow problem where an optimal solution for this algorithm is known to exist [391]. The shields offer an additional means to further reduce any coupling violations that may remain after the TSV spreading step. These two steps can produce some overlap between the TSVs and circuit cells, as the circuit cells are not allowed to move during these steps.

The next step, consequently, slightly relocates the cells to eliminate these overlaps through the use of a new force expression,

$$\mathbf{F^n} + \mathbf{F^{hold}} + \mathbf{F^{overlap}} + \mathbf{F^{KOZ}} = 0. \tag{9.36}$$

Although the nature of the constituent force does not change, these forces are applied to different objects as the optimization objective is different from the previous steps. Thus, the overlap force $\mathbf{F^{overlap}}$ is equal to the move force $\mathbf{F^{move}}$ from (9.18) where a density map of the circuit determines the amplitude. Moreover, the force $\mathbf{F^{KOZ}}$ is the previous spread force. This force however is not based on the spread vector applied to the TSVs. Rather, a KOZ vector is applied to each cell based on the position of the TSVs. The sum of these vectors is added to the current cell position to determine the target position for this cell.

These placement steps have been tested with the IWLS benchmark circuits [389]. They are compared with two other placements where TSV coupling is not considered (WL only) and only spreading the TSVs is employed (CA) without inserting TSV shields. These results are listed in Table 9.7, where combined TSV spreading and shielding (CA + SI) reduce the coupling, as described by the S parameters in units of decibels (dB). The amplitude of the S parameter between a TSV pair is modeled analytically by

$$S(d) = 8.79 \cdot 1.09^{-d} - 0.0126 \cdot d - 33.2, \tag{9.37}$$

where d is the distance between the pair of TSVs. A threshold of −28 dB indicates whether a coupling violation between neighboring TSVs exists. As these results indicate (see Table 9.7), CA + SI eliminates TSV coupling at the expense of an increase in wirelength (as compared to the wirelength objective, which is less than 5%).

Table 9.7 Comparison of Wirelength Among TSV Coupling Aware and Unaware Placement Techniques for the IWLS Benchmark Circuits [390]

Benchmark Circuit	# of Signal TSVs	WL Objective Only (mm)	CA	SI	CA + SI
aes_core	1427	303.2	318.8	306.5	319.2
wb_commax	1096	516.1	521.2	517.7	521.2
Ethernet	1501	453.1	454.1	460.3	454.3
des_perf	1114	387.5	391.6	389.4	392.1
vga_lcd	1976	339.2	340.3	348.6	340.4
Average	–	100%	101.6%	101.3%	101.6%

9.4.3 ANALYTIC PLACEMENT FOR THREE-DIMENSIONAL ICs

In analytic placers for 3-D circuits, the circuit blocks are treated as interconnected 3-D cells, in other words, as cubic blocks. This approach does not depict the discrete nature of a 3-D system as the circuit blocks can only be placed in a specific discrete number of tiers; yet allows the formulation of a continuous, differentiable, and possibly convex objective function that can be optimally solved [385]. Since this approach does not consider the discrete number of tiers available for circuit placement, more than one step—referred to as a legalization step(s)—is required to finalize the cell and intertier via placement within a 3-D circuit without overlaps among cells. Similar to other placement approaches, such as the SA or force directed method, the choice of wirelength model greatly affects the efficiency of the placement algorithm.

Early efforts described the length of a net connecting multiple cells by the Euclidean distance among the cells connected in 3-D [385]. Alternatively, the distance of the terminals of a net can be adopted as the objective function to characterize the wirelength. To consider the effects of the intertier interconnects, a weighting factor is used to increase (or, more accurately, penalize) the distance in the vertical direction, controlling the decision as to where to insert the intertier vias. This weight behaves as a control parameter that favors the placement of highly interconnected cells within the same or adjacent physical tier. In addition, the resulting placement should be without overlaps and support a design rule compliant TSV placement [387]. This requirement can lead to placement of TSVs outside the net bounding box, which increases the wirelength. To limit these longer paths for TSVs, any placement of a TSV outside the bounding box of the intertier net is penalized to limit wire overhead [387].

Modern analytic placers, however, employ an expression based on the logarithm of the sum of the exponentials (log-sum-exp) to describe the wirelength, as utilized in efficient 2-D placers [374]. Describing a placement problem with a graph $H = (V, E)$ where V are the vertices of the graph that represent the blocks and E are the edges that represent the nets of the circuit, an expression for the total wirelength is the sum of the length of each edge (net) e of every block v [392],

$$\gamma\sum_{e\in E}\left(\ln\sum_{v_i\in e} e^{x_i/\gamma} + \ln\sum_{v_i\in e} e^{-x_i/\gamma} + \ln\sum_{v_i\in e} e^{y_i/\gamma} + \ln\sum_{v_i\in e} e^{-y_i/\gamma}\right), \tag{9.38}$$

where x_i, y_i is the coordinate of the center of block v, and γ is a tuning parameter to guarantee numerical stability. This *log-sum-exp* function (note each term in (9.38)) is widely used in analytic placement algorithms [374,393] with high quality results.

A similar expression is also formulated to describe wirelength in the z-direction by considering the edges related to the intertier nets. The sum of these expressions constitutes the objective function optimized to produce placements of minimum wirelength. This function is, however, constrained by density functions that remove overlaps between cells, between TSVs, and between cells and TSVs. Consequently, the placement problem can be reformulated as a constrained optimization problem,

$$\min WL(\mathbf{x},\mathbf{y}) + aZ(\mathbf{z}), \tag{9.39a}$$

$$\text{subject to nonoverlap constraints,} \tag{9.39b}$$

where both the WL and Z wirelength terms are based on (9.38), and a is a weighting factor that trades off wirelength for the number of TSVs (typical values range from 10 to 10,000 [393,394]).

The non-overlap constraints can be described through appropriate density functions for both cells and TSVs. For 3-D systems, the stack is formed as cubes [394], and the placement of a block or TSV within this cube increases the number of cells and TSVs, which can be described by

$$D_{b,k}(\mathbf{x},\mathbf{y},\mathbf{z}) = \sum_{v \in V} (P_x(b,v,k)P_y(b,v,k)P_z(b,v,k)), \tag{9.40}$$

where $P_x(b,v,k)$, $P_y(b,v,k)$, and $P_z(b,v,k)$ describe the overlap within cube b of tier k due to block v. The sum of these terms determines the density function $D_{b,k}(\mathbf{x},\mathbf{y},\mathbf{z})$. The constrained optimization problem based on these density functions is not, in general, convex and can include nonsmooth and nondifferentiable functions. To address this problem, a bell shaped function [395] is used to yield appropriate density expressions. These differentiable functions are merged with the primary objective function, resulting in an unconstrained minimization problem of the form,

$$\min WL(\mathbf{x},\mathbf{y}) + aZ(\mathbf{z}) + \lambda \sum_{b,k} (\hat{D}_{b,k}(\mathbf{x},\mathbf{y},\mathbf{z}) + T_{b,k} - M_{b,k})^2. \tag{9.41}$$

This function is successively optimized when several methods, such as the conjugate gradient, are utilized. The coefficient λ increases until the density penalty becomes sufficiently small. The terms $\hat{D}_{b,k}$ and $T_{b,k}$ describe, respectively, the density of cells and TSVs in cube b of tier k, while $M_{b,k}$ is the area available for the cells within this cube. Since the location of the TSVs is not known during global placement, the TSV density changes during the iterations. The assumption to approximate this density is that for each intertier net, one TSV is used per tier crossing and some area is reserved for the TSVs within the bounding box of each intertier net [394]. This formulation of the placement problem has been integrated with a multilevel framework that subsequently performs placement from the coarsest level to the finest level of the framework. During this process, the placement from a coarser level is used as the initial placement during iterations at the next finer level [393].

Although the results produced by analytic placement techniques satisfy (9.38), a greater decrease in wirelength and number of TSVs is achieved through a more elaborate wirelength model.

This model is based on a weighted average (WA) of wirelength, is differentiable, and is described by [394]

$$\sum_{e \in E}\left(\frac{\sum_{v_i \in e} x_i e^{\frac{x_i}{\gamma}}}{\sum_{v_i \in e} e^{\frac{x_i}{\gamma}}} - \frac{\sum_{v_i \in e} x_i e^{-\frac{x_i}{\gamma}}}{\sum_{v_i \in e} e^{-\frac{x_i}{\gamma}}} + \frac{\sum_{v_i \in e} y_i e^{\frac{y_i}{\gamma}}}{\sum_{v_i \in e} e^{\frac{y_i}{\gamma}}} - \frac{\sum_{v_i \in e} y_i e^{-\frac{y_i}{\gamma}}}{\sum_{v_i \in e} e^{-\frac{y_i}{\gamma}}}\right). \tag{9.42}$$

This WA model has been analytically demonstrated to produce lower estimation error then the *log-sum-exp* model in (9.38) and converges to the HPWL, as parameter γ approaches zero.

Based on (9.41) and the WA model, the TSV aware placement process is illustrated in Fig. 9.15. This process consists of three major steps: (1) global placement, (2) TSV insertion and legalization, and (3) detailed placement of individual tiers. The global placement is also performed at multiple levels of granularity where a coarsening phase is followed by a refinement stage. For the first level of the coarsening phase, an initial placement is produced with the force directed method. For the subsequent levels, the placement from the previous level is used as the initial placement for the current level. During this step, the inclusion of $T_{b,k}$ in the objective function demonstrates that the TSV density is considered and whitespace is reserved. The TSVs are inserted during the second step of the process, which can cause some cells to shift, requiring some legalization to avoid significant displacement of the cells to accommodate the TSVs. In the last step, a detailed placement is required for each tier. Standard 2-D placement techniques are used since the TSV positions have been determined and any remaining overlaps can be removed while the wirelength is further decreased. Similar to this step, routing can be performed on a per tier basis with classic 2-D routing algorithms.

This placement process has been tested on the IBM-PLACE benchmark circuits [396] and a comparison with the placement method in [393] employing *log-sum-exp* has been evaluated.

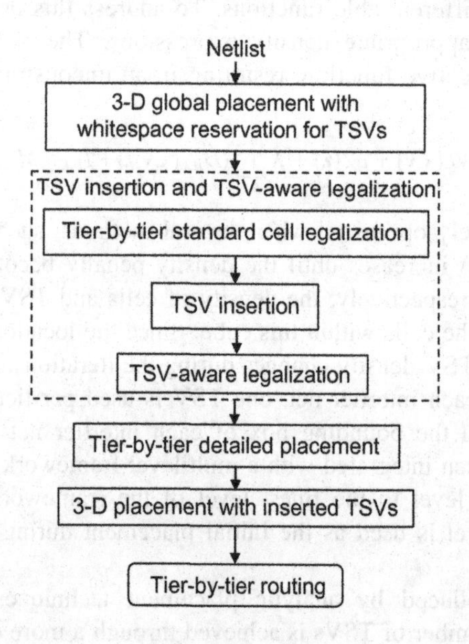

FIGURE 9.15

Analytic placement process for 3-D circuits considering number of TSVs and wirelength [394].

Table 9.8 Comparison of Wirelength and Number of TSVs Between Two 3-D Analytic Placement Algorithms

			Placer From [393]		Placer From [394]	
Circuit	# of Cells	# of Nets	Wirelength ($\times 10^7$)	# of TSVs ($\times 10^3$)	Wirelength ($\times 10^7$)	# of TSVs ($\times 10^3$)
ibm01	12K	12K	0.37	0.87	0.33	0.57
ibm03	22K	22K	0.84	2.92	0.76	2.76
ibm04	27K	26K	1.11	3.36	0.99	2.53
ibm06	32K	33K	1.45	3.40	1.23	3.97
ibm07	45K	44K	2.27	4.46	1.87	4.95
ibm08	51K	48K	2.36	4.43	2.02	4.62
ibm09	52K	50K	2.08	3.37	1.85	3.27
ibm13	82K	84K	4.14	4.37	3.34	3.83
ibm15	158K	161K	8.74	27.53	7.61	15.56
ibm18	210K	201K	12.88	38.35	11.34	12.21
Average	–	–	1.00	1.00	0.87	0.84

The results listed in Table 9.8 show, on average, a reduction of 13% and 16%, respectively, in wirelength and number of TSVs. The improvement in these metrics is attributed to the different ways that the density is controlled. In [393], filler cells with no real connections are inserted to satisfy density constraints; thereby increasing the circuit area. Alternatively, density control is achieved through (9.40) without filler cells [394].

9.4.4 PLACEMENT FOR THREE-DIMENSIONAL ICs USING SIMULATED ANNEALING

As with floorplanning, placement techniques for 3-D circuits based on SA have been developed with the same objectives such as minimizing the wirelength, number of TSVs, and circuit area. Due to the similarity of these approaches, multiobjective methods that extend beyond traditional objectives are discussed in this section. These objectives include routing congestion, circuit temperature, and power supply noise [397,398].

Each of these objectives adds an additional term to the cost function optimized by the SA engine. An example of an objective function is [397]

$$c_7 A^{\text{total}} + c_8 W^{\text{total}} + c_9 D^{\text{total}} + c_{10} T^{\text{total}}, \tag{9.43}$$

where A^{total} is the total area of the 3-D system, W^{total} is the total wirelength, D^{total} is the decoupling capacitance to satisfy a target noise margin, and T^{total} is the maximum temperature of the substrate. The c_i terms denote user defined weights that control the importance of each objective during the placement process.

To manage these different objectives, additional information describing the individual circuit blocks is required, including: (1) the current signature of each block, (2) the number of metal layers dedicated for the power distribution network, (3) the number and location of the power/ground pins, and (4) the allowed voltage ripple on the power/ground lines due to power supply noise. From this information, the required decoupling capacitance for each circuit can be determined.

This decoupling capacitance is distributed to the neighboring whitespace. This space includes those areas not only within the same tier but also on adjacent physical tiers. Since TSVs and buffers also compete for this available area, the difficulty and computational time to accommodate all of these diverse and competing objectives increase.

To determine the available whitespace within each tier, a vertically constrained graph is utilized. The upper boundary of the blocks at the ith level of the tree is compared to the lower boundary of the blocks at the $(i+1)^{th}$ level of the tree. An example of the process for determining the available whitespace is illustrated in Fig. 9.16. Although the whitespace can be extended to the adjacent planes, the decoupling capacitance allocated to these spaces may not be sufficient to suppress the local power supply noise. In these cases, the whitespace is expanded in the x and y directions to accommodate additional decoupling capacitance. The expansion procedure is depicted in Fig. 9.17.

An efficient representation of the circuit blocks is required where, as in floorplanning, the SP technique [338] can be adopted. To reach the SA freezing temperature, the solution generated at each iteration of the SA algorithm is perturbed by swapping operations between pairs of blocks. These perturbations include both intraplane and interplane swapping, as previously discussed in Section 9.2.

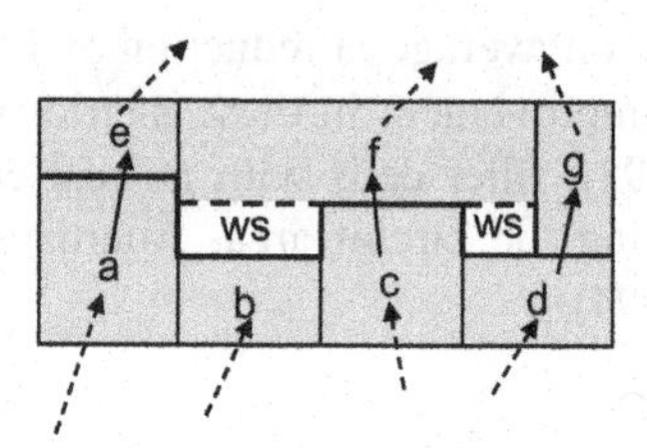

FIGURE 9.16

Process for determining available whitespace (WS), which is illustrated by the white regions.

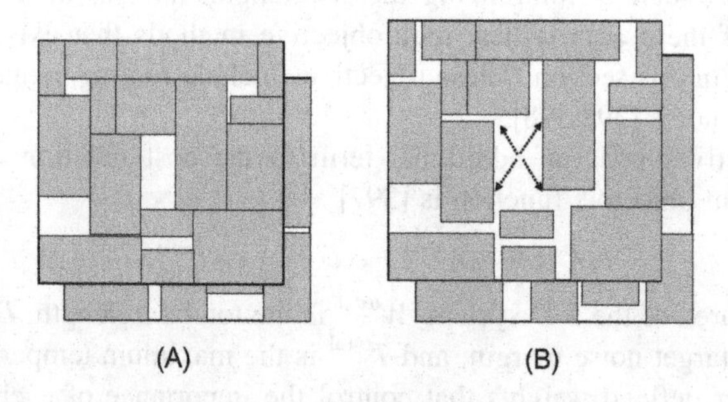

FIGURE 9.17

Block placement of an SOP. (A) Initial placement, and (B) increase in the total area in the x and y directions to extend the area of the whitespace.

Although SA is a robust optimization technique, the effectiveness of a multi-objective placement technique depends greatly on the accuracy of the physical model(s) used to describe the various objectives including the area and length of the circuit interconnections. For example, inaccuracy in the model characterizing the power distribution network of a 3-D system can either insufficiently reduce the noise or the redundant decoupling capacitance, excessively increasing the physical area, leakage current, or total wirelength. Important traits of a power distribution network to suppress power supply noise are the impedance characteristics of the paths to the current load, number and location of the power supply pins, and the decoupling capacitors. The accuracy can be further improved by including the parasitic effective series resistance and inductance of the decoupling capacitors and the inductive impedance of the interconnect paths [275,399,400]. These multi-objective cost functions have been applied to the GSRC [355] and GT benchmarks [401]. Some of these results are listed in Table 9.9, demonstrating the tradeoffs among the different design objectives [397,402].

The versatility of SA supports a multitude of objective functions similar to (9.43). These multi-objective functions, however, require careful handling of the solution space to maintain efficient computational times. A significant portion of the computational time may be spent in determining the individual terms that compose the cost function.

Multi-step SA is another means to accelerate multi-objective placement [398]. Multi-objective placement for 3-D circuits consists of three key elements in handling the third dimension. These elements include routing congestion to improve the quality of the placement of a 3-D circuit.

9.4.5 SUPERCELL-BASED PLACEMENT FOR THREE-DIMENSIONAL CIRCUITS

The notion of supercells is analogous to standard cells. Supercells can be conceptualized where the top level of a system contains several thousand supercells, and the layout of the overall circuit is adjusted to the size of a standard supercell [398]. The supercells are macros with a specific height and varying width, as illustrated in Fig. 9.18. Placing the supercells in rows requires some silicon area (whitespace) between successive rows, which can be used for TSV and buffer placement. Conversely, with other techniques discussed in this chapter, this method presets the whitespace within a circuit. The implication of this practice is a more canonical design process, and a greater flexibility in placing the buffers and TSVs. The resulting area (and wirelength) of the overall circuit may, however, be larger than standard placement approaches. Each supercell is treated as a 2-D block. The layout of supercell-based circuits can be produced with standard commercial design flows, including pins for power, ground, and I/Os.

The circuit volume (as there are multiple tiers) of supercells is gridded. Each bin within this 3-D grid is associated with an area density and an available routing density (described by a triplet). The area density indicates the number of supercells placed within this bin without overlap, while the routing density in the z-direction is the number of TSVs routed through this bin. Furthermore, the wirelength among the interconnections of the supercells is estimated with Steiner and minimum-spanning trees [403].

As the placement is also congestion driven, the objective function is

$$\text{cost} = \text{OBJ}_{\text{wirelength}} + \text{OBJ}_{\text{TSV}} + \text{OBJ}_{\text{congestion}}, \tag{9.44}$$

Table 9.9 Placement of a Four Tier SOP With Diverse Design Objectives [397]

Circuits		Area/Wire Driven (mm^2, m, nF, °C)				Decap Driven (mm^2, m, nF, °C)				Multi-objective (mm^2, m, nF, °C)			
Name	Size	Area	Wire	Decap	Temp	Area	Wire	Decap	Temp	Area	Wire	Decap	Temp
n50	50	221	26.6	18.0	87.2	232	30.5	5.2	85.2	294	35.5	9.3	76.2
n100	100	315	66.6	78.2	86.5	343	73.1	69.2	81.7	410	77.0	77.9	78.5
n200	200	560	17.1	226.3	96.4	693	20.5	223.2	96.2	824	21.3	229.1	85.4
n300	300	846	28.6	393.8	100.1	843	28.6	393.8	100.1	844	28.6	393.8	100.1
gt100	100	191	13.2	60.8	71.0	207	16.8	42.5	70.9	264	18.6	55.2	59.2
gt300	300	238	19.6	342.5	93.2	248	22.3	334.9	99.5	256	22.3	343.9	85.3
gt400	400	270	28.1	493.1	114.0	268	32.5	482.0	111.6	282	34.6	492.6	91.1
gt500	500	316	30.3	645.3	99.7	321	35.4	632.4	98.0	321	34.8	635.8	95.8
Ratio		1.00	1.00	1.00	1.00	1.07	1.13	0.97	0.99	1.18	1.19	0.99	0.90

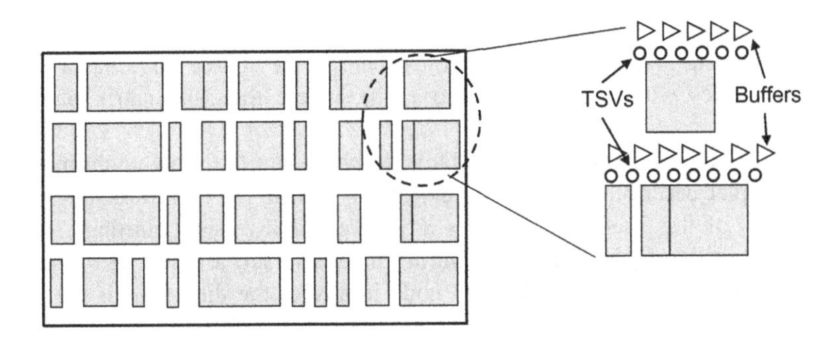

FIGURE 9.18

Layout of supercells. Supercells have the same heigth and varying width. The space around the supercells is used for buffers and TSVs [398].

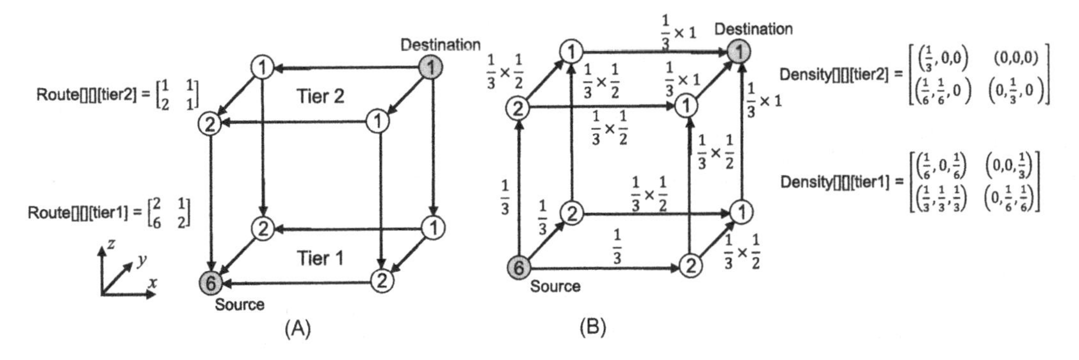

FIGURE 9.19

An example of computing the matrices of a two tier grid, (A) route counts, and (B) routing density.

where the objective targeted by each term is evident by the related subscript. Thus, the first term considers the wirelength, which in this technique is the weighted HPWL. The length of each segment is multiplied by a weighting factor, penalizing long wires (leading to shorter lengths). These weighting factors are empirically determined, which may limit the flexibility of the technique [398]. The second term considers the number of TSVs for every intertier wire where a different weight is applied to each net. Finally, the third term is the routing density, the most computationally expensive term. Probabilistic congestion models such as described in [404] and [405] can be used to compute this term. A faster alternative however is discussed below.

The model forms a 3-D matrix to route each node with grid coordinates x, y, and z. The entries of this 3-D matrix are determined by enumerating the number of routes to reach a node from other neighboring nodes. An example of a two tier grid is depicted in Fig. 9.19. Assuming a pair of source and destination nodes (shown in gray), directed edges towards the source node are drawn through all possible routes. The route count for the destination node is set to one. The source node is reached by following the arrows along each route, where the route count for each node along the path is computed by adding all of the weights of the incoming arrows. This weight is equal to the route count of the starting node of this edge. Consequently, the source node has a route count of six (three possible routes with a weight of two each), where the 3-D matrix of routes is also depicted in Fig. 9.19.

The routing density matrix, which is also 3-D, is derived from the route count matrix. An entry in this matrix is a triplet describing the routing density of a node for each direction x, y, and z. To determine the entries of this matrix, the routes are traversed inversely, starting from the source node and terminating at the destination node. During this traversal, a triplet of routing densities is assigned to each node. The routing density of a node in any of the directions is the density of the outgoing edge in that direction, assuming an outgoing edge exists. This density is the sum of densities of the incoming edges D_T (see (9.46)) multiplied by the ratio of the route counts between the start and end nodes of the outgoing edge. For all but the source and destination nodes of a path, this density along an outgoing edge, for instance, the x-direction (a similar expression applies for the other directions) is

$$\text{density } [x][y][z] \cdot x = D_T \frac{\text{route}[x+1][y][z]}{\text{route}[x][y][z]}, \tag{9.45}$$

where D_T is the sum of densities of the incoming edges,

$$D_T = \text{density}[x-1][y][z] + \text{density}[x][y-1][z] + \text{density}[x][y][z-1]. \tag{9.46}$$

Note that the density of the destination node is zero as no outgoing edges exist, and the density of the outgoing edges from the source node is the ratio of the route counts from the corresponding matrix. An example of determining the routing densities for the grid shown in Fig. 9.19A is illustrated in Fig. 9.19B.

The congestion term in (9.44) can be determined from these matrices. If these matrices are calculated during each iteration of the SA, the computational time can be significant. To avoid this situation, the matrix is constructed only once at the beginning of the placement process. During later iterations, only the density of the nets affected by the latest iteration is calculated. To further speed up the placement process, a two step SA is employed.

The placement algorithm described in [398] produces a random placement that satisfies area, TSV capacity, and routing density constraints. Although this placement may satisfy design constraints, the SA temperature corresponding to this placement may not be the lowest, and therefore, the placement has yet to be finalized. This placement is consequently modified by changing the location of two supercells, producing new candidates. As updating the length of the wires and routing densities for each candidate is computationally expensive, even if these quantities are only reevaluated to account for the relocated supercells, the third term in (9.44) is ignored for several iterations to avoid increasing the computational time. After a specific number of placement candidates that lower the temperature of the SA are produced, the second phase of the SA begins. All of the terms of the cost function in (9.44) are included during this phase to balance the placement in terms of routing density. The boundary between the two phases is user defined. In [398], a maximum number of states for this step is the square of the number of supercells or when one-tenth of the number of successful moves has been reached. Alternatively, the algorithm terminates when, during the second step, either a predefined number of iterations is reached or no change in temperature is observed for three consecutive supercell moves.

Table 9.10 Comparison of Wirelength, Number of TSVs, and Routing Density of 3-D Placements With Different Optimization Objectives [398]

Circuit	# of Supercells	# of Global Multipin Nets	Characteristic	Obj(0)	Obj(1 and 2)	Obj(1 and 2 and 3)
ibm18	1000	42,985	Total length	315,957	431,211	433,905
			Avg. length	2.34	3.20	3.20
			Max length	28	20	23
			# of TSVs	80,670	24,135	24,140
			Max density	436.10	510.96	381.33
			Avg. density	162.78	222.16	223.55
			CPU time (s)	1792	1989	1.17×10^5

The outcome of the SA-based algorithm is a supercell placement where the TSVs are yet to be assigned to the intertier nets. Consequently, another step is required to determine the physical location of the TSVs to minimize the intertier density. This objective is achieved by formulating the problem as a network flow and employing a minimum cost maximum flow algorithm to relate the TSVs with the nets.

Supercell placement has been applied with different objectives to the ISPD98 benchmark circuits [396]. These objectives, obj(0), obj(1), obj(2), and obj(3), correspond, respectively, to wirelength, wirelength and longest interconnect, TSV density, and routing density. The results listed in Table 9.10 demonstrate that decreasing the longest interconnect increases the total interconnect length, which is further aggravated when the routing density term is added. Maintaining a lower routing density results in an increase in the number of supercells, leading to longer wirelength. Moreover, if the TSV density is added to the cost function, a significant decrease in the number of TSVs occurs. The computational time increases with the number of terms in the cost function. The technique has also been demonstrated on an industrial circuit containing a low density parity check (LDPC) decoder. The LDPC decoder utilizes a 0.18 μm CMOS process with the MIT Lincoln Laboratories (MITLL) 3-D technology (discussed in Chapter 16, Case Study: Clock Distribution Networks for Three-Dimensional ICs) and is interconnect limited. As reported in [398], the 3-D LDPC exhibits superior performance by improving the area-delay-power product by $2.5 \times 1.75 \times 2.5 = 10.9$ times. Another interesting result is the improvement in computational speed achieved by incrementally computing the routing density. Not determining the global routing density for each iteration yields a two orders of magnitude improvement in computational time.

9.5 ROUTING TECHNIQUES

One of the first routing approaches for 3-D ICs demonstrated the complexity of the problem, as compared to a simpler single device layer with multiple interconnects [406]. Recent investigations related to channel routing in 3-D ICs have shown the problem as NP-hard [407]. Consequently, different heuristics have been considered to address routing in the third dimension [408,409].

The most straightforward approach to address the routing problem for 3-D systems is to first place the 3-D circuit including the TSVs and, subsequently, perform routing separately for each tier, where each tier is routed with 2-D routing tools. Some placement methodologies discussed in Section 9.4 include a 2-D routing step for each tier, producing a fully routed multitier system. Once the TSV locations are fixed, the start and landing pads of these vias are treated as net pins, allowing 2-D routing techniques to be applied.

Based on this observation, a useful approach for routing 3-D circuits converts the routing intertier interconnect problem into a 2-D channel routing task, as the 2-D channel routing problem has been efficiently solved [410,411]. A number of methods can be applied to transform the problem of routing the intertier interconnects into a 2-D routing task, which requires utilizing some of the available routing resources for intertier routing. Intertier interconnect routing is composed of five major stages including (1) intertier channel definition, (2) pseudo-terminal allocation, (3) intertier channel creation (channel alignment), (4) detailed routing, and (5) final channel alignment [408]. Additional stages route the 2-D channels, both the intertier and intratier interconnects, and order the channels to determine the wire routing order for the 2-D channels.

Each of these stages includes certain issues that should be separately considered. For example, a 3-D net should have two terminals in the intertier channel, and therefore, inserting pseudo-pins may be necessary for certain nets. In addition, aligning the channels may be necessary due to the different widths of the 2-D channels. Aligning the 2-D channels with adjacent tier of the 3-D system forms an overlapping region, which serves as an intertier routing channel. An example of this alignment is shown in Fig. 9.20. Since channel alignment can be necessary, however, at a later stage of the algorithm, the width of the 2-D channels is based on a detailed route of these channels without wires. Detailed channel routing with safe ordering follows all of the 2-D channels for both the intratier and intertier interconnects, typically utilizing a SA algorithm. The technique is completed, if necessary, with a final channel alignment. This approach has been applied to randomly generated circuits and has produced satisfactory results [408].

Alternatively, intertier routing can be implemented without decomposing these interconnects into several intratier wires. Instead, these wires are constructed as 3-D Steiner trees where several intratier (2-D) nets are connected by TSVs. To construct these 3-D trees, the following inputs are required: (1) a set of m nets, where each net is associated with a number of pins p_i, (2) a 3-D

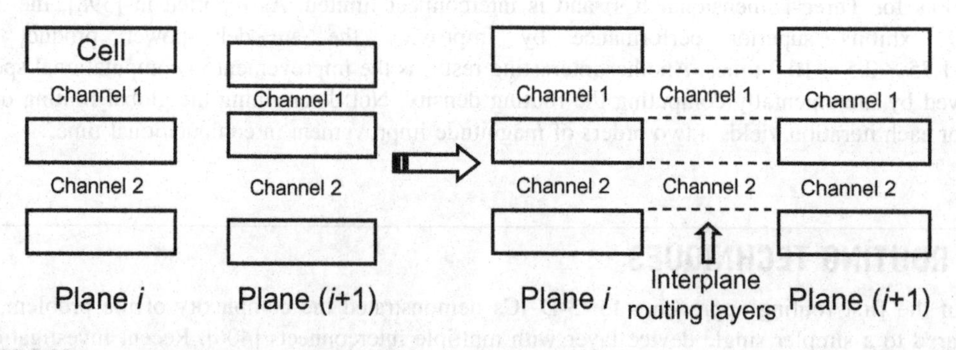

FIGURE 9.20

Channel alignment procedure to create intertier routing channels.

routing grid that describes the available routing resources of the vertical stack, (3) capacity information relating to each edge of this grid, where the capacity of the vertical edges (z-direction) corresponds to the number of TSVs for this edge, (4) the location of the pins p_i within the 3-D grid, (5) the thermal behavior of the circuit, and (6) a 3-D thermal grid different from the routing grid including thermal resistances along the edges and power sources at the nodes.

With this information, the routing technique proceeds in two stages. The 3-D trees are constructed and the TSVs are relocated to improve the thermal profile of the circuits without degrading circuit timing [412]. Pseudocode of the algorithm is provided in Fig. 9.21. The tradeoff for this thermal driven TSV relocation is an increase in wirelength; however, this increase is inevitable if the thermal behavior of the circuit is considered during the physical design process. This situation is explained by noting that if the signal TSVs are not relocated to lower the temperature of the circuit, dummy thermal TSVs are added to achieve this objective. Thermal TSVs, however, also require area, which increases the wirelength. The effect of this overhead depends upon the efficiency of the different techniques.

The tree construction produces intertier trees, where the maximum delay of the sinks of the existing branches of a tree increases least. A new branch is added each time to the tree, connecting a new pin from the set p_i. The tree, therefore, gradually grows based on the Steiner Elmore routing tree (SERT) algorithm [413]. The requirement for a minimum increase in the maximum sink delay means that the physical connection that produces the optimum delay must be determined. To achieve this objective, the optimization methodology employed in the techniques presented in Chapter 10, Timing Optimization for Two-Terminal Interconnects, and Chapter 11, Timing Optimization for Multiterminal Interconnects, is followed where a two variable problem is optimized. The function being optimized is the Elmore delay [414] of the sinks of the tree, and the two

3-D Steiner tree construction algorithm
input: netlist NL, routing grid, thermal profile Z
output: 3-D Steiner tree for each net T_n

1.	for (each net $m \in NL$)
2.	$T_n = p_0(m)$
3.	Q_n = set of pins of m except p_0
4.	while ($Q_n \neq 0$)
5.	for (each pin $a \in Q_n$)
6.	for (each edge $e \in T_n$)
7.	x = connection point for a -> e;
8.	y = TSV location $e(x,a)$;
9.	update $dly(p)$ for all $p \in T_n \cup a$
10.	$X(a,e) = \max\{dly(p)\}$
11.	(a_{min}, e_{min}) = pin + edge pair with min X
12.	$T_n = T_n \cup e_{min}$
13.	remove a_{min} from Q_n =
14.	update Z periodically
15.	for (each non-timing critical T_n violating capacity)
16.	rip-up-and reroute T_n under Z

FIGURE 9.21

Pseudocode of 3-D routing algorithm targeting reductions in both performance and temperature [412].

variables include the linking point along an existing branch of the tree (for example, the *x*-direction) and the TSV location along the other horizontal direction. An assumption, however, is made if two pins of the tree are located in more than one tier. A stacked TSV is considered. Strictly speaking, this assumption may not be valid as successive vertical edges of the routing grid may not have the same capacity or may have been assigned a different number of nets during the routing process. In this case, a detour of the vertical routing process is required to reach a vertical edge that can accept additional TSVs. Beyond these subtle limitations, 3-D Steiner tree construction in [412] demonstrates a significant decrease in delay when circuits from the ISCAS89, ITC99, and ISPD98 [396] benchmark circuits are routed for 3-D circuits with four physical tiers. Reported results exhibit an average decrease in delay of 52% and 11% over, respectively, a 3-D maze router [415] and a 3-D A-tree router [416] while variations in the number of TSVs required by each router is about 6%.

Although these techniques offer a routing solution for standard cell and gate array circuits, alternative techniques that support different forms of vertical integration, for example, SOP, are also required. In an SOP, the routing problem can be described as connecting the I/O terminals of the blocks located on the tiers of the SOP through the interconnect and pin layers. These layers, which are called "routing intervals," are sandwiched between adjacent tiers of an SOP. The structure of an SOP is illustrated in Fig. 9.22, where each routing interval consists of pin redistribution layers and x–y routing layers between adjacent tiers. Communication among blocks located on nonadjacent tiers is achieved through vias that penetrate the active device layers, notated by the thick solid lines shown in Fig. 9.22.

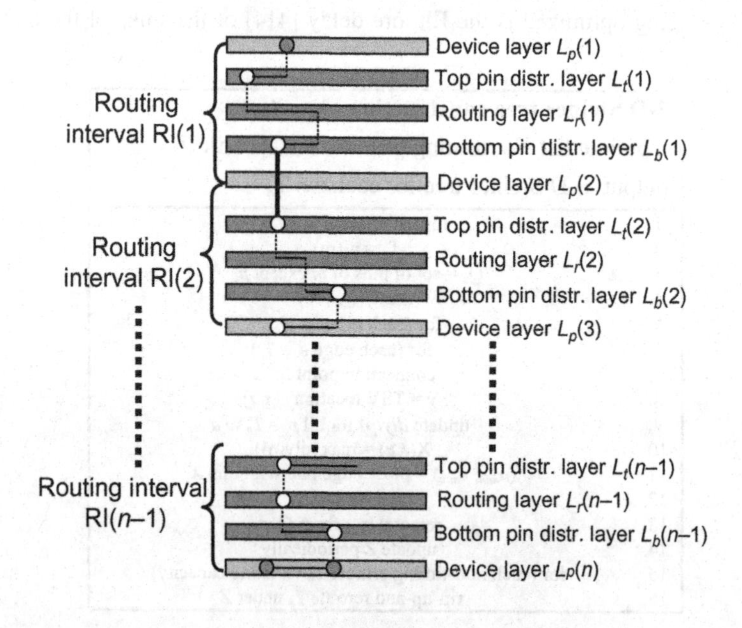

FIGURE 9.22

An SOP consisting of n tiers. The vertical dashed lines correspond to vias between the routing layers, and the thick vertical solid lines correspond to through silicon vias that penetrate the device layers.

For those systems where the routing resources, such as the number of pin distribution layers, are limited, multiobjective routing is required to achieve a sufficiently small form factor. Other factors, such as integrating passive and active components, further enhance the demand for multi-objective routing approaches. A multi-objective approach can consider, for example, wirelength, crosstalk, congestion, and routing resources [409].

A multivariable function that accurately characterizes each of these objectives is necessary to produce an efficient route of each wire net_i in an SOP. The wirelength can be described by the total Manhattan distance in the x, y, and z directions, where the z-direction describes the length of the intertier vias. The crosstalk produced from the neighboring interconnects is

$$xt_{net_i} = \sum_{s \in NL, s \neq net_i} \frac{cl(net_i, s)}{|z(net_i) - z(s)|}, \tag{9.47}$$

where $cl(net_i,s)$ is the coupling length between two interconnects, net_i and s, and $z(net_i)$ denotes the routing layer in which a wire net_i is routed. The netlist that describes the connections among the nets of the blocks is denoted as NL. The delay metric used in the objective function for an interconnect net_i is the maximum delay of a sink of wire net_i,

$$D^{\max} = \max\{d_{net_i} | net_i \in NL\}. \tag{9.48}$$

Finally, the total number of layers used to route an SOP consisting of n tiers is

$$L^{\text{tot}} = \sum_{1 \leq i \leq n} (|L_t(i)| + |L_r(i)| + |L_b(i)|). \tag{9.49}$$

For each routing interval i, $L_t(i)$, $L_r(i)$, and $L_b(i)$ denote, respectively, the top pin distribution layer, routing layers, and bottom pin distribution layer. Combining expressions (9.47) through (9.49), the global route of an SOP proceeds by minimizing the following objective function,

$$c_{11}L^{\text{tot}} + c_{12}D^{\max} + \sum_{s \in NL}(c_{13}xt_s + c_{14}wl_s + c_{15}via_s), \tag{9.50}$$

where via_s is the number of vias included in wire s and the factors, c_{11}, c_{12}, c_{13}, c_{14}, and c_{15}, correspond to the weights that characterize the significance of each of the objectives during the routing procedure.

The steps of the global routing algorithm optimizing the objective function described by (9.50) is illustrated in Fig. 9.23. To distribute the pins to each circuit block, two different approaches can be followed; coarse (CPD) and detailed (DPD) pin distribution. The difference between these two

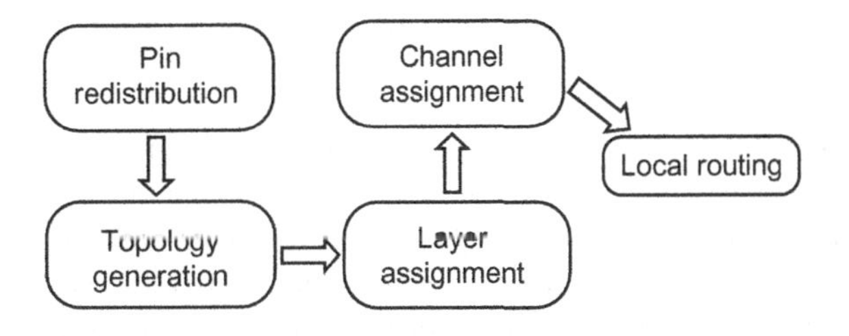

FIGURE 9.23

Stages of a 3-D global routing algorithm [409].

methods lies in the computational time. The complexity for CPD is $O(p \times u \times v)$, where p is the number of pins and $u \times v$ is the size of the grid on which the pins are distributed, while the complexity for DPD is $O(p^2 \log p)$, exhibiting a quadratic dependence on the number of pins.

A topology (i.e., Steiner tree) is generated for each of the interconnects in the routing interval to optimize the performance of the SOP. During the layer assignment process, the assignment of the routed wires is chosen to minimize the number of routing layers. The complexity of the layer assignment is $O(m \log m)$, where m is the total number of interconnects. The complexity of the channel assignment step is $O(|P| \cdot |C|)$, where $|P|$ is the number of pins and $|C|$ is the number of channels. The algorithm terminates with a local route between the pins at the boundary of the block and the pins within the routing intervals. Application of this technique to the GSRC and GT benchmark circuits exhibits an average improvement in routing resources of 35% with an average increase in wirelength of 14% as compared to a route that only minimizes the wirelength. The maximum improvement can reach 54% with an increase in wirelength of 24% [409].

9.6 LAYOUT TOOLS

Beyond physical design techniques, sophisticated layout tools are a crucial component for the back end design of 3-D circuits. A fundamental requirement of these tools is to effectively depict the third dimension and, in particular, the intertier interconnects. Different types of cells for several 3-D technologies have been investigated [417]. Layout algorithms for these cells demonstrate the benefits of 3-D integration. Other traditional features, such as impedance extraction, design rule checking, and electrical rule checking, are also necessary.

Visualizing the third dimension is a difficult task. The first attempt to develop tools to design 3-D circuits at different abstraction levels was introduced in 1984 [418], where symbolic illustrations of 3-D circuit cells at the technology, mask, transistor, and logic level are offered. A recent effort in developing an advanced toolset for 3-D ICs is demonstrated in [419,420], where the Magic layout tool has been extended to 3-D (i.e., 3-D Magic). To visualize the different tiers of a 3-D circuit, each tier is illustrated on separate windows, while special markers are introduced to notate the intertier interconnects, as depicted in Fig. 9.24. Additionally, impedance extraction is supported where the technology parameters are retrieved from predefined technology files. The 3-D Magic tool is also equipped with a reliability analysis design tool called ERNI 3-D [419]. This tool provides the capability to investigate certain reliability issues in 3-D circuits, such as electromigration, bonding strength, and interconnect joule heating. ERNI-3D is, however, limited to a two tier 3-D circuit structure.

A process design kit has been constructed [177] for designing 3-D circuits based on the MITLL 3-D fabrication technology [307]. This kit is based on the commercial Cadence Design Framework, and offers several unique features for 3-D circuits, such as visualizing circuits on the individual tiers, highlighting features for intertier interconnects, and 3-D design rule checking. The design rules for 3-D circuits have been largely extended to include aligning the intertier interconnects with the backside vias, which is an additional interconnect structure within the MITLL fabrication technology.

Although these tools consider important issues in providing a layout environment for 3-D circuits, there is significant room for improvement, as existing capabilities are limited to a specific

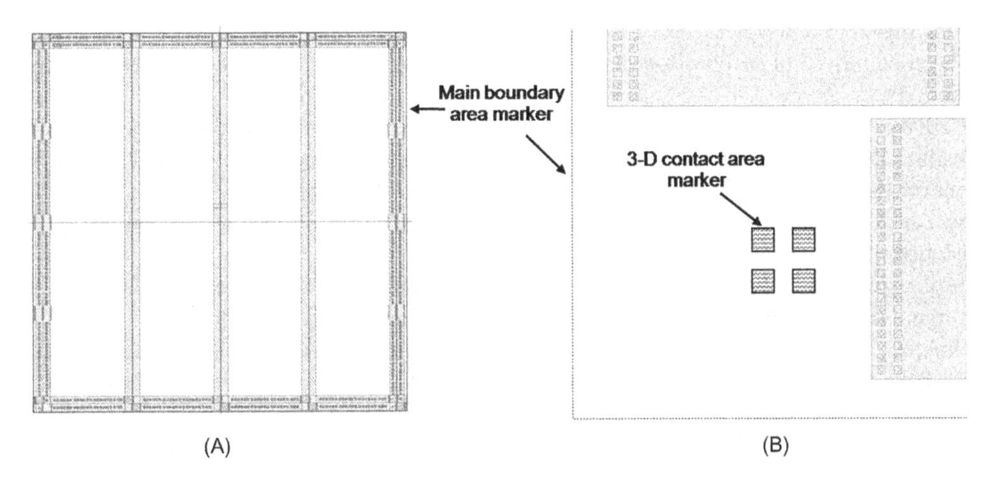

FIGURE 9.24

Layout windows with different area markers; (A) layout window for tier 1, and (B) layout window for tier 2 (the windows are not on the same scale).

3-D technology and number of tiers. In addition, a complete front and back end design flow for 3-D circuits does not yet exist. This situation is greatly complicated by the lack of a standardized fabrication technology for 3-D circuits. The complexity and heterogeneity of 3-D integration pose significant obstacles in providing an effective design flow. Managing thermal effects, which is discussed in the next chapter, requires thermal analysis, an inseparable element of the physical design process of 3-D integrated systems.

9.7 SUMMARY

The physical design techniques used during different stages of the design flow for 3-D circuits are discussed, emphasizing the particular traits of 3-D ICs. The primary concepts discussed in this chapter are summarized as follows:

- A variety of partitioning, floorplanning, and routing algorithms for 3-D circuits has been developed that consider the unique characteristics of 3-D circuits. In these algorithms, the third dimension is either fully incorporated as a continuous variable or represented as an array of tiers.
- The objective function(s) within 3-D physical design algorithms has been extended to include routing congestion, power supply noise, and decoupling capacitance allocation in addition to traditional objectives, such as wirelength and area.
- The main challenges for developing 3-D physical design algorithms include the efficient representation of blocks in three physical dimensions, an increase in the solution space, an accurate wirelength model, and management of the vertical interconnects.

- Among different representations of the circuit blocks in three dimensions, the SP representation has been employed in several floorplanning (and placement) efforts to represent blocks within a vertical stack.
- The HPWL model used for 2-D circuits is typically inadequate for estimating the length of wires in 3-D circuits due to the additional interconnect required to connect to the TSVs whenever the interconnects are placed beyond the bounding box of a net.
- TSVs can be treated either as another type of "circuit cell" or as a separate resource that consumes silicon area.
- Floorplanning and placement techniques can be distinguished between methods that incorporate partition as a first step and those techniques where tier partition is not considered.
- Techniques that incorporate partitions prohibit intertier moves and are typically faster but can suffer from a limited solution space, thereby degrading the quality of the resulting solution.
- In the second stage of the floorplanning algorithms after partitions have been applied, solution perturbations are realized by different intratier moves among the blocks, such as swapping two blocks within a tier.
- The notion of spatial slacks can be used to guide the movement of blocks to ensure that sufficient whitespace exists to insert all (or the majority) of the TSVs while not greatly increasing the total area of the circuit.
- TSVs can be simultaneously inserted into the whitespace regions formed during floorplanning or at a later stage of the physical design process.
- These whitespaces can also be used for buffers, decoupling capacitors, and thermal and/or shielding TSVs, further constraining the placement of TSVs within a 3-D system.
- Force directed methods, SA, and analytic and heuristic expressions have been developed for placing 3-D systems.
- In force directed methods, the TSVs require additional forces to avoid both overlaps between TSVs and cells and to guarantee a safe distance (keep out zone) from the active devices to reduce noise coupling between TSVs and nearby devices.
- Routing in 3-D ICs is less developed as compared to placement and floorplanning due to the ability to individually route each tier as a 2-D circuit assuming the position of the TSVs are known.
- As the position of the TSVs is approximately known during the placement process, full 3-D routing can decrease wirelength and signal delay although care must be taken due to the non-polynomial complexity of these algorithms.

CHAPTER

TIMING OPTIMIZATION FOR TWO-TERMINAL INTERCONNECTS* 10

CHAPTER OUTLINE

The three-dimensional (3-D) interconnect prediction models discussed in Chapter 7, Interconnect Prediction Models, indicate a considerable reduction in interconnect length for 3-D circuits, which can improve the speed and power characteristics of modern integrated circuits. In addition to this important advantage, 3-D integration demonstrates many opportunities for heterogeneous systems-on-chip [305]. Integrating circuits from diverse fabrication processes into a single multitier system can result, however, in substantially different impedance characteristics for the interconnects of each physical tier within a 3-D circuit. These particular interconnect traits present new challenges and opportunities for 3-D circuits. This situation is further complicated by the intertier through silicon vias (TSVs), which can exhibit different impedance characteristics as compared to the horizontal (or intratier) interconnects. By considering the disparate interconnect impedance characteristics of 3-D circuits, the performance of the intertier interconnects can be significantly improved.

Based on the interconnect prediction models discussed in Chapter 7, Interconnect Prediction Models, these intertier interconnects are typically the longest lines within a 3-D circuit. Improving the performance of these global interconnects is therefore of significant importance, since the overall performance of the system is also significantly enhanced. In this chapter, a technique to decrease the delay of two-terminal intertier interconnects by optimally placing the intertier vias is described.

*The terms tier and plane are used interchangeably in this chapter.

Three-Dimensional Integrated Circuit Design. DOI: http://dx.doi.org/10.1016/B978-0-12-410501-0.00010-1

In the following section, the characteristics of the intertier interconnects, which require a different interconnect model as compared to two-dimensional (2-D) circuits, are discussed.

A closed-form solution for the minimum delay of a two-terminal intertier net that includes only one intertier via is provided in Section 10.2. The performance degradation that can occur by not optimally placing the intertier via is also discussed. Two-terminal nets that comprise more than one intertier via are investigated in Section 10.3. Additionally, an algorithm for placing intertier vias is described along with simulation and analytic results. A comparison of the proposed technique with a wire sizing algorithm is also presented. A short summary of the chapter is presented in the last section.

10.1 INTERTIER INTERCONNECT MODELS

The impedance characteristics of the metal layers that belong to different physical tiers of a 3-D circuit can vary significantly. This variation can be attributed to several causes. For example, consider a 3-D circuit where a processor is integrated with a few memory tiers. The fabrication process used for the processor and the memory tier can result in different impedances among the metal layers of the bottom and upper tiers. In addition, the magnitude of the process variations, which can affect, for example, the metal layer and interlayer dielectric thickness, of each of those processes is different. Consequently, the nominal impedance of the lines in each tier is affected differently. As a more specific example, consider the fabrication process for 3-D circuits manufactured by the MIT Lincoln Laboratories (MITLL) [177,307]. In this process, the third tier of the 3-D circuit can be optimized for RF circuits. In this case, the resistance of the metal layers in the RF tier is approximately an order of magnitude lower than that of the metal line resistance in the other digital tiers.

If the same manufacturing process, however, is used to fabricate all of the tiers of a 3-D circuit, variations in the impedance characteristics of the lines in a 3-D circuit can also originate from other causes, such as the bonding style and technology. Consider the interconnect structure, which is extracted to determine the capacitance on the topmost metal layer, shown in Figs. 10.1A and B for, respectively, a 2-D and 3-D circuit [422]. In a 2-D circuit, the metal layer located immediately below the parallel lines is treated as a ground tier. In a 3-D circuit where face-to-face bonding is employed, the topmost metal layer of the second tier, which is flipped and bonded onto the first

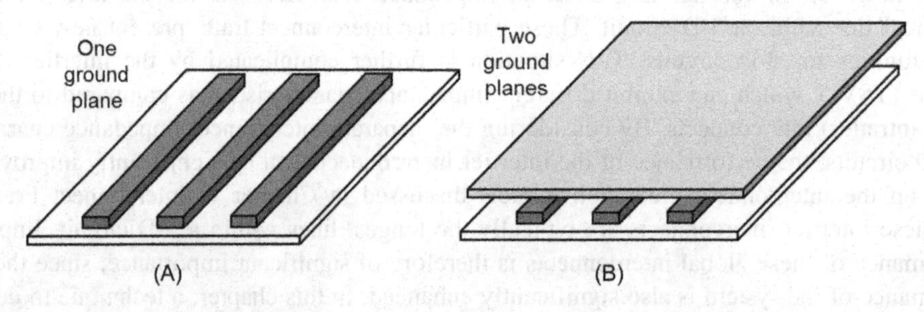

FIGURE 10.1

Global interconnect structures for impedance extraction. (A) Three parallel metal lines over a ground plane in a 2-D circuit, and (B) three parallel metal lines sandwiched between two ground planes in a 3-D circuit.

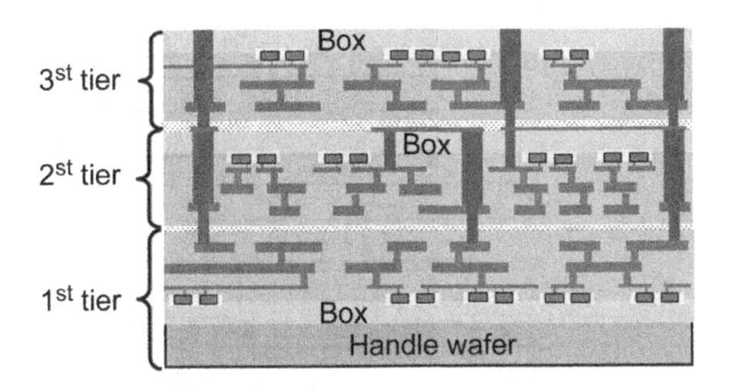

FIGURE 10.2

A three-tier FDSOI 3-D circuit [177,307]. Tiers one and two are face-to-face bonded, while tiers two and three are face-to-back bonded.

tier, behaves as a second ground plane for the topmost metal layer of the first physical tier of the 3-D circuit. Alternatively, in the case of face-to-back bonding, a ground plane can result from the substrate of the second tier if bulk CMOS technology is used to fabricate the tiers of the 3-D circuit. This second ground plane can significantly alter the capacitance of the interconnects. As an example, consider the MITLL 3-D process, where the 3-D system is illustrated in Fig. 10.2. In this technology, the lines of the topmost metal layer of the first and second tier exhibit a different capacitance as compared to the capacitance of the global metal layer on the third tier.

Another important factor in determining the performance of the intertier interconnects is the impedance characteristics of the intertier vias, which depend upon the bonding style and technology. Face-to-back bonding will likely result in intertier vias with a greater length as compared to face-to-face bonding since in the latter case the vertical distance among the tiers is greater. Furthermore, the type of technology can have a profound effect on the impedance of the intertier vias.

To justify this argument, the capacitance of different intertier via structures, illustrated in Figs. 10.3 through 10.5, has been extracted with a commercial impedance extraction tool [423]. The structure shown in Fig. 10.3 corresponds to that segment of the intertier via that is surrounded by the horizontal metal levels of a tier within a 3-D system, and is independent of the target technology (i.e., bulk CMOS or silicon on insulator (SOI)). The number of metal levels surrounding the via, however, depends upon the fabrication process. Alternatively, the structure shown in Fig. 10.4 corresponds to an intertier via surrounded by other vias, which traverse the layers of the dielectric and bonding material (assuming for simplicity to have identical dielectric constants). Such a via structure can exist in an SOI technology. For 3-D circuits where each tier is fabricated with mainstream CMOS processes or with a combination of SOI and CMOS processes, the structure in Fig. 10.5 can be utilized for capacitance extraction.

The intertier interconnects are therefore modeled as an assembly of horizontal interconnect segments with different impedance characteristics connected by intertier vias. This model is considerably different from the typical interconnect model used in a 2-D circuit, where a two-terminal net is modeled by a single segment with the same uniform distributed impedance throughout the interconnect length. An intertier interconnect connecting two circuits located on different physical tiers is illustrated in Fig. 10.6. In the following sections, it is shown that the delay of this type of 3-D interconnect can be improved by appropriately placing the intertier vias.

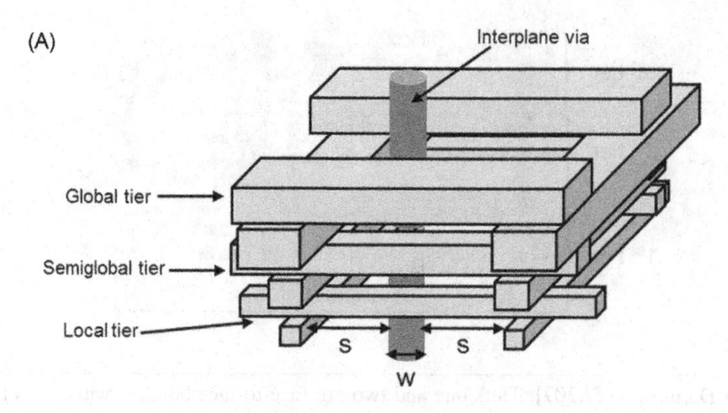

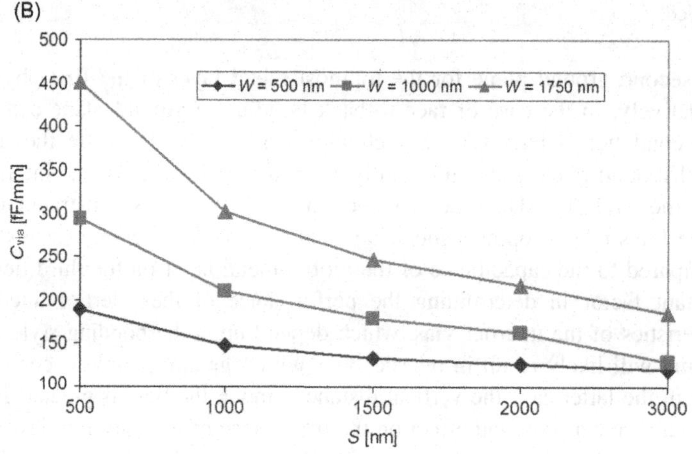

FIGURE 10.3

Capacitance extraction for an intertier via structure, (A) intertier via surrounded by orthogonal metal layers, and (B) capacitance values for different via sizes and spacing values. The same dielectric material is assumed for all of the layers (i.e., $\varepsilon_d = \varepsilon_i = \varepsilon_{SiO_2}$).

10.2 TWO-TERMINAL NETS WITH A SINGLE INTERTIER VIA

In this section, the simple case of an intertier two-terminal net connecting two circuits located on different physical tiers including only one intertier via is considered. The delay model and the dependence of the delay on the via location are discussed in Section 10.2.1. A physical explanation of the variation in delay with the via location is provided in Section 10.2.2, whereas the optimum via location for different impedance characteristics of the interconnect segments is determined in Section 10.2.3. The improvement in delay for this type of interconnect is demonstrated in Section 10.2.4.

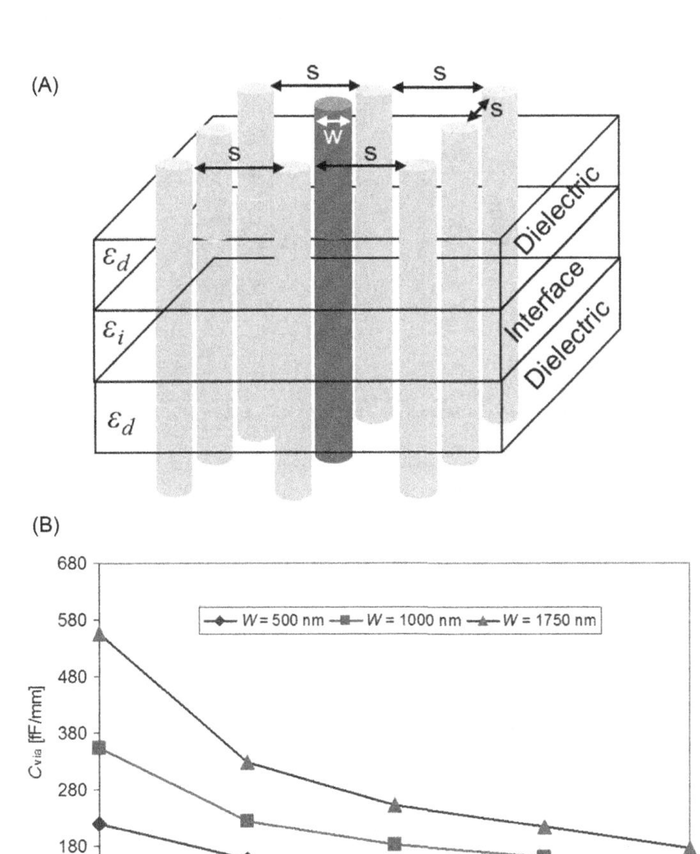

FIGURE 10.4

Capacitance extraction for an intertier via structure, (A) intertier via through layers of dielectric and the bonding interface, surrounded by eight intertier vias, and (B) capacitance values for different via sizes and spacings.

10.2.1 ELMORE DELAY MODEL OF AN INTERTIER INTERCONNECT

As previously mentioned, due to the nonuniformity of the interconnects, each segment is modeled as a distributed *RC* line with different impedance characteristics, as shown in Fig. 10.6. The driver is modeled as a step input voltage source with a linear output resistance R_S, and the interconnect is terminated with a capacitive load C_L. The total resistance and capacitance of the horizontal segments 1 and 2 are $R_1 = r_1 l_1$, $R_2 = r_2 l_2$, and $C_1 = c_1 l_1$, $C_2 = c_2 l_2$, where r_1, r_2 and c_1, c_2 denote, respectively, the resistance and capacitance per unit length. The length of the two horizontal segments is l_1 and l_2, and the via length is l_{v1}. Since the via may span more than one physical tier, l_{v1} can be expressed as

$$l_{v1} = (n-1)l_v, \tag{10.1}$$

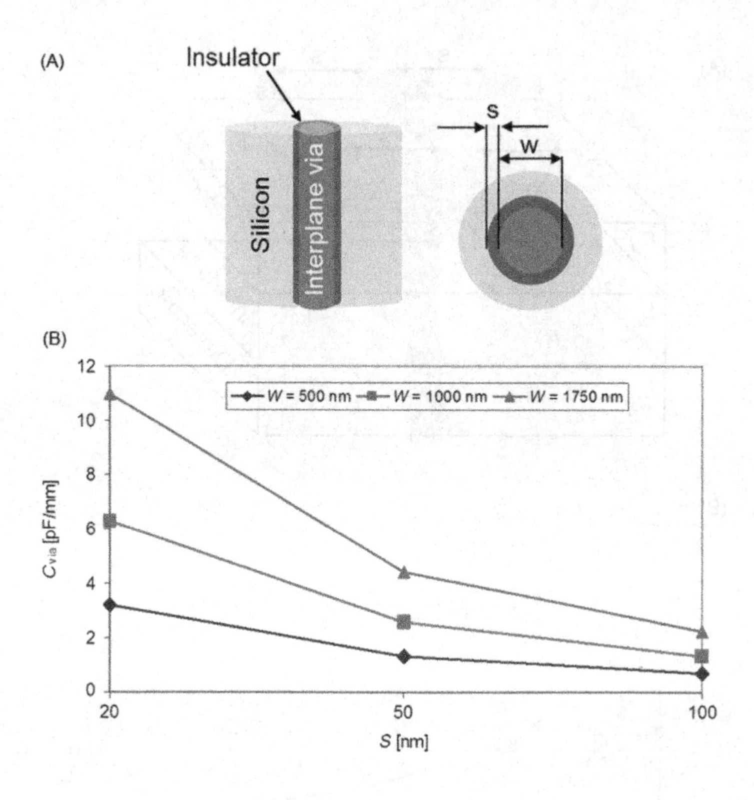

FIGURE 10.5

Capacitance extraction for an intertier via structure, (A) intertier via through silicon substrate, surrounded by a thin insulator layer, and (B) capacitance for different via sizes and thicknesses of the insulator layer.

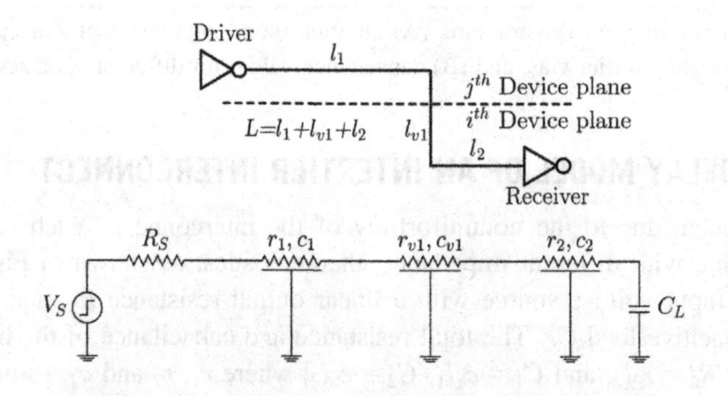

FIGURE 10.6

Two terminal intertier interconnect with single via and corresponding electrical model.

where n is the number of physical tiers among the connected circuits and l_v is the length of the via connecting two metal layers located on two adjacent physical tiers. This via length is determined by the fabrication process and can typically range from 10 to 200 μm [51,155,424]. The total length of the line can be expressed as

$$L = l_1 + l_{v1} + l_2. \tag{10.2}$$

Due to the nonuniform impedance characteristics of the line, the via location or, alternatively, the length of each horizontal wire segment, affects the delay of the line. Consequently, the objective is to place the via to ensure the interconnect delay is minimum. To analyze the delay of a line, the distributed Elmore delay model has been adopted due to the simplicity and high fidelity of this model [425]. The accuracy of the model can also be further improved, as discussed in [426]. However, unlike a single tier, more than one set of fitting coefficients is required in a 3-D system. The objective is to determine the via location or, alternatively, the length of segments 1 and 2, to minimize the distributed Elmore delay of the line.

The distributed Elmore delay for the system shown in Fig. 10.6 is

$$T_{el} = \frac{R_1C_1}{2} + C_1R_s + \frac{R_{v1}C_{v1}}{2} + C_{v1}(R_s + R_1) + \frac{R_2C_2}{2} + C_2(R_s + R_1 + R_{v1}) + C_L(R_s + R_1 + R_{v1} + R_2). \tag{10.3}$$

Substituting the total resistance and capacitance with the per unit length parameters, and using (10.1) and (10.2), the Elmore delay described in (10.3) is a function of the length of the first segment l_1,

$$T_{el}(l_1) = A_1l_1^2 + A_2l_1 + A_3, \tag{10.4}$$

where

$$A_1 = \left(\frac{1}{2}\right)(r_1c_1 - 2r_1c_2 + r_2c_2), \tag{10.5}$$

$$A_2 = R_s(c_1 - c_2) + (n-1)l_v(r_1c_{v1} - r_{v1}c_2 + r_2c_2 - r_1c_2) + L(r_1c_2 - r_2c_2) + C_L(r_1 - r_2), \tag{10.6}$$

$$\begin{aligned} A_3 = {} & ((n-1)l_v)^2\left(\frac{r_1c_{v1}}{2} - r_{v1}c_2 + \frac{r_2c_2}{2}\right) + (n-1)l_v((Lc_2 + C_L)(r_{v1} - r_2) + R_s(c_{v1} - c_2)) \\ & + C_L(R_s + Lr_2) + R_sLc_2 + \frac{L^2r_2c_2}{2}. \end{aligned} \tag{10.7}$$

Eq. (10.4) describes a parabola, but the convexity is not guaranteed since A_1 can be negative. The second derivative of (10.4) with respect to l_1 is

$$\frac{d^2T_{el}}{dl_1^2} = 2A_1. \tag{10.8}$$

Depending upon the sign of A_1, the propagation delay of the line exhibits either a minimum or a maximum as l_l varies or, alternatively, as the location of the via along the line changes. The following notations are introduced to facilitate the analysis,

$$r_{21} = \frac{r_2}{r_1}, \tag{10.9}$$

$$c_{12} = \frac{c_1}{c_2}. \tag{10.10}$$

From (10.9) and (10.10), the second derivative is

$$\frac{d^2T_{el}}{dl_1^2} = r_1c_2(r_{21} + c_{12} - 2). \tag{10.11}$$

Since r_1c_2 is always positive, the sign of (10.11) and, consequently, the timing behavior of the line only depends upon the sign of the term in the parentheses. To minimize the propagation delay, the following inequality should be satisfied,

$$r_{21} + c_{12} - 2 > 0. \tag{10.12}$$

Note that the notations introduced in (10.9) and (10.10) describe resistance and capacitance ratios and, consequently, are dimensionless. If (10.12) is not satisfied, the delay of the line exhibits a maximum.

10.2.2 INTERTIER INTERCONNECT DELAY

In the previous subsection, (10.12) only depends on the impedance characteristics of the segments that constitute the line. The dependence of the delay on the impedance of the segments of the line is analyzed in this subsection. To better explain this dependence, consider initially an *RC* interconnect line of uniform impedance. An optimum wire shaping function of the propagation delay has been shown to exist for *RC* interconnects [427]. The optimum wire sizing function is a monotonically decreasing function of the width, as illustrated in Fig. 10.7. This decrease in width occurs because the downstream capacitance is small close to the receiver and, therefore, this capacitance is charged by a larger resistance due to the reduction in line width. Alternatively, the downstream

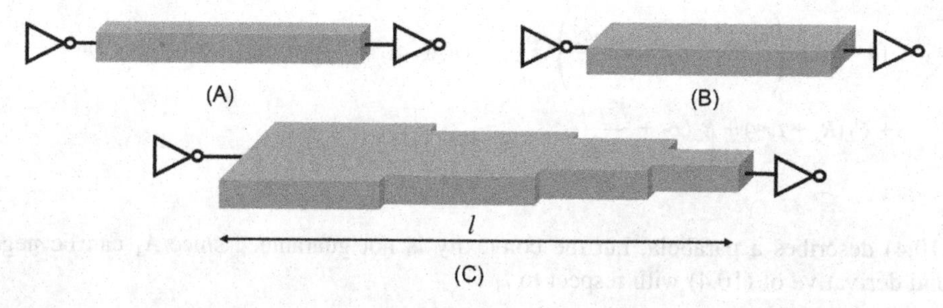

FIGURE 10.7

An example of interconnect sizing. (A) An interconnect of minimum width, W_{min}, (B) uniform interconnect sizing $W > W_{min}$, and (C) nonuniform interconnect sizing $W = f(l)$.

capacitance close to the driver is large and, consequently, a small resistance charges the capacitance, due to the widening of the interconnect towards the driver.

An intertier interconnect with nonuniform impedance characteristics can be treated as a tapered uniform line consisting of only two segments. If r_0 and c_0 denote, respectively, the resistance and capacitance per unit area, the resistance and capacitance per unit length of the segments of the intertier interconnect can be described as $r_1 = r_0/w_1$, $r_2 = r_0/w_2$, $c_1 = c_0w_1$, and $c_2 = c_0w_2$, where w_1 and w_2 are the width of the segments of the line. Thus, r_{21} and c_{12} characterize the tapering factor of the line. If $r_{21} > 1$ and $c_{12} > 1$, the tapering factor is smaller than one, and the delay of the line can be reduced by appropriately selecting the length of the line segments or, alternatively, the via location. These impedance ratios for r_{21} and c_{12} correspond to a decreasing shaping function of the interconnect, which decreases the interconnect delay [427]. Consequently, the via location should be selected to ensure the shape of the intertier interconnect approaches the sizing function of a uniform interconnect comprised of only two segments. Note that the optimum placement of the via does not necessarily achieve the optimum tapering factor, as described by the optimum wire sizing function, yet a significant reduction in the delay of the line is possible. In addition, as discussed in Section 10.3.3, the optimum via placement can achieve a greater improvement in delay as compared to wire sizing for intertier interconnects. This behavior is mainly due to the nonuniform characteristics of the interconnect and the discontinuities that the intertier vias create in the interconnect.

If $r_{21} > 1$ and $c_{12} < 1$, both the resistance and the capacitance per unit length of segment 2 are greater than the *RC* impedance of segment 1 and, typically, the optimum via location ensures the length of the segment with the higher impedance is minimized. The same relationship applies to the case where $r_{21} < 1$ and $c_{12} > 1$. If $r_{21} < 1$ and $c_{12} < 1$, a longer l_1 increases the resistance of the line, while a longer l_2 increases the capacitance of the line. The tapering factor becomes greater than one and the delay of the line exhibits a maximum. Consequently, the nonuniform impedance characteristics of the intertier interconnects, which are increased by the impedance of the intertier vias makes the process of placing a via a non-convex function, unlike for wire sizing [428]. This argument is analytically proven in Section 10.3.

10.2.3 OPTIMUM VIA LOCATION

In Section 10.2.1, the timing behavior of the line is shown to depend on the impedance characteristics of the segments that comprise the line (see expression (10.12)). Placing a via such that the delay is minimized based on the impedance of the interconnect segments is presented in this subsection. Note that while the timing behavior of the line with respect to the via location does not depend on the interconnect length, the via location that achieves the minimum delay does depend upon the interconnect length.

From (10.4)–(10.7), the value of l_1 for which the delay exhibits an extremum, either minimum or maximum, is

$$l_1 = -\left[\frac{(n-1)l_v(r_1c_{v1} - r_{v1}c_2 + r_2c_2 - r_1c_2) + R_s(c_1 - c_2) + L(r_1c_2 - r_2c_2) + C_L(r_1 - r_2)}{r_1c_1 - 2r_1c_2 + r_2c_2}\right]. \tag{10.13}$$

Since no constraints have been applied to the value of l_1, the extreme point can occur for values other than within the physical domain of l_1, i.e., $l_1 \in [l_d, l_r]$, where $l_d = l_{min}$ and $l_r = L - l_{vl} - l_{min}$.

The minimum distance between a via and a cell is determined by $l_{\min}$, as constrained by the design rules of the fabrication process. The lemma in Appendix B characterizes the optimum via location for different values of r_{21}, c_{12}, and l_1.

Depending upon the sign of (10.11) and the value of l_1 in (10.13), the optimum via location is determined for each possible case:

i. $d^2T_{el}/dl_1^2 > 0$. If $l_1 \in [l_d, l_r]$, the propagation delay is minimum when l_1 is the value described in (10.13). Consequently, the via should be placed at a distance l_1 from the driver. SPICE measurements of the 50% propagation delay for a 600 μm line are illustrated in Fig. 10.8 versus the via location l_1. Two observations can be made. First, that the delay exhibits a minimum and, second, that the minima shift to the right as r_{21} increases. If $l_1 < l_d$, the via should be placed closest to the driver, while if $l_1 > l_r$, the via should be placed closest to the receiver.

ii. $d^2T_{el}/dl_1^2 < 0$. In this case, the delay of the line reaches a maximum for the value of l_1 as described in (10.13). If $l_1 \in [l_d, l_r]$, according to the lemma in Appendix B, for $l_1 < (l_r + l_d)/2$, the via should be placed closest to the receiver, while for $l_1 > (l_r + l_d)/2$, the via should be placed closest to the driver. In Fig. 10.9, SPICE measurements of the 50% propagation delay for a 600 μm line is shown as a function of the via location l_1. Note that the delay reaches a

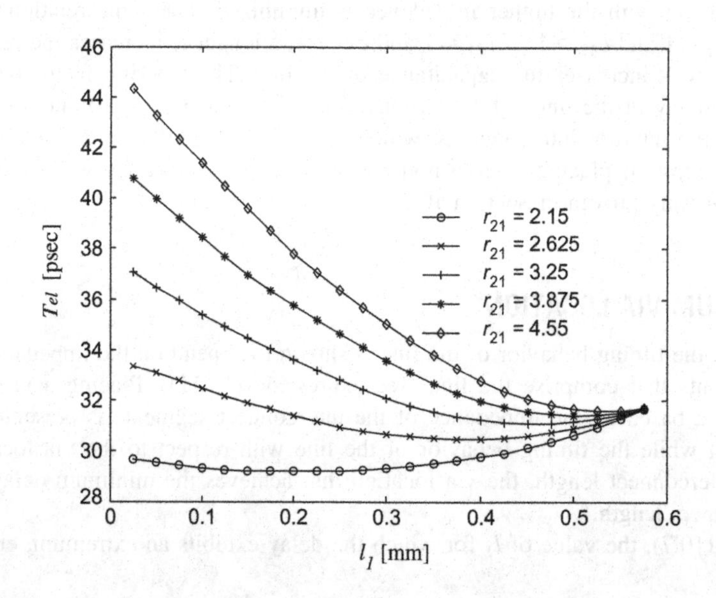

FIGURE 10.8

SPICE measurements of 50% propagation delay of a 600 μm line versus the via location l_1 for different values of r_{21}. The interconnect parameters are $r_1 = 79.5$ Ω/mm, $r_{v1} = 5.7$ Ω/mm, $c_{v1} = 6$ pF/mm, $c_2 = 439$ fF/mm, $c_{12} = 1.45$, $l_v = 20$ μm, and $n = 2$. The driver resistance and load capacitance are, resepctively, $R_S = 50$ Ω and $C_L = 50$ fF.

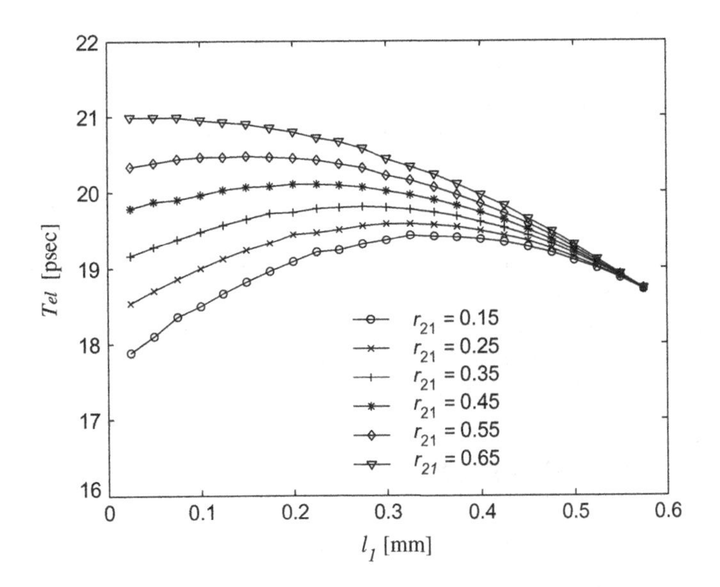

FIGURE 10.9

SPICE measurements of the 50% propagation delay for a 600 μm line versus the via location l_1 for different values of r_{21}. The interconnect parameters are $r_1 = 79.5\ \Omega$/mm, $r_{v1} = 5.7\ \Omega$/mm, $c_{v1} = 6$ pF/mm, $c_2 = 439$ fF/mm, $c_{12} = 0.46$, $l_v = 20$ μm, and $n = 2$. The driver resistance and load capacitance are, respectively, $R_S = 50\ \Omega$ and $C_L = 50$ fF.

maximum and that the maximum shifts to the right as r_{21} increases. If $l_1 < l_d$, the via should be placed closest to the receiver, while if $l_1 > l_r$, the via should be placed closest to the driver.

iii. $d^2T_{el}/dl_1^2 = 0$. If the second derivative equals zero, (10.4) becomes

$$T_{el}(l_1) = A_2 l_1 + A_3, \tag{10.14}$$

which is a linear function of l_1. The first derivative of (10.14) is equal to A_2. For $A_2 < 0$, (10.14) is strictly decreasing, and the delay is minimum by placing the via closest to the receiver. For $A_2 > 0$, (10.14) is strictly increasing, and the delay is minimum by placing the via closest to the driver. Note that the fundamental minimum distance of a via from a cell is technology dependent. In the special case where $r_{21} = c_{12} = 1$, from (10.6) and (10.7), $A_2 \ll A_3$ and the delay is independent of l_1. However, as n increases, A_2 also increases and the choice of n affects the rate of change in the delay. This particular case corresponds to the model introduced in [298] to evaluate the performance of 3-D ICs, which does not capture the timing behavior of intertier interconnects in 3-D circuits as analyzed in cases A and B.

As illustrated in Figs. 10.8 and 10.9, the optimum via location shifts to the right (left) when r_{21} increases (decreases). To explain this behavior, consider the definitions of r_{21} and c_{12} in (10.9) and (10.10), where r_{21} (c_{12}) describes the resistance (capacitance) ratio of the corresponding horizontal segments. Referring to Fig. 10.8, both r_{21} and c_{12} are greater than one, which means that segment 1 is less (more) resistive (capacitive) than segment 2. Assuming that $r_{21} = 1$, the delay of the line decreases as the length of the more capacitive segment l_1 (i.e., C_1) decreases. However, l_1 does not vanish because C_2 increases as l_1 decreases, approaching C_1. Consequently, as l_1 is decreased beyond a certain distance, described in (10.13), the delay starts to increase. As r_{21} becomes greater than one, the optimum point occurs with increasing values of l_1, although segment 1 is more capacitive than segment 2 ($c_{12} > 1$). This behavior occurs because the delay depends not only on the capacitance but also on the current, which is controlled by the resistance of each segment. Due to the distributed nature of the impedance of the line, an increase in the resistance of the second segment (increasing r_{21}) can be compensated, ensuring that the delay remains minimum, by reducing the capacitance that this resistance sees near the receiver (the upstream resistance). In the case where $c_{12} \gg 1$, $r_{12} \ll 1$, and segment 1 is both more capacitive and resistive than segment 2, l_1 becomes small to reduce the overall delay. In this case, (10.13) yields negative values for l_1 and is essentially zero (the via should be located next to the driver).

10.2.4 IMPROVEMENT IN INTERCONNECT DELAY

SPICE simulations of the 50% propagation delay of two terminal nets with a single via are listed in Table 10.1 for various impedance characteristics. The interconnect propagation delay listed in columns 8, 9, and 11 corresponds to the cases where the via is optimally placed, the via is placed at the center of the interconnect, and at the maximum delay point, respectively. In column 10, the improvement in delay over the case where the via is placed at the center of the line is reported, while the improvement in delay over placing the via at the endpoints of the line is listed in column 12. If the delay exhibits a minimum according to (10.11), the maximum delay point coincides with one of the end points of the interconnect, otherwise the maximum delay point is described by (10.13). The delay analysis considers relatively short interconnect lengths where repeaters cannot reduce the interconnect delay, and therefore, optimally placing the vias is the primary method to decrease the delay of these interconnects.

Note the considerable improvement in delay achieved by optimally placing the via in those cases where the optimum location coincides with the end points of the interconnect. This large

Table 10.1 Spice Simulations for Two Terminal Interconnects With a Single Intertier Via

Length (μm)	R_S (Ω)	C_L (fF)	r_1 (Ω/mm)	c_2 (fF/mm)	r_{21}	c_{12}	T_{opt}	T_{center}	Impr (%)	T_{max}	Impr (%)
200	100	50	159.5	239	3.50	1.55	19.79	20.02	1.16	20.76	4.90
200	100	50	159.5	239	2.85	1.85	19.89	20.08	0.96	20.44	2.77
300	120	100	93.8	387	2.88	1.67	36.03	36.30	0.75	37.62	4.41
300	120	100	93.8	387	2.88	1.35	34.56	34.80	0.69	35.92	3.94
400	100	100	75.8	287	2.15	2.87	29.30	34.98	19.39	42.34	44.51
400	100	100	75.8	287	3.70	0.77	24.74	28.94	16.98	33.34	34.76
500	30	100	23.8	287	1.50	2.57	54.02	63.37	17.31	72.58	34.36
500	30	100	23.8	287	3.35	2.37	58.18	58.69	0.88	59.77	2.73
500	30	100	23.8	287	3.15	0.87	51.09	54.83	7.32	59.02	15.52
Average improvement									7.27		16.43

improvement occurs because in these cases the minimum delay point, as described by (10.13), occurs at values between the end points of the interconnect, i.e., [0, L]. For these instances, the propagation delay has an almost linear dependence (with large slope) on the via location and, therefore, the improvement in delay is significant. Alternatively, for those instances where the optimum occurs for positive locations smaller than the interconnect length L, the variation in the delay improvement is small as the delay exhibits a parabolic form, and consequently, the rate of change in the delay with via location is small.

For several interconnect instances, the optimum location is either close to the receiver or the driver. Routing congestion therefore causes some of the intertier vias to be placed at nonoptimal locations. In addition, the improvement in delay that can be achieved varies considerably with the interconnect parameters. In Figs. 10.10 and 10.11, the decrease in the improvement in delay resulting from the nonoptimal placement of the vias is illustrated for a 500 μm intertier interconnect. The ratios r_{21} and c_{12} range from 0.1 to 5 and the intertier via is placed 50 μm from the optimum location.

A nonnegligible reduction in the improvement in delay can occur from the nonoptimal placement of the via for a 50 μm departure from the optimal location. In addition, the decrease in improvement is greater for those values of r_{21} and c_{12}, where the optimum is at the end points of the line. For the optimum location corresponding to points along the interconnect length, the decrease in the delay improvement is small as depicted by the approximately flat portion of the surface in Figs. 10.10 and 10.11. In addition, the dependence of the delay improvement on the capacitance ratio c_{12} is in general greater than the dependence on the resistance ratio r_{12}. For low values of the driver resistance, the dependence of the delay improvement on r_{12} is significant, as shown in Fig. 10.10. Increasing the driver resistance lowers this dependence, as depicted in Fig. 10.11, as the term $R_S \sum_i C_i$ dominates the delay of the line.

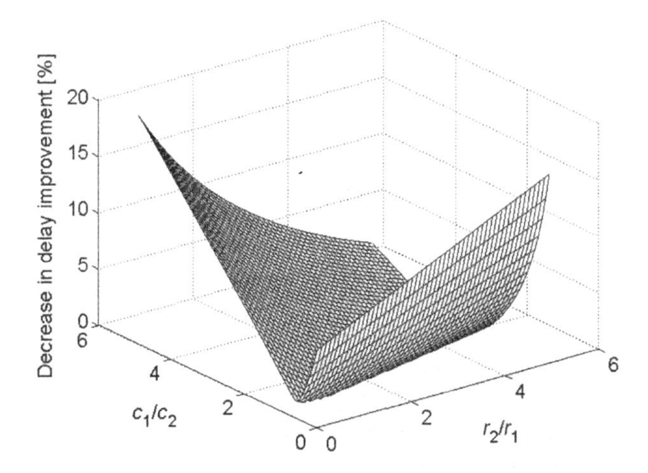

FIGURE 10.10

Decrease in the delay improvement caused by the nonoptimal placement of the intertier via for a 500 μm interconnect. The interconnect parameters are $r_1 = 23.5$ Ω/mm, $r_{v1} = 270$ Ω/mm, $c_{v1} = 270$ fF/mm, $c_2 = 287$ fF/mm, $l_v = 15$ μm, and $n = 2$. The driver resistance and load capacitance are, respectively, $R_S = 30$ Ω and $C_L = 100$ fF.

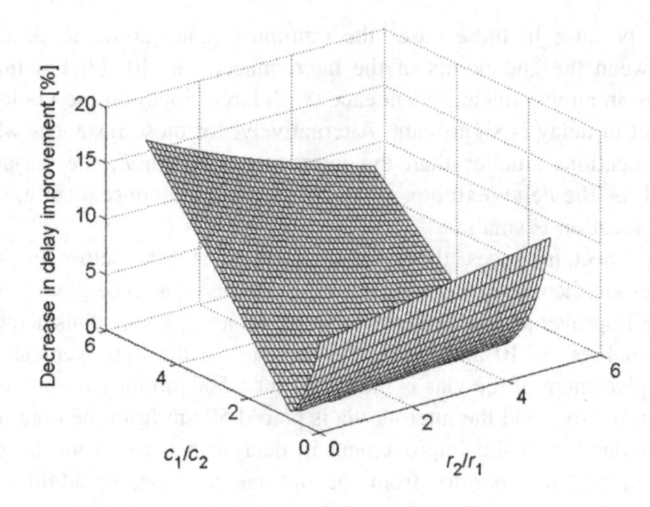

FIGURE 10.11

Decrease in the delay improvement due to the nonoptimal placement of the intertier via for a 500 μm interconnect. The interconnect parameters are $r_1 = 23.5$ Ω/mm, $r_v1 = 6.7$ Ω/mm, $c_{v1} = 270$ fF/mm, $c_2 = 287$ fF/mm, $l_v = 15$ μm, and $n = 2$. The driver resistance and load capacitance are, respectively, $R_S = 100$ Ω and $C_L = 100$ fF.

This dependence shows that for large drivers, even if the interconnect is almost uniform ($r_{21} \approx 1$), considerable savings in delay can be achieved by optimally placing the via. Furthermore, as larger drivers are used to reduce the interconnect delay, the improvement in delay becomes greater. If small drivers are utilized, the savings in delay is primarily obtained from the difference in the capacitance of the segments, even if $c_{12} \approx 1$. Additionally, a slight decrease in the maximum degradation of the improvement in delay that occurs from a nonoptimal via placement can be observed between Figs. 10.10 and 10.11. This decrease occurs due to the increase in the driver resistance, which is considered fixed and can greatly affect the interconnect delay, while the delay variation due to the impedance characteristics of the segments is reduced. A similar behavior applies for the capacitance ratio c_{12} where the load capacitance is increased. In the following section, two-terminal nets with multiple intertier vias are considered.

10.3 TWO TERMINAL INTERCONNECTS WITH MULTIPLE INTERTIER VIAS

In the previous section, the case where a two terminal interconnect includes only one intertier via is discussed. Due to routing obstacles and placed cells, however, a stacked intertier via that spans multiple physical tiers may not be possible. In addition, in the previous analysis, no constraints on the location of the via is imposed. This situation is rather impractical since existing interconnects and active devices may not permit an intertier via to be placed at a particular location along the path of the interconnect. In addition, since these vias penetrate the device layers, certain regions are reserved for intertier via placement to minimally disrupt the location of the transistors.

Due to these reasons, a two terminal intertier interconnect in a 3-D circuit consists of multiple intertier vias, where each one of these vias is placed within a certain physical interval

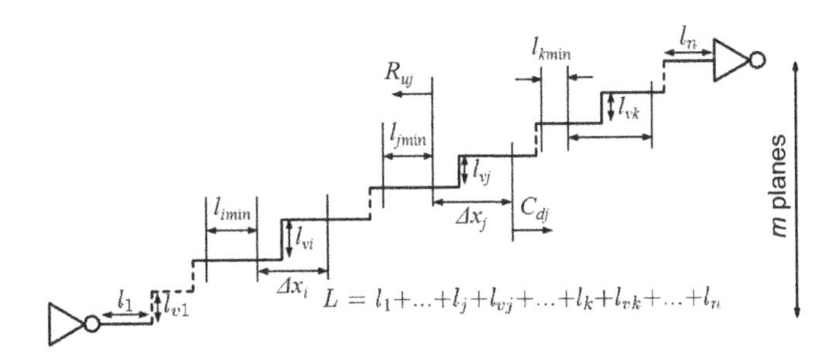

FIGURE 10.12

Intertier interconnect consisting of m segments connecting two circuits located n tiers apart.

(i.e., white space). A schematic of an intertier interconnect connecting two circuits located n tiers apart is illustrated in Fig. 10.12. The horizontal segments of the line are connected through the vias, which can traverse more than one tier. Consequently, the number of horizontal segments within the interconnect m is smaller or at most equal to the number of physical tiers between the two circuits, i.e., $n \geq m$, where the equality only applies when each of the vias connects metal layers from two adjacent physical tiers.

Each horizontal segment j of a line is located on a different physical tier with length l_j. The vias are denoted by the index of the first of the two connected segments. For example, if a via connects segment j and $j+1$, the via is denoted as v_j with length l_{vj}. Note that tiers j and $j+1$ are not necessarily physically adjacent. The total length of the line L is equal to the summation of the length of the horizontal segments and vias,

$$L = l_1 + l_{v1} + \cdots + l_j + l_{vj} + \cdots l_n. \tag{10.15}$$

The length of each horizontal segment of a line is bounded,

$$l_{j\min} \leq l_j \leq l_{j\min} + \Delta x_j \quad \text{for } j = 1, \tag{10.16a}$$

$$l_{j\min} \leq l_j \leq \Delta x_{j-1} + l_{j\min} + \Delta x_j \quad \text{for } j \in [2, m-1], \tag{10.16b}$$

$$l_{j\min} \leq l_j \leq l_{j\min} + \Delta x_{j-1} \quad \text{for } j = m, \tag{10.16c}$$

and the via placement is also constrained,

$$0 \leq x_j \leq \Delta x_j, \tag{10.17}$$

where $l_{j\min}$ denotes the minimum length of the interconnect segment on tier j, and Δx_j is the length of the interval in which the via that connects tiers j and $j+1$ is placed. This interval length is called the "allowed interval" here for clarity. x_j is the distance of the via location from the edge of the allowed interval.

$l_{j\min}$ is the length of an interconnect segment connecting two allowed intervals or an allowed interval and a placed cell. These lengths are considered fixed. Alternatively, the routing path of a net is not altered except for the via location within the allowed interval. Each horizontal segment is assumed to be on a single metal layer within the physical tier. In the case where a horizontal segment is on more than one layer, as the outcome of a layer assignment algorithm [429], the problem can be approached in two different ways. The intratier vias can be treated as additional variables where the location of these vias needs to be determined. This formulation requires, however, the generation of

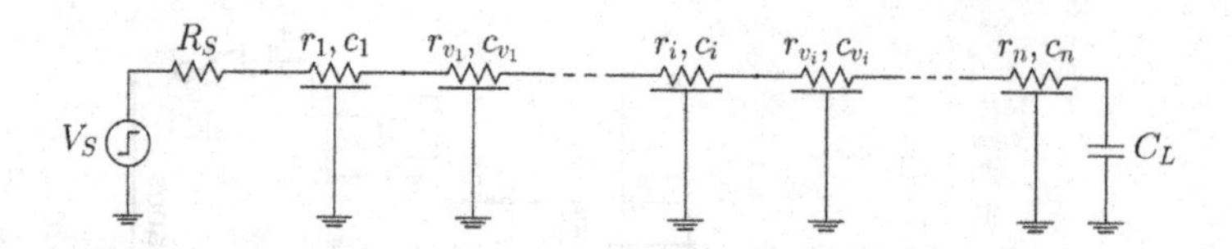

FIGURE 10.13

Intertier interconnect model composed of a set of nonuniform distributed *RC* segments.

Table 10.2 Notation for Two Terminal Intertier Interconnects

Notation	Definition
R_S	Driver resistance
C_L	Load capacitance
r_j (c_j)	Resistance (capacitance) per unit length of horizontal segment j
r_{vj} (c_{vj})	Resistance (capacitance) per unit length of intertier via v_j
R_j (C_j)	Total interconnect resistance (capacitance) of horizontal segment j
R_{vj} (C_{vj})	Total interconnect resistance (capacitance) of intertier via v_j
R_{u_j}	Upstream resistance of the allowed interval for via v_j
C_{d_j}	Downstream capacitance of the allowed interval for via v_j

additional allowed intervals, specifically for the intratier vias. Alternatively, the first and last section of the segment connected to the intertier vias remains as a variable while the remaining sections of the horizontal segment constitute the minimum length of segment $l_{j\min}$, which is constant.

The distributed Elmore delay model is used to determine the delay of these interconnects. The corresponding electrical model of the line is depicted in Fig. 10.13. The related notation is listed in Table 10.2. The distributed Elmore delay of a two terminal interconnect in matrix form is

$$T(\mathbf{l}) = 0.5\mathbf{l}^{\mathrm{T}}\mathbf{A}\mathbf{l} + \mathbf{b}\mathbf{l} + D, \tag{10.18}$$

$$\mathbf{l} = \begin{bmatrix} l_1 & l_2 & \cdots & l_{m-1} & l_m \end{bmatrix}^{\mathrm{T}}, \tag{10.19}$$

$$\mathbf{A} = \begin{bmatrix} r_1c_1 & r_1c_2 & \cdots & r_1c_m \\ \vdots & \vdots & \cdots & \vdots \\ r_1c_m & r_2c_m & \cdots & r_mc_m \end{bmatrix}, \tag{10.20}$$

$$\mathbf{b} = \begin{bmatrix} r_1\left(\sum_{i=1}^{m-1} c_{vi}l_{vi} + C_L\right) + c_1R_S \\ \vdots \\ r_mC_L + c_m\left(R_s + \sum_{i=1}^{m-1} r_{vi}l_{vi}\right) \end{bmatrix}, \tag{10.21}$$

$$D = R_s \sum_{i=1}^{m-1} c_{vi} l_{vi} + C_L \sum_{i=1}^{m-1} r_{vi} l_{vi} + \frac{1}{2} \sum_{i=1}^{m-1} r_{vi} c_{vi} l_{vi}^2 + R_s C_L. \tag{10.22}$$

Since (10.22) is a constant quantity, the optimization problem can be described as follows:

(P) **minimize** $T(\mathbf{l}) = 0.5\mathbf{l}^{\mathrm{T}}\mathbf{A}\mathbf{l} + \mathbf{b}\mathbf{l}$ $T(\mathbf{l}) = 0.5\mathbf{l}^{\mathrm{T}}\mathbf{A}\mathbf{l} + \mathbf{b}\mathbf{l}$,
subject to (10.15)–(10.17).

As described by the following theorem, the primal problem (P) is typically not convex and, therefore, convex quadratic programing optimization techniques are not directly applicable.

Theorem 1: The primal optimization problem (P) is convex *iff*

$$r_{j+1}c_j - r_j c_{j+1} > 0. \tag{10.23}$$

Proof: **A** is a positive definite matrix if all subdeterminants are positive. By elementary row operations, the subdeterminants of **A** are positive *iff* (10.23) applies. If (10.23) applies, **A** is positive definite and (P) is a convex optimization problem.

Note that condition (10.23) must be satisfied for every segment j to make (P) into a convex optimization problem. Due to the heterogeneity of 3-D circuits, (10.18) typically is of indefinite quadratic form. Therefore, convex quadratic programing techniques may not produce the global minima. Certain transformations can be applied to convert (P) into a convex optimization problem [430]; the objective functions, however, are no longer quadratic. Alternatively, (P) can be treated as a geometric programming problem. Geometric programs include optimization problems for functions and inequalities of the following form,

$$g(\mathbf{y}) = \sum_{j=1}^{M} s_j y_1^{a_{1j}} y_2^{a_{2j}} \ldots y_n^{a_{nj}}, \tag{10.24}$$

$$s_j y_1^{a_{1j}} y_2^{a_{2j}} \ldots y_n^{a_{nj}} \le 1, \tag{10.25}$$

where the variables y_j and coefficients s_j must be positive and the exponents a_{ij} are real numbers. Although equality constraints are not allowed in standard geometric problems, (P) can be solved as a generalized geometric program, as described in [431], where a globally optimum solution is determined. Alternatively, by considering the particular characteristics of the optimization problem, an efficient heuristic is proposed for placing the intertier vias.

10.3.1 TWO TERMINAL VIA PLACEMENT HEURISTIC

In this section, a heuristic for the near optimal intertier via placement of two terminal nets that include several intertier vias is described. The key concept in the heuristic is that the optimum via placement depends primarily upon the size of the allowed interval (estimated or known after an initial placement) rather than the precise location of the via. Consider the intertier interconnect line shown in Fig. 10.12, where the optimum location for via j connecting interconnect segments j and $j+1$ is to be determined. With respect to this via, the critical point (i.e., $\partial T_{el}/\partial x_j = 0$) of the Elmore delay is

$$x_j = -\left[\frac{l_{vj}(r_j c_{vj} - r_{vj} c_{j+1} + r_{j+1} c_{j+1} - r_j c_{j+1}) + R_{uj}(c_j - c_{j+1}) + \Delta x_j (r_j - r_{j+1}) c_{j+1} + C_{dj}(r_j - r_{j+1})}{r_j c_j - 2 r_j c_{j+1} + r_{j+1} c_{j+1}}\right], \quad (10.26)$$

where R_{uj} and C_{dj} are, respectively, the upstream resistance and downstream capacitance of the allowed interval for via j (see also Table 10.2), as shown in Fig. 10.12. The Elmore delay of the line with respect to x_j can be either a convex or a concave function. The remaining discussion in this section applies to the case where the Elmore delay of the line is a convex function with respect to x_j. A similar analysis can be applied for the concave case.

In (10.26), the optimum via location x_j^* is a monotonic function of R_{uj} and C_{dj}. The sign of the monotonicity depends upon the interconnect impedance parameters of the segments j and $j+1$ connected by via j. As the size of the allowed intervals for all of the vias is constrained by (10.17), the minimum and maximum value of R_{uj} and C_{dj} can be readily determined, permitting the evaluation of the extrema of x_j^*, $x_{j\min}^*$ and $x_{j\max}^*$. Due to the monotonic dependence of x_j on R_{uj} and C_{dj}, the optimum location for via j, x_j^* lies within the range delimited by $x_{j\min}^*$ and $x_{j\max}^*$. The following four cases can be distinguished

i. if $x_{j\max}^* \leq 0$, $x_j^* = 0$, and the optimum via location coincides with the lower bound of the interval as defined by (10.17).

ii. if $x_{j\min}^* \geq \Delta x_j$, $x_j^* = \Delta x_j$, and the optimum via location coincides with the upper bound of the interval as defined by (10.17).

iii. if $\Delta x_j \geq x_{j\min}^* \geq 0$ and $\Delta x_j \geq x_{j\max}^* \geq 0$, the bounded interval as defined by (10.17) reduces to

$$0 \leq x_{j\min}^* \leq x_j \leq x_{j\max}^* \leq \Delta x_j. \quad (10.27)$$

In this case, the via location cannot be directly determined. However, by iteratively decreasing the range of values for x_j^*, the optimal location for via j can be determined.

The following example is used to demonstrate that the physical domain for x_j^* iteratively decreases to a single point, the optimum via location. A detailed analysis of the heuristic is provided in Appendix C. Consider segments i, j, and k depicted in Fig. 10.12, where segments i and k are, respectively, located upstream and downstream of segment j. From (10.16) and (10.17), the minimum and maximum values of R_{ui}^0, R_{uj}^0, R_{uk}^0, C_{di}^0, C_{dj}^0, and C_{dk}^0 are determined, where the superscript represents the number of iterations. Assume that $x_{\min}^{*\,0}$ and $x_{\max}^{*\,0}$ are obtained from (10.26) to satisfy (10.27) for all three segments i, j, and k. As the range of values for the via location of segments i and k decreases according to (10.27), the minimum (maximum) value of the upstream resistance and downstream capacitance of segment j increases (decreases), i.e., $R_{uj\min}^0 < R_{uj\min}^1$, $C_{dj\min}^0 < C_{dj\min}^1$, $R_{uj\max}^1 < R_{uj\max}^0$, and $C_{dj\max}^1 < C_{dj\max}^0$. Due to the monotonicity of x_j^* on R_{uj} and C_{dj}, $x_{j\min}^{*0} < x_{j\min}^{*1}$ and $x_{j\max}^{*\,1} < x_{j\max}^{*\,0}$. The range of values for x_j^* therefore also decreases and, typically, after two or three iterations, the optimum location for the corresponding via is determined. In Fig. 10.14, the convergence of the heuristic at the optimum via location is depicted for one via of a two terminal interconnect. For this example, $c_j > c_{j+1}$ and $r_j > r_{j+1}$, and, therefore, $x_{j\min}^* = f(R_{j\min}, C_{j\min})$ and $x_{j\max}^* = f(R_{j\max}, C_{j\max})$. As shown in Fig. 10.14, the heuristic converges to the optimum via location within several iterations.

iv. if $x_{j\min}^* \leq 0$ and $x_{j\max}^* \geq \Delta x_j$, the via location cannot be directly determined. Additionally, the bounding interval cannot be reduced. Consequently, some loss of optimality occurs. This

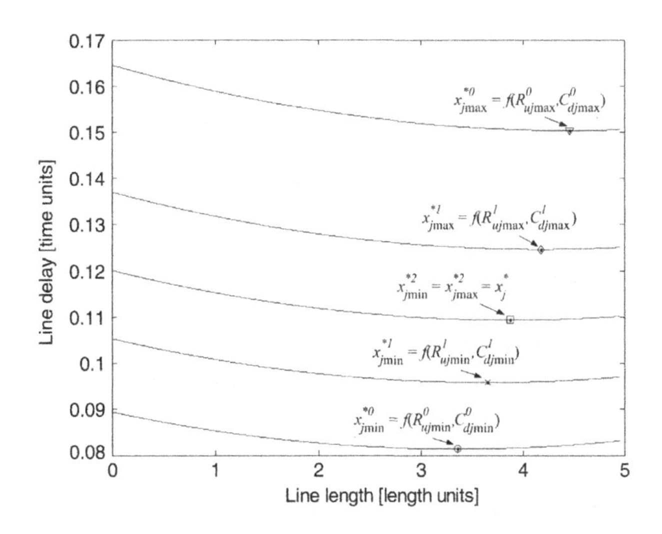

FIGURE 10.14

Case (iii) of the two terminal net heuristic. The allowed interval is iteratively decreased ensuring the optimum via location is eventually determined.

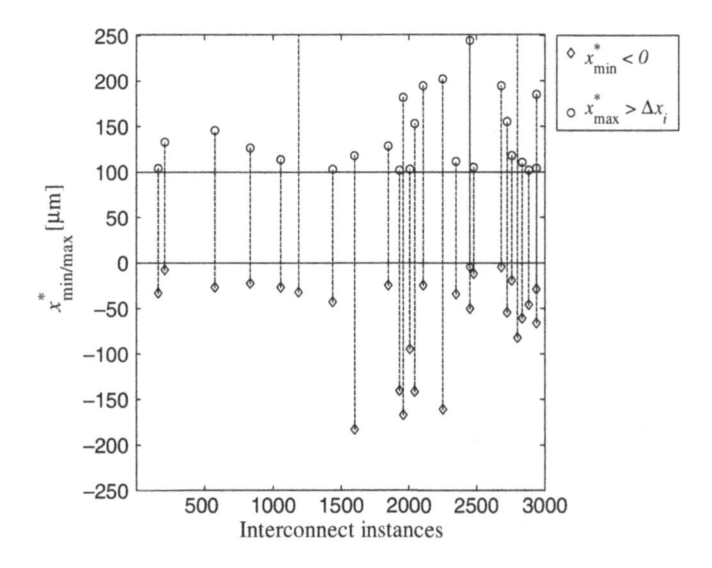

FIGURE 10.15

A subset of interconnect instances depicted by the dashed lines for case (iv) of the via placement heuristic. The interconnect traverses eight tiers and has a length $L = 1.455$ mm. The resistance r_j and capacitance c_j of each interconnect segment range, respectively, from 10 to 50 Ω/mm and 100 to 500 fF/mm.

departure from the optimal location, however, is typically smaller than a few per cent for all of the tested conditions, as shown by the results presented in Section 10.3.3.

The variation between the extrema of the upstream resistance and downstream capacitance due to the relatively small size of the allowed interval of the via placement ensures that a significant variation between the values $x^*_{j\min}$ and $x^*_{j\max}$ for most interconnects does not occur. To justify this argument, the extreme values of x^*_j are plotted in Fig. 10.15 for the fourth intertier via of a two-terminal net traversing eight tiers. This example represents a significant variation in the extreme values of x^*_j, due to the location of the via and the large number of vias along the interconnect. Indeed, the large number of vias (seven in this example) implies a considerable variation in the value of the downstream capacitance and upstream resistance, particularly for the fourth via. Note that the last via of the interconnect, for instance, only sees a variation in the upstream resistance. As shown in Fig. 10.15, less than 1% of the interconnect instances yield such boundary values for x^*_j, ensuring the inequalities $x^*_{j\min} \leq 0$ and $x^*_{j\max} \geq \Delta x_j$ are satisfied. This small per cent (1% in this case) is similar or smaller for the other vias of the interconnect due to the smaller variation in the upstream resistance or downstream capacitance. In addition, the inequalities, $x^*_{j\min} \leq 0$ and $x^*_{j\max} \geq \Delta x_j$, are typically satisfied, where the size of the allowed interval for a via is relatively small as compared to the size of the allowed intervals for the remaining vias. A nonoptimal via placement for this interconnect segment does not significantly affect the overall delay of the line.

Furthermore, for case (iv) to occur, there should be at least two vias for which an optimum location cannot be found. Indeed, if all but one of the vias are placed, the exact value of the upstream (downstream) resistance (capacitance) for the nonoptimized via can be obtained and, therefore, the location of that via can be determined from (10.26). The nonoptimal placement of a via does not necessarily affect the optimal placement of the remaining vias. For example, any via placed according to the criteria described in (i) and (ii) is not affected by the placement of the remaining vias. Therefore, as noted earlier, the size of the allowed intervals rather than the exact location of the vias is the key factor in determining the optimum via locations. Based on this heuristic, an algorithm for placing a via in two terminal nets is presented in the following subsection, along with simulation and analytic results for these interconnects.

10.3.2 TWO TERMINAL VIA PLACEMENT ALGORITHM

The heuristic described in the previous subsection has been used to implement an algorithm that exhibits an optimal or near-optimal via placement for two terminal intertier interconnects in 3-D ICs, exhibiting significantly lower computational time as compared to geometric programming solvers. The pseudocode of the two-terminal via placement algorithm (TTVPA) is illustrated in Fig. 10.16. The input to the algorithm is a description of the intertier interconnect where the minimum length of the segments and the size of the allowed intervals are provided. In the first step of the algorithm, the array of the maximum and minimum upstream (downstream) resistance (capacitance) for every allowed interval is determined. In the following steps, for each unprocessed via, the range of values for the optimum via location, as given by (10.26), is evaluated. In step five, these values are compared to the inequalities described in the previous subsection. If an optimal via location is determined in this step, the via is marked as processed and the capacitance and

```
Two-terminal via placement algorithm:(lmin, Δx)
1.   Determine Cdmin, Cdmax, Rumin, Rumax
2.   while S ≠ Ø
3.      if iter < max_iter
4.        sj ← an unprocessed via
5.        obtain x*jmin and x*jmax from (10.26)
6.        check for the inequalities in (i) – (iv)
7.        if sj is optimized (cases i-ii)
8.          store optimum via location
9.          S ← S - {sj}
10.         update Cdmin, Cdmax, Rumin, Rumax
          elseif Δxj decreases (case iii)
11.         update lminj, Cdmin, Cdmax, Rumin, Rumax
          else (case iv)
12.         go to step 3
        else (the non-optimized vias)
13.       place via at the center of the allowed interval
14.       store via location
15.       S ← S - {sj}
16.  exit
```

FIGURE 10.16

Pseudocode of the proposed two terminal net via placement algorithm.

resistance arrays are updated. If, after a number of iterations, there are nonoptimal vias, in step 14 these vias are placed at the center of the corresponding allowed intervals and the algorithm terminates. Other criteria, such as routing congestion, can alternatively be applied to place nonoptimal vias rather than placing these vias at the center of the corresponding allowed intervals. Further criteria can be considered to search for the optimum location of the nonoptimal vias, trading off runtime with accuracy.

10.3.3 APPLICATION OF THE VIA PLACEMENT TECHNIQUE

The via placement algorithm has been applied to several interconnect instances to validate the accuracy and efficiency of the heuristic. Both SPICE simulations and optimization results are provided. Two terminal intertier interconnects for different numbers of physical tiers are evaluated. The impedance characteristics of the horizontal segments and vias are extracted for several interconnect structures using a commercial impedance extraction tool [423]. Copper interconnect is assumed with an effective resistivity of 2.2 μΩ-cm. Based on the extracted impedances the resistance and capacitance of the horizontal segments range, respectively, from 25 to 125 Ω/mm and 100 to 300 fF/mm for a 90 nm technology node [252,432]. The cross section of the vias is 1 μm × 1 μm with 1 μm spacing from the surrounding horizontal metal layers, assuming a SOI process as described in [307]. For all of the interconnect structures, the total and minimum length of each

Table 10.3 SPICE Simulations Demonstrating the Delay Savings Achieved by a Near-Optimal Via Placement

Length (μm)	T_{center} (ps)	T_{rnd} (ps)	T_{min} (ps)	Improvement (%)	n
1017	12.35	12.64	11.42	8.14 (10.68)	4
1180	13.37	14.42	12.33	8.43 (16.95)	4
849	11.00	11.71	10.27	7.11 (14.02)	4
969	13.52	14.96	12.12	11.55 (23.43)	4
967	12.38	12.59	11.72	5.63 (7.42)	4
1612	18.54	19.85	17.24	7.54 (15.14)	5
1537	20.80	19.47	19.37	7.38 (0.52)	5
1289	17.78	18.43	16.45	8.09 (12.04)	5
1443	18.77	19.54	18.07	3.87 (8.14)	5
1225	16.97	18.33	15.62	8.64 (17.35)	5
2118	30.52	34.81	26.44	15.43 (31.66)	7
2130	27.92	27.32	25.94	7.63 (5.32)	7
1961	28.49	30.67	26.16	8.91 (17.24)	7
2263	35.58	40.11	31.31	13.64 (28.11)	7
2174	32.31	30.34	29.16	10.80 (4.05)	7
Average improvement				8.85 (14.14)	

The resistance and capacitance per unit length of the vias are, respectively, $R_{VI} = 6.7\ \Omega/mm$ *and* $c_{VI} = 6\ pF/mm$*. The length of the vias is* $L_{VI} = 20\ \mu m$*. The driver resistance is* $R_s = 15\ \Omega$ *and the load capacitance is* $C_L = 100\ fF$*. The length of the allowed intervals is* $\Delta x_i = 200\ \mu m$*, and the number of tiers is given by* n.

horizontal segment is randomly generated. For simplicity, all of the vias connect the segments of two adjacent physical tiers (i.e., $m = n$).

Delay simulations are reported in Table 10.3. The delay of the line T_{center}, where the vias are placed at the center of the allowed intervals, is listed in column 2. The delay T_{rnd}, listed in column 3, corresponds to the line delay for a random via placement. The minimum interconnect delay T_{min}, where the vias are optimally placed, is listed in column 4. The via locations or, equivalently, the length of the horizontal segments, are determined from the via placement algorithm. The improvement in delay as compared to the case where the vias are placed at the center of the line is listed in column 5. The number in parentheses corresponds to the improvement in delay over a random via placement. Note that the variation in the improvement in delay changes significantly for those listed instances, although the interconnect lengths are similar and the load capacitance and driver resistance are the same. This considerable variation demonstrates the strong dependence of the line delay on the impedance characteristics of the segments of the line and supports modeling the intertier interconnect as a group of nonuniform segments. Additionally, depending upon the impedance characteristics of the line segments, placing a via at the center of the allowed intervals is, for certain instances, near-optimal, explaining why the improvement in delay is not significant in these instances. The same characteristic applies to those cases where a random placement is close to the optimum placement. Nevertheless, as listed in Table 10.3, an improvement of up to

Table 10.4 Optimization Results for Different Two Terminal Intertier Interconnects and Numbers of Physical Tiers n

			Delay Improvement (%)						
			Vias Placed at the Center		Random Via Placement				
n	Average Interconnect Length (μm)	Δx_is (μm)	Avg	Max	Avg	Max	YALMIP/TTVPA Runtime Ratio × Times	Max. Error (%)	Instances
3	270	50	3.36	11.10	5.88	22.24	141.7	0.005	10,000
3	520	100	4.59	17.63	5.02	17.63	148.6	0.008	10,000
3	1020	200	5.90	23.12	10.27	47.08	145.6	0.013	10,000
4	405	50	4.02	13.01	6.00	25.97	209.4	0.006	10,000
4	781	100	5.26	16.95	7.91	34.10	574.3	0.003	10,000
4	996	100	3.45	9.87	5.37	20.05	95.5	0.008	5000
4	1155	150	5.94	21.61	8.89	44.46	125.4	0.011	10,000
4	1302	200	6.31	22.76	9.87	46.91	241.6	0.021	5000
4	1600	300	8.73	26.85	12.58	55.91	111.4	0.025	5000
5	540	50	4.48	13.73	6.16	27.49	112.14	0.005	10,000
5	1040	100	5.69	17.97	7.79	35.82	335.40	0.012	10,000
5	1541	150	6.35	22.36	8.63	46.26	549.35	0.017	10,000
5	1277	100	3.91	10.84	5.57	22.04	93.1	0.003	5000
5	1684	200	7.06	19.65	10.09	41.28	226.4	0.009	5000
5	2076	300	9.77	26.74	14.27	55.45	90.1	0.009	5000
7	1840	100	4.64	12.81	6.21	25.16	84.6	0.008	5000
7	2440	200	8.26	23.77	11.07	47.56	89.0	0.002	5000

32% is observed for relatively short interconnects, demonstrating that an optimum via placement can significantly enhance the speed of 3-D circuits (in addition to the primary benefit of reduced wire length and therefore lower power).

The TTVPA is compared both in terms of optimality and efficiency to two optimization solvers. The first solver, Yalmip [433], is a general optimization solver that supports geometric programing, while GLOPTIPOLY [434] is an optimization solver for nonconvex polynomial functions. Due to the excessive computational time of GLOPTIPOLY (greater than three orders of magnitude slower than YALMIP), only comparisons with YALMIP are reported. Optimization results are listed in Table 10.4 for different values of Δx ranging from 50 to 300 μm.

As reported in column 9 of Table 10.4, TTVPA exhibits high accuracy as compared to YALMIP. These results are independent of the number of tiers that comprise the 3-D interconnect, demonstrating that TTVPA yields optimum solutions for most interconnect instances. In addition, for those cases where some of the vias are not optimally placed, the loss of optimality is insignificant (as previously discussed). In column 8, the runtime ratio of YALMIP to TTVPA is listed. TTVPA is approximately two orders of magnitude faster than YALMIP. The complexity of TTVPA has an almost linear dependence on the number of intertier vias. As depicted in Fig. 10.16, each via is typically processed once; otherwise, a maximum of two to five iterations is required to place a via. In all of the experiments, the algorithm terminates within a number of iterations smaller than five times the total number of vias within the interconnect.

As listed in Table 10.4, the savings in delay from the near-optimal via placement strongly depends upon the length of the allowed intervals. For example, doubling the length of the allowed intervals for via placement increases considerably the maximum improvement in delay, exhibiting a twofold improvement for specific interconnections. As the length of the allowed intervals increases, the constraints in (10.16) are relaxed and a greater benefit from optimally placing the vias is achieved.

The effect of the nonuniformity of the intertier interconnects on the improvement in delay is graphically illustrated in Fig. 10.17, where the improvement in delay for intertier interconnects spanning four and five physical tiers is depicted. The average savings in delay of highly nonuniform interconnects (i.e., $r_{(j+1)}/r_j$ and $c_j/c_{(j+1)}$ ratios of 1 to 10) can be significant, approaching 10% and 13% for a moderately sized length, where the vias are placed, respectively, at the center of the

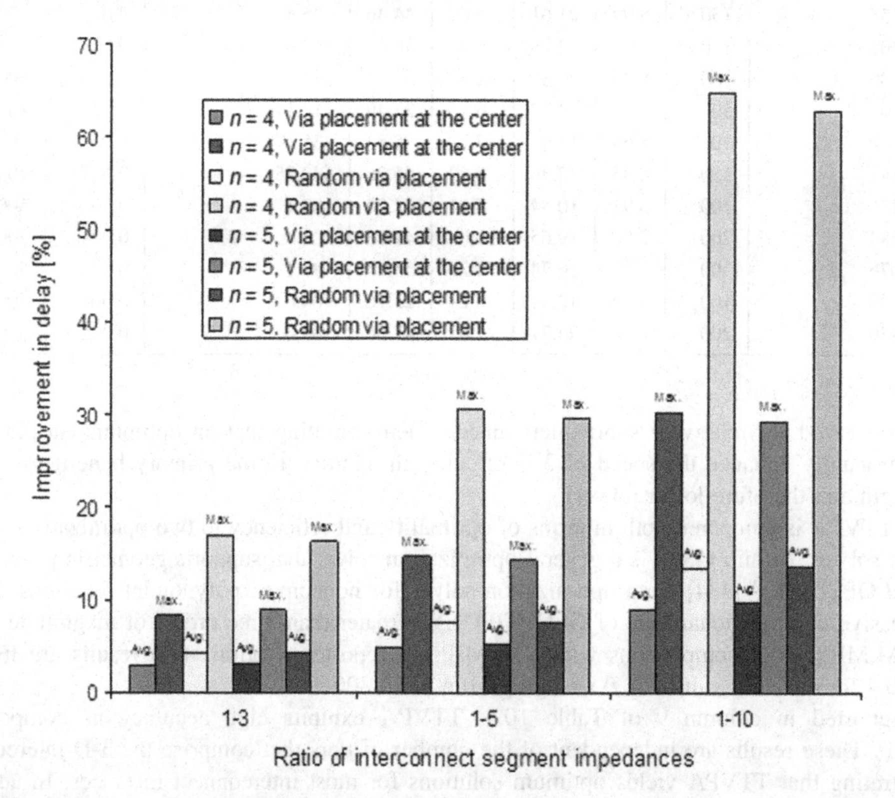

FIGURE 10.17

Average and maximum improvement in delay for different range of interconnect segment resistance and capacitance ratios. The vias are placed either at the center of the allowed intervals or randomly, as explained in the legend of the diagram.

allowed intervals and are randomly placed. The maximum improvement can exceed 60%, as shown in Fig. 10.17.

The length of most of the interconnects listed in Tables 10.1 through 10.4 is such that repeater insertion cannot improve the performance of these interconnects. Alternatively, as previously mentioned, wire sizing or a via placement technique can be utilized to improve the speed of these interconnects. Wire sizing is a well known technique to reduce interconnect delay. Wire shaping, however, is not always feasible due to routing congestion or obstacles such as placed cells. Additionally, as the interconnect is widened to lower the interconnect resistance, the capacitance and, consequently, the power consumption of the interconnect increases.

The wire sizing algorithm described in [435] has been applied to several interconnects to improve the line delay. The interconnect length is divided equally among the horizontal segments that constitute the interconnect. For the same interconnects, the line delay where the width is minimum and the vias are optimally placed is also determined. In Fig. 10.18, the average interconnect delay for both the via placement and wire sizing technique is shown. The instance where the optimum via placement outperforms wire sizing (and vice versa) is depicted. The average delay improvement ranges from 6.23% for $n = 4$ to 17.8% for $n = 5$, justifying that via placement can be an effective delay reduction technique for intertier interconnects in 3-D circuits without requiring additional area. The primary reason wire sizing does not achieve a significant reduction in delay is due to the via impedance characteristics and because the vias cannot be sized as aggressively as the

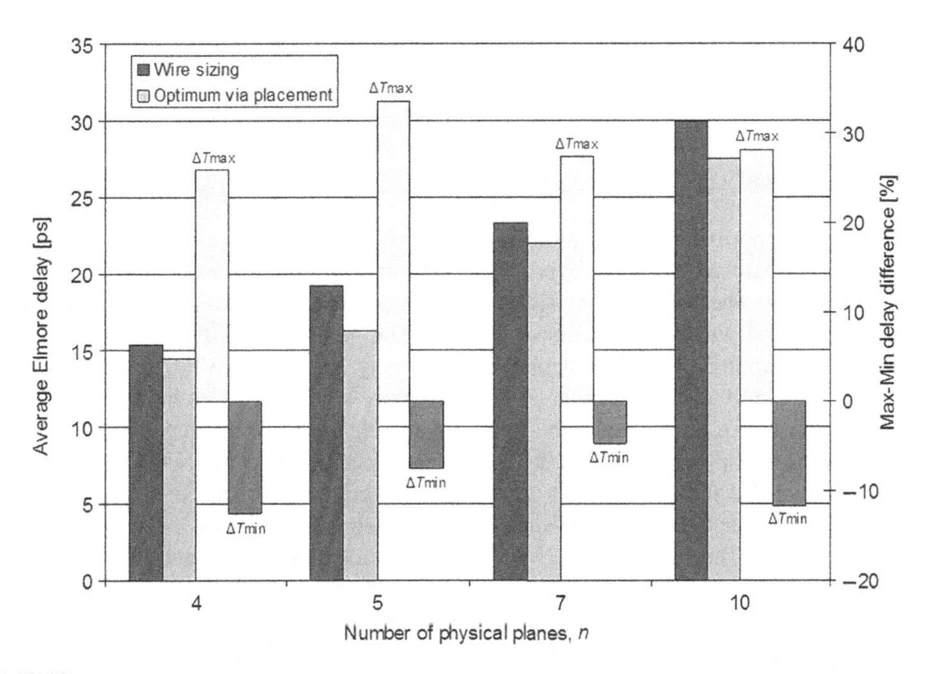

FIGURE 10.18

Comparison of the average Elmore delay based on wire sizing and optimum via placement techniques. The instance where the optimum via placement outperforms wire sizing (and vice versa) is also depicted.

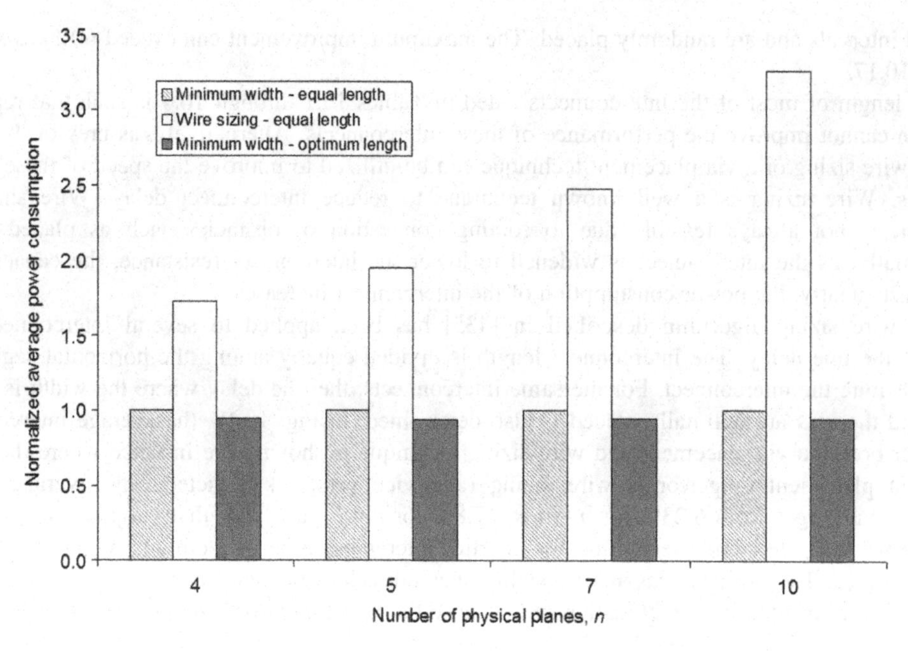

FIGURE 10.19

NAPC for minimum width and wire segments of equal length, wire sizing, and wire segments of equal length, and minimum width and optimum via placement, yielding segments of different length.

horizontal segments. Furthermore, via sizing is not particularly desirable as wider vias decrease the via density or, equivalently, the number of intertier interconnects that can be routed within a 3-D system.

In Fig. 10.19, the normalized average power consumption (NAPC) for several interconnects is illustrated. The crosshatched bar corresponds to minimum width interconnects and horizontal segments of equal length. The white bar considers those interconnects where the horizontal segments are of equal length and wire sizing has been applied. The NAPC where the line segments are of minimum width and optimum length is depicted by the gray bar. The TTVPA technique dissipates lower power as compared to the other two approaches. Wire sizing increases the capacitance of the line, resulting in an NAPC greater than the optimum via placement technique. Note that although the NAPC is lowest for optimally placed vias, the capacitance and thus the NAPC is not minimum as the time constant for the interconnect also depends upon the resistance of the line. Due to this dependence, the capacitance is not designed to be minimum width to avoid an increase in the interconnect resistance. The reduction in NAPC is of considerable importance due to pronounced thermal effects in 3-D circuits, as discussed in Chapter 12, Thermal Modeling and Analysis, and Chapter 13, Thermal Management Strategies for Three-Dimensional ICs.

From the results presented in this section, exploiting the nonuniform impedance characteristics of the intertier interconnects when placing the vias can improve the delay of the lines. This improvement in delay can decrease the number of repeaters required to drive a global line or eliminate repeaters in semiglobal lines. In addition, wire sizing can be avoided, thereby saving significant power. Decreasing the number of repeaters and avoiding wide lines reduces the overall power consumption, which is an important issue in 3-D circuits due to thermal effects.

The overhead due to the placing of the vias is the additional effort for placement and routing to generate an allowed interval for the vias, which increases the routing congestion. Other techniques, however, also require similar if not greater overhead. Repeaters, for example, consume silicon area, dissipate power, and block the metal layers within a physical tier when driving nets on the topmost metal layer. Wire sizing requires routing resources reserved for wider interconnect segments. A discussion on placing vias for the important class of multiterminal nets is deferred to the following chapter.

10.4 SUMMARY

A technique for timing driven placement of intertier vias in two terminal 3-D interconnects is presented in this chapter. The key points of this technique can be summarized as follows:

- Intertier interconnect models of 3-D circuits vary considerably from traditional 2-D interconnect models. This deviation is due to several reasons, such as the heterogeneity of 3-D circuits, diverse fabrication technologies, and the variety of bonding styles.
- For an intertier interconnect that includes only one intertier via, a closed-form solution is provided for placing a via that minimizes the Elmore delay.
- For intertier interconnect comprised of multiple segments, conditions are provided that minimize the delay of a line by placing intertier vias.
- Geometric programing can produce the globally optimum solution of the via placement problem.
- An accurate and efficient heuristic as compared to geometric programing solvers is described. The fundamental concept of the heuristic is that via placement depends primarily on the size of the allowed intervals where the vias can be placed, not on the precise location of the other vias along the line.
- The improvement in delay depends upon the size of the allowed intervals and the interconnect impedance characteristics.
- The proposed via placement technique is compared to a wire sizing technique and exhibits a greater average improvement in interconnect performance. Wire sizing cannot effectively handle the nonuniformity of the intertier interconnects and the discontinuities due to the intertier vias.
- The via placement technique decreases the dynamic power consumed by the interconnects as compared to wire sizing techniques, since wider lines are not necessary.
- Timing driven via placement can be an alternative to repeater insertion to improve the speed of 3-D interconnect systems.

CHAPTER

TIMING OPTIMIZATION FOR MULTITERMINAL INTERCONNECTS*

11

CHAPTER OUTLINE

A significant improvement in the performance of two-terminal intertier interconnect in three-dimensional (3-D) circuits is demonstrated in the previous chapter. A technique which accurately determines the via location that minimizes the delay of a two-terminal interconnect is described and applied to several interconnect systems. Multiterminal interconnects, however, constitute a significant portion of the interconnects in an integrated circuit. Improving the performance of these nets in 3-D circuits is a challenging task as the sinks of these interconnects can be located on different physical planes. In addition to decreasing the delay, a timing optimization technique should not significantly affect the routed tree. In this chapter, a via placement technique for intertier trees is presented. The task of placing the vias in an intertier tree to decrease the delay of a tree is described in the following section. A heuristic solution to this problem is described in Section 11.2. Algorithms based on this heuristic are presented in Section 11.3. The application of these algorithms to several intertier trees is discussed in Section 11.4. Finally, a summary is provided in Section 11.5.

*The terms tier and plane are used interchangeably in this chapter.

Three-Dimensional Integrated Circuit Design. DOI: http://dx.doi.org/10.1016/B978-0-12-410501-0.00011-3

11.1 TIMING DRIVEN VIA PLACEMENT FOR INTERTIER INTERCONNECT TREES

The problem of placing vias in intertier trees to decrease the delay of these trees is investigated in this section. A simple intertier interconnect tree (also called for simplicity an interconnect tree) is illustrated in Fig. 11.1A, while related terminology is listed in Table 11.1. The sinks of the tree are located on different physical tiers within a 3-D stack. Subtrees not directly connected to the intertier vias, which do not contain any intertier vias (i.e., intratier trees), are also shown. The interconnect segments from each physical tier are denoted by a solid line of varying thickness. Different objective functions can be applied to optimize the performance of an interconnect structure. In this chapter, the weighted summation of the distributed Elmore delay of the branches of an interconnect tree is considered as the objective function,

$$T_w = \sum_{\forall s_{pq}} w_{s_{pq}} T_{s_{pq}}, \tag{11.1}$$

where $w_{s_{pq}}$ and $T_{s_{pq}}$ are, respectively, the weight and distributed Elmore delay of sink s_{pq}. The weights are assigned to the sinks according to the criticality of the sinks. For a via connecting multiple interconnect segments, or equivalently, for a via with degree greater than two, there are several candidate directions d_i along which the delay can be decreased. The placement of vias along these directions is constrained by the lengths l_{d_i}, as shown in Fig. 11.1B, where the l_{d_i} terms are not generally equal. In addition, vias can span more than one physical tier. For example, consider the via connecting sinks s_{23} and s_{33}. This via traverses two physical tiers, where the allowed interval for placing the via can be different for each tier.

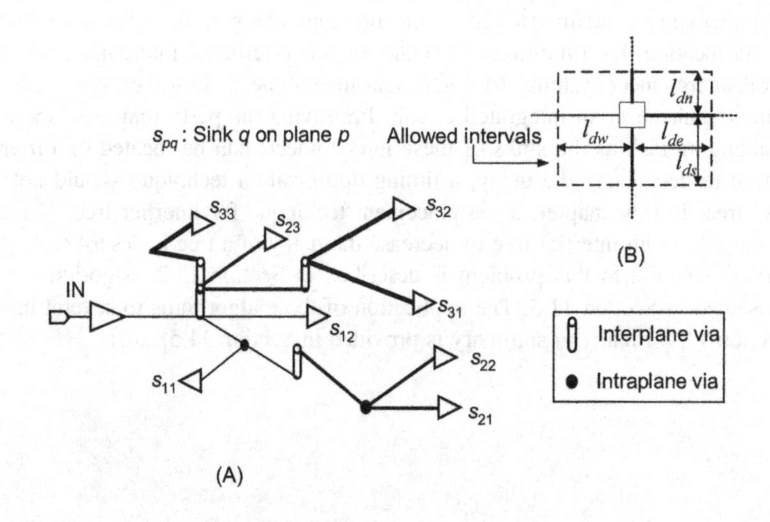

FIGURE 11.1

Intertier interconnect tree. (A) Typical intertier interconnect tree, and (B) intervals and directions that the intertier via can be placed.

Table 11.1 Notation for Two-Terminal Nets and Interconnect Trees

Notation	Definition
R_S	Driver resistance
C_L	Load capacitance
r_j (c_j)	Resistance (capacitance) per unit length of horizontal segment j
r_{vj} (c_{vj})	Resistance (capacitance) per unit length of intertier via v_j
R_j (C_j)	Total interconnect resistance (capacitance) of horizontal segment j
R_{vj} (C_{vj})	Total interconnect resistance (capacitance) of intertier via v_j
R_{u_j}	Upstream resistance of the allowed interval of via v_j
$R_{u_{ij}}$	Common upstream resistance of the allowed interval of via v_i and v_j
d_i	Candidate direction for a *type*-2 move
C_{dv_j}	Total downstream capacitance of the allowed interval of via v_j (in every direction d_i)
$P_{s_{pq}}$	Path from root of the tree to sink s_{pq}
$P_{s_{pq}v_j}$	Path to sink s_{pq} including v_j in every candidate direction
U_{kj}	Set of vias located upstream of v_j up to and including v_k, and belonging to at least one path $P_{s_{pq}v_j}$
$\overline{P_{s_{pq}v_j}}$	Path to sink s_{pq} that does not include v_j
$\overline{P_{s_{pq}U_{kj}}}$	Path to sink s_{pq} that does not include any of the vias in the set U_{kj}
$P_{s_{pq}v_jd_i}$	Path to sink s_{pq} that includes v_j and belongs to direction d_i
$\overline{P_{s_{pq}v_jd_i}}$	Path to sink s_{pq} including v_j in every candidate direction except for d_i
$C_{dv_jd_i}$	Downstream capacitance of the allowed interval of via v_j for the paths $P_{s_{pq}v_jd_i}$
$C_{dv_j\overline{d_i}}$	Downstream capacitance of the allowed interval of via v_j for the paths $\overline{P_{s_{pq}v_jd_i}}$

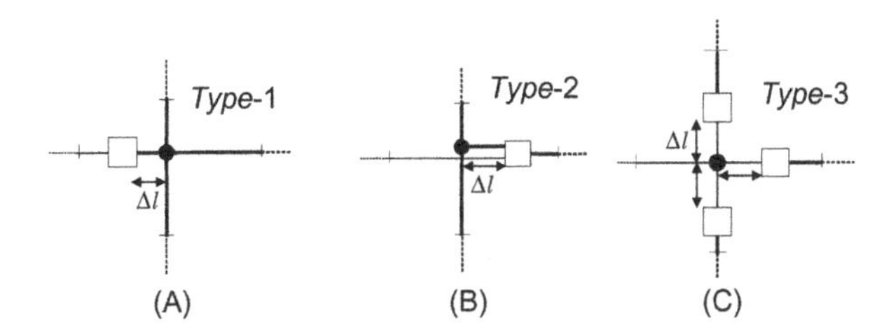

FIGURE 11.2

Different intertier via moves. (A) *Type*-1 move (allowed), (B) *type*-2 move (allowed), and (C) *type*-3 move (prohibited).

Three different types of moves for an intertier via are defined. A *type*-1 move is shown in Fig. 11.2A. This type of move requires the insertion of an intratier via (to preserve connectivity), as depicted by a dot in Fig. 11.2A. In the following analysis, the effect of these additional intratier vias on the delay of the tree is assumed to be insignificant. The impedance characteristics of the intratier vias are assumed to be considerably lower than the impedance characteristics of the intertier vias [436], particularly if bulk CMOS technology is used for the upper tiers. Alternatively, this effect can be included by appropriately shrinking the length of the allowed interval of the intertier via.

A *type*-2 move is shown in Fig. 11.2B. A *type*-2 move differs from a *type*-1 move in that an additional interconnect segment of length Δl is inserted. Although an additional interconnect segment is required for this type of move, a reduction in the delay of the tree can occur. The segments of length Δl illustrated in Fig. 11.2B are located on the same *y*-coordinate but on different physical tiers, yet are shown on different coordinates for added clarity.

Another type of move is illustrated in Fig. 11.2C, where additional intertier vias are inserted, and is denoted as a *type*-3 move. This type of move is not permitted for two reasons. The additional intertier vias outweigh the reduction in delay resulting from optimizing the length of the connected segments due to the high impedance characteristics of the intertier vias. Additional intertier vias also increase the vertical interconnect density, which is undesirable. The routing congestion also increases as these vias typically block the metal layers within a tier, adversely affecting the length of the allowed intervals for the remaining nets.

As with two-terminal nets, certain constraints on the total length of each source to sink path apply to interconnect trees

$$L_{s_{pq}} = l_1 + l_{v1} + \ldots + l_j + l_{v_j} + \ldots + l_{v_{n-1}} + l_n, \tag{11.2}$$

where l_j and l_{vj} are, respectively, the length of the horizontal segment j and via v_j which describe the path from the root of the tree to the sink s_{pq}. The number of segments that constitute the path to this sink is denoted as n. The constraint in (11.2) is adapted to consider any increase in wirelength that can result from a *type*-2 move for some branches of the tree. In addition, the length of each segment of the tree and the via placement are also constrained by

$$l_{j\min} \le l_j + l_{di,\ j-1} + l_{j\min} + l_{di,\ j}, \forall i \in \{w, e, s, n\}, \forall j \in [2, n-1] \text{ and } l_{di,0} = l_{di,n} = 0, \tag{11.3}$$

$$0 \le x_j \le l_{di,\ j}, \forall i \in \{w, e, s, n\}. \tag{11.4}$$

Consequently, the constrained optimization problem for placing a via within an intertier interconnect tree can be described as

(P1) minimize T_w,
subject to (11.2) through (11.4), $\forall$ sink s_{pq}, and via v_j.

With similar reasoning as for two-terminal nets, (P1) typically includes an indefinite quadratic form $\mathbf{l}^{\mathbf{T}}\mathbf{A}\mathbf{l}$, where $\mathbf{A}$ is the matrix described in Eq. (7.20) adapted for interconnect trees. Certain transformations can be applied to convert (P1) into a convex optimization problem [430]; the objective functions, however, are no longer quadratic. Alternatively, accurate heuristics are described in the following section to determine the location of the vias that minimizes the delay of the intertier interconnect trees.

11.2 MULTITERMINAL INTERCONNECT VIA PLACEMENT HEURISTICS

Near-optimal heuristics for placing vias in interconnect trees in 3-D circuits are presented in this section. Initially, a heuristic for minimizing the weighted delay of the sinks of an intertier tree is described in Section 11.2.1. A second heuristic for optimizing the delay of a critical path in an intertier tree is discussed in Section 11.2.2.

11.2.1 INTERCONNECT TREES

In this section, placing an intertier via within an interconnect tree in a 3-D circuit to minimize the summation of the weighted Elmore delay of the branches of the tree is described. Since several moves for intertier vias are possible, as discussed in Section 11.1, the expressions that determine the via location are different in multiterminal nets. To determine which type of move for those vias with a degree of connectivity greater than two will yield a decrease in the delay of the tree, the following conditions apply.

Condition 1: If $r_j > r_{j+1}$, only a *type*-1 move for v_j can reduce the delay of the tree.

Proof: The proposition is analytically proven in Appendix D. The condition can also be intuitively explained. A *type*-2 move increases by Δl the length of segment j. The reduction in l_{j+1} is counterbalanced by the additional segment with length Δl on the $j+1$ tier (see Fig. 11.2B). Consequently, the total capacitance of the tree increases. If *condition 1* is satisfied, a *type*-2 move also increases the total resistance of the tree and, therefore, the delay of the tree will only increase by this via move.

Condition 2: For a candidate direction d_i, if $r_j < r_j + 1$ and

$$\sum_{\forall s_{pq} \in \overline{P_{s_{pq} v_j d_i}}} w_{s_p} \left(r_j + r_{j+1}\right) C_{d_{vj}\overline{d_i}} \leq \sum_{\forall s_{pq} \in P_{s_{pq} v_j d_i}} w_{s_p} \left(r_{j+1} - r_j\right) C_{d_{vj} d_i}, \tag{11.5}$$

is satisfied, a *type*-2 move can reduce the delay of the tree.

Proof: This condition is also intuitively demonstrated. All of the interconnect segments located upstream from v_j see an increase in the capacitance by $c_j \, \Delta l$, increasing the delay of each downstream sink v_j. Consequently, only a reduction in the resistance can decrease the delay of the tree. Alternatively, the sinks located downstream of via v_j on the candidate direction d_i see a reduction in the upstream resistance by $(r_j - r_{j+1})\Delta l < 0$, while the sinks downstream of via v_j on the other directions see an increase in the upstream resistance by $(r_j + r_{j+1})\Delta l$. For a *type*-2 move, both the weighted sum of these two components as determined by the weight of the sinks and the downstream capacitances must be negative, resulting in a decrease in the delay of the tree.

Condition 2 is evaluated for each via of a tree with degree greater than two. If (11.5) is satisfied for more than one direction, the direction that produces the greatest value of the right hand side (RHS) of (11.5) is considered the optimum direction for that via. Finally, note that both *conditions 1* and *2* are only necessary and not sufficient conditions. Following the notation listed in Table 11.1, the critical point for a via connecting two segments on tiers j and $j+1$ and satisfying *condition 1* is

$$x_{type\text{-}1} = \left[\frac{\left(\sum_{v_i \in U_{1j}} \sum_{s_m \in \overline{P_{s_m U_{ij}}}} w_{s_m} R_{u_{ij}} + \sum_{s_p \in P_{s_p v_j}} w_{s_p} R_{u_j} \right) \left(c_{j+1} - c_j\right) - l_{v_j} \left(r_j c_{v_j} - r_{v_j} c_{j+1}\right)}{\sum_{s_p \in P_{s_p v_j}} w_{s_p} \left(r_j c_j + r_{j+1} c_{j+1} - 2 r_j c_{j+1}\right)} + \frac{\left(r_j - r_{j+1}\right)\left(c_{j+1} l_{d_w} + C_{d v_j}\right)}{\sum_{s_p \in P_{s_p v_j}} w_{s_p} \left(r_j c_j + r_{j+1} c_{j+1} - 2 r_j c_{j+1}\right)} \right]. \tag{11.6}$$

For a *type*-2 move along a candidate direction d_i, the critical point for a via connecting two segments on tiers j and $j+1$ is

$$x_{type\text{-}2} = \left[\frac{\sum_{s_p \in P_{spv_jd_i}} w_{s_p} r_{j+1}\left(C_{dv_jd_i} + c_{j+1} l_{d_i}\right) - \sum_{v_i \in U_{1j}} \sum_{s_m \in \overline{P_{s_m U_{ij}}}} w_{s_m} R_{u_{ij}} c_k}{\sum_{s_p \in P_{spv_j}} w_{s_p}\left(r_j c_j + r_{j+1} c_{j+1}\right)} - \frac{\sum_{s_p \in P_{spv_j}} w_{s_p}\left(r_j c_{v_j} - c_{j+1} l_{d_i} + C_{d_j} + r_{j+1} C_{dv_j\overline{d_i}} + R_{u_j} c_k\right)}{\sum_{s_p \in P_{spv_j}} w_{s_p}\left(r_j c_j + r_{j+1} c_{j+1}\right)} \right]. \tag{11.7}$$

11.2.2 SINGLE CRITICAL SINK INTERCONNECT TREES

Cases exist where the delay of only one branch of a tree requires optimization. Although the heuristic presented in the previous section can be used for this type of tree, a computationally simpler, yet accurate, optimization procedure for single critical net trees is described here. Denoting the critical sink of the tree by s_c, the weight for this sink w_{sc} is one, while the assigned weight for the remaining sinks are zero. Consequently, the expression that minimizes the delay is significantly simplified. In addition, the approach is different as compared to the optimization problem discussed in the previous section. More specifically, those intertier vias that belong to the critical branch (the on path vias) are placed according to the heuristic for two-terminal nets. There is no need to test *conditions* 1 and 2 for these vias, as any *type*-2 move only occurs in the direction that includes the critical sink. Regarding those vias that are not part of the critical path (the off path vias), these vias are placed to minimize the capacitance of the tree. A simple interconnect tree depicting this terminology is illustrated in Fig. 11.3. This situation occurs since the noncritical sinks of the tree only contribute as capacitive loads to the delay of the critical sink.

The location of the off path vias is readily determined since the impedance characteristics of the interconnect segments are known. Note that in this case, the placement of the off path vias is always optimal. Any loss of optimality is due to the location of the on path vias. As the near-optimal

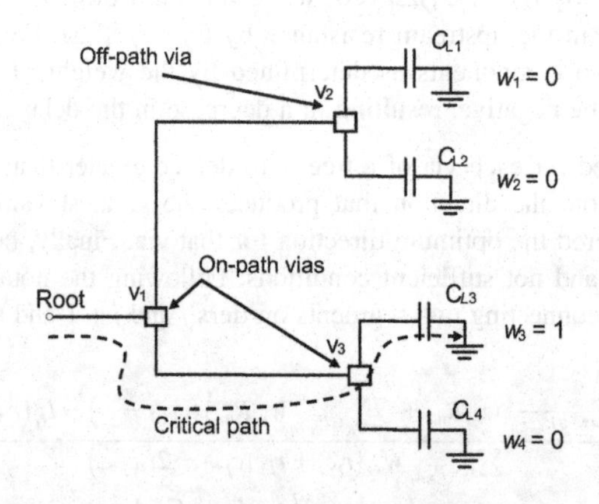

FIGURE 11.3

Simple interconnect tree, illustrating a critical path ($w_3 = 1$) and on path and off path intertier vias.

two-terminal net heuristic is used to place the on path vias, the loss of optimality is negligible. In the following section, these heuristics are used to develop efficient algorithms for placing vias in multiterminal nets in 3-D ICs.

11.3 VIA PLACEMENT ALGORITHMS FOR INTERCONNECT TREES

Efficient near-optimal algorithms for placing vias among intertier interconnects are presented in this section. Based on the aforementioned heuristic described in Section 11.2.1, an efficient algorithm for intertier interconnect trees is presented in Section 11.3.1. A second algorithm that places intertier vias to minimize the delay for the particular case of interconnect trees with a single critical branch is discussed in Section 11.3.2.

11.3.1 INTERCONNECT TREE VIA PLACEMENT ALGORITHM (ITVPA)

The via placement optimization algorithm for multiterminal nets is presented in this section. The input to the algorithm is an intertier interconnect tree where the minimum length of the segments, the weight of the sinks, and the length of the allowed intervals are provided. Pseudocode of the algorithm is shown in Fig. 11.4. Due to the different types of possible moves in intertier interconnect trees, the candidate direction for via placement is initially determined in steps one through five. The *move_type* routine operates from a leaf to the root, where the type and direction of the move of each via of the tree have a degree greater than two. *Conditions* 1 and 2 are tested for each via and direction. For those vias at the last level of the tree (close to the sinks), the downstream capacitance is determined. Alternatively, for those vias that belong to the next level closer to the root, the downstream capacitance cannot be accurately determined, and (11.5) is evaluated for the extreme values of C_{dvj}. If (11.5) is satisfied for only one of these extrema, a via can be placed along a nonoptimal direction, resulting in loss of optimality. Such instances, however, are not typically encountered, as discussed in the following section. In step six, the *optimize_tree_delay* routine places the vias within a tree to minimize (11.1). This routine is a slight modification of the algorithm used for two-terminal nets.

Interconnect tree via placement algorithm: (l_{min}, Δx, l_{di}, w_{si})

```
1. foreach physical tier i, i = n → 1
2.    foreach inter tier via j on tier i
3.       if via_degree > 2
4.          move_type(j)
         else
5.          goto step 2
6. optimize_tree_delay()
7. exit
```

FIGURE 11.4

Pseudocode of the Interconnect Tree Via Placement Algorithm (ITVPA).

```
Single critical sink via placement algorithm: (lmin, Δx, ldi, Wspq)
1.  foreach off path via j
2.        set via j to min. capacitance location
3.  foreach on path via i
4.     direction_move(i)
5.  optimize_tree_delay()
6.  exit
```

FIGURE 11.5

Pseudocode of the near-optimal Single Critical Sink interconnect tree Via Placement Algorithm (SCSVPA).

11.3.2 SINGLE CRITICAL SINK INTERCONNECT TREE VIA PLACEMENT ALGORITHM

Although the heuristic presented in Section 11.2.2 can be used to improve the delay of trees with a single critical path, a simpler optimization procedure for single critical net trees is described in this section. The input to the single critical sink via placement algorithm (SCSVPA) is a description of the intertier interconnect tree where the minimum length of the segments, the weight of the sinks, and the length of the allowed intervals are provided. Pseudocode of the algorithm is shown in Fig. 11.5. In steps one to three, each of the off path vias is placed at the location that minimizes the capacitance within the corresponding allowed interval. The direction that includes the critical sink of the tree is set by the *direction_move* routine as the direction along which the on path vias can be placed. In step five, the *optimize_tree_delay* routine is utilized to determine the location of the on path vias. As previously mentioned, any loss of optimality for this type of tree results from the heuristic that places the via in two-terminal nets. As described in Chapter 10, Timing Optimization for Two-Terminal Interconnects, this heuristic produces results similar to optimization solvers. SCSVPA naturally exhibits significantly lower computational time as compared to general purpose solvers.

11.4 DISCUSSION OF VIA PLACEMENT RESULTS

These algorithms are applied to several example intertier interconnect trees to evaluate efficiency and accuracy. Trees for a different number of tiers and sinks are analyzed. A discussion of the limitations of these algorithms is provided in this section. The impedance characteristics of the horizontal segments and vias are extracted for several interconnect structures using a commercial impedance extraction tool [423]. Copper interconnect is assumed with an effective resistivity of 2.2 μΩ-cm. Based on the extracted impedances, the resistance and capacitance of the horizontal segments range, respectively, from 25 to 125 Ω/mm and 100 to 300 fF/mm, respectively, for a 90 nm CMOS technology node [252,432]. The cross-section of the vias is 1 μm × 1 μm, with 1 μm spacing from the surrounding horizontal metal layers, assuming a silicon on insulator (SOI) process, as described in [307]. The total and minimum length of each horizontal segment is randomly generated for each of the interconnect structures. For simplicity, all of the vias connect the segments of two adjacent physical tiers. The savings in delay that can be achieved by optimally placing the vias is listed in Table 11.2 for different via placement scenarios.

Table 11.2 Optimization Results for Various Intertier Interconnect Trees for Different Number of Sinks and Physical Tiers n

n	Number of Sinks	Avg. Branch Length (μm)	Avg. Maximum Branch Length (μm)	l_{di} (μm)	Delay Improvement (%)				Instances
					$x_i^* = l_{di}/2$		$x_i^* = 0$		
					Avg	Max	Avg	Max	
3	4	153	186	50	2.72	9.33	3.79	11.25	10,000
3	4	307	376	100	4.23	15.17	6.03	17.94	10,000
4	4	208	273	50	1.11	3.53	2.49	5.63	5000
4	4	828	1100	200	3.12	10.29	6.42	13.50	5000
4	4	1243	1650	300	4.07	14.15	7.76	19.38	5000
4	8	431	569	100	3.90	13.24	7.71	19.68	10,238
5	4	264	362	50	1.25	3.83	2.40	5.89	5000
5	4	1054	1452	200	3.62	11.55	6.56	12.04	5000
5	4	791	1089	300	3.90	11.61	6.95	19.34	5000
5	8	454	660	50	0.90	2.69	2.27	4.98	5000
5	8	521	738	100	1.78	5.55	4.33	8.40	5000
5	8	779	1111	150	2.38	7.44	5.67	11.90	5000
5	8	1038	1481	200	2.91	8.71	6.74	12.58	5000
6	8	306	455	50	1.11	3.17	2.36	4.89	5000
6	8	615	913	100	2.00	5.44	4.09	9.85	5000
6	8	922	1373	150	2.72	7.01	5.43	11.72	5000
6	8	921	1371	200	3.32	10.02	6.61	14.21	5000
6	16	555	845	50	0.86	2.74	2.52	4.95	4970
6	16	637	934	100	1.68	4.82	4.84	9.26	5059
6	16	953	1404	150	2.28	6.10	6.32	12.96	5021

In Table 11.2, the optimization results for interconnect trees with different number of sinks and tiers are reported. The accuracy and efficiency of ITVPA are similar to that of TTVPA (discussed in Chapter 10, Timing Optimization for Two-Terminal Interconnects) as the optimization routine for ITVPA is the same as in TTVPA, after a move type and direction have correctly been determined. As mentioned in Section 11.3.1, a nonoptimal direction can be selected to place a via. A nonoptimal direction for via placement, however, can only be chosen if the connected branches are slightly asymmetric (i.e., have similar impedance characteristics and criticality). In these cases, the value of the downstream capacitance and weight of these branches are close, making the weighted delay of these sinks similar. For these slightly asymmetric branches, (11.7) yields $x^*_{type\text{-}2} = 0$ for *type*-2 moves, meaning that if a nonoptimal direction is chosen, the delay of the tree is not affected. Alternatively, a *type*-2 move usually occurs for highly asymmetric trees where the delay of a branch dominates the delay of the tree. As described in (11.5), the type of move for an intertier via depends upon the weight of the branches and the impedance characteristics of the interconnect segments. Consider the symmetric tree shown in Fig. 11.6. The difference in the criticality and impedance of the branches is captured from the value of the assigned weights.

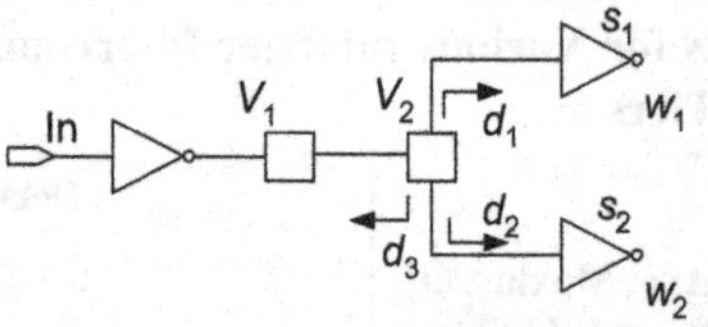

FIGURE 11.6

A symmetric tree including two intertier vias. The interconnect parameters per tier are $r_1 = 10.98$ Ω/mm, $r_2 = 11.97$ Ω/mm, $r_3 = 96.31$ Ω/mm, $c_1 = 147.89$ fF/mm, $c_2 = 202$ fF/mm, and $c_3 = 388.51$ fF/mm, and the allowed interval $l_{di \cdot v2} = 75$ μm.

Table 11.3 Optimal Via Location, Direction of Move, and Type of Move for via V_2 Shown in Fig. 11.6, as Determined From ITVPA for Various Values of W_1 and W_2

w_1	w_2	x^*_{v2} (μm)	Move	Direction, d_i	w_1	w_2	x^*_{v2} (μm)	Move	Direction, d_i
0.50	0.05	Δx_{v2}	Type-1	d_0	0.45	0.55	Δx_{v2}	Type-2	d_1
0.55	0.45	Δx_{v2}	Type-1	d_0	0.40	0.60	Δx_{v2}	Type-1	d_0
0.60	0.40	Δx_{v2}	Type-2	d_1	0.35	0.65	Δx_{v2}	Type-2	d_2
0.65	0.35	Δx_{v2}	Type-2	d_1	0.30	0.70	Δx_{v2}	Type-2	d_2
0.70	0.30	Δx_{v2}	Type-2	d_1	0.25	0.75	Δx_{v2}	Type-2	d_2
0.75	0.25	Δx_{v2}	Type-2	d_1	0.20	0.80	Δx_{v2}	Type-2	d_2
0.80	0.20	Δx_{v2}	Type-2	d_1	0.15	0.85	Δx_{v2}	Type-2	d_2
0.85	0.15	Δx_{v2}	Type-2	d_1	0.10	0.90	Δx_{v2}	Type-2	d_2
0.90	0.10	Δx_{v2}	Type-2	d_1	0.05	0.95	Δx_{v2}	Type-2	d_2
0.95	0.05	Δx_{v2}	Type-2	d_1	0.03	0.97	68.70	Type-2	d_2

In Table 11.3, the optimum location for via v_2 is listed for different weights. From these values, note that a *type*-2 move for via v_2 only occurs when branch s_2 dominates the delay of the tree. When the assigned weight of the branches of the tree is of similar value, a *type*-2 move does not occur as this move would only increase the interconnect delay of the tree, as determined by the ITVPA (i.e., $x_j^* = \Delta x_j$). Consequently, for a *type*-2 move, if the weight of the sinks, which originate from a via, are similar, the algorithm does not allow the via to be relocated.

The improvement in delay of the interconnect trees is listed in columns 6 through 9 of Table 11.2. The results are compared to the case where the vias are initially placed at the center of the allowed interval (i.e., $x_i = l_{di}/2$), and to the case where the vias are placed at the lower edge of the allowed interval (i.e., $x_i = 0$). The improvement in delay depends upon the length of the allowed interval. This dependence, however, is weak as compared to two-terminal nets. In addition, the improvement in delay is lower than point-to-point nets for the same allowed length intervals. This reduction in delay improvement occurs for two reasons.

For those vias with degree greater than two, which constitute the majority of intertier vias in interconnect trees, after the type of move for each via is determined, the actual interval length that these vias are allowed to move is $l_{di}/2$ and not l_{di} (see Fig. 11.2). Furthermore, in ITVPA, any modification to the routing tree is strictly confined within the allowed interval which least affects the routing tree. This constraint requires an additional interconnect segment for *type*-2 moves. If this constraint is relaxed, an additional interconnect segment is not necessary and the length of the interconnect segments can be further reduced, resulting in a considerably greater improvement in speed. Maintaining fixed paths limits the efficiency of the algorithms; however, these algorithms are applicable to placement and routing tools for 3-D ICs as long as these tools provide an allowed interval for via placement. In addition, interconnect routing can consider other important design objectives such as thermal effects or routing congestion. These algorithms for placing vias in multi-terminal nets can be applied as a subsequent post-processing step without significantly affecting the initial layout produced by existing tools.

In Table 11.4, placement results for single critical branch interconnect trees are reported. The improvement in the delay of these trees is listed in columns 6 through 9 of Table 11.4, as compared to the situation where the vias are initially placed at the center of the allowed interval (i.e., $x_i = l_{di}/2$) and where the vias are placed at the lower edge of the allowed interval (i.e., $x_i = 0$). This improvement is lower than for those interconnect trees listed in Table 11.2 as only *type*-1 moves can occur for the off path vias. Indeed, any *type*-2 move for the off path via only increases the off path capacitance and, in turn, the delay of the critical leaf. A smaller number of vias can, therefore, be relocated to reduce the delay of the single critical sink trees. Alternatively, for off path vias, placing the via at the center of the allowed intervals usually produces an optimum placement, resulting in a

Table 11.4 Optimization Results for Various Single Critical Sink Interconnect Trees for Different Number of Sinks and Physical Tiers *n*

					Delay Improvement (%)				
					$x_i^* = l_{di}/2$		$x_i^* = 0$		
n	Number of Sinks	Avg. Branch Length (μm)	Avg. Maximum Branch Length (μm)	l_{di} (μm)	Avg	Max	Avg	Max	Instances
4	4	341	453	50	2.72	8.95	3.70	14.63	5000
4	4	1021	1363	150	1.61	5.52	2.18	11.01	5000
4	4	1368	1821	200	1.36	5.49	1.92	8.59	5000
5	4	433	595	50	3.09	9.37	4.55	19.44	5000
5	4	1299	1790	150	1.80	5.55	2.85	13.42	5000
5	4	1734	2391	200	1.51	5.66	2.44	11.35	5000
5	8	427	612	50	2.53	8.10	4.09	16.20	5000
5	8	853	1227	100	1.94	7.22	2.98	12.63	5000
5	8	1282	1845	150	1.57	6.55	2.46	14.04	5000
5	8	1711	2461	200	1.33	4.81	2.25	10.44	5000
6	8	505	753	50	2.88	8.90	4.39	17.8	5000
6	8	1009	1512	100	2.14	6.10	3.45	15.20	5000
6	8	1511	2265	150	1.71	5.46	2.83	12.16	5000
6	16	523	779	50	2.52	8.29	4.54	13.25	4963
6	16	1045	1564	100	1.91	6.69	3.10	10.53	4977
6	16	1563	2351	150	1.55	5.36	2.96	12.37	4976

smaller overall improvement in the delay of this type of tree, as is demonstrated by the test structures. This situation occurs since *type*-2 moves for the off path vias are not allowed as a *type*-2 move increases the off path capacitance.

Typically, the larger the allowed interval, the greater the improvement in delay. Consequently, efficient placement tools for 3-D circuits that generate sufficiently large allowed intervals are desired. These intervals can be available space reserved for intertier interconnect routing (i.e., white space). For interconnect trees, the improvement in delay is smaller than for two-terminal nets. This decreased improvement in delay is due to the constraint of placing the vias within the allowed intervals to minimally affect local routing congestion. If the placement of the vias is permitted within an entire region (e.g., the bounding box of the net), a greater decrease in delay can occur. Assigning a region for placing vias, however, increases the congestion within a 3-D circuit as the same number of vias compete for sparser routing resources.

Despite the considerably lower computational time of these algorithms, further speed improvements can be achieved if more than one net are simultaneously processed. Although these algorithms support multiple net optimization without significant modification, a single net at a time approach likely yields improved results as the most critical nets are routed first. Net ordering algorithms [26] can be used to prioritize the routing of these interconnects, permitting a considerable reduction in the delay of these nets. Additionally, since the number of intertier interconnects is small as compared to the number of intratier interconnects [291], processing these interconnects one net at a time will not significantly increase the total computational time.

Thermal issues are important in 3-D ICs, as discussed in Chapter 13, Thermal Management Strategies for Three-Dimensional ICs, where additional dummy vias are utilized to control the average and peak temperature of the upper tiers within a 3-D system. Additionally, thermally aware cell placement improves the heat distribution and removal characteristics. These two techniques are decoupled from the via placement problem, which is a later step in the design process. Consequently, thermal issues are not strongly connected with the via placement approach.

In the techniques discussed in Chapter 13, Thermal Management Strategies for Three-Dimensional ICs, the thermal vias are placed in the available space among the blocks with either uniform or nonuniform densities. Alternatively, some of the thermal vias within this available space can be replaced with signal vias connecting circuits on different tiers. Placing the signal vias prior to placing the thermal vias within these regions results in large allowed intervals, thereby improving the effectiveness of this via placement technique. In this case, where the via placement can significantly enhance the thermal profile of a physical tier, the allowed interval for some vias can be decreased or removed. This practice, however, trades off performance for thermal management.

11.5 SUMMARY

Algorithms for timing driven placement of intertier interconnect trees are presented in this chapter. The primary characteristics of these algorithms are:

- Several types of moves exist for intertier vias, requiring the insertion of either intratier or intertier vias. Moves that require additional intertier vias are excluded to maintain a low intertier via density.

- The weighted summation of the delay at the sinks of a tree is the objective function that is minimized.
- A two step heuristic is presented. The direction of the move for each via is initially determined followed by the second step, placement of the via to minimize the delay.
- Another heuristic for placing intertier vias within trees to minimize the delay of the critical path is provided.
- Both of these heuristics exhibit linear complexity with the number of intertier vias.
- The improvement in delay depends upon the size of the allowed intervals for placing each via.

CHAPTER

THERMAL MODELING AND ANALYSIS

12

CHAPTER OUTLINE

A primary advantage of three-dimensional (3-D) integration, significantly greater packing density, is also the greatest threat to this emerging technology as aggressive thermal gradients can form among the tiers within the 3-D IC. Thermal problems, however, are not unique to vertical integration. Due to scaling, elevated temperatures and hotspots within traditional two-dimensional (2-D) circuits can greatly decrease the maximum achievable speed and significantly degrade the reliability of a circuit [437,438]. In addition, projected peak temperatures greatly deviate from International Technology Roadmap for Semiconductor predictions of maximum operating temperatures in next generation ICs [18]. Thermal awareness has, therefore, become another primary design issue in modern integrated circuits [439,440].

In 3-D integration, controlling the operating temperature is a prominent design objective [441]. Peak temperatures within a 3-D system can exceed the thermal limits of existing packaging technologies. Two key elements are required to effectively mitigate thermal issues in ICs: (1) accurate thermal models integrated with an effective thermal analysis methodology, and (2) an efficient thermal management strategy. In 3-D systems, the primary objectives of this strategy are to manage thermal gradients among the physical tiers while maintaining the operating temperature within tolerable levels. Cost constraints can further limit these systems as high temperatures require expensive packaging and cooling methods. Thermal models and analysis techniques for 3-D circuits are reviewed in this chapter, while a discussion of thermal management methodologies is deferred to the following chapter. Since many thermal analysis techniques have been extended from earlier approaches applied to 2-D circuits, these 2-D approaches are also discussed when appropriate, to offer a better understanding as well as distinguish the different requirements of vertically integrated systems.

Three-Dimensional Integrated Circuit Design. DOI: http://dx.doi.org/10.1016/B978-0-12-410501-0.00012-5

The primary demand for thermal models is high accuracy with low complexity while thermal analysis techniques should have affordable storage requirements, be as fast as possible, and capable of handling complex 3-D systems. A preliminary discussion of the heat transfer process within the context of ICs is offered in Section 12.1 to better understand the rationale behind the development of the various models, including the case of intertier cooling, where specific features of 3-D ICs are also described. Closed-form expressions for first-order thermal analysis of 3-D circuits are discussed in Section 12.2. Advanced thermal modeling of 3-D circuits requires a mesh structure of the volume of the circuit before applying numerical analysis methods. These fine scale models are reviewed in Section 12.3. An important element in modeling the thermal behavior of 3-D circuits is the through silicon via (TSV) which constitute a primary medium of heat flow within these circuits. Consequently, TSV models both for signal and thermal TSVs (TTSVs) are also reviewed. Additionally, models appropriate for liquid cooled 3-D circuits are presented in this section. Several analysis techniques that include these models to determine the temperature within 3-D circuits are discussed in Section 12.4. A synopsis of the primary issues presented throughout this chapter is provided in Section 12.5.

12.1 HEAT TRANSFER IN THREE-DIMENSIONAL ICs

In integrated circuits, heat originates from the transistors that behave as heat sources and also from self-heating of both the devices and the interconnects (joule heating), which can significantly elevate the circuit temperature [189,192]. The primary heat transfer mechanism within the volume of an integrated circuit is conduction, while different forms of convection are considered at the package boundaries depending upon the cooling mechanism. Examples include natural or forced air cooling through fans. This situation also applies to 3-D circuits; however, aggressive cooling mechanisms have been proposed that enable convective heat transfer between tiers (or layers) [442,443]. The heat transfer process for 3-D systems, where intertier cooling is employed, is discussed in Section 12.1.1, including a description of the technological implications of this cooling mechanism. Radiation is traditionally not considered as a heat transfer mechanism within integrated circuits.

The heat diffusion equation can be used to model the heat transfer process of conduction. Consequently, for a Cartesian coordinate system, heat conduction in the volume of a solid is described by [444]

$$\frac{\partial}{\partial x}\left(k\frac{\partial T}{\partial x}\right)+\frac{\partial}{\partial y}\left(k\frac{\partial T}{\partial y}\right)+\frac{\partial}{\partial z}\left(k\frac{\partial T}{\partial z}\right)+\dot{q}=\rho c_p\frac{\partial T}{\partial t}, \tag{12.1}$$

where k is the thermal conductivity (W/m-K), ρ is the density (kg/m^3), c_p is the specific heat (J/kg-K), and $\dot{q}$ is the rate of the generated heat per unit volume (W/m^3). Solving (12.1) produces the temperature at each point within the volume of the medium over time $T(x, y, z, t)$. Depending upon the characteristics of the system and the specific scenario, (12.1) can be further simplified. For example, if the thermal conductivity is modeled as constant with temperature (the typical case), (12.1) is

$$\left(\frac{\partial^2 T}{\partial x^2}\right)+\left(\frac{\partial^2 T}{\partial y^2}\right)+\left(\frac{\partial^2 T}{\partial z^2}\right)+\frac{\dot{q}}{k}=\frac{1}{a}\frac{\partial T}{\partial t}, \tag{12.2}$$

where $a = k/\rho c_p$ is the thermal diffusivity (m^2/s). If steady-state conditions are assumed, (12.2) further reduces to the Poisson equation,

$$\left(\frac{\partial^2 T}{\partial x^2}\right) + \left(\frac{\partial^2 T}{\partial y^2}\right) + \left(\frac{\partial^2 T}{\partial z^2}\right) + \frac{\dot{q}}{k} = 0. \tag{12.3}$$

If heat transfer takes place only in certain directions, additional terms can be dropped, further simplifying (12.3). The transfer of heat through natural or forced convection is likely to occur in at least one surface of an IC. The heat transfer rate through convection at this surface is [444]

$$q_{\text{conv}} = hA(T_s - T_\infty), \tag{12.4}$$

where h is the convection coefficient (assuming an average value for the entire surface), A is the area of the surface, T_s is the temperature at the surface, which is assumed to be uniform across the entire surface, and T_∞ is the temperature of the coolant (gas or fluid).

Within a 3-D system, heat flows through disparate materials with considerably different thermal properties. These materials include semiconductor, metal, dielectric, and possibly polymer layers used for tier bonding, where a cross-section of a 3-D system illustrating some of these materials is depicted in Fig. 12.1A. Additionally, the package, solder bumps, thermal interface materials, heat

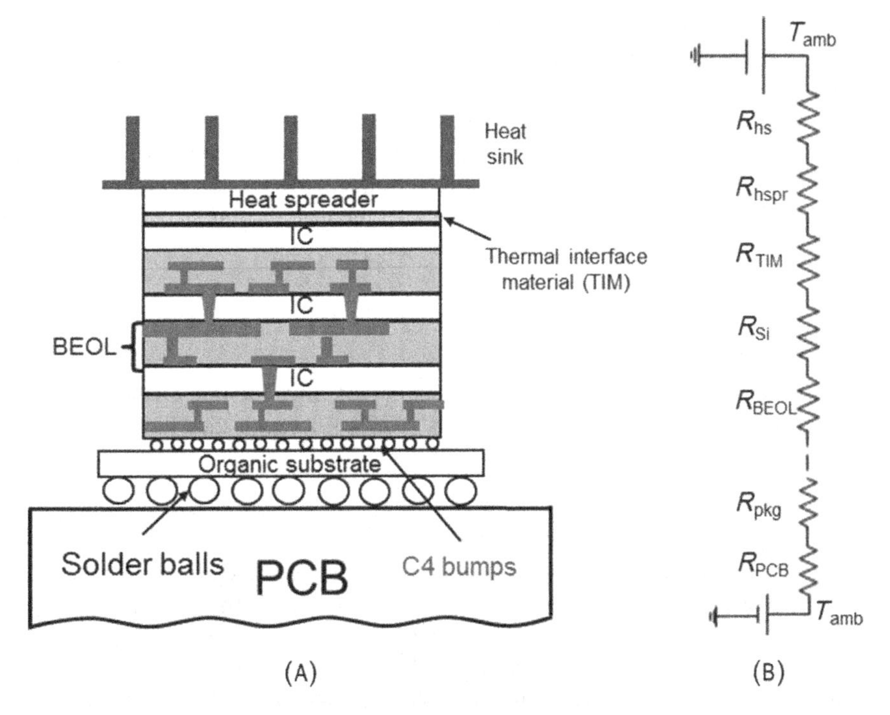

FIGURE 12.1

Cross-session of a 3-D stack illustrating (A) the variety of materials, including the package and heat sink, which increase the complexity of the thermal analysis process, and (B) a thermal circuit used to model the flow of heat along the z-direction.

Table 12.1 Thermal Properties of Materials within an Integrated Circuit [445–450]

Material	Thermal Conductivity (W/m-k)
Silicon	110–148
Copper	400
Thermal interface material (TIM)	1.6–4
Heat spreader	400
Back end of the line (BEOL)	0.3–5.2
Silicon dioxide	1.2
FR4 board	4.3
Tungsten	174

spreaders, and heat sinks all exhibit different thermal properties. Consequently, from a thermal perspective, a 3-D circuit constitutes a highly heterogeneous system, the analysis of which at fine granularity challenges most commercial tools that support multiphysics analysis. An exemplary list of the materials used in the thermal modeling of 3-D ICs is provided in Table 12.1, highlighting the greatly different thermal conductivity of these materials. Although this list is by no means exhaustive, layers of these materials are typically considered in the thermal analysis of most ICs. A range of values are provided due to the different materials for certain layers [445–450].

These aforementioned expressions describe the transfer of heat within a solid medium. However, fluidic cooling has been investigated specifically for 3-D systems due to the higher power densities that these systems encounter. To better understand the thermal models and analysis techniques discussed later in this chapter, an overview of the manufacture and design issues relating to fluidic cooling is provided in the following subsection.

12.1.1 LIQUID COOLING

One of the primary metrics that determine the quality of any cooling medium is the thermal resistance of the junction (measured in degree/Watt (°C/W)) between the silicon substrate and the ambient, described by

$$R_{th} = \frac{T_{\max} - T_{\text{amb}}}{Q}, \tag{12.5}$$

where $T_{\max}$ and T_{amb} are, respectively, the maximum temperature of the system and the temperature of the ambient, and Q is the power dissipated by the system. The thermal resistance in (12.5) may not correspond to a single material or heat transfer process but can include different materials between the substrate and ambient as well as the resistance due to conductive and convective heat flow. There are, therefore, several ways to interpret this expression.

A thermal resistance R_{th} can be determined for any cooling mechanism. The smaller the resistance, the lower the maximum temperature within the system. Alternatively, the smaller the resistance, the higher the power of the system that can be supported for a specific maximum temperature. Traditional heat sinks and forced air cooling (i.e., fans) lead to a thermal resistance of approximately 0.5°C/W, which is sufficient for the power levels of integrated systems to ensure the maximum temperature ranges between 85 and 110°C. This thermal resistance, however, must be further reduced to accommodate expected increases in power densities [18]. Assuming, for example, a maximum allowable temperature of 85°C and an ambient temperature of 27°C, a typical heat sink can accommodate a power density of 100 W/cm^2. The requirements become more stringent for vertically integrated systems as the power densities increase due to the reduced volume.

The use of liquids for cooling computing systems has been proposed for several decades as a more effective means to remove heat as compared to forced air cooling in high performance integrated systems [451]. Liquid cooling can support extremely low thermal resistances, <0.1 °C/W, effectively removing the heat for power densities up to 790 W/cm^2 within planar circuits, as shown for a case study in [452]. More recently, liquid flow cooling has been applied to processors in server systems within data centers [443], where (de-ionized) water is used as coolant with satisfying results. The additional benefit of using water as a coolant is that the heated water at the outlet of the cooling system can be used to heat buildings. Considering that cooling mechanisms can dissipate up to nearly half of the power budget of a data center [454], reusing the removed heat from the computing systems for other purposes greatly improves the overall energy efficiency of these data centers. Other refrigerants, such as R123 and R245ca, have been utilized as coolants, and two-phase cooling has also been explored [455]. The discussion herein, however, is limited to single-phase cooling (the temperature remains below the boiling temperature of the coolant) with the use of water as the coolant.

These case studies for 2-D ICs have historically been applied to silicon test vehicles [456,457] and more recently to server systems composed of commercial IBM and Intel processors [443]. A liquid cooling scheme used in these case studies may, however, prove inefficient for multi-tier circuits as the maximum temperature of the system may not appear at the tier attached to the heat sink but, rather, at another tier. Consequently, liquid cooling for 3-D systems should be applied to each individual tier, requiring microchannels within the substrate of each tier. Manufacturing processes that support the construction of microfluidic channels and ensure interconnectivity between tiers have, therefore, been investigated by several research groups [442,443]. The key idea is to etch microchannels, where the number and dimensions of these microchannels affect the efficiency of the heat transfer process. A liquid coolant flows through the channels to remove the heat from each tier. With multiple channels, rather than one wide channel, the flow of heat is improved as the total surface is significantly larger [444,452].

A schematic of the cross-section of a 3-D system with microchannels is illustrated in Fig. 12.2 based on the prototypes presented in [442,453]. A number of off-chip components relating to the design of this cooling scheme exists, such as the pump and heat exchanger, the design and optimization of which are beyond the scope of this book. Instead, the intent here is to offer an overview of the manufacturing features and expressions describing the flow of heat and thermal resistance of such a system. A discussion relating to thermal models of the microchannels for liquid cooled 3-D circuits is presented later in the chapter.

In Fig. 12.2A, a typical structure for microchannel-based cooling is shown, where channels of rectangular cross-section with width W_{ch} and depth (or height) H_{ch} are depicted. The length of the channel is equal to the width of the integrated circuit, denoted here as L_{chip}. The channels are separated with walls of thickness W_{fin} in which the TSVs are formed and connect the adjacent tiers. Thus, the pitch of the channels is $p_{ch} = W_{ch} + W_{\text{fin}}$. Consequently, for a system with width W_{chip},

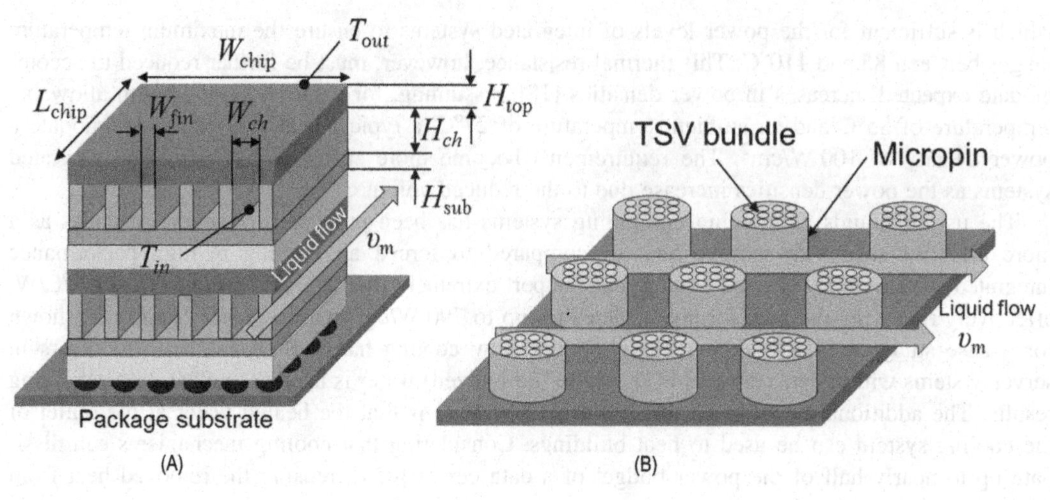

FIGURE 12.2

Schematic of a cross-section of a 3-D system with intertier liquid cooling through (A) microfluidic channels, and (B) through a micropin array [442].

the number of channels is $n_{ch} = \left\lceil \frac{W_{chip}}{p_{ch}} \right\rceil$. A different manufacturing process, shown in Fig. 12.2B, produces an array of micropins, where each pin can contain a bundle of TSVs. A prototype circuit where each pin with a 150 μm diameter contains a bundle of 4 × 4 TSVs with a TSV diameter of 13 μm resulted in a TSV density of about 31,000 TSVs/cm^2 [442], a density sufficient for vertically interconnecting several 3-D systems.

12.1.1.1 Design considerations for liquid cooled heat sinks

As shown in Fig. 12.2, a large number of issues exist that can affect the physical design of a microchannel-based heat sink. Other important parameters that do not relate to the geometry of a heat sink are not shown in the figure but should be considered in the design process and include the pressure drop, flow rate of the liquid, type of flow, and pumping power. Each of these parameters is constrained by limitations related to manufacturing yield, heat transfer efficiency, and electrical performance. For example, manufacturing limitations affect the width of the fins W_{fin}, thickness of the cover H_{top}, and depth of the channels H_{ch} (implicitly constrained by the aspect ratio of the TSVs). From a heat transfer perspective, the usual assumption of only laminar and fully developed flow can restrain the pumping power, cross-sectional area of the channel, and the number of channels [453]. Finally, from an electrical perspective, the increase in delay (and power) of the vertical interconnects passing through the high aspect ratio fins poses an additional set of constraints.

The multitude and diverse nature of the design parameters and constraints make the development of an effective methodology for designing a microchannel heat sink a formidable task. As a result, several published works address the issue of minimizing the thermal resistance of the microchannel heat sink by determining the optima for a small set of design parameters.

As the intent here is to enhance the understanding of the importance of the different design variables, the focus is on the effects of these different parameters, rather than a design solution for a specific heat sink.

Starting from the thermal resistance of a microchannel heat sink, three components can be distinguished,

$$R_{th} = R_{\text{cond}} + R_{\text{conv}} + R_{\text{heat}} = \frac{\Delta T}{Q} = \frac{T_{\text{out}} - T_{\text{in}}}{Q} = \frac{H_{\text{sub}}}{k_{\text{sub}} L_{\text{chip}} W_{\text{chip}}} + \frac{1}{hA_{ch}} + \frac{1}{\rho c_p \dot{V}}, \tag{12.6}$$

where R_{cond}, R_{conv}, and R_{heat} are, respectively, the conductive, convective, and thermal capacity resistance of the liquid. The heat produced within each tier due to circuit switching and interconnect joule heating is conducted to the boundaries of the channels (see also Fig. 12.2) through the substrate (typically silicon) with thermal conductivity k_{sub} and thickness H_{sub}. The liquid flowing in the channels removes heat by convection, resulting in R_{conv}, where h is the convection coefficient (as also mentioned in (12.4)) and A_{ch} is the total surface of the channels within a tier. The third component of the thermal resistance is due to heating of the liquid which absorbs energy flowing downstream from the channels. This last component depends upon the thermophysical properties of the liquid, which are the density, specific heat (see also (12.1)), and flow rate of the coolant. Water as a coolant has been considered in several studies, as water has a high volumetric capacity ρc_p of 4.18 J/°C-cm^3 and, with a flow of 10 cm^3/s, results in $R_{\text{heat}} = 0.024$ °C/W, a small contribution to the overall thermal resistance [452]. From (12.6), the thermal resistance can be decreased by (1) reducing the thickness of the substrate H_{sub} after etching the microchannels, (2) increasing the total surface of the channels, (3) increasing the heat transfer convection coefficient, and (4) increasing the flow rate. However, due to manufacturing constraints and the interdependency of these parameters, the process of minimizing the thermal resistance is a complex multivariable constrained optimization problem.

As the conductive resistance is bounded by the minimum thickness of the substrate H_{sub}, and the area of the circuit is fixed, emphasis has been placed on the effect of the surface of the channels, which can conveniently be described by the ratio $\beta = W_{ch}/p_{ch}$. To determine the convection coefficient h, some fundamental dimensionless numbers, commonly used in heat transfer and fluid dynamics, are required. These constants are the Reynolds Re, Nusselt Nu, and Prandtl Pr numbers (note that these numbers can be described in several equivalent forms by simple transformations),

$$Re = \frac{u_m D_h}{v}, \tag{12.7}$$

$$Nu = \frac{hD_h}{k_f}, \tag{12.8}$$

$$Pr = \frac{v}{\alpha}, \tag{12.9}$$

where, respectively, v is the kinematic viscosity, u_m is the mean velocity, α is the thermal diffusivity, and k_f is the thermal conductivity of the fluid. The hydraulic diameter of the channel is

$$D_h = \frac{2H_{ch}W_{ch}}{(H_{ch} + W_{ch})}. \tag{12.10}$$

In addition, the pumping power, volumetric flow rate, and pressure drop of the fluid through the microchannels can be written as [457]

$$\overline{P} = \dot{V}\Delta P = n_{ch} u_m H_{ch} W_{ch} \Delta P. \tag{12.11}$$

By solving these expressions, the minimum thermal resistance of a microchannel heat sink can be determined for different channel geometries and number of channels. The simultaneous solution of these expressions, however, requires the use of empirical relationships or curve fitting as in [456] or numerical approaches as in [457]. In addition, certain solutions for the cross-section of the channels may be rejected if these values violate the assumptions of the problem. For example, a choice of ratio β and channel aspect ratio $\gamma = H_{ch}/W_{ch}$ may result in a Reynolds number which does not correspond to fully developed laminar flow. Alternatively, the resulting pressure drop may not be acceptable or can require a prohibitive increase in pumping power $\overline{P}$, thereby yielding a low thermal resistance.

Thus, rather than solving for a specific heat sink design, insight is offered here describing the effects of the geometry and number of channels on the thermal resistance given by (12.6), assuming heat generated from the integrated system. Furthermore, the flow is assumed to be laminar and fully developed, which implies that $20 < Re < 2000$ [457]. The thermodynamic and hydrodynamic properties of the materials are also assumed to be constant with temperature. The dimensionless ratios β, γ, and number of channels n_{ch} are considered in this discussion.

For a specific β and number of channels, a higher channel aspect ratio γ is preferred to lower the thermal resistance, which favors deeper microchannels. Simultaneously, H_{sub} should be minimum thickness to ensure the generated stress does not exceed the bending strength of the substrate material [457]. In addition, deep channels require high aspect ratio TSVs, although the process yield is a challenge and can counteract the delay and power benefits of 3-D integration.

Alternatively, if the height and number of channels are maintained constant, thereby increasing the cross-section of the channels (i.e., increasing β), the cross-sectional area of the channels increases which is included in the convective component of the thermal resistance. As this component tends to be dominant, increasing the total surface to transfer heat by increasing β reduces the thermal resistance. Additionally, h depends upon the mean velocity if the properties of the liquid and geometry of the channel do not change. The mean velocity, in turn, increases if the aspect ratio decreases [457]. Consequently, as the height is held constant, the increase in β for constant n_{ch} reduces the aspect ratio, thereby increasing the mean velocity and h. As the number of channels remains the same, this situation requires the fins to be narrower, which reduces the TSV density of the stack.

Another tradeoff appears when varying the number of channels within the heat sink. Additional channels increase the total surface area available to transfer heat, but a large n_{ch} requires smaller channels, thereby increasing the resistance of the flow through the channel, which reduces the mean velocity, and therefore, reduces h [457]. Furthermore, the number of channels also affects the TSV density and location, since more channels (or alternatively more fins) provide greater flexibility in placing the TSVs, but fewer TSVs (per fin) may be available if the channel aspect ratio is decreased to counterbalance the increase in n_{ch}. Overall, the design of the microchannel heat sink requires careful tradeoffs of several physical and material parameters and can have serious implications on the electrical performance of the resulting system, which often is treated as a secondary concern during the design process of the heat sink.

The heat sink should exhibit a thermal resistance that maintains the temperature of the circuit within a given power limit. This power is often assumed to be constant, which is not typically the case as the power is dependent on the temperature of the circuit. Additionally, the temperature profile of the circuit is required for an effective reliability analysis. Producing a temperature map of a circuit requires solving the appropriate heat transfer expressions (e.g., (12.1) through (12.4)) which depend upon the design variables and operating and boundary conditions of the circuit.

Solving these expressions for a 3-D system is not a straightforward task due to the inhomogeneous (or heterogeneous) mixture of materials as well as the different shapes and features which comprise a 3-D system. Due to the physical structure of 3-D systems, specific assumptions relating to the flow of heat within the volume of the system are made to ensure that the thermal analysis process is tractable. A typical assumption is that the heat flows primarily in the vertical direction, whereas the lateral walls are considered adiabatic (no heat is exchanged with the environment across these surfaces).

Despite these assumptions, a complete thermal analysis remains a highly complex issue. Most analysis techniques focus on the steady-state behavior of the circuits, while a smaller number of methods consider the transient thermal behavior of 3-D ICs. For either type of analysis, thermal conductivities, for example in (12.1), have to be accurately determined. Extracting this information for an entire system is a difficult and time consuming task. A multitude of models and related techniques are employed to determine the temperature of a 3-D stack for different granularities and accuracies. In the following subsections, thermal models of increasing complexity developed over the past few years are reviewed.

12.2 CLOSED-FORM TEMPERATURE MODELS

First order analysis of the thermal behavior of 3-D systems can be performed through the use of one-dimensional (1-D) thermal models. An example is depicted in Fig. 12.1B. This thermal model is accurate if the flow of heat moves exclusively along the z-direction. The assumption of 1-D heat flow is justified by the short height of the 3-D stack and because the lateral boundaries of a 3-D IC are considered to behave adiabatically. Although thermal models based on analytic expressions exhibit the lowest accuracy, these models provide a coarse estimate of the thermal behavior of a circuit. This estimate may be of limited value at later stages of the design process where more accurate models are required; however, analytic models are useful during the early stages of the design process where physical information describing the circuit does not yet exist. These first order models are used to determine several design characteristics, such as packaging and cooling strategies and estimates of overall system cost [459].

In a 1-D thermal model, each material layer is modeled as a thermal resistor, heat sources as current sources, and temperature differences as voltage differences. An example of the relevant expressions describing the transfer of heat based on this model for a three tier 3-D circuit is shown in Fig. 12.3. The thermal equations resemble Kirchhoff's voltage law (KVL) expressions, as shown in the right half of the figure. This duality between the flow of current and of heat is extensively employed for thermal analysis both in first order analytic expressions and more elaborate models [444], as discussed in later sections of this chapter. To determine the temperature of the tiers based

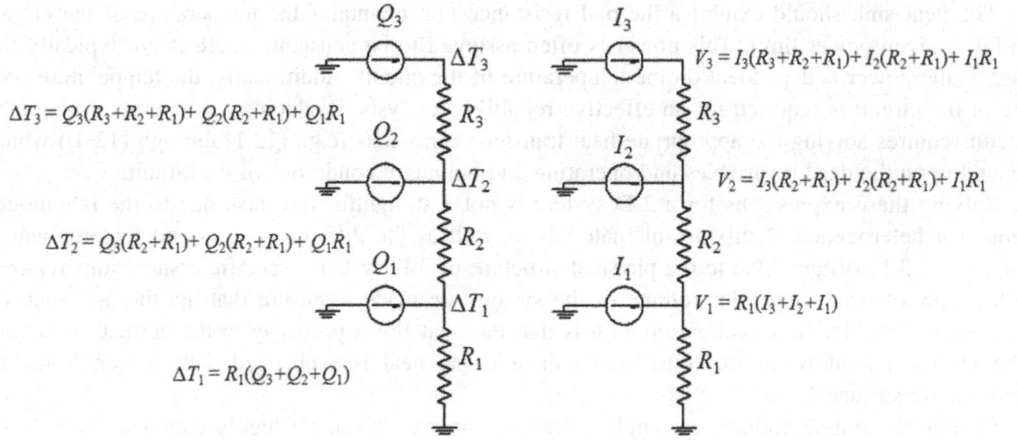

$$\Delta T_3 = Q_3(R_3+R_2+R_1)+Q_2(R_2+R_1)+Q_1R_1$$

$$\Delta T_2 = Q_3(R_2+R_1)+Q_2(R_2+R_1)+Q_1R_1$$

$$\Delta T_1 = R_1(Q_3+Q_2+Q_1)$$

$$V_3 = I_3(R_3+R_2+R_1)+I_2(R_2+R_1)+I_1R_1$$

$$V_2 = I_3(R_2+R_1)+I_2(R_2+R_1)+I_1R_1$$

$$V_1 = R_1(I_3+I_2+I_1)$$

FIGURE 12.3

Example of the duality of thermal and electrical systems.

on this model, as shown in Fig. 12.3, the heat generated within each tier and corresponding thermal resistances need to be determined.

A simple approach to model a 3-D system, with a structure as shown in Fig. 12.1A, is a cube consisting of multiple layers of silicon, aluminum, silicon dioxide, and polyimide. As depicted in Fig. 12.4, each layer is a homogeneous layer with a constant thermal conductivity. The devices on each tier are treated as isotropic heat sources and modeled as a thin layer on the top surface of each silicon layer. Either the top or bottom surface of the 3-D circuit is assumed to be adiabatic, although a non-negligible portion of the generated heat flows to the ambient through this surface. More detailed models also include the flow of heat through this surface. Alternatively, the other side of the 3-D IC, which is typically connected to a heat sink, is treated as an isothermal surface. These conditions simplify the analysis, as only a small set of important parameters needs to be investigated. Closed-form expressions also support fast design exploration as the size of the problem is reduced to only a few design parameters. The objective of these models is not to address circuit performance issues due to high temperatures but rather system level decisions; for example, the package, die stacking order, cooling mechanism, heat spreading materials, package level interconnects, and other system wide parameters that affect the design and cost of the overall system.

The heat generated within the silicon layers is primarily due to the transistors. Self-heating of the metal oxide semiconductor field effect transistor (MOSFET) devices can also cause the temperature of a circuit to rise significantly. Certain devices can behave as hotspots, causing significant local heating. For a two tier 3-D structure, an increase of 24.6°C is observed due to the silicon dioxide and polyimide layers acting as thermal barriers for the flow of heat towards the heat sink [441]. Although the dielectric and bonding layers behave as thermal barriers, the silicon substrate of the upper tiers spreads the heat, reducing the self-heating of the MOSFETs. Simulation results indicate that by reducing the thickness of the silicon substrate from 3 to 1 μm in a two tier 3-D IC, the temperature rises from 24.6 to 48.9°C [441]. Thicker silicon substrates, however, decrease the packaging density and increase the length of the intertier interconnects. In addition, high aspect ratio vias can be a challenging fabrication task, as discussed in Chapter 3, Manufacturing

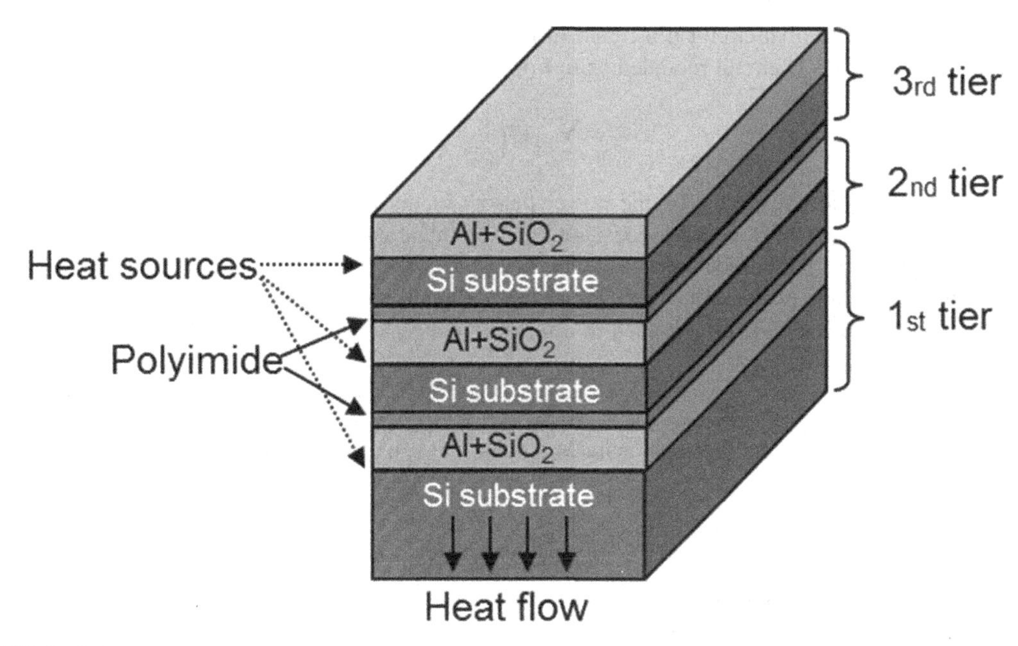

FIGURE 12.4

Thermal model of a 3-D circuit where 1-D heat transfer is assumed. Each layer is assumed homogeneous with a single thermal conductivity [441].

Technologies for Three-Dimensional Integrated Circuits. If the silicon substrate is completely removed, as in the case of 3-D silicon-on-insulator (SOI) circuits, self-heating can cause the temperature to rise to 200°C, which can catastrophically affect the operation of the ICs. In this model, the interconnects (BEOL) are implicitly included by considering a specific aluminum density within a dielectric layer, as depicted in Fig. 12.4. This situation is described by

$$K_{\text{eff}} = (1 - d_w)k_{ox} + d_w k_{\text{metal}}, \tag{12.12}$$

where k_{ox} and k_{metal} are, respectively, the thermal conductivity of the intralayer dielectric and interconnect metal and d_w is the interconnect density. This expression does not consider the several thermal paths that exist within the BEOL within each tier and does not differentiate between horizontal and vertical wires and metal contacts. Rather, an average thermal conductivity is determined by recognizing that a part of the volume of each BEOL layer consists of a dielectric $(1-d_w)$ while the rest of the volume consists of metal d_w. This notion of an effective thermal conductivity is extensively utilized throughout the literature. For example, the TSV density within the silicon substrate is included to determine an average thermal conductivity within the substrate. This averaging is a convenient way to simplify the thermal analysis process, although in certain cases averaging can lead to significant inaccuracy, as discussed in later sections of this chapter.

To estimate the maximum rise in temperature on the upper tiers of a 3-D circuit, similar to that shown in Fig. 12.4, a simple closed-form expression based on 1-D heat flow has been developed, consistent with the thermal circuit illustrated in Fig. 12.1B. Consequently, the temperature increase ΔT_j at the tier j of a 3-D circuit modeled as in Fig. 12.4 can be described by

$$\Delta T_j = \sum_{i=1}^{j} \left[R_i \left(\sum_{k=i}^{n} \frac{P_k}{A} \right) \right], \tag{12.13}$$

where P_k/A and R_k are, respectively, the power density of tier k and the thermal resistance from k to the ambient. The power density does not include interconnect joule heating, while the heat removal properties of the interconnects are only implicitly included in R_k.

Assuming the same power consumption and thermal resistance for all but the first tier, which is valid for a homogeneous 3-D circuit such as a memory cube, the increase in temperature is [458]

$$\Delta T_n = P\left[\frac{R}{2}n^2 + \left(R_{ps} - \frac{R}{2}\right)n\right]. \tag{12.14}$$

The thermal resistance of the first tier includes the thermal resistance of the package and the silicon substrate,

$$R_{ps} = \frac{t_{\text{Si1}}}{Ak_{\text{Si}}} + \frac{t_{pkg}}{Ak_{pkg}}, \tag{12.15}$$

where t_{Si1} and t_{pkg} are the thickness and k_{si} and k_{pkg} are the thermal conductivity of, respectively, the silicon substrate of the first tier and the package. The thermal resistance of an upper tier k is

$$R_k = \frac{t_{\text{sik}}}{Ak_{\text{Sik}}} + \frac{t_{\text{dielk}}}{Ak_{\text{dielk}}} + \frac{t_{\text{ifacek}}}{Ak_{\text{ifacek}}}, \tag{12.16}$$

where t_{sik}, t_{dielk}, and t_{ifacek} are the thickness and k_{Sik}, k_{dielk}, and k_{ifacek} are the thermal conductivity of, respectively, the silicon substrate, dielectric layers, and bonding interface for tier k. From (12.14) to (12.16), the increase in temperature on the topmost tier for different number of tiers and power densities of a 3-D system is illustrated in Fig. 12.5 for a typical thickness and thermal conductivity of the substrates, dielectrics, and bonding materials. As shown in Fig. 12.5, the increase in temperature exhibits a square dependence on the number of tiers and a linear relationship with the power density. Note that the thermal resistance of the package (or, equivalently, the junction thermal resistance in (12.5)) provides the greatest contribution to the increase in temperature. Further to this point, more recent results have shown that the choice of heat sink and package can change the purely monotonic increase in temperature with the number of tiers. If the boundary condition of the top surface being adiabatic is relaxed, the heat flows through both the top and bottom surfaces of a 3-D stack (i.e., package and heat sink) and, depending on the thermal resistance of the package and heat sink, the tier with the highest temperature is not always the tier farthest from the heat sink (or, alternatively, closest to the package). Rather, the temperature within the stack increases monotonically up to some tier and decreases for the remaining planes. For a two tier 3-D system, for the structure shown in Fig. 12.1, the second tier can exhibit a higher temperature if the following condition applies [459],

$$\frac{Q_2}{Q_1} > \frac{k_{pkg}}{k_{hs}}, \tag{12.17}$$

where k_{hs} is the thermal conductivity of the heat sink. From Fig. 12.5, the temperature within a 3-D circuit is exacerbated even for a small number of tiers. The effect of the interconnects on removing

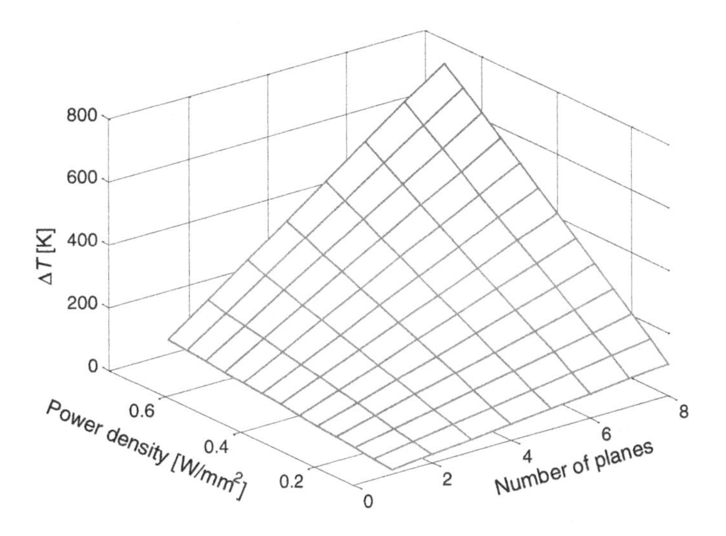

FIGURE 12.5

Increase in temperature in a 3-D circuit for different number of tiers and power densities.

the heat is not explicitly described, and interconnect joule heating has not been incorporated into these expressions. The increase in temperature on a specific tier k within a 3-D circuit considering the heat removal properties of the interconnect and the rise in temperature due to interconnect joule heating is described by [460]

$$T_{\mathrm{Si}_k} - T_{\mathrm{amb}} = \sum_{i=1}^{k-1}\left[\sum_{r=1}^{k_i}\frac{t_{\mathrm{ILD},ir}}{k_{\mathrm{ILD},ir}s f_{ir}}\eta_{ir}\left(\sum_{s=r}^{k_i} j_{\mathrm{rms},ir}^2 \rho_m H_{ir} + \sum_{j=i+1}^{n}\Phi_j\right)\right] + \sum_{i=1}^{k} R_i\left(\sum_{m=i}^{n}\Phi_m\right), \tag{12.18}$$

where the first term represents the temperature increase from the interlayer dielectrics (ILDs), while the second term yields the rise in temperature caused by the package, the bonding materials, and the silicon substrate(s). The notations in (12.18) are defined in Table 12.2. Expression (12.18) considers a 1-D model of the heat flow within a 3-D system similar to that based on Fig. 12.4 but applies a more accurate model of the different thermal conductivities and heat sources as compared to (12.14) to (12.16).

By including intertier vias and interconnect joule heating in the thermal model of a 3-D system, the thermal behavior of a 3-D circuit can be more accurately modeled. The rise in temperature of a two tier 3-D system is evaluated for two scenarios. In one scenario, interconnect joule heating and intertier vias are not considered, while in the second scenario, interconnect thermal effects are included. A decrease of approximately 40°C in the temperature of the bottom Si substrate is observed in the second scenario as compared to the first scenario. This result is an early indication of the important role that intertier vias play on reducing the overall temperature of a 3-D system by decreasing the effective thermal resistance in the vertical direction.

Table 12.2 Definition of the Symbols Used in (12.18)

Notation	Definition
T_{amb}	Ambient temperature
n	Total number of tiers
N_i	Number of metal layers in the ith tier
ir	r^{th} interconnect layer in the ith tier
t_{ILD}	Thickness of ILD
k_{ILD}	Thermal conductivity of ILD materials
sf	Heat spreading factor
η	Via correction factor, $0 \leq \eta \leq 1$
j_{rms}	Root mean square value of current density for interconnects
ρ_m	Electrical resistivity of metal lines
H	Thickness of interconnects
Φ_m	Total power density on the mth tier, including the power consumption of the devices and interconnect joule heating
R_1	Total thermal resistance of package, heat sink, and Si substrate (bottom tier)
R_i $(i > 1)$	Thermal resistance of the bonding material and the Si substrate for each tier

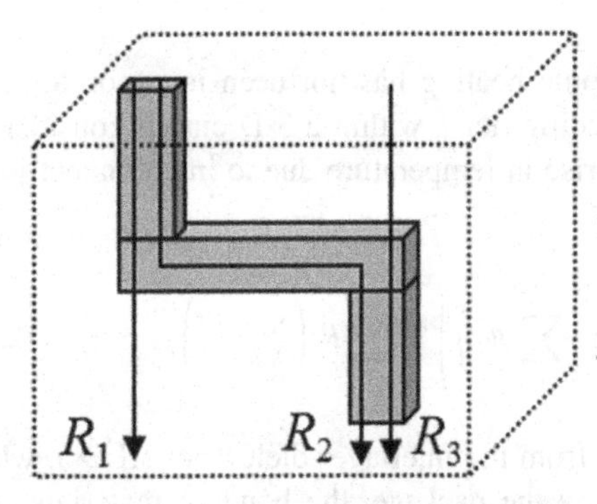

R_1 : Metal-dielectric

R_2 : Metal (horizontal and vertical)

R_3 : Dielectric-metal

FIGURE 12.6

Different vertical heat transfer paths within a 3-D IC [461].

Although (12.18) includes the effect of the interconnect on the heat flow process (as a BEOL layer within the stack), the several heat transfer paths that can exist within the interconnect structures require investigation. For example, assuming 1-D heat flow, heat is only transferred vertically within the intratier metal layers. Due to physical obstacles such as circuit cells or routing congestion, a continuous vertical path may not be possible for certain interconnections. This situation is depicted in Fig. 12.6 where different thermal paths are illustrated. As with current flow, heat flow also follows the path of the highest thermal conductivity. Consequently, interconnections consisting of horizontal segments in addition to intertier vias cause the heat flow to deviate from the vertical direction and spread laterally over a certain length, depending upon the length and thermal conductivity of each thermal path. By considering several thermal paths that exist within the BEOL layers, as shown in Fig. 12.6, the effective thermal conductivity of the buried interconnect layer consisting of a dielectric and metal is [461]

$$k_{\text{eff}} = k_{ox} + k_{\text{metal,eff}} = k_{ox} + \frac{t_{bi}}{A_{\text{int}}}\left[\frac{1}{R_1} + \frac{1}{R_2} + \frac{1}{R_3}\right], \tag{12.19}$$

where t_{bi} and A_{int} are, respectively, the thickness of the interconnect layer and the area of the buried interconnect layer. The thermal resistance of the paths is given by R_i, where these paths are considered in parallel, similar to electrical resistors connected in parallel. This duality implies that the presence of multiple thermal paths in a region produces a change in the total thermal conductivity of that region as compared, for example, with the thermal conductivity described by (12.12). By considering the different parallel thermal paths that can exist along the metal layers of each physical tier, a more accurate model of the flow of heat within the BEOL is achieved, although a single thermal resistor is utilized to characterize the entire layer.

As high temperatures can affect the reliability of a circuit, early publications on thermal analysis of 3-D circuits employing these first order models investigated self-heating of these devices [461]. Primary candidates include those devices that exhibit a high switching activity, such as clock drivers and buffers [461], which can suffer greatly from local heating, resulting in degraded performance. By considering the various thermal paths that can exist within interconnect structures, the effect of a rise in temperature on these devices has been investigated [461]. The increase in peak temperature as a function of the power density of the clock drivers placed above different interconnect structures is illustrated in Fig. 12.7. The existence of a thermal path with a horizontal metal segment exhibits inferior heat removal properties as compared to an exclusively vertical thermal path. In addition, the resulting increase in temperature in a 3-D IC is higher than bulk CMOS but not necessarily worse than SOI, as illustrated in Fig. 12.7.

Another factor that can affect the thermal profile of a circuit is the physical adjacency of the devices. Thermal coupling among neighboring devices within a 3-D circuit is greater as proximity increases, further increasing the temperature of the circuit [441,461]. The temperature is shown to exponentially decrease with gate pitch. This behavior implies that the area consumed by certain circuit elements, such as a clock driver, should be greater in a 3-D circuit than in a 2-D circuit to guarantee reliable operation by lowering thermal degradation.

The assumption of 1-D heat flow permits a circuit to be modeled by a few serially connected resistors. Additionally, by including the interconnect power and the different thermal paths by appropriately adapting the thermal conductivity of some layers, the accuracy of the thermal models of 3-D ICs is significantly improved. The major assumption, and simultaneously, primary drawback of the closed-form expressions describing the temperature of a 3-D circuit is that each physical tier is characterized by a single heat source. This assumption implies that all of the heat sources that can exist within a tier collapse to a single heat source, as depicted in Fig. 12.3. Consequently, phenomena such as thermal coupling and intratier thermal gradients cannot be captured. Thus, although this approach is sufficiently accurate during early steps in the design process, knowledge of the actual power density and temperature within each physical tier is essential to thermal design methodologies to maintain thermally tolerant circuit operation. More accurate models required for these techniques are presented in the following subsection.

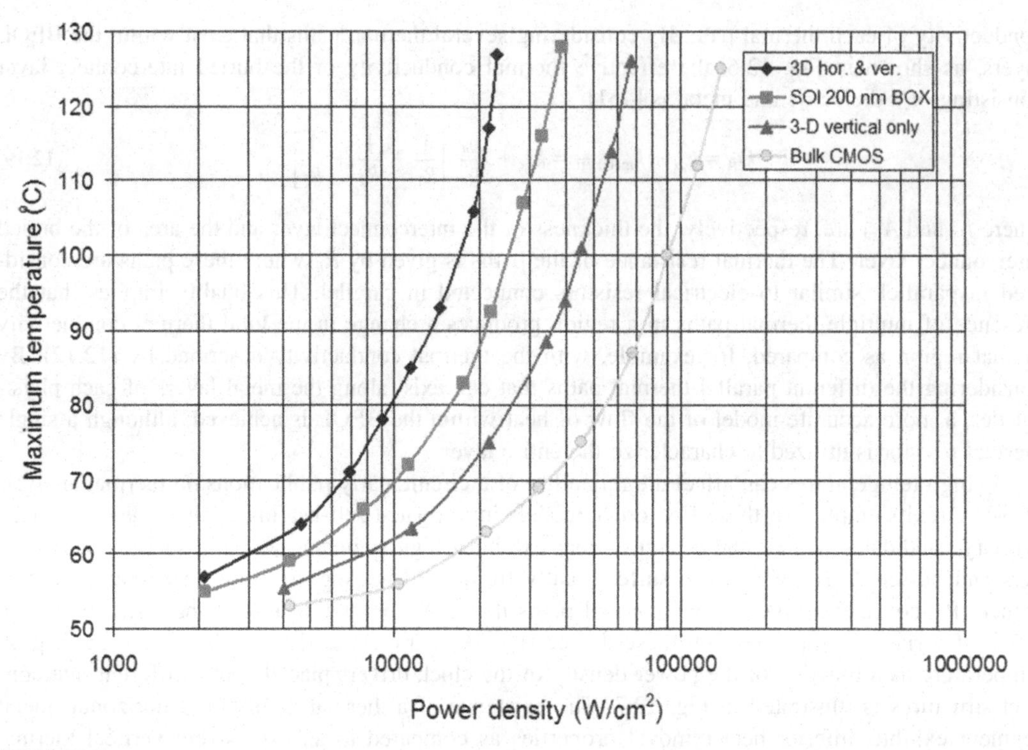

FIGURE 12.7

Maximum temperature versus power density for 3-D ICs, SOI, and bulk CMOS [461]. The difference among the curves for the 3-D ICs is that the first curve (3-D horizontal and vertical) includes thermal paths with a horizontal interconnect segment, while the second curve includes only continuous vertical flow of heat through the wires.

12.3 MESH-BASED THERMAL MODELS[1]

In the previous subsection, thermal models based on analytic expressions to evaluate the temperature of a 3-D circuit are discussed. In all of these models, the heat generated within each physical tier is represented by a single value. Consequently, the power density of a 3-D circuit is assumed to be a vector in the vertical direction (i.e., *z*-direction). In addition, the thermal network is represented as a 1-D resistive network, as shown in Fig. 12.1B.

The temperature and heat flow within each tier of a 3-D system, however, can fluctuate considerably, yielding temperature and power density vectors that vary in all three directions. Mesh-based thermal models capture this critical information by representing the volume of a circuit with a set of tiles. Each tile is thermally modeled with a small number of resistors (and capacitors if the thermal transient behavior is also analyzed), as illustrated in Fig. 12.8. The tiles are connected through nodes at the tile boundaries forming a 3-D thermal network, where the temperature at each node is

[1]The terms tile, cell, and control volume are interchangeably used throughout the chapter.

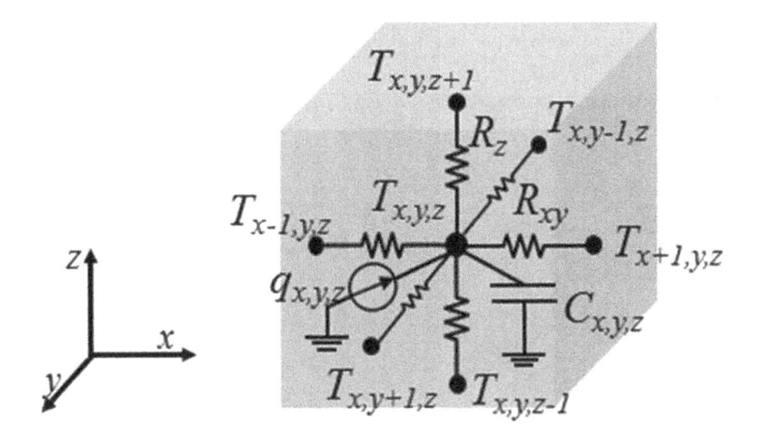

FIGURE 12.8

Unit tile (or cell) including a thermal resistor in each x, y, z-direction. A thermal capacitor models the heat capacity of the tile and a heat source $q_{x,y,z}$ for the power consumed by the devices or the joule heating of the wires within this cell.

determined. Although in this figure only two different thermal resistances are indicated, R_z and R_{xy}, which relate to the transfer of heat in, respectively, the vertical and horizontal directions, all resistances can be different. In addition, some elements may not be included in each tile. For example, if no heat is generated within a cell, the current source can be omitted. If only a steady-state analysis is intended, the capacitors are not required. Furthermore, a resistor in the vertical direction is not included in the case of the topmost/bottommost tier of the stack. Therefore, which elements are present in each tile depends not only on the components of the 3-D stack within each cell but also on the intended analysis. Similar to the *R*(*L*)*C* extraction process for IC layouts, the thermal components within the volume of each tile must be extracted. This process, however, is not straightforward as typically the volume of a tile contains different materials. In other words, the tiles are not in general homogeneous. For example, a tile may include some segment of a wire, interlayer dielectric, metal contacts, some diffusion area, a TSV, and/or silicon. The volume of the tiles can become arbitrarily small yet finite, ensuring that each tile contains only one material, making the thermal elements more easily determined. This approach, however, greatly increases the number of cells and, consequently, the number of nodes that needs to be analyzed, rendering the computational time impractical. A characteristic example is the use of multiphysics solvers, which can practically analyze only the smallest 3-D structures [462,463]. Thus, researchers have resorted to approximations to reduce the number of tiles needed to characterize a 3-D circuit. Based on these experimental results and comparisons between multiphysics solvers and proposed approximations, tiles with dimensions on the order of tens of micrometers are computationally practical while offering reasonable accuracy [449,464].

Tiles of this scale can contain interconnects, dielectrics, and/or silicon, as the feature size of modern technology nodes is on the scale of nanometers. Consequently, several methods have been developed to determine the thermal components of the tiles. The majority of the methods emphasize thermal resistance as most of these techniques focus on steady-state analysis. The current

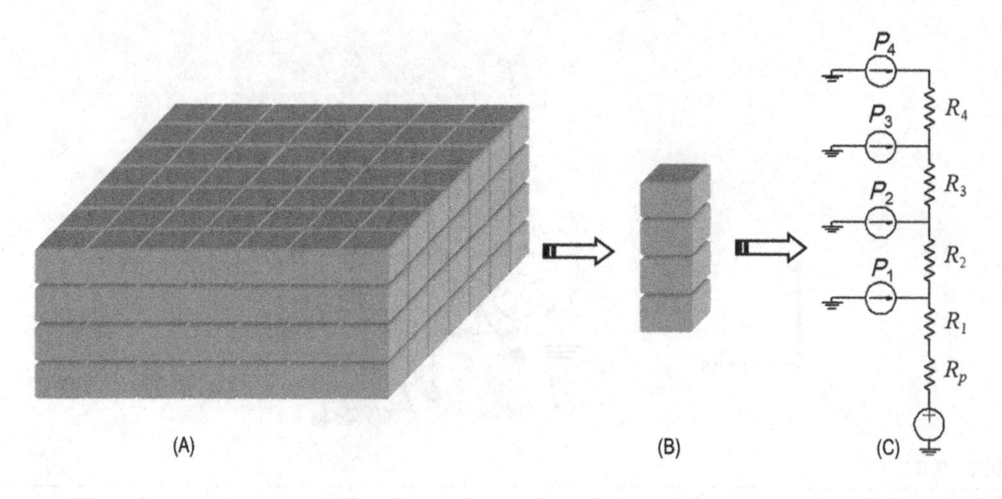

FIGURE 12.9

Thermal model of a 3-D IC. (A) A 3-D tile stack, (B) one pillar of the stack, and (C) an equivalent thermal resistive network. R_1 and R_p correspond, respectively, to the thermal resistance of the thick silicon substrate of the first tier and the thermal resistance of the package [466].

source is typically the current passing through the transistors within the cell or equivalently, the heat generated by the current flowing through the wires for those tiles within the BEOL layers.

An early model, a compromise between a 1-D thermal circuit and a full mesh, models a 3-D system as a thermal resistive stack, as shown in Fig. 12.9. The discretized volume of the system shown in Fig. 12.9A is segmented into single pillars, as illustrated in Fig. 12.9B [465–467]. Each pillar is successively modeled by a 1-D thermal network including thermal resistors and heat sources, as shown in Fig. 12.9C. The heat sources include all of the heat generated by the devices contained within each tile. Resistances related to the TSV are also included in the pillar. The absence of a TSV between two tiers is incorporated by removing the via resistances, ensuring that heat will not flow through those resistors. The voltage source at the bottom of the network models the isothermal surface between the heat sink and the bottom silicon substrate. Additional resistors, not shown here, can be used to incorporate the flow of heat among neighboring pillars.

Comparing the compact model of a single pillar of the stack with the simpler 1-D model used to produce the closed-form solution presented in Section 12.2, several similarities exist. Both of these models use resistors and heat sources modeled as current sources. Note that although the voltage source included in Fig. 12.9C, which considers the heat sink, does not appear in Fig. 12.3, this element of the model is implicitly included in (12.18) as the closed-form expression describes the rise in temperature in tier k of a 3-D system (i.e., $\Delta T = T_{si\_N} - T_{amb}$) rather than the absolute temperature generated by a compact 1-D model.

Merging several cells into a pillar reduces the number of nodes at which the temperature needs to be determined, thereby reducing the computational complexity of the problem. The accuracy, however, may be degraded. In addition, this model has been developed for a specific technology and cannot be used to explore different technologies where physical characteristics differ. The usefulness of a model depends not only on the complexity which is related to the number of parameters that need to be determined but also on the capability of the model to support different geometries and fabrication parameters since 3-D integration is manifested in diverse manufacturing processes. In addition, this model does not describe the disparate materials that can exist within a cell other than a TSV.

In 3-D systems, the primary direction where the heat flows is vertical. Accurately modeling the direction of the thermal behavior of the TSVs is highly important, as the TSVs provide a path of high thermal conductivity along this vertical direction. Thus, a non-negligible number of works has been developed to thermally model the TSVs [465–471]. Due to enhanced heat conduction properties, TSVs have also been used solely for the purpose of facilitating the flow of heat. These heat conduits are called thermal TSVs (TTSVs) and several thermal management techniques have been developed to allocate these resources across the volume of a 3-D system, ensuring the resulting system satisfies the temperature specifications. Consequently, different models exist for signal and TTSVs, where the TTSVs are not subject to joule heating as no current flows through these vias.

An alternative cooling approach that makes obsolete the need for TTSVs and supports more efficient heat removal is that of integrated liquid cooling, as discussed in Section 12.1.1. As the presence of fluid flow adds a convective component of heat transfer, the modeling process for microchannels is not the same as within solids where conduction is the primary mechanism of heat transfer. Consequently, the unit cells within a discretized 3-D system describing a microchannel need to be modeled differently as compared to the volume of a solid, as shown in Fig. 12.8. In the next subsections, thermal models of varying complexity for different types of TSVs and fluidic channels are discussed.

12.3.1 THERMAL MODEL OF THROUGH SILICON VIAS

The simplest thermal model of a TSV is a resistor (similar to (12.15)) equal to the inverse of the thermal conductivity k_m of the metal used for the TSVs (typically copper or tungsten) and the area of the TSV A_{TSV}, and multiplied by the length of the TSV t_{TSV} or, alternatively, the length of the cell (where the cell contains only part of the TSV),

$$R_{\text{TSV}} = \frac{t_{\text{TSV}}}{A_{\text{TSV}} k_m}. \tag{12.20}$$

This model, however, neglects several important aspects which affect the model accuracy. For example, joule heating is excluded and the non-negligible lateral flow of heat through the TSV liner is not considered. In addition, the case where multiple TSVs are included within a single cell requires a different modeling approach. These aspects are discussed in the following subsections.

12.3.1.1 Thermal through silicon vias

As previously mentioned, TTSVs act exclusively as heat pipes, allowing heat to flow to the heat sink, alleviating hotspots within a 3-D stack. Early thermal management techniques modeled

TTSVs as a single resistance, ignoring lateral heat transfer effects. However, lateral flow of heat should not be neglected as this mechanism affects the overall heat transfer process. Although the thermal conductivity of the surrounding dielectric of the TSV liner is considerably lower than that of silicon and metal, the liner thickness is typically about a micrometer. Due to this short thickness, heat flows laterally towards the thermally less resistive metallic TSV, facilitating the heat removal process through the 3-D stack. Assuming a cell includes a TSV within the silicon substrate, as shown in Fig. 12.10, the different physical parameters of the TSV typically used in thermal models are considered. The simulation setup to determine the thermal conductivity along the path of the flow of heat is depicted in Fig. 12.11. The structure illustrated in Fig. 12.11A applies a heat source at the left boundary surface, while the top and bottom surfaces are adiabatic. In a similar way, a heat source is applied at the top surface in Fig. 12.11B while the lateral walls are considered adiabatic. Auxiliary blocks (see Fig. 12.11B) are added to ensure that the heat spreads uniformly before reaching the target cell, and a small section ΔH is evaluated to determine the local thermal conductivity. Note that this small segment includes only part of the TSV, liner, and silicon substrate, which is not necessarily consistent with larger cells since these larger cells often include other

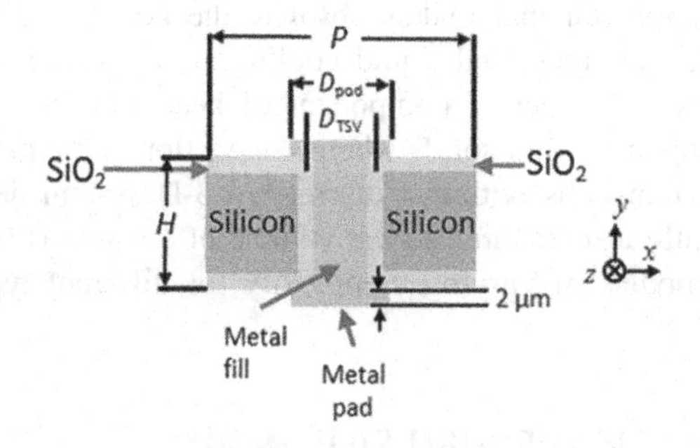

FIGURE 12.10

Cross-section of a cell including a TSV within the silicon substrate [468].

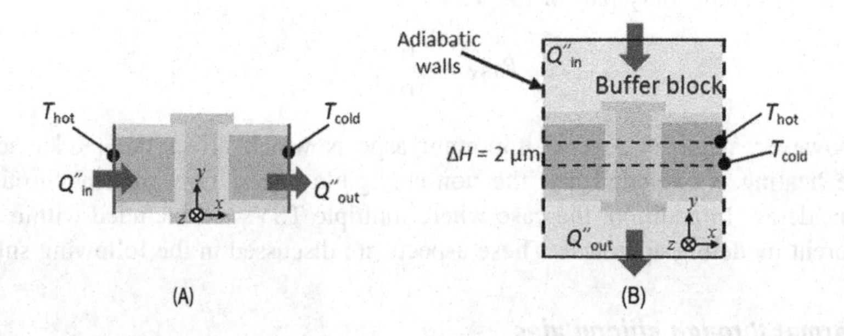

FIGURE 12.11

Simulation setup for determining the thermal conductivity of the cell shown in Fig. 12.10 along (A) the *xy*-plane, and (B) along the *z*-direction [468].

materials. From this perspective, two distinct thermal conductivities are determined along the *xy*-plane and *z*-direction through the following expressions [468],

$$k_{xy} = \left(90t_{\mathrm{SiO_2}}^{-0.3} - 148\right)\left(\frac{D_{\mathrm{TSV}}}{P}\right)H^{0.1} + 160t_{\mathrm{SiO_2}}^{0.07} \tag{12.21}$$

$$k_z = 128e^{\left(\frac{D_{\mathrm{TSV}}}{P}\right)} \text{ for } 0.002 \le \frac{t_{\mathrm{SiO_2}}}{H} \le 0.01, \tag{12.22}$$

$$k_z = 130e^{\left(1.1\frac{D_{\mathrm{TSV}}}{P}\right)} \text{ for } 0.01 < \frac{t_{\mathrm{SiO_2}}}{H} \le 0.02, \tag{12.23}$$

$$k_z = 260\left(\frac{D_{\mathrm{TSV}}}{P}\right) + 115 \text{ for } 0.02 < \frac{t_{\mathrm{SiO_2}}}{H} \le 0.04, \tag{12.24}$$

$$k_z = 300\left(\frac{D_{\mathrm{TSV}}}{P}\right) + 120 \text{ for } 0.04 < \frac{t_{\mathrm{SiO_2}}}{H} \le 0.1, \tag{12.25}$$

$$k_z = 135\ln\left(\frac{D_{\mathrm{TSV}}}{P}\right) + 380 \text{ for } 0.1 < \frac{t_{\mathrm{SiO_2}}}{H} \le 0.2, \tag{12.26}$$

where $t_{\mathrm{SiO_2}}$ is the thickness of the silicon dioxide (or another dielectric) layer surrounding the TSV, and D_{TSV}, P, and H are, respectively, the diameter, pitch, and height of the TSV. These expressions are applicable for a range of these parameters: liner thickness of 0.2 to 2.0 μm, TTSV diameter of 10 to 50 μm, TTSV length greater than 20 μm, and $0.1 \le D_{\mathrm{TSV}}/P \le 0.77$. These expressions are compared with simulations from the Icepak solver [463] and the error is less than ± 10%.

This model, however, only considers that segment of the TSV within the silicon substrate, which is not the case for TSV-last processes. In addition, the thermal conductivity in these expressions is either an exponential, parabolic, or logarithmic function depending upon the value of certain physical parameters, without offering any intuitive insight. To consider those parts of a TTSV through the bonding and BEOL layers, for example, and to offer a more intuitive model of a TTSV, another model that employs three thermal resistances per TTSV has been developed. The rationale behind this model is based on the three major heat transfer paths witihin a volume of a 3-D stack that includes a stacked set of TTSVs. A stacked TTSV is a better option for removing heat as a continuous structure has the least thermal resistance and removes heat more efficiently. Consequently, employing stacked TTSVs is a reasonable practice to facilitate the vertical flow of heat.

For the structure shown in Fig. 12.11, a small volume of a 3-D system is evaluated with COMSOL [462]. This volume corresponds to a segment of a three tier 3-D circuit with a single TTSV, which can be extended to an *n*-tier circuit. The physical structure of this stack is illustrated in Fig. 12.12A. The cross-section of the circuit and the temperature distribution determined from COMSOL multiphysics are illustrated in Fig. 12.12B. Although for different fabrication technologies the materials and geometries of the circuit can vary, the underlying structure remains the same. This model is based on a 3-D technology employing wafer bonding. As labeled in Fig. 12.12A, each tier of the circuit consists of three layers describing, respectively, the silicon substrate (Si), the interlayer dielectric (ILD) and metal interconnects (i.e., BEOL), and the bonding layer. The heat sources include the power generated by the active devices on the top surface of the Si substrates and Joule heating due to the interconnects surrounded by the ILD.

As shown in Fig. 12.12, three major paths of heat flow are illustrated. Heat flows vertically through silicon (path 1) and the TTSV (path 3) and laterally through the liner of the TSV (path 2)

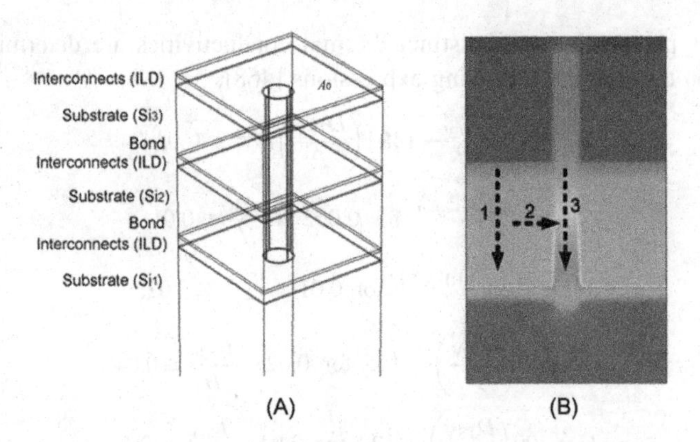

FIGURE 12.12

A segment of a three tier 3-D IC with a TTSV, where (A) is the geometric structure, and (B) is the cross-section of a TTSV of this segment. The area of the circuit is denoted by A_0. The three main paths of heat transfer are depicted by the dashed lines.

towards the more conductive metal fill within the TTSV. The flow of heat along each of these paths can be modeled by a resistance. If the model is intended to support design exploration, the model should be linked with the physical characteristics of the TSV including, for example, the thickness of the liner and diameter of the TSV. As the heat transfer process can be more complicated (there are many more paths in addition to these three paths), some fitting coefficients are used to improve the accuracy of the model. Based on these heat flow paths, the following expressions describe the thermal resistance of each TSV,

$$R_1 = \frac{1}{k_1 A}\left(\frac{t_{\text{BEOL}}}{k_{\text{BEOL}}} + \frac{l_{\text{ext}}}{k_{\text{Si}}}\right), A = A_0 - \pi\left(\frac{D_{\text{TSV}}}{2} + t_{\text{SiO}_2}\right)^2, \tag{12.27}$$

$$R_2 = \frac{t_{\text{BEOL}} + l_{\text{ext}}}{k_1 k_{\text{TSV}} \pi \left(\frac{D_{\text{TSV}}}{2}\right)^2}, \tag{12.28}$$

$$R_3 = \frac{\ln\left(\frac{D_{\text{TSV}}}{2} + t_{\text{SiO}_2}\right) - \ln\frac{D_{\text{TSV}}}{2}}{2\pi k_2 k_{\text{SiO}_2}(t_{\text{BEOL}} + l_{\text{ext}})}, \tag{12.29}$$

$$R_4 = \frac{1}{k_1 A}\left(\frac{t_{\text{BEOL}}}{k_{\text{BEOL}}} + \frac{t_{\text{Si2}}}{k_{\text{Si}}} + \frac{t_b}{k_b}\right), \tag{12.30}$$

$$R_5 = \frac{t_{\text{BEOL}} + t_{\text{Si}_2} + t_b}{k_1 k_{\text{TSV}} \pi \left(\frac{D_{\text{TSV}}}{2}\right)^2}, \tag{12.31}$$

$$R_6 = \frac{\ln\left(\frac{D_{\text{TSV}}}{2} + t_{\text{SiO}_2}\right) - \ln\left(\frac{D_{\text{TSV}}}{2}\right)}{2\pi k_2 k_{\text{SiO}_2}(t_{\text{BEOL}} + t_{\text{Si}_2} + t_b)}, \tag{12.32}$$

$$R_7 = \frac{1}{k_1 A}\left(\frac{t_{\text{BEOL}}}{k_{\text{BEOL}}} + \frac{t_{\text{Si}_3}}{k_{\text{Si}}} + \frac{t_b}{k_b}\right), \quad (12.33)$$

$$R_8 = \frac{t_{\text{Si}_3} + t_b}{k_1 k_{\text{TSV}} \pi \left(\frac{D_{\text{TSV}}}{2}\right)^2}, \quad (12.34)$$

$$R_9 = \frac{\ln\left(\frac{D_{\text{TSV}}}{2} + t_{\text{SiO}_2}\right) - \ln\left(\frac{D_{\text{TSV}}}{2}\right)}{2\pi k_2 k_{\text{SiO}_2}(t_{\text{Si}_3} + t_b)}, \quad (12.35)$$

$$R_s = \frac{t_{\text{Si}_1} - l_{\text{ext}}}{k_1 k_{\text{Si}} A_0}, \quad (12.36)$$

where t_b and $t_{\text{Si}n}$ are the thicknesses of, respectively, the bonding layer and the substrate for tier n. The thermal conductivity of the TSV and the liner are notated, respectively, by k_{TSV} and k_{SiO_2}. In (12.27)–(12.36), the resistances, R_2, R_5, and R_8, are the thermal resistances of the filling material (e.g., copper) of the TTSV. R_2 describes the resistance of the TSV for the first or last tier, where the TSV may be "blind" (i.e., enclosed within the substrate) and, consequently, l_{ext} is employed to capture this case. In addition, there is no bonding layer for this tier, therefore, this term is omitted from (12.28). The resistances R_3, R_6, and R_9 denote the lateral thermal resistance of the insulator liner (e.g., SiO_2) of the TTSV. The resistances R_1, R_4, and R_7 denote the thermal resistance of the TTSV surroundings (see Fig. 12.13) for each of the three physical tiers. The thermal resistance of the silicon substrate of the first tier is denoted by R_s due to the considerably different thickness of the substrate (note that a separate resistance is also added for the tier with the thick silicon substrate in the model shown in Fig. 12.9C). Although this model captures more of the technological characteristics of a 3-D system, two fitting coefficients k_1 and k_2 are required for the model to provide sufficient accuracy with an average and maximum error of, respectively, 3% and 6% for the investigated scenarios [469]. The k_1 and k_2 coefficients adjust the lumped model representation of the thermal resistance of the TTSV when used with a distributed model of heat flow.

A method to eliminate fitting coefficients while not degrading accuracy is to model the TTSV with additional resistors. The stack (or single) TTSV is modeled as a distributed resistive wire. This situation means that additional cells are used to model the target volume, which naturally improves the precision without the need for fitting coefficients but increases the computational complexity. Thus, for the volume shown in Fig. 12.12A, a comparison between the distributed and lumped models is performed, where the reference temperatures are the results reported by an finite element method (FEM) solver [462]. The results listed in Table 12.3 indicate that more than a hundred segments can have an undesirable overhead on the computational time [469]. Alternatively, the overhead for producing the fitting coefficients for specific structures is performed only once and may be a more computationally efficient method for thermal analysis without sacrificing accuracy (model A in Table 12.3). An example of the usefulness as well as accuracy of this model is given in Fig. 12.14, where the maximum rise in temperature for a three tier circuit is plotted as a function of the thickness of the liner (e.g., t_{SiO_2}). Note that the 1-D model cannot capture this temperature increase, while the accuracy of the distributed model marginally improves when more than 100 segments are utilized.

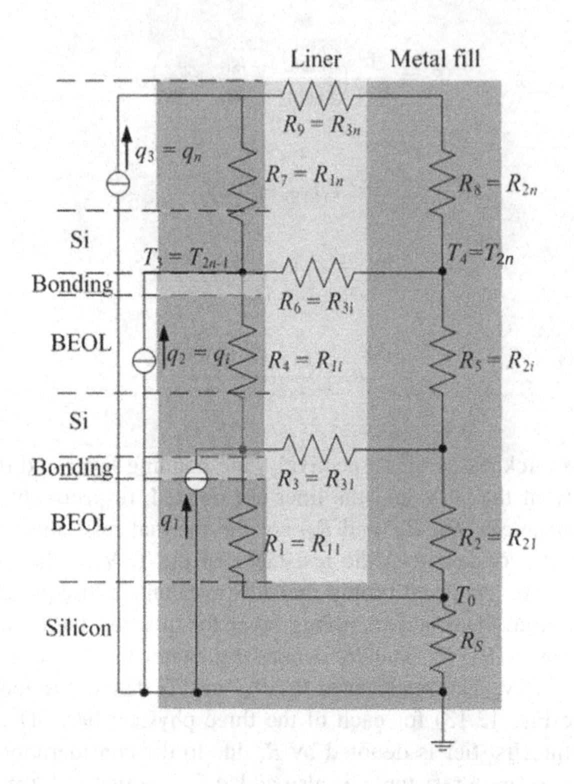

FIGURE 12.13

Thermal model of a TTSV in a three tier circuit, extendible to n tiers, where double notation is used to demonstrate that the model can be extended to a 3-D stack of n tiers.

Table 12.3 Error and Computational Time Versus Number of Segments in the Distributed Model

Model (# of Segments)	*B* (1)	*B* (20)	*B* (100)	*B* (500)	A	1-D
Max. error	23%	12%	6%	5%	4%	30%
Avg. error	19%	11%	4%	3%	2%	23%
Time (ms)	1	3	32	2474	–	–

Inclusion of the lateral path more accurately describes the flow of heat and reduces overdesign. If excluded, typically higher temperatures are predicted, which, in turn, require more demanding thermal management schemes. To illustrate this situation, a DRAM memory-processor system has been analyzed employing this model and a 1-D model (only one resistor models a TSV as in [459]), and is compared with FEM simulation. The maximum rise in temperature for the more complex model, 1-D model, and FEM is, respectively, 12.8, 20, and 12°C, exhibiting the high inaccuracy introduced by ignoring the lateral flow of heat through the liner.

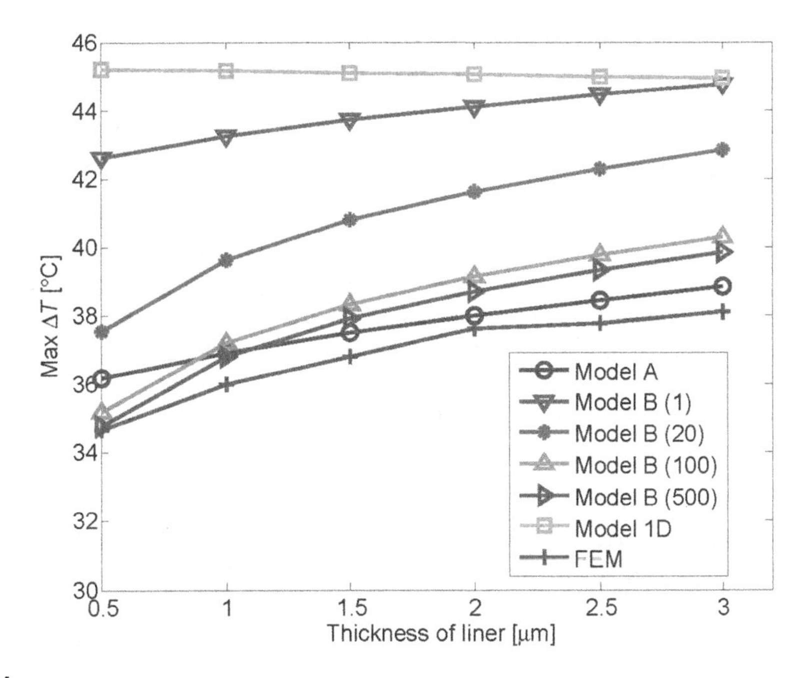

FIGURE 12.14

Maximum rise in temperature in a three tier 3-D circuit for different dielectric liner thicknesses, where $D_{TSV} = 10$ μm. The other parameters are $t_{SiO_2} = 7$ μm, $t_b = 1$ μm, $t_{Si2} = t_{Si3} = 45$ μm, $k_1 = 1.3$, and $k_2 = 0.55$.

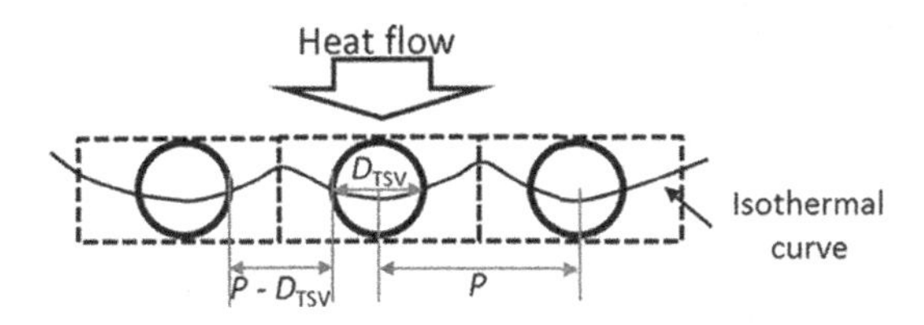

FIGURE 12.15

Neighboring cells bending the isothermal curves due to the TSVs [470].

Both of the models discussed in this section require some fitting to enhance accuracy. In the first model, this fitting is achieved by changing the form of the function describing the thermal conductivity. This approach, however, offers limited insight into the dependence of the flow of heat on the physical parameters, such as the radius of the TSV. The second model uses certain fitting coefficients to improve precision. The resulting inaccuracy depends upon the ratio of the TSV radius to the TSV pitch. This ratio causes bending in the isothermal curves among adjacent TSV cells, as shown in Fig. 12.15. For a similar range of D_{TSV}/P (i.e., [0.125 to 0.25]) as in [468], a correction factor models the bending of the isothermal curves under the presence of identical cells. Thus, the model used in [470] magnifies the thermal resistance of a TTSV with a corrective factor θ, which is a linear function of the TSV space-to-radius ratio, $\delta = 2(P\text{-}D_{TSV})/D_{TSV}$. This function is

$$\theta = \beta_1\delta + \beta_2, \tag{12.37}$$

where β_1 and β_2 are fitting coefficients which can be determined from either simulated or measured data with linear fitting techniques [470].

In general, a structure with juxtaposed TSV cells is not applicable, as the adjacent cells may not contain any TSVs and, consequently, the heat will flow in different ways. Arrays of TTSVs are typically used to mitigate thermal hotspots. The rationale behind the use of these TTSV arrays is that the effective thermal conductivity increases in this local region, lowering the high temperatures. Since these TTSV arrays require non-negligible area, models that include more than one vertical path for the heat to flow lead to fewer TTSVs and are therefore a useful means to limit this area overhead while also satisfying thermal limits [469].

12.3.1.2 Signal through silicon vias

Although some models have been published for TTSVs, there are few models that capture the thermal behavior of signal TSVs, where the primary difference is that heat is generated within these TSVs (joule heat) due to the current flowing through those wires. In addition, as the current flowing through the TSV is not constant (even for power/ground TSVs), the transient thermal behavior of the target cell must be considered. For a basic structure, such as the tapered TSV shown in Fig. 12.16, commercial multiphysics solvers, such as COMSOL, can be employed to analyze the thermal behavior of this cell. Another reasonably efficient approach for multiphysics characterization of TSVs is the hybrid time-domain finite-element method [471].

The thermal resistance of a tapered TSV including the lateral flow of heat through the liner and the contribution of the starting and landing pads is described by

$$R_p = 10^{12} \times \left\{ \pi k_{\text{SiO}_2} H \left[\frac{0.5(D_T + D_B)}{t_{\text{SiO}_2}} + 1 \right] + \left(\pi k_{\text{SiO}_2} / h_{\text{SiO}_2} \right) \left[0.5D_{\text{pad}}^2 - (0.5D_T + t_{\text{SiO}_2})^2 - (0.5D_B + t_{\text{SiO}_2})^2 \right] \right\}^{-1}, \tag{12.38}$$

where the notation of the different physical traits are illustrated in Fig. 12.16. Note that this expression includes the contribution of the resistance of the starting and landing pads of the TSV in

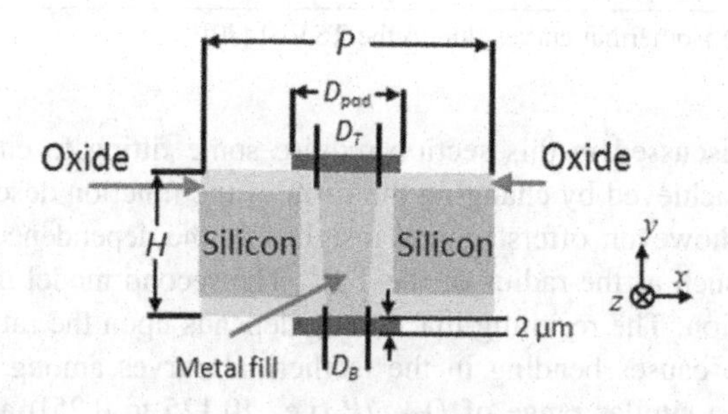

FIGURE 12.16

Schematic of a tapered TSV.

addition to the metal fill of the TSV. Hence, the diameter of the pads D_{pad} is included in the second term of the expression. To determine the transient thermal behavior, a periodic trapezoidal signal pulse is applied at the top of the TSV (the opposite situation can also occur, should the signal propagate upwards through the stack), and the transient temperature of the TSV is observed. The oxide thickness is an important factor in evaluating the transient thermal response of the TSV. This situation also applies for the case of stacked TSVs [471]. Although stacked signal TSVs are infrequent due to routing constraints, TSVs for power and ground distribution networks are usually placed at the periphery of a tier (or circuit blocks within a tier) stacked across several physical tiers [472]. In addition, a thinner oxide increases the lateral flow of heat, although the electrical capacitance of the TSV increases. In addition, the tapering of a TSV affects the temperature distribution across the length of the TSV, where the highest temperature appears at the bottom of the TSV with the smallest cross-section. Finally, interactions between the electrical and thermal characteristics of the TSV must be considered due to the dependence of the electrical resistance on temperature.

Before completing the discussion of TTSV models, note that this analysis is based on the assumption that specific material properties, such as the thermal conductivity, are constant. This assumption facilitates solving the related expressions but can overestimate the rise in temperature within a TSV. To investigate the deviation of the resulting temperatures due to the assumed independence of the thermal conductivity, an analysis of a single TSV has been performed assuming material properties independent of temperature. The temperature is overestimated by up to 5.7% for $t_{SiO_2} = 20$ nm, while this divergence increases to 15.2% for $t_{SiO_2} = 100$ nm [471]. The overestimate in temperature is attributed to the increased conductivity of the liner. Although the conductivity of the substrate decreases, the overall effect is a decrease in temperature [471]. In this analysis, the thermal conductivity of the metal fill, liner, and silicon substrate is approximated as a fourth order polynomial of temperature T,

$$x(T) = \sum_{i=0}^{4} c_n T^n, T_0 \leq T \leq T_1, \tag{12.39}$$

where $x(T)$ is the material property approximated by the polynomial, c_i is the corresponding coefficients, and T_0 and T_1 limit the temperature range for which this approximation is valid. These coefficients are listed in Table 12.4 for different material and temperature ranges.

All of the models discussed in this section include a different number of elements as compared to the typical unit cell structure shown in Fig. 12.8. Standard transformations can be used to map these models to a six resistor cell model [473]. Thermal models of channels for liquid cooling in 3-D systems are described in the following subsection.

12.3.2 THERMAL MODELS OF MICROCHANNELS FOR LIQUID COOLING

A conceptual drawing of intertier integrated cooling with channels etched within the substrate of the tiers is illustrated in Fig. 12.2. Once the structure of the heat sink is determined, the thermal behavior of the liquid flowing through the microchannels can be modeled. The important difference for thermal modeling of a 3-D system that employs liquid cooling is that the greatest portion of the generated heat is removed through convection. Considering (12.6) for the thermal resistance of the heat sink, a thermal model of one channel includes all three resistive components, where fully developed hydrodynamic and thermal flow is assumed.

Table 12.4 Polynomial Coefficients for Temperature Dependent Material Parameters [471]

	Metal Fill			Liner		Bonding
Material Coefficients	Cu $k_{Cu}(T)$	Tungsten (W) $k_W(T)$	Poly-Si $k_{Poly\text{-}Si}(T)$	Silicon $k_{Si}(T)$	SiO_2 $k_{SiO_2}(T)$	BCB $k_{BCB}(T)$
c_0	420.33208	191.23977	441.10556	332.14097	0.54335	0.08511
c_1	−0.06809	−0.07538	−4.71735	−1.07848	0.00105	6.96767×10^{-4}
c_2	0	0	0.02008	0.00158	0	0
c_3	0	0	-3.76157×10^{-5}	-1.08505×10^{-6}	0	0
c_4	0	0	2.60417×10^{-8}	2.81425×10^{-10}	0	0
Applicable temperature range (K)	(200, 1200)	(300, 1000)	(300, 400)	(300, 1300)	(273, 1000)	(297, 339)

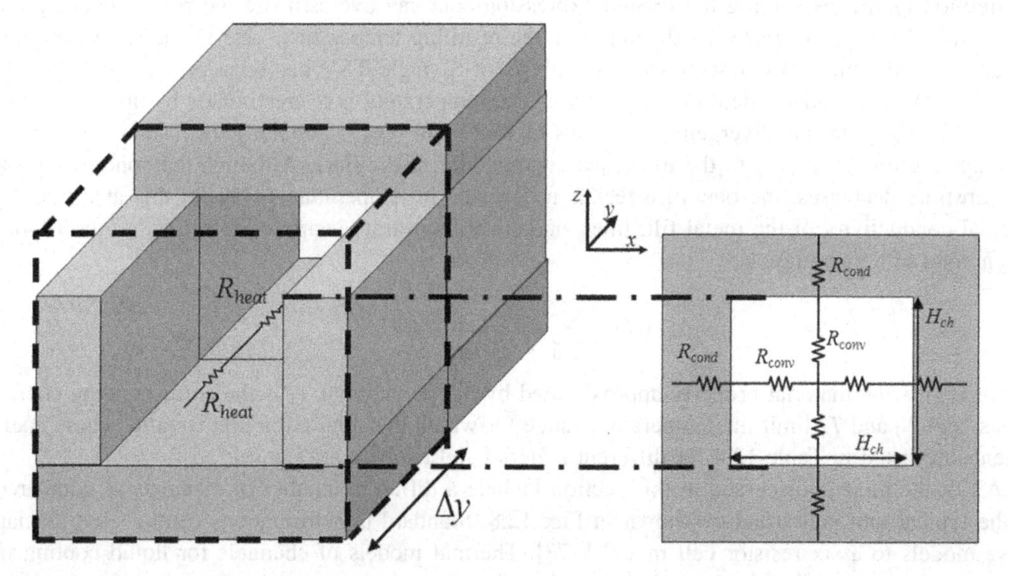

FIGURE 12.17

Thermal model of microchannel with conductive and convective thermal resistances.

The partial volume of a discretized microchannel along with the surrounding walls is shown in Fig. 12.17, where the flow of the coolant is along the y-axis. The different thermal resistances are notated where, depending upon the fineness of the mesh, the resistances within one cell can vary. For example, if the size of a grid cell is equal to the cross-section of the channel, only convective and thermal capacity resistances should be considered. If the boundaries of a cell extend within the sidewalls, the conductive resistance should also be included. The magnitude of the resistances is determined by the expression, $R_{cond} = t/(k_{sub}A_{wall})$. The area of the solid which corresponds to the sidewall of the channel is A_{wall}, and the thickness t is equal to the length of the edge of the solid segment in either the x- or z-direction. The resistance due to convection for cell i is

$$R_{\text{conv}_{ij}} = \frac{(T_{ij}^s - T_{ij}^l)}{Q_{ij}} = \frac{1}{h \sum A_{\text{wall}}}, \tag{12.40}$$

where T_{ij}^s and T_{ij}^l are, respectively, the temperature of the solid sidewall j within cell i and the temperature of the fluid. Q_{ij} is the heat transferred from the solid to the fluid. For full thermal flow, the temperature of the fluid within a cell can be considered constant in addition to assuming the same temperature for all of the walls of the channel within the cell i. To determine the convection resistance, the transfer coefficient h is determined from the Nusselt number (see (12.8)), which, in turn, can be determined by empirical relationships where the aspect ratio of the cross-section of the channel is treated as a variable [456,474]. Thus, the convection coefficient depends upon this aspect ratio. The total area of the walls is determined by the geometry of the channel. Once the Nusselt number is determined, the convection resistance is [475]

$$R_{\text{conv}_{ij}} = \frac{W_{ch} H_{ch}}{N_u k_f \Delta y (W_{ch} + H_{ch})^2}, \tag{12.41}$$

where the cell is assumed to contain the entire cross-section of the channel.

The third type of thermal resistance, which is due to the heat absorbed by the fluid flowing downstream from the channel, is [475]

$$R_{\text{heat}_i} = \frac{1}{\rho c_p u_m W_{ch} H_{ch}}, \tag{12.42}$$

which is considered constant throughout the length of a channel as a mean fluid velocity u_m is assumed and the cross-section is the same for all cells. The same resistance may be used for all of the channels if the assumption of a constant average fluid velocity is appropriate.

If the assumption of a constant heat flux is removed, these expressions no longer accurately describe the different resistances. This behavior is usually known as the "thermal wake effect" (i.e., the thermal trace due to the fluid flowing through the channel) [476]. For certain grid cells, the heat generated within the solid walls is transferred through the fluid to other cells downstream from the flow, adding to the thermal resistance of those cells. This phenomenon is qualitatively illustrated in Fig. 12.18. In addition, within the same cell, a heated segment of the substrate under the channel transfers heat to the sidewalls of the channel transverse to the fluid flow. To model this effect, the thermal model shown in Fig. 12.17B is augmented with a voltage controlled current source [477]. To better understand this situation, consider the rise in temperature within a cell i_d due to the thermal wake effect, as described by

$$\Delta T_{l,id} = P_{tc,i_s} a_{i_d - i_s}, \tag{12.43}$$

where P_{tc,i_s} denotes the heat transferred to the fluid through transverse convection from a source cell i_s. The thermal wake function is treated as a transconductance denoted by $a_{i_d - i_s}$. For those cells located farther downstream from the inlet location of the channel, more of these components are included to describe the cumulative nature of the thermal wakes. This situation implies that the added inaccuracy of not including the thermal wake effect in predicting the temperature of the fluid within the channel varies along the length of the channel. The temperature is, typically,

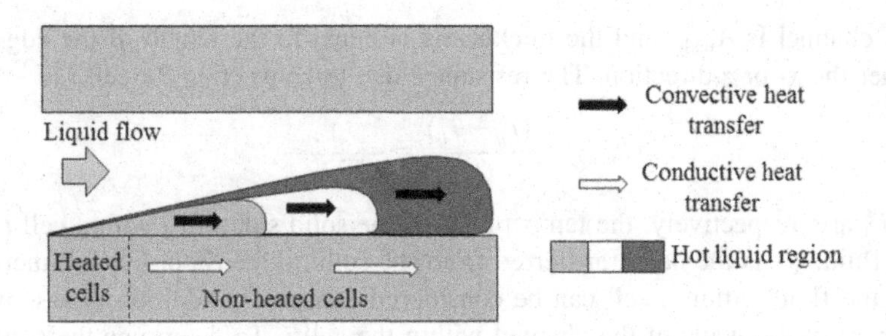

FIGURE 12.18

Schematic illustration of the thermal wake effect, which leads to an exponential decay of the temperature downstream from the channel due to the heated cells located upstream. The transfer of heat occurs both downstream and transverse to the flow within the channel [477].

overestimated for those cells located closer to the inlet of the channel and is underestimated for those cells located closer to the outlet of the channel (since the components of the thermal wake function accumulate downstream). Whether this effect should be included depends upon the status of the flow (developed vs. developing) and the desired accuracy of the model, as simulations exhibit an inaccuracy of up to ~10%, if the thermal wake effect is excluded from the thermal model.

Irrespective of the presence of liquid cooling, once the thermal elements of each cell are determined, the entire volume of a 3-D stack is converted into a mesh and appropriate analysis techniques can be used to determine the temperature at each node within the mesh. These techniques are presented in the following section.

12.4 THERMAL ANALYSIS TECHNIQUES

Methods for analyzing different heat transfer mechanisms have been investigated in the past and remain an active scientific topic. Thermal issues in integrated circuits have also been extensively studied over the past decades, where many well established techniques for thermal analysis have been explored and adapted to the specific traits of integrated circuits [478].

As discussed in Section 12.3, the entire volume of a 3-D system is discretized into a mesh. A 3-D mesh consisting of finite hexahedral elements (i.e., parallelepipeds), called cells, tiles, or control volumes, is illustrated in Fig. 12.19 (note that elements with different shapes can be utilized). The mesh can also be nonuniform for those regions with complex geometries or nonuniform power densities to provide either high accuracy or improve the computational time. A much finer mesh is often required at the interface of materials with greatly different conductivities. Alternatively, for a large volume of constant thermal conductivity, a coarser grid is a better choice as the analysis is computationally more efficient without affecting accuracy.

The volume of the circuit can be discretized with the use of several disparate methods, such as the FEM [479], finite difference [480,481], finite volume [482], and boundary element methods [439]. The thermal elements (e.g., heat generators and thermal resistors) connecting the vertices of the grid cells are modeled as discussed in the previous section. The vertices of the cells, typically

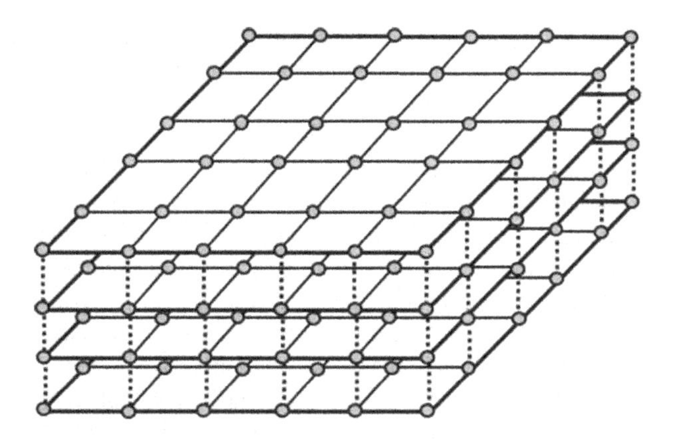

FIGURE 12.19

A four tier 3-D circuit discretized into a mesh.

called nodes or degrees of freedom, correspond to the unknown temperatures that are determined during the analysis process. Since most of the physical features within an integrated system are at the nanometer scale and each circuit includes millions of these cells, meshes with hundreds of millions of nodes can be produced. Once the volume of the system has been discretized and the thermal elements of the grid cells have been extracted, a matrix system describing the differential thermal expressions is obtained. This matrix system for a steady-state analysis (the objective of most methods) is a linear system of matrices of the form,

$$\mathbf{GT} = \mathbf{P}, \tag{12.44}$$

where $\mathbf{G}$ is the matrix of thermal conductance, $\mathbf{T}$ is the vector of temperature nodes that needs to be computed, and $\mathbf{P}$ is the vector containing the power sources of the system. Note that this matrix system is similar to that formed when solving for *IR* drops within power distribution networks [275]. Consequently, the duality of the thermal and electrical networks means that methods applied to the analysis of power distribution networks can also be adapted for thermal analysis. The resulting matrices are rather sparse. Appropriate techniques for solving sparse matrices are employed to decrease the computational time.

A number of techniques exist to solve the system of (12.44), where these methods are usually described as either direct or iterative solvers. Although the direct methods can produce high quality results for relatively small systems [482] (e.g., hundreds of thousands of nodes), the efficiency decays for large scale systems (e.g., multimillion nodes). Thus, scalability is a major issue for direct methods. Additionally, as the matrices of the system in (12.44) must be explicitly formed, the memory requirements can be significant.

An efficient means to improve the performance of direct or single level iterative solvers is to employ a nonuniform mesh to represent a 3-D system. A nonuniform mesh reduces the number of nodes to be solved and can be particularly helpful where different physical scales are required. This situation is encountered when the thermal behavior of a system simultaneously includes the 3-D circuit, package, and printed circuit board (PCB). The use of a nonuniform grid, however, must be carefully realized as the different cells can cause discontinuities at the interfaces, delaying convergence or causing the solver to diverge. Consequently, the interfaces between the different meshes must be treated carefully to establish continuity for the flow of heat throughout the entire system. The application of nonuniform grids has been compared to the use of a uniform mesh for a packaged integrated system mounted on a PCB [450]. The number of nodes decreases from 206,000 to 35,000 (6×) and the analysis time decreases by a factor of 91 [450].

Although the use of variable size meshes decreases the number of nodes or unknowns, direct solvers are not scalable. Alternatively, iterative methods can be more efficient, reducing both the computational time and the memory usage (12.44), where the most common (and powerful) iterative solvers are those based on multigrid methods [483]. In general, the computational complexity of multigrid techniques grows linearly (multiplied by a small constant) with the number of unknowns [483]. Another advantage as compared to other iterative schemes, such as the Gauss-Siedel method, is that multigrid solvers can efficiently remove the high frequency errors that can slow or hinder convergence in other methods. Due to these superior properties, multigrid methods have proven particularly efficient in solving these systems and, consequently, have been utilized with high success in a variety of different fields, such as fluid and molecular dynamics [484], in addition to electronic design automation.

Multigrid methods 1) construct coarser grids starting from an original fine grid, 2) solve the coarse grid, and 3) map and correct the coarse grid solutions with respect to the fine grid, where the process is repeated until the residue error reaches a predefined lower limit. This procedure is typically called a V-cycle [485,486] and is illustrated in Fig. 12.20. As depicted in this figure, a multigrid solving method requires a coarse grid operator, interpolation (for mapping the solution back to the fine grid), and restriction operators (for producing the coarse grid at the next level). A detailed treatment of multigrid methods can be found in several related publications [483,485,486]. Multigrid methods are divided into two categories: algebraic and geometric.

Although both methods have been employed in a large number of problems related to integrated circuits, geometric multigrid techniques have exhibited better performance. This enhanced performance is achieved since algebraic methods are applicable to more general problems and are, therefore, not readily adapted to the traits and requirements of integrated circuits. A specific difference between algebraic and geometric methods is that the former method utilizes fixed simple smooth functions and ensures convergence through complicated coarsening procedures, while the latter method uses specific smooth functions to implement effective smoothing and coarse grid correction as the hierarchy of the grid is fixed [487]. The choice of these operators plays a significant role in producing a robust and effective technique.

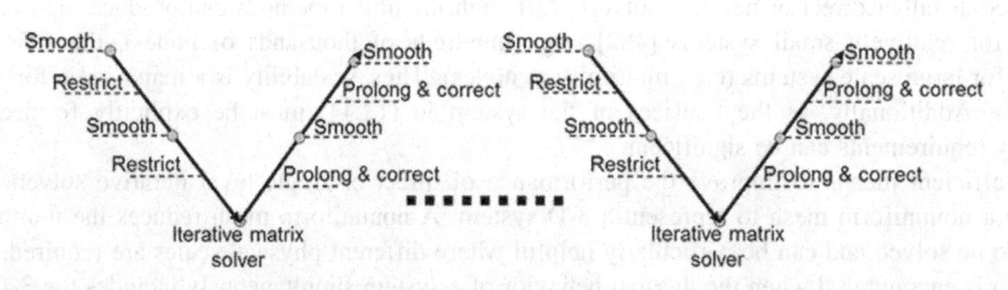

FIGURE 12.20

Traditional V-cycles of multigrid methods with coarsening and refining stages [488].

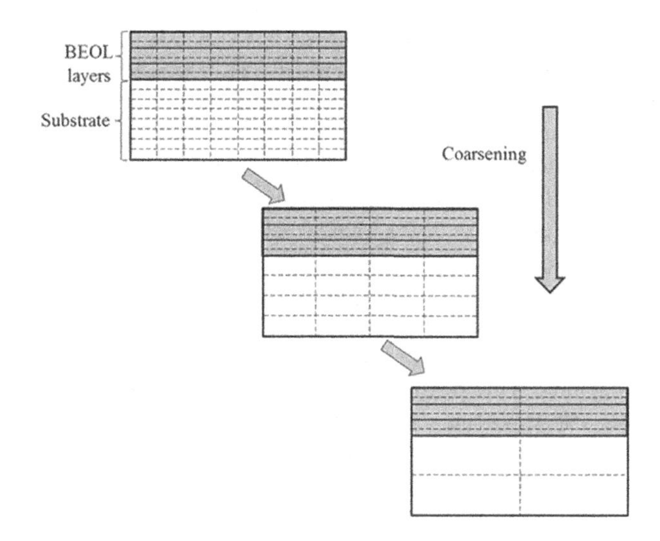

FIGURE 12.21

Coarsening process excluding the BEOL layers in the z-direction to ensure that valuable physical information is not lost, improving the overall efficiency and accuracy of the multigrid technique [487].

In addition, the grid hierarchy can be selected to ensure that the important characteristics of the system are not lost, thereby not degrading precision. In the case of integrated systems, irrespective of 2-D or 3-D, several thin layers exist, where the thermal conductivities can be quite different. This situation is typically relevant to the BEOL layers, which include dielectric and metal layers about 1 μm thick, while the silicon substrate is much thicker. A grid hierarchy based on coarsening all of the layers in all directions (x, y, z) often ignores important physical information describing these thin layers, delaying convergence. An appropriate choice, therefore, is to not coarsen the portion of the grid along the z-direction within the coarsening stages of the V-cycles [487]. This approach is shown in Fig. 12.21, and, although initially applied to a 2-D system, the same notion is applicable to a multi-tier 3-D structure. Another reason that also justifies this selective coarsening is the flow of the heat, as the primary heat conduction path within an integrated circuit is primarily along the z-direction. Since microchannel cooling is within the substrate, there is another major path where heat flows and, consequently, the grid hierarchy as well as the related coarsening and smoothing operators must be adapted to accommodate the existence of this second important thermal path [488].

The application of multi-grid methods has produced impressive reductions in both computational time and storage needs for both steady-state and transient state analysis [487,488]. As compared to the conjugate gradient method, thermal analysis of a planar circuit with ~2.5 million nodes has demonstrated a decrease in runtime and memory of, respectively, 420× and 32× while large circuits with ~12 million nodes are simply not solvable with non-multigrid methods [487].

The computational efficiency of multigrid techniques has been further improved through the use of GPU architectures [488]. The primary issue is to partition the solution process onto the underlying hardware architecture to best exploit the available computing resources (e.g., cores). The gains achieved from the resulting analysis procedure may, however, not be portable to other GPU architectures. Recent applications of multigrid methods to the thermal analysis of 3-D circuits including microchannel cooling of the nVIDIA Geforce GTX 285 GPU have demonstrated a 35× speed up over iterative solvers running on a quad core processor [488].

Other techniques have also been investigated for the thermal analysis of integrated circuits [489]. These methods are based on the Green's function, the discretized form of which can be solved by spectral transformations such as the discrete cosine transform [490]. The use of these transformations, however, is suitable only for homogeneous layers. In the case of 3-D circuits, the TSVs within the silicon substrate produce discontinuities in the thermal conductance within the volume of the substrate. One approach is to employ an average thermal conductivity (an approach typically used in the closed-form models). This homogenization process however degrades accuracy irrespective of the efficiency of the analysis technique [491]. Moreover, the thermal effect of the TSVs is not accurately captured. One possible way to address this issue is to treat the TSVs as virtual power sources. Since the TSVs are replaced with virtual power sources, the substrate of each tier only consists of silicon; hence, the transformation used for 2-D circuits can also be utilized for a multi-tier stack. Modeling the TSVs as virtual power sources replaces the thermal conductivity of the TSVs in matrix **G** with the silicon substrate. Consequently, the thermal conductance matrix described by (12.44) is converted into two matrices which describe the homogeneous substrate without the TSVs and the difference between the TSV and substrate thermal conductivities [489]. The application of this method to a 3-D circuit has performed well over a finite difference solver, but the number of grid cells is on the order of several thousands [489]. Thus, scalability remains an issue for thermal analysis techniques based on the Green's function.

Another technique that resembles solutions based on the Green's function but removes the limitation for operating on homogeneous circuit regions is the "power blurring" technique, which originates from the field of image processing [492]. A digital image undergoes spatial filtering for certain operations, such as sharpening or blurring. To implement the blurring operation, the image f is convoluted with a matrix w, called a mask, as described by [493]

$$g(x,y) = \sum_{s=-a}^{a} \sum_{t=-b}^{b} w(s,t) \cdot f(x+s, y+t), \tag{12.45}$$

where $a = (m-1)/2$ and $b = (n-1)/2$ for a $m \times n$ matrix. Similarly, in the case of thermal analysis of integrated circuits, the power map of a circuit is convoluted with the response mask to obtain the thermal profile of the circuit. In the frequency domain, convolution corresponds to simple multiplication. The response mask represents a heat spreading function. For a unit source applied to a location within the circuit [492], the impulse response of this source provides one response mask. An illustration of the power blurring method is provided in Fig. 12.22. The application of (12.45)

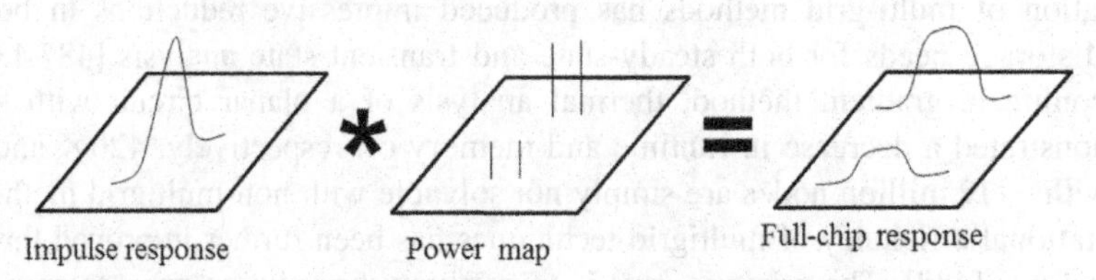

FIGURE 12.22

Principle of power blurring method [491].

to a single response mask corresponds to the temperature distribution due to a single power source (whether this source corresponds to a single device or a circuit block depends upon the desired granularity of the analysis). Superposition is used to determine the overall temperature distribution across the entire power map of the circuit.

The performance of this method largely depends on the quality as well as number of response masks. The response masks correspond to the impulse response of the power sources and are determined prior to thermal analysis. Although any thermal analysis technique is applicable to this task, these methods use commercial tools, such as ANSYS, to produce the response masks. This setup task has proven to be computationally expensive and requires careful treatment to ensure that the computational load is appropriately balanced with the desired accuracy. An important issue is how many response masks are required. Early application of this method to planar circuits employs only a single response mask, where one heat source is present at the center of the circuit [492]. Another response mask can be applied at the edge of the circuit to include boundary effects, but the error introduced due to these effects can, alternatively, be eliminated by an added correction factor. Extraction of the response masks becomes more complicated when applied to 3-D circuits.

To produce a response mask, the rise in temperature in every tier due to the presence of a power source in one tier of the stack must be captured. As an example, if a two tier system is assumed, four response masks are required, which are noted as $Mask_{11}$, $Mask_{12}$, $Mask_{21}$, and $Mask_{22}$, where $Mask_{xy}$ denotes the response mask in tier x due to the presence of a heat source in tier y. The temperature profiles, T_1 and T_2, respectively, in tiers 1 and 2 are described by

$$T_1 = Mask_{11} * P_1 + Mask_{12} * P_2, \tag{12.46}$$

$$T_2 = Mask_{21} * P_1 + Mask_{22} * P_2, \tag{12.47}$$

where P_1 and P_2 are, respectively, the power map in tiers 1 and 2 [494].

Power blurring has been applied to a three tier circuit at a fine level of granularity (0.1 μm), where the circuit area is 3 mm × 3 mm including about 714,000 power sources [491]. The computational time is significantly lower than analyzing the same system with a network of thermal resistors. Based on this case study, the crucial elements to broadly apply this technique relate to the computational resources and time required to produce the response masks, which can be a tedious process if applied for each transistor size and orientation of the transistor channel. To remedy a potentially intractable condition, the response mask for a group of transistors is utilized. This group accurately represents a set of standard cells if the physical characteristics of these groups are not very different [491]. This grouping process, however, is performed manually and consequently the efficiency of this practice depends strongly on design experience. Alternatively, the response masks can be determined only once for each thermal cell, and this information can be integrated within the design library.

Note that the discussion of these techniques does not favor one method over another method but, rather, highlights the advantages and limitations of these techniques. These techniques can also be used in a complementary and integrated manner, a path that has yet to be adequately explored. For example, multigrid thermal analysis at the circuit block level can indicate hotspots, and power blurring can be applied to small regions of a circuit with transistor level granularity to pinpoint (and alleviate) the most thermally fatigued devices.

12.5 SUMMARY

Thermal models and related analysis techniques for 3-D ICs are discussed in this chapter. The primary concepts discussed in the chapter are

- High temperatures deteriorate circuit reliability and lifetime, increase leakage power, and decrease circuit performance.
- Higher temperatures and thermal gradients are predicted for 3-D ICs due to increased power densities and greater distances between the circuits on the upper tiers and the heat sink(s).
- Thermal coupling can further increase self-heating of devices, which leads to hotspots.
- Liquid cooling has been shown to effectively remove heat from 3-D circuits. The design of a microchannel heat sink requires careful design tradeoffs between the geometry of the channels and the characteristics of the off-chip cooling system. Manufacturing and electrical issues can further constrain the design of the heat sink.
- Thicker silicon substrates facilitate the conductive heat transfer process; however, high aspect ratio intertier vias reduce performance and are difficult to fabricate.
- Thermal models of 3-D circuits include analytic expressions, compact thermal resistive networks, and 3-D grids for numerical analysis. The models are listed in ascending order of increasing computational complexity and accuracy.
- A first order analysis of thermal behavior of 3-D systems can be performed with 1-D thermal circuits, where a thermal circuit is appropriate if the flow of heat is assumed to occur exclusively in the z-direction. In a 1-D model, each material layer is modeled as a thermal resistor, heat sources as current sources, and temperature differences as voltage differences.
- The objective of 1-D models is not to address circuit performance issues due to high temperatures but, rather, system level exploratory decisions. For example, the package, die stacking order, cooling mechanism, heat spreading materials, package level interconnects, and other system wide parameters can affect the cost of the overall system.
- The boundary conditions for thermal models of 3-D circuits typically assume adiabatic walls for the lateral and top surfaces and an isothermal surface for the bottom surface attached to the heat sink.
- The lateral flow of heat through the TSV liner is not negligible and if ignored can lead to overestimation of the temperature. Moreover, the oxide thickness is an important factor in determining the transient thermal response of the TSVs.
- Mesh-based thermal models typically include thermal resistors, capacitors, and current sources to model the thermal properties of a single material (homogeneous) or several materials (heterogeneous) within a cell.
- Thermal analysis methods typically require converting the volume of the target system into a mesh. The mesh can be nonuniform for regions with complex geometries or nonuniform power densities, often a requirement to provide either high accuracy or computational efficiency. The volume of the circuit can be discretized with the use of several disparate methods, such as the finite element, finite volume, finite difference, and boundary element methods.
- The differential thermal expressions can be solved with several techniques, where these methods are usually distinguished as direct or iterative. Scalability is a major issue for direct methods.

- Iterative methods reduce both the computational time and the use of memory for solving thermal differential expressions, where the most common (and powerful) iterative solvers are those based on multigrid methods. In general, the computational complexity of multigrid techniques grows linearly with the number of unknowns.
- The process of multigrid methods is composed of constructing coarser grids starting from a fine grid, solving the coarser grid, and mapping and correcting the coarse grid solutions by comparing to the original fine grid. The process is repeated until the residue error has reached a predefined limit. This procedure is typically called a V-cycle.
- Other techniques based on the Green's function have also been investigated for thermal analysis of integrated circuits. The discretized form of these methods are solved by spectral transformations, such as the discrete cosine transformation.
- Another technique that resembles solutions based on Green's function but removes the limitation for only operating on homogeneous regions is the "power blurring" technique, which originates from the field of image processing.
- The performance of the power blurring method depends on the quality as well as number of response masks. The response masks correspond to the impulse response of the power sources and are determined prior to thermal analysis. In 3-D circuits, a response mask is produced for each tier due to the presence of a power source in another tier within the stack.

CHAPTER

THERMAL MANAGEMENT STRATEGIES FOR THREE-DIMENSIONAL ICs

13

CHAPTER OUTLINE

In the previous chapter, a number of thermal models within three-dimensional (3-D) systems are reviewed. These models, employed in the thermal analysis process, provide a temperature map of a circuit. Based on this information, thermal management techniques can, in turn, be applied to mitigate excessive temperatures (i.e., "hot spots") or thermal gradients within and across tiers. These temperature related phenomena are expected to become more pronounced due to the increasing power densities in 3-D systems. This situation is also exacerbated by the greater distance of the heat sources from the heat sinks. The reduced volume of 3-D systems also allocates smaller area for the heat sink, reducing the heat transferred to the ambient [495].

Thermal management methodologies can be roughly divided into two broad categories: (1) approaches which control power densities within the volume of the 3-D systems, and (2) techniques that target an increase in the thermal conductivity of the 3-D stack. Methods that consider both objectives have also recently appeared and are discussed in this chapter. Note that this categorization of techniques is not based on the methods for achieving the target objective or the stage of the design flow. Rather, the thermal management methods discussed in this chapter are driven by the design objective, ensuring that the resulting 3-D circuits exhibit a low thermal risk. The subsections within each section are, alternatively, presented based on the means utilized to achieve the target objective. Consequently, those methodologies that manage the power density throughout the volume of the 3-D stack are discussed in Section 13.1. Strategies that accelerate the transfer of the

Three Dimensional Integrated Circuit Design. DOI: http://dx.doi.org/10.1016/B978-0-12-410501-0.00013-7

generated heat within a 3-D circuit to the ambient are described in Section 13.2. Hybrid approaches, such as active cooling, where both objectives are simultaneously addressed, are discussed in Section 13.3. These concepts are summarized in Section 13.4.

13.1 THERMAL MANAGEMENT THROUGH POWER DENSITY REDUCTION

Careful control of the peak power density within a 3-D integrated system is a primary means to lower the peak temperature of a 3-D stack while reducing thermal gradients across each physical tier as well as among tiers. Methods in this category can be divided into two types. Those techniques applied during the design process that prudently distribute the power within a multi-tier system [351,496], and online (or real-time) techniques that adapt the spatial distribution of temperature within the stack over time by controlling the computational tasks executed by the system [497,498]. Both types of techniques are discussed in the following subsections.

The elimination of hot spots and reduction in thermal gradients within 3-D circuits requires the extension of physical design techniques to include temperature as a design objective. Several floorplanning, placement, and routing techniques for 3-D ICs and systems-on-package (SoP) have been developed that consider the high temperatures and thermal gradients throughout the tiers of these systems in addition to traditional objectives such as area and wirelength minimization. Emphasizing temperature can result in significant penalties in area—by placing for example, high power blocks far from each other—and performance due to a potentially significant increase in wirelength. Consequently, most techniques balance the various design objectives, producing systems that provide significant performance while satisfying temperature constraints. These techniques are applied to several steps of the design flow. In Section 13.1.1, thermal driven floorplanning techniques are discussed, while thermal driven placement techniques are reviewed in Section 13.1.2. Thermal management methodologies during circuit operation are discussed in Section 13.1.3.

13.1.1 THERMAL DRIVEN FLOORPLANNING

Traditional floorplanning techniques for two-dimensional (2-D) circuits typically optimize an objective function that includes the total area of the circuit and the total wirelength of the interconnections among the circuit blocks. Linear functions that combine these objectives are often used as cost functions where, for 3-D circuits, an additional floorplanning requirement may be minimizing the number of intertier vias to decrease the fabrication cost and silicon area, as discussed in Chapter 9, Physical Design Techniques for Three-Dimensional ICs.

Different issues with thermal aware floorplanning can also lead to a number of tradeoffs. The techniques discussed in this section highlight the advantages and disadvantages of the different choices to produce highly compact and thermally safe floorplans.

A thermal driven floorplanning technique for 3-D ICs includes the thermal objective,

$$\cos t = c_1 wl + c_2 \text{area} + c_3 iv + c_4 g(T), \tag{13.1}$$

where c_1, c_2, c_3, and c_4, are weight factors and *wl*, *area*, and *iv* are, respectively, the normalized wirelength, area, and number of intertier vias [351]. The last term is a cost function to represent the temperature. An example of this function is a ramp function of the temperature, as shown in

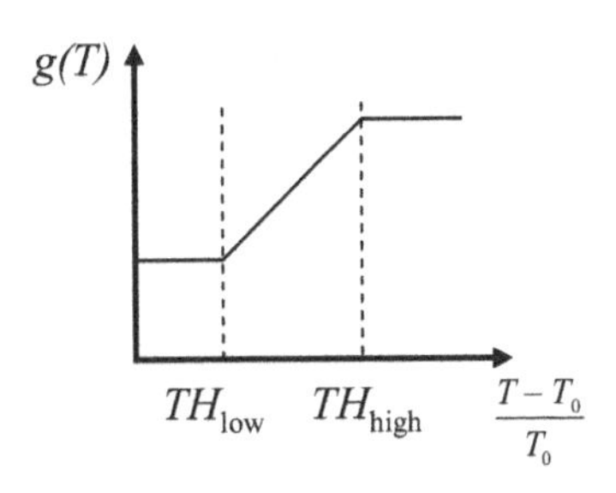

FIGURE 13.1

Cost function of the temperature [351].

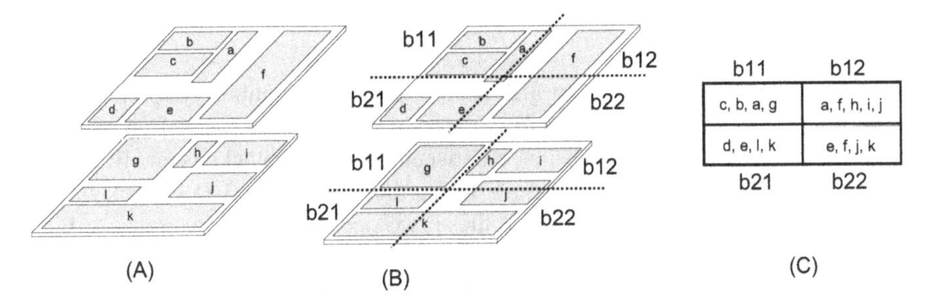

FIGURE 13.2

A bucket structure example for a two tier circuit consisting of 12 blocks. (A) A two tier 3-D IC, (B) a 2 × 2 bucket structure imposed on a 3-D IC, and (C) the resulting bucket index.

Fig. 13.1. Note that the cost function does not intersect the abscissa but, rather, the plateaus. Consequently, this objective function does not minimize the temperature of the circuit but, rather, constrains the temperature within a specified level. Indeed, minimizing the circuit temperature may not be an effective objective, leading to prohibitively long computational times or failure to satisfy other design objectives.

As with thermal unaware floorplanning techniques, the choice of floorplan representation also affects the computational time. Sequence pair and corner block list representations have been used for 3-D floorplanning, as discussed in Chapter 9, Physical Design Techniques for Three-Dimensional ICs. In addition to these approaches, a low overhead scheme is realized by representing the blocks within a 3-D system with a combination of 2-D matrices that correspond to the tiers of the system and a bucket structure that contains the connectivity information for the blocks located on different tiers (a combined bucket and 2-D array (CBA)) [351]. A transitive closure graph describes the intratier connections of the circuit blocks. The bucket structure can be envisioned as a group of buckets imposed on a 3-D stack. The indices of those blocks that intersect a bucket are included, irrespective of the tier on which a block is located. A 2 × 2 bucket structure applied to a two tier 3-D IC is shown in Fig. 13.2, where the index of the bucket is also depicted. To explain the bucket index notation, consider the lower left tile of the bucket structure shown in

Fig. 13.2C (i.e., b21). The index of the blocks that intersects with this tile on the second tier is d and e, and the index of the blocks from the first tier is l and k. Consequently, b21 includes d, e, l, and k.

Simulated annealing (SA) is employed to optimize an objective function, as in (13.1) for thermal floorplanning of 3-D circuits. The SA scheme converges to the desired freezing temperature through several solution perturbations. These perturbations include one of the following operations, some of which are unique to 3-D ICs:

1. block rotation;
2. intratier block swapping;
3. intratier reversal of the position of two blocks;
4. move of a block within a tier;
5. intertier swapping of two blocks;
6. *z*-neighbor swap;
7. *z*-neighbor move.

The last three operations are unique to 3-D ICs, while the *z*-neighbor swap can be treated as a special case of intertier swapping of two blocks. Therefore, two blocks located on adjacent tiers are swapped only if the relative horizontal distance between these two blocks is small. In addition, the *z*-neighbor move considers the move of a block to another tier of the 3-D system without significantly altering the *x-y* coordinates. Examples of these two operations are illustrated in Fig. 13.3.

Moreover, every time a solution perturbation occurs, the cost function is reevaluated to gauge the quality of the new candidate solution. Computationally expensive tasks, such as wirelength and temperature calculations, are therefore invoked. To avoid this exhaustive approach, incremental changes in wirelength for only the related blocks and interconnections are evaluated, as applied to the techniques described in Chapter 9, Physical Design Techniques for Three-Dimensional ICs. Note that the thermal profile of the heat diffusion can change the temperature across an area that extends beyond the recently moved blocks.

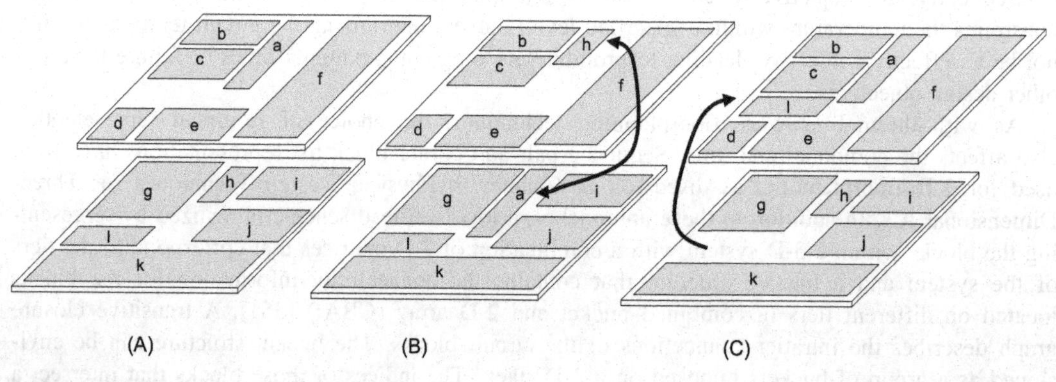

FIGURE 13.3

Intertier moves. (A) An initial placement, (B) a *z*-neighbor swap between blocks a and h, and (C) a *z*-neighbor move for block l from the first tier to the second tier.

Consequently, each block perturbation requires the thermal profile of a 3-D circuit to be determined. This strict requirement, however, can increase the computational time, becoming a bottleneck for temperature related physical design techniques. A thermal profile, therefore, is invoked only after a specific operation or after a specified number of iterations. The reason behind this practice is that not all operations significantly affect the temperature of an entire system; rather, only a portion of the system. For example, intratier moves of two small area blocks or the rotation of a block are unlikely to significantly affect the temperature of a system, whereas other operations, such as a *z*-neighbor swap or a *z*-neighbor move, can significantly affect the temperature of the tiers and, more broadly, the entire system.

Thermal analysis techniques to determine the temperature of a 3-D circuit, each with different levels of precision and efficacy, can be applied, as discussed in Chapter 12, Thermal Modeling and Analysis. To ascertain the effects of different thermal analysis approaches on the total time of the thermal floorplanning process, thermal models with different accuracy and computational time have been applied to MCNC benchmarks in conjunction with this floorplanning technique. These results are reported in Table 13.1, where a compact thermal modeling approach is considered. A significant tradeoff between the computational runtime and the decrease in temperature exists between these thermal models. With thermal driven floorplanning, a grid of resistances is utilized to thermally model a 3-D circuit, exhibiting a 56% reduction in temperature [351]. The computational time, however, is increased by approximately an order of magnitude as compared to conventional floorplanning algorithms. Alternatively, if a closed-form expression is used for the thermal model of a 3-D circuit, the decrease in temperature is only 40%. The computational time is, however, approximately doubled in this case. Other design characteristics, such as area and wirelength, do not significantly change between the two models.

As the block operations allow intertier moves, exploring the solution space becomes a challenging task [352]. To decrease the computational time, floorplanning can be performed in two separate phases. In the first step, the circuit blocks are assigned to the tiers of a 3-D system to minimize area and wirelength, effectively ignoring the thermal behavior of the circuit during this first stage. This phase, however, can result in highly unbalanced power densities among the tiers. A second step that limits these unbalances is therefore necessary. An objective function to accomplish this balancing process is [496]

$$\cos t = c_5 \cdot wl + c_6 \cdot \text{area} + c_7 \cdot \text{dev}(F) + c_8 \cdot P + c_9 \cdot \text{TOP}, \tag{13.2}$$

Table 13.1 Decrease in Temperature Through Thermal Driven Floorplanning [351]

	CBA W/O Thermal Objective		CBA-T		CBA-T-Fast	
Circuit	T (°C)	Runtime (s)	T (°C)	Runtime (s)	T (°C)	Runtime (s)
ami33	471	23	160	466	204	56
ami49	259	86	151	521	196	144
n100	391	313	158	4322	222	446
n200	323	1994	156	6843	242	4474
n300	373	3480	167	17,484	208	4953
Avg.	1	1	0.44	9.71	0.6	1.82

where c_5, c_6, c_7, c_8, and c_9 notate weighting factors. Beyond the first two terms that include the area and wirelength of the circuit, the remaining terms consider other possible design objectives for 3-D circuits. The third term minimizes the imbalance that can exist among the dimensions of the tiers within the stack, based on the deviation dimension approach described in [350]. Tiers with particularly different areas or greatly uneven dimensions can result in a significant portion of unoccupied silicon area on each tier.

The last two terms in (13.2) consider the overall power density within a 3-D stack. The fourth term considers the power density of the blocks within the tier as in a 2-D circuit. Note that the temperature is not directly included in the cost function but is implicitly captured through management of the power density of the floorplanned blocks. The cost function characterizing the power density is based on a similarly shaped function as the temperature cost function depicted in Fig. 13.1. Thermal coupling among the blocks on different tiers is considered by the last term and is

$$\mathrm{TOP} = \sum \left(\sum_i P_i + P_{ij} \right), \tag{13.3}$$

where P_i is the power density of block i, and P_{ij} is the power density due to overlapping block i with block j from a different tier. The summation operand adds the contribution from the blocks located on all of the other tiers other than the tier containing block j. If a simplified thermal model is adopted, an analytic expression as in (13.3) captures the thermal coupling among the blocks, thereby compensating for some loss of accuracy originating from a crude thermal model.

This two step floorplanning technique has been applied to several Alpha microprocessors [499]. Results indicate a 6% average improvement in the maximum temperature as compared to 3-D floorplanning without a thermal objective [496]. In addition, comparing a 2-D floorplan with a 3-D floorplan, an improvement in area and wirelength of, respectively, 32% and 50% is achieved [496]. The peak temperature, however, increases by 18%, demonstrating the importance of thermal issues in 3-D ICs.

The reduction in temperature is smaller than the one step floorplanning approach. Alternatively, for a two step approach, the solution space is significantly smaller, resulting in decreased computational time. The interdependence, however, of the intratier and intertier allocation of the circuit blocks is not captured, which can yield inferior solutions as compared to one step floorplanning techniques.

SA methods have also been employed for floorplanning modules in SoP (see Section 2.2), which is another variant of 3-D integration with coarse granularity. A cost function similar to (13.2) includes the decoupling capacitance and temperature in addition to area and wirelength. The modules in each tier of the SOP are represented by a sequence pair to capture the topographical characteristics of the SOP. To avoid the computational overhead caused by thermal analysis of the SOP during SA iterations, an approximation of the thermal profile of the circuit is used.

The temperature for an initial floorplan of the SoP is produced using the method of finite differences. The finite difference approximation given by (12.3) can be written as $RP = T$, where R is the thermal resistance matrix. The elements of the thermal matrix contain the thermal resistance (or conductance) between two nodes in a 3-D mesh, while the temperature T and power vector P contain, respectively, the temperature and power dissipation at each node. Any modification to the placement of the cells causes all of these matrices to change. To determine the resulting change in temperature, the thermal resistance is updated and multiplied with the power density vector. This approach, however, leads to long computational times. Consequently, assuming the modules in the SoP exhibit similar thermal conductivities dominated by the volume of silicon, the thermal resistance matrix is not updated, although any module move results in some change in the matrix.

Only local changes in the power densities due to a move of a module are considered. Consequently, for each modification of the block placement, the change in the power vector ΔP is scaled by R, and the change in the temperature vector is evaluated. A new temperature vector is obtained after the latest move of the blocks within a 3-D system. Results from applying this thermally aware SoP floorplanning technique are listed in Table 13.2, where the results from a placement based on traditional area and wirelength objectives are also provided for comparison [397].

Although SA is the dominant optimization scheme used in most floorplanning and placement techniques for 3-D ICs [351,397,409], thermal aware floorplanners based on the force directed method have also been investigated. The motivation for employing this method stems from the lack of scalability of the SA approach. The analytic nature of force directed floorplanners, however, formulates the floorplanning problem into a continuous 3-D space. A transition is required between placement in the continuous volume and assignment without overlaps (i.e., legalization) within the discrete tiers of a 3-D system, potentially leading to a nonoptimal floorplan. In addition, as block level floorplanning includes components of dissimilar sizes, a different solution process is required as compared to floorplanning at the standard cell level [500].

This process includes more stages as compared to a traditional force directed method, as discussed in Chapter 9, Physical Design Techniques for Three-Dimensional ICs, due to legalization issues that can result during tier assignment. The different stages of the method are illustrated in Fig. 13.4, which are distinguished as (1) temperature aware lateral spreading, (2) continuous global optimization, and (3) optimization and tier assignment among the tiers within the 3-D stack. Assuming that the floorplan is a set of blocks $\{m_1, m_2, \ldots, m_n\}$, the method minimizes (1) the peak temperature $T_{\max}$ of the circuit, (2) the wirelength, and (3) the circuit area, the product of the maximum width and height of the tiers within the 3-D stack. Each block m_i is associated with

Table 13.2 Thermal Driven Floorplanning for Four Tier 3-D ICs [397]

	Area/Wire Driven (mm², m, nF, °C)				Thermal Driven (mm², m, nF, °C)			
Circuit	**Area**	**Wire**	**Decap**	**Temp**	**Area**	**Wire**	**Decap**	**Temp**
n50	221	26.6	18.0	87.2	377	84.1	29.7	68.9
n100	315	66.6	78.2	86.5	493	24.5	93.6	69.8
n200	560	17.1	226.3	96.4	1077	38.8	243.6	76.2
gt100	846	28.6	393.8	100.1	1310	20.4	405.3	86.6
gt300	191	13.2	60.8	71.0	474	28.0	92.7	52.3
gt400	238	19.6	342.5	93.2	528	37.0	392.1	72.1
gt500	270	28.1	493.1	114.0	362	38.5	512.0	89.2
gt600	316	30.2	645.3	99.7	541	76.5	684.4	80.3
Ratio	1.00	1.00	1.00	1.00	1.75	1.51	1.08	0.80

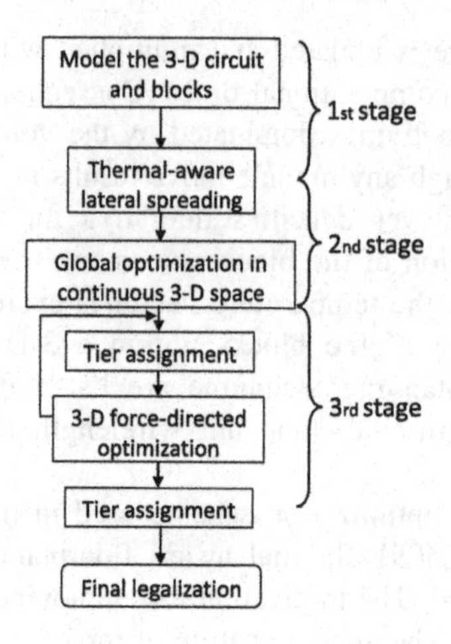

FIGURE 13.4

Three stage floorplanning process based on the force directed method [500].

dimensions W_i and H_i, area $A_i = W_i \times H_i$, aspect ratio H_i/W_i, and power density P_{mi}. The height of the blocks is a multiple of the thickness of the tiers, which is assumed to be D for all of the physical tiers and L tiers are assumed to comprise the 3-D stack. A valid floorplan is an assignment of non-overlapping blocks within a 3-D stack, where the position of each block is described by (x_i, y_i, l_i), denoting the horizontal coordinate of the lower left corner of the block and tier l_i.

The continuous 3-D space within which the blocks are allowed to move and rotate consists of homogeneous cubic bins. The height of these bins is set to $D/2$, and the other two dimensions are half the size of the minimum block size. To determine the length of the connections between blocks, the half perimeter wirelength (HPWL) model is utilized. Based on this structure, two different forces are exerted on the blocks, where the aim of a filling force F^f is to remove overlaps, while a thermal force F^{th} reduces the resulting peak temperature. A filling force is formed for each bin by considering the density of the blocks within this bin. This density is determined for each bin based on the sum of the blocks covering a bin. Having determined all of the forces for each bin, the filling force applied to each block is equal to the summation of the forces related to all of the bins occupied by this block. Alternatively, the thermal force is based on the thermal gradient within the 3-D space. Obtaining the thermal gradients requires thermal analysis of the system, which in [500] is achieved through a spatially adaptive thermal analysis package [501].

The combination of these forces is the total force exerted on each block in each physical direction. Similar to (9.14), a system of equations is solved, for which the total force applied to the blocks in the x-direction is

$$Cx = a_x[\beta_x F_x^f + (1 - \beta_x)F_x^{th}], \tag{13.4}$$

where a_x and β_x are weighting parameters. Parameter a_x controls the significance of the wirelength, area, and thermal objectives. Equivalently, β_x characterizes the relative importance between the two forces in each direction. All of these parameters are empirically determined.

Having determined the forces on each block, the floorplanning process begins by spreading the blocks laterally within an *xy*-plane rather than the entire 3-D space. This spreading is at odds with traditional force directed methods where the blocks collapse at the center of the floorplan with high overlaps (see Chapter 9, Physical Design Techniques for Three-Dimensional ICs). As this situation results in strong filling forces, allowing the blocks to scatter throughout the 3-D space causes some large blocks to move to the boundary of the 3-D space, for example, close to the tier adjacent to the heat sink. A consequence of this practice can be a poor initial floorplan subjected to hot spots. Alternatively, spreading the blocks only in the *xy*-plane initially avoids this situation while simultaneously reducing the strong filling forces. Furthermore, this initial lateral spreading more evenly distributes the thermal densities, offering an initial distribution of the interconnect.

The second stage follows with global placement of the blocks within the volume of the system based on the filling and thermal forces. An issue that arises during this step is determining the thermal forces, which requires a thermal analysis of the circuit. As the thermal tool to perform this task is based on a tiered structure [501], a continuous floorplan is temporarily mapped into a discrete space. The temperature map for each tier is produced, and the thermal forces ensure that the global placement in continuous space can proceed. The global placement iterates until the overlap between the blocks is reduced to 5 to 10%.

An approach to this interim tier mapping can be implemented stochastically, where the goal is to allocate the power density of the blocks to a specific tier to thermally analyze the system [500]. Considering that the center of block m_i with coordinates (x_i, y_i, z_i) is located between tiers q and $q-1$ during an iteration of the global placement, this block is placed within tier q or $q-1$ with a probability of, respectively,

$$P(m_i, q) = \left(\frac{z_i + \frac{1}{2}D - Z_{q-1}}{D}\right), \tag{13.5}$$

$$P(m_i, q-1) = \left(\frac{Z_q - (\frac{1}{2}D + z_i)}{D}\right). \tag{13.6}$$

Accordingly, the probability for a power density of block m_i to be allocated to tier q or $q-1$ is, respectively,

$$\text{MPD}(m_i, q) = PD_i \cdot P(m_i, q), \tag{13.7}$$

$$\text{MPD}(m_i, q-1) = PD_i \cdot P(m_i, q-1). \tag{13.8}$$

Having produced a floorplan in a continuous 3-D space, tier assignment is realized. If this task takes place as a postprocessing step, however, inferior results can be produced. Instead, a third stage is introduced, as shown in Fig. 13.4, where tier assignment is integrated with floorplanning in a 2.5-D domain. As shown in Fig. 13.5 illustrating a continuous floorplan, a tier assignment of block 2 in either the first or second tier results in a different level of overlap in blocks 1 and 3. This overlap guides the force directed method with the tier assignment to ensure that the global placement produced by the previous stage is not significantly degraded.

This approach reduces the significant mismatches that can occur during tier assignment. Although these mismatches are often resolved as a postprocessing step, the heterogeneity of the shapes and sizes of the blocks can lead to significant degradation from the optimum placement produced during the second stage. Consequently, by integrating the tier assignment with the force directed method, these disruptive changes can be avoided.

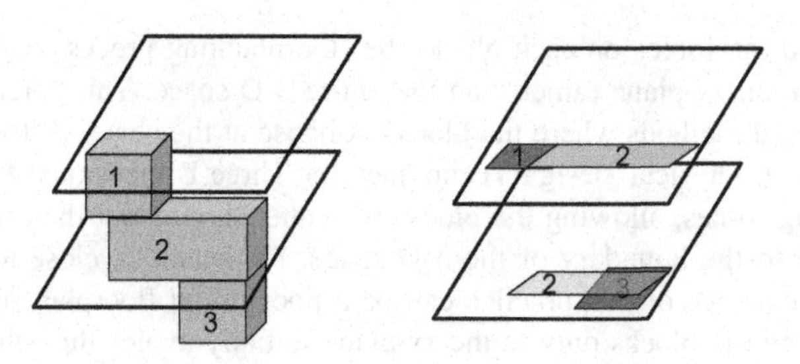

FIGURE 13.5

Transition from a continuous 3-D space to discrete tiers. Block 2 is assigned to either the lower or upper tier, which results in different overlaps.

The final step of the technique removes any remaining minor overlaps between blocks, where rotating the blocks has been demonstrated to improve the results as compared to moving the blocks within each tier. The topographical relationship among the blocks is captured in addition to the orientation of the blocks [502]. Although this technique is not based on a fixed outline, a boundary is assumed during legalization to detect an increase in the area of a tier, which can, in turn, cause an undesirable area imbalance. These violations are treated as additional overlaps. Additional moves and rotations are performed to remove these overlaps.

This force directed method has been compared to the SA based approach where CBA is employed. Some results are listed in Tables 13.3 and 13.4. In Table 13.3, the two methods are compared without considering thermal issues. The results indicate that the force directed method produces comparable results with CBA in area and number of through silicon vias (TSVs) but exhibits a decrease in wirelength. More importantly, the computational time is reduced by 31% [500]. If the thermal objective is added to the floorplanning process, the force directed method performs better in all of the objectives with a greater reduction in computational time than reported in Table 13.4. Note, however, that if the dependence between power and temperature is included in the thermal analysis process, the savings in time is significantly lower.

In addition to analytic techniques, other less conventional approaches to floorplan 3-D circuits have been developed. These approaches include genetic algorithms where, as an example, the thermal aware mapping of 3-D systems that incorporate a network-on-chip (NoC) architecture [503]. Merging 3-D integration with NoC is expected to further enhance the performance of interconnect limited ICs (the opportunities that emerge from combining these two design paradigms are discussed in Chapter 20, 3-D Circuit Architectures).

Consider the 3-D NoC shown in Fig. 13.6. The goal is to assign the tasks of a specific application to the processing elements (PEs) of each tier to ensure that the temperature of the system and/or communication volume among the PEs is minimized. The function that combines this objective characterizes the fitness of the candidate chromosomes (i.e., candidate mappings), described by

$$S = \frac{1}{a + \log(\max_{temp})} + \frac{1}{\log(\mathrm{comm}_{cost})}. \tag{13.9}$$

As with traditional genetic algorithms, an initial population is generated [504]. Crossover and mutation operations generate chromosomes, which survive to the next generation according to the

Table 13.3 Comparison for Area and Wirelength Optimization [500]

Circuit	CBA				3-D Scalable Temperature Aware Floorplanning (STAF) (No Temperature)			
	Area (mm^2)	HPWL (mm)	# of TSVs	Time (s)	Area (mm^2)	HPWL (mm)	# of TSVs	Time (s)
ami3	35.30	22.5	93	23	37.9	22.0	122	52
ami49	1490.00	446.8	179	86	1349.1	437.5	227	57
n100	5.29	100.5	955	313	5.9	91.3	828	68
n200	5.77	210.3	2093	1994	5.9	168.6	1729	397
n300	8.90	315.0	2326	3480	9.7	237.9	1554	392
Aggregate relative to CBA					+4%	−12%	−1%	−31%

Table 13.4 Comparison for Temperature Optimization [500]

Circuit	CBA					3-D STAF				
	Area (mm^2)	HPWL (mm)	# of TSVs	Temp. (°C)	Time (s)	Area (mm^2)	HPWL (mm)	# of TSVs	Temp. (°C)	Time (s)
ami3	43.2	23.9	119	212.4	486	41.5	24.2	116	201.3	227
ami49	1672.6	516.4	251	225.1	620	1539.4	457.3	208	230.2	336
n100	6.6	122.9	1145	172.7	4535	6.6	91.5	753	156.8	341
n200	6.6	203.7	2217	174.7	6724	6.2	167.8	1356	164.6	643
n300	10.4	324.9	2563	190.8	18475	9.3	236.7	2173	168.2	1394
Aggregate relative to CBA						−6%	−16%	−12%	−6	−75%

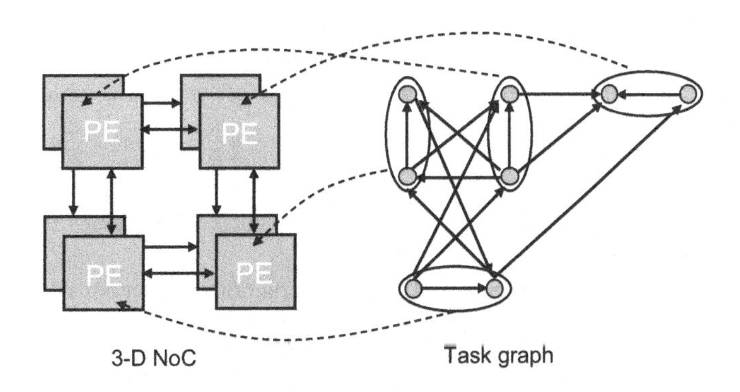

FIGURE 13.6

Mapping of a task graph onto physical PEs within a 3-D NoC [503].

relative chromosomal fitness. The floorplan with the highest fitness is selected after a number of iterations or if the fitness cannot be further improved.

13.1.2 THERMAL DRIVEN PLACEMENT

Placement techniques can also be enhanced by the addition of a thermal objective. The force directed method used in cell placement [371] and discussed in Section 9.3.1 has been extended to incorporate the thermal objective during the placement process [505] of standard cells in 3-D systems. In this approach, repulsive forces are applied to those cells that exhibit high temperatures (i.e., "hot blocks") to ensure that the high temperature cells are placed at a greater distance from each other. The applied forces comprise both thermal forces and overlap forces. Since the objective is to reduce the temperature, the thermal forces are set equal to the negative of the thermal gradients. This assignment places the blocks far from the high temperature regions.

Determining these gradients, as discussed in the previous section, is a critical step of any temperature aware technique. Another method to obtain the temperature of a 3-D circuit is the finite element method [505]. The 3-D stack is discretized into a mesh consisting of unit cells, as discussed in Section 12.3. The thermal gradient of a point or node of an element is determined by differentiating the temperature vector along the different directions,

$$\mathbf{g} = \left[\frac{\partial \mathbf{T}}{\partial x} \quad \frac{\partial \mathbf{T}}{\partial y} \quad \frac{\partial \mathbf{T}}{\partial z}\right]^{\mathrm{T}}. \tag{13.10}$$

Exploiting the thermal electric duality, the modified nodal technique in circuit analysis [473], where each resistor contributes to the admittance matrix (i.e., matrix stamps), is utilized to construct element stiffness matrices for each element. These elemental matrices are combined into a global stiffness matrix for the entire system, notated as $\mathbf{K_{global}}$. This matrix is included in a system of equations to determine the temperature of the nodes that characterizes the entire 3-D circuit. The resulting expression is

$$\mathbf{K_{global}} \mathbf{T} = \mathbf{P}, \tag{13.11}$$

where $\mathbf{P}$ is the power consumption vector of the grid nodes. To determine this vector, the power dissipated by each element of the grid is distributed to the closest nodes. From solving (13.11), the temperature of each node is determined during each iteration of the force directed algorithm. The thermal forces can be determined from the thermal gradient between grid nodes, requiring the following expression to be solved,

$$\mathbf{Ci} = \mathbf{f_i}, \quad \forall \mathbf{i} \in \{\mathbf{x}, \mathbf{y}, \mathbf{z}\}, \tag{13.12}$$

where $\mathbf{f_i}$ is the force vectors in the *x*, *y*, and *z* directions. Matrix $\mathbf{C}$ describes the cost of a connection between two nodes, as defined in [371].

After the stiffness matrices are constructed, an initial random placement of the circuit blocks is generated. Based on this placement, the initial forces are computed, permitting the placement of the blocks to be iteratively determined. This recursive procedure progresses as long as an improvement above some threshold value is exhibited. The procedure includes the following steps [505]:

1. the power vector resulting from the new placement is determined;
2. the temperature profile of the 3-D stack is calculated;

3. the new value of the thermal and overlap forces is evaluated;
4. the matrices of the repulsive forces are updated;
5. a new placement is generated.

After the algorithm converges to a final placement, a postprocess step follows. During this step, the circuit blocks are positioned without any overlap within the tiers of the system. If one tier is packed, the remaining cells, initially destined for this tier, are positioned onto an adjacent tier. A similar process takes place in the y-direction to ensure that the circuit blocks are aligned into rows. A divide and conquer method is applied to avoid any overlap within each row. A final sorting step in the x-direction includes a postprocessing procedure, after which no overlap among cells should exist.

The efficiency of this force directed placement technique has been evaluated on MCNC [396] and IBM-PLACE benchmarks [506], demonstrating a 1.3% decrease in average temperature, a 12% reduction in maximum temperature, and a 17% reduction in average thermal gradient. The total wirelength, however, increases by 5.5%. This technique achieves a uniform temperature distribution across each tier, resulting in a significant decrease in thermal gradients as well as maximum temperature. The average temperature throughout a 3-D IC, however, is only slightly decreased. This technique, consequently, focuses on mitigating hot spots across a multi-tier system.

In all of the techniques presented in this section, the heat is transferred from the upper tiers to the bottom tier primarily through the power and signal lines and the thin silicon substrates of the upper tiers. No additional means other than redistributing the major heat sources throughout the 3-D stack lessens any significant thermal gradients. Furthermore, these techniques typically assume the power density associated with each module, block, or cell is temporally fixed. However, thermal management is also applicable in real-time, where power densities are monitored and adjusted to prevent the appearance of hot spots that affect circuit performance and contribute to aging, degrading the reliability of the system. These methods are described in the following section.

13.1.3 DYNAMIC THERMAL MANAGEMENT TECHNIQUES

Thermal management techniques during circuit operation—typically used in processors—have received significant attention since around 2000 due to increasing power densities. The advent of multi-core architectures has further fueled interest in these techniques as each core within a processor is managed separately, offering several ways to thermally manage these complex computing systems. Although techniques for dynamic power management exist [507], simply reducing power is not sufficient due to the thermal coupling of circuits and spatially varying thermal conductivities across a system. Decreasing power can lower the peak or average temperature of an integrated system, yet the thermal gradients may be higher if appropriate thermal management techniques are not applied. The appearance of hot spot(s) in some region(s) of a circuit can be attributed to the inability of power reduction techniques to eliminate these thermal gradients despite the average temperature of the circuit being maintained within thermal limits.

Dynamic thermal management (DTM) methods can be applied to both software and hardware, where a combination of techniques is often used. In hardware, dynamic frequency voltage scaling (DVFS) and clock gating (or throttling) are commonplace techniques, while in software, workload (or thread) migration is usually employed to cool down cores [508]. The granularity at which thermal policies are applied can also vary, in particular, for multi-core processors. For example, a single voltage/frequency pair can be chosen for an entire processor (a global policy) or each core may have a separate voltage/frequency depending upon the temperature of the core (a distributed policy).

A number of tradeoffs exist among these choices, leading to a relatively broad design space for dynamic thermal management. Software methods (driven by the operating system) are usually coarse grained methods and are less effective than hardware techniques since the latter can respond faster to alleviate steep transient thermal loads. Software methods can, however, be implemented relatively easily by scheduling tasks among the different parts of a system. In a multi-core system, threads are swapped among processors or assigned to different processors depending upon the temperature of each core, where these changes typically take place at tens of milliseconds [509,510]. Several techniques use a 10 ms interval for job scheduling as this interval is also used in the Linux kernel for timer interrupts [509]. DVFS mechanisms are more complex to implement. This complexity increases if a distributed (per core) DVFS scheme is employed. Despite the greater complexity, however, these techniques are widely used in modern processors to maintain the temperature within specified limits [511,512].

Determining the temperature of a circuit during operation is another important aspect of dynamic thermal management as slow responses can render these techniques inefficient or excessively strict. Different means are used to determine the temperature of a circuit including thermal sensors [512] and performance counters that characterize a core, such as accesses to register files, cycle counts, and number of executed instructions [508]. Information from thermal sensors is used directly, where the primary issue for sensors is the response time to changes in temperature, recognizing that thermal constants within integrated systems are typically on the order of milliseconds. Thermal sensors constitute a reliable means for measuring circuit temperature to guide a thermal management policy. Alternatively, the information from performance counters is loosely connected to temperature and, although widely explored in the literature, outputs from these components represent a proxy and should be used with caution [509].

Less complex dynamic thermal policies are also available, where the online workload schedule is based on the thermal profile of the target system obtained offline for the expected combination of workloads [513]. This strategy reduces the overhead of thermal management; the efficiency, however, may be lower than fully online methods. These methods are particularly effective for those scenarios where the workload combinations differ from those workloads employed during offline energy profiling of the system.

Several efforts employing dynamic thermal management approaches, both in software and hardware, have been published [497,508, 510–512]. Similarly, several works pay attention to the design and allocation of thermal sensors within 2-D circuits [514]. Although these techniques can also be applied to 3-D circuits, the resulting efficiency may not be similar. This drop in efficiency can be attributed to the strong thermal coupling between adjacent circuits. This coupling is more pronounced in the vertical direction due to the significantly smaller physical distance between circuits and that heat primarily flows in the vertical direction. Other reasons for developing novel thermal policies for 3-D systems include the case where memory (e.g., DRAM) tiers are stacked on top of a processor tier [515]. Although the processor tier is located next to the heat sink, care must be placed to ensure processor operation does not cause the temperature to rise beyond the strict thermal limits of DRAM [516]. Higher overall power can occur due to more frequent refreshing of the stored data. The remainder of this section reviews the evolution of thermal management techniques for 3-D systems, emphasizing either multi-tier processor architectures or a combination of a processor tier vertically integrated with tiers of memory.

13.1.3.1 Dynamic thermal management of three-dimensional chip multi-processors with a single memory tier

Although automatic control theory has been used to guide workload scheduling for thermal management [508], most techniques are based on heuristics due to low complexity characteristics. A heuristic OS-level scheduling algorithm targeting 3-D processors has been proposed in [498]. The key concept of this technique is to consider the strong thermal coupling along the vertical direction in the scheduling process to balance temperatures across the stack, thereby reducing thermal gradients and decreasing the frequency in those cores with hot spots. The occurrence of hot spots requires aggressive thermal measures which degrade system performance. Consider a 3-D multi-core system. The assumption is that irrespective of the floorplan, which can also be thermally driven, leading to the stacking of energy demanding cores with low power caches, stacking of cores may be unavoidable. In this case, careful scheduling lowers the temperature of the system, where the scheduling process considers the entire stack rather than individual tiers. Assigning tasks to those cores situated in the tier next to the heat sink should, therefore, not be based on the temperature of a specific tier (which is lower than in other tiers) but, rather, the temperature of those cores in tiers farther away.

Due to the vertical thermal coupling, rather than assigning a single task to one core, a stack of cores (for example, an entire 3-D system split into several pillars of cores) is considered, where a task is assigned to every processor within this stack. The terms, "super-core" and "super-tasks," are used to describe the main features of the heuristic. Assuming a 3-D system with n tiers and m cores per tier, the tasks on $n \times m$ cores require scheduling. With this notation, a super-core consists of n cores, where a task is assigned to each core to distribute a balanced temperature. The assignment process forms super-tasks that require comparable power. These super-tasks are assigned to each super-core.

To balance the power across the super-cores, the power of all of the $(n \times m)$ tasks is sorted. Each task and associated power are assigned to the available m bins (of super-tasks). The assignment proceeds iteratively by inserting power values, in descending order, to that bin with the smallest total power during this iteration. The process continues until every task has been assigned to a bin. An example assignment is shown in Fig. 13.7, where sorting of power within tasks and the resulting task assignment to four bins are depicted for a twotier 3-D system containing four cores per tier. As the main operation of this heuristic is sorting, the time complexity is $O(mn\log(mn))$ [498].

In addition to task scheduling, DVFS within each super-core is assumed, although DVFS for a core is not necessarily straightforward. Thus, if a core exceeds a thermal threshold, DVFS scales the voltage supply of the core with the highest power (which may or may not be the overheated core within a super core). The objective is to penalize the task that consumes the highest power, which may cause overheating in a vertically adjacent core. DVFS is applied to only one core within a super-core, which limits the benefits of this technique.

Job scheduling is applied to the benchmark suite of applications, SPEC2000 [517], where several workloads are combined to form benchmark sequences with diverse power profiles and

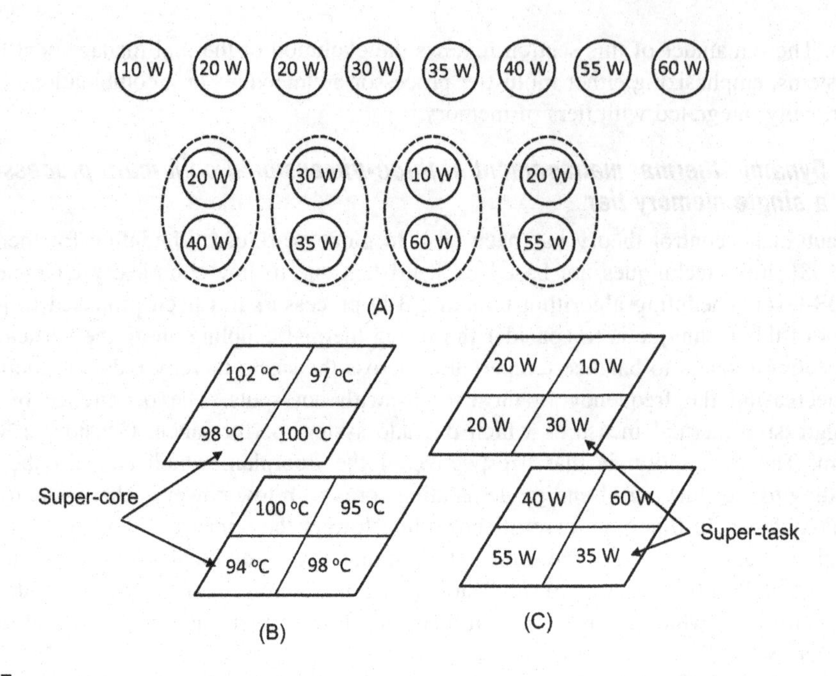

FIGURE 13.7

Temperature balancing heuristic where (A) the tasks are sorted in descending power and assigned to super-tasks, (B) the temperature of each core, and (C) the super-tasks assigned to the super-cores [498].

thermal loads. For example, thermal analysis of the *crafty* and *mcf* workloads yields a HC (hot-cold) scenario. These workloads are applied to a 3-D multi-core architecture with two tiers and four cores per tier, where a P4 Northwood architecture operating at 3 GHz is assumed for each core.

To evaluate the effectiveness of the heuristic, other task assignments have also been considered. The baseline for comparison is a random assignment of tasks to cores based on the Linux 2.6 scheduler [509]. The scheduler works similarly to Linux but with a slightly smaller scheduling interval of 8 ms. A round robin scheduler is also utilized for comparison. Finally, another scheduling approach, where the objective is to balance temperature by assigning a high power task to a cool core but without considering the temperature of the vertically adjacent cores, is also evaluated for the purposes of comparison.

A comparison in terms of peak temperature among these task scheduling methods for the target workload scenarios demonstrate that the heuristic assignment reduces the peak temperature by up to 24°C as compared to a random assignment. Other techniques similarly decrease the peak temperature. Thermal coupling in task scheduling supports the removal of a large number of thermal problems when DVFS is employed. DVFS, however, often counteracts the benefits of job scheduling as DVFS requires a greater performance overhead.

In contrast to the method described in [498], where frequent use of DVFS is avoided, other techniques consider combining both software and hardware techniques. The objective of these methodologies is to balance the effectiveness of each technique against the overhead without favoring one technique over the other. Thus, a thermal management framework has been developed for 3-D chip multi-processors (CMP), where clock gating, DVFS, and workload scheduling are all applied to satisfy thermal limits without degrading performance [497]. This goal is achieved by applying these techniques both in a distributed (or local) and global (or centralized) manner. A distributed approach provides greater versatility when applying a thermal policy as allowing DVFS only for a core within each super-core is a restricted approach. The structure and operation of this framework are discussed in the following subsection.

13.1.3.2 Software/hardware (SW/HW) thermal management framework for three-dimensional chip multi-processor

The development of this framework requires a set of tools for architectural, power, and thermal modeling. These tools include the M5 architectural simulator [518], a Wattch-based EV6 model [519], CACTI [520], the approach described in [521] for power modeling of the cores, caches, and leakage power, and a tool based on [522] for thermally analyzing these systems. Irrespective of the specific tools, architectural simulators, power models and/or simulators, and thermal analysis tools are all required to design and evaluate a dynamic thermal management policy for CMP or multi-processor system-on-chip (MPSoC) systems. The framework is based on specific guidelines, which are based on a first order thermal analysis of a two tier CMP, as illustrated in Fig. 13.8. In this example, the electrical-thermal duality is used to analyze the thermal behavior of a CMP, where each core is represented by a single node in an equivalent thermal model. Only a simple path for the heat to flow is assumed to exist within the stack. At first glance, the thermal conductivity of the cores in the upper tier (e.g., core I) is lower than the thermal conductivity of those cores in the lower tier, which implies that the temperature of core I is higher than in cores J and K. Additionally, thermal coupling between cores in the same tier (cores J and K) is considerably lower, implying comparable cooling efficiencies. These observations suggest that workload scheduling should consider the different cooling efficiencies of the cores and the effect that a workload can have on other cores within a system.

To extend this analysis to a CMP with m cores, specific data that affects the thermal behavior of the cores need to be determined. Consequently, the cooling efficiency of each core determines the workload schedule. The schedule is extracted from the steady-state heat conduction expression,

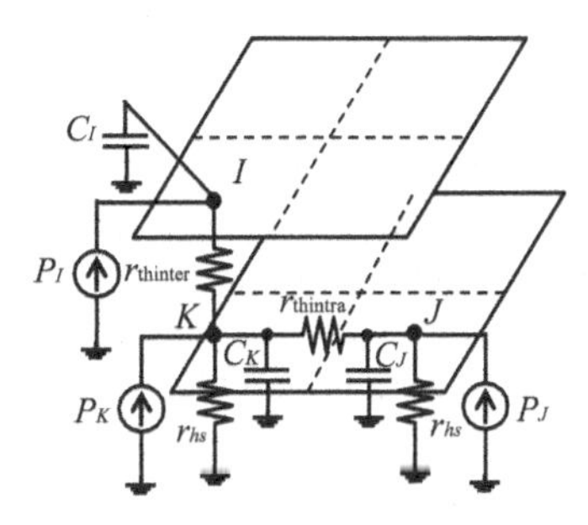

FIGURE 13.8

First order thermal model, where each core is thermally modeled by a node with power P_i, specific heat C_i, and inter- and intratier thermal resistances [497].

$\boldsymbol{T} = \boldsymbol{P}\boldsymbol{R}_{th}^{-1}$, where matrix R_{th} includes the thermal resistance connecting the nodes (i.e., cores) within the thermal model. Notating the temperature of core i as T_i, the temperature is

$$T_i = \sum_{j=0}^{m-1} r_{thi,j} \cdot p_j, \tag{13.13}$$

where $r_{thi,j}$ is the thermal resistance between cores i and j, and p_j is the power consumed by core j. The row i,j of matrix R_{th} describes the effect of the cores on the temperature at core i. Furthermore, considering the quadratic relationship between dynamic power consumption and voltage, and that the operating frequency exhibits a roughly linear dependence on voltage, power dissipation can be described as $p_j = s_{i,j} f_j^3$ where f_j is the operating frequency of core j. The term $s_{i,j}$ is the product of the switching activity of the core multiplied by the switched capacitance, which is linearly proportional to the number of instructions per cycle (IPC) of the job being executed on core j. The figure of merit, thermal impact per performance (TIP), is introduced to formulate guidelines for the thermal control of 3-D CMPs,

$$\mathrm{TIP}_{i,j}^{f} = \frac{dT_i}{df_j}, \tag{13.14}$$

$$\mathrm{TIP}_{i,j}^{\mathrm{IPC}} = \frac{dT_i}{d\mathrm{IPC}_j}, \tag{13.15}$$

which denotes the effect of core j on the temperature of core i, where the frequency f_i (and voltage) through DVFS and the workload IPC_j are assigned to core j.

Robust thermal management of a CMP is intended to improve performance while satisfying specific thermal constraints, which can vary among cores. A metric to describe the performance of a CMP is the total number of instructions per second executed by a CMP,

$$\mathrm{CMP}_{\mathrm{IPS}} = \sum_{i=0}^{m-1} \mathrm{IPC}_i \cdot f_i. \tag{13.16}$$

The thermal constraints that apply to a CMP are determined by the thermal limit of each core, which is assumed to be equal for all cores. Consequently, the requirement is that $\forall_{i=0}^{m-1} T_i \leq T_{\max}$. Satisfying this constraint requires equating the thermal impact per performance of all cores. This decision leads to assigning different frequencies among those cores with different cooling efficiencies, and executing jobs with different IPCs. This method results in two approximate design guidelines. For intertier processors, the frequencies and IPCs should be assigned based on the cooling efficiency of the cores, given by (13.13) to (13.16), where both the frequency and IPC are, in general, different. This guideline is in accordance with the heuristic described in [498], where the assignment of a workload to a core considers both the temperature as well as the location of a core within the stack. The capability to transfer heat to the ambient is affected by the power dissipated during an assigned workload. Alternatively, among cores situated within the same tier, where the cooling efficiency is roughly the same based on the thermal model of Fig. 13.8, the same frequency and workloads with similar IPC are assigned.

These guidelines underpin the thermal management policy developed at the operating system level [497], where the temperature is obtained from the thermal sensors. Performance counters gather information for workload monitoring and to estimate the IPC. With this information and the aforementioned guidelines, this framework applies distributed workload migration and real-time thermal control, globally adapting the power and thermal attributes across CMPs.

At the global CMP (global) level, the power and thermal budgets are determined for each core using a hybrid online/offline technique. For each workload the optimal voltage/frequency pair is determined. The temperature of each workload is computed, and the power is updated to consider the dependence on temperature. The voltage/frequency (V/F) pair is chosen to satisfy the thermal limit. After several iterations, the temperature and temperature dependent power converge. The resulting V/F pair is stored in a look-up table. This table is integrated with the OS and is periodically invoked.

Thermal balancing is also leveraged by a distributed policy where the IPC of each core is monitored and, if required, adjusted to guarantee thermal safety. This adjustment is primarily between vertically adjacent cores, since greater thermal heterogeneity is noted. The workload migration swaps jobs to assign those workloads with high IPCs to those cores with higher cooling efficiencies. This migration of workloads takes place every 20 ms. If thermal transients, however, occur at faster rates, other thermal control measures are considered, such as DVFS and clock gating. These techniques are applied locally, thereby providing better control of thermal gradients across the stack as compared to a centralized approach. Moreover, due to the considerable impact on performance of clock gating, DVFS is the primary method. Clock gating is only used for thermal emergencies.

This framework is applied to a 3-D CMP consisting of three physical tiers. Two of these tiers host eight Alpha21264 cores (four in each tier) assuming a 90 nm CMOS node, and the third tier contains the L2 cache. A comparison of the framework with a strictly distributed thermal policy [508] for this 3-D CMP, where the simulated workloads are based on applications from SPEC2000 [517] and Media benchmark suites, exhibits an improvement of, on average, 30%. This situation is due to the strong thermal coupling of the vertically adjacent cores. Local control cannot capture this coupling since the workload and V/F pair are chosen for this core. Employing power and thermal budgeting at the global level can, however, address this limitation, improving the overall throughput of the CMP.

In addition to the cooling efficiency, an analysis of the characteristics of the workloads can yield greater performance of the CMP, assuming the same thermal limits [523]. To assess these improvements, the workloads are classified as compute bound or memory bound, depending upon the memory requirements of the workloads. Analyzing memory transfers between the core and in-stack memory provides useful information that can be included within the thermal policies. Considering that instructions are executed at a clock rate f_{CPU} and the off-chip memory transfers due to L2 cache misses are performed at a clock rate $f_{\text{off-chip}}$, the time to execute a task is [523],

$$t_{ex}(f_{\text{CPU}}) = \frac{w_{\text{on-chip}}}{f_{\text{CPU}}} + \frac{w_{\text{off-chip}}}{f_{\text{off-chip}}}, \tag{13.17}$$

where $w_{\text{on-chip}}$ is the number of clock cycles to execute CPU instructions without a cache miss, and $w_{\text{off-chip}}$ is the number of clock cycles for external transfers if the core stalls. This simple expression describes whether a workload is compute bound (e.g., high $w_{\text{on-chip}}$) or memory bound (e.g., high $w_{\text{off-chip}}$). Both of these quantities depend upon the type of application and the time needed for a core to execute an application. The off-chip clock rate is also considered constant, while the clock cycles for off-chip memory transfer are modeled as a function of the number of cache misses. $w_{\text{off-chip}} = (aN_{\text{miss}} + b) \cdot f_{\text{CPU}}$ is the number of L2 misses denoted by N_{miss} [524]. The coefficients a

and b are fixed and depend on the target architecture, while the number of misses is determined by performance monitors, such as PAPI [525].

Collection of this information can quantify the speed up required to execute a workload due to an increase in the core clock frequency (which affects the temperature). This speed up for each core and workload at two different frequencies is

$$SU = \frac{t_{ex}(f_{CPU}^{new})}{t_{ex}(f_{CPU}^{ref})}. \tag{13.18}$$

This simple metric computes the gains in IPS for a workload with different core clock frequencies. These speed ups, however, cannot be exploited unless the temperature of the core is lower than the maximum temperature at the target frequency. Application of these higher frequencies is performed effectively if the *instantaneous* temperature rather than the steady state temperature (SST) of a core is employed. The use of the *instantaneous* temperature tracks the closeness of each core to the maximum allowed temperature and core frequency. This approach is not considered in [497] where the SST of a core determines the thermal effect on the cores and drives the workload allocation policy. Note that the term *instantaneous* temperature refers to the temperature of a core over several milliseconds, since thermal constants are orders of magnitude greater than the clock frequency.

A considerably lower instantaneous temperature, notated as T_{inst} from the maximum allowed temperature T_{max}, provides a temperature slack [523] which can be exploited by operating the core at a higher frequency. Performance improvements can, therefore, be determined by SU for each workload. Thus, not only the cooling efficiency of each core but also the SU can improve the IPS of a 3-D CMP [523], as described by (13.16). These frequency adjustments of finer granularity should, however, be carefully performed due to the strong thermal coupling in the vertical direction. This coupling has a detrimental effect on the temperature of the other cores within the 3-D CMP, particularly those cores located in other tiers at the same 2-D position.

The optimization problem of maximizing (13.16) is solved subject to the constraint that the temperature of those cores farthest from the heat sink is equal to T_{max}. Solving this optimization problem with the use of Laplace multipliers leads to the following expression applied to each core i [523],

$$\frac{R_{thi}}{IPC_i} \cdot \frac{dP(f_i)}{df_i} = M, \tag{13.19}$$

where M is a constant. Determining the precise frequency for each core that satisfies (13.19), however, is not feasible since the clock frequency changes discretely. Only a small set of clock frequencies is supported by a CMP. Consequently, an approximate solution is offered to ensure that the frequency assigned to each core deviates the least from (13.19).

Solving (13.19) requires the thermal resistance R_{thi}, power P_i, and IPC_i of each core i. The thermal resistance of each core in the CMP is illustrated in Fig. 13.8, while the power is determined by the IPC of the core. Allocation of the IPC to the cores for each workload proceeds according to the heuristic of Section 13.1.3.1, where an example is depicted in Fig. 13.7. Rather than explicit power values, the IPC of each workload matches the super-tasks (i.e., a set of tasks) to the super-cores of the CMP. This procedure is repeated and threads are migrated at regular intervals (of 100 ms) [523].

Table 13.5 Average Power Dissipation (P_{avg}) and Average Temperature Slacks (T_{slack}) of 3-DIS-IT and 3-DI-SST [523]

Benchmark Combination	3-DI-SST		3-DIS-IT	
	P_{avg} (W)	T_{slack} (°C)	P_{avg} (W)	T_{slack} (°C)
hipc-hm	91.63	7.29	123.40	2.46
hipc-mm	113.86	5.08	118.97	4.29
hipc-lm	99.00	6.95	104.63	6.02
mipc-hm	104.30	7.32	150.53	2.30
mipc-mm	105.94	5.17	115.32	4.25
mipc-lm	84.56	7.53	87.77	6.99
lipc-hm	69.98	10.44	70.14	10.40
lipc-mm	85.03	7.37	85.70	6.94
lipc-lm	119.34	5.74	124.64	4.85

Several scenarios based on the benchmark applications SPEC2000/2006 [517] and ALPbench [526] are used to evaluate the efficiency of thread migration based on the *instantaneous* temperature of each core. These scenarios include workloads with high, low, and mixed IPCs denoted, respectively, as HIPC, LIPC, and MIPC. The *SU* of these benchmarks is also classified as high, low, and mixed, where the frequency of each core switches between 1 and 2 GHz. The resulting IPS for a 3-D CMP comprising two core tiers and one memory tier are compared for two different thread management approaches. The first approach follows from [497]. The HIPCs are assigned to those cores closer to the heat sink (3-DI) and to cores with a lower SST, setting the clock frequency of each core to ensure that the SST does not exceed T_{max} (3-DI-SST). Alternatively, the thread allocation assigns core frequencies based on the temperature slack of each core ($T_i(t) \leq T_{max}$) (IT), and assigns the super-task with the highest sum of IPCs to the coolest super-core. The thread with the highest *SU* (3-DIS) is assigned to the core of each super-core located closest to the heat sink (3-DIS-IT). A comparison between these two thread allocation policies results in an average IPS improvement of 18.5% for those scenarios listed in Table 13.5. The temperature slack for 3-DIS-IT is much lower, demonstrating that most cores operate close to T_{max} due to the higher frequency, yielding both a higher total IPS and power for each scenario.

13.1.3.2.1 Dynamic thermal management of three-dimensional chip multi-processors with multiple memory tiers

These techniques relate to CMPs when a single memory tier is considered. Adding more memory tiers improves performance as fewer cache misses occur; however, thermal management becomes a more acute issue as the distance of those tiers from the heat sink increases. The use of nonconventional memories, such as magnetic RAM [527], can alleviate this situation since a nonvolatile memory tier does not leak current, reducing the overall power of the stack. The introduction of these memory technologies requires different approaches for dynamic thermal management of 3-D CMPs since these memory technologies exhibit substantially different characteristics as compared to

SRAM-based cache. For example, nonvolatile memories are slower and require higher energy to write, while the endurance is lower than SRAM. The main traits of different memory technologies are reported in Table 13.6.

As magnetic memory is slower than SRAM, these memory tiers should be used to store less immediate data, while frequently accessed data or frequent write accesses should be placed within the SRAM tier. Considering a 3-D CMP that includes a mixture of memory tiers, an objective to improve the IPS of the CMP while satisfying thermal limits can be satisfied by power gating a cache at the "way-level" and applying DVFS for the cores of the CMP. An example of an architecture combining a mixture of memory technologies is shown in Fig. 13.9, where one processing tier with four cores is stacked with three tiers of SRAM and one tier of MRAM for L2 cache. L1 cache is integrated within the processing tier. The memory organization in each tier is also shown in the figure along with the vertical bus connecting the cache to the cores. The cores are connected

Table 13.6 Parameters of Different Memory Technologies Fabricated in 65 nm Technology [529]

	SRAM	MRAM	PCRAM
Cache size	128 kb	512 kb	2 MB
Area (mm^2)	3.62	3.30	3.85
Read latency (ns)	2.252	2.318	4.636
Write latency (ns)	2.264	11.024	23.180
Read energy (nJ)	0.895	0.858	1.732
Write energy (nJ)	0.797	4.997	3.475
Static power at 80°C (W)	1.131	0.016	0.031
Write endurance	10^{16}	4×10^{12}	10^9

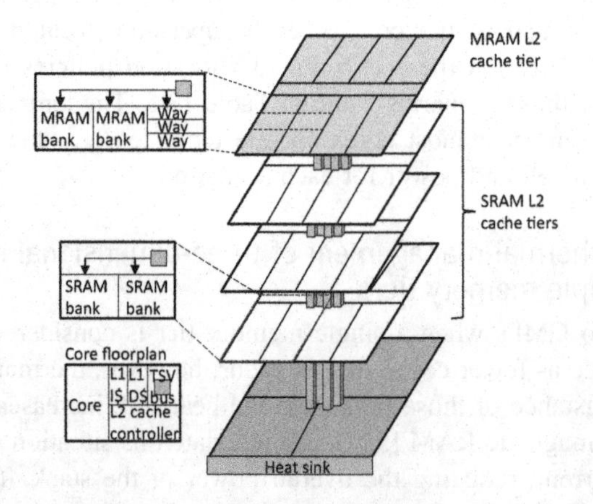

FIGURE 13.9

3-D CMP consisting of a single four core tier with three tiers of SRAM and one tier of MRAM [527].

through a crossbar switch (a large number of cores would require a NoC topology [528]). Allocation of the cache ways[1] to each core lowers the temperature [527]. This allocation occurs dynamically both for the core and cache tiers where the leakage power of the memory tiers is also considered within the power budget. Key to this allocation strategy remains the notion of heterogeneous thermal coupling in the vertical direction throughout the entire stack, including the memory tiers.

To illustratively explain the concept behind the cache way and clock frequency allocation process, consider the example shown in Fig. 13.10. Five different schemes are applied to maximize IPS. In Fig. 13.10A, a low frequency clock (1 GHz) is chosen for both cores and cache ways allocated to each core from all of the SRAM tiers. As core 1 executes a memory demanding

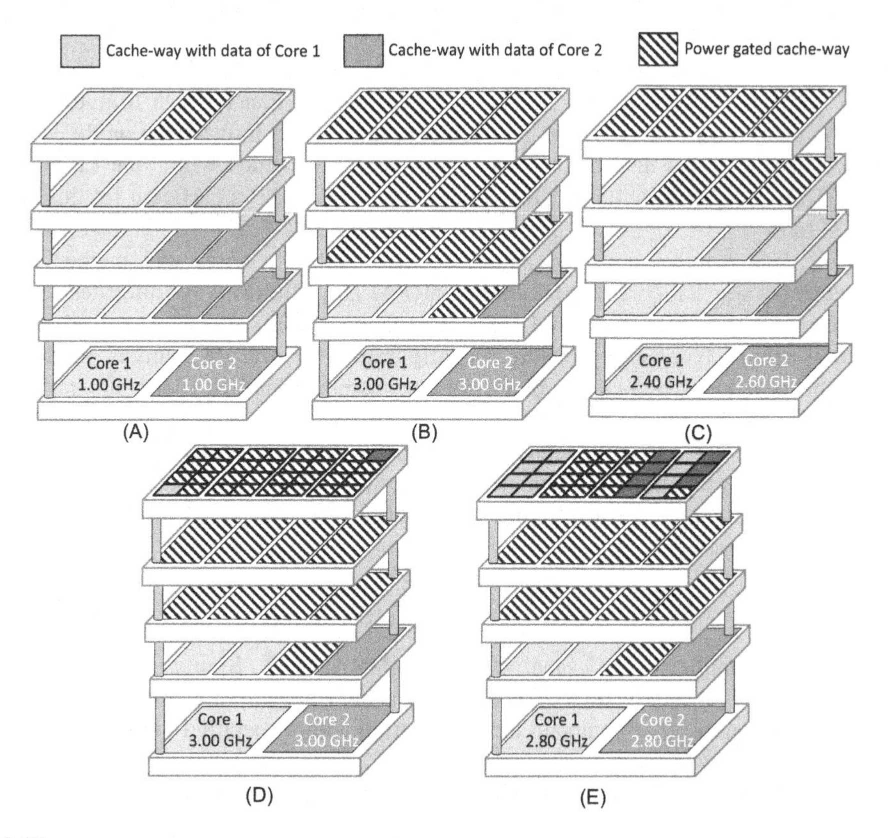

FIGURE 13.10

Dynamic thermal management schemes for a 3-D CMP employing a mixture of SRAM, MRAM, and DVFS, (A) SRAM-1 GHz core, (B) SRAM-3 GHz core, (C) SRAM-core DVFS, (D) hybrid-3 GHz core, and (E) hybrid-core DVFS [527].

[1]A "cache way" is a cache block within each set of a set associative cache memory. An *n*-way set associative cache indicates that each set of the memory is composed of two cache ways (i.e., blocks).

benchmark (*Art*), additional cache ways are provided to this core. The frequency cannot be substantially increased due to the greater power required by the L2 cache. In Fig. 13.10B, the clock frequency of both cores rises to 3 GHz but the cache available for each core is decreased to avoid excessive temperatures. Data would be lost and should therefore be transferred from main memory. In Fig. 13.10C, the clock of each core is adjusted through a DVFS mechanism, supporting a tradeoff between the capacity of the available L2 cache and the clock frequency. Replacing a tier of SRAM with MRAM (see Fig. 13.10D) and turning off some SRAM tiers supports a clock frequency of 3 GHz for both cores. Some data can be stored in the MRAM tiers, avoiding slow off-chip memory transfers, thereby achieving a better IPS than the system shown in Fig. 13.10B. Depending upon the workload, this hybrid-DVFS approach also employs DVFS to tradeoff the number of activated cache ways with clock frequency to ensure the resulting IPS of the 3-D CMP is maximized.

In the schemes shown in Figs. 13.10D and E, data stored in the MRAM tier are transferred into the faster SRAM tiers to limit the on-chip energy expended to transfer data from/to the memory. Data migration takes place with counters for each core that collect information on the least recently used and most recently used data blocks. This policy is implemented within the L2 memory controller [527]. Thus, a cache miss in the SRAM tier is compensated if a cache hit for this data occurs in the MRAM tier.

Maximizing IPS requires a set of parameters for both the cores and memory, including the clock frequency of each core, the number of SRAM and MRAM ways allocated to each core, and the number of activated SRAM and MRAM ways physically allocated on top of each core. This allocation occurs dynamically, where a configuration interval of 50 ms is employed [527]. Partition of the power gated memory blocks is also demonstrated in [530] but not to increase CMP throughput.

The IPS for the target 3-D architecture is analytically described in [527]. Two additional concepts are introduced to determine the number of cache ways allocated to each processor as well as which ways are activated. Consequently, the performance improvement (PI) of IPS in terms of the SRAM cache ways w_i^S assigned to core i is

$$\mathrm{PI}_i = \frac{\partial \mathrm{IPS}_i}{\partial w_i^S}. \tag{13.20}$$

A similar expression applies to the cache ways of the MRAM, w_i^M. Similar to PI, the performance loss (PL) incurred by the activation of one more cache way within a super-core l_i^M, entails a decrease in clock frequency f_i to maintain the temperature within specified limits,

$$\mathrm{PL}_i = \frac{\partial \mathrm{IPS}_i(l_i^S)}{\partial l_i^S}. \tag{13.21}$$

A similar expression applies for the cache ways of the active MRAM, l_i^M. With PI, PL, and the Lagrange multipliers, the IPS reaches a maximum for a specific temperature if the SRAM and MRAM cache ways for each core are selected to ensure that PI and PL are equal among all cores. Greedy heuristic algorithms and the bisection method are typically utilized to determine the SRAM and MRAM cache ways allocated to each core as well as the number of active ways within a configuration interval. The number of active ways depends upon the clock frequency of the core and the maximum tolerated temperature.

Table 13.7 Combinations of Benchmark Applications to Compare the Performance of Different Thermal Management Schemes Based on SRAM/MRAM L2 Cache [527]

Scenarios of Benchmark Applications	Benchmark Applications for Each Scenario
HIPC-LM	equake, parser
LIPC-LM	lbm, mcf06, sjeng, ammp
HIPC-HM	gcc, bzip
LIPC-HM	gcc, art

Table 13.8 Average Clock Frequency f_{avg} (GHz) and Allocated L2 Cache Capacity C_{tot} (MB) for Each Thermal Management Scheme [527]

Benchmark Application	SRAM-3 GHz	SRAM-DVFS		Hybrid-3 GHZ	Hybrid-DVFS	
	C_{tot}	f_{avg}	C_{tot}	C_{tot}	f_{avg}	C_{tot}
HIPC-LM	0.250	2.20	0.375	1.375	2.60	7.875
LIPC-LM	0.250	3.00	0.250	1.250	3.00	1.250
HIPC-HM	0.250	2.60	0.875	1.375	2.80	7.875
LIPC-HM	0.375	2.40	0.875	5.750	2.80	7.875

To demonstrate the efficiency of hybrid-memory and the performance benefits of the cache way allocation, a set of benchmarks is evaluated on the target architecture. The features of the memory tiers are listed in Table 13.6. The processor tier is based on the Intel Core i5 technology operating at 3 GHz. The benchmark applications are based on the SPEC2000/2006 and include combinations of workloads with HIPCs and low IPCs (LIPCs) as well as high memory (HM) and low memory (LM) demands (see Table 13.7). The schemes depicted in Figs. 13.10C–E are compared to the scheme where the cores operate at 3 GHz, SRAM is only used, and most of the cache ways are power gated to satisfy temperature requirements (Fig. 13.10B). The results of this comparison are reported in Table 13.8, where SRAM-DVFS (Fig. 13.10C) improves IPS by 26.7% and hybrid-DVFS exhibits an increase in IPS of, on average, 55.3%. Moreover, the energy-delay product (EDP) is also improved by 78.2% for SRAM-DVFS. This improvement is achieved by activating additional cache ways and using a lower clock frequency. The fewer cache misses allow the execution to finish earlier than SRAM-3 GHz, yielding improved EDP. Similarly, the hybrid-DVFS yields a lower EDP as compared to the hybrid-3 GHz of, on average, 32.1%.

In all of these dynamic thermal management (DTM) techniques, thermal control is achieved by recognizing the thermal heterogeneity (e.g., cooling efficiency) among cores in a 3-D processor and carefully moving the heat generated within each region of the system. The thermal conductivity of these regions is fixed, similar to the physical design techniques discussed in Sections 13.1.1 and 13.1.2. However, as discussed in Section 12.3.1.1, the intertier interconnects can carry significant heat toward the heat sink, reducing the temperature and the thermal gradients within a 3-D IC. Consequently, these structures enhance the flow of heat to the ambient in addition to connecting circuits located on different physical tiers within the stack. Efficiently placing the available vertical connections or adding more TSVs to increase the thermal conductivity of the 3-D circuits facilitates the flow of heat towards the ambient, providing another method to control the thermal behavior of these circuits. These techniques are discussed in the following section.

13.2 THERMAL MANAGEMENT THROUGH ENHANCED THERMAL CONDUCTIVITY

Methods that facilitate the removal of heat from a 3-D stack are discussed in this section. A 3-D system is typically designed to ensure that the thermal conductivity within each tier is increased and the thermal resistance towards the heat sink is as low as possible. As integrated systems consist primarily of layers of dielectric, metal, and silicon, emphasis is placed on increasing or redistributing the volume of metal within the 3-D stack to increase the thermal conductivity of specific regions within each tier. Furthermore, as the primary direction of heat flow is vertical, the density of the (metallic) TSVs plays a significant role in lowering the thermal resistance along this path. Alternatively, liquid cooling techniques for 3-D ICs can be employed as these techniques are more efficient in mitigating thermal issues. The fluid flows between adjacent tiers, enabling faster removal of heat through each tier, avoiding highly thermal resistive paths to the heat sink.

Several techniques exist to insert thermal vias to decrease the temperature in those tiers located farthest from the heat sink. The insertion of thermal intertier vias entails an area and/or wiring overhead, which depends upon both design and technological parameters. Furthermore, techniques exist that determine the number of these thermal vias applied to diverse stages of the design flow, resulting in different efficiencies. If thermal via insertion cannot satisfy the temperature constraints, auxiliary horizontal wires are used to facilitate the lateral spreading of heat towards the TSVs, another means to lower the temperature of a 3-D stack.

The use of both vertical and horizontal interconnections to move heat within a stacked system has been considered in systems-in-package (SiP) technologies [531]. Communication among tiers in SiP passes through vertical off-chip wires connected with wide metal stripes to the I/O pad area of each tier. An example of this technology is illustrated in Fig. 13.11. Prototype structures of this

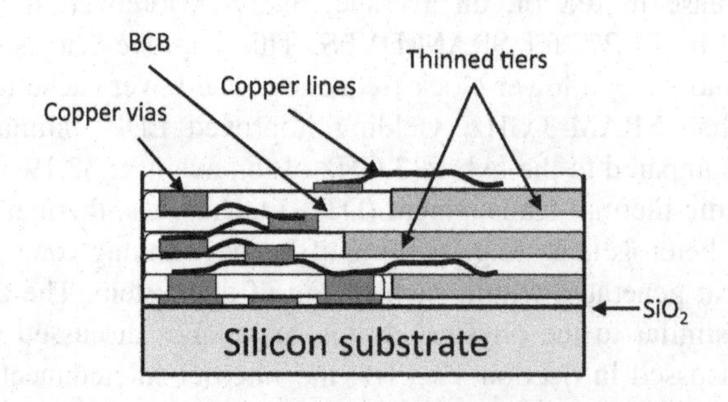

FIGURE 13.11

Cross-sectional view of a 3-D ultra-thin system with peripheral copper TSVs [531].

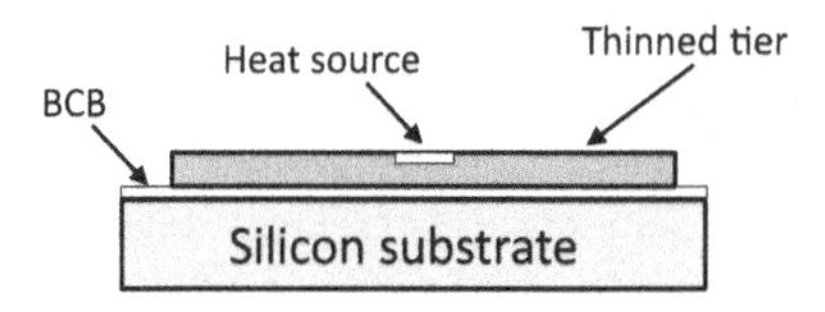

FIGURE 13.12

Cross-sectional view of a two tier structure with a spatial heat source to evaluate the effects of the metal grid/ plate and thickness of the adhesive materials on the thermal behavior of the structure (not to scale) [531].

ultra-thin tier technology demonstrate several issues and tradeoffs related to thermal management of vertically integrated systems. Benzocyclobutene (BCB), for example, can be used as an adhesive layer with a thermal conductivity of 0.18 W/m-K. This material hinders the flow of heat in the vertical direction. Vertical flow of heat can be averted if the silicon substrate for each tier other than the first tier is thinned to reduce the length of the thermal path to the heat sink. This practice is beneficial; however, extreme substrate thinning in the range of 10 μm degrades thermal flow since the volume of the silicon is too small, preventing the heat from spreading laterally within the tiers, leading to hot spots [531]. Simulations have demonstrated that for the structure shown in Fig. 13.12, decreasing the BCB layer from 3 to 2 μm lowers the vertical thermal resistance of a two tier system by 22% (where the heat source area is 2×10^5 μm^2). For the same sized heat source, the thermal resistance increases by 17% when the silicon substrate is thinned from 15 to 10 μm [531].

To further facilitate the flow of heat, an intermediate layer with embedded metal structures, such as a grid, is utilized. These structures allow the heat to spread laterally into the highly conductive vertical vias, lessening the rise in temperature. These structures, however, are only efficient for a certain physical distance from the vias, termed the effective transverse thermal transfer length,

$$L_T = \sqrt{\frac{k_{Cu} t_{Cu} t_{BCB}}{k_{BCB}}}, \tag{13.22}$$

where k_{Cu} and k_{BCB} are, respectively, the thermal conductivity of copper and BCB, and t_{Cu} and t_{BCB} are, respectively, the thickness of the copper grid/plate and BCB layer. As described in this expression, the effective thermal length does not change linearly with the transverse thermal resistance of the copper grid/plate [531].

Early investigation of thermal issues [531] demonstrated the potential and limitations of enabling faster flow of heat within 3-D systems. As fabrication processes for TSVs have evolved, the use of thermal TSVs (TTSVs) as heat conduits has gained popularity and TTSVs have been included in several physical design techniques. Alternatively, TSVs are employed as a means to shield a circuit from both rises in temperature and electrical noise generated by adjacent circuit blocks within the same tier [532]. The objective is not to facilitate the flow of heat but rather to block the rise in temperature of a circuit block from adjacent blocks dissipating significant power. The efficiency of this approach, where a metal guard ring is replaced by a ring of uniformly spaced TSVs, improves as the TSV diameter grows, as illustrated in Fig. 13.13 [532].

Although this practice is beneficial, hot spots developed within the block cannot be completely removed. This situation is more pronounced if heat is generated from the adjacent blocks in the

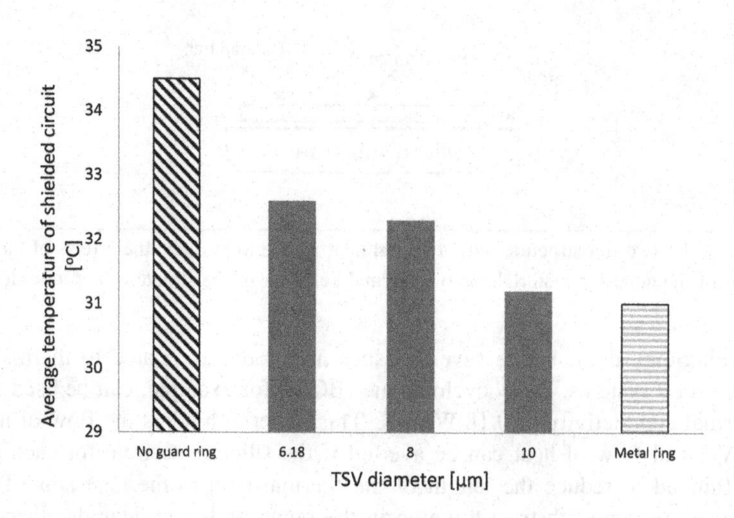

FIGURE 13.13

Average temperature of a circuit surrounded by resistors used as heating elements where different means such as a TSV or metal ring are used to thermally isolate the circuit [532].

vertical direction or if the effective thermal length limits efficient heat transfer to the periphery of the blocks. Consequently, integrating the TSV insertion process into the physical design process of a 3-D system can lead to more thermally robust and reliable solutions. Issues related to including TSVs within the physical design process include allocation or planning of the TTSVs, the system granularity to insert TSVs (for example, standard cell or block) whether TTSVs (and more broadly, temperature) is an objective or constraint, and the overhead of the TTSV on other design objectives such as performance, wirelength, and area. Several TSV planning techniques are discussed in the following subsection.

13.2.1 THERMAL VIA PLANNING UNDER TEMPERATURE OBJECTIVES

In Chapter 9, Physical Design Techniques for Three-Dimensional ICs, the available space among the cells or circuit blocks in each tier of a 3-D system is employed to allocate signal TSVs to least affect the placement of the components and the length of the interconnections. Similarly, the available space can also mitigate thermal issues by placing thermal vias within the whitespace. As thermal gradients differ across and among tiers, two different types of TTSV densities are typically computed. Vertical TTSV densities among tiers and a horizontal TTSV density for each tier are determined. Thermally driven methods for floorplanning, placement, and routing have been developed to determine these densities to achieve a broad repertoire of objectives.

The temperature or thermal gradients within a 3-D stack is interestingly treated as an objective and the TTSV density as a design constraint and *vice versa*, leading to different problem formulations. Another issue that arises is whether TTSV planning should be integrated with the floorplanning or placement steps or be applied as a postprocessing step. Integrating TTSV planning within a physical design process rather than as a postprocessing step can lead to better results, but the need for thermal analysis during each iteration can adversely affect computational time.

Extending floorplanning methods to 3-D systems, as discussed in Chapter 9, Physical Design Techniques for Three-Dimensional ICs, requires adapting the objective cost function to consider the allocation of the TTSVs. An additional term is added to the cost functions, such as in (9.9), to include the thermal objective. This added function is typically either the maximum temperature within the substrate [533] or the normalized maximum temperature with respect to the original peak temperature of the circuit [534]. Extending the cost function can, however, adversely affect both the solution time to produce a floorplan that satisfies a fixed outline and any temperature constraints [534]. To mitigate this issue, a two phase SA algorithm is applied. In this process, the SA algorithm proceeds without initially considering the thermal objective, producing a floorplan that satisfies a target outline while optimizing the wirelength and area. In the second phase, the SA algorithm includes the temperature objective. Furthermore, the SA algorithm also commences not with the lowest cost floorplan from the first phase, but rather with a higher cost floorplan to provide greater flexibility when the temperature objective is included. Both intertier and intratier moves are applied to the circuit blocks, where the sequence pair method represents the position of the blocks. A sequence pair is produced for each tier within a 3-D system. A change in the sequence pair captures a move of a block or a swap between two blocks.

TTSV assignment takes place during the second phase of the SA, where the blocks are moved to ensure that sufficient whitespace exists to accommodate any TTSV requirements. This task is driven by two guidelines, where the whitespace of the adjacent tiers overlap to allocate the TTSVs. Furthermore, the algorithm ensures greater whitespace around those blocks with higher temperatures since these blocks demand larger TTSV densities. In [534], to save computational time, TTSV assignment is typically applied during the last several iterations of the second phase. Variants of this technique, for example, single versus two phase SA based floorplanning and simultaneous versus postprocess TTSV assignment, are applied to standard benchmark circuits implemented as four tier stacks. Results on wirelength, maximum temperature, and success rates are reported in Table 13.9. The success rate indicates the percentage of valid floorplans that satisfy the fixed outline constraint out of the total number of runs. In Table 13.9, A.R. notates the aspect ratio of the outline of the floorplan, and WS denotes the area of the whitespace as a per cent of the total floorplan area.

The results listed in Table 13.9 demonstrate that the maximum temperature of a 3-D circuit is reduced when TTSVs are inserted. A greater decrease in temperature is observed when TTSV planning is an integral part of the floorplanning process. Although the 3-D stacks exhibit a higher temperature than a 2-D version of the circuits, the TTSVs offer a non-negligible decrease in temperature. The computational time of the thermal aware floorplanning method including TSV planning increases by three times as compared to the case where TTSV planning is not considered.

To mitigate the increased computational times, a thermal analysis of the circuit can be performed less frequently or, alternatively, faster (yet reasonably accurate) techniques can be used rather than solving the temperature matrix in (12.44). Unlike the matrix manipulation methods in [533], thermal analysis of 3-D circuits uses random walks [535]. Moreover, the TSV planning in [533] assumes a non-fixed outline. Relaxing the outline constraint allows for a less complex objective function. A weighted term for the maximum temperature is also utilized. Having as inputs (1) a set of circuit blocks, (2) the dimensions of these blocks along with the related power

Table 13.9 Reduction in Temperature for Thermal Driven Floorplanning With TTSV Allocation [534]

Circuit	A.R.	WS	No TSV Assignment			TSV Assignment During SA			TSV Assignment as Postprocessing		
			Succ. Rate	HPWL	Max. Temp.	Succ. Rate	HPWL	Max. Temp.	Succ. Rate	HPWL	Max. Temp.
n100	1	10	100	201,252	286.337	100	203,408	227.840	100	202,448	266.073
n100	1.5	10	100	201,928	287.679	100	206,102	259.817	100	203,505	268.282
n100	2	10	90	207,032	284.848	100	207,750	241.208	90	208,537	257.537
n100	2	15	100	207,838	241.368	100	211,837	196.453	100	209,839	222.829
n200	1	10	100	368,415	311.592	100	380,961	271.824	100	370,457	289.217
n200	1.5	10	100	376,818	292.612	100	380,787	249.314	100	378,891	266.282
n200	2	15	90	387,580	312.756	90	389,621	278.527	90	391,264	273.262
Avg.	–	–	–	1.0 ×	1.0 ×	–	1.015 ×	0.85 ×	–	1.007 ×	0.92 ×

consumption, (3) a connectivity netlist for the circuits blocks, and (4) certain parameters of the 3-D system, for example, the number of tiers and traits of the TTSVs, allows SA algorithms to be employed to optimize a cost function.

This function includes a weighted linear combination of wirelength and area in addition to temperature, similar to (13.1). Thus, SA progresses in a single phase but with faster thermal analysis to counterbalance the longer time needed to produce a high quality floorplan. As the TTSVs are allocated close to the high temperature blocks, the area of these blocks is enlarged to include the TTSVs, incurring an increase in area. To capture the area overhead of the TTSVs, a thermal via map is used where each entry describes the density of the TTSVs for a region of a tier. For each of these regions, space is created by vertically shifting overlapping blocks to fit the TTSVs in all of the tiers of the system, where a maximum TSV density v_{max} is permitted.

At every iteration of the technique, a new TTSV density is determined, described by the thermal conductivity of those regions with inserted TSVs. The updated thermal conductivity k_{new} is

$$k_{new} = k_{old} \frac{T_{cur}}{T_{target}}, \tag{13.23}$$

where k_{old} is the existing thermal conductivity (i.e., TTSV density), and T_{cur} and T_{target} are, respectively, the current and target temperature of a block. The thermal conductivities are updated in descending temperatures of the blocks as the low temperature blocks require a lower TTSV density. This density for each entry of the TTSV matrix is

$$v = \min\left(v_{max}, c\frac{k_{new} - k_{old}}{k_{via} - k_{old}} \right), \tag{13.24}$$

where k_{via} is the thermal conductivity of a single TTSV and c is a user defined constant.

The tradeoff of treating TSV planning as either an integral part of floorplanning or as a postprocessing step can again be used for runtime savings. In [534], the results are of lower quality if TTSV planning is a postprocessing step. Moreover, the use of random walks for thermal analysis allows the temperatures to be computed locally where the TTSVs are allocated to improve a hot spot. Limiting thermal analysis to a smaller area further saves computational time.

A comparison among the different approaches for TTSV planning applied to the GRSC benchmarks is listed in Table 13.10. Interestingly, if a floorplan targeting only wirelength and area is utilized as the baseline, and TTSVs are added, the resulting temperatures may be higher for certain circuits as compared to the case where a thermal driven floorplan without TTSVs is generated using the methods discussed in Section 13.1.1. This situation must not lead to the conclusion that TTSVs do not offer any benefit. Rather, omission of the temperature term in the objective function leads to highly compact floorplans that include overlaps of high temperature blocks from adjacent tiers. Adding TTSVs can cause certain blocks to shift to make space for the TTSVs. These new positions, however, may cause new overlaps that can yield higher temperatures. Instead, thermal aware floorplanning without TTSVs alleviates high temperatures by shifting blocks from each other, avoiding the creation of these hot spots. Summarizing, TTSV planning with floorplanning results in the lowest temperatures, similar to [534], yet with a non-negligible increase in area and wirelength that can reach, respectively, 47% and 22%.

This significant increase can be attributed to the large number of TTSVs at the periphery of those blocks with a high temperature. This number is large if the area of these blocks is significant,

Table 13.10 Comparison of Thermal Aware Floorplanning Approaches [533]

	Area/Wirelength Driven			Area/Wirelength Driven With TTSVs			Thermal Driven W/O TTSVs			Integrated Approach With TTSVs		
Benchmarks	**Area**	**Wirelength**	**Temp.**	**Area**	**Wirelength**	**Temp.**	**Area**	**Wirelength**	**Temp.**	**Area**	**Wirelength**	**Temp.**
n50	58,491	91,521	136.6	59,309	91,986	126.5	62,517	91,363	120.7	86,093	102,425	94.1
n50b	66,490	87,838	145.1	72,564	90,886	115.7	68,694	85,173	118.1	82,925	94,088	108.6
n50c	63,666	92,418	129.2	64,521	92,900	122.1	64,532	91,808	110.6	80,303	100,013	86.8
n100	57,664	135,970	123.6	61,431	138,729	92.3	68,480	142,521	82.0	83,311	155,972	87.7
n100b	49,950	120,431	112.9	51,095	121,297	98.0	61,490	127,801	84.8	81,893	148,806	71.7
n100c	53,040	132,142	128.8	54,135	133,800	95.4	63,745	138,324	85.7	81,596	152,045	76.4
n200	50,190	215,549	135.6	52,472	218,601	105.2	62,220	270,123	97.9	74,414	310,017	75.8
n200b	55,385	226,447	125.9	57,579	228,792	103.7	70,596	250,672	69.1	82,599	284,590	98.7
n200c	52,877	250,970	123.6	53,601	251,855	110.8	66,150	250,582	74.4	77,465	304,035	68.7
n300	81,340	313,680	186.9	83,801	316,041	146.9	117,600	334,304	51.4	136,907	468,086	56.4
Avg. ratio	1.000	1.000	1.000	1.035	1.012	0.831	1.195	1.054	0.678	1.473	1.220	0.625

prohibiting the allocation of TTSVs close to the hot spot. Thus, thermal driven placement where the TTSVs are allocated as standard cells may be more beneficial, offering greater opportunity to reduce temperature. Contrary to temperature unaware methods for TSV planning, where inserting TSVs at the standard cell level incurs unacceptable overhead, this approach may be viable when thermal issues are considered during the design process.

Allocating TTSVs at the standard cell level decreases the physical distance between a low resistivity thermal path (i.e., the TTSV) and the heat sources (i.e., the standard cells), offering an efficient method to move heat towards the ambient. TTSV insertion can satisfy a variety of design objectives (not simultaneously), such as [536]

maximum or average thermal gradient ($g_{\max}$ or g_{ave})
maximum or average temperature ($T_{\max}$ or T_{ave})
maximum or average thermal via density ($d_{th\max}$ or $d_{th\text{ave}}$)

The design objective is to identify those regions where thermal vias are most needed (the hot spots) and place thermal vias within those regions at the appropriate density. This assignment, however, is mainly restricted by two factors: the routing blockage caused by these vias and the area of the whitespace that exists within each tier. Note that although the density of the thermal vias can vary among different whitespace allocations, the thermal vias within each whitespace are uniformly distributed.

To determine the number of thermal vias for a 3-D circuit, the temperature at specific nodes within the volume of the circuit needs to be evaluated. The finite element method combined with (13.10) and (13.11) is utilized to determine the temperature of the nodes within a 3-D grid in [536], although other methods discussed in the previous chapter can also be used. An iterative approach is applied to determine the thermal conductivity of certain elements, similar to that shown in Fig. 12.8, to minimize the thermal objective. More specifically, when initializing the optimization procedure, an ideal thermal gradient is selected when a moderate number of TTSVs are assumed. Additionally, the initial temperature profile characterizes the thermal gradients within the whitespace regions where thermal vias can be added. Furthermore, the minimum thermal conductivity is assumed for these regions, coinciding with no thermal vias in these whitespace regions. The thermal conductivity of the whitespace is iteratively modified, updating the temperature of the nodes. The algorithm terminates when any further change in the thermal conductivity does not significantly improve the desired objective.

Inserting thermal vias can significantly affect the thermal conductivity of the tiers. Since these vias facilitate the transfer of heat in the z-direction, the thermal conductivity in the z-direction significantly differs from the conductivity in the x and y directions. To quantify the effects of the thermal vias in terms of a change in the thermal conductivity within a 3-D grid, the following expressions describe the change in (13.11),

$$K_z^{\text{eff}} = d_{th} K_{\text{via}} + (1 - d_{th}) K_z^{\text{tier}}, \tag{13.25}$$

$$K_x^{\text{eff}} = K_y^{\text{eff}} = \left(1 - \sqrt{d_{th}}\right) K_{\text{lateral}}^{\text{tier}} + \frac{\sqrt{d_{th}}}{\frac{1-\sqrt{d_{th}}}{K_{\text{lateral}}^{\text{tier}}} + \frac{\sqrt{d_{th}}}{K_{via}}}. \tag{13.26}$$

d_{th} is the density of the thermal vias and $K_{\text{lateral}}^{\text{tier}}$ and K_z^{tier} are the thermal conductivity of a physical tier of a 3-D system in, respectively, the horizontal and vertical directions when no thermal vias are employed. Values of 2.15 and 1.11 W/m-K are used for, respectively [536], $K_{\text{lateral}}^{\text{tier}}$ and K_{Z}^{tier} for the MIT Lincoln Laboratories 3-D process technology [307]. The thermal conductivity of the vias is denoted as K_{via}. The thermal conductivity of copper is 398 W/m-K. With these expressions, the vertical and lateral thermal conductivity or, alternatively, the density of the thermal vias in each region is iteratively determined until the target thermal gradient or temperature is reached. Note that (13.25) is identical to (12.12).

In Fig. 13.14, (13.25) and (13.26) are plotted for a variable density of thermal vias. The thermal conductivity in the z-direction is about two orders of magnitude greater than in the horizontal direction. Only the vertical thermal conductivity is, therefore, updated during the optimization process, while the thermal conductivity in the horizontal directions are determined from the thermal via density obtained from the algorithm and (13.26).

This method has been evaluated on MCNC and IBM-PLACE benchmark circuits [396], where interconnect power consumption is not considered. Some results are listed in Table 13.11. Note that the target objectives are decreased by increasing the density of the thermal vias. A considerable reduction in all of the aforementioned objectives (see, p. 491) is observed. In addition, as compared to a uniform distribution of thermal vias, fewer thermal vias are inserted to reduce the temperature. Furthermore, by analyzing the distribution of the thermal vias throughout a four tier 3-D circuit, the density of the vias is smaller in the upper tiers. This behavior is explained by noting that in a 3-D IC the thermal gradients substantially increase the temperature in the upper tiers. By placing additional thermal vias in the lower tiers, these thermal gradients are mitigated, reducing the temperature in the upper tiers.

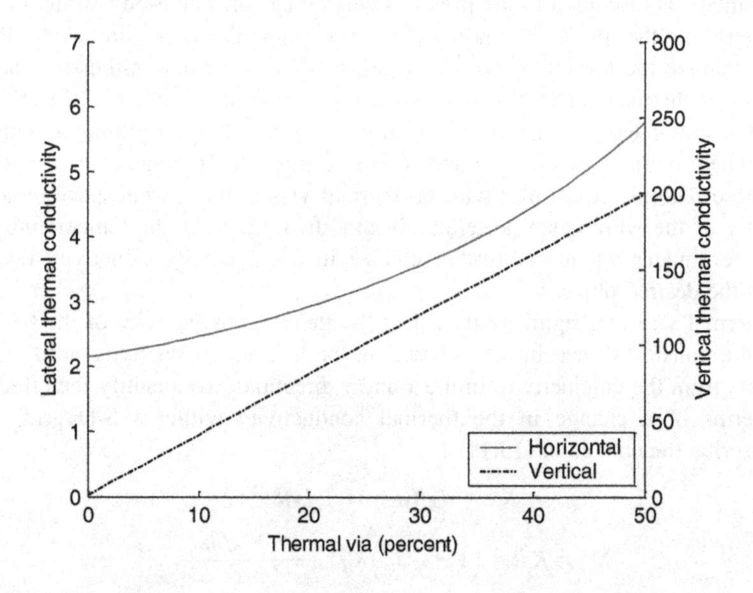

FIGURE 13.14

Thermal conductivity versus thermal via density.

Table 13.11 Average Per cent Change of Thermal Objectives as Compared to the Case With no Thermal Vias [536]

	Average Percent Change					
Objective	g_{max}	g_{ave}	T_{max}	T_{ave}	d_{thmax}	d_{thave}
g_{max}	−68.1	−60.8	−44.5	−25.9	44.9	10.2
g_{ave}	−75.7	−70.7	−51.6	−29.5	50	17.6
T_{max}	−71.1	−64.5	−47.3	−27.3	50	12.3
T_{ave}	−73.2	−67.4	−49.2	−28.3	50	14.3
d_{thmax}	−55.5	−43.3	−31.4	−19.2	25	4.2
d_{thave}	−79.2	−75.3	−54.7	−31.0	50	23.9

13.2.2 THERMAL VIA PLANNING UNDER TEMPERATURE CONSTRAINTS

Although TTSVs incur area and wirelength overhead during both the floorplan and placement steps, the density constraints for TTSVs are typically not highly restrictive. The situation is more delicate if TTSV planning is applied to a later step of the design process (e.g., routing) where there is less flexibility in moving circuits blocks within the 3-D stack. In these cases, a reasonable approach is to make the TTSV density an objective where temperature is a constraint. This approach is orthogonal to the techniques presented in the previous subsection. Consequently, thermal via planning can be described as the problem of minimizing the number of thermal vias while constraining the temperature and capacity of the thermal vias. Based on these observations, the problem of determining the minimum number of TSVs to satisfy a specific temperature constraint can be described by the nonlinear programing (NLP) problem,

$$\min \sum_{i=2}^{n} d_{thi}, \tag{13.27}$$

where d_{thi} is the TTSV density per tier for a 3-D circuit comprising n physical tiers. In addition, a number of constraints apply to (13.27), such as temperature (i.e., the temperature of the circuits cannot exceed a specified value), capacity of the TSVs in each tile, a lower bound on the number of TSVs to ensure that the wirelength of the circuit does not increase, and the heat flow equality (i.e., the incoming and outgoing heat flow for every tile should be equal).

Compact thermal models, such as the model described in Section 12.3, are preferred due to the lower computational time required to obtain the thermal profile of 3-D circuits as compared to finite element and finite difference methods [351,537–539]. Using the compact thermal model as a baseline, described in Section 12.3, a 3-D circuit is discretized into tiles. The tiles located at the same x–y coordinates but on different tiers constitute a pillar modeled by a group of serially connected resistors and heat sources (see Fig. 12.9).

Relating the thermal conductivity of the TSVs with the serially connected resistors in a single pillar, (13.27) can be rewritten as

$$\min. \sum_{k \geq 2} \left(\frac{R_{via} I_{i,j,k}}{T_{i,j,k} - T_{i,j,k-1}} - n_{TSV} \right), \tag{13.28}$$

where R_{via} is the thermal resistance of one TSV [465] and n_{TSV} is the number of TSVs that exhibit the same thermal resistance as a tile within a 3-D grid. $I_{i,j,k}$ and $T_{i,j,k}$ are, respectively, the heat flow in the z-direction and temperature of the tiles in the grid, and i, j, and k are the indices of the tiles.

Efficiently solving this NLP is a formidable task. The thermal via planning process is, therefore, divided into a two stage problem: determining the intratier TTSV density within each tier of a 3-D circuit and determining the intertier TTSV density among the tiers within the stack. Depending upon the formulation of these problems and the applied constraints, different intratier and intertier distributions of TTSVs are produced.

13.2.3 MULTI-LEVEL ROUTING

The technique of multi-level routing [539,540] is extended to 3-D ICs including thermal via planning. Multi-level routing with thermal via planning can be treated as a three stage process, illustrated in Fig. 13.15, which includes a coarsening phase, initial solution generation at the coarsest level of the grid, and subsequent refinement process until the finest level of the grid is reached. Before the coarsening phase is initiated, the routing resources, capacity of the TSVs, and power density in each tile are determined. The power density and routing resources are determined during each coarsening step. At the coarsest level (level k), an initial routing tree is generated. At this

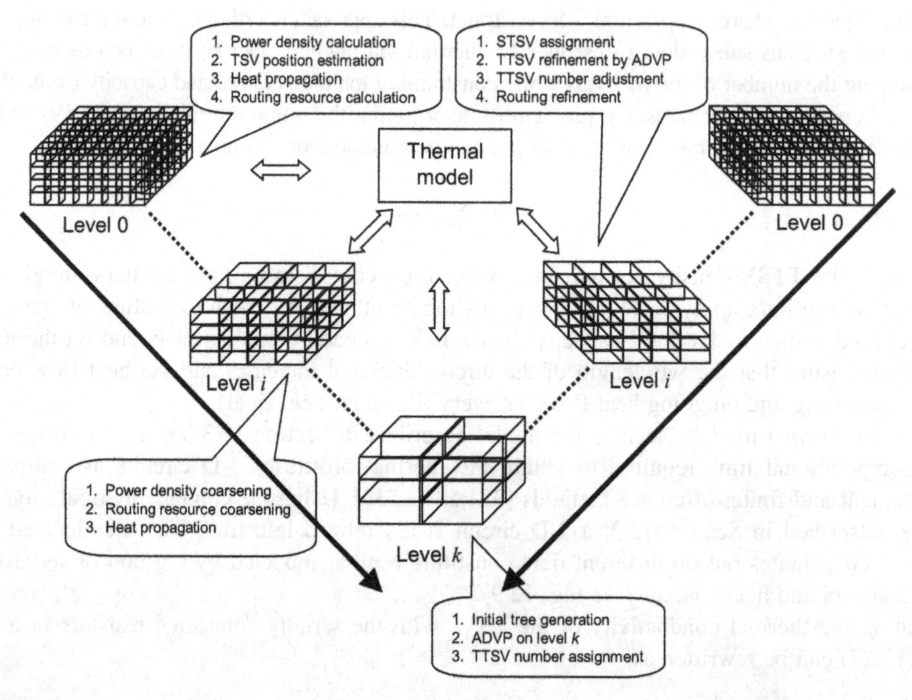

FIGURE 13.15

Multi-level routing process with thermal via planning [538].

point, the TSV planning step is invoked, assigning TSVs to each tile within a coarse grid. During the refinement phase, the TSVs are distributed to preserve the solution produced during the previous level. If the final temperature at the end of the refinement phase does not satisfy a specified temperature, the TSVs are further adjusted to achieve the target temperature.

The TSV planning step is based on the alternating direction TSV planning (ADVP) algorithm, which distributes the TSVs in alternate directions. The TSVs are distributed during the first step among the tiers of the 3-D IC and, during the second step, within each tier of the circuit. This algorithm reduces overall runtime since the thermal profile during the multi-level routing process increases the execution time. The problem of distributing the TSVs among the tiers of a 3-D system can be described as a convex problem. An analytic solution is determined if the capacity bounds for the TSVs are removed,

$$a_n : a_{n-1} : \ldots : a_3 : a_2 = \sqrt{P'_n} : \sqrt{P'_n + P'_n} : \ldots : \sqrt{\sum_{k=3}^{n} P'_k} : \sqrt{\sum_{k=2}^{n} P'_k}, \tag{13.29}$$

where a_i and P'_i are, respectively, the number of TSVs and the power density of each tile of a grid consisting of n tiers.

A corresponding analytic solution cannot be easily determined, however, for the horizontal or intratier distribution of TSVs. Alternatively, *heat propagation* and *path counting* replace the thermal profiling step. *Heat propagation* considers the propagation of the heat flow among the tiles of the grid and is determined by evaluating the different paths for transferring heat to the lower tiers of the grid. Different *heat propagation* paths are illustrated in Fig. 13.16.

The multi-level routing and ADVP algorithm are applied to MCNC benchmarks and compared to both the TSV planning approach described in [537] and to a uniform distribution of TSVs. As listed in Table 13.12, the ADVP algorithm achieves a significant decrease in the number of TSVs to maintain the same temperature, likely resulting in lower fabrication cost and less routing congestion. From Table 13.12, a considerable reduction in the number of TSVs is achieved by the ADVP algorithm without an increase in computational time.

Although the ADVP algorithm reduces the number of TTSVs, a further decrease in the TTSV density is achieved by applying a different approach to determining the local TSV densities.

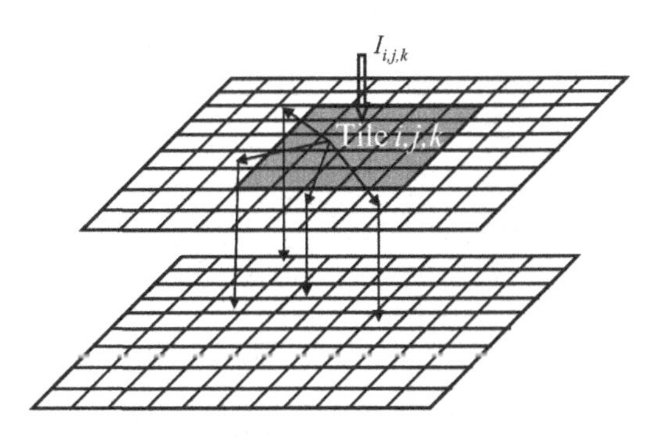

FIGURE 13.16

Heat propagation paths within a 3-D grid.

Table 13.12 Comparison of TSV Planning Techniques

	m-ADVP [538]				m-VPPT [537]				Uniform TSV			
Circuits	*T* (°C)	#TSV	Area Ratio	Runtime (s)	*T* (°C)	#TSV	Area Ratio	Runtime (s)	*T* (°C)	#TSV	Area Ratio	Runtime (s)
ami33	77.0	1282	2.5%	1.55	77.1	1801	3.5%	1.76	77.1	2315	4.5%	1.62
ami49	77.0	20,956	0.9%	13.5	77.1	43,794	1.8%	12.15	76.9	166,366	6.8%	16.17
n100	77.0	11,885	1.5%	7.66	77.0	22,211	2.8%	8.31	76.8	30,853	3.9%	7.54
n200	77.0	13,980	1.8%	12.24	77.2	18,835	2.4%	10.89	77.1	30,346	3.9%	12.21
n300	77.0	17,646	1.3%	20.44	77.1	30,161	2.2%	21.73	76.9	57,342	4.2%	22.42
Avg.		1.0	1.6%	1.0		1.68	2.6%	1.01		3.55	4.6%	1.06

Both the intertier and intratier via density problem are converted into a convex problem through several simplifications [541]. The primary assumption of this approach is that the silicon substrate of each of the upper physical tiers and the bonding material at the interface between two tiers are treated as a single material with homogeneous material properties. The thickness of this material is the summation of the thickness of the silicon layer and the bonding layer, with a thermal conductivity K_{avg} equal to the average thermal conductivity of the silicon and bonding material. The solution of these convex problems produces a different distribution of intertier and intratier TSVs [541], as compared to the distribution produced by ADVP and a uniform TSV distribution. These differences are summarized in Table 13.13, where K_{via} is the thermal conductivity of the TSVs, and S is the area of the circuit blocks within each tier of a 3-D system.

This modified thermal via planning step has been applied to MCNC and GSRC benchmark circuits and compared with the ADVP and other TSV distribution algorithms [537,538]. Some results are reported in Table 13.14, where the solution of the approximate convex problems reduces the number of thermal vias required to reach a prespecified temperature.

The improved thermal via planning step is integrated into a hierarchical floorplanning technique [541] for 3-D circuits, where the circuit blocks are initially partitioned onto the tiers of the circuit. Since no intertier moves are allowed after the partitioning step is completed, the partitioning step is crucial in determining the overall quality of the final result. This partitioning problem is treated as a sequence of knapsack problems [542], where the hottest blocks are placed on the lower tiers of a 3-D circuit to prevent steep thermal gradients. The heat generated by these blocks is transferred to the heat sink. Furthermore, overlap among these high power density blocks is avoided. The integrated thermal via planning and floorplanning approach is compared with a non-integrated

Table 13.13 Different Solutions for Distributing TSVs in 3-D ICs

Algorithm	Intertier Planning	Intratier Planning
m-VPPT [537]	$d_{thi}:d_{thj} = I_i:I_j$	$d_{thik}:d_{this} = I_{ik}:I_{is}$
m-ADVP [538]	$\frac{d_{thi}+\alpha}{d_{thj}+\alpha} = \sqrt{I_i}:\sqrt{I_j}, \quad a = \frac{K_{via}}{K_{avg}S}$	$d_{thik}:d_{this} = I_{ik}:I_{is}$
TVP [541]	$\frac{\lambda d_{thi}+1}{\lambda d_{thj}+1} = \sqrt{I_i}:\sqrt{I_j}, \quad \lambda = \frac{K_{via}}{K_{avg}} - 1$	$d_{thik}:d_{this} = \sqrt{I_{ik}}:\sqrt{I_{is}}$

Table 13.14 Comparison Among the Required Numbers of TSVs

	m-ADVP [538]		m-VPPT [537]		TVP [541]	
Circuit	T_{max}	#*T*-via	T_{max}	#*T*-via	T_{max}	#*T*-via
ami33	76.8	1109	76.7	1360	77.5	981
ami49	77.0	21,668	77.1	28,793	77.2	19,857
n100	77.2	16,731	76.9	25,205	77.0	14,236
n200	77.1	14,273	76.4	17,552	77.1	12,566
n300	76.8	19,337	76.5	25,995	76.9	17,853
Avg.		1.12		1.51		1.00

approach, where the floorplan (initially omitting the thermal objectives) is generated first, followed by thermal via planning during a postprocessing step. The integrated technique requires 16% fewer thermal vias to achieve the same temperature, with a 21% increase in computational time and an almost 3% reduction in total area.

Note that in these techniques the dependence between temperature and power is not considered, which produces a pessimistic result with higher TTSVs densities. This pessimism is due to the reduction in leakage power as the initial temperature is decreased towards the target low (or minimum) temperature. Thus, linking the power of the circuit with the decrease in temperature gained by the TTSVs can produce a target temperature with fewer TTSVs that, in turn, lowers the area and wirelength overhead while more easily satisfying the performance objectives [543].

13.2.4 THERMAL WIRE INSERTION

In addition to the benefits of the added thermal vias, thermal wires can enhance the heat transfer process. These thermal wires are horizontal wires that connect regions with different thermal via densities through TTSVs. These thermal wires are treated as routing channels wherever there are available tracks. Both TTSVs and wires can be integrated into the routing process [544]. Given a placement of cells within a 3-D IC, the technology parameters, and a temperature constraint, sensitivity analysis and linear programming methods are utilized to route a circuit. For routing purposes, a 3-D grid is imposed on a 3-D circuit, as shown in Fig. 13.17 for a two tier 3-D circuit. The thermal model of the circuit is based on a resistive network, as discussed in Section 12.3. Note that interconnect power is not considered in this thermal model [544].

Placing TTSVs and wires to decrease the circuit temperature adversely affects the available routing resources while increasing the routing congestion. Each vertical edge of the routing grid is, therefore, associated with a specific capacity of intertier vias. A similar constraint applies for the horizontal edges, which represent horizontal routing channels. The width of the routing channel is equal to the edge width of the tiles. As shown in Fig. 13.18, the thermal wire and vias affect the routing capacity of each tile.

The 3-D global routing flow is depicted in Fig. 13.19, where thermal vias and wires are inserted to achieve a target temperature under congestion and capacity constraints. A 3-D minimum Steiner

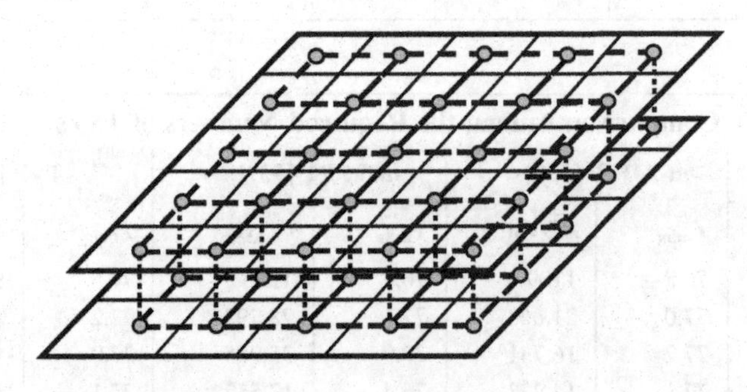

FIGURE 13.17

Routing grid for a twotier 3-D IC. Each horizontal edge of the grid is associated with a horizontal wire capacity. Each vertical edge is associated with an intertier via capacity.

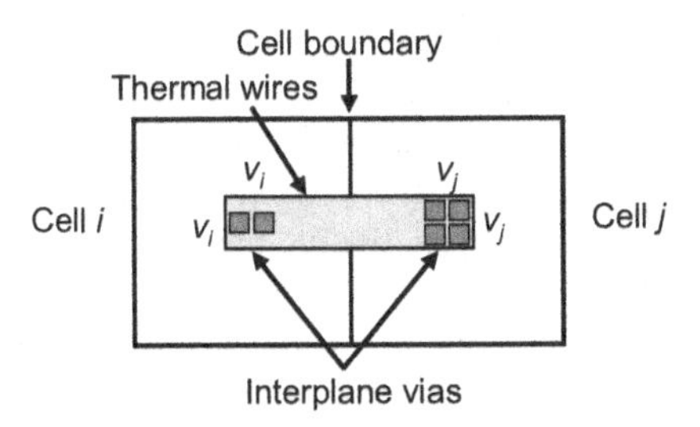

FIGURE 13.18

Effect of a thermal wire on the routing capacity of each grid cell. v_i and v_j denote the capacity of the intertier vias for, respectively, cell i and j. The horizontal cell capacity is equal to the width of the cell boundary.

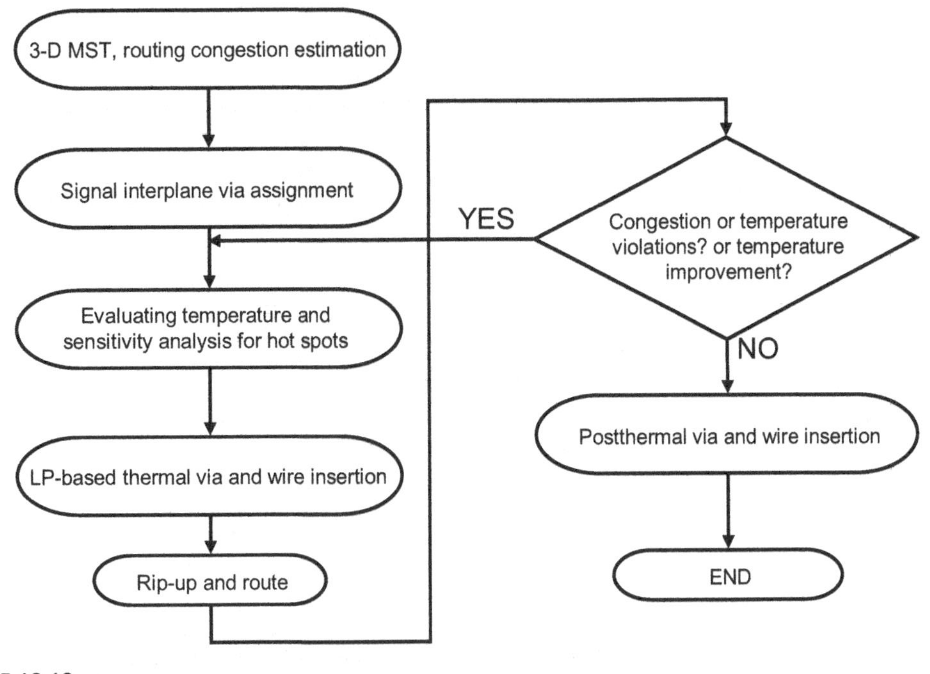

FIGURE 13.19

Flowchart of a temperature aware 3-D global routing technique [544].

tree is initially generated, followed by an intertier via assignment. A 2-D maze router produces a thermally driven route within each tier of the circuit. In the following steps [544], an iterative procedure is applied to insert the thermal vias and wires and complete the physical routing process. A sensitivity analysis is used to perform a linear programing based thermal via and wire insertion process during the first step of the iterative procedure. If an insertion violates any congestion constraints or causes overflow in the routing channels, a rip-up and route step is used to resolve these conflicts [544]. The thermal profile and sensitivity analysis steps are repeated to determine whether the target temperature has been achieved. The iterations are terminated when no remaining violations exist or the temperature of the circuit cannot be improved. The complexity of the algorithm is bounded by the complexity of the iterative procedure. The complexity of each iteration is $O(NG\log G + G^3)$ where N and G are, respectively, the number of nets in the circuit and the number of cells in the routing grid.

To determine the efficiency and overhead of inserting thermal wires, a number of benchmark circuits have been routed with several routing approaches that employ different thermal management methods. The traits of the benchmark circuits are listed in Table 13.15 and are based on the MCNC and IBM placement benchmark suites [506]. All of the benchmark circuits are composed of a small number of circuit blocks, and the largest routing grid includes 1,600 unit grid cells.

The thermal via and wire insertion process integrated with the routing process (TA) is compared to routing with thermal via and wire insertion as a postprocessing step after global routing (P), and simultaneous routing and thermal via insertion but without the use of thermal wires (V). Some results describing the peak temperature, wirelength, and computational time based on these methods are reported in Table 13.16. Note the excessively high temperatures when no thermal management is employed. This situation is due to the assumption of 0°C as the ambient temperature in [544], while the target maximum temperature is 80°C. As shown by these results, inserting auxiliary wires is another means to further lower temperature within a 3-D system with a small increase in wirelength.

Table 13.15 Benchmark Circuits and Routing Grid Size [544]

Circuit	# of Circuit Blocks	# of Nets	Grid Size
biomed	6417	5743	28 × 28
industry2	12,149	12,696	31 × 31
industry3	15,059	21,939	35 × 35
ibm06	32,185	33,521	40 × 40

Table 13.16 Comparison of Various Metrics Among Different 3-D Global Routing Approaches [544]

		Peak Temperature			Wirelength ($\times 10^5$)			Computational Time (s)		
Circuit	T_{init} (°C)	TA	P	V	TA	P	V	TA	P	V
biomed	237.1	81.9	105.6	115.3	1.82	1.77	1.78	255	137	188
industry2	207.5	82.4	106.6	116.3	6.04	5.92	6.01	855	473	591
industry3	202.0	79.2	99.1	112.5	9.85	9.75	9.71	1807	1405	1686
ibm06	236.4	81.2	99.2	131.4	18.08	18.19	18.12	2956	1585	2274

13.3 HYBRID METHODOLOGIES FOR THERMAL MANAGEMENT

Methods that either redistribute the power densities within a 3-D system (either dynamically or during design time) or increase the thermal conductivity of the tier, allowing heat to be removed faster, are presented in the previous two sections. Cooling methods, however, are not considered. More complete thermal management solutions based on external cooling strategies are the focus of this section. These techniques consider the features of the cooling mechanism and combine thermal management both at runtime and during design time as an efficient manner to control thermal issues. Methods to best combine liquid cooling with dynamic thermal management and thermal aware design to satisfy temperature constraints while limiting the effects of high temperatures on system reliability are discussed in this section.

Liquid cooling systems and microchannels etched in the substrate between tiers are discussed in Chapter 12, Thermal Modeling and Analysis. Those parameters affect the heat removal capacity of liquid cooling, including the type of coolant, flow rate of the fluid, and energy dissipated by the cooling mechanism. Combining this mechanism with dynamic thermal management, for example, task migration, task scheduling, and DVFS, results in a large solution space that can be further increased if thermal aware floorplanning is added as another thermal management option. All of these methods make thermal optimization challenging.

Another important aspect of hybrid thermal management is the response time to address thermal events (i.e., exceeding a temperature or thermal gradient threshold). Liquid cooling includes mechanical parts and, therefore, changes in the cooling system, for example, the flow rate, require a relatively long time as compared to the thermal constant of the system and the response time of other dynamic thermal management techniques. To illustrate the interplay among the different thermal management approaches, a 3-D MPSoC is discussed where liquid cooling, workload scheduling, DVFS, and thermal aware floorplanning are employed to address the peak temperature and thermal gradients.

Two and four tier MPSoCs are considered where, respectively, one and two tiers include an UltaSPARC T1 processor [545] consisting of eight cores and caches. The other tiers contain the L2 cache and part of the crossbar switch connecting the cores and caches within the MPSoC. A floorplan of the tiers are depicted in Fig. 13.20A. A 90 nm CMOS node is assumed as the manufacturing technology for these 3-D systems. Cavities are etched into the substrate of each tier, forming microchannels with a cross-section of 100 μm × 100 μm and a pitch of 100 μm. Intertier interconnects with a uniform distribution of TSVs, a TSV diameter of 50 μm, and a TSV pitch of 100 μm are

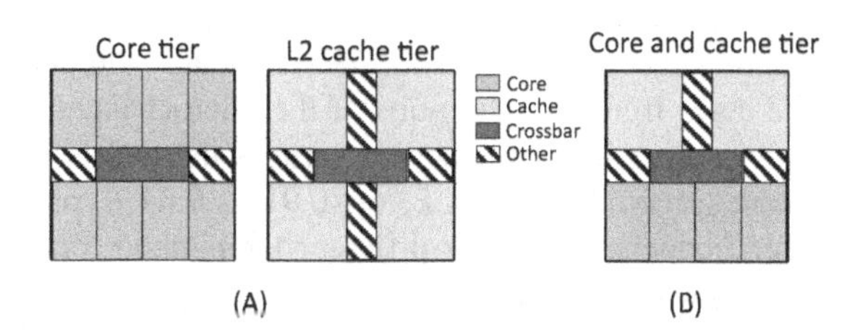

FIGURE 13.20

Floorplan of a 3-D MPSoC, (A) cores and L2 caches are placed in separate tiers, and (B) cores and caches share the same tier [546].

assumed. A uniform distribution of TSVs is a restrictive assumption, where a nonuniform distribution of TSVs is more common and has a lower impact on wirelength. Based on these traits and the area of the tiers within the MPSoC, 66 microchannels are used within each tier [546]. Furthermore, convective single phase cooling is assumed, where the coolant is water and the cooling system is supplied by a pump and regulated by valves. The flow rate within the microchannels ranges from 0.01 to 0.0323 L/min. A single pump supplies 60 3-D MPSoCs, and the power can reach 180 watts, demonstrating that the energy requirements of the cooling system are not negligible. Exploiting other less power demanding techniques can be combined with liquid cooling to lower the power budget of the cooling mechanism in addition to the power consumed by the MPSoC [546].

Liquid cooling removes heat faster from those cores located close to the inlet port since the temperature of the coolant is lower at this point, allowing more heat to be removed. As the temperature of the coolant increases towards the outlet port at a fixed rate along the flow direction (i.e., the flow is thermally developed), less heat is removed and, therefore, those cores located close to the outlet port exhibit higher temperatures. Varying the flow rate within the microchannels improves the heat removal properties, although a similar thermal behavior across the tier is observed. The higher flow rate allows more heat flux to be absorbed as the fluid more quickly exits the microchannels, leading to lower temperatures at the outlet port of the microchannel. The difference in temperature of the cores for different flow rates is significant and can reach 40°C for those cores located close to the outlet port. Alternatively, this temperature variation is much lower for the cores located at the inlet port of the microchannels. The temperature of the fluid is low at the inlet port and does not significantly vary with the flow rate, exhibiting similar heat removal capabilities over the range of flow rates.

High flow rates can decrease the core temperature below typical operating conditions while simultaneously dissipating high power. This observation has motivated the introduction of other less power demanding (and faster) methods to lower the temperature within 3-D MPSoCs. In addition, liquid cooling is susceptible to spatial thermal gradients due to the thermal development of the fluid in the microchannel. Task scheduling and migration are, therefore, combined with liquid cooling since the temperature of those cores closer to the inlet port exhibit a lower temperature. Employing different utilization rates, e.g., 90%, 70%, 40%, and 10%, each core is assigned a utilization ratio to satisfy the temperature limits [546]. Among the different core workloads, the smallest difference in temperature among the cores is observed when a "flow descending" allocation of tasks is performed. This allocation implies that higher utilization rates are assigned to those cores close to the inlet port of the fluid. This assignment avoids the creation of hot spots while also reducing thermal gradients across the tiers. To consider thermal gradients due to liquid cooling, the task utilization method employs a queue of tasks adjusted to include the position of the cores with respect to the distance of the cores from the inlet ports of the microchannel [546].

DVFS is also explored as another DTM method to address thermal issues. Two different voltage/frequency pairs are assumed, the nominal (V, F) of $(0.91\ V, 0.84\ F)$ and $(0.83\ V, 0.67\ F)$. The cores switch between voltage/frequency pairs for different temperature ranges, where a core temperature below the lower bound leads to an increase in frequency, while a slower voltage/frequency pair is selected if the upper bound is exceeded. The temperature ranges are (73, 77) °C, (78, 80) °C, and (82, 85) °C. The application of DVFS is affected by the thermal gradients caused by the liquid cooling. Thus, those cores located closer to the outlet port switch more frequently between voltage/frequency pairs as compared to the cores close to the inlet port.

These results demonstrate that the different thermal management techniques are interdependent and should not be applied separately. Since each of the techniques is characterized by a different overhead type and magnitude, each technique is applied only as long as required.

A complex system is often controlled by both cyber physical [547] and rule-based fuzzy controllers [546]. A set of rules are listed in Table 13.7, where three factors, the distance (*D*) from the inlet port, temperature (*T*), and utilization (*U*) of each core, can be categorized as low (L), medium (M), and high (H). The rules indicate the fuzzy controller decisions based upon the voltage/frequency pair of each core and the flow rate. The fuzzy controller can be implemented as a software routine, which is invoked each time a thermal map of the system is obtained through thermal sensors. In [546], this sampling occurs every 100 ms. Additionally, the core utilizations are computed within the same interval based on the workload of the queue for each core. With this information and the set of rules listed in Table 13.17, the fuzzy controller selects the appropriate voltage/frequency pair for each core and the flow rate of the fluid.

This thermal management framework is compared with several other thermal management policies. A case where standard air cooled packaging is used for the stack is compared to the liquid cooling solution with a variable flow rate. The air cooled system leads to prohibitive temperatures for a four tier system, ranging up to 178°C. The thermal gradients, however, do not exceed 10°C across tiers for a four tier system. The peak temperatures are much lower for the liquid cooled system ranging between 56°C to 85°C for different liquid flow policies. If only liquid cooling is employed for the bottom tier, the resulting thermal gradients within the stack are higher, reaching 15°C for a two tier system. If liquid cooling is applied to each tier, however, the thermal gradients are less than 5°C for two tiers.

Table 13.17 A Set of Rules for the Fuzzy Controller (X is a "don't care") [546]

IF			THEN	
D Is	**AND T Is**	**AND U Is**	**V/F Is**	**AND Flow Rate Is**
L	X	X	H	L
M	L	X	H	L
M	M	L	L	L
M	M	M	M	M
M	M	H	M	M
M	H	L	L	L
M	H	M	M	M
M	H	H	M	H
H	L	X	H	L
H	M	L	L	L
H	M	M	M	L
H	M	H	H	M
H	H	L	L	M
H	H	M	L	H
H	H	H	M	H

Consequently, liquid cooling offers higher efficiency in reducing temperatures if combined with position aware workload scheduling and DVFS. High flow rates lower the temperature, yet the power of the pump and other components of the system lessen the advantages of this approach. To avoid this situation, lower flow rates are selected to produce thermally efficient 3-D stacks. Small deviations from the target temperature can be addressed by DVFS and/or task migration. Both methods require lower overhead and shorter response time. A fuzzy controller affects the flow and utilization rates, achieving a 63% decrease in the energy of the cooling system when the maximum flow rate is used.

A last noteworthy point is that the inherent traits of liquid cooling can be considered in the design flow to further improve the thermal behavior of 3-D systems. As an example, the floorplan shown in Fig. 13.20B is used when a core is replaced with a portion of an L2 cache. This mixed floorplan reduces the heat absorbed at the outlet port of the microchannels. Placing the low power caches in these locations balances the temperature across the tiers, employing fewer thermal techniques, each of which requires a performance and/or power overhead.

13.4 SUMMARY

Thermal management techniques for 3-D ICs both during design time and online are discussed in this chapter. The important concepts of this chapter are summarized as follows:

- Thermal management methodologies can be roughly distinguished into: (1) approaches which control the power densities within the volume of the 3-D systems, and (2) those techniques that target an increase in the thermal conductivity of a 3-D stack.
- Physical design techniques, such as floorplanning, placement, and routing, use a thermal objective to manage thermal issues in 3-D ICs. These techniques decrease thermal gradients and temperatures in 3-D circuits by redistributing the blocks among and within the tiers of a 3-D circuit.
- Thermal management during design time requires frequent thermal analysis of the 3-D stack. The time required for this task depends on the thermal model and the analysis method. More accurate thermal models achieve improved results but often require unacceptably high computational time. Alternatively, simpler models reduce runtime but are often inaccurate.
- A compromise between the accuracy and computational requirements of the thermal model is necessary.
- The heat in 3-D ICs primarily flows vertically rather than laterally. Consequently, the thermal resistance in the horizontal direction can be treated as a *constant* to improve the computational time for thermal profiling.
- The third dimension can greatly affect the computational time by significantly increasing the solution space of the thermal design techniques.
- A multitude of optimization methods have been employed to design a 3-D circuit with respect to temperature and thermal gradient constraints, including SA, force directed methods, genetic algorithms, and convex, linear, and non-linear programming.
- In SA based techniques, multi-phase techniques are faster if only temperature is considered during certain stages.

- Dynamic thermal management has also been applied to 3-D systems. These methods consider the strong vertical thermal coupling to adapt established techniques employed for 2-D circuits, such as task scheduling and migration, DVFS, clock throttling, and power gating.
- Multi-level dynamic thermal management operating at both global (3-D stack) and local (per core) levels offer greater flexibility to control temperature at lower overhead.
- Formal control methods can be applied but often low overhead heuristics integrated as software routines in the operating system kernel are preferred.
- Dynamic thermal management techniques assign similar operating frequencies and tasks to cores within a tier but disparate frequencies and tasks for cores in different physical tiers.
- Intertier vias that do not carry an electrical signal are called thermal or dummy vias (TTSVs). These thermal vias are utilized in 3-D circuits to transfer heat to the ambient.
- Techniques for TTSV planning should consider several issues such as the system granularity at which TSV insertion is applied—for example, standard cell or block—whether TTSVs (and more broadly temperature) is an objective or a constraint, and the overhead of the thermal TSV, such as performance, wirelength, and area.
- TTSVs create routing obstacles. These vias should, therefore, be judiciously inserted and the allocation performed within the available whitespace within the tiers.
- Thermal via planning describes the problem of either minimizing the number of thermal vias while satisfying temperature and intertier via capacity constraints or optimizing a multiobjective function including temperature under TTSV density constraint(s). Thermal via planning techniques can significantly decrease the temperature and thermal gradients within a 3-D circuit.
- Thermal wires in the horizontal direction are equivalent to thermal vias and can lower thermal gradients in 3-D circuits by facilitating the flow of heat within tiers.
- Combining active cooling with dynamic and physical design thermal management techniques offers the greatest temperature reduction in 3-D systems.
- Active (liquid) cooling can drastically decrease the temperature throughout a 3-D stack but suffers from a long response time, increased power requirements for the cooling mechanism, and spatial thermal gradients.
- Augmenting thermal management with task scheduling and/or migration alleviates some of these issues.
- DVFS supports finer temperature control, leading to superior solutions as compared to individually applying each of these techniques.
- Adapting the floorplan of a 3-D system to the features of the cooling system provides another opportunity to further regulate temperature and thermal gradients, offering a holistic thermal management methodology.

CHAPTER

CASE STUDY: THERMAL COUPLING IN 3-D INTEGRATED CIRCUITS

14

CHAPTER OUTLINE

The importance of thermal issues in 3-D ICs is discussed in previous chapters. Prior work, some of which is also discussed in Chapter 12, Thermal Modeling and Analysis, has focused on simulation [548] and modeling [459, 552,553] of hot spot formation and propagation within 3-D ICs. Tools such as 3D-ICE [551] and a 3-D extension to HotSpot [552,553] to model thermal profiles of 3-D circuits have been developed and provide a visual interpretation of hot spot formation based on the power requirements of each device plane. Models and simulations have been extended to block level floorplanning including global wire congestion, permitting the location of the highly active blocks within a 3-D IC to be adjusted based on the location of the thermal hot spots [351,554–556]. Additional mitigation techniques, including the use of passive techniques such as thermal through silicon vias (TTSVs) [489] and active techniques such as microchannel or micro-fluidic cooling [455,477], have been proposed to address heat removal in 3-D ICs.

Although extensive theoretical work has provided an understanding of heat flow in 3-D ICs, there has been limited experimental work quantifying the flow of heat between device planes. Meindl et al. experimentally characterized the effects of microfluidic cooling techniques on both 2-D and 3-D circuits [557–559]. Additional experimental results have characterized microfluidic

Three-Dimensional Integrated Circuit Design. DOI: http://dx.doi.org/10.1016/B978-0-12-410501-0.00014-9

cooling methods [455]. Experiments characterizing an intertier cooling system for a vertically stacked DRAM/multiprocessor system-on-chip have been described [560]. Numerical and experimental characterization of thermal hot spots within a packaged DRAM-on-logic 3-D IC has also been described [561]. Similar to this work, intertier thermal propagation is investigated. The primary purpose of the results described in this chapter, however, is to characterize intra and intertier thermal coupling to improve design methodologies and techniques for stacked ICs. The experimental results discussed herein provide insight into the effects of the location of the heat source and active cooling on thermal gradients within 3-D ICs.

A test circuit has been fabricated by Tezzaron Semiconductor in a 130nm CMOS technology with 1.2 μm diameter TSVs. A face-to-face bonding technique to vertically stack the two logic device tiers is used. This test circuit is designed to also evaluate the effects of inter and intratier thermal resistance on hot spot formation.

This chapter is composed of the following sections. The thermal propagation test circuit is described in Section 14.1. Experimental characterization of thermal coupling for a set of test configurations is presented in Section 14.2. A discussion of the experimental results including the effects of hot spot formation and mitigation techniques is provided in Section 14.3. A comparison of experimental results with simulations is provided in Section 14.4. Some conclusions are offered in Section 14.5.

14.1 THERMAL PROPAGATION TEST CIRCUIT

A 3-D test circuit comprised of two silicon layers with back metal on one of the device planes has been fabricated to experimentally analyze horizontal and vertical thermal coupling in 3-D ICs. The experimental results are useful for evaluating thermal propagation paths within TSV-based 3-D structures. An illustration of this test circuit is provided in Fig. 14.1.

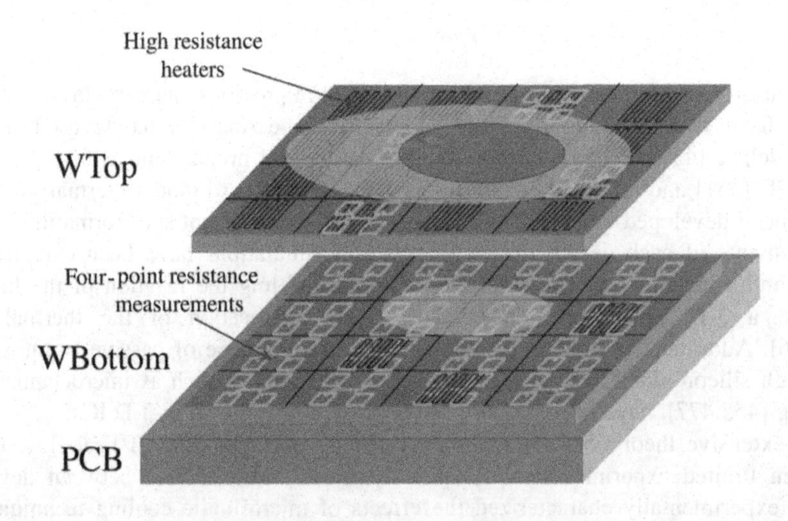

FIGURE 14.1

Heat propagation from one tier spreading into a second stacked tier.

14.1.1 3-D IC FABRICATION TECHNOLOGY

Each device plane is individually processed in a 130 nm CMOS technology, provided by Chartered Semiconductor, before 3-D bonding, TSV fabrication, and wafer thinning by Tezzaron Semiconductor. The Chartered fabrication process includes low power 1.5 volt and 2.5 volt transistors, six metal layers per device plane, a single polysilicon layer, dual gates for the 2.5 volt transistors, and low and nominal threshold voltage devices [562]. The sixth metal level on each die is allocated for face-to-face bonding to vertically stack the two logic device planes.

14.1.2 3-D TEST CIRCUIT

The test structures are designed to investigate thermal coupling between adjacent tiers and include both resistive thermal sources and thermal sensors. The thermal sensors use four point voltage measurements. Each thermal source is paired with a resistive thermal sensor on an adjacent metal level, and these pairs are distributed throughout each tier within a two tier 3-D stack. The thermal sources are heater resistors with a maximum applied voltage of 28 volts, producing a maximum current of 1.5 amperes. The heaters are 200 μm × 210 μm, similar to the dimensions of the heaters in [561], and are in metal 2. Within this area, the total length of the heater is 2,120 μm and the width is 6 μm. Metal 2 is 0.42 μm thick and has a nominal sheet resistance of 0.053 Ω/□. The resistance of the heaters is therefore 18.7 Ω. Joule heating through the resistive heater is adjusted by controlling the current flow, and therefore, the I^2R power consumed within the 200 μm × 210 μm area. The thermal sensors, with dimensions of 200 μm × 86 μm, are placed directly above the heaters in metal 3. The total length of the sensors is 4,442 μm. The resistance of the sensors is 117.7 Ω for a width of 2 μm, metal 3 thickness of 0.42 μm, and nominal sheet resistance of 0.053 Ω/□. The temperature sensor provides a calibrated four point measurement tested at a low current to avoid joule heating. The resistive heater, thermal sensor, and combined heater and sensor are shown in Fig. 14.2.

Similar resistive heaters and thermal sensors are included in the aluminum back side metal layer, as shown in Figs. 14.3A and B. The differences between the heaters and sensors located in the logic tiers and the back side metal include: (1) the heaters and sensors are not vertically stacked on adjacent metal layers as in the logic tiers, as there is only a single back side metal layer, (2) the back side metal layer is almost three times the thickness of either metal 2 or 3 (1.2 μm as compared to 0.42 μm), and (3) due to the greater width and thickness, larger currents pass through the back side metal. The heaters and sensors on the back side metal support thermal coupling through the thinned silicon to the thermal sensors on the internal logic layer, providing enhanced understanding of the effects of thermal spreading through the silicon to the neighboring device planes.

The location of the on-chip thermal sensors and resistive heaters with respect to the backside sensors is shown in Fig. 14.4. A microphotograph of the 5 mm × 5 mm 3-D IC depicts two locations from which thermal data are collected. The center-to-center distance between the back metal sensors is 1.1 mm, while the on-chip sensors are 1.3 mm apart. A cross cut view of the complete 3-D IC stack is shown in Fig. 14.5. Each device plane, labeled as WTop and WBottom in the figure, includes a thermal sensor on metal 3 and a resistive heater on metal 2. The backside metal heaters and sensors are at the top of the stack, as shown in Fig. 14.5. In addition, the thickness of both the WBottom silicon and the active portion of the test structure is included in the illustration shown

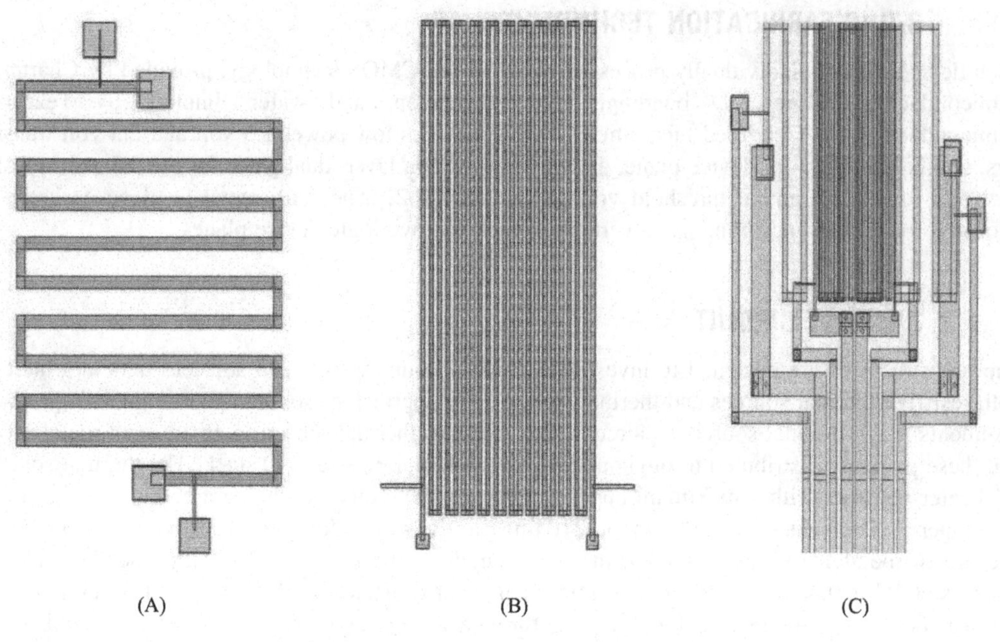

FIGURE 14.2

Physical layout, (A) on-chip resistive heater, (B) on-chip four-point resistive thermal sensor, and (C) overlay of the resistive heater and resistive thermal sensor.

in Fig. 14.5, indicating a significantly smaller thermal resistive path to the top of the 3-D stack than to the board below.

Through silicon vias with 1.25 μm diameter are placed 100 μm apart across both WTop and WBottom. The effects of the TSVs on the thermal propagation process are minimal. The focus of the test circuit is to investigate horizontal and vertical thermal coupling in stacked ICs. In addition, more recent work has indicated that placing an excessive number of TSVs, particularly TSVs with small diameters where a larger portion of the area is occupied by insulating material (i.e., SiO_2), actually increases the on-chip temperature as the insulating material has poor heat conducting properties as compared to the silicon being replaced [563]. The more critical parameter influencing the heat dissipation characteristics is the physical nature of the silicon substrate [564].

14.2 SETUP AND EXPERIMENTS

The on-chip and back metal thermal sensors require calibration prior to experimental analysis of thermal coupling between logic tiers. Calibration is performed by setting the die temperature through a thermal chuck and measuring the resistance of the sensors at each temperature ranging from room temperature (27°C) to 120°C. Resistance data for calibrating the on-chip and back metal fourpoint thermal sensors are provided in Table 14.1. A temperature controlled hot plate is used to

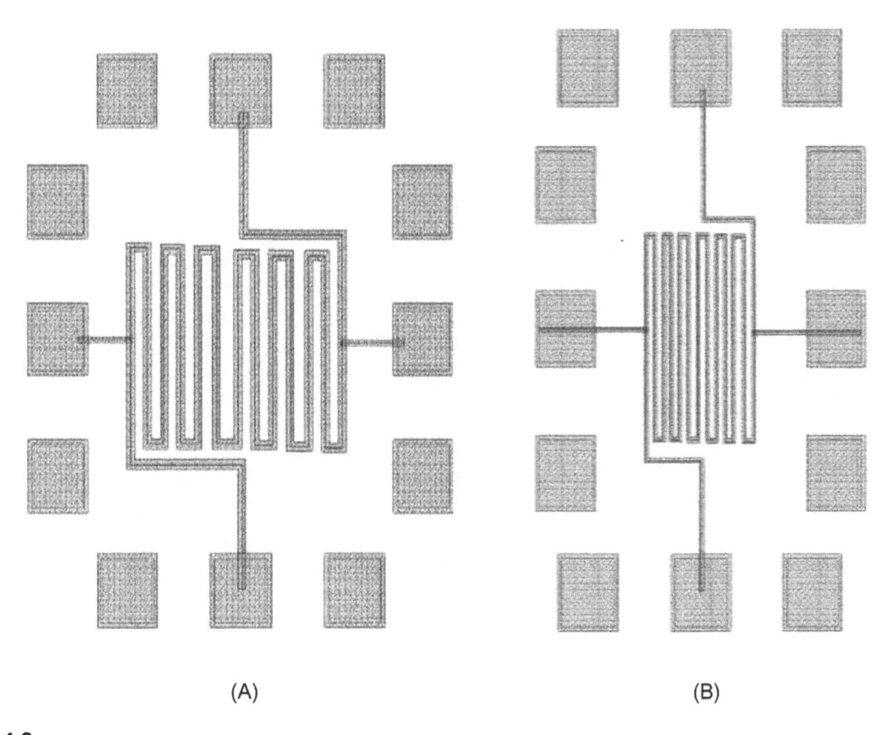

FIGURE 14.3

Physical layout, (A) back metal resistive heater and (B) back metal four-point resistive thermal sensor.

measure resistance as a function of temperature for the thermal sensors on WTop, WBottom, and the back metal at sites 1 and 2. The data are fitted to a second order polynomial expression to determine the temperature from the resistance measurements.

The resistance as a function of temperature for each calibrated sensor is shown in Fig. 14.6. All of the sensors exhibit a linear response to temperature. The on-chip sensors, shown in Fig. 14.6A, produce consistent results on the same logic layer. A difference in resistance does, however, exist between the top and bottom logic layers, as shown in Fig. 14.6A, due to process variation between the two tiers. Within a specific die, the sensors produce consistent results, demonstrating that the thermal sensors can be calibrated from a single sensor. This behavior, however, is not the case with the back metal sensors, as shown in Fig. 14.6B. The difference in resistance between the two sites on the back metal reveals greater process variations on the back metal layer than the on-chip metal layers, requiring each thermal sensor to be individually calibrated and normalized at room temperature.

The experimental setup includes the use of an HP 4145B Semiconductor Parameter Analyzer and an HP 16058-60003 Personality Board. In addition, Interactive Characterization Software from Metric Technology Inc. is used to determine the settings for the parametric analyzer. A Keithley 2420 SourceMeter is used as a current source to supply the on-chip and back metal heaters with

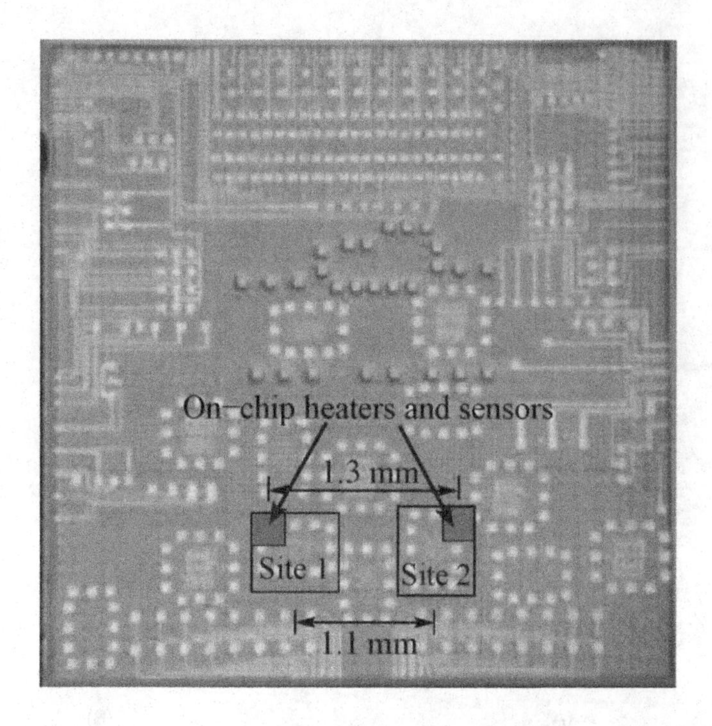

FIGURE 14.4

Microphotograph of the test circuit depicting the back metal pattern with an overlay indicating the location of the on-chip thermal test sites.

0 to 110 mA of current (130 mA for the back metal). The parametric analyzer sweeps the voltage on the sensor from 0.1 volts to 0.6 volts in 0.01 volt increments, permitting the average resistance to be determined across this voltage range. Measurements are made at each site and each sensor (a total of six data locations) for each current level. In addition, there are six different heater locations, as shown in Fig. 14.5. The resistance measurements are converted to temperature based on the results depicted in Fig. 14.6.

The resistive heaters are controlled to provide different test conditions to emulate common on-chip devices and to investigate the effects on the temperature profile and thermal conduits within the 3-D stack. A current is individually supplied to the heaters on WBottom and WTop. These results are shown in Figs. 14.7A and B. The power density of the resistive heaters ranges up to 17.4 watt/mm^2 for the heaters placed on metal 2 of each die, and up to 24.3 watt/mm^2 for the heaters placed on the back metal (determined from the currents, resistances, and effective areas). The power density of common on-chip devices is listed in Table 14.2. Note that the power density generated in these experiments is consistent with practical integrated circuits [565–567].

The results from different test conditions are also depicted in Fig. 14.7. The effects of the metal heat spreaders (metal interconnect) on the in-tier thermal profile are examined by removing the metal heat spreaders that surround the sensors and heaters located at WTop site 2. The results from this experiment are shown in Fig. 14.7C and are compared to the experimental results from WTop

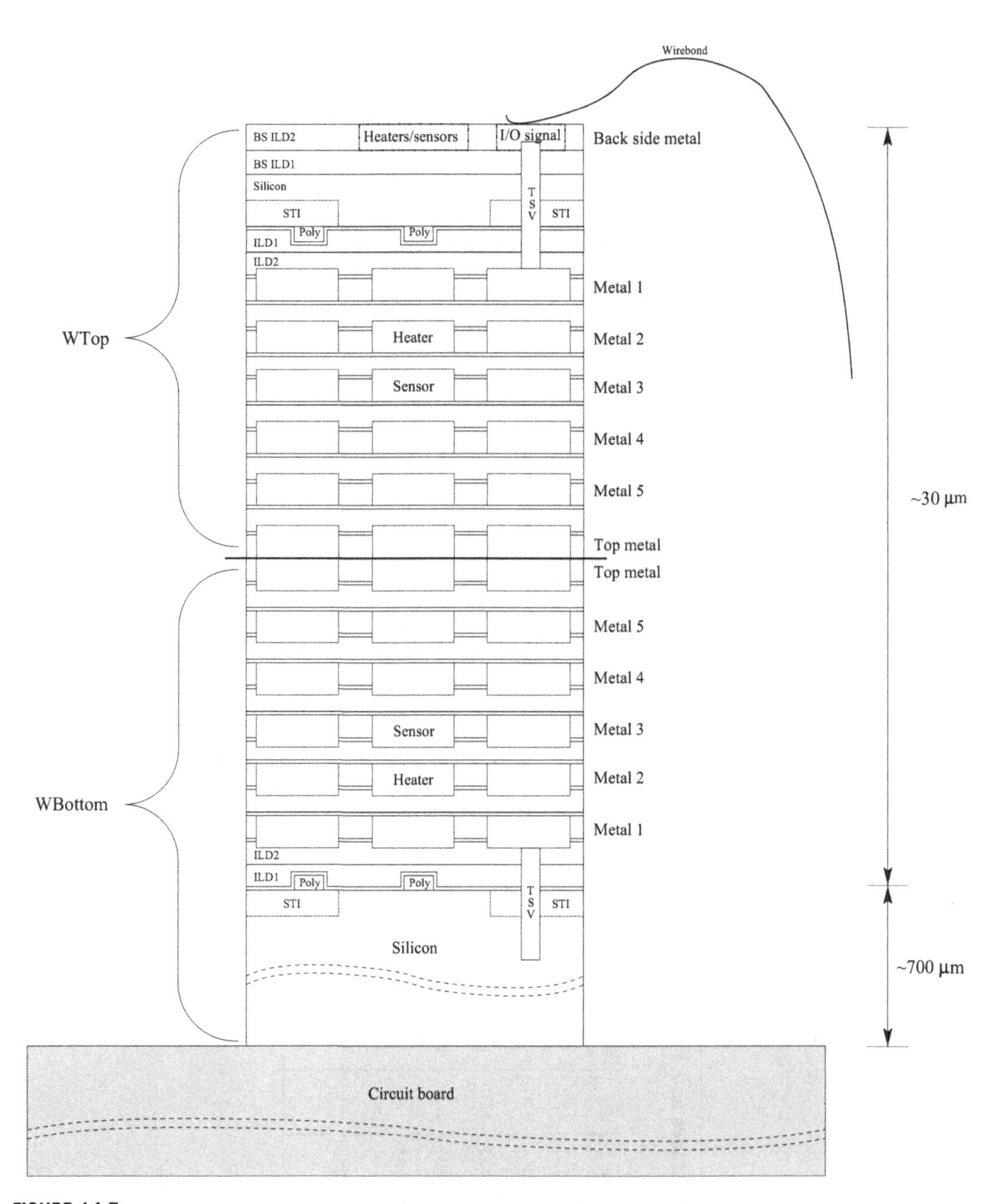

FIGURE 14.5

Placement of thermal heaters and sensors, respectively, in metals 2 and 3 in the two stacked device planes. The placement of the back metal heaters and sensors is also shown.

Table 14.1 Resistance as a Function of Temperature for Calibrating the On-Chip and Back Metal Thermal Sensors

Heater Location	Temperature (°C)																			
	27	30	35	40	45	50	55	60	65	70	75	80	85	90	95	100	105	110	115	120
WTop, Site 1	123.1	125.0	127.2	130.0	132.5	135.4	137.3	139.7	142.0	144.7	147.0	149.3	151.7	154.0	156.4	158.3	160.5	163.3	165.3	167.0
WTop, Site 2	123.3	124.4	127.0	130.4	133.3	135.4	137.7	140.1	142.9	145.0	147.5	150.0	152.0	154.7	156.9	158.9	161.5	164.8	166.8	168.7
WBottom, Site 1	119.5	120.5	123.0	125.5	127.9	130.3	132.7	135.1	137.5	139.9	142.0	144.7	147.0	149.0	151.0	152.8	155.3	157.5	159.8	162.0
WBottom, Site 2	119.7	120.4	123.1	125.1	127.4	129.9	132.3	134.4	136.5	139.2	141.8	143.9	146.3	148.3	150.3	152.4	154.7	156.8	158.8	161.0
Back Metal, Site 1	24.25	24.55	25.10	25.55	26.10	26.60	27.05	27.65	28.15	28.70	29.15	29.75	30.15	30.60	31.05	31.55	32.00	32.55	32.95	33.45
Back Metal, Site 2	20.15	20.60	21.15	21.60	21.90	22.30	22.80	23.20	23.55	24.05	24.50	24.85	25.25	25.65	26.15	26.55	26.95	27.30	27.60	28.00

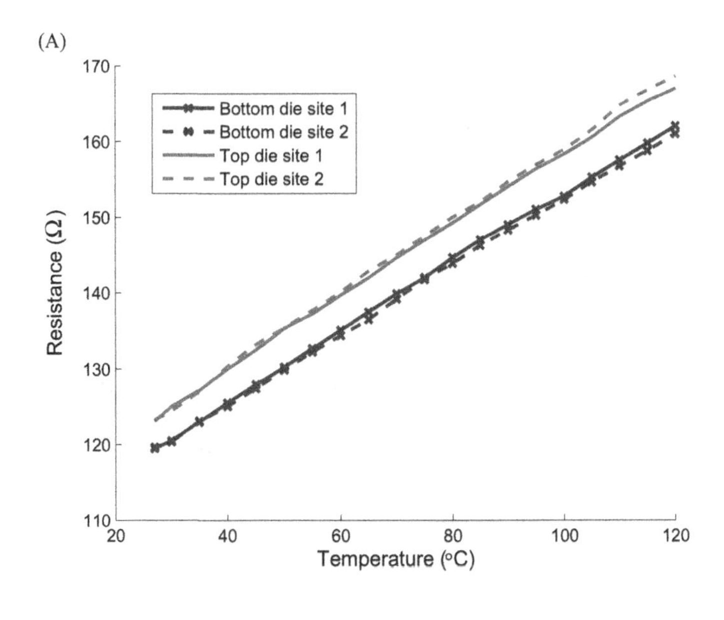

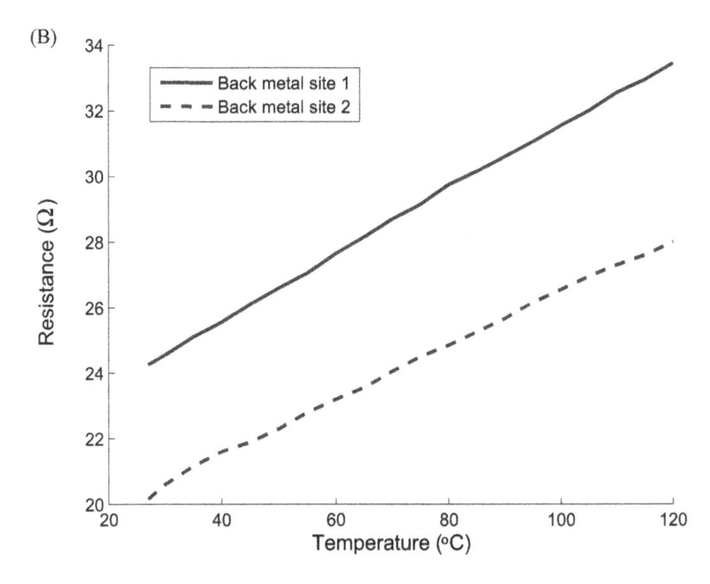

FIGURE 14.6

Calibration of (A) on-chip thermal sensors, and (B) back metal thermal sensors.

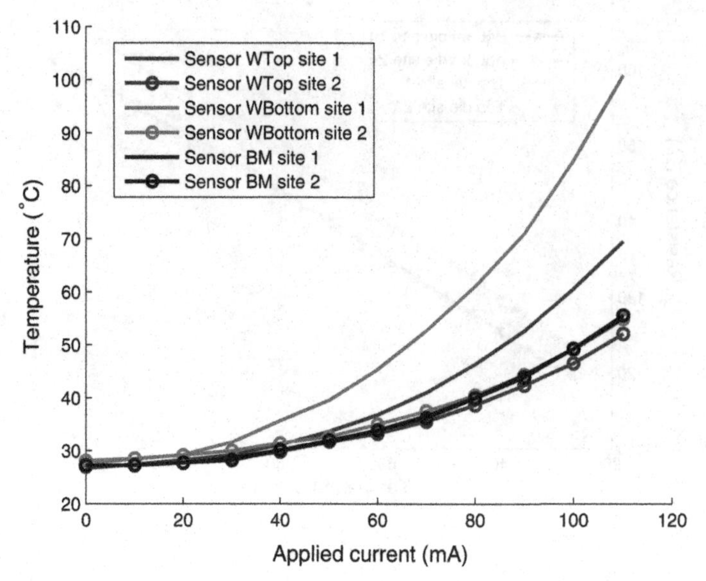

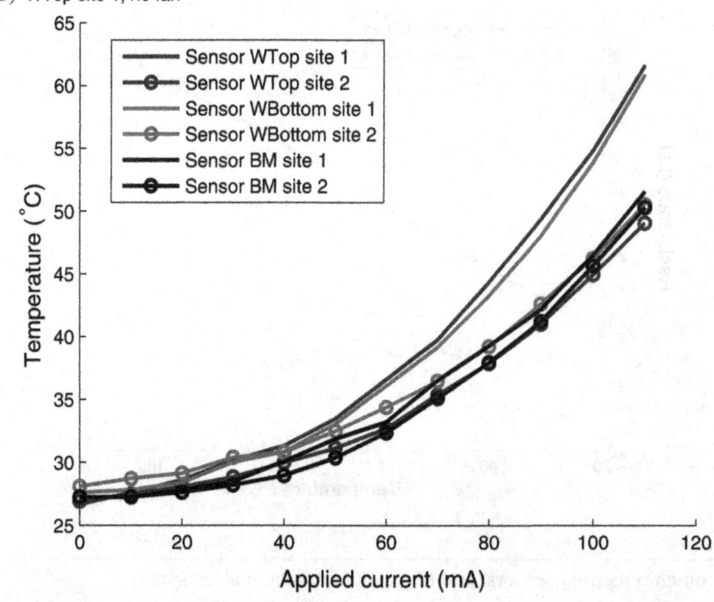

FIGURE 14.7

Experimental results for the different test conditions. Each label describes the device plane, site location of the heater, and whether active cooling is applied.

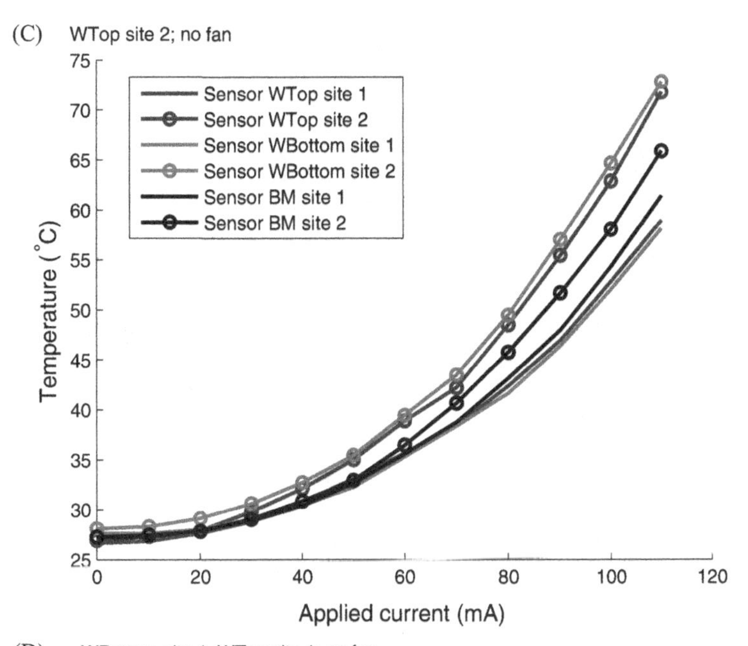

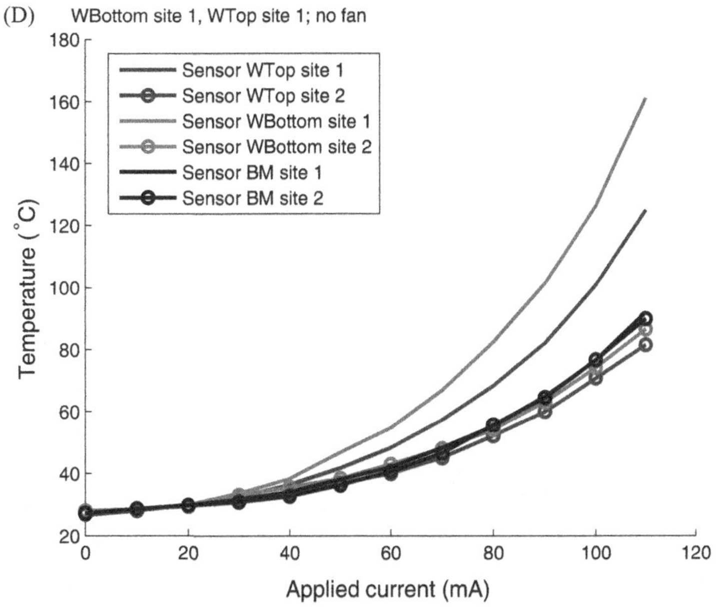

FIGURE 14.7

(Continued)

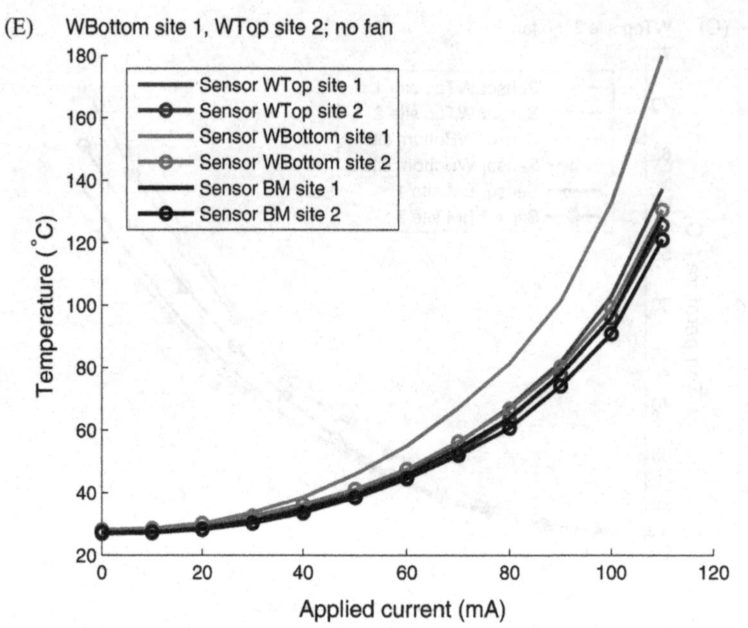

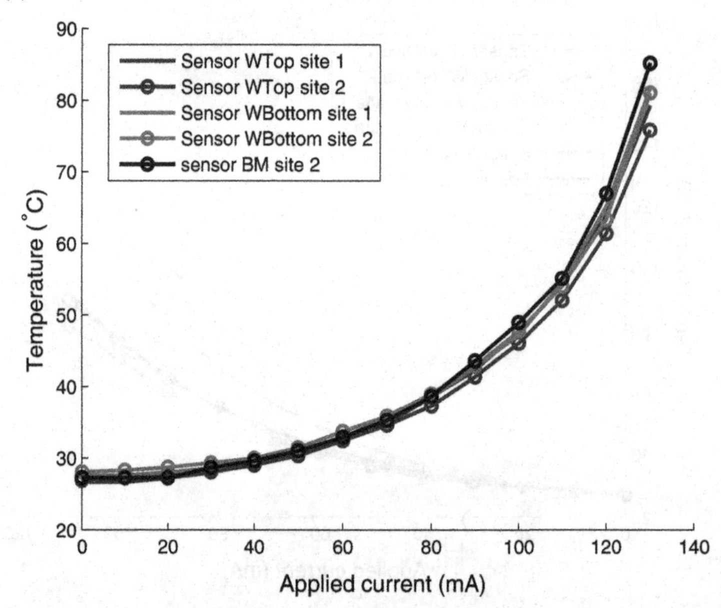

FIGURE 14.7

(Continued)

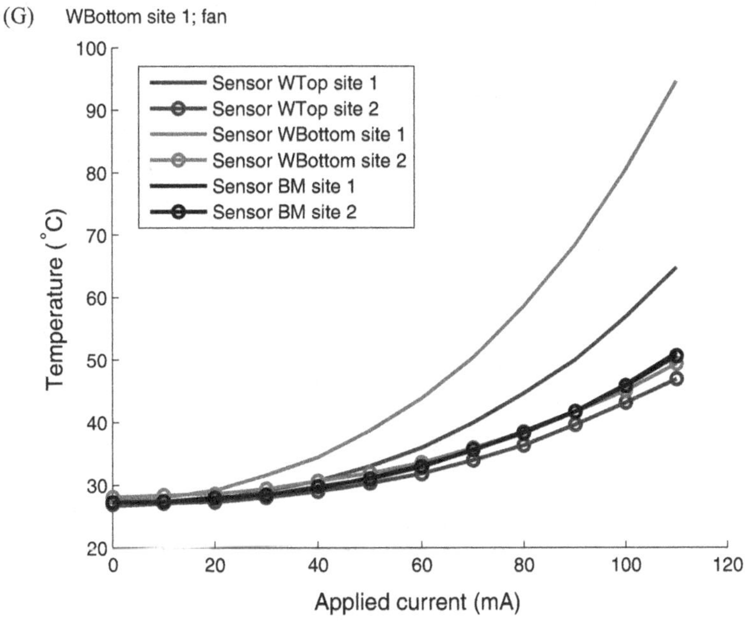

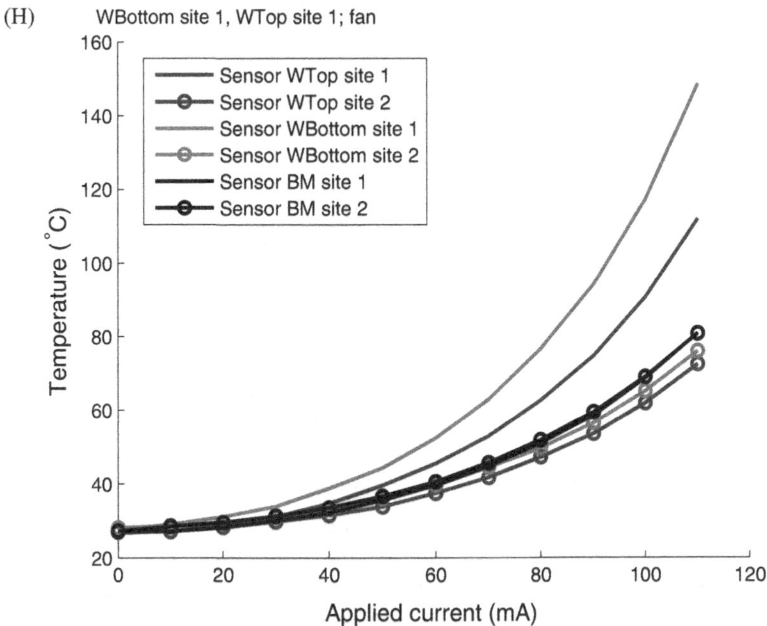

FIGURE 14.7

(Continued)

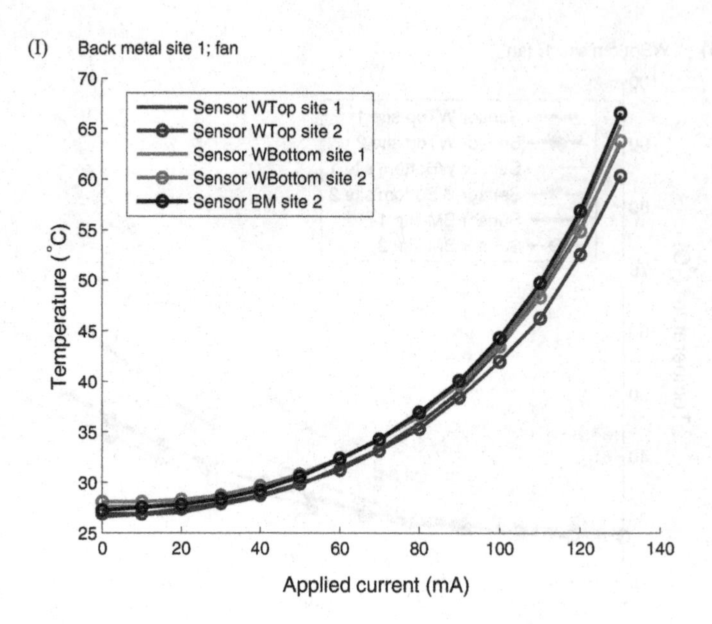

FIGURE 14.7

(Continued)

Table 14.2 Power Density of Common Integrated Circuits

Circuit	Description	Power Density (watt/mm^2)	References
Power transistor	Output stage of an LDO voltage converter	20	[565]
SC	Switched capacitor converter for different technologies	0.77–4.6	[566]
PWM	Pulse width modulator	1.39	[567]

site 1 shown in Fig. 14.7B. The effects of placing two highly active device blocks on the thermal gradients are explored. In this case, two separate conditions are examined: (1) the heaters on WBottom and WTop are simultaneously active and stacked directly above each other, and (2) the heaters on WBottom and WTop are simultaneously active and physically nonaligned. These results are depicted in Figs. 14.7D and E. A fourth test condition is used to evaluate the flow of heat from a back metal heat source to the on-chip thermal sensors. One of the two back metal sensors operates as a heater while the other sensor detects the temperature. The effects of placing the CMOS 3-D IC in a location where heat may couple from the backside of a thinned silicon substrate are shown in Fig. 14.7F. The final test condition examines the effects of active cooling on heat dissipation within a stacked IC. A 12 volt, 0.13 ampere, 6,500 RPM, and 8 CFM (cubic feet per minute) fan is placed 1 inch above the 3-D stack for convective heat removal. The fan is used for three different resistive heater conditions: (1) active heater in WBottom site 1, (2) active heater on WBottom and WTop site 1, and (3) active heater on back metal site 1. The results of this cooling experiment are illustrated in, respectively, Figs. 14.7G through 14.7I.

14.3 DESIGN CONSIDERATIONS BASED ON EXPERIMENTAL RESULTS

The experimental test conditions described in Section 14.2 provide insight into thermal propagation paths and hot spot formation within 3-D ICs. All of the test configurations include heaters placed in metal 2, which emulate the heat generated by the active devices within a device plane. Temperature data for all test conditions are determined from the four on-chip and two back metal resistive thermal sensors. The location of the resistive heaters determines the specific test condition. The temperature at the six thermal sensors as a result of activity from the single heater, double heaters, and active cooling is listed in Table 14.3. The data are derived from converted measured resistances. Design considerations to minimize hot spot formations are discussed in the subsections below.

14.3.1 EFFECT OF BLOCK PLACEMENT ON HOT SPOT FORMATION AND MITIGATION TECHNIQUES

The effects of stacking two dies on the temperature profile of a 3-D IC are significant, as shown in Figs. 14.7A and B. The maximum observed temperature on WBottom site 1 increases by 65.7% when the resistive heater is active on WBottom as compared to an active heater on WTop (100.9°C from 60.9°C), as indicated by the data listed in Table 14.4. The maximum observed temperatures occur for the maximum applied current of 110 mA. For heater currents less than 110 mA, the percent increase in temperature for WBottom as compared to WTop decreases due to the exponential drop in temperature as the current is reduced. Data from the remaining five thermal sensors reveal a temperature increase of 6.1% to 13.0% (see Table 14.4), with the 13.0% increase occurring on the sensor located directly above the active heater on WTop. Although the measured temperature is lower at the thermal sensors 1.3 mm from the hot spot, the drop is not as significant as reported in [561], where over a 50% drop in temperature is measured at a distance of 500 μm. The observed temperature on WBottom when a resistive heater is active on WBottom site 1 is within 20% of the measured temperature at the same WBottom sensor when an active heater is on WTop site 1, for all heater currents less than 50 mA. Placing a highly active device block in WBottom requires special consideration as the thermal resistance along the path to the top of the 3-D stack is higher than to a block placed in WTop.

Three potential mitigation techniques can be considered. During placement and routing, blocks with high current loads can be placed in those locations that minimize the thermal resistance between the block and the heat sink. The current load greatly affects the thermal profile (as revealed by less than a 20% change in temperature for currents less than 50 mA). Another approach to control the activity of the block is to ensure that the average current load remains below an acceptable value, thereby not forming hot spots. Passive and active cooling techniques such as TTSVs and microfluidic channels can also be used to remove heat from the internal device planes by changing the thermal resistive paths within the highly active circuit blocks.

14.3.2 HORIZONTAL AND VERTICAL THERMAL CONDUITS

The experiment examining in-tier thermal spreading (Figs. 14.7B and C) requires further discussion. By removing the in-tier metal heat spreaders surrounding the thermal sensors and resistive

Table 14.3 Temperature Measurements From the Four On-chip and Two Back Metal Thermal Sensors for Different Heater Activities

Heater Location	Convective Cooling	Sensor Location	Current (mA)													
			0	10	20	30	40	50	60	70	80	90	100	110	120	130
WBottom, Site 1	No	WTop, Site 1	26.7	27.4	28.2	29.2	31.1	33.7	36.8	41.0	46.4	52.6	60.6	69.6	–	–
		WTop, Site 2	26.9	27.3	27.7	28.5	29.8	31.7	33.3	35.6	38.7	42.4	46.5	52.1	–	–
		WBottom, Site 1	27.6	28.4	29.2	31.6	35.7	39.6	45.4	52.7	61.1	71.0	84.8	100.9	–	–
		WBottom, Site 2	28.1	28.5	29.1	30.0	31.5	32.7	35.0	37.6	40.5	44.3	49.1	54.9	–	–
		Back Metal, Site 1	27.3	27.3	27.7	28.7	30.1	31.6	33.9	36.6	40.1	44.4	49.2	55.1	–	–
		Back Metal, Site 2	27.3	27.3	27.7	28.2	30.1	32.0	33.9	36.3	40.0	43.9	49.2	55.6	–	–
WBottom, Site 1	Yes	WTop, Site 1	26.7	27.0	27.6	29.2	30.7	33.1	36.0	40.0	44.7	50.0	56.9	64.8	–	–
		WTop, Site 2	26.8	27.1	27.3	28.1	29.0	30.4	31.9	34.0	36.4	39.7	43.2	46.9	–	–
		WBottom, Site 1	27.8	28.0	29.2	31.6	34.5	38.8	43.9	50.3	58.6	68.4	80.4	94.5	–	–
		WBottom, Site 2	27.9	28.4	28.6	29.5	30.7	32.0	33.7	36.0	38.5	41.7	45.1	49.4	–	–
		Back Metal, Site 1	27.2	27.4	27.7	28.6	29.5	31.2	33.2	35.6	38.5	41.8	46.0	51.1	–	–
		Back Metal, Site 2	27.0	27.4	28.0	28.5	29.8	31.0	33.0	35.7	38.4	41.7	45.9	50.6	–	–
WTop, Site 1	No	WTop, Site 1	27.0	27.6	28.6	30.2	31.3	33.5	36.6	39.8	44.4	49.4	54.9	61.6	–	–
		WTop, Site 2	26.8	27.5	28.1	28.8	30.0	31.1	32.7	35.4	37.9	41.0	44.9	49.1	–	–
		WBottom, Site 1	27.5	27.8	28.4	30.0	30.8	33.1	36.1	39.2	43.3	48.1	53.9	60.9	–	–
		WBottom, Site 2	27.8	28.7	29.1	30.4	30.8	32.5	34.4	36.5	39.2	42.6	46.3	50.6	–	–
		Back Metal, Site 1	27.1	27.3	27.6	28.2	29.0	30.4	32.4	35.1	38.0	41.2	45.7	50.3	–	–
		Back Metal, Site 2	27.2	27.3	27.8	28.5	30.1	31.9	33.2	36.5	39.3	42.2	46.5	51.6	–	–
WTop, Site 2	No	WTop, Site 1	26.5	26.9	27.6	28.8	30.7	33.1	35.6	38.4	42.4	47.0	52.8	59.0	–	–
		WTop, Site 2	26.7	27.3	27.9	29.8	32.1	35.0	38.9	42.2	48.5	55.4	62.8	71.8	–	–
		WBottom, Site 1	27.5	27.6	28.0	29.0	30.4	32.3	35.3	38.4	41.6	46.4	52.0	58.1	–	–
		WBottom, Site 2	27.4	28.3	29.1	30.6	32.7	35.4	39.5	43.5	49.5	57.0	64.7	72.8	–	–
		Back Metal, Site 1	27.1	27.3	27.9	28.9	30.4	32.6	35.6	38.7	43.2	48.0	54.3	61.4	–	–
		Back Metal, Site 2	26.9	27.5	27.8	29.1	30.8	32.9	36.4	40.7	45.8	51.7	58.1	65.9	–	–

WBottom, Site 1 WTop, Site 1	No	WTop, Site 1	27.1	28.0	30.2	32.5	36.4	41.8	48.4	57.3	68.3	82.1	100.7	124.8	–	–
		WTop, Site 2	27.3	28.1	29.8	31.9	34.0	36.9	40.1	45.1	52.2	60.0	70.8	81.4	–	–
		WBottom, Site 1	27.9	28.6	30.0	33.7	38.4	46.8	54.8	66.9	82.4	101.4	126.1	161.0	–	–
		WBottom, Site 2	28.0	28.5	29.4	33.1	35.7	38.4	43.1	48.2	54.2	63.1	74.2	86.5	–	–
		Back Metal, Site 1	27.2	28.5	29.9	31.7	34.1	38.2	42.1	48.4	55.3	64.9	76.5	91.9	–	–
		Back Metal, Site 2	27.6	28.7	29.7	30.9	32.7	36.4	40.8	46.6	55.7	64.5	76.6	89.9	–	–
WBottom, Site 1 WTop, Site 2	No	WTop, Site 1	27.2	27.8	29.6	31.7	35.6	40.6	47.0	56.1	67.5	81.9	103	137.2	–	–
		WTop, Site 2	27.0	27.1	28.5	30.6	34.2	38.9	44.9	52.8	63.7	77.7	96.1	125.6	–	–
		WBottom, Site 1	27.8	28.2	30.0	33.7	38.6	46.0	55.0	67.1	81.5	101.4	132.9	179.9	–	–
		WBottom, Site 2	28.0	28.5	30.2	32.5	36.3	41.2	47.6	56.4	66.9	80.4	99.3	130.6	–	–
		Back Metal, Site 1	27.3	27.5	28.3	30.3	33.6	38.5	44.6	52.1	60.7	74.1	91.0	121.0	–	–
		Back Metal, Site 2	27.2	27.3	28.6	30.9	34.6	39.3	45.7	54.3	64.0	78.1	96.1	128.4	–	–
WBottom, Site 1 WTop, Site 1	Yes	WTop, Site 1	26.9	27.2	28.6	31.1	34.7	39.6	45.6	53.0	62.7	74.8	90.8	112.0	–	–
		WTop, Site 2	26.8	27.1	28.3	29.8	31.5	33.8	37.5	41.8	47.3	53.6	62.0	72.6	–	–
		WBottom, Site 1	27.9	29.0	31.0	33.9	38.8	44.3	52.4	63.0	76.7	94.3	117.3	148.6	–	–
		WBottom, Site 2	28.1	28.7	29.6	31.0	33.1	36.1	39.7	44.6	49.7	56.6	65.3	76.2	–	–
		Back Metal, Site 1	27.1	27.2	28.1	30.1	32.2	35.6	39.7	44.9	51.0	58.7	68.9	81.0	–	–
		Back Metal, Site 2	27.4	28.5	29.5	31.2	33.5	36.6	40.6	45.8	51.9	59.5	69.2	80.9	–	–
Back metal, Site 1	No	WTop, Site 1	26.6	26.7	27.2	28.0	29.2	30.5	32.5	34.9	38.6	42.4	47.6	54.1	64.1	79.3
		WTop, Site 2	27.1	27.3	27.5	28.1	29.0	30.4	32.5	34.6	37.3	41.4	46.1	52.1	61.4	75.9
		WBottom, Site 1	27.5	27.6	28.0	28.6	29.8	31.2	32.9	35.7	38.6	42.7	47.2	54.5	65.2	81.7
		WBottom, Site 2	28.0	28.3	28.7	29.4	30.0	31.5	33.8	35.9	39.0	42.9	47.8	54.0	63.4	81.1
		Back Metal, Site 1	–	–	–	–	–	–	–	–	–	–	–	–	–	–
		Back Metal, Site 2	0.00	5.18	7.13	9.70	8.80	13.66	20.28	28.60	39.29	46.81	55.63	61.65	–	–
Back metal, Site 1	Yes	WTop, Site 1	27.0	26.9	27.0	27.8	28.6	29.8	31.5	33.3	35.8	39.0	43.4	48.6	55.1	63.5
		WTop, Site 2	27.1	26.9	27.4	28.0	28.7	29.9	31.2	33.2	35.3	38.4	42.0	46.2	52.5	60.3
		WBottom, Site 1	27.4	27.6	28.0	28.8	29.5	30.8	32.3	34.3	36.5	40.0	43.9	49.1	56.2	65.4
		WBottom, Site 2	27.9	28.1	28.3	28.7	29.7	30.8	32.4	34.2	36.7	39.7	43.5	48.3	54.9	63.8
		Back Metal, Site 1	–	–	–	–	–	–	–	–	–	–	–	–	–	–
		Back Metal, Site 2	27.1	27.5	27.8	28.4	29.2	30.6	32.4	34.3	36.9	40.1	44.3	49.7	56.9	66.7

Table 14.4 Per cent Increase in Temperature When Active Circuit Is Located in Internal Tier (WBottom)

Heater Location	Sensor Location	Current (mA)											
		0	10	20	30	40	50	60	70	80	90	100	110
WBottom, Site 1 % increase as compared to WTop, Site 1	WTop, Site 1	0.00	−0.72	−1.39	−3.31	−0.64	0.60	0.55	3.02	4.50	6.48	10.38	12.99
	WTop, Site 2	0.00	−0.73	−1.42	−1.04	−0.67	1.93	1.83	0.56	2.11	3.41	3.56	6.11
	WBottom, Site 1	0.00	2.16	2.82	5.33	15.91	19.64	25.76	34.44	41.11	47.61	57.33	65.68
	WBottom, Site 2	0.00	−0.70	0.00	−1.32	2.27	0.62	1.74	3.01	3.32	3.99	6.05	8.50
	Back Metal, Site 1	0.00	0.00	0.36	1.77	3.79	3.95	4.63	4.27	5.53	7.77	7.66	9.54
	Back Metal, Site 2	0.00	0.00	−0.36	−1.05	0	0.31	2.11	−0.55	1.78	4.03	5.81	7.75

heaters, the ability of the heat to spread horizontally is diminished. The maximum temperature from a sensor directly above an active heater in WTop site 2 is 16.4% higher than an active heater in WTop site 1 for a heater current of 110 mA. All of the other sensors exhibit temperatures 14.9% to 27.7% higher than when metal heat spreaders are present and a 110 mA heater current is applied, as listed in Table 14.5. The increase in temperature at site 2 when the heater on WTop site 2 is active is expected; however, the lower temperature indicated by the negative change for the on-chip sensors at site 1 is not as great as expected. The thermal resistance from site 2 to the heat sink is larger as the horizontal metal heat spreaders have been removed, resulting in higher on-chip temperatures. Placing thermal spreaders to horizontally distribute the heat and thereby lower the effective thermal resistance reduces the maximum temperature experienced by those blocks 1.3 mm and a device plane away. It is therefore critical to reduce both the inter and intratier thermal resistances by providing paths for the heat to flow from the hot spots to the heat sink.

14.3.3 MULTIPLE ALIGNED ACTIVE BLOCKS

The placement of two highly active circuit blocks aligned directly above one another has a significant effect on the thermal profile of a 3-D IC. Placing an active block on WBottom produces a higher thermal resistive path than a block placed on WTop. An analysis of two vertically aligned active circuit blocks has been performed by comparing the temperature for a single heater placed on WBottom with the increase in temperatures caused by placing two active heaters located at WTop and WBottom site 1. The largest temperature increase occurs at WTop site 1, where a 79.4% increase in temperature is observed (from 69.6°C to 124.8°C) for an applied current of 110 mA. The maximum on-chip temperature occurs on WBottom site 1, where the maximum temperature increases from 100.9°C to 161.0°C when a current of 110 mA flows through, respectively, one active heater and two active heaters, producing a 59.6% increase in temperature. The remaining four sensors reveal an increase in temperature of 56.5% to 66.9%, corresponding to absolute temperatures of 81.4 to 91.9°C, from original temperatures of 52.1 to 55.6°C. The per cent increase in temperature when two active circuit blocks are vertically stacked as compared to a single active block is listed in Table 14.6, and the corresponding temperatures are listed in Table 14.3.

The magnitude of the current also has a significant effect on the thermal profile within a 3-D IC. When applying 40 mA through both heaters, the increase in temperature remains below 20% for all thermal sensors as when applying 40 mA to just the WBottom heater. Two mitigation techniques include limiting the current flow, or deactivating the circuit block to allow heat to flow from the hot spot. Deactivation is particularly useful in high activity circuit blocks. Active and passive heat removal techniques, such as microfluidic channels and thermal TSVs, also apply.

14.3.4 MULTIPLE NONALIGNED ACTIVE BLOCKS

The effect of increased spacing on the maximum temperature between vertically nonaligned active circuits is discussed in this subsection. A comparison is made between two vertically aligned resistive heaters (both of the heaters in site 1) and two heaters separated by 1.3 mm (WBottom heater site 1 and WTop heater site 2). A 0.0% to 5.4% reduction in temperature for all six thermal sensors up to a current of 30 mA is observed when shifting the WTop heater from site 1 to site 2. The on-chip thermal sensors at site 1 reveal temperature changes ranging from −2.9% to 0.5% for currents up to 90 mA. A maximum increase in temperature at site 1 of 11.8% at WBottom and 9.9% at WTop is measured for a peak current of 110 mA, as listed in Table 14.7. The on-chip thermal

Table 14.5 Per cent Increase in Temperature from Intratier Thermal Spreading

Heater Location	Sensor Location	Current (mA)											
		0	10	20	30	40	50	60	70	80	90	100	110
WTop, Site 2 % increase as compared to WTop, Site 1	WTop, Site 1	0.00	− 2.81	− 3.39	− 4.52	− 1.87	− 1.17	− 2.69	− 3.48	− 4.50	− 4.89	− 3.71	− 4.35
	WTop, Site 2	0.00	− 0.70	− 0.68	3.33	7.06	12.39	18.97	19.24	27.89	35.09	39.83	46.23
	WBottom, Site 1	0.00	− 0.72	− 1.41	− 3.35	− 1.31	− 2.45	− 2.25	− 2.09	− 3.81	− 3.46	− 3.50	− 4.53
	WBottom, Site 2	0.00	− 1.45	0.00	0.69	6.12	9.06	14.73	19.17	26.08	33.75	39.77	44.05
	Back Metal, Site 1	0.00	0.00	0.34	1.32	0.94	2.38	7.15	6.01	9.77	13.73	16.74	19.05
	Back Metal, Site 2	0.00	0.69	0.68	3.01	6.52	8.41	12.62	15.78	20.50	25.35	27.14	30.94

Table 14.6 Per cent Increase in Temperature When Two Active Circuit Blocks Are Vertically Aligned

Heater Location	Sensor Location	Current (mA)											
		0	10	20	30	40	50	60	70	80	90	100	110
WTop, Site 2 % increase as compared to WTop, Site 1	WTop, Site 1	0.00	2.12	6.89	11.35	17.01	24.05	31.40	39.88	47.29	55.98	66.19	79.43
	WTop, Site 2	0.00	2.81	7.62	12.15	14.22	16.45	20.41	26.85	35.03	41.58	52.10	56.47
	WBottom, Site 1	0.00	0.71	2.76	6.40	7.43	18.24	20.72	27.04	34.85	42.77	48.64	59.55
	WBottom, Site 2	0.00	0.00	0.72	10.48	13.36	17.40	22.97	28.33	33.81	42.36	51.18	57.61
	Back Metal, Site 1	0.00	4.49	7.82	10.53	13.22	20.76	24.25	32.03	37.98	46.12	55.42	66.86
	Back Metal, Site 2	0.00	5.18	7.13	9.70	8.80	13.66	20.28	28.60	39.29	46.81	55.63	61.65

Table 14.7 Per cent Increase in Temperature When Two Active Circuit Blocks Are Not Vertically Aligned

Heater Location	Sensor Location	Current (mA)											
		0	10	20	30	40	50	60	70	80	90	100	110
WTop, Site 2 and WBottom, Site 1 % increase as compared to WTop, Site 1 and WBottom, Site 1	WTop, Site 1	0.00	− 0.69	− 1.94	− 2.41	− 2.16	− 2.86	− 2.91	− 2.14	− 1.23	− 0.26	2.22	9.94
	WTop, Site 2	0.00	− 3.41	− 4.51	− 4.22	0.57	5.26	12.21	17.06	21.83	29.47	35.85	54.27
	WBottom, Site 1	0.00	− 1.40	0.00	0.00	0.53	− 1.77	0.38	0.32	− 1.08	0.00	5.41	11.75
	WBottom, Site 2	0.00	0.00	2.85	− 1.90	1.77	7.17	10.41	16.99	23.32	27.35	33.85	51.01
	Back Metal, Site 1	0.00	− 3.63	− 5.36	− 4.47	− 1.39	1.00	5.95	7.65	9.80	14.22	18.90	31.63
	Back Metal, Site 2	0.00	− 4.92	− 3.81	0.00	5.80	8.14	12.05	16.50	14.90	21.06	25.41	42.87
WTop, Site 1 and WBottom, Site 1 % increase as compared to WTop, Site 1 and WBottom, Site 1	WTop, Site 1	0.00	1.41	4.82	8.68	14.48	20.51	27.57	36.89	45.49	55.57	69.89	97.28
	WTop, Site 2	0.00	− 0.70	2.77	7.42	14.87	22.58	35.11	48.49	64.51	83.32	106.63	141.39
	WBottom, Site 1	0.00	− 0.71	2.76	6.40	8.00	16.14	21.18	27.45	33.39	42.77	56.68	78.29
	WBottom, Site 2	0.00	0.00	3.59	8.38	15.37	25.82	35.77	50.14	65.02	81.31	102.34	138.01
	Back Metal, Site 1	0.00	0.69	2.04	5.59	11.64	21.97	31.64	42.12	51.49	66.90	84.81	119.64
	Back Metal, Site 2	0.00	0.00	3.05	9.70	15.11	22.92	34.77	49.82	60.04	77.73	95.18	130.95

sensors at site 2 detect an exponentially increasing temperature from an applied current of 40 mA (0.6% for WTop and 1.8% for WBottom) to a peak current of 110 mA (54.3% for WTop and 51.0% for WBottom). The exponential increase in temperature at the back metal sensors for currents above 30 mA indicates strong thermal coupling through the back metal layer. A maximum on-chip temperature of 179.9°C occurs at WBottom site 1. The maximum temperature at WTop site 1 is 137.2°C, while the remaining sensors exhibit a maximum temperature ranging between 121.0°C and 130.6°C.

Three primary design issues are noted: (1) moving highly active circuit blocks on the same tier apart reduces the maximum temperature as compared to vertically aligned circuits. (2) Although nonaligned circuit blocks reduce the maximum temperature, the effective thermal resistance of the path to the heat sink (air in this case) significantly affects the maximum on-chip temperature, as indicated by the increase in temperature for currents above 30 mA. (3) Proper floorplanning within 3-D ICs requires analysis of the heat generated by each circuit block and the appropriate placement of *hot* blocks in both the vertical and horizontal directions, as placement algorithms must consider these issues to minimize peak on-chip temperatures.

14.3.5 ADDITIONAL DESIGN CONSIDERATIONS

The 3-D structures are particularly beneficial for heterogeneous systems. Some of these circuits commonly have blocks bonded to the back side of the silicon [568–570] (e.g., VCSELs within an optical interconnect system [571]). The thermal heaters on the back metal emulate these blocks and support the analysis of the thermal profile and conduits from the back side of the silicon to the rest of the 3-D structure.

The fourth and fifth test conditions evaluate, respectively, heat flow from a source on the back side of the silicon to the on-chip thermal sensors and the reduction in the maximum temperature when a fan is used to convectively cool a 3-D IC. All of the thermal sensors produce results within 6% of each other for all current load conditions through the back metal heater. The heat flows from a source on the back metal, evenly dispersing heat to the rest of the 3-D IC. Thermal spreading from a hot spot on the back metal is more pronounced. It is, however, difficult to isolate a circuit block from thermal effects caused by a back metal heat source. For the fifth test condition, a fan 1 inch above the 3-D IC reduces the temperature by 0.7% for a 10 mA current through a back metal heater to 21.9% for a current load of 130 mA. In the case where both on-chip heaters in site 1 are active, the maximum temperature is reduced by 0.7% and 11.9% for, respectively, currents of 10 mA and 110 mA. The per cent reduction in temperature for three different heater configurations and heater currents of 0 mA to 110 mA is listed in Table 14.8. Despite an approximately 12% reduction in peak temperature with the use of a fan, additional heat removal techniques like TTSVs and microfluidic channels are necessary for those hot spots located deep within the 3-D IC.

14.4 VERIFICATION OF EXPERIMENTAL RESULTS WITH SIMULATIONS

Simulations of the fabricated 3-D test circuit have been compared with experimental measurements. Temperatures are extracted from the simulations, and the thermal resistance per unit length is determined. The experimental data exhibit good agreement with these simulations.

Table 14.8 Per cent Decrease in Temperature with Convective Cooling

Heater Location	Sensor Location	Current (mA)													
		0	10	20	30	40	50	60	70	80	90	100	110	120	130
WTop, Site 2 and WBottom, Site 1 % increase as compared to WTop, Site 1 and WBottom, Site 1	WTop, Site 1	0.00	− 1.41	− 2.06	0.00	− 1.25	− 1.75	− 2.14	− 2.42	− 3.67	− 5.00	− 6.10	− 6.91	–	–
	WTop, Site 2	0.00	− 0.70	− 1.38	− 1.35	− 2.58	− 4.25	− 4.06	− 4.35	− 6.02	− 6.45	− 7.17	− 9.87	–	–
	WBottom, Site 1	0.00	− 1.41	0.00	0.00	− 3.42	− 2.07	− 3.19	− 4.37	− 4.17	− 3.66	− 5.23	− 6.34	–	–
	WBottom, Site 2	0.00	− 0.37	− 1.79	− 1.74	− 2.33	− 2.25	− 3.90	− 4.22	− 4.97	− 6.00	− 8.07	− 10.01	–	–
	Back Metal, Site 1	0.00	0.34	0.00	− 0.33	− 1.88	− 1.20	− 1.96	− 2.86	− 3.82	− 5.85	− 6.50	− 7.27	–	–
	Back Metal, Site 2	0.00	0.34	1.02	1.00	− 0.94	− 2.96	− 2.52	− 1.58	− 4.07	− 5.04	− 6.89	− 8.97	–	–
WTop, Site 1 and WBottom, Site 1 % increase as compared to WTop, Site 1 and WBottom, Site 1	WTop, Site 1	0.00	− 2.77	− 5.16	− 4.21	− 4.86	− 5.23	− 5.81	− 7.48	− 8.23	− 8.85	− 9.88	− 10.23	–	–
	WTop, Site 2	0.00	− 3.41	− 5.15	− 6.63	− 7.37	− 8.39	− 6.31	− 7.38	− 9.46	− 10.65	− 12.31	− 10.87	–	–
	WBottom, Site 1	0.00	1.41	3.36	0.60	1.07	− 5.31	− 4.22	− 5.77	− 6.96	− 6.98	− 7.03	− 7.67	–	–
	WBottom, Site 2	0.00	0.73	0.71	− 6.33	− 7.08	− 6.05	− 7.89	− 7.54	− 8.34	− 10.35	− 11.91	− 11.94	–	–
	Back Metal, Site 1	0.00	− 4.62	− 5.99	− 5.07	− 5.57	− 6.75	− 5.70	− 7.20	− 7.78	− 9.47	− 10.01	− 11.88	–	–
	Back Metal, Site 2	0.00	− 0.66	− 0.63	0.92	2.32	0.79	− 0.47	− 1.87	− 6.85	− 7.69	− 9.74	−9.96	–	–
WTop, Site 1 and WBottom, Site 1 % increase as compared to WTop, Site 1 and WBottom, Site 1	WTop, Site 1	0.00	0.73	− 0.71	− 0.69	− 2.00	− 2.55	− 3.01	− 4.50	− 7.16	− 7.97	− 8.84	− 10.12	− 14.12	− 19.92
	WTop, Site 2	0.00	− 1.40	− 0.35	− 0.34	− 0.99	− 1.58	− 3.85	− 4.19	− 5.45	− 7.29	− 8.93	− 11.19	− 14.47	− 20.48
	WBottom, Site 1	0.00	0.00	0.00	0.70	− 1.01	− 1.29	− 1.85	− 3.99	− 5.30	− 6.26	− 7.02	− 9.99	− 13.72	− 20.00
	WBottom, Site 2	0.00	− 0.74	− 1.45	− 2.13	− 1.05	− 2.00	− 4.04	− 4.70	− 5.96	− 7.43	− 8.94	− 10.56	− 13.39	− 21.32
	Back Metal, Site 1	–	–	–	–	–	–	–	–	–	–	–	–	–	–
	Back Metal, Site 2	0.00	1.03	2.07	− 0.99	− 1.27	− 1.52	− 1.73	− 2.96	− 4.45	− 8.15	− 9.49	− 9.75	− 15.09	− 21.88

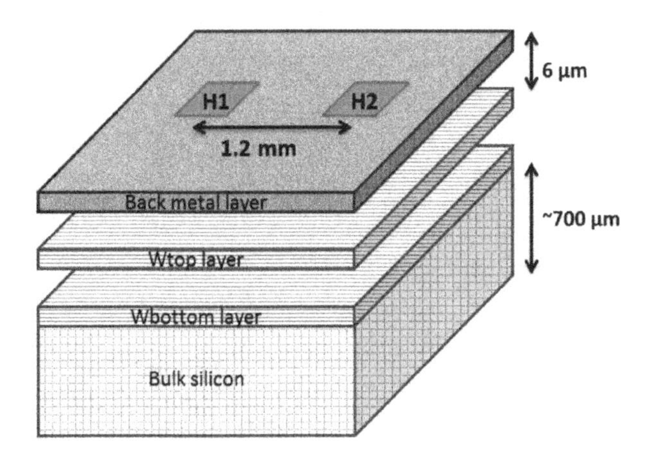

FIGURE 14.8

Structure of the 3-D test circuit consisting of two silicon tiers and one back metal layer. Each tier has two separately controlled heaters (H1 and H2). The back metal is connected to WTop using thermal through silicon vias.

14.4.1 SIMULATION SETUP AND TOOLS

The HotSpot simulator [552,553,572] is used to analytically investigate thermally conductive paths within 3-D structures. The structure shown in Fig. 14.8 is used to analyze heat propagation paths within a 3-D stack, including the dependence of thermal conductivity on temperature. This stack consists of two silicon layers and a single aluminum back metal layer (i.e., WTop, WBottom, and BackMetal layers). The back metal is connected to WTop using TTSVs, modeled as 6 μm high tungsten vias. Thermally passive (no heat is generated) layers are included in the simulation to more accurately model the 3-D test circuit (e.g., silicon dioxide, bulk silicon, and the metal layers). Two heaters, modeled as heat dissipating blocks, are placed 1.1 mm apart on the back metal, and 1.3 mm apart on metal 2 of WTop and WBottom. Six heater/sensor sites are placed across the structure to evaluate the flow of heat in both the horizontal and vertical dimensions. Different heaters are turned on to model different on-chip power dissipating blocks and related thermal paths. The temperatures are determined at each of the six sites.

14.4.2 COMPARISON TO EXPERIMENTAL RESULTS

A comparison of the measured temperatures with the simulations is presented in Figs. 14.9 to 14.11 for, respectively, the horizontal, vertical, and diagonal paths. A diagonal path is a path from WBottom layer site 1 to WTop layer site 2. The worst case temperature difference between experiment and simulation with a constant thermal conductivity is 25%, while the worst case difference with a temperature dependent thermal conductivity is 7%.

The thermal resistance per unit length is compared in Figs. 14.12 to 14.14 for, respectively, horizontal, vertical, and diagonal paths. The first datum shown in Figs. 14.13 and 14.14 deviates from the expected trend of the experimental data and the simulated results, hence measurement error is assumed, and is therefore not considered when evaluating the difference between the simulated and experimental

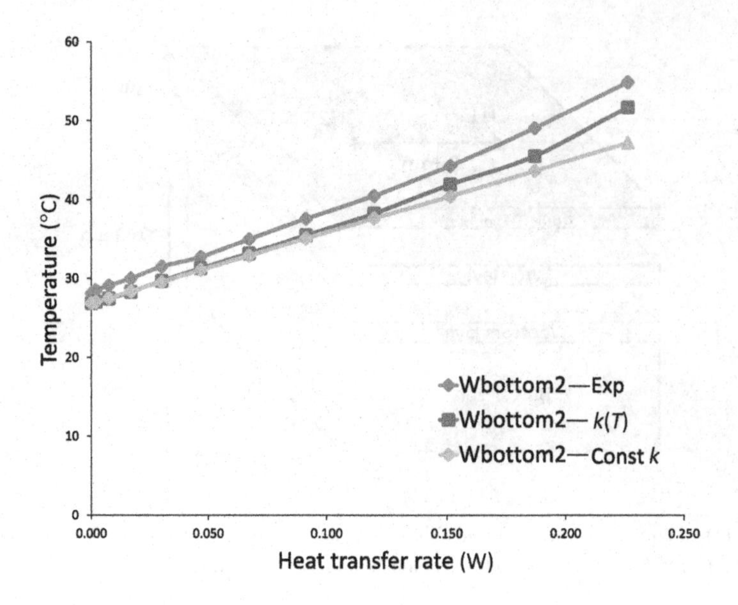

FIGURE 14.9

Comparison of temperatures for a horizontal path (length = 1,300 μm).

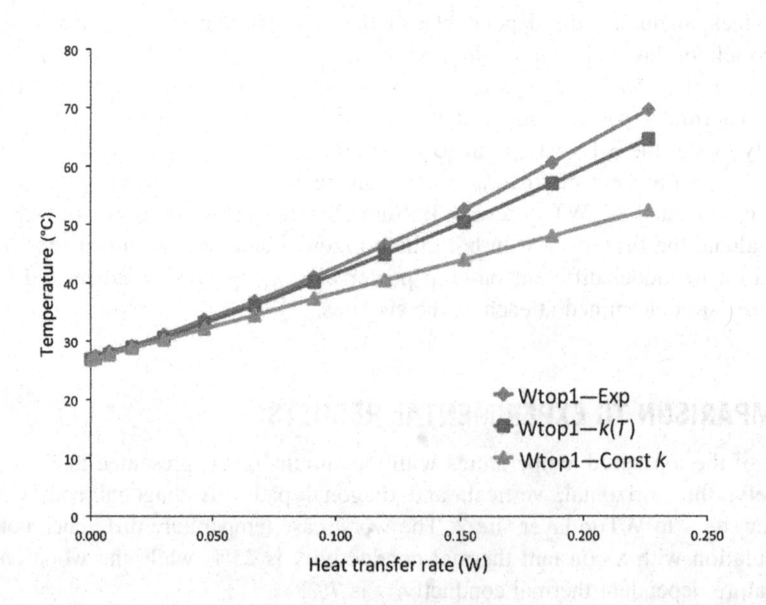

FIGURE 14.10

Comparison of temperatures for a vertical path (length = 10 μm).

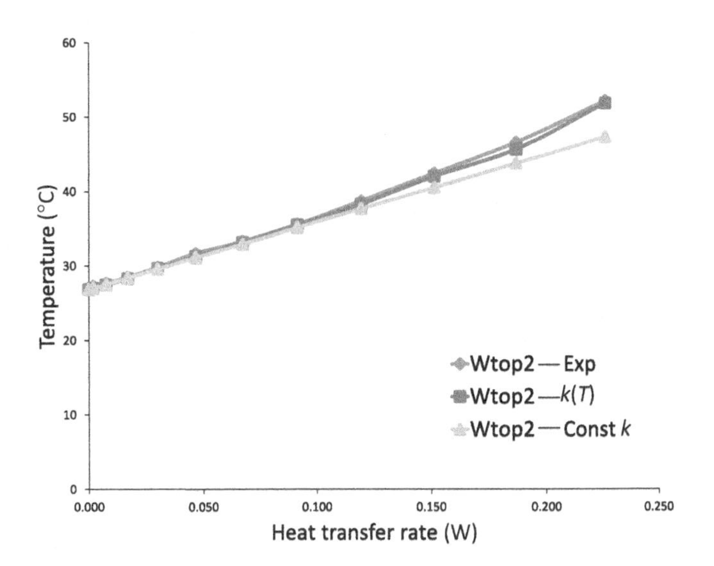

FIGURE 14.11

Comparison of temperatures for a diagonal path (length = 1,300 μm).

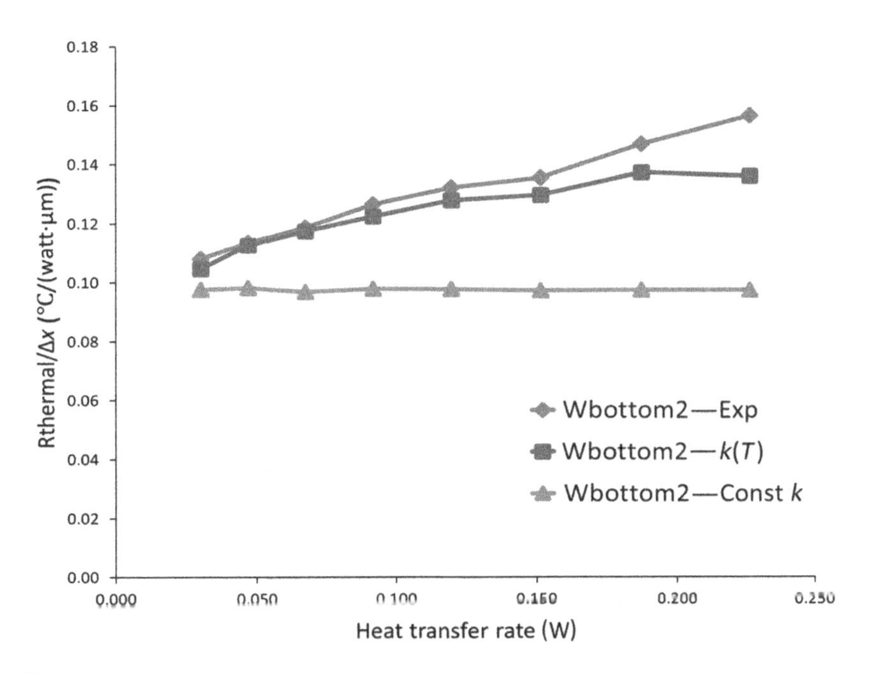

FIGURE 14.12

Comparison of thermal resistance per unit length for a horizontal path (length = 1,300 μm).

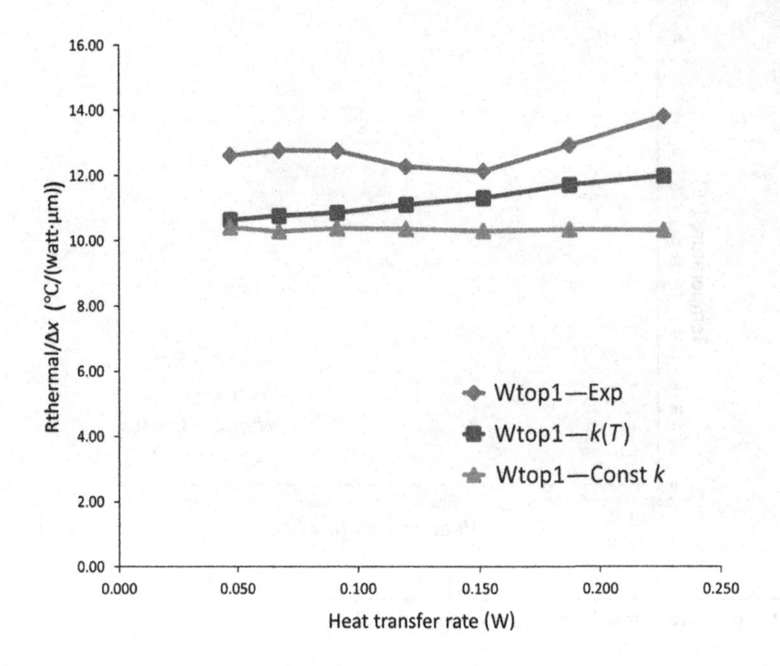

FIGURE 14.13

Comparison of thermal resistance per unit length for a vertical path (length = 1,300 μm).

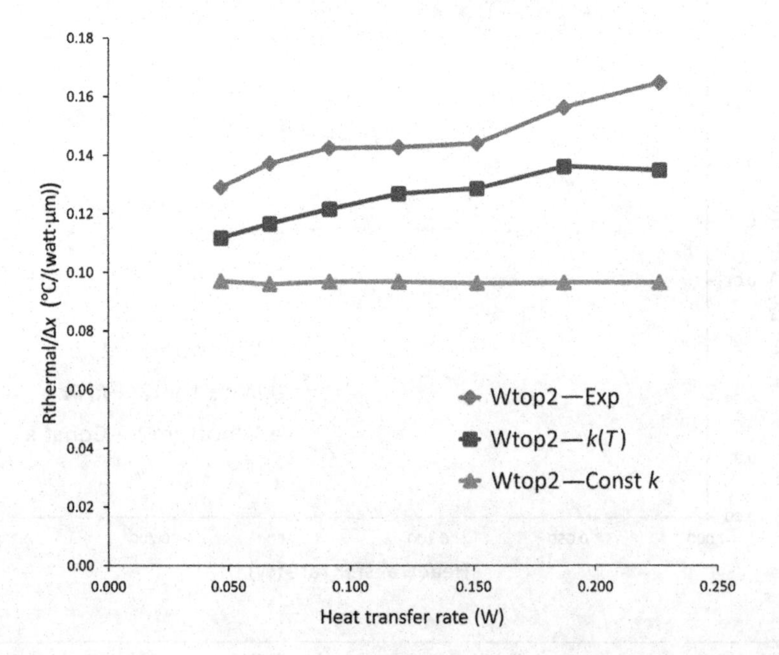

FIGURE 14.14

Comparison of thermal resistance per unit length for a diagonal path (length = 1,300 μm).

data. The worst case difference between the experimental results and simulation assuming a constant thermal conductivity is 38%, while the worst case difference assuming a temperature dependent thermal conductivity is 19%. It is therefore important to model the temperature dependence of the thermal conductivity. Both the experiments and simulations indicate that the lateral thermal paths conduct more heat than the vertical thermal paths. The thermal resistance per unit length of the vertical path is two orders of magnitude larger than the thermal resistance per unit length of the horizontal path, since the thermal conductivity of SiO_2 is much lower than the thermal conductivity of silicon.

14.4.3 EFFECT OF DENSITY OF TSVs ON THERMAL COUPLING

The number of TSVs has a significant effect on thermal coupling between tiers. TSVs are commonly fabricated using copper or tungsten, materials with higher thermal conductivity as compared to the insulating layer [127,173]. The TSVs, therefore, form thermal paths to conduct heat from the on-chip hot spots to the heat sink. The extension to HotSpot presented by Coskun et al. [573] has been used to capture the effects of the TSV density on the temperature profile. This ability of the tool to characterize materials on a per block basis supports different simulation setups with an increasing number of TSVs. The WTop site 1 heater/sensor pair has been analyzed in this simulation for four different setups: (1) no TSVs, (2) 2% TSVs, (3) 10% TSVs, and (4) 20% TSVs. The simulated TSVs are characterized with a 1.25 μm diameter, 100 μm pitch, and thermal resistivity of 0.005 mK/watt, similar to the TSVs in the fabricated 3-D test circuit. Each test structure has been simulated over the full power range of the heater in WTop site 1. The simulation results are illustrated in Fig. 14.15.

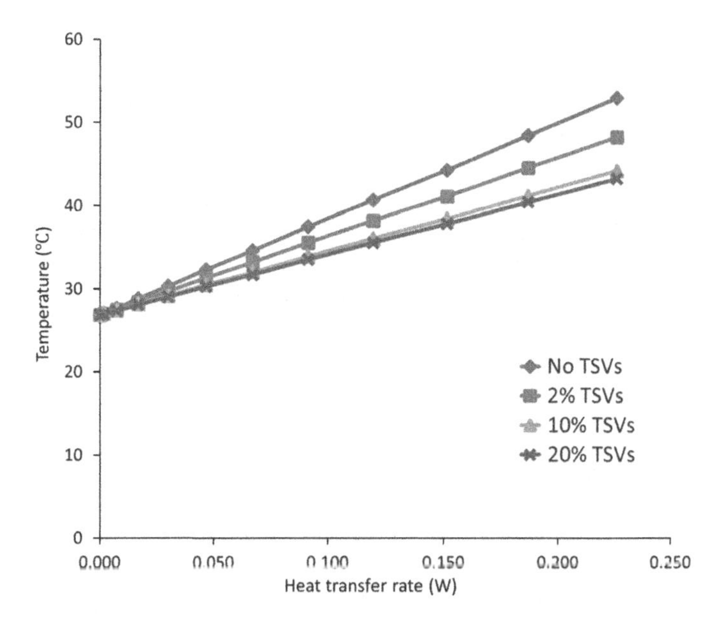

FIGURE 14.15

Simulated temperature at the WTop site 1 sensor for four densities of TSVs placed between the WTop site 1 heater/sensor pair and the back metal.

These results demonstrate that the temperature measured at WTop site 1 decreases with increasing density of TSVs. This result confirms the claim that additional TSVs form thermal paths from the on-chip hot spots to the heat sink. An interesting finding is that the decrease in temperature is not constant. This behavior is due to the number of parallel thermal resistors with increasing TSV density (similar to electrical resistors placed in parallel).

14.5 SUMMARY

The performance and reliability of a 3-D integrated circuit are greatly affected by significant heat gradients. The major contributions of this chapter are summarized as follows:

- Thermal effects can potentially alter the performance of the clock and power networks due to hot spot formation.
- Proper block placement and active and passive heat removal techniques are critical to ensure a 3-D IC operates within the specified thermal design power (TDP) envelope.
- A three-dimensional test circuit examining thermal propagation within a 3-D stack has been designed, fabricated, and tested.
- Five test conditions are examined to characterize thermal propagation in 3-D integrated circuits: (1) the location of the active circuit, (2) the impact of intratier thermal spreading, (3) the relative vertical alignment of the two active circuits, (4) heat flow from a thermal source at the back side of the silicon, and (5) convective cooling of the 3-D IC.
- The experimental data is confirmed by simulations conducted on a model of the fabricated 3-D test circuit.
- Based on experimental and simulated results, design suggestions are provided to better manage hot spot formation while reducing the effects on neighboring circuit blocks.
- Two important considerations for heat reduction are described: (1) the position of a block relative to a heat sink significantly affects the thermal resistance and therefore the flow of heat from the hot spots, and (2) the effective design of both the inter and intratier thermal conduits provides an important means for removing heat.
- The described test circuit provides enhanced understanding of thermal hot spot formation, propagation, and modeling of 3-D integrated circuits.

CHAPTER

SYNCHRONIZATION IN THREE-DIMENSIONAL ICs

15

CHAPTER OUTLINE

The clock signal constitutes the cornerstone of synchronous integrated systems as this signal provides the timing reference to maintain correct temporal operation. An important step in the delivery of the clock signal is the design of the clock distribution network, which for the majority of modern integrated systems includes a hierarchical clock network topology with several distribution levels [574]. Higher integration densities, larger silicon area, increasing power consumption, and process and environmental variability are some of the primary reasons that have led to multiple clock domains within modern integrated systems [264]. Consequently, the problem of distributing the clock signal in these complex systems can be described as the task of delivering a robust and reliable clock signal to all clock sinks (e.g., flip flops) across a circuit block, which can be as large as a processing core, or across a clock domain under skew, slew, power, and wiring constraints. Depending upon the performance specifications of the system, the significance of these constraints can vary, meaning that the design space changes according to the system specifications.

In hierarchical clock networks, the uppermost level of the hierarchy spans a large area and is often achieved with a symmetric topology requiring equidistant paths. These paths have the same traits (e.g., latency and slew) from the source to the roots of the next level of the clock network hierarchy. The global networks consist of several different topologies, such as H-tree, X-tree, ring, and mesh/grid topologies [575,576]. These global topologies often end at predetermined and reserved locations distributed throughout the system. Alternatively, the lower levels of the hierarchy connect to the clock pins of the circuit cells, where there can be several thousand clock sinks at each hierarchical level. A synthesized clock tree with specific timing characteristics typically

Three-Dimensional Integrated Circuit Design. DOI: http://dx.doi.org/10.1016/B978-0-12-410501-0.00015-0

provides the clock signal to all of these sinks. A multitude of clock tree synthesis (CTS) algorithms have been published over the past few decades that are supported by design automation tools for 2-D circuits.

With the advent of 3-D ICs, several efforts have focused on clock networks in three physical dimensions. The requirements for these 3-D clock synthesis techniques stem from the vertical interconnects (i.e., the through silicon vias (TSV)). The impedance characteristics of the TSV differ considerably from those of the horizontal wires and the partial clock networks that exist in each tier of a 3-D stack make pre-bond test a challenging task.

The design of 3-D clock network topologies is the topic of this chapter where both global topologies along with 3-D CTS algorithms are reviewed. Certain algorithms for the synthesis of two-dimensional (2-D) clock trees have been employed in the development of 3-D CTS methods. These algorithms are discussed in Section 15.1. The design and performance of 3-D global clock distribution networks are presented in Section 15.2. Synthesis techniques for 3-D clock trees are compared in Section 15.3. Enhanced techniques that stress practical issues relating to the synthesis of 3-D clock trees are reviewed in Section 15.4. The primary concepts of this chapter are summarized in Section 15.5. Note that most of the results described in this chapter are based on simulations and models while silicon measurements from an example test circuit characterizing the performance of global 3-D clock distribution networks are presented in Chapter 16, Case Study: Clock Distribution Networks for Three-Dimensional ICs.

15.1 SYNTHESIS TECHNIQUES FOR PLANAR CLOCK DISTRIBUTION NETWORKS

Synthesis techniques for clock distribution networks have evolved significantly over the past several decades, producing clock network topologies that satisfy several design objectives and constraints [577,578–582]. Among these methods, some early techniques, including the methods of means and medians (MMM) [578] and deferred-merge embedding (DME) [581], have been extended to 3-D clock distribution networks. Consequently, these two techniques are discussed in this section, providing the necessary background to better follow the sections dedicated to the synthesis of 3-D clock network topologies.

CTS techniques generate routed paths for a large number of clock sinks while ensuring a target clock skew is satisfied along with wirelength, clock slew, and power constraints. More precisely, clock skew is the difference between the clock signal arrival time of *sequentially-adjacent* registers [574]. Two registers are *sequentially-adjacent* if these registers are connected with combinatorial logic (i.e., no additional registers intervene between *sequentially-adjacent* registers). An illustration of this data path is provided in Fig. 15.1. An expression for the clock skew T_{skew} between these registers is also shown in Fig. 15.1, where T_{C_i} and T_{C_j} are, respectively, the arrival time of the clock signal at register R_i and R_j.

Other design objectives, such as robustness and reliability, are also used to enhance manufacturing technologies at the nanometer scale [584,585]. Minimizing any of these objectives typically entails tradeoffs with other design requirements due to the interdependence of the clock network characteristics. Consequently, synthesis methods ensure that design specifications are satisfied for all of the targeted objectives. For a large number of CTS methods, clock skew is the primary

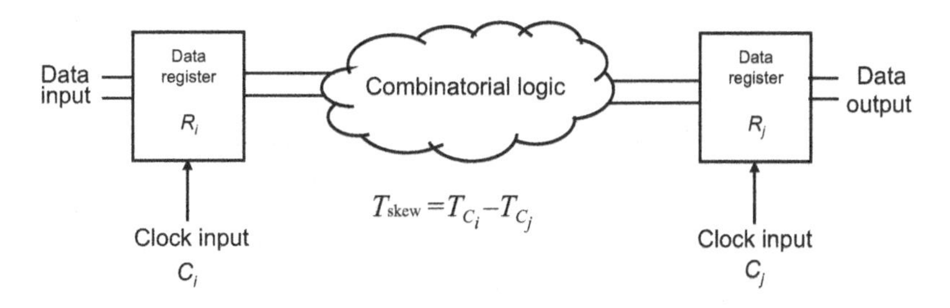

FIGURE 15.1

A data path depicting a pair of *sequentially-adjacent* registers [577,583].

objective, where both zero skew and bounded skew algorithms have been developed [579,586]. Both the MMM and DME techniques discussed in this section focus on clock skew.

15.1.1 METHOD OF MEANS AND MEDIANS

The MMM algorithm iteratively divides the clock sinks within a routing grid (or plane) with the use of *cuts* (i.e., bipartitions) in the x and y directions, producing two new sets of clock sinks for each *cut*. Assuming a set of m clock sinks, each clock sink is notated as s_i with coordinates (x_i, y_i) in the routing plane. The clock sinks can be described by the finite set $S = \{s_1, s_2, s_3, \ldots s_m\} \subset \Re^2$. The location of the center of mass for this set is

$$x_c = \frac{\sum_{i=1}^{m} x_i}{m}, \tag{15.1}$$

$$y_c = \frac{\sum_{i=1}^{m} y_i}{m}, \tag{15.2}$$

where these coordinates constitute the root of the clock tree. To proceed with the *cuts* along the physical directions, an ordering of the clock sinks is utilized. Two different orderings of the clock sinks are formed by separately considering the x and y coordinates of the clock sinks in increasing order, meaning that for each element of a subset $S_x(S)$ of the set, s_i precedes s_j in $S_x(S)$ if $x_i \leq x_j$. A similar rule applies to the subset $S_y(S)$ with respect to the y-coordinate. With these orderings, four different subsets are defined around the median of each ordering,

$$S_L(S) \equiv \left\{s_i \in S_x | i \leq \left\lceil \frac{m}{2} \right\rceil \right\}, \tag{15.3}$$

$$S_R(S) \equiv \left\{s_i \in S_x | \left\lceil \frac{m}{2} \right\rceil < i \leq m \right\}, \tag{15.4}$$

$$S_B(S) \equiv \left\{s_i \in S_y | i \leq \left\lceil \frac{m}{2} \right\rceil \right\}, \tag{15.5}$$

$$S_T(S) \equiv \left\{s_i \in S_y | \left\lceil \frac{m}{2} \right\rceil < i \leq m \right\}. \tag{15.6}$$

These four subsets are produced by applying a *cut* at the median of each ordering along each direction. All of these subsets contain approximately the same number of clock sinks as $| |S_L| - |S_R| | \leq 1$. The same inequality applies to the subsets S_B and S_T. The center of mass is determined for each of these new subsets. The algorithm connects these newly determined points with the center of the mass of the initial set S, which is the root of the tree. This process is repeated, producing two sets, left (L) and right (R), for each x-cut and bottom (B) and top (T) sets for each y-cut. The process terminates when each new set includes only a single clock sink.

A simple example of this algorithm is shown in Fig. 15.2, where routes between the center of mass for different sets are noted. Although routes between a center of mass and a subsequent center of mass, determined by a *cut* in the next iteration, are of equal length, different route lengths can occur as the algorithm progresses, as depicted in Fig. 15.2A. To address this situation, two pairs of *cuts* are applied. The pair of *cuts* leads to lower skew. This look-ahead procedure yields superior routings. As shown in Fig. 15.2B, the tree resulting from the y- and x-cuts exhibits a lower skew as compared to the tree shown in Fig. 15.2A. The complexity of the algorithm, including the look-ahead function, is $O(m \log m)$.

The MMM method ensures that after each iteration where *cuts* are performed, the routing path to the center of mass of the newly formed subsets is of equal length, leading to low (or zero) skew. This routing, however, is typically longer than the length of the minimum rectilinear Steiner tree (RMST) with the same set of clock sinks [403]. As described in [578], the wirelength of the generated clock tree increases proportionally to $3/2\sqrt{m}$, while for RMST, the length of the tree grows as $(\sqrt{m}+1)$ for m points (i.e., clock sinks) uniformly distributed across a routing grid. The ratio between the two routing algorithms increases by a constant factor of approximately 1.5×, demonstrating that the wirelength overhead of the MMM method is not excessive.

To limit this additional overhead, the MMM method is used in two phases. In the first phase, where the generated *cut* subsets include a large number of clock sinks, the routing proceeds with the MMM method. In the second phase, where the number of clock sinks within each subset produced at later iterations is small, standard routing algorithms are preferable to guarantee minimum wirelength with a low and acceptable increase in clock skew. The depth of the tree (or number of iterations) after which a routing technique other than MMM is applied depends upon the skew that can be tolerated.

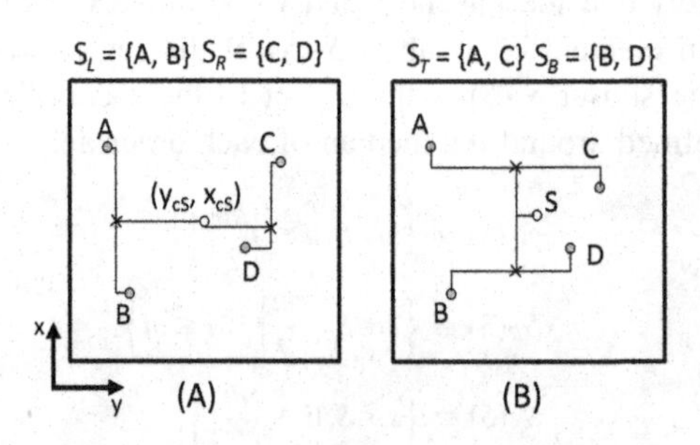

FIGURE 15.2

Simple example of the MMM clock synthesis method where a clock tree is generated. (A) Without look-ahead, and (B) with look-ahead. An *xy*-cut leads to larger skew in (A) than a *yx*-cut in (B).

15.1.2 DEFERRED-MERGE EMBEDDING METHOD

Another popular technique that has been extended to support the synthesis of 3-D clock networks is the DME method, which is a two phase CTS technique [581,582]. The input to the method is the connectivity of the sinks within some topology (typically a tree), where the location of the sinks is known and the internal nodes are placed to ensure that zero skew is achieved at minimum wirelength (assuming a linear delay model) [581]. Consequently, the DME technique produces optimal zero skew clock networks assuming a linear delay model, while suboptimal clock trees are produced if other more complex delay models, such as the Elmore delay model, are utilized. In this case, although the results are suboptimal, these results are in general of high quality, yielding zero skew clock trees with low wirelength. The advantage of the DME method is that as compared to MMM, the wirelength is minimized in addition to constructing a zero skew clock network.

The DME technique is not constrained by the topology of the clock network and, therefore, can be applied as a post-processing step once the internal nodes of a clock tree have, for example, been generated by the MMM method. Note, however, that the choice of topology or the choice of technique to generate the clock network affects the quality of the results produced by the DME method (i.e., the total wirelength since zero skew is guaranteed).

The DME method consists of a bottom-up and a top-down phase. In the bottom-up phase the feasible placement locations of each internal node are determined. Each internal node is a parent node for the root of (two) merged subtrees. In the top-down phase, the internal nodes are embedded into one of these feasible locations, ensuring that minimum wirelength is achieved (for the linear delay model) in addition to guaranteeing zero skew for all of the sinks within the clock network.

To explain this method, some related terminology is introduced. In the MMM method, a finite set S containing m clock sinks and the corresponding location of these sinks are considered as input. A clock tree notated as $T(S)$ is embedded by the DME method in the Manhattan plane, where the location of each internal node of the tree is determined to ensure that the skew is zero and the wirelength of the clock tree is minimum. The placement of each node is denoted as $pl(u)$. Assuming that u is the parent node of another node w, e_w denotes the directed edge from the parent to the child node. The cost of this edge is notated as $|e_w|$. As the Manhattan plane embeds the clock network, this cost is equal (or larger) to the Manhattan distance between these two nodes. Accordingly, the cost of the entire tree $T(S)$ is the sum of the wirelength of all of these edges.

Similarly, the delay of any path from source node u to node w is notated as $t_d(s_u,s_w)$. The clock skew between two paths from the same source node is described by $|t_d(s_u,s_w)-t_d(s_u,s_v)|$. Assuming the root of the clock network is s_r, the clock skew of $T(S)$ is the maximum difference $|t_d(s_r,s_i)-t_d(s_r,s_j)|$ among all pairs of sinks s_i, s_j belonging to S. If the linear delay model is used to estimate the delay of each source-sink path, this delay is the summation of the edges constituting the path from a source node u to a sink node w,

$$t_{ld}(u,w) = \sum\nolimits_{e_v \in \text{path}(u,w)} |e_v|. \tag{15.7}$$

In addition to this notation, some other definitions are required to describe the DME method. The loci for each internal node determined during the bottom-up step are called *merging segments*. The merging segments are line segments with slope ± 1 or, alternatively, with an angle of $\pm 45°$ from the x–y routing directions. Similarly, a set of points on the Manhattan plane within a fixed distance from a Manhattan arc forms a *tilted rectangular region* (TRR), which is inscribed within

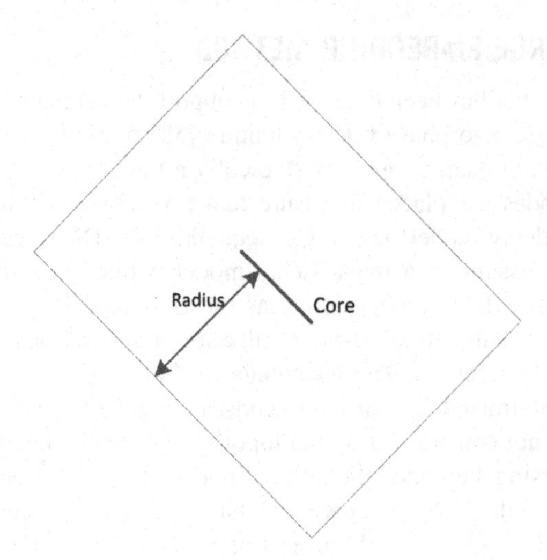

FIGURE 15.3

TRR where the *core* of the region is a Manhattan arc and the boundary points are at a *radius* distance from the *core*.

Manhattan arcs. An example of a TRR is illustrated in Fig. 15.3, where the *core* arc and *radius* of the TRR are also shown. The *core* arc consists of all points located farthest from the boundary of the TRR (always assuming a Manhattan distance).

Considering a node u, the merging segment $ms(u)$ for this node is determined as follows. If u belongs to S, the merging segment is the location of the clock sink; otherwise, u is an internal node and the length and position of the merging segment depend on the merging segments of the children nodes of u. Assuming that nodes a and b are the children nodes of u and since this step proceeds in a bottom-up manner, the merging segments of nodes a and b, notated, respectively, as $ms(a)$ and $ms(b)$ are known and are also Manhattan arcs [581]. This situation is illustrated in Fig. 15.4. The merging segment of node u is also a Manhattan arc and is obtained by intersecting two TRRs that have as core the segments $ms(a)$ and $ms(b)$. Consequently, $ms(u)$ is formally described as $ms(u) = \text{TRR}_a \cap \text{TRR}_b$. The radius of TRR_a and TRR_b is, respectively, $|e_a|$ and $|e_b|$, as also shown in the figure. The merging cost for the subtrees rooted at nodes a and b is $|e_a| + |e_b|$. The objective is to determine these radii (i.e., $|e_a|$ and $|e_b|$) such that the merging cost is minimized and the constraint of zero skew is satisfied.

Assuming a linear delay model, the lower bound for the merging cost is equal to the minimum Manhattan distance between $ms(a)$ and $ms(b)$, denoted as $MD_{\min} = d(ms(a), ms(b))$. If the delay of nodes a and b is highly unbalanced, the merging cost deviates from this minimum to satisfy the zero skew requirement and may require additional wiring. The requirement of zero skew at node u can be satisfied from

$$t_{ld}(a) + |e_a| = t_{ld}(b) + |e_b|, \tag{15.8}$$

where $t_{ld}(a)$ and $t_{ld}(b)$ denote, respectively, the (linear model) delay to the sinks of the subtrees rooted at nodes a and b. Note that since the zero skew requirement is satisfied at each level of the

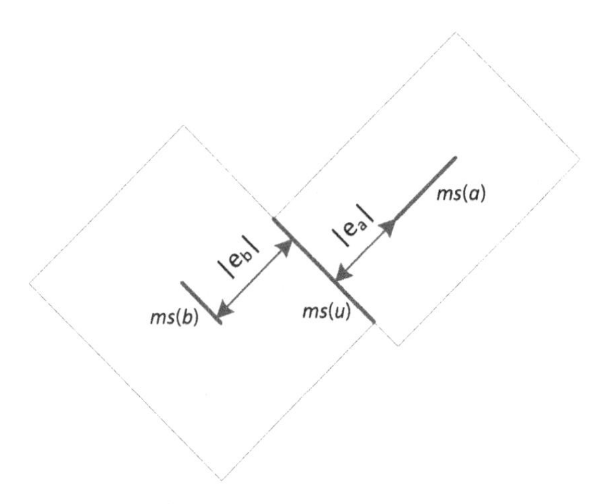

FIGURE 15.4

Merging segment $ms(u)$ for node u that is the parent node of nodes a and b based on TRR_a and TRR_b.

tree, these delays are the same at each sink of the subtrees from these nodes. The radius of TRR_a and TRR_b is obtained by letting $|e_a| + |e_b| = MD_{\min}$, and if $|t_{ld}(a) - t_{ld}(b)| \leq MD_{\min}$,

$$|e_a| = \frac{MD_{\min} + t_{ld}(b) - t_{ld}(a)}{2}, \tag{15.9}$$

and

$$|e_b| = MD_{\min} - |e_a|. \tag{15.10}$$

If the condition $|t_{ld}(a) - t_{ld}(b)| \leq MD_{\min}$ is not satisfied, additional wirelength is required to balance the skew from node u to nodes a and b, increasing the merging cost.

The bottom-up phase recursively merges the nodes of $T(S)$ until the root node s_r is reached. The clock driver is often placed at a specific location which is not intercepted by the merging segment of the root node. An additional segment connects to s_r without degrading the performance of the resulting tree. The bottom-up phase (i.e., *deferred-merge*) terminates when the merging segment for the root node is obtained.

The top-down phase (i.e., *embedding*) follows. The exact placement of each node $pl(u)$ is successively determined starting from the root node. For s_r, any point along $ms(s_r)$ can, in general, be chosen as $pl(s_r)$. For any other internal node u, any point on $ms(u)$ can be selected as $pl(u)$, which is at a distance $|e_u|$ or less from the placement $pl(p)$ of the parent node of u. From the construction of the merging segments, this location must exist [582]. An example of this situation is illustrated in Fig. 15.5.

Operation of the DME technique can be better understood with a practical example. A simple clock topology with eight sinks (e.g., s_1 to s_8) is depicted in Fig. 15.6. The assumption of the same (or zero) delay for each sink permits the TRR for each sink to be determined. As the location of each sink is known and zero delay is assumed, $t_{ld}(s_i) = 0 \forall s_i$ (15.9) and (15.10) are equal to half the

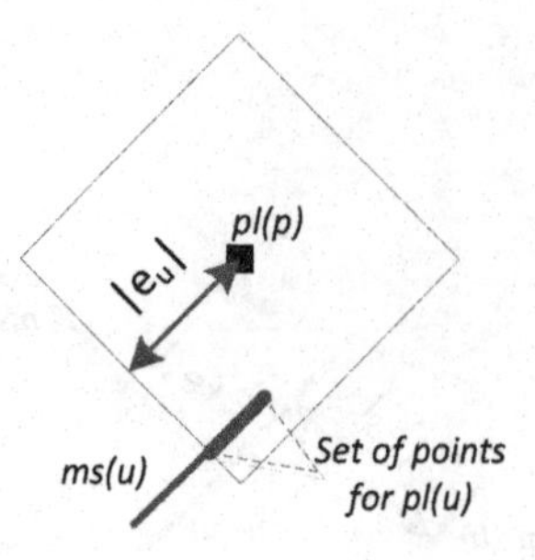

FIGURE 15.5

TRR with a *core* point. The placement location of the parent node p, $pl(p)$ (which is known from the previous iteration) and *radius* equal to the wirelength of edge e_u. The segment of $ms(u)$ within the TRR is the thick line and represents the set of valid placement locations for node u.

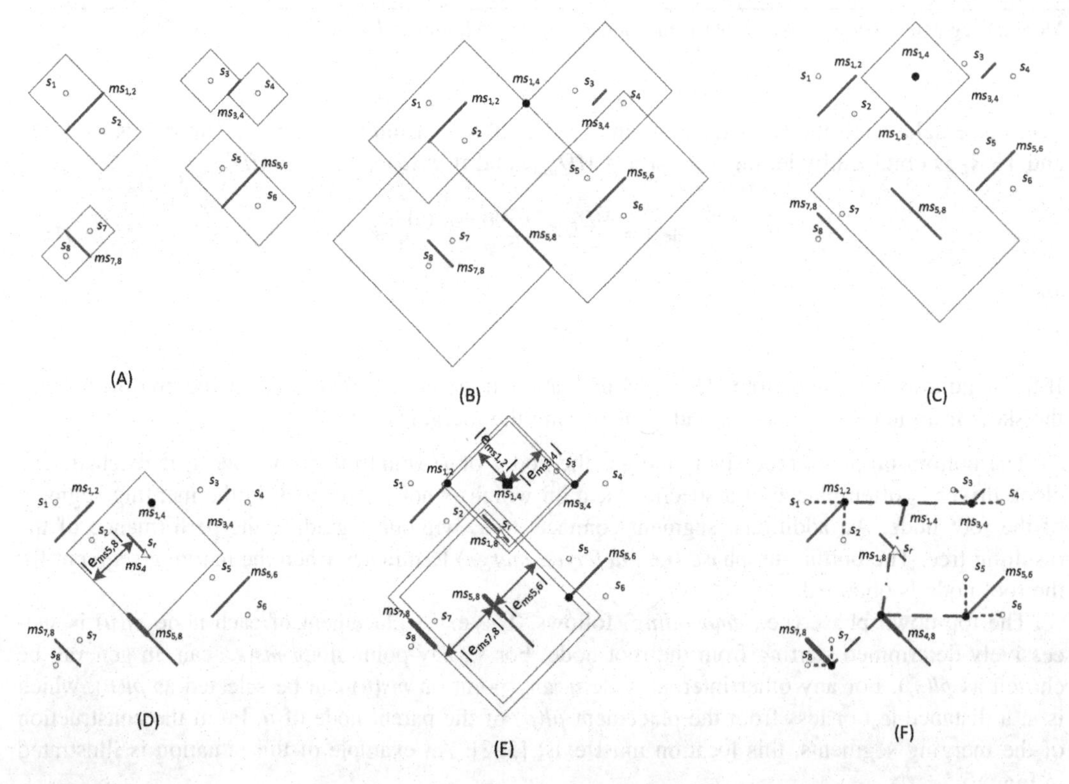

FIGURE 15.6

Example of the DME method for a tree with eight sinks. (A) to (C) Bottom-up phase where the recursive derivation of the merging segments is accomplished and (D) to (F) top-down phase where the exact placement of each internal node is determined.

Manhattan distance between each pair of sinks s_i and s_j (see Fig. 15.6A). The radius of the TRRs for each pair of sinks is the same, and the merging segment for each parent node $ms_{i,j}$ is illustrated by the thick lines. In Fig. 15.6B, the merging segments of the nodes at the next upper level of the tree are obtained. Note that in this specific example, $ms_{1,4}$ is a single point as the TRRs for nodes $s_{1,2}$ and $s_{3,4}$ only intersect at one point. In Fig. 15.6C, the merging segment $ms_{1,8}$ for the root node s_r is at the intersection of the TRRs for nodes $s_{1,4}$ and $s_{5,8}$.

Since the merging segments of all nodes have been determined, the top-down phase determines the placement of the internal nodes within the tree. As previously mentioned for the root of the tree, any point along $ms_{1,8}$ can be chosen. This point is indicated by a triangle in Fig. 15.6D. To determine $pl(s_{1,4})$ and $pl(s_{5,8})$, the TRRs are shown with, respectively, radius $|e_{ms1,4}|$ and $|e_{ms5,8}|$ from the placement of the root node $pl(s_r)$. As $ms_{1,4}$ corresponds to only a single point, only the TRR for node $s_{1,4}$ is plotted in Fig. 15.6D. The thickest line segment depicts the valid placement points for node $s_{5,8}$, where any point on this segment can be selected as $pl(s_{5,8})$. Similarly, the TRRs are drawn in Fig. 15.6E to determine the valid placement location for nodes $s_{1,2}$, $s_{3,4}$, $s_{5,6}$, and $s_{7,8}$. Note that for nodes $s_{1,2}$ and $s_{3,4}$, valid placements are possible only at a single point. The resulting clock tree is shown in Fig. 15.6F, where the solid lines indicate merging segments and all other lines depict the branches of the tree. The dots depict the placement of the internal nodes within the tree.

15.2 GLOBAL THREE-DIMENSIONAL CLOCK DISTRIBUTION NETWORKS

A number of clock network topologies have been developed for 2-D circuits. These topologies can be symmetric, such as H- and X-trees, highly asymmetric, such as buffered trees and serpentine-shaped structures [575,583], and grid like structures, such as rings and meshes [587]. Clock distribution networks are typically structured as a global network with multiple smaller local networks. Within the global clock network, the clock signal is distributed to specific locations across the circuit. These locations are the source of the local networks that pass the clock signal to the registers or other circuit components.

A symmetric structure, such as an H-tree, is often utilized in global clock networks [583], as shown in Fig. 15.7. The most attractive characteristic of symmetric structures is that the clock signal ideally arrives simultaneously at each leaf of the clock tree.

The traversed distances are, however, substantially longer as compared to minimum length trees to preserve symmetry. Due to the extremely long interconnects in global clock networks operating at gigahertz frequencies, clock networks are typically modeled as a transmission line, since inductive behavior is likely to occur [588]. Inductive behavior can cause multiple reflections at the branch points, directly affecting the performance and power consumed by the clock network. To lessen the reflections at the branch points of H-trees, the interconnect width of the segments at each branch point is halved (for a $2\times$ change in the line width) to ensure that the total impedance seen at that branch point is maintained constant (matched impedance).

Due to several reasons, however, such as load imbalances, process variations, and crosstalk, the arrival time of the clock signal at various locations within a symmetric tree can be different, producing clock skew. To mitigate this situation, a combination of an H-tree and a mesh is often employed, as shown in Fig. 15.8 [587]. There are several ways to implement these networks in 3-D circuits. A straightforward extension of a two tier circuit is illustrated in Fig. 15.9. Both of

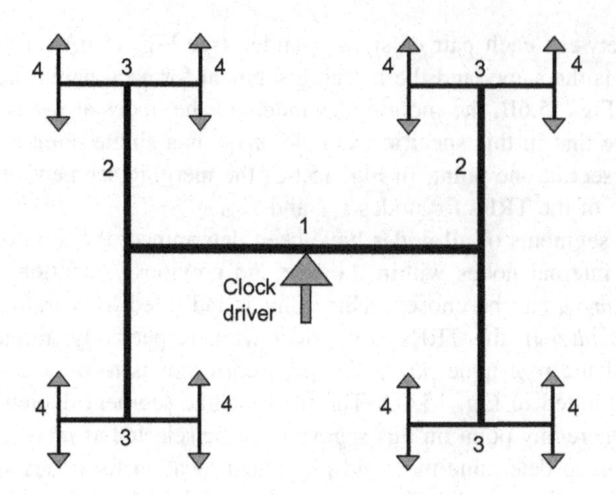

FIGURE 15.7

Two-dimensional four level H-tree.

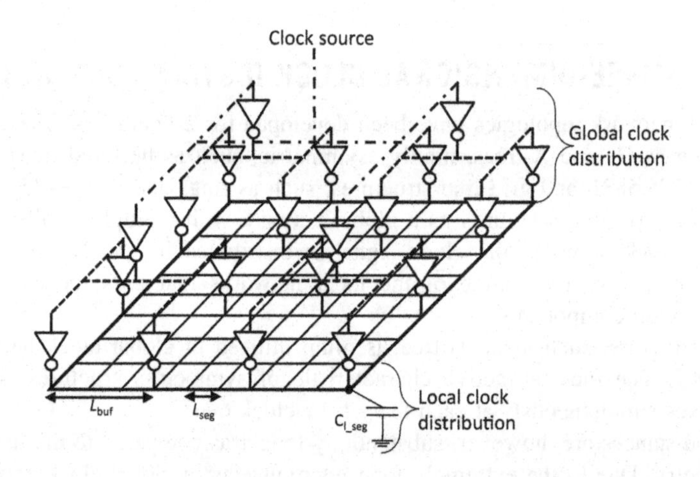

FIGURE 15.8

Buffered and symmetric clock tree that drives a grid, where each unit grid constitutes a local clock network modeled as a lumped capacitor C_{l_seg}.

these topologies include two H-trees, yet the behavior of these topologies is quite different. The network shown in Fig. 15.9A utilizes only one H-tree network to provide the clock signal in both of the tiers. The clock signal is propagated to the lower tier by TSVs connected to the leaves of the H-tree in the upper tier. Local clock networks are rooted at the TSVs in the lower tier, distributing the clock signal to the clock pins in that tier. The H-tree shown with the dashed line is only used for pre-bond test and is disconnected once the two tiers are bonded. Thus, some redundancy in this topology exists but much less wiring is required in normal post-bond operation as compared to the

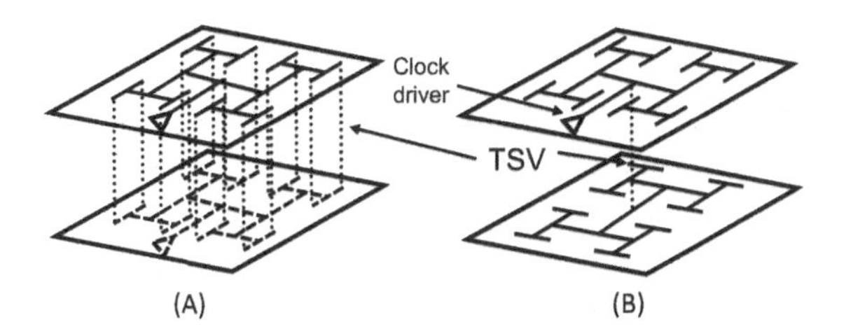

FIGURE 15.9

Global 3-D clock distribution networks based on planar symmetric H-trees, where during normal operation (A) one H-tree and multiple TSVs distribute the clock signal, and (B) two H-trees and a root TSV distribute the clock signal.

topology shown in Fig. 15.8. Indeed, the topology depicted in Fig. 15.9B requires two full H-trees in both tiers during normal (post-bond) operation, dissipating greater power. A group of TSVs (resembling a vertical trunk) connects the roots of the two H-trees, simplifying, in general, the clock distribution within the stack.

Another difference between the two topologies is the number and size of the clock buffers that drive the clock load. As the clock load shown in Fig. 15.9A is driven by a network placed within a single tier, more and larger clock buffers are used as compared to the number and size of the clock buffers in each tier of the topology depicted in Fig. 15.9B. Note that the total number of buffers for the network shown in Fig. 15.9B can be greater. Extending these topologies to a larger number of tiers exacerbates this tradeoff. More buffers are needed to drive the clock load within all of the tiers through stacked TSVs connected to the leaves of the H-tree in only one of the tiers. Alternatively, replicating the H-tree in all of the tiers, as shown in Fig. 15.9B, is a reasonable approach as fewer TSVs are used. The large amount of wiring required for this topology may, however, be prohibitive. An important issue that also arises is the robustness of these two topologies and, more generally, of 3-D clock networks to process variations. This issue is discussed in Chapter 17, Variability Issues in Three-Dimensional ICs.

An alternative topology that is more robust to process variations and load imbalances is the combination of a tree and mesh, as shown in Fig. 15.8. In a case study, the option to dedicate an entire tier to the clock circuits and global networks has been investigated [589]. Separating the task of clock delivery from the logic provides opportunities to individually optimize the clock network and any related circuits that generate the clock signal.

The key concept is to place a mesh within each tier and a global symmetric tree (such as an H-tree) to drive the meshes through TSVs. This dedicated clock tier is placed at the middle of a 3-D stack for the purpose of symmetry, as depicted in Fig. 15.10 for a network spanning four logic tiers. In this configuration, the clock tier is connected to the other tiers from both faces, which means that both face-to-back and face-to-face interconnections are necessary. The intertier interconnects between these two bonding styles are different (e.g., microbumps vs. TSVs), an issue which needs to be considered during the design process.

To assess the performance of this approach, the Alpha 21264 processor has been deployed within a 3-D system [590]. A 2-D floorplan of this processor is split into quadrants, each quadrant placed onto a physical tier, forming a five tier stack including the dedicated clock tier. The cache is

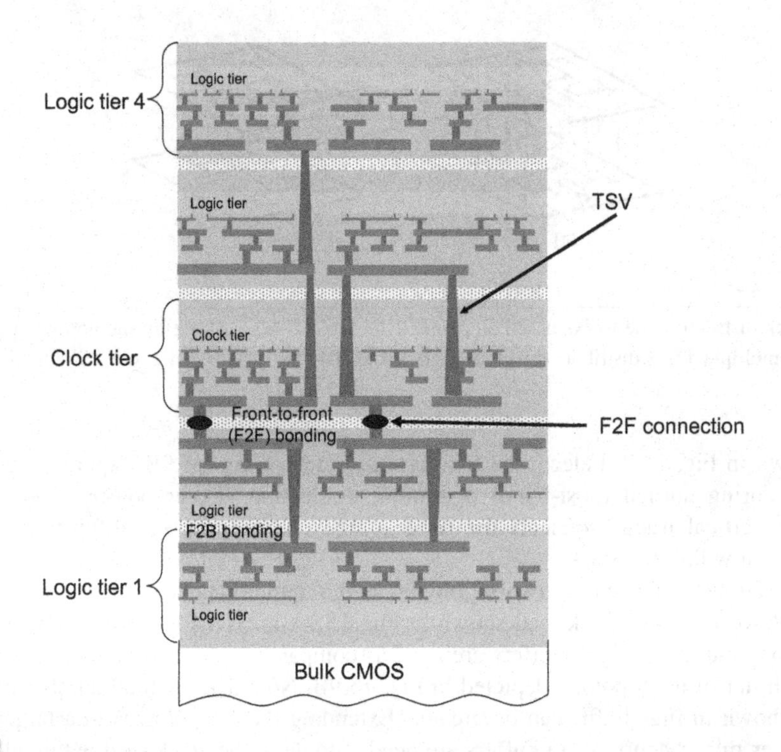

FIGURE 15.10

Cross-section of a 3-D stack of five tiers with one dedicated clock tier and four logic tiers [589].

split into two tiers and the integer and floating point units are each placed onto two separate tiers. The instruction fetch unit is shared between one of the cache tiers and the tier with the floating point unit. Although this floorplan is rather crude, it offers a reasonable test case for evaluating the performance of a clock tier. The capacitance of each tier is estimated and assigned to each grid unit, where the clock grid has a granularity of $L_{seg} = 100$ μm. Clock buffers are inserted at intervals of $L_{buf} = 300$ μm (see Fig. 15.8).

Based on this partition for the planar floorplan of the microprocessor, which is also assumed to be driven by a combined tree and grid network structure, the footprint of the circuit is greatly reduced. More specifically, the area of the clock grid in the 2-D processor is 2.5 mm × 2.5 mm, which decreases to 1.25 mm × 1.25 mm for the 3-D processor since four tiers are utilized. This decrease in area leads to fewer final stage buffers. The number of these buffers drops from 81 to 25. The buffers drive the grid in all four tiers, which has a total area similar to the 2-D grid since the initial planar floorplan is "folded" to produce the 3-D processor. Consequently, the size of these buffers in a 2-D system is twice as large as a 3-D system. However, as both the total number of buffers and the area of the clock tree driving the grids within the logic tiers have decreased, the power consumption of the clock network has also been reduced. For a 32 nm process node [252], the maximum skew and power consumed by the planar and 3-D system are reported in Table 15.1.

Table 15.1 Comparison of Clock Skew and Power Dissipation Between 2-D and Four Tier 3-D Microprocessor With Dedicated Clock Tier [589]

	Floorplan	Skew (ps)	Power (W)
Slice 1	2-D	9.86	3.15
	3-D	7.9	2.67
Slice 2	2-D	3.04	3.04
	3-D	2.51	2.51

The smaller dimensions of the clock network and the fewer buffers decrease both the skew and the power by 15% to 20%.

The drawback of this global clock distribution network is the requirement for a dedicated tier for the clock circuitry and network. This approach suffers from the same difficulties related to pre-bond test as the topology shown in Fig. 15.9A. Consequently, a more practical approach would combine a clock tree with a mesh in each tier sharing a common clock source [591]. In Fig. 15.11, the basic elements of a synchronization system for a two tier 3-D circuit is illustrated. An external reference clock is provided prior to bonding the clock generating circuitry. After bonding, a tristate multiplexer in tier 2 disconnects the intratier phase locked loop (PLL) and selects the output from the PLL in tier 1. In this configuration, both networks are driven from a single PLL and are connected at the root of the global tree. The phase detector and DLL in tier 2 ensure that the clock signals within the two tiers are temporally aligned. The clock skew between the two tiers primarily depends upon the process variations and bandwidth of the DLL [591].

An alternative approach that can suppress the additional skew due to process variations increases the number of connections between the clock networks of the two tiers. With this topology, the clock phase alignment circuit can be removed, simplifying the clock circuitry. Furthermore, as several more nodes are shorted between the clock networks, the process variations can be compensated, reducing the overall intertier skew. Interestingly, the power does not decrease, which, in principle, is counterintuitive. This contradictory behavior is due to the use of electrostatic discharge (ESD) diodes for each TSV to protect the driving circuits during manufacturing, which considerably increases the overall power consumption to deliver the clock. Another drawback of this approach is that the physical design of the network is constrained to provide appropriate TSV connections at the shorting nodes.

These topologies have been evaluated on an eDRAM. The test circuit is 5.6 mm × 10.9 mm for an IBM 45 nm silicon-on-insulator (SOI) technology [592] and operates at 2.5 GHz. The TSV shorts add a power overhead of only 1%, but the overhead due to the capacitive load of the ESD diodes increases the power consumption by 10% to 30% [591]. Corner analysis has also shown that the approach with shorted grids and typical process corners for both tiers exhibits twice as low a clock skew as compared with the case where the tiers are in different corners (e.g., fast-slow) [591].

These results not only demonstrate the important role of selecting the appropriate number of TSV to manage skew but also show the importance of process variations in 3-D ICs where both

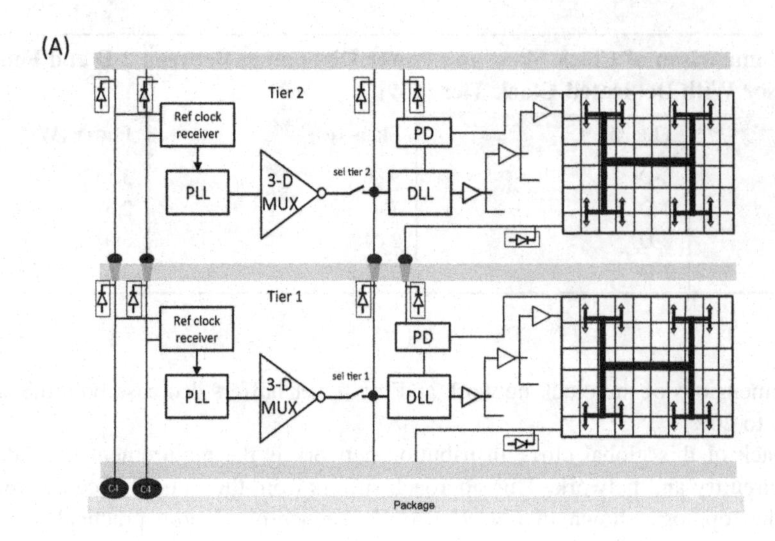

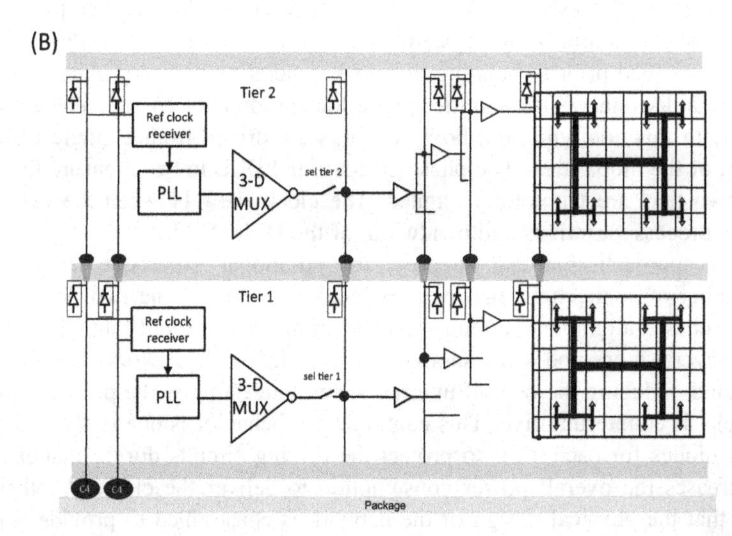

FIGURE 15.11

Two clock delivery networks, (A) the networks are shorted only at the initial stages of the clock distribution, and (B) TSVs connect the clock networks at the lower levels of the clock network hierarchy [591].

intertier and intratier variability must be considered when determining the performance of a 3-D clock network. Due to the importance of this issue, the effects of process and environmental variability on the design of global clock distribution networks are discussed in Chapter 17, Variability Issues in Three-Dimensional ICs.

As a final remark for this section, note that the existing clock synthesis techniques discussed in the following sections have not sufficiently investigated this issue. Consequently, significant effort remains for further improving the present art in 3-D CTS.

15.3 SYNTHESIS OF THREE-DIMENSIONAL CLOCK DISTRIBUTION NETWORKS[1]

Similar to previously discussed techniques, tackling other physical design issues for 3-D circuits such as floorplanning, planar CTS methods have also been adapted to manage the multi-tier nature of these systems and the vertical connections. The majority of 3-D CTS algorithms consist of two primary stages. In the first stage, a suitably modified version of the MMM algorithm, as discussed in Section 15.1, produces a connection topology for the clock sinks located on several tiers. Different variants of this topology generation algorithm have been developed depending upon the target objective(s) of the synthesis technique. In the second stage, another popular technique, DME (also discussed in Section 15.1), is used after appropriate adjustments to embed the 3-D connection topology into a multi-tier stack.

The diversity of 3-D CTS methods can be attributed to the disparate performance objectives as well as the evolution of these methods over the past few years [593–598]. As listed in Table 15.2, 3-D CTS techniques produce clock networks that satisfy a number of objectives in addition to zero (or close to zero) skew and (near-)minimum wirelength. An important objective is lower power consumed by the 3-D clock trees, which, as reported in several works [593–597], entails several tradeoffs among the number of TSVs, wirelength of the 3-D tree, and the TSV capacitance.

Considering the global topology and the structures shown in Fig. 15.9, similar tradeoffs have been observed [593,595] where multi-TSV clock trees dissipate, in general, less power than single TSV trees. Limiting the number of TSVs within a 3-D clock tree does not increase power if the TSV capacitance is high, as this capacitive load is counter balanced by the savings due to the lower capacitance of the horizontal interconnects.

Additionally, pre-bond test, which has been shown to be a useful means to improve the yield of 3-D systems [599], requires a different treatment of the clock synthesis process, a situation unique to 3-D systems. To better understand the added complexity of pre-bond testing to the clock synthesis process, consider the example shown in Fig. 15.12 of a two tier clock tree with several sinks and a multi-TSV topology.

Table 15.2 CTS Methods for 3-D Circuits and Related Objectives Satisfied by Each Technique

	Objective/Constraint							
CTS Technique	Zero Skew	Clock Slew	Power	Number of TSVs	Pre-bond Testability	Buffer Insertion	Whitespace Aware	TSV Defect Tolerance
[593]	✓	✓	✓	✓	✗	✓	✗	✗
[594]	✓	✓	✓	✓	✗	✗	✗	✗
[595]	✓	✓	✓	✓	✓	✓	✗	✗
[596]	✓	✓	✓	✓	✓	✓	✗	✗
[597]	✓	✓	✓	✓	✗	✓	✓	✗
[598]	✓	✗	✗	✗	✗	✓	✗	✓

[1]To enhance readability, different naming conventions are used to describe the techniques presented herein as compared to those in the original source publications.

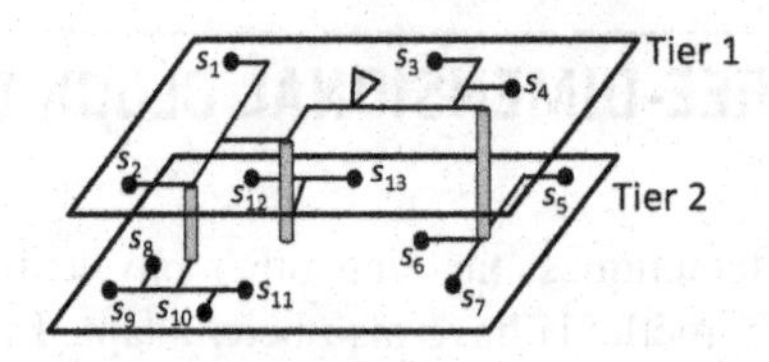

FIGURE 15.12

Multi-TSV clock tree with 13 sinks and three TSVs spanning two tiers.

The sinks in tier 2 form three disconnected subtrees connected with the subtree in tier 1, forming a 3-D tree connecting all of the 13 sinks. To support pre-bond test for both tiers, a clock tree should exist in either tier and satisfy all of the objectives as compared to a fully connected clock tree of the bonded two tier system. This requirement cannot be satisfied with the illustrated topology unless a redundant network is utilized in tier 2. This redundant network should ensure that the same clock skew and slew are maintained as for the post-bond 3-D tree. In addition, a pre-bond test for tier 1 does not require a redundant tree; however, since the clock loads are different than the full tree, the characteristics of the clock network also differ. Clock synthesis techniques that consider this situation have been developed, where these techniques consider routing and placement obstacles [600] and the placement of TSVs within the available whitespace.

Another issue considered in the CTS process is fault tolerance. As TSVs can suffer from fabrication or assembly defects [601], fault tolerance is an important issue.

The characteristics of several clock synthesis methods are summarized in Table 15.2, where the different objectives addressed by each technique are listed. In the following subsections, techniques with and without pre-bond test are described separately due to the different approaches and resources.

15.3.1 STANDARD SYNTHESIS TECHNIQUES

Classic synthesis techniques for 2-D clock trees are discussed in Section 15.1. These approaches can produce clock trees for 3-D circuits, assuming that each tier contains a tree connecting all of the clock sinks within that tier. The roots of all of these trees are connected through a single vertical connection (e.g., a few TSVs connected in parallel), providing a complete tree spanning the entire 3-D circuit. This single TSV topology, however, is a rather naïve approach as the advantages of the third dimension are not adequately exploited [593]. Consequently, a multi-TSV topology is a superior approach, offering a considerable reduction in wirelength, which typically translates to lower power.

The 3-D clock synthesis problem has been casted in several ways, where the primary objectives are to produce a zero-skew multi-TSV tree with minimum (or near-minimum) wirelength and/or power and the minimum [594] or a bounded number of TSVs [593] for a set of clock sinks $S = \{s_1, s_2, s_3, \ldots s_m\} \subset \Re^3$. The clock sinks span several physical tiers. The location of these sinks is described by a triplet of coordinates (x_i, y_i, z_i) where z_i indicates the tier of s_i. In addition, the slew should satisfy a target constraint that is typically less than 5% of the clock period.

Techniques for multi-TSV clock tree synthesis proceed in two phases. A topology connecting the clock sinks in all of the tiers of a 3-D system is first generated, followed by an extension of the

DME method to embed the topology throughout the tiers of the 3-D system. Topology generation extends the traditional MMM algorithm. As described in Section 15.1, the MMM algorithm operates recursively by bipartitioning a set of clock sinks using an x- or y-cut until only one sink belongs to each set. In 3-D circuits, the clock sinks within a subset produced by a *cut* are located in different tiers, where those sinks are connected by TSVs. Additionally, a z-cut divides S in several ways, resulting in different demands for TSVs.

The classic MMM algorithm is, therefore, indifferent to TSVs, which can lead to undesirable situations either due to insufficient usage of TSVs to reduce the total wirelength or excessive utilization of TSVs, resulting in lower savings (or even increasing) in power. To address these issues, the MMM algorithm has been extended to control the number of TSVs required by the synthesized tree topologies [593,594].

The MMM-TSV-Bound algorithm, called MMM-TB, generates topologies based on satisfying a user specified upper limit on the number of TSVs. The TSVs are distributed across the area of the tiers [593]. An example of the effect that different bounds of TSVs have on the resulting topologies is shown in Fig. 15.13. As observed in this figure, for larger TSV bounds, additional TSVs are used, where these TSVs appear closer to those sinks connecting a smaller number of internal nodes and/or sinks. This behavior decreases the total wirelength since the horizontal interconnections are shorter. The MMM-TB algorithm begins with a set of sinks S and a TSV bound for each tier within the 3-D system. If *stack*(S) denotes the number of tiers spanning the sinks of S, at each partitioning iteration, two sets S_1 and S_2 are produced based on two conditions.

If the TSV bound is one, the existing set S is partitioned to ensure that sinks with the same z_i (i.e., from the same tier) are assigned to the same subset. This partition is straightforward as long as *stack*(S) = 2. If, however, the sinks span more than two tiers and since *cuts* bipartition set S, there are (*stack*(S)−1) iterations that partition the sinks into the z-direction to ensure that sinks with

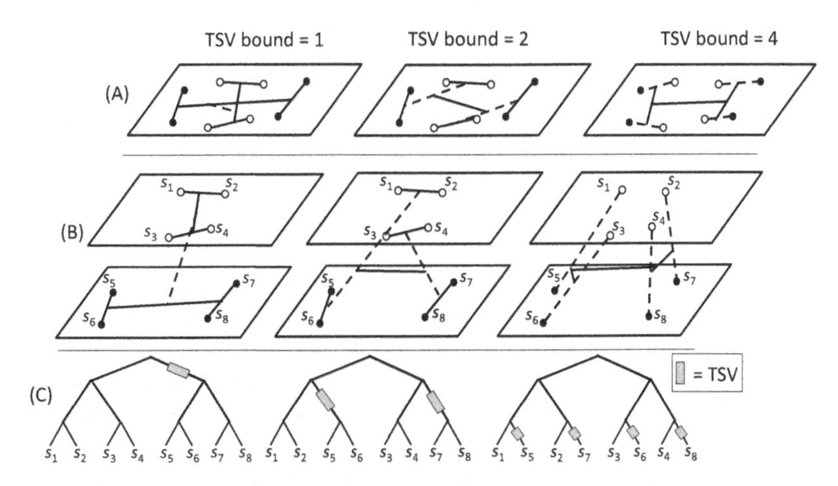

FIGURE 15.13

Several abstract trees for a set of eight sinks generated by the MMM-TB algorithm for different bounds of TSVs, (A) is the 2-D view of these trees and the dashed lines denote TSVs, (B) a 3-D view of the same trees, and (C) the resulting connection topologies where the gray rectangles refer to a TSV [593].

```
Z-cut( set S, subset S_T, subset S_B)
Input: set S = {s_1, s_2,...,s_k}, root tier index z_r
Output: subsets S_T and S_B

1:   z_min = min(z_1,..., z_i,...,z_k), s_i = (x_i, y_i, z_i) ∈ S
2:   z_max = max(z_1,..., z_i,...,z_k), s_i = (x_i, y_i, z_i) ∈ S
3:   if (z_r ≤ z_min) then
4:     S_T = {s_1,..., s_i,...,s_kl}, z_i ∈ [z_min + 1, z_max]
5:     S_B = {s_kl+1,..., s_j,...,s_k}, z_j = z_min
6:   else if (z_r ≥ z_max) then
7:     S_T = {s_1,..., s_i,...,s_kl}, z_i = z_max
8:     S_B = {s_kl+1,..., s_j,...,s_k}, z_j ∈ [z_min, z_max-1]
9:   else
10:    S_T = {s_1,..., s_i,...,s_kl}, z_i = z_r
11:    S_B = {s_kl+1,..., s_j,...,s_k}, z_j ≠ z_r
```

FIGURE 15.14

Pseudocode of the z-cut procedure for the MMM-TB algorithm [593].

the same z_i are on the same tier. The procedure, illustrated in Fig. 15.14, demonstrates the process in which the partitions in the z-direction occur. In this figure, $z_{\min}$ ($z_{\max}$) is the lowest (highest) index of the tier that contains the sinks of set S at a specific iteration of the MMM-TB algorithm. In addition, the root of the clock tree is assumed to be located at z_r. Indexing the tiers is achieved in ascending order from the bottom tier to the uppermost tier.

As shown in Fig. 15.14, if $z_r \leq z_{\min}$, which means the root is located at a lower tier than any of the sinks in S, all sinks with $z_i = z_{\min}$ form the (bottom) subset S_B and the remaining sinks (e.g., $z_i \in [z_{\min} + 1, z_{\max}]$) constitute the (top) subset S_T. Alternatively, if $z_r \geq z_{\max}$, which means the root is located at a higher tier than any of the sinks in S, all sinks with $z_i = z_{\max}$ form the (top) subset S_T and the remaining sinks (e.g., $z_i \in [z_{\min}, z_{\max} - 1]$) constitute the (bottom) subset S_B. In any other case, the (top) subset S_T includes those sinks in the same tier as the root of the tree, while the (bottom) subset S_B contains the remaining sinks in the other tiers. An example of this procedure for a small set $S = \{a, b, c\}$ is depicted in Fig. 15.15, where the number of TSVs resulting from different z-cuts are illustrated by the thick line segments.

In the second condition, where the TSV bound is greater than one, a *cut* along the horizontal direction is applied to the median of the x- or y-coordinate of the sinks as in MMM and the z-dimension is not considered. The resulting subsets require several TSVs if sinks with a different z-coordinate are contained in these subsets. At the end of each partitioning step, a new TSV bound is assigned to each subset to ensure that the TSV bound is maintained. This new TSV bound is determined by estimating the number of TSVs required by the newly formed subsets and dividing the bound by the estimated number of TSVs. The TSV estimates are determined from the minimum number of sinks within each tier resulting from the horizontal *cuts*. The complexity of the MMM-TB algorithm remains the same as that of MMM, which is $O(m \log m)$.

The MMM-TB algorithm guarantees that the number of TSVs does not exceed the TSV bound. Any number of TSVs can, however, be employed as long as this bound is satisfied. The number of TSVs greatly affects the total wirelength and, therefore, the power consumed by the clock network. An exhaustive sweep is used to determine the optimum number of TSVs. This approach, however,

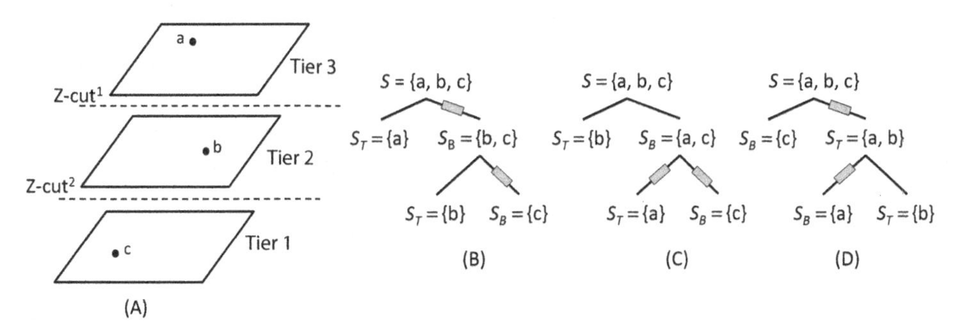

FIGURE 15.15

A set of sinks $S = \{a, b, c\}$ where the effect of the recursive z-cuts in the MMM-TB algorithm is exemplified. (A) Two z-cuts are successively applied, (B) the source is in tier 3 and z-cut^1 is followed by z-cut^2, (C) the source is in tier 2 and the sinks in this tier are first extracted, and (D) the source is in tier 1 and z-cut^2 is followed by z-cut^1.

is computationally expensive, particularly for large TSV bounds. The MMM-TB can be further enhanced to determine the optimal number of TSVs in terms of power. Note that simply using as many TSVs as possible may not yield the least power if the capacitance of the TSVs is high as this trait can adversely affect the savings in power by the decrease in the horizontal wirelength. To determine the number of TSVs, look ahead operations similar to those operations used in [578] are employed [593]. These operations proceed by (a) a z-cut followed by xy-cuts, and (b) xy-cuts followed by a z-cut. The distance of the newly formed centers of mass is compared for cases (a) and (b). A cost function is applied in both cases that relates the resulting number of TSVs and wirelength to estimate the power. The sequence of steps that yields the lower estimate of power is selected during the following iteration.

The second phase of the synthesis process is a variant of the DME method, adapted to consider TSVs and a slew constraint. To satisfy slew constraints, a general practice is to not allow the capacitance seen by any node of the tree to exceed a specific limit, C_{max} [602]. This constraint is met by adding clock buffers to ensure that the load seen by these buffers remains below C_{max}. The merging segments produced from the bottom-up pass of the DME method (see Section 15.1), computed in [593] based on the Elmore delay, are also determined for the buffers whenever the downstream capacitance of the internal nodes exceed C_{max}. During the top-down phase, DME is extended to evaluate the type of merging (e.g., merging of an internal node and a buffer from the root of the subtrees or a standard merging of an internal node with the children nodes) and determines the placement of the internal node and buffer. An example of the two different merging types encountered when buffer insertion is integrated with the DME method is illustrated in Fig. 15.16.

Application of the MMM-TB algorithm and the integrated buffer insertion with the DME technique to several benchmarks [603] has demonstrated that multi-TSV clock trees can greatly reduce power, ranging from 16.1% to 18.8%, 10.3% to 13.7%, and 6.6% to 8.3%, as compared to a single TSV clock network where the TSV capacitance is, respectively, 15, 50, and 100 fF, for a two tier circuit. The corresponding decrease in total wirelength is, respectively, 24.0% to 26.5%, 23.9% to 26.6%, and 16.6% to 18.9%. Another advantage of multi-TSV clock trees is that the average slew and distribution of slews are better managed due to the reduced wirelength. Although the

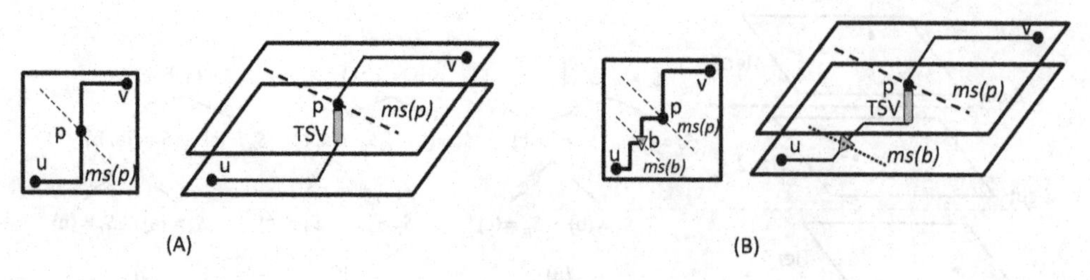

FIGURE 15.16

Examples of merging segments for two intertier nodes *u*, *v* merged with node *p*. (A) An unbuffered tree, and (B) a buffered tree.

combination of the MMM-TB algorithm and the DME method produces efficient multi-TSV clock trees, a greater reduction in wirelength for unbuffered trees is demonstrated by other variants of the MMM and DME methods, which are presented below to offer greater insight.

15.3.1.1 Further reduction of through silicon vias in synthesized clock trees

A simple extension of MMM in three physical dimensions, called MMM-half perimeter wirelength (HPWL), is introduced in [594], where a user defined parameter, ρ $(0 \leq \rho \leq 1)$ determines the direction of a *cut*. If $\rho = 0$, the standard MMM is employed, and the *z*-dimension is ignored. Alternatively, if $\rho = 1$, the sinks are partitioned among the tiers comprising the 3-D circuit, and the resulting per tier subsets are bipartitioned using *xy*-cuts as in MMM. For any other ρ, a horizontal *cut* is performed along the geometric median in a 2-D Manhattan plane. The HPWL for these subsets of sinks is determined. If the HPWL of a subset of sinks is smaller than $\rho \cdot$ HPWL of all of the sinks, the sinks are partitioned among the tiers (i.e., a *z*-cut) rather than along the *xy*-directions. This condition shares some concepts with the MMM-TB algorithm, where the TSV bound is used to determine the direction of the sink partition. Furthermore, in the MMM-TB algorithm, constraining the number of TSVs does not guarantee that the total vertical wirelength is minimized.

The MMM-TB algorithm provides the number of TSVs required to connect the sinks during the recursive top-down partition. However, TSV usage also depends upon the tier in which the internal nodes of the connection topology of $T(S)$ are placed. This information, however, is not determined by the MMM-TB algorithm. In addition, the standard DME technique is applied to a 2-D plane where the sinks are projected and does not minimize the number of TSVs.

To demonstrate the different number of TSVs that can originate from diverse placements of the internal nodes of a tree topology, consider the example shown in Fig. 15.17. Depending upon the placement of nodes x_1, x_2, and the root node s_r, the number of TSVs can differ significantly for a two tier tree, as depicted in Fig. 15.17. As illustrated in this example, embedding the nodes for cases (A) and (H) yields the fewest number of TSVs. An algorithm that embeds the tiers to ensure a minimum vertical length is presented in [594]. The vertical length for a 3-D clock tree is

$$L_{V_{\text{total}}} = \sum_{\forall i, j \in G(S)} L_{\text{TSV}}(z_i - z_j), \qquad (15.11)$$

where L_{TSV} is the length of one TSV connecting adjacent tiers, and the summation applies to every pair of nodes within the connection topology of tree $T(S)$. Any pair of nodes placed within the

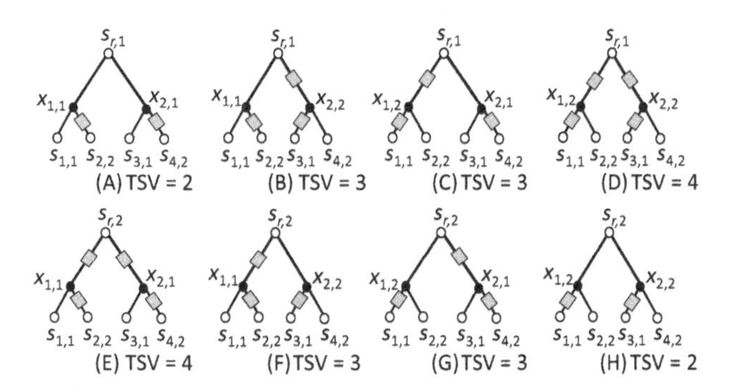

FIGURE 15.17

A tree with four sinks embedded in two tiers for different cases of embedding the internal nodes x_1 and x_2 and the root node s_r and the resulting number of TSVs for each case. The notation $x_{i,j}$ ($s_{i,j}$) implies the placement of node x_i (s_i) in tier j. (A) TSV = 2, (B) TSV = 3, (C) TSV = 3, (D) TSV = 4, (E) TSV = 4, (F) TSV = 3, (G) TSV = 3, and (H) TSV = 2.

same tier does not contribute to the vertical wirelength since $z_i = z_j$. To determine the number of TSVs for a node i of $T(S)$ and, subsequently, the vertical wirelength, the following expressions are used

$$n_{\mathrm{TSV}}(i) = \begin{cases} 0, & \text{if } i \text{ is a sink node,} \\ n_{\mathrm{TSV}}(T_{\mathrm{left}}(i)) + n_{\mathrm{TSV}}(T_{\mathrm{right}}(i)) \\ \quad + n_{\mathrm{TSV}}(i, T_{left}(i)) + n_{\mathrm{TSV}}(i, T_{right}(i)), \\ \text{if } i \text{ is an internal node,} \end{cases} \tag{15.12a–b}$$

where $n_{\mathrm{TSV}}(T_{\mathrm{left}}(i))$ and $n_{\mathrm{TSV}}(T_{\mathrm{right}}(i))$ are, respectively, the number of TSVs contained in the subtrees rooted at the children of node i and $n_{\mathrm{TSV}}(i, T_{\mathrm{left}}(i))$ and $n_{\mathrm{TSV}}(i, T_{\mathrm{right}}(i))$ are, respectively, the number of TSVs connecting node i to subtrees $T_{\mathrm{left}}(i)$ and $T_{\mathrm{right}}(i)$.

If the children of node i are sinks and the location of the sinks is known, the tier(s) where node i can be embedded is notated as $ET(i)$ to ensure that the minimum number of TSVs is determined by the tier of the children nodes. If, however, the children nodes are internal nodes, the embedding tiers for these nodes are sets of embedding tiers. In this case, from these sets of embedding tiers, the set of tiers $ET(i)$ that lead to the minimum number of TSVs $n_{\mathrm{TSV}}(i)$ for this node can be determined. An example of embedding an internal node x based on the embedding tiers of the children nodes x_1 and x_2 is shown in Fig. 15.18. If the children nodes are sinks, which means that $x_1 = s_1$ and $x_2 = s_2$, the length of the vertical connection for merging these nodes at node x is $|t_2 - t_1|$, where $t_2 = z_2$ and $t_1 = z_1$ are the tier of the sinks, as depicted in Fig. 15.18A.

If the children nodes are also internal nodes, two cases can be distinguished. If the embedding tiers $ET(x_1)$ and $ET(x_2)$ have some common tier(s), the minimum vertical length to merge these nodes with node x is $n_{\mathrm{TSV}}(T_{\mathrm{left}}(x)) + n_{\mathrm{TSV}}(T_{\mathrm{right}}(x))$, and the tier where the node is embedded is between tier t_2 and t_3, as shown in Fig. 15.18C. Choosing any other tier for embedding x results in an increase in $n_{\mathrm{TSV}}(x)$, as depicted in Fig. 15.18D. In the second case, where the set of embedding tier(s) for the children nodes x_1 and x_2 do not share any common embedding tier, as shown in Fig. 15.18E, node x is embedded between tier t_2 and t_3. The difference between these two cases, as

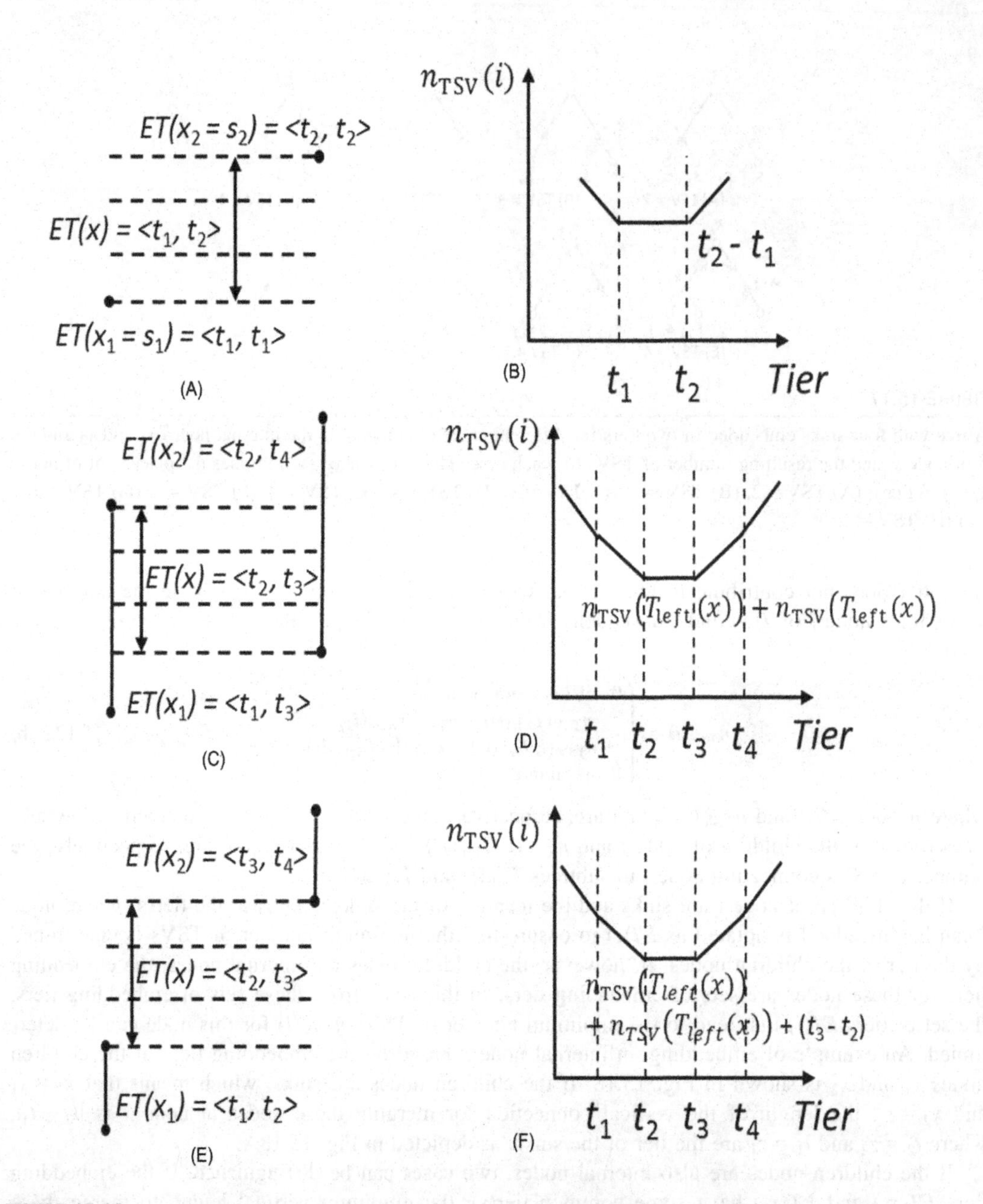

FIGURE 15.18

Different cases to determine the number of embedding tiers for node x, where the children nodes x_1 and x_2 are (A) clock sinks, and (C) and (E) are internal nodes. The minimum number of TSV for (A), (C), and (E) are shown, respectively, in (B), (D), and (F) [594].

shown in Figs. 15.18C and 15.18E, a common embedding tier for the children nodes exists in Fig. 15.18C and, consequently, merging with node x does not require a TSV. As shown in Fig. 15.18E, however, $|t_3 - t_2| > 1$ and at least one TSV is required to embed node x.

The set of embedding tiers for the internal nodes (required by (15.12a)) is iteratively determined in a bottom-up process [594],

$$ET(x) = <\min(t_1, t_2), \max(t_1, t_2)>, \tag{15.13}$$

where

$$t_1 = \max(\min(ET(T_{\text{left}}(x))), \min(ET(T_{\text{right}}(x)))), \tag{15.14}$$

and

$$t_2 = \min(\max(ET(T_{\text{left}}(x))), \max(ET(T_{\text{right}}(x)))). \tag{15.15}$$

The optimality of these expressions has been proven through a series of lemmas [594]. The process of embedding the nodes of a tree to ensure that the minimum vertical length is employed exhibits a complexity of $O(m)$ where m is the number of nodes of tree $T(S)$.

Once the embedding tiers for the nodes within a tree are determined, the placement of the nodes of the tree follows based on the DME technique, where the Elmore or a linear delay model is employed. In a linear delay model, the TSVs are modeled differently due to a disimilar capacitance as compared to the horizontal wires. As with the application of DME to 2-D clock trees with a linear delay model, zero skew with minimum wirelength is also produced for 3-D clock trees. The situation is more subtle since the Elmore delay is used for e_a and e_b for merging nodes a and b (see Section 15.1). This approach can lead to a suboptimal wirelength due to the lack of a specific position along a merging segment to balance the delay from the merging node x to the root of subtrees T_a and T_b. This behavior requires wire snaking, which results in detouring from the minimum wirelength.

15.3.2 SYNTHESIS TECHNIQUES FOR PRE-BOND TESTABLE THREE-DIMENSIONAL CLOCK TREES

The previously discussed synthesis techniques produce 3-D clock trees for any number of tiers, yet suffer from a drawback that limits usability. As pre-bond test can guarantee high yield of 3-D systems [604], the correct functionality and expected performance of each tier should be verified prior to assembly of the stack. This demand places considerable constraints on the design of 3-D systems. For pre-bond test methodologies of 3-D circuits, see [605]. For the design of clock trees, specific changes must be made to provide pre-bond test. The fundamental difference of pre-bond test related to CTS techniques is the additional circuitry. The salient features of these techniques are presented in this section.

A method for providing a 3-D clock distribution network that can be tested during pre-bond is to utilize single TSV topologies as shown, for example, in Fig. 15.19, where these topologies include fully connected trees in each tier of the 3-D system. One clock input for each tier provides the clock signal to all of the clock pins within that tier. This approach, however, yields minor improvements over 2-D circuits, reducing the advantages of 3-D integration. Instead, adapting multi-TSV clock networks to support pre-bond test is more advantageous. The main challenge is in multi-TSV topologies.

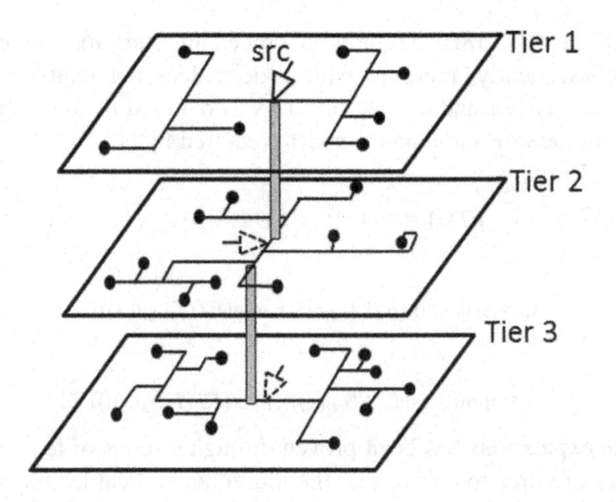

FIGURE 15.19

Three tier clock tree using a single TSV for the intertier connections. This topology is pre-bond testable as each tier includes a network connecting all of the sinks.

There is typically only one tier that contains a fully connected tree (typically the tier where the primary clock driver resides), while the other tiers comprise several (or a forest) of trees connected to the full tree through TSVs. These several trees within each tier are however disconnected from each other during pre-bond test operation. A non-negligible number of clock inputs is required to clock these disconnected trees prior to bonding, complicating the test process and increasing cost.

Additionally, even for the tier that includes the fully connected clock network, test is not a simple process as this network drives the entire clock load. During pre-bond test, only part of this load exists, which means that clock skew and slew can deviate significantly from the target skew and slew during normal operation. This network can, therefore, cause several timing violations that can incorrectly be interpreted as faults. To mitigate this situation, two measures are applied: (1) insert a buffer at each node of the complete network that connects to a TSV, and (2) add an appropriate number of redundant networks that connect the local trees of a tier with each other during pre-bond test. During normal operation, that is post-bond, these redundant trees are disconnected from the rest of the clock network [606]. Buffer insertion from (1) decouples the fully connected tree from the rest of the network. Consequently, the capacitance at each node of this clock tree remains the same during both pre- and post-bond operation.

An example of a pre-bond testable clock tree in two tiers is shown in Fig. 15.20. Drawn with dashed lines, the buffers before the TSVs, the redundant tree, and the transmission gates (TGs) that disconnect this tree during normal operation, are noted in this figure. Also note that an additional control signal is required for this structure to switch the TGs in tier 2.

Having described the characteristics of pre-bond testable clock networks, the synthesis problem is formulated as follows: for a set of sinks $S = \{s_1, s_2, s_3, \ldots s_m\} \subset \Re^3$ distributed across n tiers, a root node s_r, and a TSV bound, synthesize a 3-D clock network with a minimum or zero clock skew both in post-bond (for the entire stack) and pre-bond (for each tier) operation. The wirelength and power of the 3-D clock network are minimized while simultaneously satisfying the clock slew constraints.

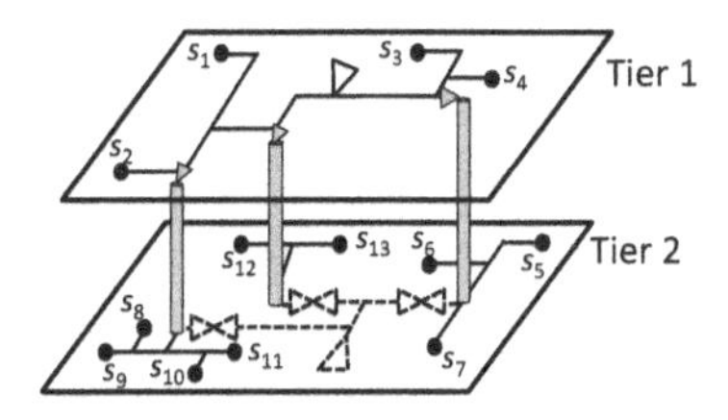

FIGURE 15.20

Pre-bond testable clock tree with multiple TSVs. The buffers are inserted before the TSVs, thereby not changing the capacitance of the tree in tier 1. TGs in tier 2 connect the redundant tree (shown as a dashed line) with the subtrees during pre-bond test. The TGs are switched off after bonding, disconnecting the redundant tree.

Employing the MMM-TB method, as described in the previous section, a connection topology for the sinks of S, denoted as $T(S)$, is generated. The MMM-TB method produces a topology where the sinks are located on the same tier as the source, called herein the *source tier*. These sinks are connected by a tree. Connections to other tiers are implemented with TSVs. To decouple the subtrees in these tiers from the network in the source tier, the following steps are followed [596]: (1) if a TSV connects the source tier with another tier, a TSV buffer is inserted in the source tier, (2) if two nodes of the 3-D clock tree are connected by a TSV traversing the source tier, a TSV buffer is added to the source tier, and (3) if one TSV interconnects two nodes without traversing the source tier, no TSV buffer is inserted. In this case, the absence of this TSV during pre-bond test has no effect on the capacitance of the source tier and, therefore, a TSV buffer is not needed.

The next phase of the synthesis process includes a modified DME, which is a two stage technique. In this pre-bond aware CTS (PBA-CTS) technique, a zero skew tree is produced that considers merging segments both for the TSV buffers and the nodes of the connection topology originating from the MMM-TB algorithm. The TSV buffers, however, often require additional wirelength to maintain the zero (or low) skew requirement [596]. This situation can result in the minimum wirelength objective to not be satisfied. The insertion of common clock buffers avoids this situation. Consider the example shown in Fig. 15.21 where two subtrees, ST_1 and ST_2, are, respectively, in tiers one and two. A TSV buffer is used to merge these subtrees as the trees are located on separate tiers. This TSV buffer adds to the delay from node E to ST_2. If this delay is much greater than the delay from E to ST_1, before the insertion of the TSV buffer, this delay increases. Wire "snaking" is added to maintain the zero skew objective, which departs from the minimum wirelength objective. A clock buffer can be inserted to avoid wire snaking. Case studies show that these imbalances occur infrequently; the use of a clock buffer for this purpose is, therefore, not a significant issue.

Another reason exists for adding clock buffers during the bottom-up process where the merging of subtrees takes place. Clock buffers are widely used to satisfy slew constraints. Inserting these clock buffers ensures that the load that each clock buffer drives does not exceed a downstream capacitance threshold C_{max} [602]. This practice has been shown to control slew. In the case of pre-bond testable 3-D clock trees, examples where the clock and TSV buffers are inserted are illustrated in Fig. 15.22. The simple case is shown in Fig. 15.22A, which is common in both 2-D and 3-D (either pre-bond testable or not) synthesized clock trees if the delay of the merged branches is highly imbalanced (e.g., $t_{dA} < t_{dB}$). In Fig. 15.22B, multiple clock buffers are utilized due to the excessive length of the wires and/or if the downstream capacitance exceeds C_{max}.

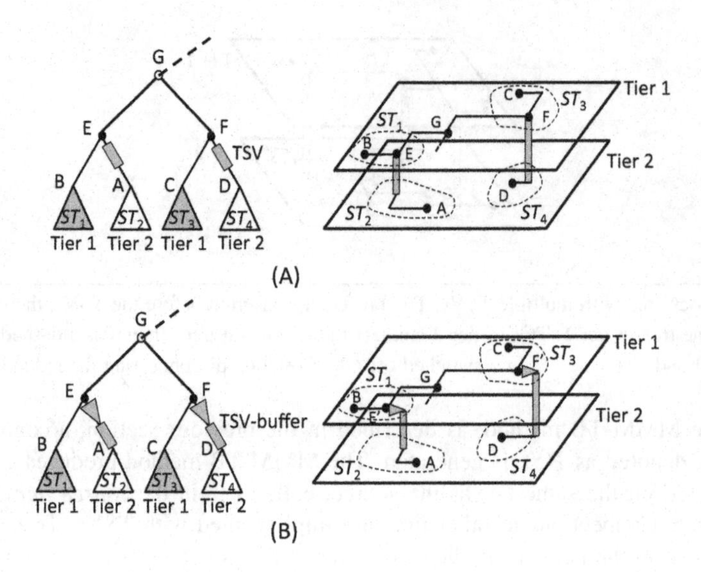

FIGURE 15.21

Portion of a 3-D clock tree consisting of several subtrees ST_i. The TSVs in (A) are replaced with TSV buffers in (B) to decouple the clock tree in tier 1 from the clock tree in tier 2 [596].

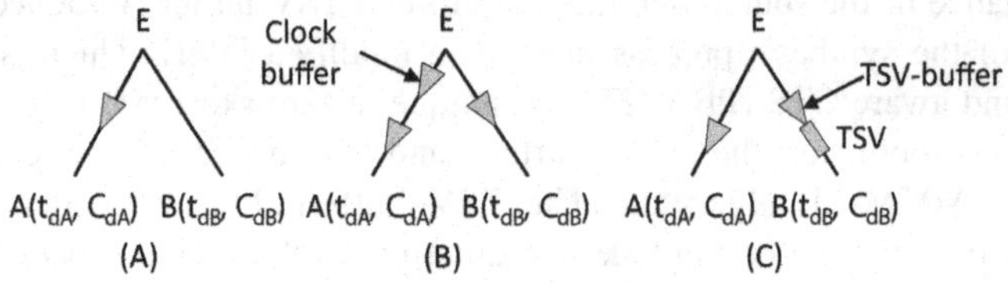

FIGURE 15.22

Different cases where TSV and/or clock buffers are inserted. (A) A clock buffer is inserted to balance the delay between the two branches where $t_{dA} < t_{dB}$, (B) multiple clock buffers are inserted due to long wires or high downstream capacitance, and (C) a TSV buffer is inserted to decouple the downstream clock tree, and a clock buffer is added to counterbalance the delay imbalance caused by the TSV buffer [596].

Another case is shown in Fig. 15.22C, which requires the addition of a clock buffer to counteract the delay imbalance caused by inserting a TSV buffer. The insertion of the TSV buffer also decouples the subtrees, thereby facilitating pre-bond test.

The output of the PBA-CTS method is a 3-D clock tree used for post-bond operation and a tree that spans all of the clock sinks lying in the source tier. Similar trees are generated for the remaining tiers, which is achieved by redundant trees connecting the unconnected trees in these tiers. This process is similar to a conventional CTS technique for planar circuits [596] comprising: (1) construction of a binary connection topology $T(S)$ generated in a top-down manner, (2) insertion of a TG at each sink node, and (3) embedding and buffering of $T(S)$ in a tier under skew and minimum wirelength constraints. Note that the sinks are the roots of the subtrees connected by the redundant tree in this tier. Insertion of the TGs, which connect (pre-bond operation) or disconnect (post-bond operation) the redundant tree, requires a control signal spanning a large portion of the tier. To limit the wirelength of this signal, the rectilinear minimum spanning tree (RMST) algorithm can be used [403,607].

IBM [603] and ISPD [608] benchmark circuits are utilized to ascertain the efficiency of this pre-bond CTS technique, where the clock sinks are randomly assigned within the tiers of the 3-D circuits. A 45 nm process node based on the Predictive Technology Model (PTM) model is assumed [252]. The clock frequency is 1 GHz and the power supply is 1.2 volts. A via-last TSV technology is used where the area, length, and liner thickness of a TSV are, respectively, 10 μm × 10 μm, 20 μm, and 0.1 μm. The clock skew and slew constraints are, respectively, 10% and 3% of the clock period. Other electrical and physical characteristics assumed for these circuits are listed in [596]. A comparison of a single TSV, which is inherently pre-bond testable, and multi-TSV clock trees, which require redundant trees and TSV buffers, has demonstrated that multi-TSV clock trees decrease wirelength from 14.8% to 24.4% and 39.2% to 42.0%, for circuits with, respectively, two and four tiers. The savings in power vary from 10.1% to 15.9% and from 18.2% to 29.7%, respectively, for circuits comprising two and four tiers.

The departure between the wirelength and power savings, as wirelength and power depend linearly on wire capacitance, is attributed to the additional circuits, such as the TSV buffers and TGs, and the TSV capacitance, which can be greater than the capacitance of the horizontal wires on a per unit length basis. Furthermore, the control signal for the TGs adds to the overall wirelength congestion of the clock trees. The breakdown of the wirelength components for several benchmark circuits implemented in two tiers is reported in Table 15.3. The wirelength breakdown is reported for the post-bond 3-D clock trees, the pre-bond trees in the source tier (tier 1), and the pre-bond trees in tier 2. Notably, the length of the control signal of the TGs constitutes a significant portion of the total wirelength (on average, 29%).

Another potential source of overhead is the TSV buffers. The change in wirelength and power from the case where no TSV buffers are used in the source tier to decouple that tree from the downstream capacitance in the other tiers is listed in Table 15.4. The change in wirelength is small; for specific benchmark circuits, a decrease in the interconnect length is observed. Alternatively, an increase in power is always observed, verifying that inserting TSV buffers narrows the power gains for multi-TSV clock trees that are pre-bond testable.

Lastly, the effect of the TSV capacitance on the performance of a 3-D clock trees is assessed by the results reported in Table 15.5 for the IBM benchmark circuit r_5. These results include disparate TSV bounds, which constrain the number of TSVs in the clock tree. Two interesting observations are noted from these results. First, the power worsens with increasing TSV capacitance irrespective of the TSV bound. Second, increasing the TSV bound lowers power as more TSVs drive a shorter interconnect length. This trend is reversed if the TSV capacitance is much larger, increasing the power by 17.7% for 2,469 TSVs and a TSV capacitance of 100 fF.

Although the results listed in Tables 15.3 to 15.5 demonstrate considerable improvements, two limitations of the CTS technique in [596] are noted. One of these limitations relates to the effect of the added TSV buffers, which can produce delay imbalances that are controlled by additional clock buffers. The outcome is increased power. Another limitation stems from the use of TGs in all but the source tier to (dis)connect the redundant trees according to the operational mode (pre- or post-bond). Indeed, as listed in column 13 of Table 15.3, the interconnect overhead due to the control signal of the TGs (although the routing of this signal is embedded with RMST) reaches, on average, 29% of the total wirelength of the post-bond clock tree. This overhead adds to the wirelength

Table 15.3 Post-bond Testable 3-D Clock Trees and Pre-bond Testable 2-D Clock Trees of the Individual Tiers for a Two Tier 3-D Circuit [596]

			Post-bond 3-D			Pre-bond Testable Tier 1			Pre-bond Testable Tier 2					
									Wirelength (μm)					
Circuit	# of *Sinks*	# of TSVs	Wirelength (μm)	Power (mW)	Skew (ps)	Wirelength (μm)	Power (mW)	Skew (ps)	Total	Subtrees	Redundant	TG Control Signal	Power (mW)	Skew (ps)
r_1	267	57	227,141	128.4	13.7	166,691	103.0	13.5	150,219	60,450	89,769	62,732	68.2	13.0
r_2	598	95	488,987	274.1	14.2	328,914	196.0	14.1	302,023	160,073	141,950	109,031	148.6	11.8
r_3	862	183	616,077	361.6	15.5	444,156	280.5	15.5	429,950	171,921	258,029	161,561	201.9	16.2
r_4	1903	265	1,311,290	763.2	15.5	889,460	536.4	14.9	846,980	421,830	425,151	259,442	422.1	15.1
r_5	3101	269	1,998,950	1115.0	29.1	1,255,760	715.9	29.1	1,236,417	743,190	493,227	310,885	615.9	20.9
f_{11}	121	44	129,391	73.3	9.4	99,393	64.1	9.2	99,169	29,998	69,171	51,214	44.3	6.3
f_{12}	117	36	127,763	71.2	6.8	96,093	60.4	6.2	93,625	31,669	61,956	42,134	42.0	5.7
f_{21}	117	42	136,676	75.6	5.0	107,834	67.0	4.7	101,968	28,841	73,127	52,241	45.0	7.3
f_{22}	91	30	80,977	46.8	15.3	61,504	40.4	15.2	59,870	19,473	40,397	29,449	26.4	14.9
Ratio			1.00	1.00	1.00	0.72	0.79	0.97	0.69	0.28	0.41	0.29	0.57	0.94

Table 15.4 TSV Buffer Insertion on Wirelength and Power [596]

		Increase (%)	
Circuit	# of TSVs	Wirelength	Power
r_1	248	−0.77	7.95
r_2	434	3.89	6.27
r_3	718	3.11	9.19
r_4	1651	2.72	11.28
r_5	2469	4.56	11.17
f_{11}	129	−1.69	8.28
f_{12}	114	0.06	7.47
f_{21}	102	−1.66	3.34
f_{22}	81	−2.68	2.30

required by the clock distribution network or, alternatively, severely curtails the benefits of pre-bond testable multi-TSV trees.

These limitations can be circumvented through more elaborate CTS methods and a suitable self-configured circuit that can substitute for the TGs [595]. This approach commences with an advanced methodology for topology generation, called herein MMM-HPWLmax, which is based on the MMM-HPWL algorithm followed by the embedding of the topology in the tiers of the stack to minimize the vertical wirelength [594]. These steps provide a post-bond 3-D clock tree. To generate the redundant clock trees for pre-bond test, the same methods are applied again to all but the source tier, where the TGs are replaced by a self-configured circuit that adapts to the operating mode.

There are two primary differences between this method, called herein low overhead pre-bond aware CTS (LO-PBA-CTS) [595], and PBA-CTS [596]. These differences are (1) relaxation of the constraint that a buffer should be placed before a TSV to decouple the tree in the source tier from the downstream subtrees, and (2) elimination of the control signal for the TGs along with the associated interconnect overhead.

Consequently, in LO-PBA-CTS, a buffer can be placed anywhere along the branch of the tree that connects to a TSV. Greater flexibility is, therefore, offered for placing the buffers, lowering the likelihood of inserting a TSV buffer that produces a delay imbalance during the node merging process. This flexibility may, however, be insufficient to prevent delay imbalances, in particular when a TSV buffer is inserted at a branch with a small capacitive load. To cope with this situation, the TSV buffers are characterized as *stable* or *unstable* [595]. A TSV buffer is stable if inserting this buffer isolates the tree of the source tier and does not require insertion of another clock buffer at another edge to satisfy delay imbalances. Otherwise, a TSV buffer is unstable.

The LO-PBA-CTS technique manages the unstable TSV buffers through the MMM-HPWLmax algorithm [594], which builds on the MMM-HPWL. Whereas the latter algorithm handles the density of TSVs by the parameter ρ during the recursive bipartition, the MMM-HPWLmax checks whether the HPWL of the sinks included in a candidate subset is less than or equal to HPWLmax,cap,

$$\mathrm{HPWL}_{\mathrm{max,cap}} = \frac{C_{\mathrm{max}} - 2C_{\mathrm{max,sink}}}{c_w}, \tag{15.16}$$

Table 15.5 TSV Capacitance and TSV Bound for Different Metrics as Compared to Single TSV Clock Trees [596]

TSV Capacitance (fF)	# of TSVs = 183						# of TSVs = 2469					
	# of Buffers	Wirelength (μm)	Power (mW)	Skew (ps)	Reduction (%)		# of Buffers	Wirelength (μm)	Power (mW)	Skew (ps)	Reduction (%)	
					Wirelength	Power					Wirelength	Power
0	2788	2,012,360	1154.9	20.5	13.0	9.3	2970	1,337,980	972.4	23.3	42.1	23.6
15	2803	2,014,790	1159.1	20.3	12.9	8.9	3134	1,368,370	1041.0	20.3	40.8	18.2
25	2814	2,021,640	1167.4	20.7	12.6	8.3	3237	1,404,560	1087.3	18.6	39.3	14.5
50	2834	2,033,640	1180.4	19.9	12.1	7.3	3603	1,489,930	1220.9	21.0	35.6	4.1
100	2890	2,071,800	1215.0	16.0	10.5	4.7	4249	1,719,590	1499.7	25.7	25.7	−17.7

where C_{max} is the maximum capacitance constraint, c_w is the unit length capacitance of the horizontal interconnects, and $C_{max,sink}$ is the maximum capacitive load of all of the sinks. MMM-HPWLmax employs the threshold max(HPWL$_{max,cap}$, $\rho \cdot$ HPWL) to decide whether partitioning the sinks occurs in the z-direction or along the xy-directions. This threshold is rather conservative as the TSV capacitance is not included; however, simulation results demonstrate that the number of unstable TSV buffers is reduced, particularly in those circuits with a high number of clock sinks [595]. Note that this treatment resembles the practice of [596] to satisfy slew constraints but targets a different design objective.

The other key difference of LO-PBA-CTS [595] as compared to PBA-CTS [596] is the use of a self-configured circuit to switch the TGs connecting the redundant clock tree in all but the source tier. This circuit is depicted in Fig. 15.23, where, in addition to the TG, only a handful of transistors are required. The circuit operates as follows. During pre-bond test, the clock signal is supplied to the root of the redundant tree and reaches the root of every subtree preceded by a TG. A rising edge of the clock signal T_{clk} is inverted by *INV1* and the device P_2 turns on, charging node *PBTEST* to V_{dd}. This high voltage drives the output of *INV3* low, turning off N_2. The output signal of inverters *INV3* and *INV4* produces the control signals for the TG.

Alternatively, if the T_{clk} signal is low (no pre-bond test), P_2 turns off. The capacitance of node *PBTEST* slowly discharges through the different leakage current paths. At some time, the output of *INV3* is driven high, turning on N_2, completely discharging node *PBTEST*. The TG is also turned off, disconnecting the redundant tree. The turn-off time of the circuit can be regulated by sizing P_2 and N_2, which are along, respectively, the charge and discharge paths of node *PBTEST*. The circuit in [595] is based on a 45 nm predictive technology [252], resulting in a turn-off time of about 100 ns.

Application of LO-PBA-CTS to the same benchmark circuits as in PBA-CTS and the same technology parameters demonstrates comparable results in terms of wirelength, clock skew, slew rate, and power. However, for specific benchmark circuits, such as *ispd09fnb1* and *ispd09fnb2*, reductions of 56% to 88% in the number of TSVs, 53% to 67% in the number of buffers,

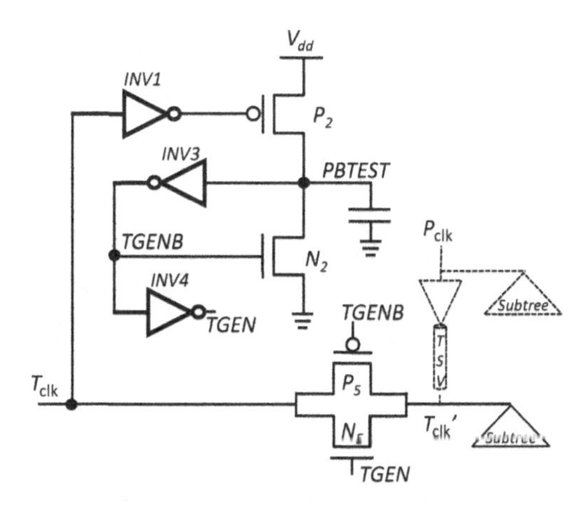

FIGURE 15.23

Self-configured circuit controlling the operation of the TG (N_5 and P_5) [595].

22% to 65% in total wirelength, and 26% to 43% in the power of the post-bond clock distribution network are observed. Additionally, the wirelength overhead of the control signal for the TGs no longer exists. Further developments in CTS techniques for 3-D clock trees emphasize other important issues, such as limitations on the TSV placement. These techniques are reviewed in the following section.

15.4 PRACTICAL CONSIDERATIONS OF THREE-DIMENSIONAL CLOCK TREE SYNTHESIS

The methods presented in Section 15.3 have significantly advanced the state-of-the-art for 3-D CTS. More recent approaches have enhanced the efficiency of these techniques by considering more practical issues, such as whitespace availability and obstacle avoidance for the TSVs (which can affect important metrics, such as clock skew, slew rate, wirelength, and power) as well as fault tolerance.

The majority of the physical design techniques reviewed in Chapter 9, Physical Design Techniques for Three-Dimensional ICs, highlight the importance of whitespace in 3-D circuits where TSVs and buffers (or repeaters) are inserted. Due to the nature of TSVs, which disrupt cell placement, the placement of the clock signal TSVs can also affect the whitespace. An approach to address this practical constraint is, after applying one of the techniques discussed in Section 15.3, to shift the TSVs to the nearest available whitespace. This approach is illustrated in Fig. 15.24, where the dashed lines depict the additional wire segment. This shift of the TSV location can negate the (near-)optimality of the DME technique (and its variants), degrading the performance of the synthesized clock network.

Considering the constraints in placement due to the whitespace, the clock synthesis problem can be formulated as follows. For a target whitespace area (fragmented into several parts, the number and size of which depends upon the floorplan parameters, see Chapter 9, Physical Design

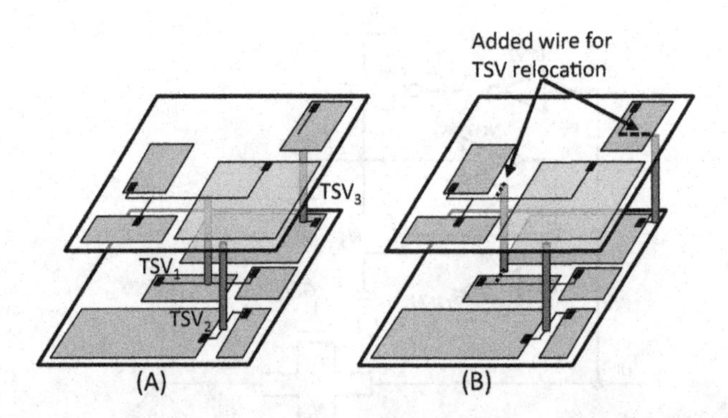

FIGURE 15.24

A two tier clock tree, (A) the initial TSV locations are shown, and (B) TSV_1 and TSV_3 are relocated within the whitespace. The relocation adds wirelength (shown by the dashed lines) which degrades the performance of the clock tree topology shown in (A).

Techniques for Three-Dimensional ICs) within each tier, a set of clock sinks S, a bound for the number of TSVs, and a slew constraint, synthesize a clock tree to ensure that (1) the number of TSVs does not exceed the TSV bound, (2) each TSV is placed within some whitespace, (3) the clock slew constraint is met, and (4) clock skew and power are minimized.

To address this problem, a three-stage method is employed [597], where these stages include: (1) pre-clustering of sinks, (2) an extension of the MMM-TB method to consider available whitespace, and (3) the application of the DME technique to place the TSVs within the whitespace. The computational complexity of this method is $O(k_{ws}m)$, where k_{ws} is the number of whitespace blocks and m is the number of clock sinks in S.

The pre-clustering step proceeds by clustering all of the clock sinks located within a specific radius from each whitespace and assigning these sinks to the corresponding whitespace. The whitespace from all of the tiers is projected onto a single plane, forming a set of whitespace regions. A parameter β controls the assignment of a sink to a whitespace. For sinks situated farther than $\beta \cdot \mathrm{HPWL}_{\mathrm{tier}}$ ($\mathrm{HPWL}_{\mathrm{tier}}$ is the HPWL of the tier) from the nearest whitespace, these sinks are clustered. This clustering means that a TSV placed in a nearby whitespace provides the clock signal to all of the sinks. To comply with the design requirements for a clock network, each cluster is treated as a subtree, the root of which is contained within a whitespace. An example of sink clustering is shown in Fig. 15.25. The traditional MMM and DME methods produce a tree connecting these sinks to the corresponding whitespace. Consequently, at the end of this stage, a new set of sinks is formed that also includes the root of these subtrees. The downstream capacitance of the roots and the delay to the sinks is, respectively, the input delay and capacitive load for these root nodes in the two successive stages of the technique. Furthermore, the clustered sinks are removed from set S.

The second stage is based on the MMM-TB algorithm (see Fig. 15.14), where the decision of the direction of bipartition also considers the whitespace. Assume that at an iteration of the

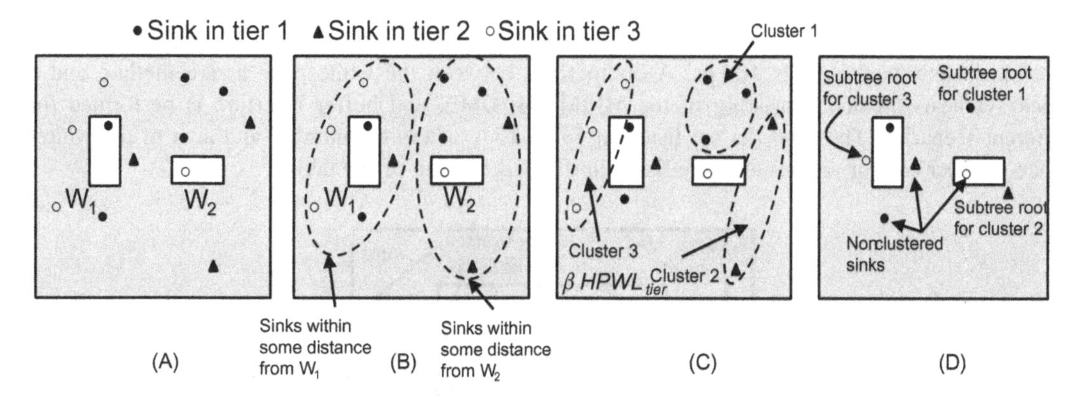

FIGURE 15.25

Pre-clustering stage of the whitespace-aware CTS method, (A) a set of sinks and whitespaces are projected onto a plane, (B) the sinks per tier are located beyond distance $\beta \cdot \mathrm{HPWL}_{\mathrm{tier}}$ from whitespaces, (C) those sinks within a cluster belong to the same tier, and (D) the root of the subtrees from the clustered sinks in each tier and some non-clustered sinks is depicted [597].

algorithm, two subsets with sinks S_A and S_B result from *xy*-cuts. If the sinks contained in these subsets are only placed within one tier, meaning that $stack(S_A) = 1$ and $stack(S_B) = 1$, no TSVs are needed and no whitespace is required to proceed with the partition. Alternatively, if $stack(S_A) > 1$ and $stack(S_B) > 1$, whitespace exists in all of the $stack(S_A) - 1$ and $stack(S_B) - 1$ tiers to ensure that the TSVs can be inserted to connect the sinks of these subsets. Otherwise, the *xy*-cut does not occur and the partition occurs in the *z*-direction (see Fig. 15.14). Generation of this connection topology that is also whitespace aware completes the second stage.

A variant of DME is employed during the last stage to produce the clock tree. During the bottom-up phase of this stage, the merging segments for all of the nodes are generated, including the nodes related to a TSV, where these nodes are placed within the whitespace. As the whitespace can contain multiple clock TSVs, crosstalk noise can develop among these TSVs. This issue emerges if the physically adjacent TSVs within some whitespace carry the clock signal at a different hierarchical level of the clock network. The latency of the clock signal from the root to these TSVs can differ. The signal transferred by these TSVs can, therefore, be out-of-phase, increasing coupling within the clock network. To alleviate this noise, modifications of the merging segments of the nodes with TSVs are relocated to the center of the nearest whitespace. The location of the TSV may be farther from this whitespace.

An example of this diversion from the initial merging segment is shown in Fig. 15.26. The whitespace is divided into several squares. The size of this space is determined by the area of a single TSV and the TSV keep-out zone. Any square that overlaps with the merging segment of that TSV is an appropriate location. If, however, this placement leads to coupling with neighboring TSVs, a more distant square is used, and wire "snaking" is required to cope with the delay imbalance. Note that as individual TSVs are placed, suboptimal placements are produced. To satisfy slew constraints, buffer insertion as in the PBA-CTS technique is applied but as a post-synthesis step.

The efficiency of this technique is evaluated on ISPD benchmark circuits for a two tier system assuming a 45 nm CMOS technology node [252]. The clock sinks and whitespace are randomly distributed across the two tiers. The clock frequency is 2 GHz and the power supply is 1.2 volts. The clock slew constraint is 100 ps. A comparison between the whitespace aware method and a clock synthesis method consisting of the MMM-TB, DME, and buffer insertion is performed for different scenarios. These results are listed in Table 15.6, where the number and area of the whitespace as a per cent of the tier area are listed in the first column of the table.

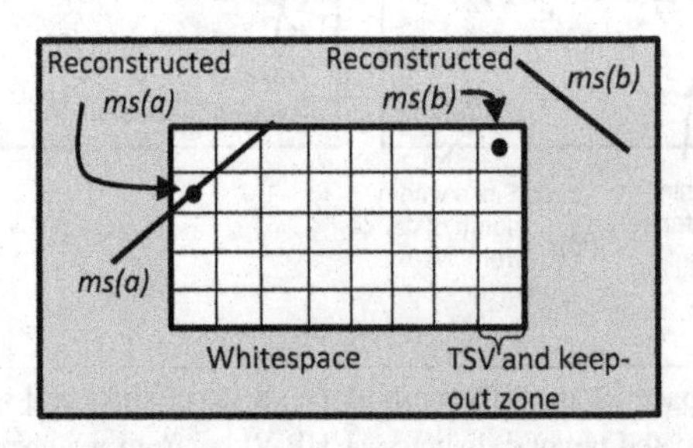

FIGURE 15.26

Reconstruction of the merging segments [597].

Table 15.6 Effect of Whitespace Area on the Number of TSVs, Skew, Power, and Slew on Different Whitespace Aware and Unaware Methods, where the Bound for TSVs is 20 [597]

	MMM-TB, DME, Buffer Insertion, W/O TSV Relocation				MMM-TB, DME, Buffer Insertion, After TSV Relocation				Whitespace Aware CTS [597]			
Whitespace No/Area (%)	# of TSVs	Skew (ps)	Power (W)	Slew Violation	# of TSVs	Skew (ps)	Power (W)	Slew Violation	# of TSVs	Skew (ps)	Power (W)	Slew Violation
4 (4.11)	20	23.81	0.299	No	20	175.78	0.358	Yes	2	28.58	0.294	No
9 (5.64)	20	23.81	0.299	No	20	148.98	0.336	Yes	9	41.28	0.314	No
16 (10.03)	20	23.81	0.299	No	20	55.14	0.314	No	14	40.09	0.308	No
29 (12.86)	20	23.81	0.299	No	20	83.06	0.309	No	18	32.45	0.309	No
36 (14.47)	20	23.81	0.299	No	20	40.94	0.304	No	19	26.41	0.310	No
55 (16.23)	20	23.81	0.299	No	20	33.22	0.302	No	20	32.01	0.302	No
131 (19.79)	20	23.81	0.299	No	20	22.98	0.301	No	20	25.33	0.302	No

Columns 6 to 9 correspond to those results after the TSVs are moved into the nearest whitespace. This practice, however, leads to slew violations, most likely due to an increase in the capacitance of the path after the TSV is relocated. Furthermore, the whitespace aware method seems to utilize fewer TSVs than the TSV bound (which, in this case, is 20). The rationale behind this behavior is that if fewer and smaller whitespaces exist, the extension of the MMM-TB method to produce the connection topology discourages *xy*-cuts that produce several TSVs since no whitespace (at least at a reasonable distance) is available to accommodate these TSVs. This lack of whitespaces exacerbates the clock skew reported in column 7, as relocating the TSVs adds significant wirelength. As the number and area of the whitespace blocks increase, however, the performance of the different methods deviates less. This result is due to less of an effect of the post-synthesis TSV relocation process than for whitespace unaware methods due to the greater availability of whitespace.

Another practical issue that has drawn attention in CTS is the presence of defects in TSVs, which can render a 3-D stack inoperable [604]. Due to the criticality of the clock networks, adding fault tolerance to a 3-D clock network is a reasonable objective. The simplest way to achieve this objective is TSV redundancy. Two or more TSVs can be utilized for each intertier connection within a clock network. The area (and power) overhead is, however, significant. Consequently, more advanced redundancy schemes have been proposed [609–612]. Some of these schemes are depicted in Fig. 15.27. The method of double (or higher) redundancy, shown in Fig. 15.27A, replicates the number of TSVs. To reduce the area overhead, for example, the 4:2 topology

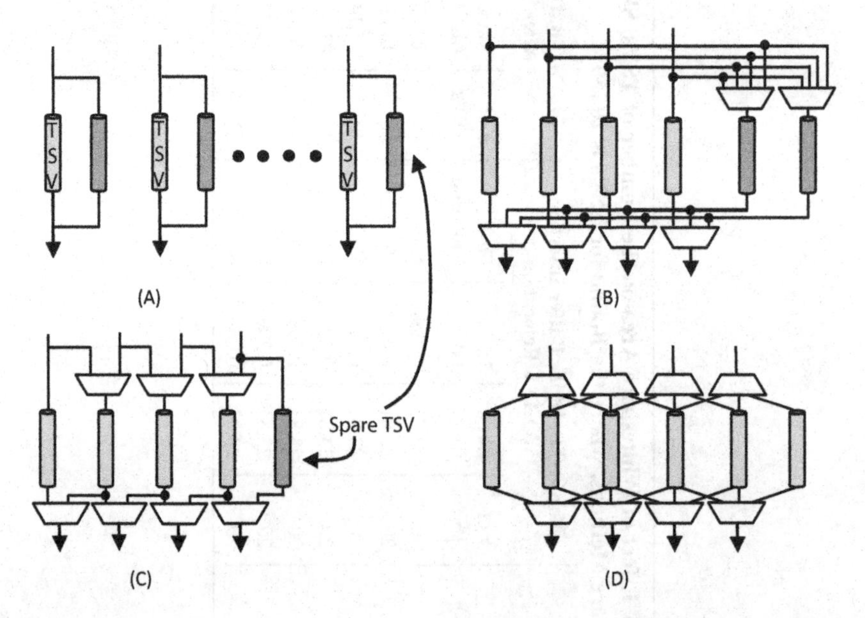

FIGURE 15.27

Different TSV redundancy schemes. (A) Double (*N*-times) redundancy [609], (B) 4:2 shared spare topology with two spare TSVs [610], (C) 4:1 shared spare topology with one spare TSV [611], and (D) 4:2 shared spare topology with no spare TSVs [612].

illustrated in Fig. 15.27B uses multiplexers to reroute the signal transferred by a defective TSV to a spare TSV. This rerouting can add significantly to the delay of a signal. In Fig. 15.27C, in the 4:1 topology, a signal is shared between two adjacent TSVs through multiplexers, where the signal to a defective TSV is shifted to a neighboring connection. A spare TSV is used, thereby tolerating a single defect. The last scheme shown in Fig. 15.27D includes no spare TSVs and shares a signal with three TSVs, which enhances the ability to tolerate TSV faults.

Introducing these defect tolerance mechanisms to a clock network is feasible but the overhead in area and power can be prohibitive. An approach that requires less overhead utilizes existing TSVs within the clock network. These TSVs are inserted by the CTS technique to tolerate faults from defective TSVs. The key concept of this approach is a TSV fault tolerant component (TFC). The operation of TFC is described in Fig. 15.28. As shown in this figure, additional circuitry is needed. This circuitry includes one TG (i.e., TGT) and multiplexers, *MUX1* and *MUX2*, while the other TGs are available from the pre-bond testable clock network. These TGs connect the redundant clock tree during pre-bond test. This scheme exhibits a lower overhead as compared to the schemes depicted in Fig. 15.27.

The essence of this technique is to form pairs of existing TSVs and use these pairs to form TFCs, which operate as follows. If the tiers are not bonded, pre-bond operation is achieved by enabling the TGs to connect the redundant clock tree and supply the clock signal across a tier. During post-bond operation without any TSV defects, all of the TGs are off and the clock signal propagates through "healthy" TSVs. If any of the two TSVs is defective, the TGs and multiplexers are configured to ensure that the clock signal delivered from the healthy TSV is propagated to another subtree through the wire segment connected by TG1 and TG2. This scheme can be further extended to include more TSVs, offering enhanced fault tolerance. The complexity and associated overheads, however, make the usage of pairs or triplets of TFCs more beneficial [598].

This method is applied as a post-synthesis technique, and should, therefore, be placed to not degrade the characteristics of the synthesized clock tree. Pairs of TSVs can form a TFC whenever a TSV is added. First, a check is performed to determine whether other TSVs are located within a

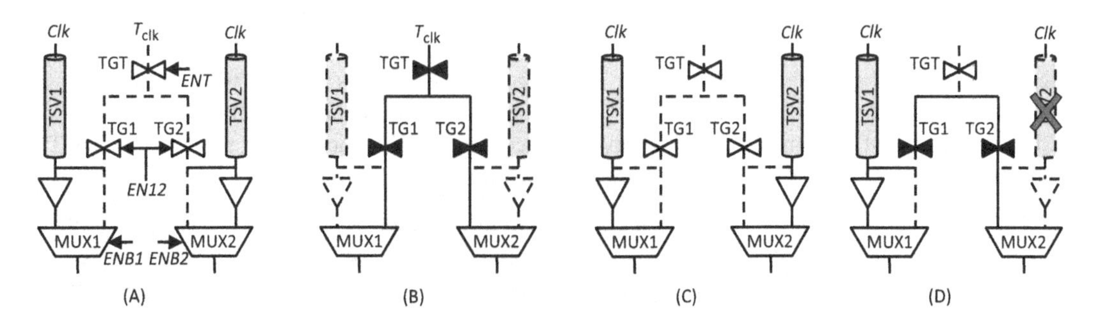

FIGURE 15.28

Operation of a TSV TFC [598], (A) a pair TFC, (B) in pre-bond operation, the redundant tree is connected (shown with solid lines) while the TSVs are not present, (C) in post-bond operation with no defects, the clock signal is transferred by the TSVs, and (D) the TSV2 is defective and part of the redundant tree is used to propagate the clock signal to an adjacent subtree through *TG2* and *MUX2*.

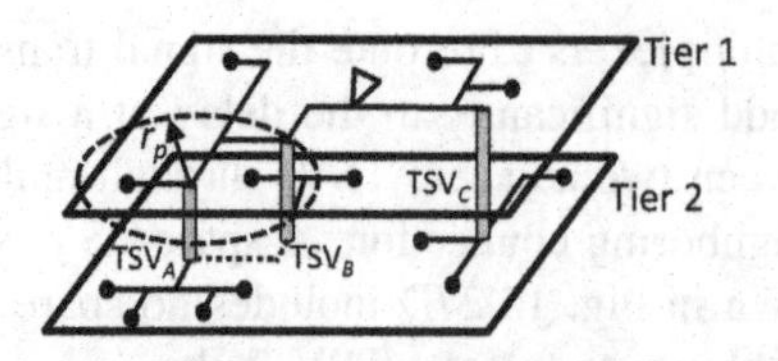

FIGURE 15.29

Example of fault tolerant CTS from adjacent TSVs. TSV_A and TSV_B are within distance r_p and form a TFC pair [598].

radius r_p from the placed TSV, as illustrated in Fig. 15.29. If several TSVs exist, the closest TSV is chosen to form a pair, yielding a TFC. If only TSVs that belong to other TFCs exist, the recently placed TSV can be used to form a triple TFC. If this possibility does not exist or if no TSVs are within the specified distance, double redundancy is employed to provide defect tolerance. A crucial component in forming a TFC is the radius r_p, which is determined from the TSV characteristics and the fabrication technology [598].

This fault tolerance scheme has been applied to several synthesized 3-D clock trees [580], where two tiers comprise the 3-D stack. A comparison among three fault tolerant techniques, double redundancy, only pair TFCs, and both pair and triple TFCs, is performed using an industrial circuit with 55.4 clock sinks. This comparison demonstrates that the pair and triple TFCs reduce the area overhead as compared to double redundancy by, respectively, 40% and 58%. The power overhead due to the use of pair and triple TFCs as compared to a clock network without fault tolerance is, respectively, 29% and 32%. The additional improvement in yield offered by the TFCs as compared to double redundancy is 1.27%, while the increase in skew and wirelength does not exceed 3%. Consequently, using the existing resources of the clock network, defect tolerance can be provided with moderate overhead.

15.5 SUMMARY

CTS techniques and global clock distribution networks for 3-D circuits are reviewed in this chapter. The majority of 3-D CTS methods are based on standard CTS methods, such as the method of MMM and the DME method. The primary enhancements, differences, and traits of 3-D CTS approaches and global clock network topologies are summarized as follows:

- Symmetric planar topologies have been extended to three dimensions, where a different number of TSVs is required to provide the clock signal to all of the tiers within a 3-D circuit. Single and multi-TSV global networks have been proposed, where the number of TSVs is traded off with power and wirelength.
- Multi-TSV topologies do not inherently support pre-bond test and, therefore, require redundant trees during pre-bond test.
- A clock network that combines in each tier a global tree with a grid is more robust to process and environmental variations, typically at the cost of increased power. The meshes can be shorted at several points through TSVs to lower intertier variations.

- A dedicated tier that includes the global clock network and related clock circuitry supports individual optimization of the clock delivery topology and circuitry. The other tiers comprise the logic and memory of the system. Each tier, except for the clock tier, includes a mesh to provide the clock signal to the clock pins in these other tiers.
- 3-D CTS techniques usually proceed in two phases. A top-down phase to generate a connection topology for a set of clock sinks, and a second phase where the generated topology is embedded into the tiers of the 3-D circuit.
- The top-down phase is based on the MMM method where a new form of partitioning, the *z*-cut, is introduced. The direction of these partitions take place recursively, resulting in different TSV requirements for the topologies. Several algorithms are described that generate topologies that either obey a TSV bound or minimize the total vertical length.
- The second phase is based on the DME technique where all of the sinks are projected onto a single Manhattan plane. Embedding of the internal nodes into tiers affects the number of required TSVs. Modifications of the DME method can embed the internal nodes while minimizing the number of TSVs.
- Multi-TSV clock trees significantly reduce wirelength and power as compared to single TSV topologies, although these improvements decrease for large TSV capacitances.
- Pre-bond test is an important requirement of 3-D clock trees. This task adds redundant trees in all but the source tier of a 3-D circuit. Unconnected local trees in those tiers are connected to the redundant tree through TGs during pre-bond test.
- Insertion of TSV buffers in the source tier decouples the clock network of this tier from subtrees in the other tiers. This practice preserves the characteristics (e.g., clock skew and slew) of the network in the source tier during both pre- and post-bond operation.
- Additional adjustments to the synthesis of clock trees are required if placement of the TSVs is constrained to limited whitespace. Clustering of sinks, modifications of the MME technique, and restructuring of the merging segments produced by the DME technique are performed to address the implications of limited whitespace and ensure that the 3-D clock trees satisfy the required specifications.
- Pre-bond testable 3-D clock trees can be augmented with fault tolerance by replacing the TGs with TFCs using the existing TSVs of the network rather than adding redundant TSVs, which require significant area overhead.
- The overhead in power and skew of the TFC is moderate as compared to 3-D clock networks that do not support fault tolerance.

CHAPTER

CASE STUDY: CLOCK DISTRIBUTION NETWORKS FOR THREE-DIMENSIONAL ICs 16

CHAPTER OUTLINE

As discussed in the previous chapter, an omnipresent and challenging issue for synchronous digital circuits is the reliable distribution of the clock signal to the many thousands to millions of sequential elements distributed throughout a synchronous circuit [574]. The complexity is further increased in three-dimensional (3-D) ICs as sequential elements belonging to the same clock domain (i.e., synchronized by the same clock signal) can be located on different tiers.

In this chapter, a variety of clock network architectures for 3-D circuits is discussed. These clock topologies have been included on a test circuit for the 3-D technology developed at MIT Lincoln Laboratories (MITLL). This fabrication process is discussed in the following section. The logic circuitry comprising the common load of the 3-D clock distribution networks is described in Section 16.2. The several clock distribution networks employed in this case study are described in Section 16.3. Models used to simulate these clock distribution networks are discussed in Section 16.4. Experimental results and a comparison of the different clock distribution networks are presented in Section 16.5. A short summary is provided in the last section of the chapter.

16.1 MIT LINCOLN LABORATORIES THREE-DIMENSIONAL IC FABRICATION TECHNOLOGY

The MITLL developed a manufacturing process for fully depleted silicon-on-insulator (FDSOI) 3-D circuits with short intertier vias (also called 3-D vias here for simplicity). The usual term through silicon via (TSV) is rather misleading for this process as the silicon substrate is fully removed and only the silicon oxide remains. The most attractive feature of this process is the high density of the 3-D vias as compared to other 3-D technologies currently under development, as

Three-Dimensional Integrated Circuit Design. DOI: http://dx.doi.org/10.1016/B978-0-12-410501-0.00016-2

reviewed in Chapter 3, Manufacturing Technologies for Three-Dimensional Integrated Circuits. The MITLL process is a wafer level 3-D integration technology with up to three FDSOI wafers bonded to form a 3-D circuit. The diameter of the wafers is 150 mm. The minimum feature size of the devices is 180 nm, with one polysilicon layer and three metal layers interconnecting the devices on each wafer. A backside metal layer also exists on the upper two tiers, providing the starting and landing pads for the 3-D vias, and the I/O, power supply, and ground pads for the entire 3-D circuit. The primary steps of this fabrication process are illustrated in Figs. 16.1 through 16.6 [177,307].

Each of the wafers is manufactured by a mainstream FDSOI process (Fig. 16.1). The second wafer is flipped and face-to-face bonded with the first wafer using oxide bonding (Fig. 16.2). The handle wafer is removed from the second wafer and the 3-D vias are etched through the oxide of both tiers. Tungsten is deposited to fill the 3-D vias, and the surface of the wafer is planarized by chemical mechanical polishing (Fig. 16.3). The backside vias and metallization are formed to provide the pads for the 3-D vias of the third wafer and the interconnection of these vias with the M1 layer of the second wafer (Fig. 16.4). The third wafer is also flipped and face-to-back bonded with

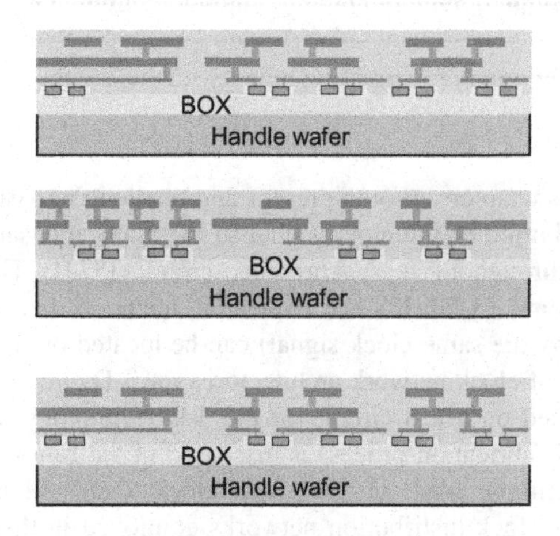

FIGURE 16.1

Three wafers are individually fabricated with an FDSOI process.

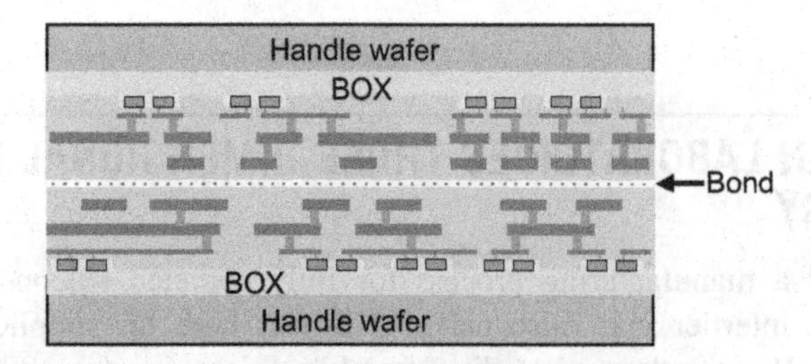

FIGURE 16.2

The second wafer is face-to-face bonded with the first wafer.

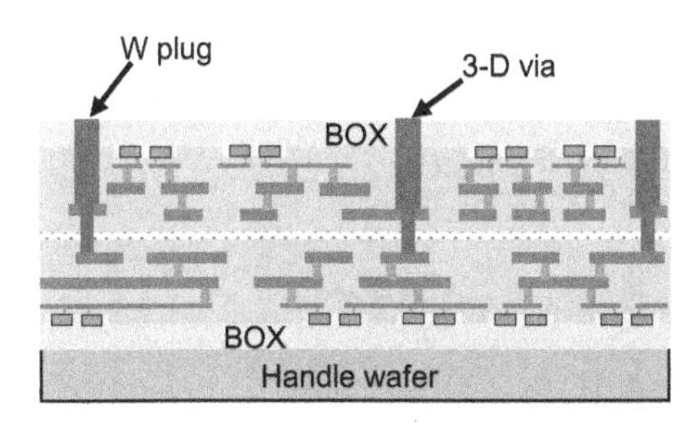

FIGURE 16.3

The 3-D vias are formed and the surface is planarized with chemical mechanical polishing.

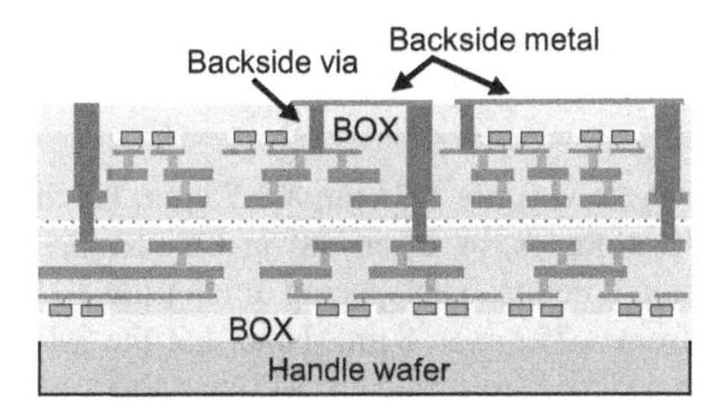

FIGURE 16.4

The backside vias are etched, and the backside metal is deposited on the second wafer.

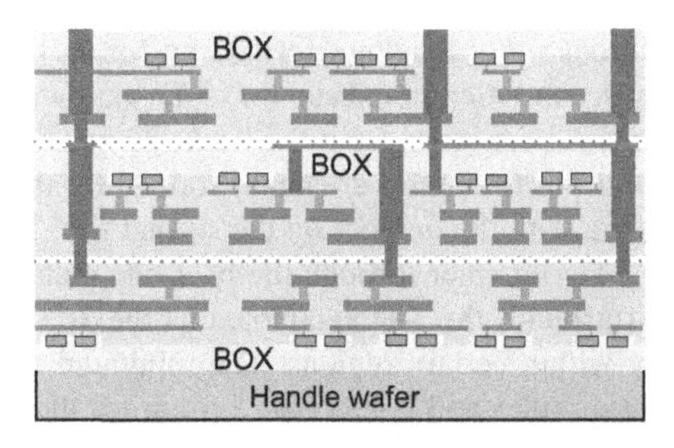

FIGURE 16.5

The third wafer is face-to-back bonded with the second wafer and the 3-D vias for that tier are formed.

the second tier (Fig. 16.5). Another etching step is used to form the 3-D vias of the third tier. The backside vias and interconnections are formed along with the I/O and power pads for the 3-D circuit. A deposited glass coating provides the passivation layer, while the overglass cuts create the necessary pad openings for the off-chip interconnections (Fig. 16.6).

A salient characteristic of this process is the short 3-D vias. As illustrated in Fig. 16.7, the total length of a 3-D via that connects two devices on the first and third tier is approximately 20 μm.

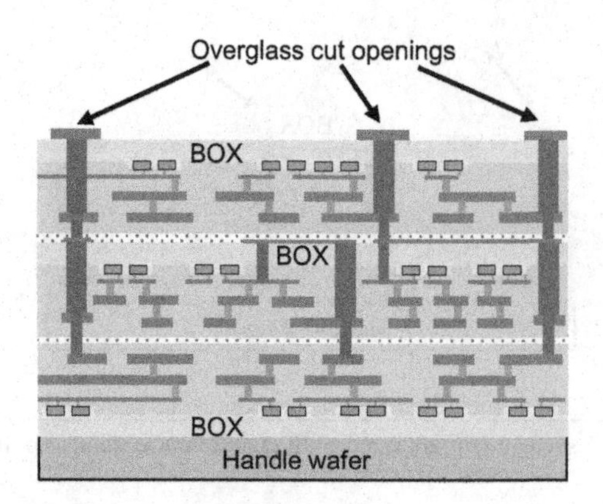

FIGURE 16.6

Backside metal is deposited and glass layers are cut to create openings for the pads.

In addition, the dimensions of these vias are 1.75 μm × 1.75 μm, much smaller than the size of the TSV in many existing 3-D technologies, as discussed in Chapter 3, Manufacturing Technologies for Three-Dimensional Integrated Circuits. The spacing among the 3-D vias depends upon the density of these vias and ranges from 1.75 μm to 8 μm. Note that the 3-D vias connecting the second and third tier can be vertically stacked, resulting in 3-D vias that directly connect devices on the first and third tiers.

A doughnut shaped structure is required for the 3-D vias in both the second and third tier to provide mechanical support for these vias. As shown in Fig. 16.7, the 3-D vias connect different metal layers in the second and third tier. A 3-D via for the second tier connects the backside metal with the M3 layer of the first tier through the doughnut formed by M3 in the second tier. The backside vias or the M3 layer of the second tier connects the 3-D via with the devices on that tier. Alternatively, a 3-D via for the third tier starts from the backside metal of the third tier and ends on the backside metal of the second tier through a doughnut formed by the M3 layer of the third tier. The backside vias connect this via with the devices on the second tier. The transistors located on the third tier are connected to the 3-D via either through the backside metal layer and backside vias, or through the M3 doughnut that surrounds the 3-D via. Note that the 3-D vias can be placed anywhere within the circuit and not only within certain regions. The minimum distance from the transistors, however, is specified by the design rules and is less than 1.5 μm for the MITLL 3-D process.

The electrical sheet resistance of the metal and diffusion layers is listed in Table 16.1 along with the bulk resistivity. The total resistance of the intratier vias and contacts is listed in Table 16.2. Since the third tier is also intended for RF circuits, a low resistance backside metal is available. In addition to the active devices, passive elements, such as resistors and metal-insulator-metal (MIM) capacitors, are also available. The resistors, however, can only be placed on the third tier where the polysilicon or active layer is utilized to form these resistors.

In support of this manufacturing process, design kits for this technology have been developed by academic institutions and supporting CAD companies. The process design kit has been developed for the Cadence Design Framework by faculty from North Carolina State University [421]. This kit includes device models for circuit simulation and a sophisticated layout tool for 3-D

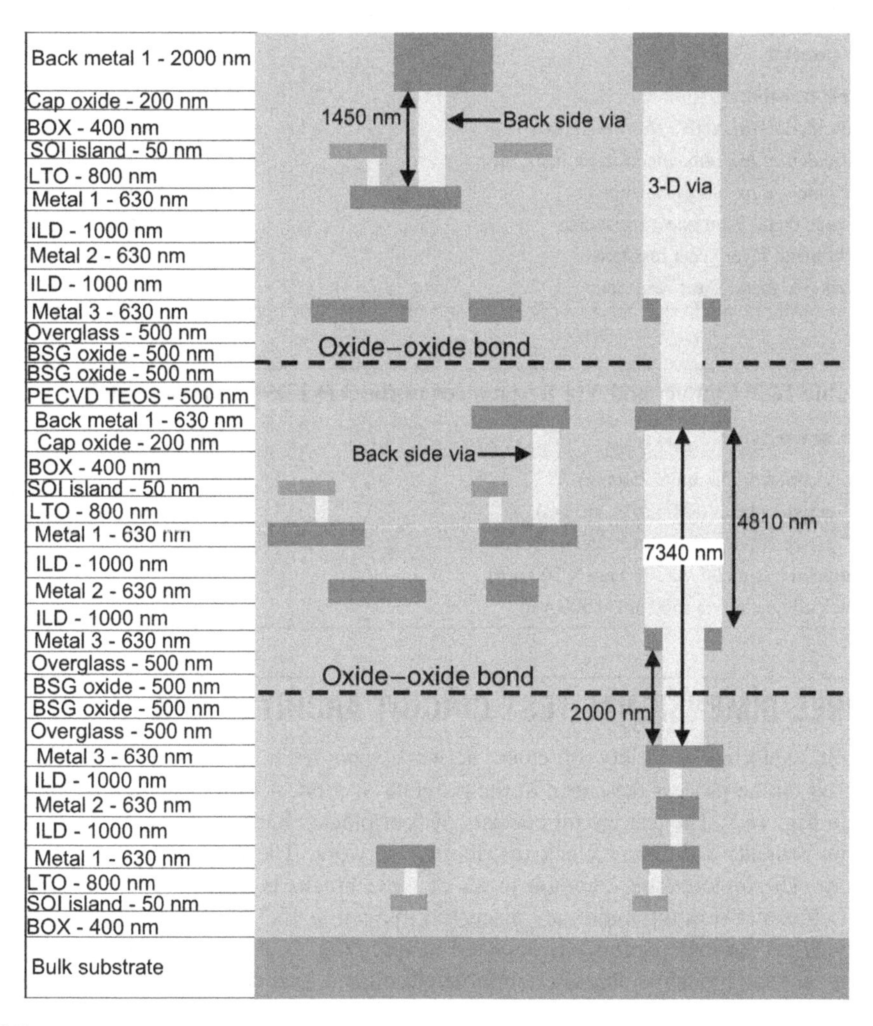

FIGURE 16.7

Layer thicknesses in the 3-D IC MITLL technology [307].

circuits. For example, circuits located on a specific tier can be separately visualized or highlighted. In addition, a circuit designed for one tier can be reproduced or transferred to another tier. Complete design rule checking and circuit extraction are also available. Electrical rule checking, however, is not included, meaning that these 3-D circuits cannot be directly checked for shorts between the power and ground lines. However, to mitigate this problem, two pins can be assigned different names on the power and ground lines. Design rule checking reports errors whenever nets with different names are crossed, thereby checking for electrical shorts.

Table 16.1 Layer Resistances of the 3-D FDSOI Process [307]

Parameter	Value
Bulk resistivity	~2,000 Ω-cm
Silicided n^+/p^+ active sheet resistance	15 ± 3 Ω/sq.
Silicided n^+/p^+ polysilicon sheet resistance	15 ± 3 Ω/sq.
Silicided n^+/p^+ sheet resistance	15 ± 3 Ω/sq.
Lower metal layer sheet resistance	~0.12 Ω/sq.
Top metal layer sheet resistance	~0.08 Ω/sq.
Backside metal sheet resistance	~0.12 Ω/sq.

Table 16.2 Contact and Via Resistances of the 3-D FDSOI Process [307]

Parameter	Value
Poly contact (250 nm × 250 nm)	10 ± 2 Ω
n^+ active contact (250 nm × 250 nm)	10 ± 2 Ω
p^+ active contact (250 nm × 250 nm)	10 ± 2 Ω
Interconnect metal via (300 nm × 300 nm)	4 Ω
Backside metal via (500 nm × 500 nm)	2 Ω

16.2 THREE-DIMENSIONAL TEST CIRCUIT ARCHITECTURE

A test circuit exploring a variety of clock network topologies suitable for 3-D ICs has been designed based on the process described in the previous section. A block diagram of the test circuit is depicted in Fig. 16.8. The test circuit consists of four blocks. Each block contains the same logic circuit but implements a different clock distribution network. The total area of the test circuit is 3 mm × 3 mm. The logic circuit common to all of these blocks is described in this section, while the different clock network topologies are discussed in Section 16.3.

An overview of the logic circuitry is depicted in Fig. 16.9. The function of this logic is to emulate different switching patterns characterizing the circuit and load conditions for the clock distribution networks under investigation. The logic is repeated in each tier and includes

Pseudorandom number generators
Crossbar switch
Control logic for the crossbar switch
Groups of four-bit counters
Current loads and an output circuit for probing.

The pseudorandom generators are based on the technique described in [613], which uses linear feedback shift registers and XOR operations to generate a random 16-bit word every clock cycle after the first few cycles required to initialize the generator. The physical layout of one random number generator is illustrated in Fig. 16.10. There are a total of nine pseudorandom generators in each circuit block, connected by groups of three to the crossbar switch within each tier.

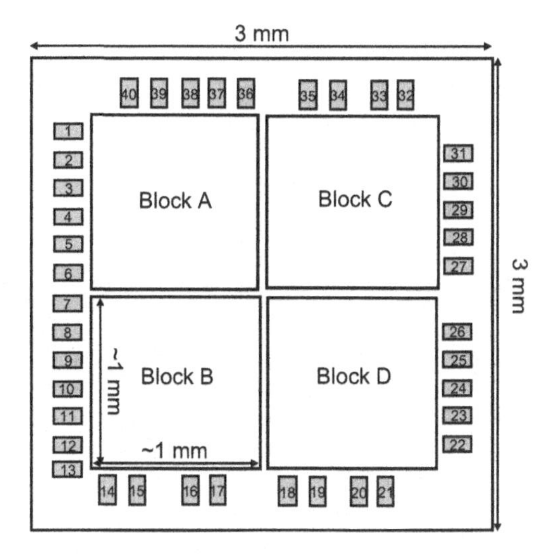

FIGURE 16.8

Block diagram of the 3-D test IC. Each block has an area of approximately 1 mm^2. The remaining area is reserved for the I/O pads (the gray shapes).

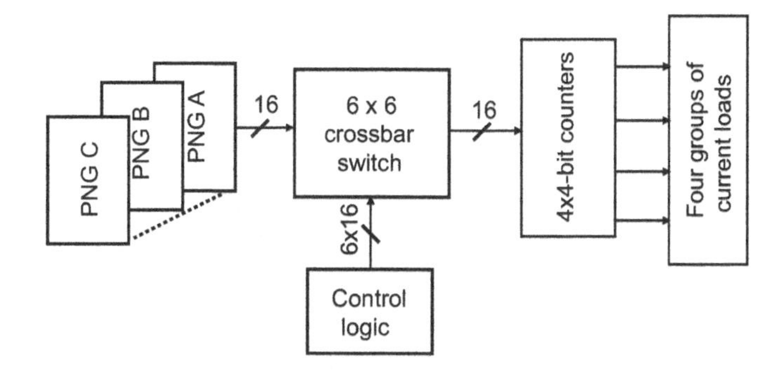

FIGURE 16.9

Block diagram of the logic circuit included in each tier of each block.

A classic crossbar switch with six input and output ports is included in each tier, where the width of each port is 16 bits. Three of the six inputs of the crossbar switch are connected to the output of the number generators, while the remaining inputs are connected to ground. The physical layout of the crossbar switch is shown in Fig. 16.11. The three output ports of the switch are connected to a group of 4-bit counters, while the remaining outputs drive a small capacitive load. Since each port is 16 bits wide, each port is connected to four 4-bit counters. These counters are, in turn, connected to current loads implemented with cascoded current mirrors. The counters and current loads are distributed across each tier. The control logic consists of an 8-bit counter that controls the connectivity among the input and output ports of the crossbar switch.

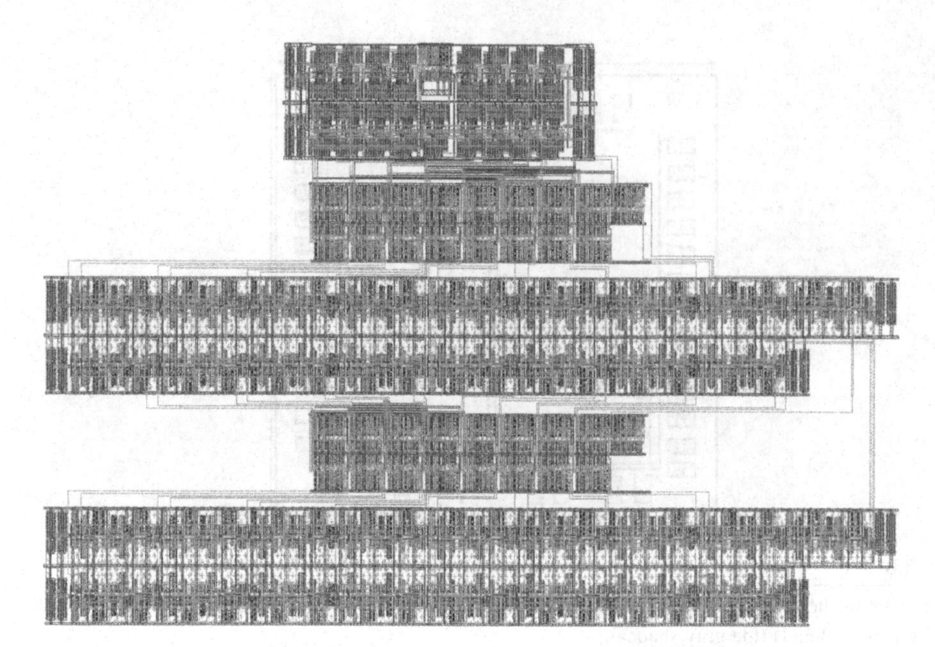

FIGURE 16.10

Physical layout of a pseudorandom number generator.

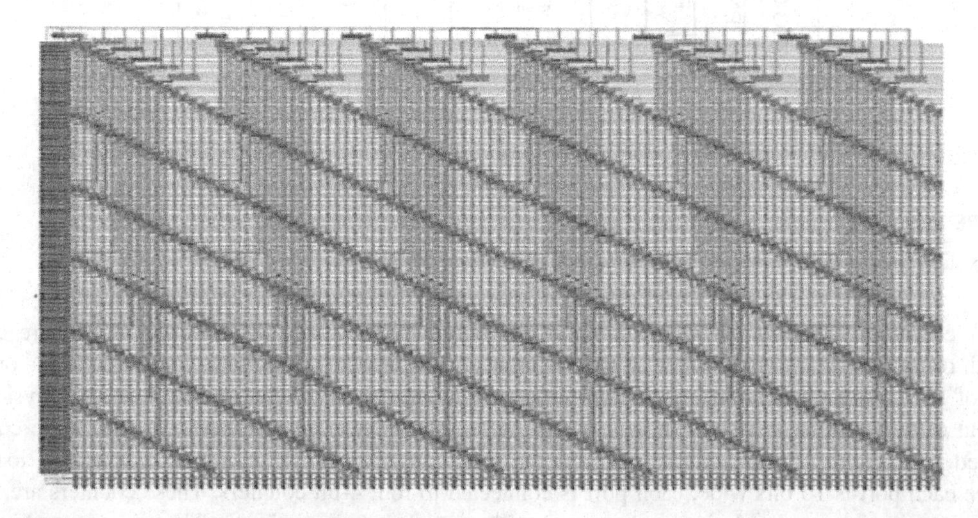

FIGURE 16.11

Physical layout of 6 × 6 crossbar switch with 16-bit wide ports.

The data flow in this circuit can be described as follows. After resetting the circuit, the pseudo-random number generators are initialized, and the control logic connects each input port to the appropriate output port. Since the control logic includes an 8-bit counter, each input port of the crossbar switch is successively connected every 256 clock cycles to each output port.

The output ports of the crossbar switch are connected to the 4-bit counters. Each of these counters is loaded with a 4-bit word, counts upwards, and loaded with a new word every time all of the bits are equal to one. The most significant bit (MSB) of each counter is connected to four current loads that are turned on when the bit is equal to one. Since the counters are loaded with random words from the random generators through the crossbar switch, the current loads draw a variable amount of current during circuit operation. This randomness mimics different switching patterns that can exist within a circuit.

The current loads are cascoded current mirrors, as shown in Fig. 16.12. In the cascoded current mirrors, the output current I_{out} closely follows I_{ref} as compared to a simple current mirror. The reference current I_{ref} is externally provided to control the amount of current drawn from the circuit. The gate of transistor M5 is connected to the MSB of a 4-bit counter, shown in Fig. 16.12 as the sel signal. This additional device switches the current sinks. The layout of a group of four current loads is illustrated in Fig. 16.13. The width of the devices shown in Fig. 16.12 is

$$W_1 = W_2 = W_3 = W_4 = 600 \text{ nm}, \quad W_5 = 2000 \text{ nm}. \tag{16.1}$$

The power supply is 1.5 volts, as set by the MITLL process.

The layout of the test circuit is illustrated in Fig. 16.14, while the connectivity of the pads to the external signals is listed in Table 16.3. Several decoupling capacitors are included in each circuit block and are highlighted in Fig. 16.14. The capacitors serve as extrinsic decoupling capacitance and are MIM capacitors. Note that the number of pads is not limited by the area of the circuit but, rather, by the maximum number of connections permitted by the probe card. The backside metal layer of the third tier is utilized for the pads. Each of the circuit blocks is supplied by

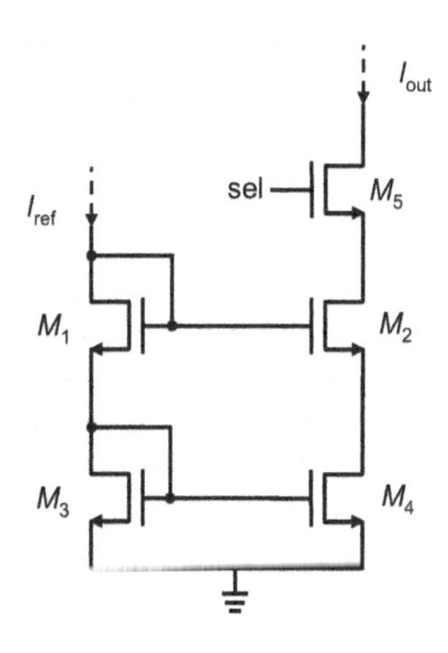

FIGURE 16.12

Cascoded current mirror with an additional control transistor.

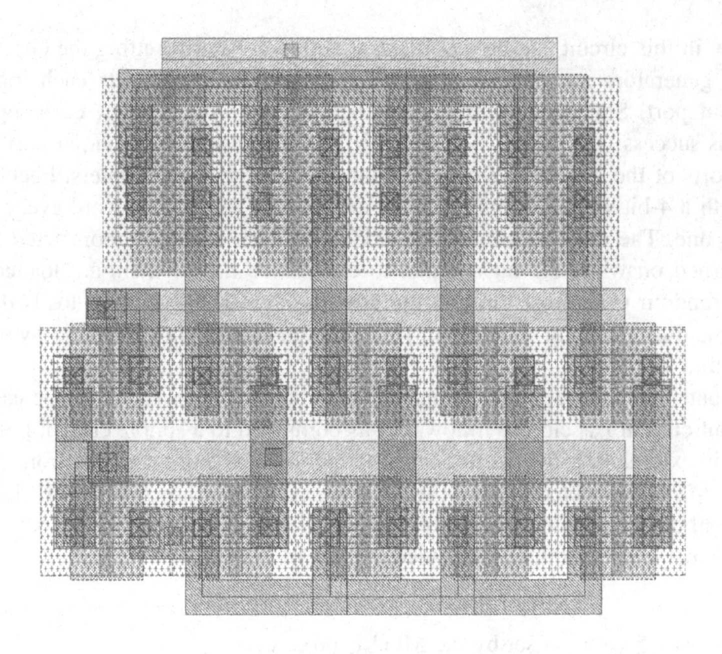

FIGURE 16.13

Four stage cascoded current mirrors.

separate power and ground pads, while only a pair of power and ground pads is connected to the pad ring to provide protection from electrostatic discharge.

16.3 CLOCK DISTRIBUTION NETWORK STRUCTURES WITHIN THE TEST CIRCUIT

Different clock distribution schemes for 3-D circuits are described in this subsection. To evaluate the specific requirements of the clock networks, consider a traditional H-tree topology. As shown in Fig. 15.7, at each branch point of an H-tree, two branches emanate with the same length. An extension of the H-tree to three dimensions does not guarantee equidistant interconnect paths from the source to the leaves of the tree. This situation is shown in Fig. 16.15, where an H-tree is replicated in each tier of a 3-D circuit. The clock signal is propagated through intertier vias from the output of the clock driver to the center of the H-tree on tiers one and three. The high impedance of these vias increases the time for the clock signal to arrive at the leaves of the tree on these tiers as compared to the time for the clock signal to arrive at the leaves of the tree on the same tier as the clock driver. Furthermore, in a multi-tier 3-D circuit, three or four branches can emanate from each branch point. The third branch propagates the clock signal to the other tiers of the 3-D circuit, as shown in Fig. 16.15. Similar to a design methodology for a two-dimensional (2-D) H-tree topology, the width of each branch is reduced to a third (or more) of the segment preceding the branch point to match the impedance at the branch point. This requirement, however, is difficult to achieve as

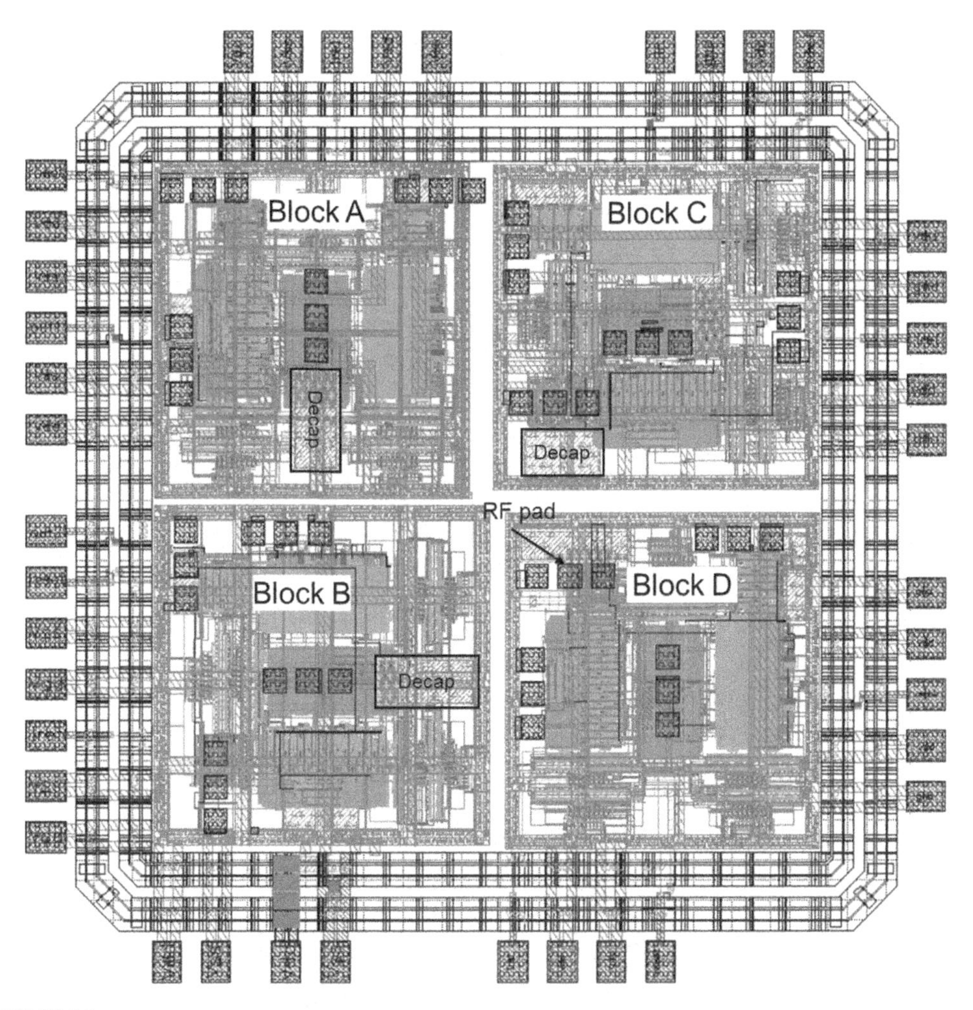

FIGURE 16.14

Physical layout of the test circuit. Some decoupling capacitors are highlighted.

the third and fourth branches are connected by an intertier via. The vertical interconnects are of significantly different length as compared to the horizontal branches and also exhibit different impedance characteristics.

Several clock network topologies for 3-D ICs are investigated in this case study. Each of the four blocks of the test circuit includes a different clock distribution structure, which are schematically illustrated in Fig. 16.16. The physical layout of these topologies for the MITLL 3-D technology is depicted in Fig. 16.17. The architectures employed in the blocks are:

Block A: All of the tiers contain a four level H-tree (i.e., equivalent to 16 leaves) with identical interconnect characteristics. The H-trees are connected through a group of intertier vias at the

Table 16.3 Pad Connectivity of the 3-D Test Circuit (Pad Index Shown in Fig. 16.8)

Index	Pad Connectivity	Index	Pad Connectivity
1	Reset	21	Reset
2	V_{dd}	22	V_{ss}
3	V_{ss}	23	V_{dd}
4	Output bit	24	Output bit
5	V_{ss}	25	V_{dd}
6	V_{dd}	26	V_{ss}
7	Output bit	27	V_{ss}
8	Reset	28	V_{dd}
9	V_{ss}	29	I_{ref}
10	V_{dd}	30	V_{ss}
11	I_{ref}	31	V_{dd}
12	V_{ss}	32	Reset
13	V_{dd}	33	V_{dd}
14	V_{dd}	34	V_{ss}
15	V_{ss}	35	Output bit
16	V_{dd}	36	V_{ss}
17	V_{ss}	37	V_{dd}
18	I_{ref}	38	I_{ref}
19	V_{dd}	39	V_{ss}
20	V_{ss}	40	V_{dd}

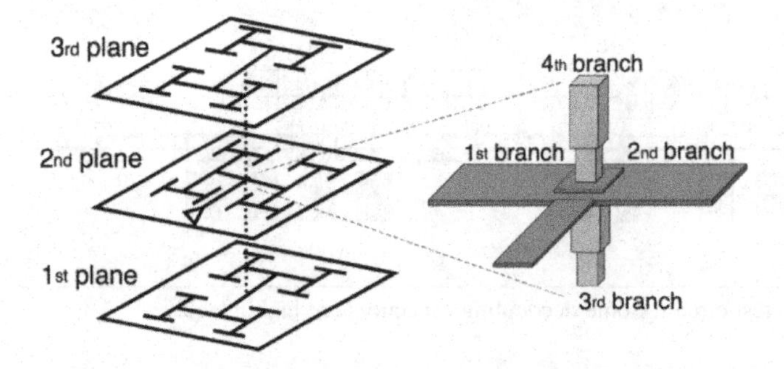

FIGURE 16.15

Two-dimensional H-trees constituting a clock distribution network for a 3-D IC.

first branching point, as shown in Fig. 16.16A. The second tier is face-to-face bonded with the first tier and both of the H-trees are placed on the third metal (M3) layer. The physical distance between these clock networks is approximately 2 μm. Note that the H-tree on the second tier is rotated by 90° with respect to the H-trees on the other two tiers. The orthogonal placement of these two clock networks effectively eliminates any inductive coupling. All of the H-trees are shielded with two parallel lines connected to ground.

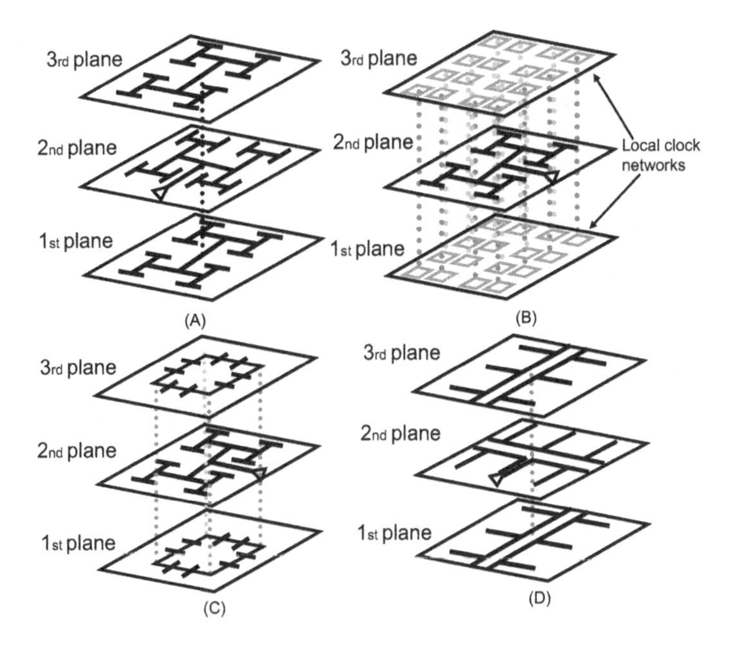

FIGURE 16.16

Different 3-D clock distribution networks within the test circuit. (A) H-trees, (B) H-tree and local rings/meshes, (C) H-tree and global rings, and (D) trunk based.

Block B: A four level H-tree is included in the second tier. Each of the leaves of this H-tree is connected through intertier vias to small local rings on the first and third tier, as illustrated in Fig. 16.16B. As in Block A, the H-tree is shielded with two parallel lines connected to ground. Additional interconnect resources are used to form local meshes. Due to the limited interconnect resources, however, a uniform mesh in each ring is difficult to achieve. The clock routing is constrained by the power and ground lines as only three metal layers are available on each tier.

Block C: The clock distribution network for the second tier is a shielded four level H-tree. Two global rings are utilized for the other two tiers, as depicted in Fig. 16.16C. Each ring is connected through intertier vias to the four branch points on the second level of the H-tree. The registers in each tier are individually connected to the ring.

Block D: The clock network on each tier consists of a trunk structure and branches that connect the registers in each tier to the trunk, as shown in Fig. 16.16D. As for Block A, the trunk for the second tier is rotated by 90° to avoid inductive coupling. Those interconnects that branch from the trunk are placed as close as possible to the registers.

Buffers are inserted at appropriate branch points within the H-trees to amplify the clock signal. In each of the circuit blocks, the clock driver for the entire clock network is located on the second tier. The clock driver on that tier is placed to ensure that the clock signal propagates through similar vertical interconnect paths to the first and third tier, resulting in the same approximate delay for

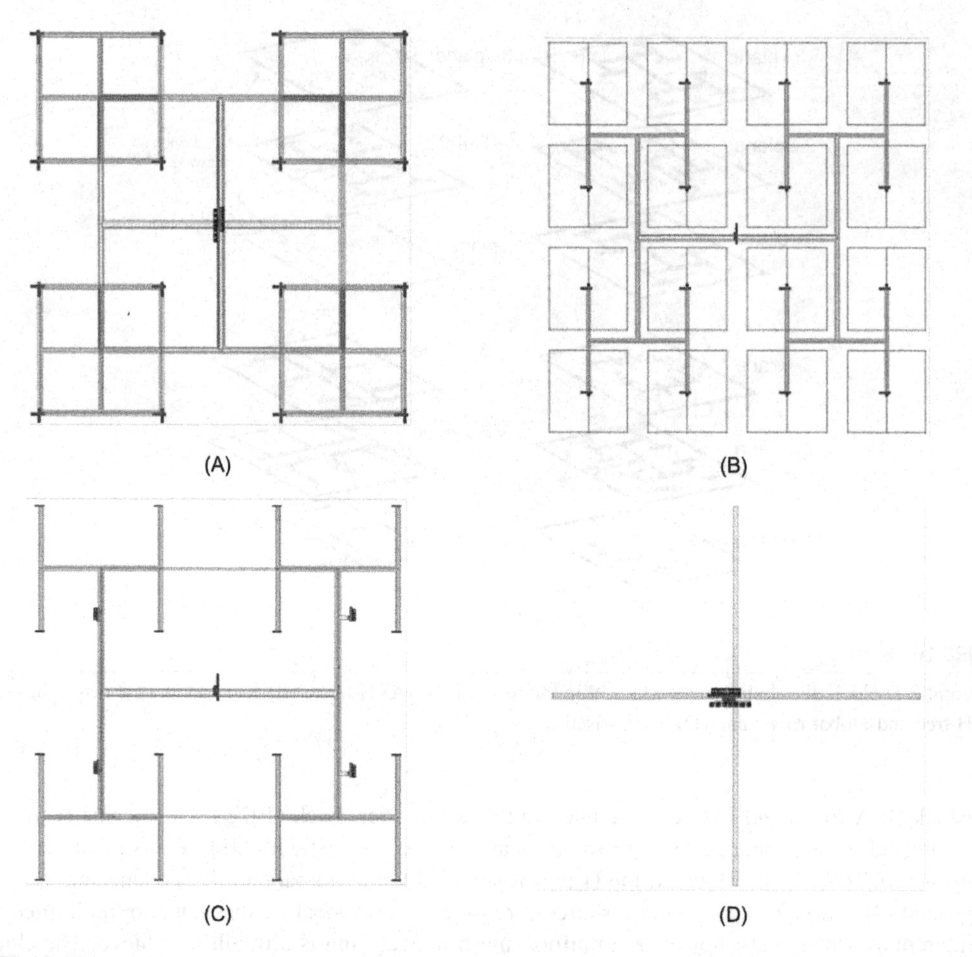

FIGURE 16.17

Physical layout of the clock distribution networks in the 3-D IC. (A) H-trees, (B) H-tree and local rings/meshes, (C) H-tree and global rings, and (D) trunk based.

the registers located on the first and third tiers. The clock driver is a traditional chain of tapered buffers [614–617].

The clock network on each tier feeds the registers located on the same tier. The off-chip clock signal is passed to the clock driver through an RF pad, as shown in Fig. 16.18. The dimensions of the RF pad are also shown in Fig. 16.18. Additional RF pads are placed at different locations on the third tier of each block for probing. These RF pads are used to measure the clock skew at different locations on different tiers within the clock network. The output circuitry is an open drain transistor connected to the RF pads by a group of intertier vias to decrease the resistance between the transistor and the probe. The probe is modeled as a series *RLC* impedance with the values shown in Fig. 16.19. The circuit depicted in Fig. 16.19 is used to determine the size of the output transistor.

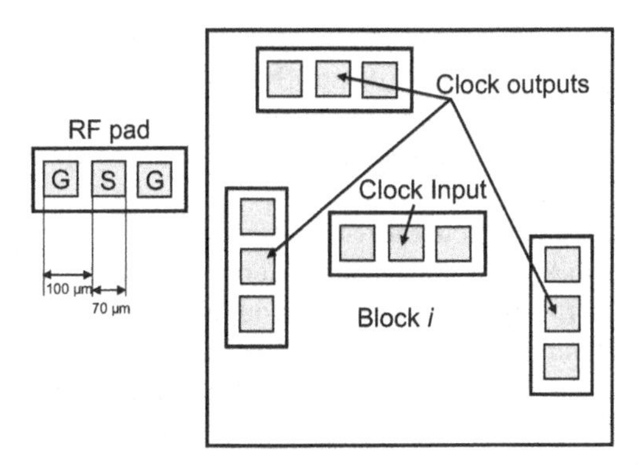

FIGURE 16.18

Clock signal probes with RF pads.

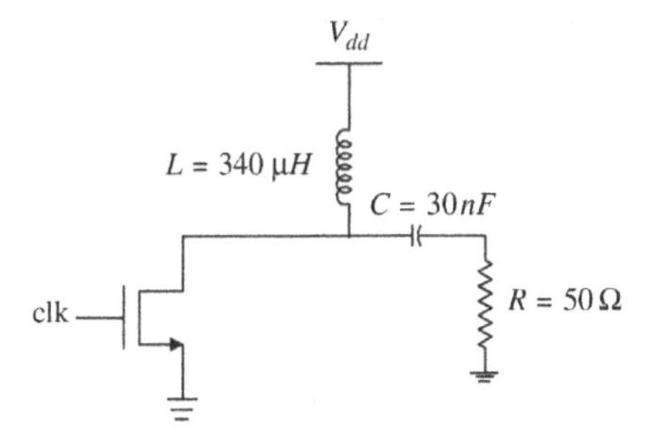

FIGURE 16.19

Open drain transistor and circuit model of the probe (includes impedance of RF pads).

In addition to the clock skew within the clock network topologies employed within the blocks of the test circuit, the power consumption of the entire clock distribution network has also been measured. Since all of the blocks include the same logic circuits, any difference in power consumption is attributed to the clock network, including the interconnect structures, clock driver, and clock buffers. The simulations characterizing the clock delay for each topology are described in the following section. The measurements of the clock skew and power dissipation of the blocks are discussed in the following Section 16.5.

16.4 MODELS OF THE CLOCK DISTRIBUTION NETWORK TOPOLOGIES INCORPORATING THREE-DIMENSIONAL VIA IMPEDANCE

Simulation of the fabricated clock distribution topologies incorporating the electrical impedance of the intertier 3-D vias is described in this section. A comparison between the simulated and experimental results is also presented here. The electrical impedance of the 3-D vias is evaluated for several diameters, lengths, dielectric thicknesses (bulk), and via-to-via spacings [58,201]. These expressions are used here to model the contribution of the 3-D vias to the delay and skew characteristics of the clock distribution topologies, and are described in detail in Chapter 4, Electrical Properties of Through Silicon Vias.

In addition to characterizing the electrical parameters of the TSVs, the electrical characteristics of the clock distribution network on each tier are determined through numerical simulation. This set of simulations has been performed for the three widths used in the fabricated test circuit, and for five different lengths. Trend lines for the capacitance, DC resistance, 1 GHz resistance, DC self- and mutual inductance, and the asymptotic self- and mutual inductance f_{asym} approximate the electrical parameters of different length interconnect segments within the clock network. These simulations include two ground return paths spaced 2 μm from either side of the clock line. These return paths behave as ground for the electrical field lines emanating from the clock line, resulting in a more accurate estimate of the capacitance.

The electrical paths of the clock signal propagating from the root to the leaves of each tier for the H-tree clock topology (see Fig. 16.16A) is depicted in Fig. 16.20. The clock network on each tier is modeled by 50 μm segments, where a π network represents the electrical properties of each segment. These 50 μm segments model the distributive electrical properties of the interconnect. Similarly, when either meshes (Fig. 16.16B) or rings (Fig. 16.16C) are used on tiers A and C (see Fig. 16.20), each 50 μm segment is replaced with an equivalent π network to more accurately represent the single mesh and ring structure within the test circuit. Note that for the mesh structures, the clock signal is distributed to tiers A and C from the leaves of the H-tree in tier B. For the rings

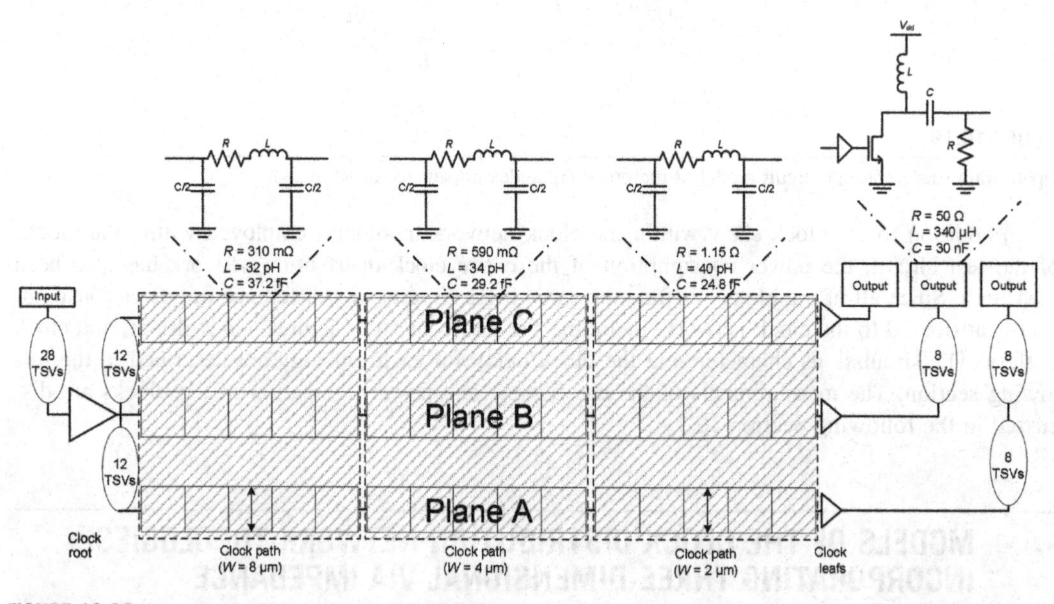

FIGURE 16.20

Structure of clock signal path from Fig. 16.16A to model the clock skew. The number within each oval represents the number of parallel TSVs between device tiers [618].

topology, the clock signal distributed within tiers A and C is driven by buffers at the second level of the H-tree. The delay from the root to the leaves of each tier is listed in Table 16.4.

Good agreement between the model and experimental data is shown. The per cent error between the model and experimental clock delays is listed in Table 16.5. A maximum error of less than 10% is achieved for the clock paths within the H-tree topology. The larger errors listed in Table 16.5 are due to the small time scale being examined.

The equivalent electrical model of a TSV used for the simulation of the clock networks is shown in Fig. 16.21. The resistance, inductance, and capacitance expressions are compared to numerical simulations for the TSV structures in the MITLL multiproject wafer (for 3-D via parameters, $r = 1$ μm, $l = 8.5$ μm, and $p = 5$ μm) in Table 16.6.

Table 16.4 Modeled Clock Delay From the Root to the Leaves of Each Tier for Each Block

Clock Distribution Network	Clock Delay (ns)		
	t_A	t_B	t_C
H-trees (Fig. 16.16A)	0.359	0.355	0.351
Local rings (Fig. 16.16B)	0.325	0.323	0.321
Global rings (Fig. 16.16C)	0.510	0.465	0.442

Table 16.5 Per cent Error Between Modeled and Experimental Clock Delay

Clock Distribution Network	Clock Delay % Error		
	t_A	t_B	t_C
H-trees (Fig. 16.16A)	0	−4.8	−7.4
Local rings (Fig. 16.16B)	36.9	8.4	−6.5
Global rings (Fig. 16.16C)	3.9	3.2	0.7

Table 16.6 Comparison of Numerical Simulations and Analytic Expressions of the TSV Electrical Parameters

Electrical Parameters	Numerical Simulation	Analytic Expressions	% Error
DC resistance (mΩ)	148	154	4.1
1 GHz resistance (mΩ)	166	177	6.6
DC self-inductance (pH)	3.9	3.9	0
f_{asym} self-inductance (pH)	2.9	3.1	6.9
DC mutual inductance (pH)	1.40	1.32	− 5.7
f_{asym} mutual inductance (pH)	1.10	1.08	− 1.8
Capacitance (fF)	1.43	–	–

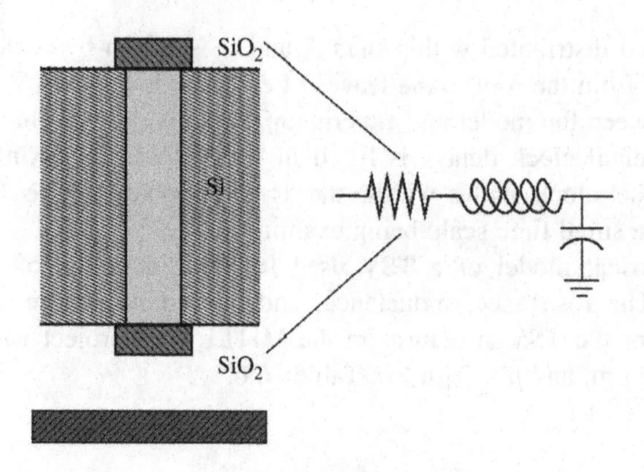

FIGURE 16.21

Equivalent electrical model of a TSV.

Table 16.7 Transistor Width of the Clock Buffers at the Root, Leaves, and Output Circuitry (all Lengths are 180 nm)

Buffer Location	W_N (μm)	W_P (μm)
Root		
Buffer 1	20	50
Buffer 2	54	136
Leaf		
Buffer 1	15	38
Buffer 2	15	38
Output circuitry		
Buffer 1	2.5	7
Buffer 2	2.5	7

The circuit parameters used to model the delay within the clock network are provided below. The dimensions of the buffer circuits at the root, leaves, and output circuitry are listed in Table 16.7. Two sets of transistor widths are provided as each location is double buffered to maintain the same signal logic level. The dimensions of the ring and a single mesh are listed in Table 16.8. These lengths are partitioned into 50 μm long segments, and each segment is modeled by an equivalent π network for a line width of 4 μm. The source follower NMOS transistor located in the output circuitry has a length of 180 nm and a width of 12 μm. The interconnect length connecting the output circuitry and pads to the leaves on each of the three device tiers varies from 0 to 150 μm depending upon the clock topology (line width of 2 μm) and is also represented by an equivalent π network.

Table 16.8 Dimensions of Local and Global Clock Rings

Topology	Length (μm)	Width (μm)
Global rings	500	500
Local rings	200	200

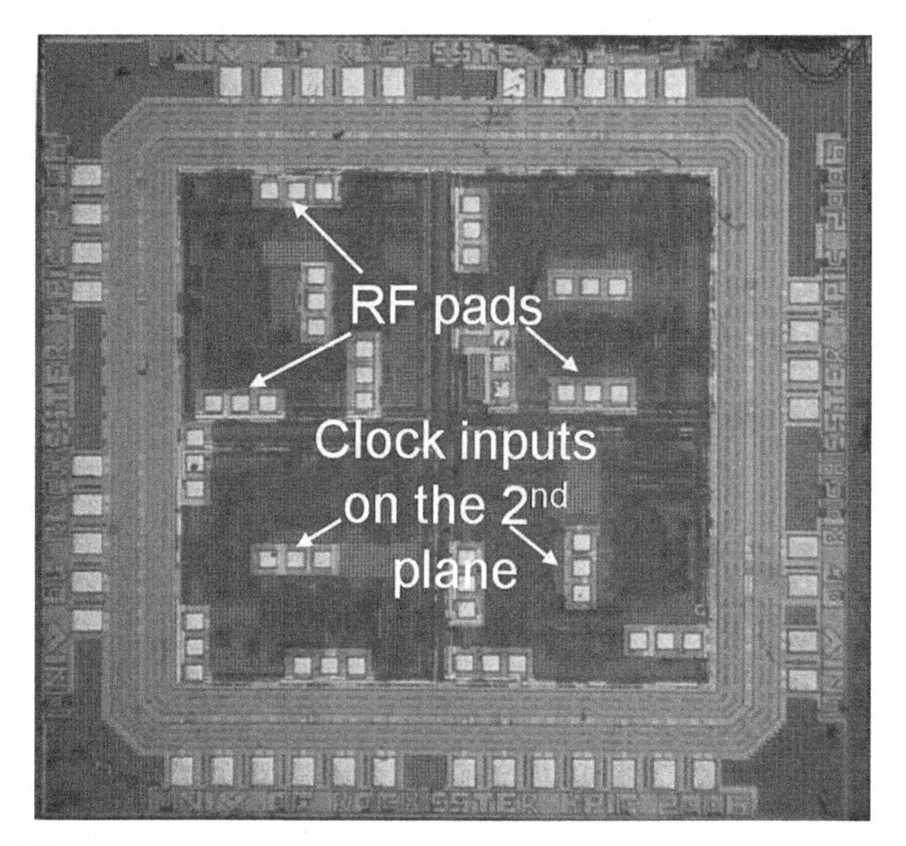

FIGURE 16.22

Top view of fabricated 3-D test circuit.

16.5 EXPERIMENTAL RESULTS

The clock distribution network topologies of the 3-D test circuit are evaluated in this section. The fabricated circuit is depicted in Fig. 16.22, where the four individual blocks can be distinguished. A magnified view of one block is shown in Fig. 16.23. Each block includes four RF pads for measuring the delay of the clock signal. The pad located at the center of each block receives the input clock signal. The clock input is a sinusoidal signal with a DC offset, which is converted to a square

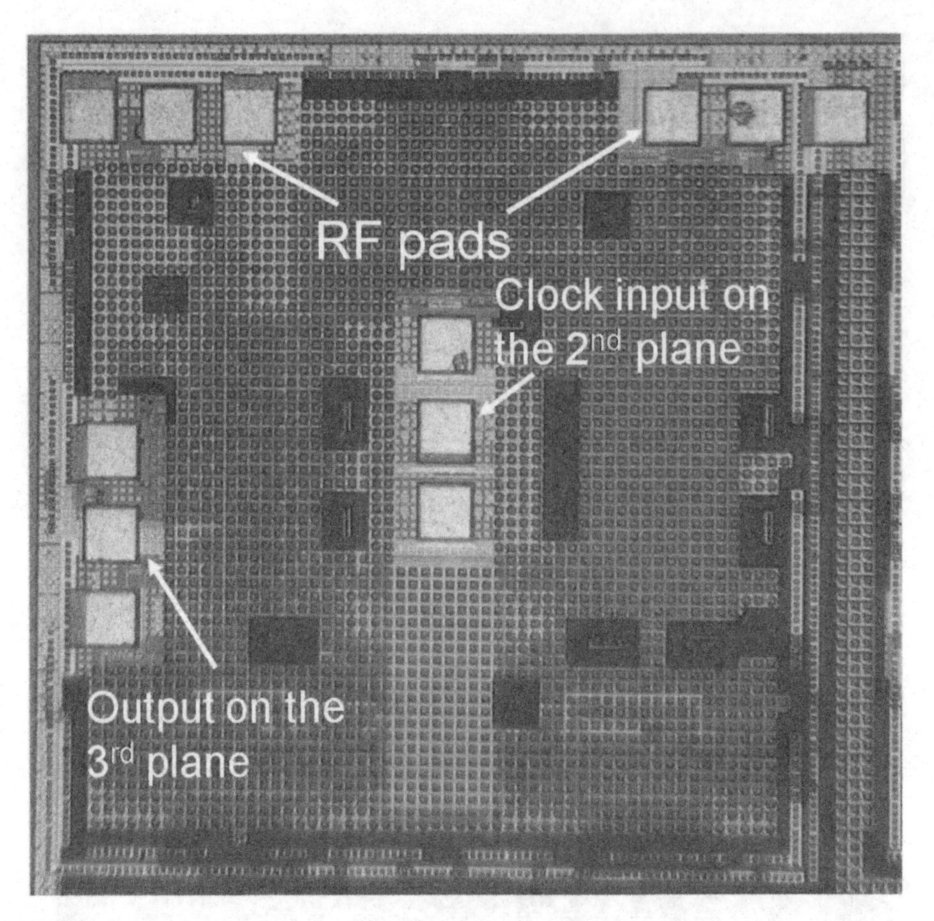

FIGURE 16.23

Magnified view of one block of the fabricated 3-D test circuit.

waveform at the output of the clock driver. The remaining three RF pads are used to measure the delay of the clock signal at specific points on the clock distribution network within each tier. A buffer is connected to each of these measurement points. The output of this buffer drives the gate of an open drain transistor connected to the RF pad. The RF probes landing on these pads are depicted in Fig. 16.24, where the die assembly on the probe station is illustrated.

A clock waveform acquired from the topology combining an H-tree and global rings, shown in Fig. 16.16C, is illustrated in Fig. 16.25, demonstrating operation of the circuit at 1.4 GHz. The clock skew between the tiers of each block is listed in Table 16.9. The topologies are ordered in Fig. 16.26 in terms of the maximum measured clock skew between two tiers. The delay of the clock signal from the RF input pad at the center of each block to the measurement point on tier i is denoted as T_{C_i} in Table 16.9. For example, T_{C_A} denotes the delay of the clock signal to the measurement point on tier A. Additionally, the skew, the difference in the delay of the clock signal between two measurement points on tiers i and j, is notated as $T_{C_{i-j}}$.

FIGURE 16.24

Die assembly of the 3-D test circuit with RF probes.

For the H-tree topology, the clock signal delay is measured from the root to a leaf of the tree on each tier, with no other load connected to these leaves. The skew between the leaves of the H-tree on tiers A and C (i.e., $T_{C_{A-C}}$) is effectively the delay of a stacked 3-D via traversing the three tiers to transfer the clock signal from the target leaf to the RF pad on the third tier. The delay of the clock signal to the sink of the H-tree on the second tier T_{C_B} is larger due to the additional capacitance within that quadrant of the H-tree. This capacitance is intentional on-chip decoupling capacitance placed under the quadrant, increasing the measured skew of $T_{C_{B-C}}$ and $T_{C_{B-A}}$. This topology produces, on average, the lowest skew as compared to the two other topologies.

In the H-tree topology, each leaf of a tree is connected to registers located only within the same tier. Allowing one sink of an H-tree to drive a register on another tier adds the delay of another 3-D via to the clock signal path, further increasing the delay. Consequently, the registers within each tier are connected to the H-tree on the same tier. Note that this approach does not imply that these registers only belong to data paths contained within the same tier.

The clock skew among the tiers is greater for the local mesh topology, as compared to the H-tree topology, primarily due to the imbalance in the clock load for certain local meshes. Indeed, this topology has only 16 tap points within the global clock distribution network; three times fewer than the H-tree topology illustrated in Fig. 16.16A. This difference can produce a considerable load imbalance, greatly increasing the local clock skew as compared to the local clock skew within the

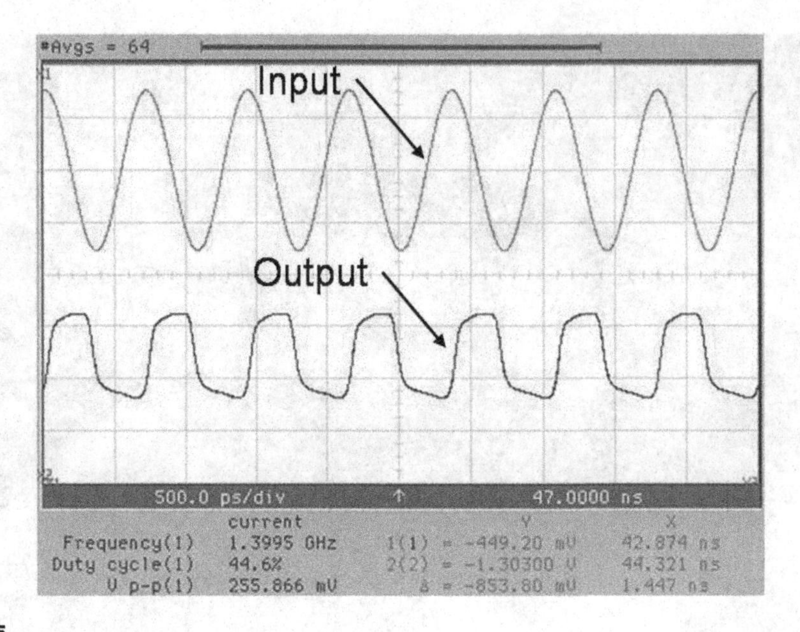

FIGURE 16.25

Clock signal input and output waveform from the topology with global rings, as illustrated in Fig. 16.16C.

Table 16.9 Measured Clock Skew Among the Tiers of Each Block

	Clock Skew (ps)		
Clock Distribution Network	$T_{C_{B-A}} = T_{C_B} - T_{C_A}$	$T_{C_{B-C}} = T_{C_B} - T_{C_C}$	$T_{C_{A-C}} = T_{C_A} - T_{C_C}$
H-trees (Fig. 16.16A)	32.5	28.3	−4.2
Local meshes (Fig. 16.16B)	−68.4	−18.5	49.8
Global rings (Fig. 16.16C)	−112.0	−130.6	−18.6

H-tree topology. By inserting the local meshes on tiers A and C, which are connected to the 16 sinks of the H-tree on the second tier, the local clock skew is smaller. The greatest difference in the load is between the measurement points on tiers A and B, which also produces the largest skew for this topology. The increase in skew, however, as compared to the H-tree topology, is moderate.

Consequently, a limitation of the local meshes topology is that greater effort is required to control the local skew. The fewer number of sinks driven by the global clock distribution network increases the number of registers clocked by each sink. To better explain this situation, consider a segment of each topology shown in, respectively, Figs. 16.27A and B. For the H-tree topology, the clock signal is distributed from three sinks, one on each tier, to the registers within the circular area depicted in Fig. 16.27A. Note that the radius of the circle on tiers A and C is slightly smaller to compensate for the additional delay of the clock signal caused by the impedance characteristics of the 3-D vias. The registers located within these regions satisfy specific local skew constraints.

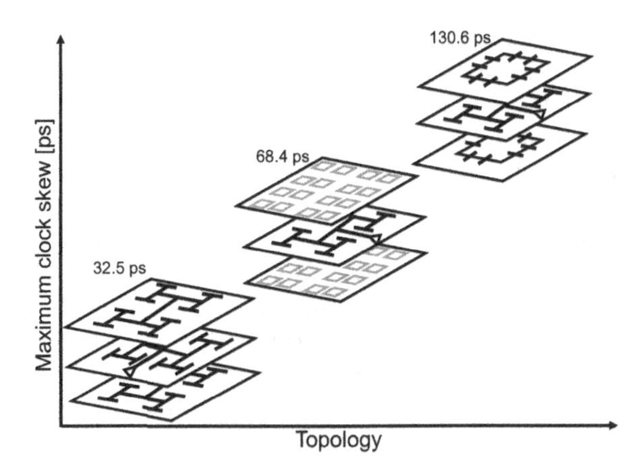

FIGURE 16.26

Maximum measured clock skew between two tiers within the different clock distribution networks.

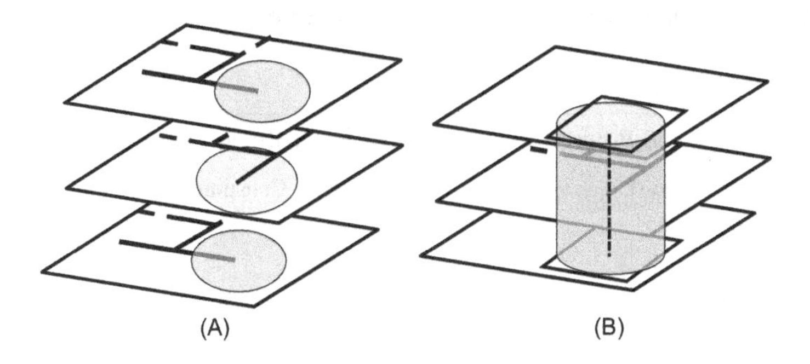

FIGURE 16.27

Part of the clock distribution networks illustrated in Figs. 16.16A and B. (A) The local clock skew is individually adjusted within each tier for the H-tree topology, and (B) the local skew is simultaneously adjusted for all of the tiers for the local mesh topology.

Alternatively, in the case of the local mesh topology, the clock signal at the sinks of the H-tree on the second tier feeds registers in each of the three tiers. Consequently, each sink of the tree connects to a larger number of registers as compared to the H-tree topology, as depicted by the shaded region in Fig. 16.27B. Despite the beneficial effect of the local meshes, load imbalances are more pronounced for this topology. Alternatively, the H-tree topology (see Fig. 16.16A) utilizes a significant amount of interconnect resources, dissipating greater power.

The clock distribution network with the global rings exhibits low skew for tiers A and C, those tiers that include the global rings. The objective of this topology is to evaluate the effectiveness of a less symmetric architecture in distributing the clock signal within a 3-D circuit. Although the clock load on each ring is nonuniformly distributed, the load balancing characteristic of the rings yields a relatively low skew between the tiers. Since the clock distribution network on the second tier is an H-tree, the skew between adjacent tiers is significantly larger than the skew between the top and bottom tiers. Note that the sinks of the H-tree are located at a great distance from the rings on tiers A and C (see Fig. 16.16C). A combination of H-tree and global rings, consequently, is not a suitable approach for 3-D circuits due to the difficulty in matching the distance that the clock signal traverses on each tier from the root of the tree or the ring to the many registers distributed across a tier.

The measured power consumption of the blocks operating at 1 GHz is reported in Table 16.10. An ordering of the blocks in terms of the measured dissipated power is illustrated in Fig. 16.28. The local mesh topology dissipates the lowest power. This topology requires the least interconnect resources for the global clock network, since the local meshes are connected at the output of the buffers located on the last level of the H-tree on the second tier. In addition, this topology requires a small amount of local interconnect resources as compared to the H-tree and global rings topologies. Most of the registers are connected directly to the local rings. Alternatively, the power consumed by the H-tree topology is the highest, as this topology requires three H-trees and additional wiring for the local connections to the leaves of each tree. In addition, the greatest number of buffers is included in this topology. This number is threefold as compared to the number of buffers used for the local mesh topology. Finally, the block with global rings consumes slightly less power than the H-tree topology due to the reduced wiring resources of the global clock network.

Table 16.10 Measured Power Consumption of Each Block Operating at 1 GHz

Clock Distribution Network	Power Consumption (mW)
H-trees (Fig. 16.16A)	260.3
Local meshes (Fig. 16.16B)	168.3
Global rings (Fig. 16.16C)	228.5

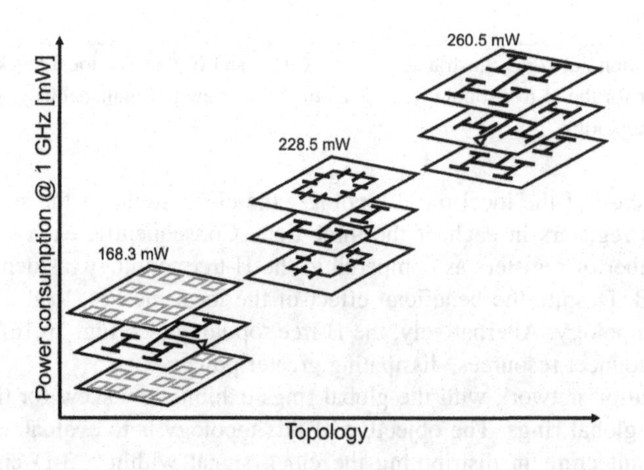

FIGURE 16.28

Measured power consumption at 1 GHz of the different circuit blocks.

Although the local mesh topology requires the least interconnect resources, a large number of 3-D vias is required for the intertier connections. Since the 3-D vias block all of the metal layers and occupy silicon area, the routing blockage increases considerably as compared to the H-tree topology. The global rings topology requires a moderate number of 3-D vias as only four connections between the vertices of the rings and the branch points of the H-tree are necessary.

Since 3-D integration greatly increases the complexity of designing a synchronization system, a topology that offers low overhead in the design process of a 3-D clock distribution network is preferable. From this perspective, a potential advantage of the H-tree topology is that each tier can be individually analyzed, since the clock distribution network in each tier is exclusively connected to registers within the same tier. Alternatively, in the local ring topology, all of the registers from all of the tiers, which are connected to each sink of the tree on the second tier, need to be simultaneously considered.

16.6 SUMMARY

A case study for investigating several clock distribution networks for 3-D ICs is described in this chapter, and measurements from a 3-D test circuit are presented. The characteristics of the circuit and related topologies are:

- A 3-D clock distribution network cannot be directly extended from a 2-D circuit due to the lack of symmetry within a 3-D structure due to the effects of the intertier vias.
- The 3-D FDSOI fabrication technology from MITLL has been used to manufacture the test circuit.
- The 3-D test circuit is composed of four independent blocks, where each block is a three tier 3-D circuit. For each block, a different clock distribution network is utilized.
- All of the blocks in each tier share the same logic circuitry to emulate a variety of switching patterns in a synchronous digital circuit.
- The maximum clock frequency of the fabricated 3-D test circuit is 1.4 GHz.
- A comparison of the clock skew and power consumption of each block is provided. A topology combining the symmetry of an H-tree on the second tier and local meshes on the other two tiers results in moderate clock skew for 3-D circuits while consuming the lowest power as compared to the other investigated topologies.

CHAPTER

VARIABILITY ISSUES IN THREE-DIMENSIONAL ICs*

17

CHAPTER OUTLINE

For several decades, the electrical behavior of both active devices and passive components in integrated circuits was primarily characterized with deterministic models. With feature size scaling well below micrometer dimensions, the increasing variability of the physical properties of these elements has rendered these models increasingly less accurate [619]. Consequently, since approximately the mid-1990s, a considerable body of research on statistical models has been developed that focus on the variability of on-chip transistors and interconnect [620–623].

Variability originates from both the manufacturing process and the environment of an integrated circuit. Environmental variability is typically attributed to fluctuations in the power supply of the circuits and temperature gradients within the ambient environment. Collectively, these variations are termed as process, voltage, and temperature or, succinctly, PVT variations.

Process variations are the result of a large and diverse number of imperfections in the manufacturing process. For example, aberrations in the stepper lens [624] or other imprecisions introduced from lithography [625] and/or during illumination [626] lead to slightly different physical properties of the transistors, thereby affecting the electrical behavior of the manufactured circuits. In addition to variations due to the optical process steps, fluctuations are also caused during fabrication of the interconnection layers. For example, the chemical mechanical polishing process utilized to smooth the interface of the deposited interconnect layers produces variations in the thickness of the metal layers [627].

*Dr. Hu Xu contributed to this chapter.

Three-Dimensional Integrated Circuit Design. DOI: http://dx.doi.org/10.1016/B978-0-12-410501-0.00017-4

Depending upon the underlying phenomena producing the manufacturing variations, these variations can be characterized as either systematic or random. For example, variations in the transistor channel length depend upon the orientation of the layout of the transistors and can be characterized in a systematic manner [628]. Alternatively, the number and distribution of the dopant atoms within the transistor channel, which determine the transistor voltage threshold, is a random phenomenon [622]. Whether a source of variation is treated as systematic or random also depends on the models that capture these variations. Therefore, if the model is of significant complexity, the assumption of a random process to model a source of physical variation may be a plausible approach from a design perspective [629].

The different scales at which process variations are manifested are illustrated in Fig. 17.1. Depending upon the stage of the fabrication process, these variations affect all of the transistors across a die in the same way but differently between dies (interdie variations), or cause the properties of each transistor within a die to differ from each other (intradie variations). Another nomenclature typically used for these types of variations that is also followed in this chapter is die-to-die (D2D) for interdie and within-die (WID) for intradie variations. The issue of variability in the manufacturing and design of integrated systems is a broad topic that by no means can be covered within this chapter (nor is this the intent). The interested reader is referred to many other excellent sources that consider this topic in much greater depth [629].

Historically, random or systematic variability was addressed by worst case design methodologies [630]; however, considering the increase in the magnitude of the variations, designing for the

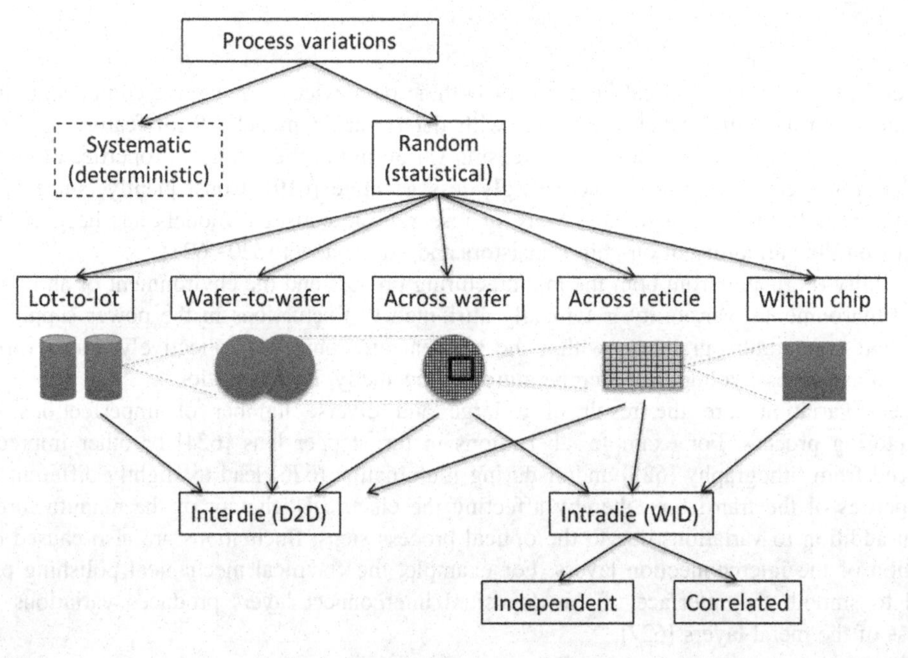

FIGURE 17.1

Classification of process variations and an illustration of the physical scale of the disparate sources of variations.

worst case can lead to a significant loss in performance at high overhead. Consequently, statistical models that reduce the pessimism of the design margins can recover some of this performance loss, yielding more competitive products. These models employ random variables to characterize WID or D2D variations, where these variables are usually assumed to be normally distributed. In this chapter, these process variations are assumed to follow a Gaussian distribution. The effects of variability on timing and, consequently, the parametric yield of 3-D circuits are discussed in the following section. An efficient model that characterizes the distribution of clock skew for 3-D clock distribution networks as compared to Monte Carlo simulations [631] is presented in Section 17.2. The combined effects of process and power noise fluctuations on the salient features of 3-D clock distribution networks are discussed in Section 17.3. Alternatively, temperature variations are not considered since thermal issues are extensively covered in previous chapters. The major concepts of this chapter are summarized in Section 17.4.

17.1 PROCESS VARIATIONS IN DATA PATHS WITHIN THREE-DIMENSIONAL ICs

Interestingly, variability in 3-D circuits has to date not been adequately investigated. The multi-tier nature of these circuits requires different statistical models to account for process variations. To exemplify this requirement, consider the datapaths illustrated in Fig. 17.2. For the datapath shown in Fig. 17.2A, one random variable can be used for each gate to describe the effects of WID variations on the delay of this gate. Furthermore, one random variable, common for all of the gates along the path, is used to capture the effect of D2D variations. Thus, a statistical model that describes the delay distribution of this path requires seven random variables.

Alternatively, for the path shown in Fig. 17.2B, which spans two physical tiers, the situation is somewhat different. The same number of random variables is required to model WID variations. Two variables are, however, needed to model D2D variations as some gates along the path are placed in another tier. Assuming that these tiers originate from different wafers (a reasonable assumption for the vast majority of systems), these variables should be modeled as independent. Furthermore, if all of the tiers are fabricated with the same process (e.g., a memory stack), these random variables are also considered to be identically distributed. Alternatively, this situation is not

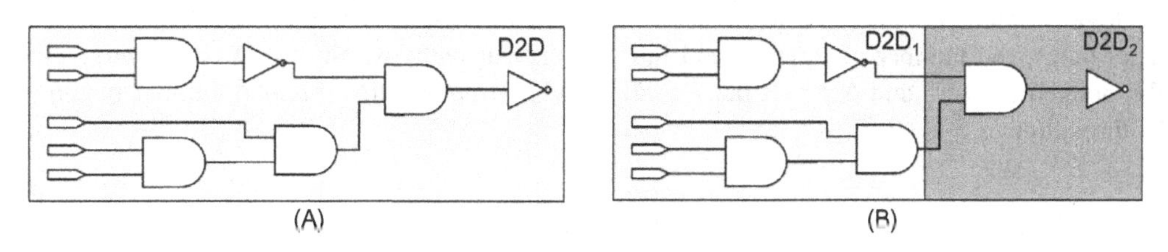

FIGURE 17.2

Example of intratier and intertier paths. (A) One random variable is required to model D2D variations, and (B) two random variables (one for each tier) are used to model D2D variations for the entire path.

the same if a different process is used for all or some of the tiers. The former case is assumed in this chapter to model the multiple sources of D2D variations that can exist within a 3-D stack. The effects of multiple D2D variations on the speed of datapaths in a 3-D circuit are described in [632]. This work constitutes the basis of this section.

A statistical delay model for 3-D circuits is based on the critical path model described in [633]. This path model describes the performance of a datapath while hiding information from the lower abstraction levels. Although this approach adds some inaccuracy, a general path model is useful in adding insight into the effects of variability on the performance of 3-D circuits. Note that the general critical path model in [633] is for planar circuits. The model utilizes two primary parameters, the number of critical paths within a circuit N_{cp} and the number of stages within each critical path n_{cp}. Another assumption of the model is that all of the n_{cp} stages within a critical path are two input NAND gates.

To model D2D variations, a single random variable G is employed for all of the gates, while the effects of WID variations are modeled by a set of independent and identically distributed random variables notated as L_{ij}, where $1 \leq i \leq N_{cp}$ and $1 \leq j \leq n_{cp}$. Consequently, L_{ij} describes the effects of WID variations on gate j of path i. Based on this notation, the maximum delay variation due to the combined effect of D2D and WID variations is [633]

$$\Delta T_{\max}^{2-D} = \max_{1 \leq i \leq N_{cp}} \left(\sum_{j=1}^{n_{cp}} a(G + L_{ij}) \right) = \max_{1 \leq i \leq N_{cp}} \left(n_{cp} aG + \sum_{j=1}^{n_{cp}} aL_{ij} \right), \tag{17.1}$$

where a is the sensitivity of the delay of the gates to process variations. Using the common assumption that both D2D and WID process variations follow a normal distribution, that is $G \sim N(0, \sigma_G)$ and $L \sim N(0, \sigma_L)$, the probability that the maximum delay variation of the critical paths is less than a specific delay τ is

$$Pr\{\Delta T_{\max}^{2-D} \leq \tau\} = F_{\Delta T_{\max}^{2-D}}(\tau) = f_G\left(\frac{\tau}{\alpha n_{cp}}\right) * \left(F_L\left(\frac{\tau}{\alpha \sqrt{n_{cp}}}\right)^{N_{cp}} \right). \tag{17.2}$$

The functions $f_K()$ and $F_K()$ are, respectively, the probability density function and the cumulative density function (cdf) of the random variable K and the sign * corresponds to the convolution operation.

The primary difference in the modeling procedure for delay variations in 3-D circuits is that a single variable G to describe D2D variations can no longer be used. This situation is due to the existence of intertier paths, as depicted in Fig. 17.2. Consequently, if a 3-D circuit includes m tiers, m random variables are required to model D2D variations. Based on the discussion of intra and intertier paths, the number of intertier and intratier critical paths within a 3-D circuit are, respectively, notated as N_{cp}^{inter} and N_{cp}^{intra}. If tier i contains N_{cp}^{i} intratier paths, the total number of intratier paths throughout a 3-D stack is

$$N_{cp}^{\text{intra}} = \sum_{i=1}^{m} N_{cp}^{i}. \tag{17.3}$$

Accordingly, the within each intratier path within any tier of a 3-D circuit is assumed to comprise n_{cp}^{intra} number of stages. Similarly, the number of stages of an intertier path is notated as

n_{cp}^{inter}, where the number of gates within this path placed in each tier i is denoted by n_{cp}^{i}. Based on this notation,

$$n_{cp}^{\text{inter}} = \sum_{i=1}^{m} n_{cp}^{i}. \tag{17.4}$$

To better understand this notation, an example of a two-dimensional (2-D) and 3-D circuit is shown in Fig. 17.3. For the 2-D circuit depicted in Fig. 17.3A, $N_{cp} = 2$ and $n_{cp} = 3$ for both paths. Alternatively, in Fig. 17.3B, there is one intertier and two intratier paths, which, respectively, means that $N_{cp}^{\text{intra}} = 2$ and $N_{cp}^{\text{inter}} = 1$. For the intratier paths, $n_{cp}^{\text{intra}} = 3$, while for the intertier path, $n_{cp}^{1} = 1$ and $n_{cp}^{2} = 2$. Another distinction in the notation of WID random variables is shown in Fig. 17.3, where the variables for the WID variations of the intratier paths are notated as L_{ij} and the variables for the WID variations of the intertier paths are notated as L'_{ij}, capturing the delay variation of gate j in the (intra or intertier) critical path j. In a manner similar to (17.1), the delay variation of a critical path i within a multi-tier circuit for the intratier paths [632] is

$$\Delta T_i^{\text{intra}} = n_{cp}^{\text{intra}} a G_{g(i)} + \left(\sum_{j=1}^{n_{cp}^{\text{intra}}} a L_{ij} \right), 1 \leq i \leq N_{cp}^{\text{intra}}, \tag{17.5}$$

and for the intertier paths [632] is

$$\Delta T_i^{\text{inter}} = \left(\sum_{j=1}^{m} \beta n_{cp}^{j} G_j \right) + \left(\sum_{j=1}^{n_{cp}^{\text{inter}}} \beta L'_{ij} \right), 1 \leq i \leq N_{cp}^{\text{inter}}, \tag{17.6}$$

where $g()$ is a mapping function that projects each intratier critical path onto one of the m tiers. The sensitivity of the gate delay to process variations for intra and intertier paths is, respectively,

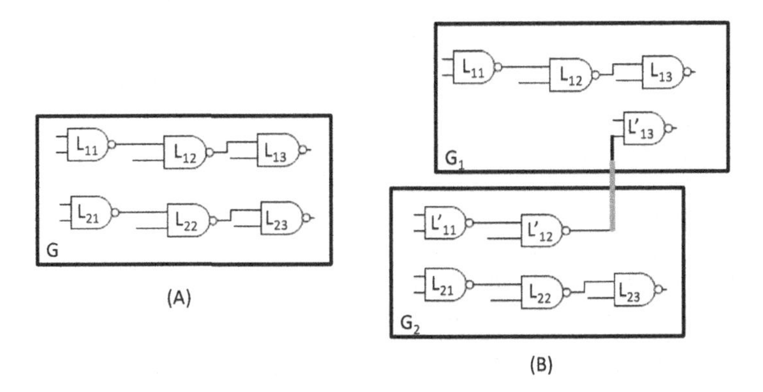

FIGURE 17.3

Notation used in the delay variability model for 2-D and 3-D circuits. (A) 2-D circuit comprising two critical paths each with three logic gates, and (B) two-tier 3-D circuit contains three critical paths each with three stages, where two paths are intratier paths and one path is an intertier path. Two random variables are required in (B) to model the D2D variations of each tier [632].

denoted by α and β. The maximum delay variation for the entire 3-D circuit is the maximum delay variation provided by (17.5) and (17.6),

$$\Delta T_{\max}^{3-D} = \max_{1 \le i \le N_{cp}^{\text{intra}}, 1 \le j \le N_{cp}^{\text{inter}}} \left(\Delta T_i^{\text{intra}}, \Delta T_j^{\text{inter}} \right). \tag{17.7}$$

Determining a closed-form expression for (17.7) is a complicated task. This intractability is due to the intertier paths [632]. Consequently, upper and lower bounds (LBs) are determined for (17.7), while for the delay variation of intratier paths, closed-form expressions are similar to (17.1). Note that in (17.5) a mapping function exists for each intratier path to another tier within the stack. Assuming that the intratier paths are divided equally among the m tiers of the stack, the cumulative distribution function (cdf) for the intratier paths in a 3-D circuit is

$$F_{\Delta T_{\max}^{3-D}}(\tau) = \left[f_G\left(\frac{\tau}{\alpha n_{cp}^{\text{intra}}}\right) * F_L\left(\left(\frac{\tau}{\alpha\sqrt{n_{cp}^{\text{intra}}}}\right)\right)^{\frac{N_{cp}^{\text{intra}}}{m}} \right]^m, \tag{17.8}$$

where * is the convolution operation. Different from a 2-D circuit the number of tiers affects the cdf, as described by (17.8). Furthermore, the critical paths are assumed to be equally split among the tiers of the stack. This mapping yields the worst case delay variability for the intratier paths [632]. These results suggest that to better manage variability in 3-D circuits, process aware physical design should carefully consider the distribution of the paths among the tiers within a stack.

Based on these analytic formulations, another conclusion is that a planar circuit has a higher likelihood to satisfy a timing constraint as compared to a 3-D circuit. The related assumptions are that both circuits contain the same number of critical paths, and that the 3-D circuit does not contain any intertier paths. Although this last assumption is rather restrictive, assuming that registers are present at the boundary of each tier for intertier nets, this argument demonstrates that process variability is an important issue for 3-D circuits, a situation that has to date been neglected. To exemplify these results, consider Fig. 17.4 where the *cdf* of different circuits with the same critical paths are considered. These curves demonstrate that a 2-D circuit exhibits a higher probability to satisfy a timing constraint (τ) as compared to a 3-D version of the same circuit. Moreover, a 3-D circuit with an even distribution of paths between the two tiers is the least likely circuit to satisfy a design constraint. Note that these results do not consider any changes that the third dimension brings into the physical design of the circuit.

To provide a complete picture of the importance of process variations on datapaths within 3-D circuits an intertier path is considered. In this case, however, the delay variation cannot be described analytically and only bounds are provided [632]. The LB is set to the greatest delay variation of the intratier paths,

$$\Delta T_{\max}^{\text{LB}} = \max_{1 \le i \le N_{cp}^{\text{intra}}} \left(\Delta T_i^{\text{intra}} \right). \tag{17.9}$$

Although this approach appears as a crude simplification, the bound in (17.9) is not loose since the intertier path delay is determined by adding the random variables for the D2D variations. This summation decreases the standard deviation of an intertier path as compared to an intratier path. Assuming therefore that all of the traits of intra and intertier paths are the same (e.g., the number

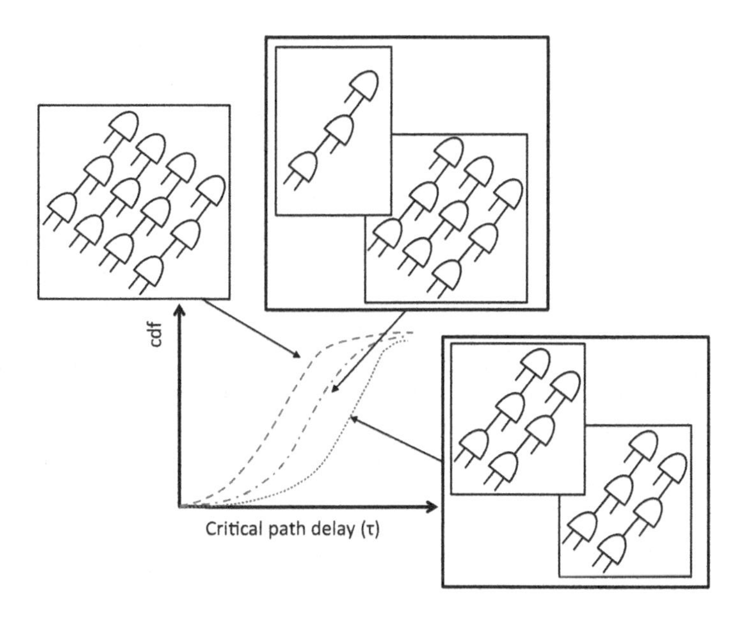

FIGURE 17.4

Cdf of a 2-D circuit (dashed line), a 3-D circuit with uneven critical path distribution between the two tiers (dashed dotted line), and a 3-D circuit with the same number of critical paths in each tier (dotted lined) [632].

of gate stages and sensitivity of each gate delay to process variations), an intertier path always exhibits a lower likelihood to limit the performance of a 3-D circuit.

Alternatively, the upper bound (UB) can be described by [632]

$$\Delta T_{\max}^{\text{UB}} = \max\left(\max_i(\Delta T_i^{\text{intra}}), \max_i \left(\Delta T_i^{\text{inter,UB}}\right)\right), \tag{17.10}$$

where

$$\Delta T_i^{\text{inter,UB}} = \sum_{j=1}^{m} \beta n_{cp}^{j} G_j^{'} + \sum_{j=1}^{n_{cp}^{\text{D2D}}} \beta L_{ij}^{'}. \tag{17.11}$$

In this expression, a set of new random variables $G_j^{'}$ is introduced which are assumed to be independent and identically distributed as G_j. Based on (17.10) and (17.11), the cdf of $\Delta T_{\max}^{\text{UB}}$ becomes [632]

$$F_{\Delta T_{\max}^{\text{UB}}}(\tau) = \left[\prod_{i=1}^{m} f_G\left(\frac{\tau}{k_1}\right) * \left(F_L\left(\frac{\tau}{k_2}\right)^{N_{cp}^{i}}\right)\right]\left[f_G\left(\frac{\tau}{k_3}\right) * \left(F_L\left(\frac{\tau}{k_4}\right)^{N_{cp}^{\text{inter}}}\right)\right], \tag{17.12}$$

where $k_1 = \alpha n_{cp}^{\text{intra}}$, $k_2 = \alpha\sqrt{n_{cp}^{\text{intra}}}$, $k_3 = \beta\sum_{i=1}^{m}(n_{cp}^{i})^2$, $k_4 = \beta\sqrt{n_{cp}^{\text{inter}}}$.

The application of this model on synthetic critical paths for a 90 nm process node demonstrates interesting tradeoffs [632]. Assuming 1,000 critical paths, each consisting of six NAND gates, and introducing the parameter $\gamma = \sigma_G^2/\sigma_{\text{tot}}^2$ to describe the importance of D2D variations (modeled as a random variable *G*) over the total variations, the mean of the maximum critical path delay is determined for different placements of the critical paths among the tiers within a 3-D system. The parameter γ has the values, 0.25, 0.5, and 0.75, where a higher γ corresponds to a greater contribution of D2D variations. Furthermore, the number of tiers changes from one (i.e., a 2-D circuit) to six tiers.

These results indicate that if only intratier paths are assumed, the mean delay increases by 9.5% for $\gamma = 0.5$, while for lower γ, the increase in the mean delay is lower. If only intratier paths are assumed, the D2D variations affect all of the gates along the critical paths within a tier in the same way, shifting the nominal mean delay. If the contribution of D2D variations is greater than the total variations, this shift is more pronounced. Alternatively, WID variations affecting each gate (or stage) of the critical paths have an averaging effect for longer (multistage) paths [634], leading to a smaller shift in the mean delay. Consequently, if the D2D variations are low, the increase in the mean delay is also low.

The placement of the critical paths among the tiers of the 3-D stack also affects the mean delay, where an uneven distribution of the critical paths among the tiers produces higher yield [632]. Although this situation suggests the placement of the critical paths in those tiers where D2D variations are lower, other design constraints may not permit such a placement. However, including variability in the physical design process for 3-D circuits can reduce the performance margins of the circuits; particularly since overall variability in 3-D systems is expected to worsen as compared to 2-D circuits [632].

Alternatively, if the critical paths span more than one tier, meaning that both intra and intertier critical paths exist in a 3-D circuit, the delay distribution of these paths decreases. This behavior is similar to the averaging effect of WID variations in intratier paths. As the gates within the intertier paths are spread over several tiers, the contribution of D2D variations from different tiers can cancel each other, yielding a narrower delay distribution of the critical paths as compared to those 3-D circuits that are comprised of only intratier critical paths. The interplay between the contribution of WID and D2D variations is applied in the following section to the design of more robust global clock distribution networks for 3-D circuits, where a highly accurate variation aware clock skew model is also described.

17.2 EFFECTS OF PROCESS VARIATIONS ON CLOCK PATHS

Several global 3-D clock distribution networks are presented in Chapter 15, Synchronization in Three-Dimensional ICs. In this section, the effects of process variations on the principal traits of 3-D clock networks, such as clock skew and power, are discussed. To consider these effects, effective statistical delay models are required to describe the distribution of the clock skew for disparate 3-D clock network topologies. A model of the delay distribution of a clock buffer including both WID and D2D variations is described in Section 17.2.1. This model produces a delay distribution of clock paths, as described in Section 17.2.2. This distribution is, in turn, utilized to describe the distribution of clock skew in 3-D clock trees, as discussed in Section 17.2.3. Based on these models, the effects of process variations on clock skew distribution for different clock networks are reviewed, and a robust 3-D clock topology is presented in Section 17.2.4.

17.2.1 STATISTICAL DELAY MODEL OF CLOCK BUFFERS

A typical buffered 3-D H-tree is illustrated in Fig. 17.5. The pairwise clock skew is defined as the skew between every pair of sinks in a 3-D clock distribution network, $S_{\text{skew}} = \{s_{i,j} | s_{i,j} = D_i - D_j, 1 \le i, j \le n_{\text{sink}}\}$. Sinks i and j can be located in any tier of the 3-D circuit. $s_{i,j}$ denotes the skew between sinks i and j. The clock signal delay to sinks i and j is denoted, respectively, by D_i and D_j. The number of clock sinks is n_{sink}. To determine the distribution of skew S_{skew} for the clock sinks, a statistical delay model for a clock buffer is required.

The overall variation in the delay of a clock buffer is dependent on the variations of the input capacitance and output resistance [635,636]. This approach considers the input slew rate to more accurately model the distribution of the buffer delay. The interconnects constituting a clock stage are modeled as distributed *RC* wires. The circuit illustrated in Fig. 17.6 is utilized to obtain the variation of the buffer delay for different input signal slew rates.

Let R_{in} denote the output resistance of a buffer driving a second buffer. The load capacitance of the buffer is denoted by C_l. Interconnects with diverse impedance characteristics are modeled by employing different R_{int} and C_{int}, where R_{int} and C_{int} denote, respectively, the resistance and

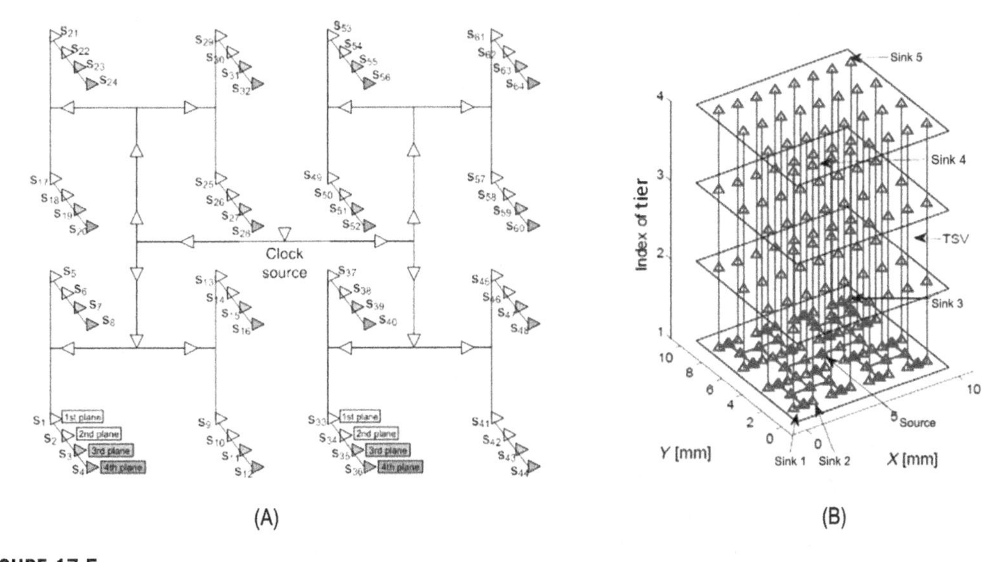

FIGURE 17.5

3-D H-tree spanning four tiers. (A) Notation for all of the 64 sinks, and (B) certain sinks used to evaluate clock skew.

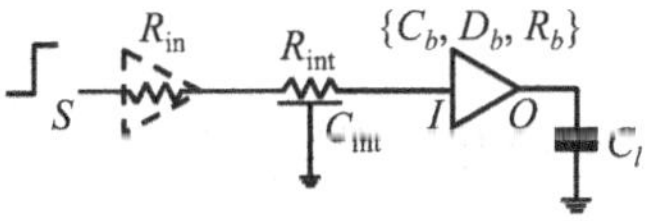

FIGURE 17.6

Elemental circuit to measure the distribution of delay due to variations in the buffer characteristics.

capacitance of the interconnects. The interconnect R_{int} and C_{int} can also be adjusted to produce different slew rates for the input signal of the buffer shown in Fig. 17.6.

For a step input signal, the Elmore delay [414] from source S to nodes I and O (in Fig. 17.6) is, respectively,

$$D_{SI} = 0.69R_{in}C_{int} + 0.38R_{int}C_{int} + 0.69(R_{in} + R_{int})C_b, \tag{17.13}$$

$$\Delta D_{SI} = 0.69(R_{in} + R_{int})\Delta C_b, \tag{17.14}$$

$$D_{SO} = D_{SI} + D_b + 0.69R_bC_l, \tag{17.15}$$

$$\Delta D_{SO} = \Delta D_{SI} + \Delta D_b + 0.69\Delta R_bC_l, \tag{17.16}$$

where C_b, R_b, and D_b are, respectively, the input capacitance, output resistance, and intrinsic delay of the buffer. Variations of C_b, R_b, and D_b are, respectively, denoted by ΔC_b, ΔR_b, and ΔD_b. For the buffer shown in Fig. 17.6, R_{in} is considered constant (for the moment).

Through several Monte Carlo simulations, the delay variation at nodes I and O is measured by setting C_l to zero (corresponding to ΔD_{SO_0}) and a different value (e.g., 200 fF, corresponding to ΔD_{SO_1}). The mean value and standard deviation of ΔC_b, ΔR_b, and ΔD_b are obtained from (17.14) and (17.16) [636]. Assuming that process variations can be described by a Gaussian distribution, the electrical characteristics of a buffer can also be approximated by a Gaussian distribution [637],

$$\Delta C_b \sim \mathcal{N}\left(0, \sigma_{C_b}^2\right), \Delta R_b \sim \mathcal{N}\left(0, \sigma_{R_b}^2\right), \Delta D_b \sim \mathcal{N}\left(0, \sigma_{D_b}^2\right). \tag{17.17}$$

σ_{C_b} is characterized from (17.14). According to (17.14) and (17.16), $\sigma_{D_{SO}}$ is dependent on σ_{C_b}, σ_{D_b}, and σ_{R_b}, and the covariance $\sigma_{D_{SO}}^2$ among these variables is

$$\sigma_{D_{SO}}^2 = \left(0.69(R_{in} + R_{int})\sigma_{C_b}\right)^2 + \sigma_{D_b}^2 + \left(0.69\sigma_{R_b}C_l\right)^2 + 1.38(R_{in} + R_{int})\mathrm{cov}(D_b, C_b)$$
$$+ 1.38C_l\mathrm{cov}(D_b, R_b) + 0.952C_l(R_{in} + R_{int})\mathrm{cov}(C_b, R_b), \tag{17.18}$$

$$\sigma_{D_b}^2 = \sigma_{D_{SO0}}^2 - \sigma_{D_{SI}}^2 - 1.38(R_{in} + R_{int})\mathrm{cov}(D_b, C_b), \tag{17.19}$$

$$\sigma_{C_b}^2 = \frac{\sigma_{D_{SI}}}{0.69(R_{in} + R_{int})}. \tag{17.20}$$

σ_{R_b} is obtained from (17.18) by substituting σ_{D_b} and σ_{C_b}, respectively, into (17.19) and (17.20).

Consider that ΔC_b, ΔR_b, and ΔD_b are used to obtain the delay variation of each buffer stage Δd_i, which is similar to ΔD_{SO}. When calculating σ_{d_i} (similar to recalculating $\sigma_{D_{SO1}}$ through (17.18)), σ_{C_b}, σ_{R_b}, and σ_{D_b} are again substituted into (17.18). In this procedure, the covariances, cov(D_b, R_b), cov(D_b, C_b) and cov(C_b, R_b), effectively cancel. Consequently, the correlation among ΔC_b, ΔR_b, and ΔD_b does not significantly affect Δd_i as long as Δd_i is based on this same correlation. Since ΔC_b, ΔR_b, and ΔD_b originate from the same source of process variations, these variables are assumed in this model to be fully correlated.

17.2.2 DELAY DISTRIBUTION OF CLOCK PATHS

The buffer delay model described in the previous subsection is used to evaluate the delay distribution of clock paths in 3-D circuits. An example of a 3-D clock path is illustrated in Fig. 17.7. The devices in different physical tiers are connected by TSVs [181], which, in turn, are modeled as *RC*

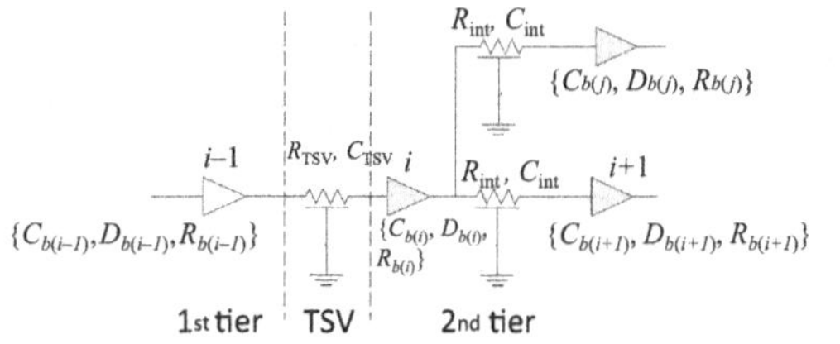

FIGURE 17.7

Electrical model of a segment of an intertier clock path.

wires of different resistance and capacitance as compared to the horizontal wires (e.g., R_{TSV} and C_{TSV} in Fig. 17.7). R_{TSV} and C_{TSV} are considered fixed.

Consider the clock path consisting of buffers $i-1$, i, and $i+1$. From (17.14) and (17.16), the delay variation Δd_i attributed to the variation of buffer i along a target path is

$$\Delta d_i = 0.69\left(R'_{in(i)} + \Delta R_{b(i-1)}\right)\Delta C_{b(i)} + 0.69\Delta R_{b(i)}\left(C'_{l(i)} + \Delta C_{b(i+1)} + \Delta C_{b(j)}\right)$$

$$+ 0.69R'_{b(i)}\Delta C_{b(j)} + \Delta D_{b(i)}, \quad (17.21)$$

$$R_{in(i)} = R_{b(i-1)} + R_{TSV}, \quad (17.22)$$

$$C_{l(i)} = 2C_{int} + C_{b(i+1)} + C_{b(j)}, \quad (17.23)$$

where the prime (′) denotes the nominal value. For buffer i, the $\Delta R_{b(i-1)}$ of the upstream buffer and $\Delta C_{b(i+1)}$ of the downstream buffer are both included in (17.21). To determine the delay of a clock path, Δd_i for all of the buffers along this path is summed. In this case, $\Delta R_{b(i-1)}\Delta C_{b(i)}$ and $\Delta R_{b(i)}\Delta C_{b(i+1)}$ are duplicated. One of these two terms therefore needs to be removed. Consequently, Δd_i is rewritten as

$$\Delta d_i = 0.69\left(R'_{in(i)}\Delta C_{b(i)} + \Delta R_{b(i)}(C'_{l(i)} + \Delta C_{b(i+1)} + \Delta C_{b(j)} + R'_{b(i)}\Delta C_{b(j)}\right) + \Delta D_{b(i)}$$

$$= 0.69\left(R'_{in(i)}\Delta C_{b(i)} + R'_{b(i)}\Delta C_{b(j)}\right) + \Delta D_{b(i)} + \delta_i, \quad (17.24)$$

where

$$\delta_i = 0.69\Delta R_{b(i)}\left(C'_{l(i)} + \Delta C_{b(i+1)} + \Delta C_{b(j)}\right). \quad (17.25)$$

The variation of ΔC_b is relatively low as compared with the nominal C_b ($\sigma/\mu < 3\%$ for both D2D and WID variations, as reported in Table 17.1). The observed delay variation of the buffers is also typically much lower than the nominal value (e.g., $\sigma/\mu \leq 5\%$ for both D2D and WID variations, as reported in [638]). δ_i can be approximated using a first order linear Taylor series expansion around zero [637],

$$\delta_i \approx \left[\frac{\vartheta\delta_i}{\vartheta\Delta R_{b(i)}}\right]_0 \Delta R_{b(i)} + \left[\frac{\vartheta\delta_i}{\vartheta\Delta C_{b(i+1)}}\right]_0 \Delta C_{b(i+1)} + \left[\frac{\vartheta\delta_i}{\vartheta\Delta C_{b(j)}}\right]_0 \Delta C_{b(j)} = 0.69C'_{l(i)}\Delta R_{b(i)}. \quad (17.26)$$

As reported in Table 17.1 and discussed in [632,638] and [639], σ/μ of the transistor characteristics is typically less than 5%. The 3σ variation is smaller than 15% of the nominal R_b and 10%

Table 17.1 Variations in the Electrical Characteristics of the Buffers

	R_b (Ω)			C_b (fF)			D_b (ps)		
Input Slew	μ	σ_{WID}	σ_{D2D}	μ	σ_{WID}	σ_{D2D}	μ	σ_{WID}	σ_{D2D}
47 (mV/ps)	371	18.8	15.3	4.9	0.04	0.03	19.9	1.04	0.85
	σ/μ	5.1%	4.1%	σ/μ	0.8%	0.7%	σ/μ	5.2%	4.3%
16 (mV/ps)	349	17.8	14.7	5.7	0.31	0.16	24.8	1.49	1.21
	σ/μ	5.1%	4.2%	σ/μ	2.3%	2.1%	σ/μ	6.0%	4.9%
6 (mV/ps)	345	16.7	13.7	7.2	0.08	0.06	30.1	2.19	1.79
	σ/μ	4.8%	4.0%	σ/μ	1.1%	0.9%	σ/μ	7.3%	5.9%

for ΔC_b. Since ΔC_b and ΔR_b are modeled as Gaussian distributions, for more than 99.7% of buffers, $\Delta C_b \Delta R_b$ is lower than 1.5% $C_b R_b$. Moreover, from the nominal value and standard deviation of C_b, R_b, and D_b, as reported in Table 17.1, $0.69\Delta C_b \Delta R_b$ and $0.69 C_b R_b$ are, respectively, much lower than ΔD_b and D_b. Consequently, approximating δ_i with (17.26) does not introduce a significant loss of accuracy.

As mentioned previously, $\Delta R_{b(i)}$, $\Delta C_{b(i)}$, and $\Delta D_{b(i)}$ are approximated as Gaussian distributions and can be assumed to be fully correlated. According to (17.24) and (17.26), Δd_i is approximated as a Gaussian distribution.

$$\Delta d_i \sim \mathcal{N}\left(0, \sigma^2_{d_i^{\text{D2D}}} + \sigma^2_{d_i^{\text{WID}}}\right), \tag{17.27}$$

$$\sigma^2_{d_i^{\text{D2D}}} = \left\{ (\sigma_1 + \sigma_2 + \sigma_3)^2 + \sigma_4^2 \textit{ if buffers i and j are in different } \text{tiers}, \right. \tag{17.28a}$$

$$\sigma^2_{d_i^{\text{D2D}}} = \left\{ (\sigma_1 + \sigma_2 + \sigma_3 + \sigma_4)^2 \textit{ if buffers i and j are in the same } \text{tier}, \right. \tag{17.28b}$$

$$\sigma^2_{d_i^{\text{WID}}} = (\sigma_5 + \sigma_6 + \sigma_7)^2 + \sigma_8^2 + 2\text{corr}(i,j)(\sigma_5 + \sigma_6 + \sigma_7)\sigma_8. \tag{17.29}$$

The terms σ_1 to σ_8 are, respectively,

$$\sigma_1 = 0.69 R'_{\text{in}(i)} \sigma_{C^{\text{D2D}}_{b(i)}}, \quad \sigma_2 = 0.69 C'_{l(i)} \sigma_{R^{\text{D2D}}_{b(i)}}, \; \sigma_3 = \sigma_{D^{\text{D2D}}_{b(i)}}, \quad \sigma_4 = 0.69 R'_{b(i)} \sigma_{C^{\text{D2D}}_{b(i)}},$$
$$\sigma_5 = 0.69 R'_{\text{in}(i)} \sigma_{C^{\text{WID}}_{b(i)}}, \quad \sigma_6 = 0.69 C'_{l(i)} \sigma_{R^{\text{WID}}_{b(i)}}, \; \sigma_7 = \sigma_{D^{\text{WID}}_{b(i)}}, \quad \sigma_8 = 0.69 R'_{b(i)} \sigma_{C^{\text{WID}}_{b(i)}}.$$

The correlation between buffers i and j is denoted by corr(i, j). A model describing this spatial correlation is discussed in Appendix E.

Consequently, for a 3-D clock path to a sink u that includes n_u clock buffers, the variation of the delay is expressed as the summation of (17.24) of each buffer along the path. The variance of the distribution of a 3-D clock path is a Gaussian distribution consisting of WID and D2D variations of the buffers,

$$\Delta D_u = \sum_{i=1}^{n_u} \Delta d_i, \tag{17.30}$$

$$\Delta D_u \sim \mathcal{N}\left(0, \sigma^2_{D_u^{\text{D2D}}} + \sigma^2_{D_u^{\text{WID}}}\right). \tag{17.31}$$

The variation of the delay of a 3-D clock path due to D2D process variations is the sum of the D2D variations of the buffer delay in all of the tiers,

$$\Delta D_u^{\mathrm{D2D}} = \sum_{j=1}^{m} \Delta D_{u(j)}^{\mathrm{D2D}}, \tag{17.32}$$

$$\Delta D_{u(j)}^{\mathrm{D2D}} = \sum_{j=1}^{n_{u(j)}} \Delta D_{u(j,i)}^{\mathrm{D2D}}, \tag{17.33}$$

where m is the number of tiers spanned by the clock tree. $\Delta D_{u(j)}^{\mathrm{D2D}}$ is the variation of the delay of the clock path from the clock source to sink u in tier j. The number of buffers located in tier j along this clock path is denoted by $n_{u(j)}$. The variation of the delay related to the ith buffer in tier j is denoted by $\Delta D_{u(j,i)}$. Since the D2D variations equally affect the buffers within the same tier, according to (17.27), (17.28), and (17.32), the distribution of $\Delta D_{u(j)}^{\mathrm{D2D}}$ is a Gaussian distribution. The D2D variations affect the buffers in different tiers independently and, therefore, $\Delta D_{u(j)}^{\mathrm{D2D}}$ is independent of $\Delta D_{u(k)}^{\mathrm{D2D}}$ for any $j \neq k$. Consequently, according to (17.32), the distribution of $\Delta D_u^{\mathrm{D2D}}$ is also a Gaussian distribution,

$$\Delta D_u^{\mathrm{D2D}} \sim \mathcal{N}\left(0, \sigma^2_{\Delta D_u^{\mathrm{D2D}}}\right), \tag{17.34}$$

$$\sigma^2_{D_u^{\mathrm{D2D}}} = \sum_{j=1}^{m} \sigma^2_{D_{u(j)}^{\mathrm{D2D}}} = \sum_{j=1}^{m} \left(\sum_{i=1}^{n_{u(j)}} \sigma_{D_{u(j,i)}^{\mathrm{D2D}}} \right)^2. \tag{17.35}$$

Alternatively, the variation of the delay of a 3-D clock path affected by WID variations is the sum of the WID variations of all of the buffers along this path. Consequently, according to (17.29), the distribution of $\Delta D_u^{\mathrm{WID}}$ is also a Gaussian distribution. The resulting variance of the delay of sink u due to WID variations is

$$\Delta D_u^{\mathrm{WID}} \sim \mathcal{N}\left(0, \sigma^2_{\Delta D_u^{\mathrm{WID}}}\right), \tag{17.36}$$

$$\sigma^2_{D_u^{\mathrm{WID}}} = \sum_{i=1}^{n_u} \sigma^2_{d_i^{\mathrm{WID}}} + 2 \sum_{1 \le i < j \le n_u} \mathrm{corr}(i,j) \sigma_{d_i^{\mathrm{WID}}} \sigma_{d_j^{\mathrm{WID}}}, \tag{17.37}$$

where corr(i, j) is the correlation between the WID variations of buffers i and j. If buffers i and j are located in different tiers, corr(i, j) = 0. The correlation of the impact of WID variations on different buffers within the same tier can be classified as systematic or random. The systematic WID variations typically exhibit a spatial correlation [638,640–642]. For those buffers located within the same tier, two types of correlations for WID variations exist, as described in Appendix E.

17.2.3 CLOCK SKEW DISTRIBUTION IN THREE-DIMENSIONAL CLOCK TREES

The clock skew between any pair of sinks in a 3-D clock tree is the difference in the clock delay between these sinks. For a 3-D clock tree with n_{sink} sinks distributed in m tiers, the nominal and variation of the clock skew $s_{u,v}$ between sinks u and v are, respectively,

$$s'_{u,v} = D'_u - D'_v, \tag{17.38}$$

$$\Delta s_{u,v} = \Delta s_{u,v}^{\mathrm{WID}} + \Delta s_{u,v}^{\mathrm{D2D}} = \Delta D_u^{\mathrm{WID}} - \Delta D_v^{\mathrm{WID}} + \Delta D_u^{\mathrm{D2D}} - \Delta D_v^{\mathrm{D2D}}. \tag{17.39}$$

The mean value of $\Delta s_{u,v}$ is $E(\Delta s_{u,v}) = E(\Delta s_{u,v}^{\text{WID}}) = E(\Delta s_{u,v}^{\text{D2D}}) = 0$. The terms $\Delta D_u^{\text{WID}} - \Delta D_v^{\text{WID}}$ and $\Delta D_u^{\text{D2D}} - \Delta D_v^{\text{D2D}}$ are independent. Consequently, $\Delta s_{u,v}^{\text{D2D}}$ and $\Delta s_{u,v}^{\text{WID}}$ are treated separately. The correlation between every two terms in the expression of $\Delta s_{u,v}^{\text{D2D}}$ is one or zero (i.e., respectively, fully correlated or uncorrelated). According to (17.32), $\Delta s_{u,v}^{\text{D2D}}$ is the sum of the terms in different tiers,

$$\Delta s_{u,v}^{\text{D2D}} = \sum_{j=1}^{m} \Delta s_{(u,v)_j}^{\text{D2D}}, \Delta s_{(u,v)_j}^{\text{D2D}} = \sum_{i=1}^{n_{u(j)}} \Delta D_{u(j,i)}^{\text{D2D}} - \sum_{i=1}^{n_{v(j)}} \Delta D_{v(j,i)}^{\text{D2D}}, \tag{17.40}$$

where $\Delta D_{u(j,i)}^{\text{D2D}}$ is the D2D delay variation related to the ith buffer in the jth tier along the clock path ending at sink u. The number of buffers in the jth tier along this path is denoted as $n_{u(j)}$.

All of the buffers in the same tier are equally affected by D2D variations, meaning that the correlation between each pair of variables in (17.40) is one. Since $\Delta D_{u(j,i)}^{\text{D2D}}$ and $\Delta D_{v(j,i)}^{\text{D2D}}$ are both modeled as a Gaussian distribution, $\Delta D_{s(u,v)_j}^{\text{D2D}}$ is also a Gaussian distribution. In (17.40), $\forall\ j_1 \neq j_2$ $(1 \leq j_1,\ j_2 \leq m)$ $\Delta D_{s(u,v)_{j_1}}^{\text{D2D}}$ is independent of $\Delta D_{s(u,v)_{j_2}}^{\text{D2D}}$. Consequently, $\Delta s_{u,v}^{\text{D2D}}$ is also described by a Gaussian distribution,

$$\Delta s_{u,v}^{\text{D2D}} \sim \mathcal{N}\left(0, \sigma_{s_{u,v}^{\text{D2D}}}^2\right), \tag{17.41}$$

$$\sigma_{D_u^{\text{D2D}}}^2 = \sum_{j=1}^{m} \sigma_{s_{u,v(j)}^{\text{D2D}}}^2 = \sum_{j=1}^{m}\left(\sum_{i=1}^{n_{u(j)}} \sigma_{D_{u(j,i)}^{\text{D2D}}} - \sum_{i=1}^{n_{v(j)}} \sigma_{D_{v(j,i)}^{\text{D2D}}}\right)^2. \tag{17.42}$$

According to (17.37), the distribution of $\Delta s_{u,v}^{\text{WID}}$ is also a Gaussian distribution,

$$\Delta s_{u,v}^{\text{WID}} \sim \mathcal{N}\left(0, \sigma_{s_{u,v}^{\text{WID}}}^2\right), \tag{17.43}$$

$$\begin{aligned}
\sigma_{s_{u,v}^{\text{WID}}}^2 = &\sum_{i=n_{u,v}+1}^{n_u} \sigma_{D_{u(i)}^{\text{WID}}}^2 + \sum_{j=n_{u,v}+1}^{n_v} \sigma_{D_{v(j)}^{\text{WID}}}^2 + 2 \sum_{\substack{i,j=n_{u,v}+1 \\ i<j}}^{n_u} \text{corr}(i,j)\sigma_{D_{u(i)}^{\text{WID}}}\sigma_{D_{u(j)}^{\text{WID}}} \\
&+ 2 \sum_{\substack{i,j=n_{u,v}+1 \\ i<j}}^{n_v} \text{corr}(i,j)\sigma_{D_{v(i)}^{\text{WID}}}\sigma_{D_{v(j)}^{\text{WID}}} - 2 \sum_{\substack{n_{u,v}+1 \leq i \leq n_u \\ n_{u,v}+1 \leq j \leq n_v}} \text{corr}(i,j)\sigma_{D_{u(i)}^{\text{WID}}}\sigma_{D_{v(j)}^{\text{WID}}},
\end{aligned} \tag{17.44}$$

where $n_{u,v}$ is the number of buffers shared by the clock paths ending at sinks u and v, as depicted in Fig. 17.8. Downstream the buffers $n_{u,v}$, the subpaths to u and v do not share a buffer.

According to (17.39) through (17.44), the variation of the clock skew $\Delta s_{u,v}$ between sinks u and v in a 3-D clock tree is modeled as a Gaussian distribution,

$$\Delta s_{u,v} \sim \mathcal{N}\left(0, \sigma_{s_{u,v}^{\text{WID}}}^2 + \sigma_{s_{u,v}^{\text{D2D}}}^2\right). \tag{17.45}$$

If the maximum tolerant skew variation is $\Delta S \geq 0$, the probability that a 3-D clock tree satisfies this constraint is

$$P\left(\left|s_{u,v}\right| \leq \Delta S\right) = \int_{-\Delta S - s'_{u,v}}^{\Delta S - s'_{u,v}} f_{\Delta s_{u,v}}(t)dt, \tag{17.46}$$

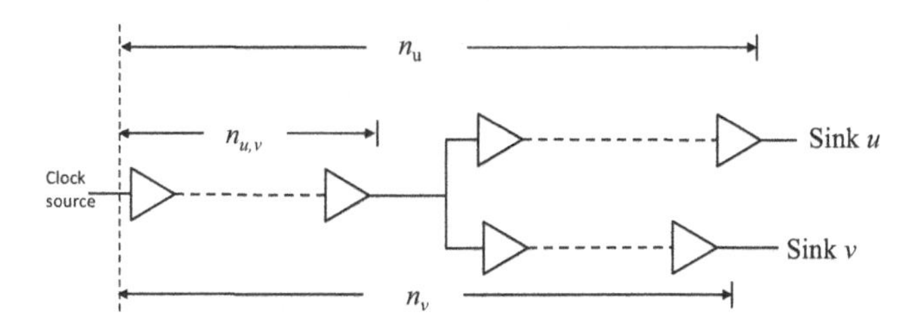

FIGURE 17.8

Clock paths to sinks u and v where the paths share $n_{u,v}$ buffers.

Table 17.2 Device and Interconnect Parameters of the Target Circuit

Parameter	W_n/L_n	W_p/L_p	V_{dd} (V)	R_b (Ω)	C_b (fF)	D_b (ps)
Value	30	60	1.0	349.0	5.7	24.8
Parameter	r_{int} (Ω/mm)	c_{int} (fF/mm)	ϕ_{TSV} (μm)	l_{TSV} (μm)	R_{TSV} (mΩ)	C_{TSV} (fF)
Value	51.2	230.2	2	20	133	52

$$f_{\Delta s_{u,v}}(t) = \frac{1}{\sqrt{2\pi\sigma^2_{s_{u,v}}}} e^{-t^2/(2\sigma^2_{s_{u,v}})}. \tag{17.47}$$

This model of skew variations is used to analyze the effects of process variations on various 3-D clock trees. This model is extended to include variations of the horizontal interconnects, as described in Appendix F.

The accuracy of the model is demonstrated through a comparison with Monte Carlo simulations. The structure used for this purpose is an H-tree clock distribution network. This H-tree is placed in a circuit with a total area of 10 mm × 10 mm. The circuit is assumed to be manufactured in a 45 nm CMOS technology. The parameters of the transistors and the interconnects are extracted from the PTM 45 nm CMOS and global interconnect models [252] and International Technology Roadmap for Semiconductors (ITRS) [643]. The clock buffers consist of two inverters connected in series. The circuit parameters are listed in Table 17.2. The ratio of the width to the channel length is denoted by W_n/L_n and W_p/L_p for, respectively, the NMOS and PMOS transistors. The interconnect resistance and capacitance per unit length are, respectively, denoted by r_{int} and c_{int}. The physical and electrical characteristics of the TSVs are also listed in Table 17.2 and are based on the data reported in [181,58] (see Chapter 4, Electrical Properties of Through Silicon Vias). The diameter and length of the TSVs are notated, respectively, as ϕ_{TSV} and l_{TSV}.

The variation of the effective channel length of transistors L_{eff} is considered in this comparison, which is the most significant source of device variations [622,633,638,640]. Note that the effects of other sources of process variations are also represented by the circuit illustrated in Fig. 17.6 and described by the analytic model in (17.13)–(17.20). The corresponding nominal L_{eff}, D2D variation ($3\sigma^{\text{D2D}}_{L\text{eff}}$), and WID variation ($3\sigma^{\text{WID}}_{L\text{eff}}$) are, respectively, 27, 2.2, and 2.7 nm [643]. The resulting variations of R_b, C_b, and D_b are listed in Table 17.1, which are obtained as discussed in Section 17.2.1

with an input slew rate of 47, 16, and 6 mV/ps. The mean value and standard deviation are denoted, respectively, by μ and σ. The ratio σ/μ indicates the importance of variations [633]. The Monte Carlo simulation is repeated 1,000 times. As reported in Table 17.1, σ of R_b, C_b, and D_b all depend on the input slew rate. Simultaneously considering both the slew rate and the load is therefore necessary when evaluating buffer delay variations.

Two H-tree topologies are used to verify the accuracy of the skew variation model. The first topology (multi-via) is illustrated in Fig. 17.5. The second topology (single via) is illustrated in Fig. 17.9. These topologies are compared in the following subsection. The H-tree spans four tiers. The clock source is located at the center of the first tier. There are 128 clock sinks in total, 32 in each tier. The clock buffers, marked with Δ are inserted following the technique described in [602]. The clock frequency is 1 GHz and the constraint on the input slew rate is 16 mV/ps (the transition time is 5% of the clock period). The number of buffers inserted into the multi-via and single via topologies are, respectively, 168 and 540. Only a few buffers are illustrated in Figs. 17.5 and 17.9 for improved readability. The wire segments between the two buffers are simulated using a standard RC π model.

Both skew variations with uncorrelated (independent) WID variations and correlated WID variations modeled by the multi-level spatial correlation approach are simulated where five levels are assumed ($l = 5$). The error of the skew variation model between any pair of sinks in the target clock tree is below 6%.

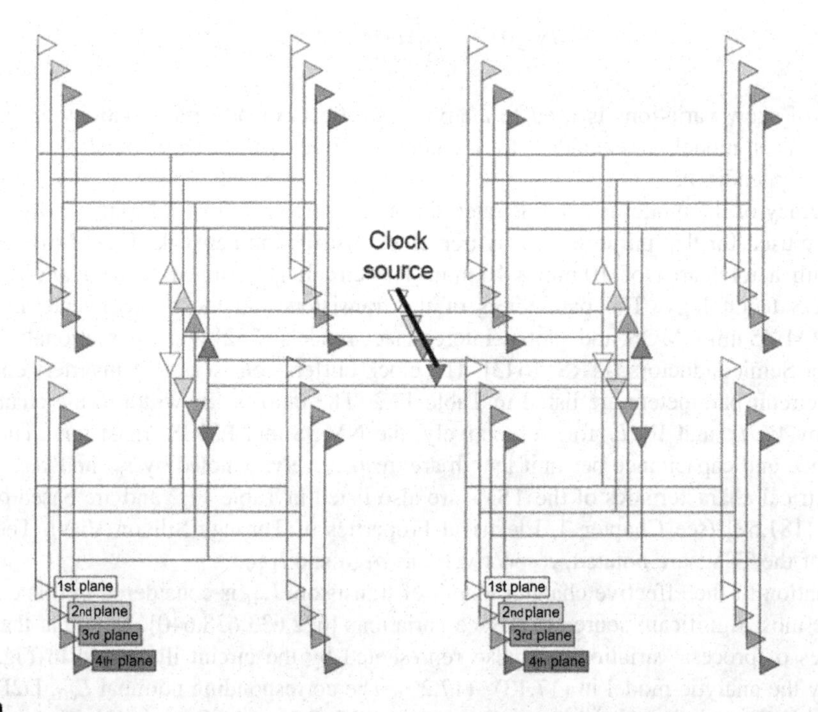

FIGURE 17.9

A single via 3-D clock H-tree.

17.2.4 SKEW VARIATIONS IN THREE-DIMENSIONAL CLOCK TREE TOPOLOGIES

The skew variation for two types of 3-D global H-trees is investigated in this subsection. The first topology (a multi-via topology) is shown in Fig. 17.5, where the clock source and buffers (except for the buffers at the last level) of a 3-D H-tree are located in a single physical tier (e.g., the first tier). In this topology, the clock signal is propagated to the sinks in other tiers by multiple TSVs. The vertical lines at each leaf correspond to a cluster of TSVs. The second clock tree topology (a single via topology) is illustrated in Fig. 17.9, where a 2-D H-tree is replicated in each tier. The clock signal is propagated by a single via (or a group of TSVs to prevent TSV failure and to lower the resistance of this vertical path), connecting the clock source to each H-tree replica.

To better understand the distribution of skew within these topologies, an intratier (pair of sinks placed in the same tier) and intertier (pair of sinks placed in different tiers) distribution of skew is considered separately. In 3-D H-trees and for intratier paths, the number, size, and location of the buffers along these paths are equal for a single tier since multi-via and single via topologies are both symmetric topologies (at least within the x and y directions). D2D variations in each tier, therefore, affect these clock paths equally. Consequently, according to (17.40), for both the multi-via and single via topologies, only WID variations affect the variation of skew between sinks located on the same tier. For both the single via and multi-via tree topologies, the variation of skew between the buffers located in the same tier exhibits the same behavior as in 2-D circuits.

For a 3-D clock tree, as described by (17.44), if R_{in} and C_l of each buffer remain unchanged, $\Delta s_{(u,v)}^{\text{WID}}$ between two sinks decreases as the number of nonshared clock buffers (e.g., the buffers after the $n_{u,v}$ buffers shown in Fig. 17.8) decreases. For a 3-D circuit with area A, the side length of each tier is $L \propto \sqrt{A/N_p}$. Consequently, the number of buffers in one tier decreases as L decreases for an increasing number of tiers forming a 3-D circuit. For the single via topology, all of the clock sinks within a tier are connected to the clock source by the same TSV. The length of this TSV and the increasing number of buffers vertically connected to this TSV do not affect the intratier skew. Consequently, for the single via topology, the distribution of skew between clock sinks in the same tier becomes narrower as the number of tiers increases.

For the multi-via topology, however, the clock sinks in the same tier connect to different TSVs. As the number of tiers increases, both the number of buffers connecting to a TSV and the length of the TSVs increase. The input slew rate decreases since the load is increasing. As reported in Table 17.1, the delay variations of the buffers after the TSVs increase. Moreover, the load of the buffers driving the TSVs increases. These changes in the topological characteristics increase $\sigma_{d(i)}$, as described by (17.29). This increase, consequently, counteracts and can surmount the decrease in variations due to the fewer clock buffers along the clock paths.

The buffers inserted into the 3-D clock trees are reported in Table 17.3. The number of inserted buffers within one tier in the single via topology is lower than the multi-via topology, which introduces a lower WID skew variation than the multi-via topology. The total number of buffers in a single via topology is, however, much higher than the multi-via topology. The numbers of buffers in each tier for both topologies decreases as the number of tiers increases. The increasing number of buffers connected to TSVs (due to the greater number of tiers) increases $\sigma_{su,v}$ in the multi-via topology but does not affect $\sigma_{su,v}$ in the single via topology. Consequently, the decrease in the number of buffers leads to a reduction in skew variation within the same tier for the single via topology, as shown by the (♦) and (■) curves in Fig. 17.10A. Nevertheless, for the multi-via topology, as

Table 17.3 Number of Buffers Inserted Into the 3-D Clock Trees

# of Tiers	1	2	3	4	5	6	7	8	9	10
Multi-via	981	588	558	264	264	242	234	138	134	134
Single via (per tier)	981	460	231	199	199	177	169	105	101	101
Single via (total)	981	920	924	995	995	1062	1183	840	909	1010

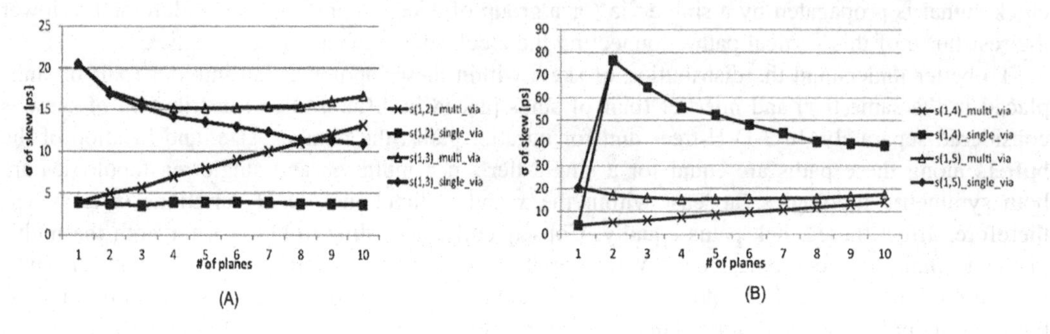

FIGURE 17.10

σ of skew for increasing number of tiers (tiers) and uncorellated WID variations for both the multi and single via topologies, (A) between sinks in the first tier, and (B) between sinks in the first and topmost tiers [644].

shown by the Δ and $\times$ curves, $\sigma_{su,v}$ within the same tier changes nonmonotonically with the number of tiers. For the short distance sinks, $\sigma_{s1,2}$ increases with the number of tiers. As a result, for multi-via 3-D H-trees, simply increasing the number of tiers does not necessarily lower the clock skew. By employing the proposed skew variation model, the number of tiers that produces the lowest skew variation can be determined. Based on this behavior, in a 3-D circuit, if the data related sinks are located mostly within the same tier, the single via topology is more efficient in reducing skew variations and can therefore support a higher clock frequency.

A different behavior is observed when considering intertier skew. As described by (17.40) through (17.42), when the target clock sinks are located in different tiers, the corresponding clock skew is also affected by D2D variations. For the single via topology, the skew variation for the intertier pairs of sinks remains approximately the same irrespective of the tiers to which the sinks belong. This behavior occurs since the paths to the sinks located in different tiers do not share any common segments (see Fig. 17.9). When the number of tiers is greater than two, the skew variation decreases as the number of tiers increases, as also shown in Fig. 17.10B. Since the paths lay in different tiers, according to (17.42), the effect of D2D variations on the 3-D single via topology is much greater than in planar H-trees.

The skew variation under the multi-via topology varies significantly from the single via topology, as illustrated in Fig. 17.10B. The skew variation between tiers depends upon the location of the related sinks. According to (17.40)–(17.42), the effects of D2D variations increase as the number of buffers located in different tiers increases. For the multi-via topology, all of the clock paths preceding the TSVs are in the first tier. The effects of D2D variations on the multi-via topology, therefore, are much smaller than the single via topology, as shown in Fig. 17.10B. Nevertheless, the skew variation of the multi-via topology changes nonmonotonically with the number of tiers. In a 3-D circuit, if the data related sinks are widely distributed on several tiers, the multi-via topology is more efficient in reducing skew variations and can therefore support a higher clock frequency.

These results indicate that the performance improvement in a 3-D clock network depends significantly on the distribution of the sinks (and, consequently, the clock paths) among the tiers. When the data related sinks are distributed on different tiers, the skew of the single via 3-D clock trees is affected more by process variations than a corresponding 2-D clock tree. This behavior is consistent with the conclusions made in Section 17.1 for data paths in 3-D circuits. The effects of process variations on 3-D clock distribution networks can be mitigated in this case by employing a multi-via topology. This topology can better exploit the traits of vertical integration (i.e., shorter wires) to significantly increase the operating frequency.

Based on these observations, a hybrid H-tree topology (multi-group topology) combining the advantages of both the single and multi-via topologies exhibits the lowest clock skew variability. The multi-group topology is illustrated in Fig. 17.11. The key concept is that the m tiers forming a

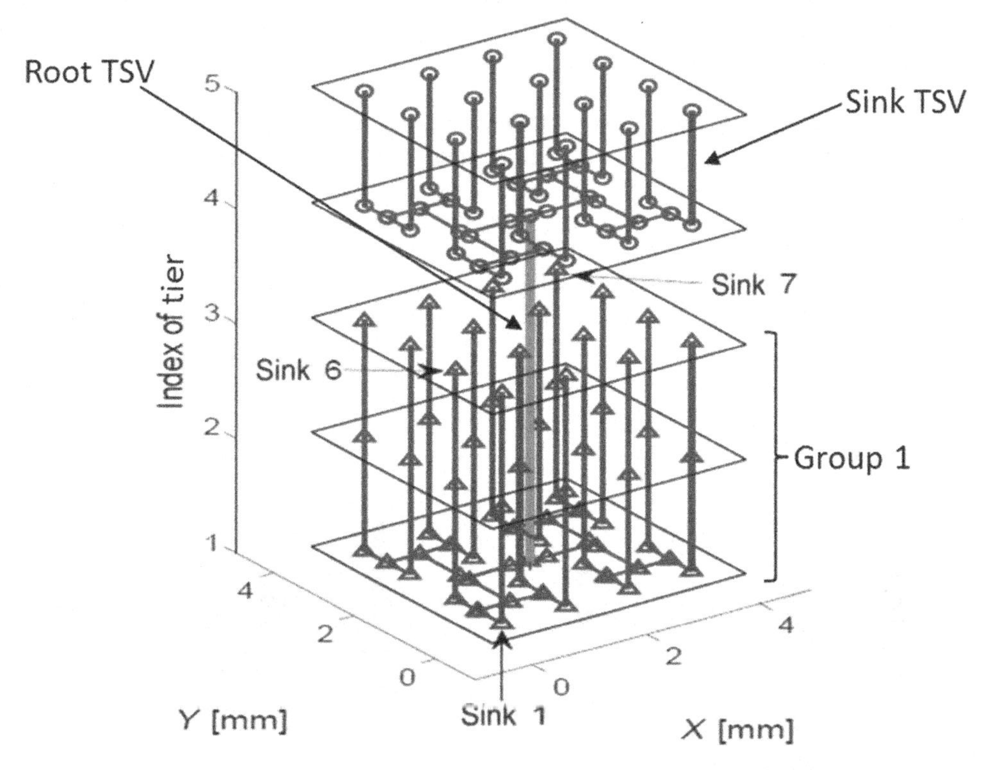

FIGURE 17.11

Example multi-group 3-D clock topology.

3-D circuit are divided into Q groups of "data related tiers." The data related tiers are the physical tiers containing data related registers. The ith group of data related tiers consists of h_i ($\leq m$) physical tiers. The clock signal is distributed within these h_i tiers by a multi-via topology.

An example of this H-tree topology is illustrated in Fig. 17.11. This H-tree includes two groups of data related tiers ($Q = 2$). The first group spans three tiers ($h_1 = 3$), and the second group spans two physical tiers ($h_2 = 2$). The buffers contained in each group of data related tiers are denoted by empty "triangles" and "dots." The TSVs connecting these buffers are called "sink TSVs." The root of the multi-via topologies is connected with a "root TSV" (or a cluster of TSVs), as illustrated by the segment at the center of the tiers.

For a 3-D circuit, if all the data related clock sinks cannot be located within the same tier but in adjacent tiers, the multi-group topology is more efficient in reducing skew variations than the other topologies. Compared with the single via topology, using Q instead of m H-trees, the multi-group topology significantly reduces skew variations between data related tiers. As compared with the multi-via topology, the multi-group topology requires fewer buffers connected to the sink TSVs than buffers connected to the TSVs of the multi-via topology. Both skew variations within a single tier and skew variations between data related tiers are therefore reduced.

To quantify the ability of a multi-group tree to reduce skew variations among the clock sinks, a 3-D circuit with eight tiers is evaluated. Two variants of the multi-group topology are considered, including two groups (hybrid 2, $Q = 2$, $h_i = 4$) and four groups (hybrid 4, $Q = 4$, $h_i = 2$) of data related tiers. Simulation results are shown in Fig. 17.12.

In Fig. 17.12, skews $s_{1,2}$ and $s_{1,3}$ (illustrated in Fig. 17.5B) and $s_{1,6}$ and $s_{1,7}$ (illustrated in Fig. 17.11) are depicted exhibiting skew variations between the nearest and farthest sinks. The results, based on independent and multi-level correlated WID variations, are denoted, respectively, by (I) and (II). $\sigma_{s1,2}$ and $\sigma_{s1,3}$ produced by the multi-group topology are lower than the multi-via topology and decrease as the number of sub-H-trees increases. For the topology with four sub-H-trees (hybrid 4), $s_{1,2}$(I), $s_{1,3}$(I), $s_{1,2}$(II), and $s_{1,3}$(II) are reduced, respectively, by 55%, 23%, 44%, and 10% as compared with the multi-via topology. Although $\sigma_{s1,3}$ within the same tier of the multi-group topology is greater (4% for hybrid 4) than the single via topology, the intertier skews $\sigma_{s1,6}$

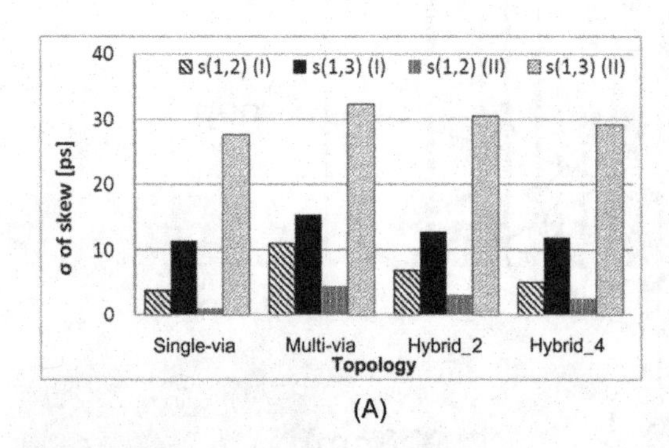

(A)

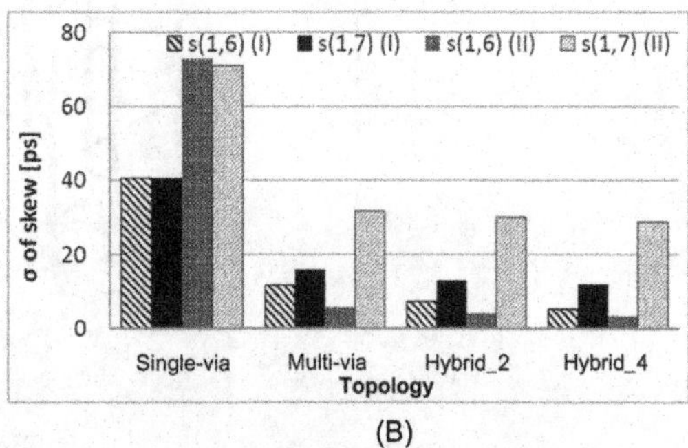

(B)

FIGURE 17.12

σ of skew for 3-D clock tree topologies. (A) Intratier skew of sink pairs $s_{1,2}$ and $s_{1,3}$, and (B) intertier skew of sink pairs $s_{1,6}$ and $s_{1,7}$ within a group of data related tiers.

and $\sigma_{s1,7}$ within a group of data related tiers of the multi-group topology are significantly reduced, as shown in Fig. 17.12B. This reduction is also greater than the multi-via topology. The number of sub-H-trees within a multi-group 3-D topology is determined by the distribution of the data related sinks. Consequently, if the data related sinks are located in adjacent tiers of a 3-D circuit, the multi-group 3-D clock tree topology is more efficient in reducing skew variations than both the single and multi-via topologies.

17.3 EFFECT OF PROCESS AND POWER SUPPLY VARIATIONS ON THREE-DIMENSIONAL CLOCK DISTRIBUTION NETWORKS

D2D variations due to the multi-tier nature of 3-D circuits increase the variation in the delay of intertier paths, as shown in Section 17.1. This greater variation in delay manifests as larger timing uncertainty in clock distribution networks that span several tiers. A greater timing uncertainty hinders the management of skew which can in turn degrade the performance of a 3-D clock distribution network, as discussed in Section 17.2. Similar to the situation where different D2D variations are assumed among the tiers of a 3-D system, each tier experiences disparate power supply noise, as discussed in Chapter 18, Power Delivery for Three-Dimensional ICs.

Power supply noise in a clock distribution network adds to the timing uncertainty, although differently than process variations. Power supply noise can produce clock jitter. Clock jitter can occur in three different ways: period jitter, cycle-to-cycle jitter, and phase jitter (or time interval error) [645]. Period jitter is the difference between the measured clock period and the ideal period, which is the most explicit form of clock jitter within a circuit. Cycle-to-cycle jitter is the variation in clock period between adjacent clock periods over a random sample of cycles [646]. For a random number of clock cycles, phase jitter is the departure in phase of a specific edge from the mean phase [646].

The jitter is produced by a phase locked loop (PLL) driving a clock distribution network. PLL jitter can be mitigated by careful design of the PLL [647]. Furthermore, power supply noise affects the clock buffers used within the distribution network, leading to higher clock jitter [648]. This increase in jitter within planar clock distribution networks is discussed in [649,650]. A similar analysis cannot however be easily applied to 3-D circuits due to the disparate power supply noise that buffers within different tiers can experience [651]. Moreover, clock distribution networks are simultaneously affected by process variations and power supply noise. Consequently, the combined effects of process variations and power supply noise on a multi-tier 3-D clock distribution network are discussed in this section where both theoretical and practical design issues are reviewed. A term introduced in [652] to describe both skew and jitter, *skitter*, is utilized throughout this section.

A methodology to determine the delay variation of a buffer stage under process and power supply variations is presented in the following subsection. With a statistical description of the delay of a clock buffer, a model that describes the combined effects of clock skew and jitter of 3-D clock trees is presented in Section 17.3.2. Based on this model, several tradeoffs among the skitter, power of the clock network, and allocation of clock buffers among tiers are considered in Section 17.3.3. Related design guidelines to mitigate skitter and lower the complexity are presented in Section 17.3.4.

17.3.1 DELAY VARIATION OF BUFFER STAGES

The distribution of the delay of a buffer stage is modeled in this subsection. The delay of a buffer stage d consists of the delay of the buffer d_b and the interconnect (horizontal wire and/or through silicon via) d_I. The variation of d is modeled in this subsection as a random variable affected by both process variations and power supply noise.

Since the variation of parameters due to process variations is typically within a small range, the delay of a buffer stage considering parameter variations can be approximated by a first order Taylor series expansion [637]

$$d(t_r, \vec{P}, C_{lw}) = d_b(t_r, \vec{P}, C_{lb}) + d_I(\vec{P}, C_{lw}) \approx \overline{d} + \sum_{p \in \vec{P}} \left(\frac{\partial d}{\partial p}\bigg|_0 \Delta p \right). \tag{17.48}$$

The input slew rate of this buffer stage is denoted by t_r. The capacitive load at the output of the buffer and wire is denoted, respectively, by C_{lb} and C_{lw}. The nominal delay is $\overline{d}$ and the subscript "0" denotes the partial derivative assuming nominal parameters. The set of parameters affected by process variations is denoted by $\vec{P}$. Each parameter is modeled as a random variable. For instance, if the variation of the channel length of three buffers is considered, $\vec{P}$ is $\{L_{b,1}, L_{b,2}, L_{b,3}\}$. The variation of a parameter Δp consists of WID and D2D variations,

$$\Delta p = \Delta p_{\text{WID}} + \Delta p_{\text{D2D}}, \tag{17.49}$$

where Δp_{D2D} is constant among the buffers and interconnects within the same die, while Δp_{WID} varies among the components within the same die [638]. The individual partial derivatives in (17.48) are

$$\frac{\partial d}{\partial p} = \frac{\partial d_b}{\partial t_r}\frac{\partial t_r}{\partial p} + \frac{\partial d_b}{\partial C_{lb}}\frac{\partial C_{lb}}{\partial p} + \frac{\partial d_b}{\partial p} + \frac{\partial d_I}{\partial C_{lw}}\frac{\partial C_{lw}}{\partial p} + \frac{\partial d_I}{\partial p}. \tag{17.50}$$

The partial derivatives in (17.50) are determined from the expression for d_b and d_I. The expression of $d_b(t_r, \vec{P}, C_{lb})$ is obtained through analytic formulas [653] or an adjoint sensitivity analysis with SPICE-based simulations [637]. To achieve higher accuracy, the latter is used. For horizontal wires, the expression of $d_I(\vec{P}, C_{lw})$ is based on the *RLC* interconnect model described in [654,655].

The variations introduced by the TSVs are discussed in [656,657], where TSV stress induced delay variations of buffers are modeled. In this chapter, the keep out zone around the TSVs is assumed to be sufficiently large ($\leq$10 μm [657,658]) to mitigate the effects of TSV stress. Consequently, the TSVs are modeled as *RLC* wires with different electrical characteristics than the horizontal interconnects

In addition to process variations, the power noise also affects the buffer delay. The supply voltage V_{dd} is affected by the power noise v, $V_{dd} = V_{dd0} + v$, where V_{dd0} is the nominal power supply voltage of the circuit. As the power noise contains several components across a large range of frequencies, the focus of this section is on the frequency noise component due to the capacitance of the on-chip power distribution network and the inductance of the package. This noise is typically between tens and several hundred MHz (e.g., 400 MHz) [649].

This noise is modeled as a sinusoidal waveform where the amplitude of the waveform is the worst case noise observed in the clock network. Assuming a clock edge arrives at the source of a clock path at time zero, t_j is the time when this clock edge arrives at buffer j. The supply noise at buffer j at time t_j is

$$v(t_j) = V_n e^{-\epsilon t_j} \sin(2\pi f_n t_j + \phi), \tag{17.51}$$

$$t_j = \sum_{i=1}^{j-1} d_i. \tag{17.52}$$

The clock frequency is much higher than the resonant noise frequency, and the clock path delay is typically lower than the period of the resonant noise. Due to the significant voltage drop, the first period of the resonant noise causes the greatest clock jitter [649]. Consequently, to investigate the effects of the worst case power noise on clock distribution networks, (17.51) is approximated by an undamped sinusoidal waveform [649]

$$v(t_j) \approx V_n \sin(2\pi f_n t_j + \phi). \tag{17.53}$$

According to (17.48) and (17.52), d_i, t_j, and $v(t_j)$ are all random variables. Since Δt_j is low as compared with $\overline{t_j}$, $v(t_j)$ can also be approximated by a first order Taylor series expansion,

$$v(t_j) = \overline{v}(t_j) + \Delta v(t_j) \approx \overline{v}(t_j) + \left.\frac{\partial v(t_j)}{\partial t_j}\right|_0 \Delta t_j, \tag{17.54}$$

$$\Delta v(t_j) \approx 2\pi V_n f_n \cos(2\pi f_n \overline{t_j} + \phi) \sum_{i=1}^{j-1} \Delta d_i. \tag{17.55}$$

The amplitude V_n and frequency f_n are determined by the switching current and the circuit characteristics. The initial phase ϕ is the phase of the resonant noise when the clock edge arrives at the source of the clock path.

In 3-D circuits, the current dissipated by the tiers can differ due to the different number and size of the devices. The amplitude and frequency of the resonant supply noise change with the current within different tiers. To demonstrate this behavior, a 1-D model of a power distribution network for a three tier circuit, shown in Fig. 17.13, is considered under different switching scenarios.

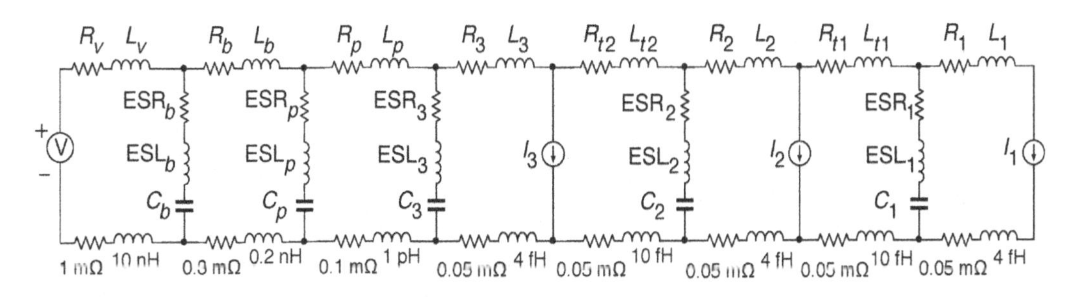

FIGURE 17.13

Simplified 1-D model of a power distribution network to evaluate global power noise. R_{ti} and C_{ti} denote, respectively, the TSV resistance and capacitance of tier i.

Table 17.4 Four Scenarios of Switching Current Within a Three Tier Circuit

Case	cur1	cur2	cur3	cur4
I_1 (A)	0	10	20	40
I_2 (A)	20	20	20	20
I_3 (A)	40	30	20	0

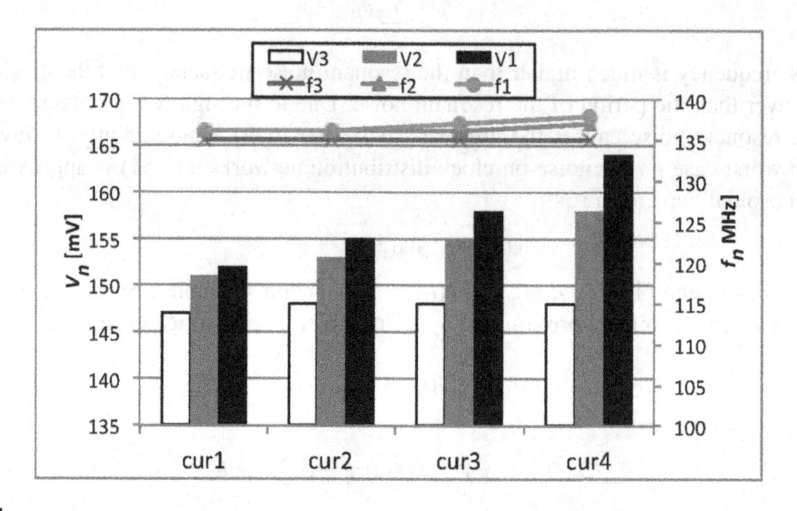

FIGURE 17.14

Amplitude and frequency of the resonant noise versus the switching current in different tiers.

Four cases of switching current are considered for this power network, which are listed in Table 17.4. The pulse width and transition times of the switching current are both 1 ns. The resulting V_n and f_n are depicted in Fig. 17.14. As illustrated in this figure, different current distributions introduce a nonnegligible difference in V_n among tiers ΔV_n. The *IR* drop and resonance impedances among the tiers both contribute to this ΔV_n. The resonant frequency is similar among tiers ($\Delta f_n \leq 3$ MHz) and does not change significantly with the current.

The electrical characteristics of the TSVs depend upon the manufacturing technology [201,181]. The change of the power noise with the total resistance of the TSVs (R_{tsv}) is illustrated in Fig. 17.15. In this figure, the amplitude and frequency of the overall supply noise are denoted, respectively, by V3–V1 and f3–f1. The DC *IR* drop is denoted by V3_dc–V1_dc. Since the resonant noise is stimulated by a current pulse, the effect of the *IR* drop is also included in V3–V1.

A larger R_{TSV} introduces a higher *IR* drop in the first and second tiers. In the third tier, the DC *IR* drop is not affected by R_{TSV} since this tier is directly connected to the package (see Fig. 17.13). Nevertheless, a higher R_{TSV} decreases the quality factor (*Q* factor) of the resonant circuit, which decreases the amplitude of the resonance. Consequently, V_n in the third tier decreases with R_{TSV}. In the first and second tiers, V_n is determined by both the resonance and the *IR* drop. Consequently, V2 and V1 increase with R_{TSV} due to the significantly higher *IR* drop. Nevertheless, the increase in V2 and V1 is not as high as the increase in the DC *IR* drop due to the lower *Q* factor.

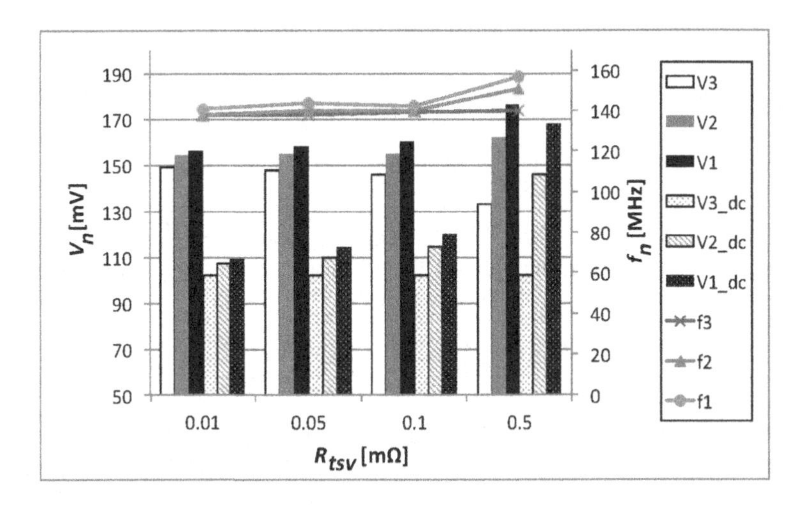

FIGURE 17.15

Resonant supply noise and *IR* drop versus the total resistance of the TSVs.

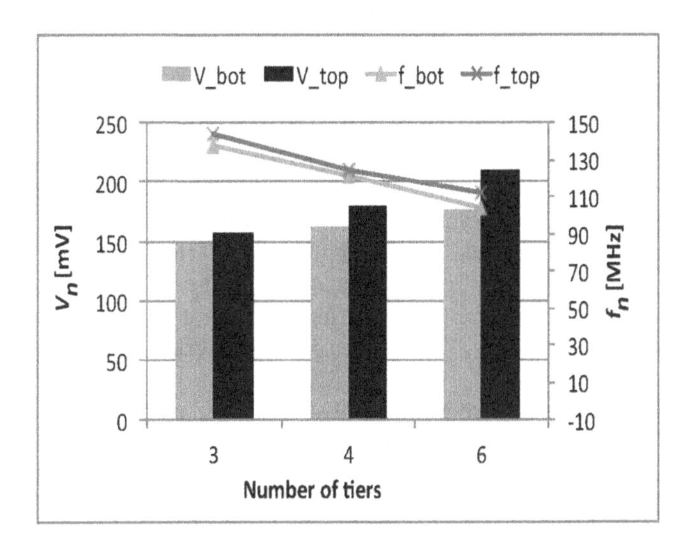

FIGURE 17.16

Resonant noise versus the number of tiers.

The resonant noise for different number of tiers in a 3-D IC is plotted in Fig. 17.16. The switching current and on-die capacitance are assumed identical for all tiers. As shown in Fig. 17.16, ΔV_n between the bottom and top tiers increases with the number of tiers. As more dies are vertically stacked, the difference in resonant noise among tiers increases.

According to (17.48), the delay variation Δd is also affected by the input slew Δt_r, which is determined by the previous buffer stage. Considering the effects of Δv and Δt_r on Δd, the delay variation of the jth buffer stage is modeled as

$$\Delta d_j \approx \sum_{p \in \vec{P}_j} \left(\frac{\partial d_j}{\partial p}\bigg|_0 \Delta p \right) + \frac{\partial d_j}{\partial v}\bigg|_0 \Delta v(t_j) + \frac{\partial d_j}{\partial tr}\bigg|_0 \Delta t_{rj}. \tag{17.56}$$

The set of statistical parameters of the jth buffer stage is denoted by $\vec{P}_j$, which is a subset of the entire parameter set, $\vec{P}_j \subseteq \vec{P}$. The input slew of the jth buffer stage Δtr_j is determined similarly to (17.56),

$$\Delta tr_j \approx \sum_{p \in \vec{P}_j} \left(\frac{\partial t_{rj}}{\partial p}\bigg|_0 \Delta p \right) + \frac{\partial t_{rj}}{\partial v}\bigg|_0 \Delta v(t_{j-1}) + \frac{\partial t_{rj}}{\partial t_{rj-1}}\bigg|_0 \Delta t_{rj-1}. \tag{17.57}$$

Substituting (17.55) and (17.57) into (17.56), Δd_j can be recursively determined considering both process variations and power noise. The coefficients in (17.56) and (17.57) are obtained through an adjoint sensitivity analysis, as previously mentioned. The resulting expression, (17.56), determines the skitter, as described in the following subsection.

17.3.2 MODEL OF SKITTER IN THREE-DIMENSIONAL CLOCK TREES

The definitions of clock skew, period jitter, and skitter are illustrated in Fig. 17.17. The clock signal is fed into the 3-D clock tree from the primary clock driver. Two flip flops are driven by this clock signal, denoted, respectively, as FF_1 and FF_2. The waveforms, clk_1 and clk_2, shown in Fig. 17.17B correspond to the clock signal driving, respectively, FF_1 and FF_2. The time when the first rising edge in Fig. 17.17B arrives at the clock input is defined as the origin. The time when this clock edge arrives at FF_1 and FF_2 is, respectively, denoted by t_1 and t_2. The arrival time of the next rising edge is t'_1 and t'_2. The number of buffers from the clock input to FF_1 and FF_2 is denoted, respectively, by $n_1 + n_2$ and $n_3 + n_4$. The skew between the first edge of clk_1 and clk_2 is $S_{1,2}$. The clock period after the first edge for FF_1 and FF_2 are, respectively, T_1 and T_2. The clock period is T_{clk}. The corresponding period jitter is $J_1 = T_1 - T_{\text{clk}}$ and $J_2 = T_2 - T_{\text{clk}}$.

Assuming the data is transferred from FF_1 to FF_2 within one clock cycle, $T_{1,2}$ is the time interval that determines the highest clock frequency supported by the circuit. The setup time requirement needs to be satisfied for the system to operate correctly [577,645]. The setup time slack notated as $\text{slack}_{\text{setup}}$ is defined as

$$\text{slack}_{\text{setup}} \equiv T_{1,2} - \max(D_{1,2}) - t_{\text{setup}}, \tag{17.58}$$

$$T_{1,2} = (t_2 - t_1) + T_2 = S_{1,2} + J_2 + T_{\text{clk}}, \tag{17.59}$$

where $\max(D_{1,2})$ denotes the longest data transfer time from FF_1 to FF_2. The setup time for FF_2 is t_{setup}. Consequently, the variation of $\text{slack}_{\text{setup}}$ is affected by the variation of $T_{1,2}$, called "setup skitter" and notated as $J_{1,2}$,

$$J_{1,2} = S_{1,2} + J_2 = t'_2 - t_1 - T_{\text{clk}}. \tag{17.60}$$

To avoid setup time violations, $\text{slack}_{\text{setup}} \geq 0$ is required under all operating conditions.

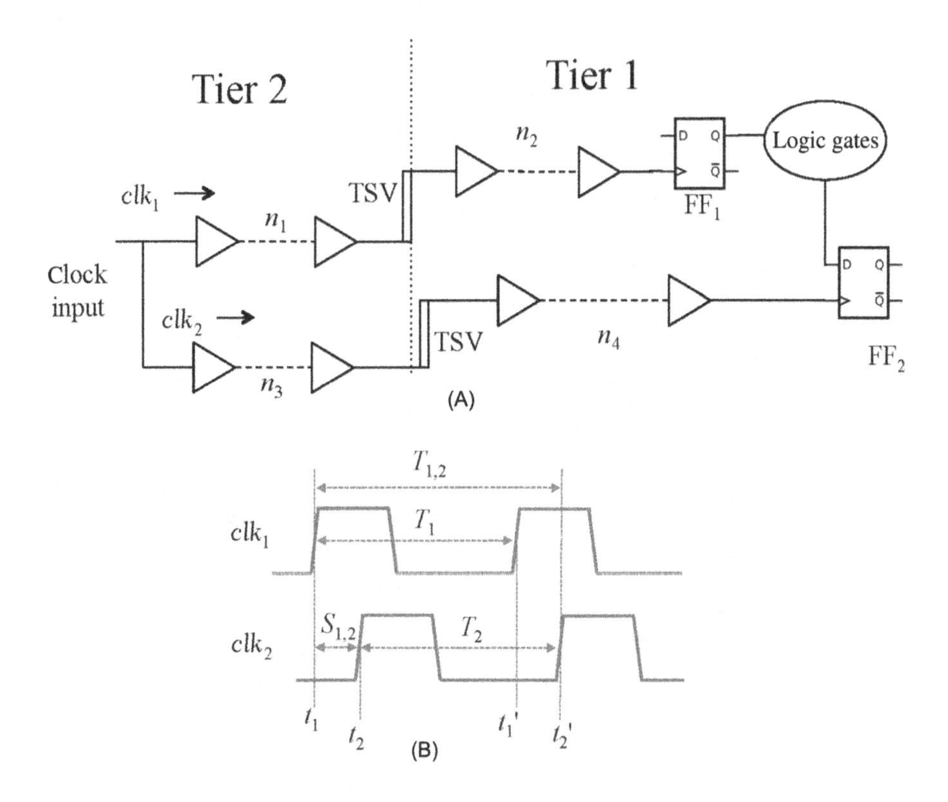

FIGURE 17.17

Clock uncertainty between 3-D clock paths. (A) Two paths and flip flops, and (B) corresponding clock signals.

According to (17.52) and (17.60), skitter $J_{1,2}$ is the linear combination of the delay of the buffer stages,

$$J_{1,2} = \sum_{k=1}^{n_3+n_4} d'_{2,k} - \sum_{k=1}^{n_1+n_2} d_{1,k}, \tag{17.61}$$

$$\overline{J}_{1,2} = \sum_{k=1}^{n_3+n_4} \overline{d}'_{2,k} - \sum_{k=1}^{n_1+n_2} \overline{d}_{1,k}, \tag{17.62}$$

$$\Delta J_{1,2} = \sum_{k=1}^{n_3+n_4} \Delta d'_{2,k} - \sum_{k=1}^{n_1+n_2} \Delta d_{1,k} \approx \sum_{p \in \vec{P}} \left(\left. \frac{\partial J_{1,2}}{\partial p} \right|_0 \Delta p \right), \tag{17.63}$$

where $d'_{2,k}$ is the delay of the kth buffer stage along the path to FF$_2$ for the second clock edge. The mean skitter $\overline{J}_{1,2}$ is the mean delay of all of the buffer stages considering the mean voltage supply noise (without process variations). Substituting (17.56) into (17.63), the partial derivative $(\partial J_{1,2}/\partial p)|_0$ is obtained. Consequently, skitter $J_{1,2}$ is approximated by a first order Taylor series expansion. Assuming all of the parameters are characterized by a Gaussian distribution, $\Delta J_{1,2}$ can also be approximated by a Gaussian distribution,

$$\Delta J_{1,2} \sim \mathcal{N}(0, \sigma_{J_{1,2}}^2), \tag{17.64}$$

$$\sigma_{J_{1,2}}^2 = \sum_{p \in \vec{P}} \left(\left. \frac{\partial J_{1,2}}{\partial p} \right|_0^2 \sigma_p^2 \right) + 2 \sum_{p,q \in \vec{P}} \left(\left. \frac{\partial J_{1,2}}{\partial p} \right|_0 \left. \frac{\partial J_{1,2}}{\partial q} \right|_0 \mathrm{cov}(p,q) \right), \tag{17.65}$$

where cov(p, q) denotes the covariance between two parameters. Assuming D2D variations are independent from WID variations [632,637], $\sigma_p^2 = \sigma_{p(\mathrm{D2D})}^2 + \sigma_{p(\mathrm{WID})}^2$. The covariance between the two parameters is determined according to the tiers to which these parameters are related and the spatial correlation between these parameters,

$$\mathrm{cov}(p,q) = \begin{cases} 0, \text{ if } p, q \text{ are of different type or belong to different tiers} \\ \mathrm{cov}(p,q)_{\mathrm{WID}} + \sigma_{p(\mathrm{D2D})}\sigma_{q(\mathrm{D2D})}, \text{ otherwise,} \end{cases} \tag{17.66a-b}$$

where the WID covariance cov$(p, q)_{\mathrm{WID}}$ is determined by the spatial correlation between parameters p and q within the same tier. Statistically, the devices (and wires) close to each other exhibit a higher correlation than those devices far from each other. This spatial correlation can be obtained from fabricated wafers [659] or through a spatial correlation model [637,660].

As shown in (17.65) and (17.66), the variance of the setup skitter $\sigma_{J_{1,2}}^2$ depends on the covariance between the process induced parameters. In 2-D circuits, the change of cov(p,q) is primarily determined by cov$(p,q)_{\mathrm{WID}}$ since the parameters are affected by the same D2D variations. The distribution of the clock paths therefore only affects $\sigma_{J_{1,2}}^2$ by changing the WID covariance. In 3-D circuits, however, D2D variations vary among tiers, and the WID covariance among tiers is zero. Consequently, the distribution of the clock paths affects the skitter variation in a highly complicated manner.

In addition to the setup time slack, the hold time slack also significantly affects circuit performance. The hold time violation can also cause the failure of the entire system [645]. Moreover, this failure cannot be removed by lowering the system wide clock frequency. As illustrated in Fig. 17.17B, the hold time slack is modeled as

$$\mathrm{slack}_{\mathrm{hold}} = \min(D_{1,2}) - S_{1,2} - t_{\mathrm{hold}}, \tag{17.67}$$

where t_{hold} is the hold time requirement. The "hold skitter" affecting slack$_{\mathrm{hold}}$ is determined by $S_{1,2}$, which is the skew between clk_1 and clk_2. Note that $S_{1,2}$ is affected by both process variations and power noise.

To correctly latch the data in FF$_2$, slack$_{\mathrm{hold}} \geq 0$ is required to avoid hold time violations under any operating condition. From Fig. 17.17B, $S_{1,2}$ is

$$S_{1,2} = t_2 - t_1 = \sum_{k=1}^{n_3+n_4} d_{2,k} - \sum_{k=1}^{n_1+n_2} d_{1,k} \approx \sum_{k=1}^{n_3+n_4} \bar{d}_{2,k} - \sum_{k=1}^{n_1+n_2} \bar{d}_{1,k} + \sum_{p \in \vec{P}} \left(\left. \frac{\partial S_{1,2}}{\partial p} \right|_0 \Delta p \right). \tag{17.68}$$

Similar to (17.52) and (17.65), the distribution of $\Delta S_{1,2}$ can be modeled as

$$\Delta S_{1,2} \sim \mathcal{N}(0, \sigma_{S_{1,2}}^2), \tag{17.69}$$

$$\sigma_{S_{1,2}}^2 = \sum_{p \in \vec{P}} \left(\left. \frac{\partial S_{1,2}}{\partial p} \right|_0^2 \sigma_p^2 \right) + 2 \sum_{p,q \in \vec{P}} \left(\left. \frac{\partial S_{1,2}}{\partial p} \right|_0 \left. \frac{\partial S_{1,2}}{\partial q} \right|_0 \mathrm{cov}(p,q) \right), \tag{17.70}$$

where the partial derivatives are obtained similarly to the coefficients in (17.63). As shown by (17.48)–(17.70), both the setup and hold time violations are simultaneously affected by process variations and power noise. This effect and the accuracy of this model are discussed in the following subsection.

17.3.3 SKITTER RELATED TRADEOFFS IN THREE-DIMENSIONAL ICs

The variation of skitter for diverse characteristics of a clock distribution network is discussed in this subsection. The number and size of the buffers change with the number of tiers spanned by the clock networks which also affects the dissipated power. The related tradeoffs are discussed in Section 17.3.3.1. The change in skitter as a function of the phase and magnitude of the power noise frequency is summarized in Section 17.3.3.2.

The electrical parameters of the transistors are based on a 32 nm PTM model [252]. The parameters of the interconnects are based on an Intel 32 nm interconnect technology [649]. The parameters of the TSVs are based on data from [181]. Both the horizontal wires and TSVs are modeled by π segments in SPICE-based simulations. The variation aware model of skitter is based on Matlab. All of the simulations are performed in a Scientific Linux server (Intel Xeon 2.67 GHz, 24 cores, 24 GB memory).

The variations considered in the simulations are listed in Table 17.5. The D2D and WID ΔL_b are extracted based on ITRS data [661]. The wire variations and ΔV_{th} are based on [637]. The variations of the TSVs are based on [656]. Note that other sources of variations can also be described by this modeling approach. For example, the TSV stress induced delay variation in [657] can be included. In this case, the distribution of d_B in (17.48) is based on the distance between the buffer and the TSVs and the stress induced buffer delay.

17.3.3.1 Skitter versus length of clock paths, number of tiers, and power dissipation

The change in setup skitter with the length of the clock paths is the topic of this subsection. The length ranges from 0.5 to 12.5 mm within two and three tier circuits. Buffers are inserted to produce a 10% T_{clk} input slew for the next stage. To emphasize the relation between skitter and the length of the clock paths, all tiers are assumed to experience similar supply noise ($V_n = 90$ mv, $f_n = 400$ MHz, and $\phi = 270^{\circ}$ [649]). Each pair of paths is distributed across different tiers, as shown in Fig. 17.17A. The resulting $\mu_{J_{1,2}}$ and $\sigma_{J_{1,2}}$ are illustrated in Fig. 17.18, where the suffixes “2” and “3” denote the results for, respectively, the two and three tier circuits.

The results from SPICE-based Monte Carlo simulations and the semi-analytic model (labeled as (m)) are both depicted in Fig. 17.18. As shown in this figure, both $\mu_{J_{1,2}}$ and $\sigma_{J_{1,2}}$ deteriorate with longer clock paths. This behavior is described by the model and exhibits reasonable accuracy. The error of the model is below 11% for $\mu_{J_{1,2}}$ and below 12% for $\sigma_{J_{1,2}}$. Not surprisingly, long clock paths introduce high skitter in 3-D clock trees. Consequently, both the mean and standard deviation of the setup skitter increase with the length of the clock paths.

The skitter has been evaluated for no TSV variations, 5% TSV variations ($\sigma/\mu = 5\%$), and 15% TSV variations. The difference in $\sigma_{J_{1,2}}$ among these three cases is around 1 ps for all of the clock paths. This situation shows that TSV variations are a second order effect, consistent with the results reported in [656].

Table 17.5 Variation of Devices, Horizontal Wires, and TSVs

Parameters	Nominal	3σ (D2D)	3σ (WID)
Channel length (nm)	32	1.5	2.5
Threshold voltage (mV)	242	24.2	24.2
Wire width (nm)	225	22.5	11.3
Wire height (nm)	388	19.4	9.7
ILD thickness (nm)	252	18.9	9.5
TSV resistance (mΩ)	133	39.9	39.9
TSV capacitance (fF)	52	15.6	15.6

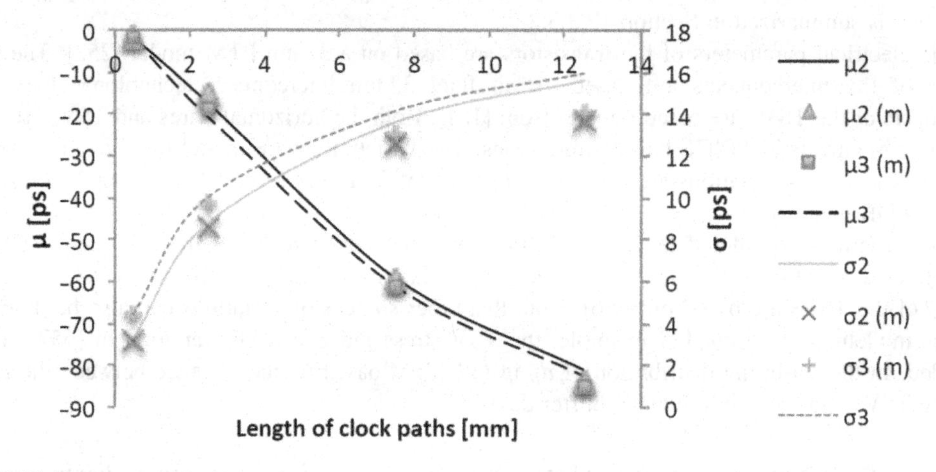

FIGURE 17.18

Skitter versus length of 3-D clock paths.

In addition to the length of the clock paths, the number of tiers spanned by these paths determines the skitter of the clock distribution network. Due to the different switching currents in power supply networks and the vertical resistance of the P/G TSVs among tiers, the devices in different tiers are subjected to different ΔV_n, as shown in Figs. 17.14 and 17.15. The tier closer to the P/G pads experiences a lower power noise, as also discussed in Chapter 18, Power Delivery for Three-Dimensional ICs.

Clock paths spanning two tiers with 20 buffers ($n_1 + n_2 = n_3 + n_4 = 20$, see Fig. 17.17A) are considered as an example. The clock source is located on tier 2. The total length of each path is 5 mm. The initial phase ϕ (270°) and frequency f_n (400 MHz) are assumed to be the same for both tiers. Two distributions of clock paths are discussed: (A) $n_1 = n_2 = n_3 = n_4 = 10$ and (B) $n_1 = n_3 = 15$, $n_2 = n_4 = 5$. Distribution (A) denotes equally divided 3-D clock paths. In distribution (B), the longest segment of the clock paths is placed in tier 2. To depict the accuracy of the model, the simulation results of the setup skitter $J_{1,2}$ for $V_{n1} = 90$ mV and different V_{n2} are shown in Fig. 17.19. As noted in this figure, $\mu_{J_{1,2}}$ changes significantly with V_{n2} while $\sigma_{J_{1,2}}$ varies only slightly with V_{n2}.

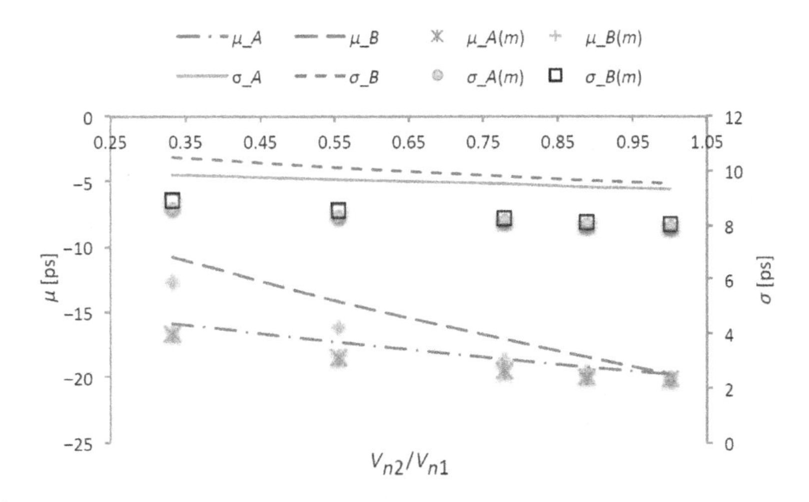

FIGURE 17.19

Skitter for $V_{n1} = 90$ mV and different V_{n2}.

The change in setup skitter $J_{1,2}$ with (V_{n2}, V_{n1}) is illustrated in Fig. 17.20. As shown in Figs. 17.20A and B, for distribution (A), $\mu_{J1,2}$ increases significantly with both V_{n2} and V_{n1}, since higher supply noise introduces greater period jitter. The clock paths of (A) are equally distributed among the tiers. $\mu_{J1,2}$ is therefore similarly affected by V_{n1} and V_{n2}. For distribution (B), however, the situation is different. As shown in Figs. 17.20C and D, $\mu_{J1,2}$ is primarily determined by V_{n2} since the longest segment of the clock paths in (B) is placed in tier 2. Consequently, for unequally distributed clock paths, the mean skitter is mainly determined by the tier containing the longest part of the clock paths.

As shown in Figs. 17.20A and B, assuming $V_{n1} = 0.09$ mV, distribution (A) produces higher $\mu_{J1,2}$ than (B) for different V_{n2}. This difference in $\mu_{J1,2}$ increases with ΔV_n ($\Delta V_n = V_{n1} - V_{n2}$), from 1% to 42% of μ_{JA}. The reason for this significant difference is that the majority of the buffers in (B) is located in tier 2, which is more sensitive to V_{n2}. More generally, assuming $V_{n1} > V_{n2}$, the mean skitter of (B) is always lower than (A).

Consequently, the distribution of the clock paths in 3-D ICs significantly affects the mean skitter due to the different V_n among tiers. However, in 2-D circuits, this mean skitter does not vary significantly with the distribution of clock paths due to the global resonant noise at low frequencies [662]. The standard deviation $\sigma_{J1,2}$ of (A) and (B) is illustrated, respectively, in Figs. 17.20E and F. Similar to $\mu_{J1,2}$, $\sigma_{J1,2}$ also increases with V_{n1} and V_{n2}. Nevertheless, $\Delta\sigma_{J1,2}$ is relatively low as compared with $\Delta\mu_{J1,2}$.

Alternatively, the mean value of the hold skitter $S_{1,2}$ is relatively low (≤ 0.5 ps) since the two clock paths have the same number, size, and distribution of buffers. Nevertheless, $\sigma_{S1,2}$ is nonnegligible for both distributions (A) and (B), as illustrated, respectively, in Figs. 17.21A and B. Similar to $\sigma_{J1,2}$, $\sigma_{S1,2}$ increases with V_{n1} and V_{n2} but $\Delta\sigma_{S1,2}$ is lower than 1.5 ps. These results demonstrate that the standard deviation of the setup and hold skitter increases with the amplitude of the resonant power noise.

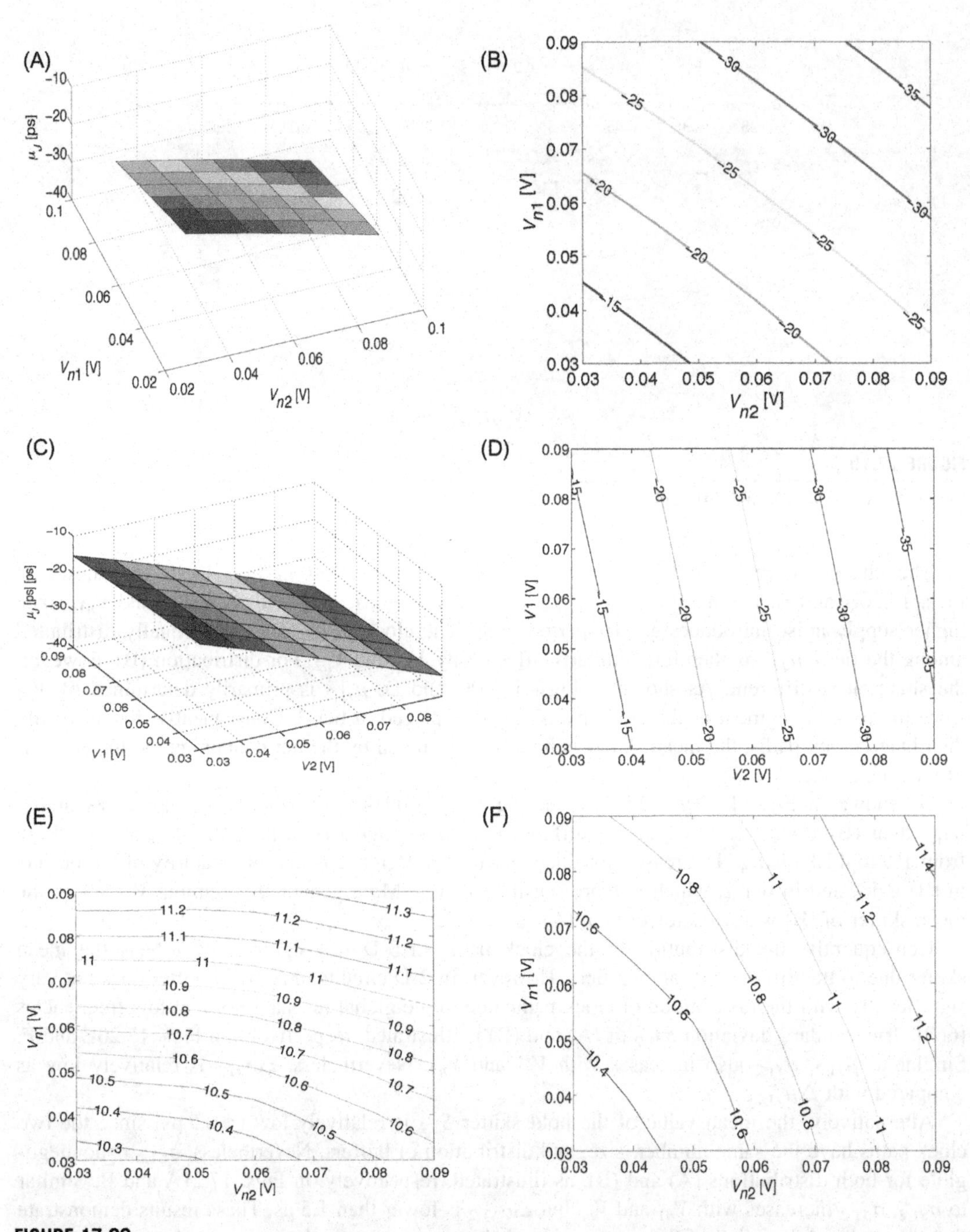

FIGURE 17.20

Setup skitter versus (V_{n2}, V_{n1}). (A) 3-D plot of μ_{JA}, (B) contour of μ_{JA}, (C) 3-D plot of μ_{JB}, (D) contour μ_{JB}, (E) contour of σ_{JA}, and (F) contour of σ_{JB}.

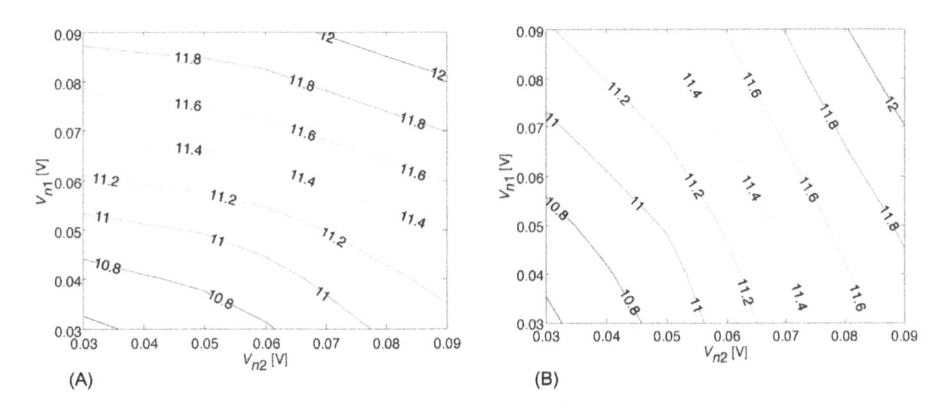

FIGURE 17.21

Hold skitter versus (V_{n1} and V_{n2}). (A) Contours for $\sigma_{SA,}$ and (B) contours for σ_{SB}.

The power consumed by the clock distribution networks constitutes a significant portion of the total power consumed by a complex integrated circuit [645]. Clock skitter also depends on the traits of the buffers, which in turn affect the power of the clock network. This power is evaluated under different constraints on the skitter. A pair of clock paths with a length of 5 mm is evaluated. These paths are both equally distributed across two tiers, where $V_{n1} = 0.09$ volts and $V_{n2} = 0.08$ volts. The skitter and power are determined from Monte Carlo simulations. A different number (14 to 40) and size (W_n) of the clock buffers are inserted along the clock paths.

Considering the Gaussian distribution of the setup skitter $J_{1,2}$ in (17.64), $J_{1,2}$ falls into the range $[\mu_{J1,2} - 3\sigma_{J1,2}, \mu_{J1,2} + 3\sigma_{J1,2}]$ with a probability of 99.7%. Within this range, $\max(J_{1,2})$ indicates the worst (maximum) skitter. For improved readability the absolute value of $\max(J_{1,2})$ is shown, where $\max(J_{1,2}) = |\mu_{J1,2}| + 3\sigma_{J1,2}$. The total power consumption under different constraints on max $(J_{1,2})$ for these clock paths is illustrated in Fig. 17.22. The shaded area depicts inferior buffer solutions. Point A denotes the lowest skitter that can be obtained for this example circuit. In the unshaded area, the skitter decreases as the buffer size and power increase. For the same constraint in skitter, those clock paths with fewer buffers are more power efficient.

As shown within the unshaded area, the clock paths with fewer buffers produce lower skitter. For the clock paths with 14 buffers, as the constraint becomes lower than 68 ps, significant power is required. For example, to decrease the $\max(J_{1,2})$ from 68 to 58 ps (15% improvement), the buffer size is increased from 4 to 10 μm. The resulting power consumption increases from 6.9 to 14.4 mW (a 109% increase). Consequently, pursuing extreme constraints on clock skitter dissipates high power.

17.3.3.2 Skitter versus phase and frequency of the power supply noise

Another parameter that skitter depends upon is the phase difference between the clock signal and the power supply noise. Assuming that the phase difference is the same between the two tiers, meaning that $\phi_1 = \phi_2$, the change in $J_{1,2}$ and $S_{1,2}$ with phase is illustrated in Fig. 17.23 where $V_{n1} = 0.09$ volts and $V_{n2} = 0.07$ volts.

As shown in Figs. 17.23A and B, the difference in ϕ produces a significant change not only in $\mu_{J1,2}$ but also in $\sigma_{J1,2}$. For instance, the highest $\sigma_{J1,2}$ is 41% greater than the lowest $\sigma_{J1,2}$ for

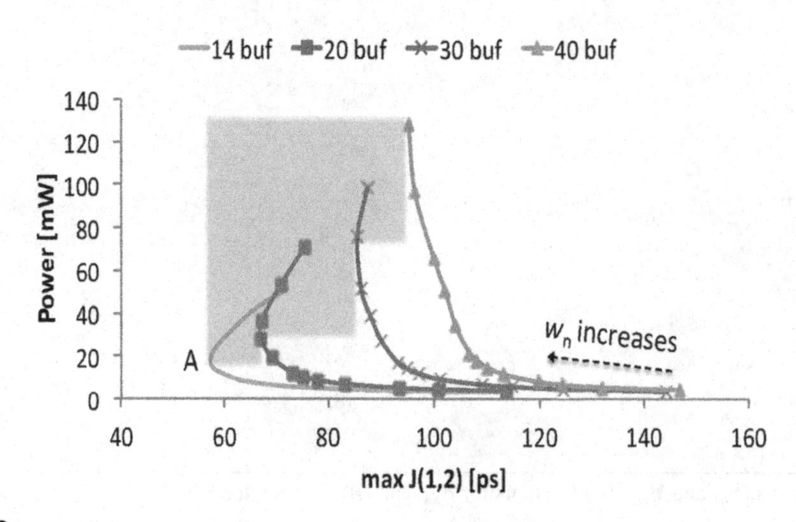

FIGURE 17.22

Tradeoff between power and maximum allowed setup skitter max($J_{1,2}$).

distribution (A) (see Fig. 17.23B). The worst case $\mu_{J1,2}$ occurs when ϕ_1 and ϕ_2 are both approximately 270°, similar to 2-D circuits [649]. The worst $\sigma_{J1,2}$, however, occurs when $\phi \approx 205°$. Therefore, if the initial phase is not 270°, the skitter can be high due to the high $\sigma_{J1,2}$. The difference in $\sigma_{J1,2}$ is low between distributions (A) and (B) since, in either case, the clock path to FF_1 and FF_2 is the same.

The behavior of the hold skitter is different. The effect of ϕ_1 and ϕ_2, on $S_{1,2}$ is shown in Fig. 17.23C. Due to the similarity between the two clock paths, the resulting $S_{1,2}$ is relatively low. The standard deviation, however, is significantly affected by ϕ. As illustrated in Figs. 17.23B and C, the change of $\sigma_{S1,2}$ is similar to $\sigma_{J1,2}$. Consequently, both for setup and hold skitter, σ changes considerably with the phase of the power supply noise. The highest σ and μ of the skitter do not occur at the same initial phase of the supply noise.

Considering the clock paths and waveforms shown in Fig. 17.17, ϕ is the time when the first clock edge arrives at the input of the clock paths. The worst case σ can be obtained by traversing all possible ϕ. Due to the excessive time required for Monte Carlo simulations, this model is highly efficient in determining the worst case skitter and the corresponding ϕ for multi-tier circuits, as compared with Monte Carlo simulations.

As the effect of the noise phase can be significant both for setup and hold skitter, several techniques, such as *RC* filtered buffers and "stacked" phase-shifted buffers [663], have been proposed to shift the ϕ seen by the clock paths. In 3-D clock distribution networks, these techniques can be applied to a portion of the clock paths in a different tier to increase $\Delta\phi$ among the tiers. The change in $\sigma_{J1,2}$ versus the shifted ($\phi_1 = \phi_2$) for distribution (A) is shown in Figs. 17.24A and B. As shown in Fig. 17.24B, the dashed line depicts $\sigma_{J1,2}$ for $\phi_1 = \phi_2$, which denotes the skitter without phase shifting. As shown by the arrow, the highest $\sigma_{J1,2}$ decreases with $\Delta\phi = \phi_2 - \phi_1$. In this case, since ϕ_2 and ϕ_1 are not simultaneously equal to 270°, the worst case $\mu_{J1,2}$ also decreases.

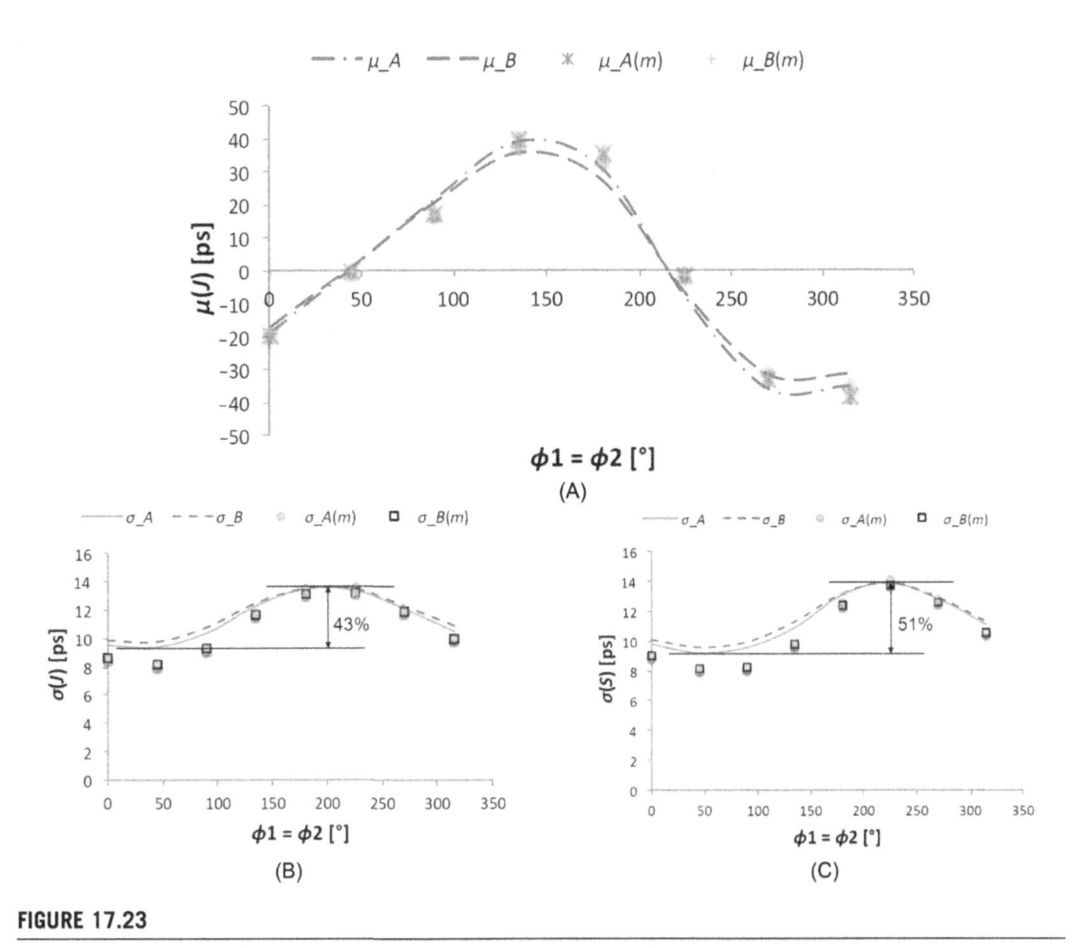

FIGURE 17.23

Skitter versus different ϕ ($\phi_1 = \phi_2$). (A) change in $\mu_{J1,2}$, (B) change in $\sigma_{J1,2}$, and (C) change in $\sigma_{S1,2}$.

In Fig. 17.24C, however, $\sigma_{J1,2}$ of distribution (B) depends strongly on ϕ_2. This behavior occurs since $\sigma_{J1,2}$ is dominated by the supply noise in the second tier. In this case, shifting ϕ among tiers provides less than a 1.5 ps decrease in $\sigma_{J1,2}$, as shown by the dashed line with arrows. Thus, for equally distributed clock paths across 3-D ICs, the worst case skitter can be decreased by shifting ϕ among those tiers with phase-shifted clock distribution networks.

Note that the proper $\Delta\phi$ is determined by traversing all of the combinations of ϕ in different tiers. The number of combinations increases exponentially with the number of tiers, requiring a large number of simulations. This unified model provides a highly efficient way to determine a valid shift in ϕ to decrease skitter in multi-tier circuits.

In addition to phase. the frequency of the power supply noise also affects the skitter. This frequency is usually considered to be the same among tiers, as shown in Figs. 17.14 and 17.15. The frequency f_n is varied to evaluate the change in skitter with the frequency of the supply noise. The

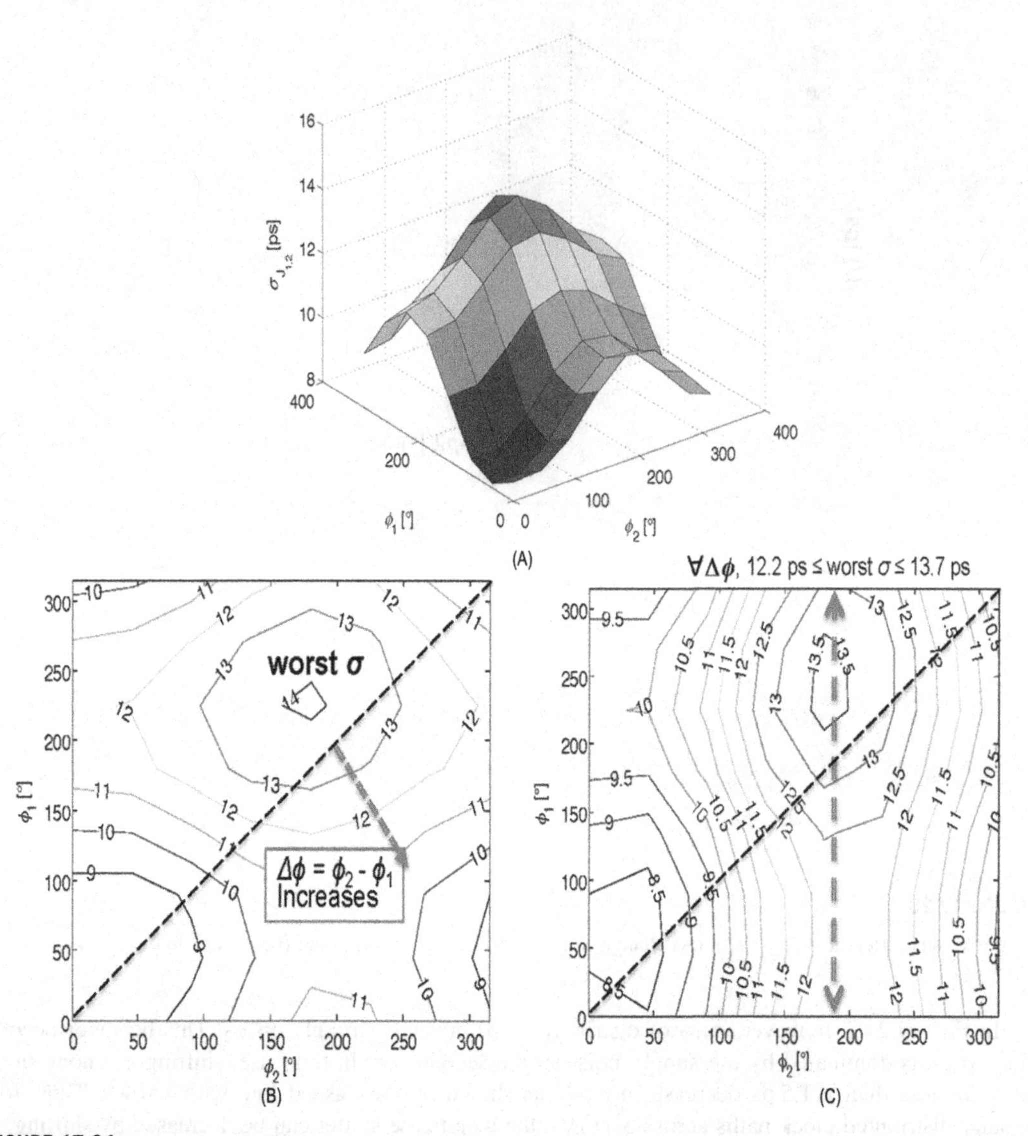

FIGURE 17.24

Skitter $J_{1,2}$ versus shifted ϕ_1 and ϕ_2. (A) 3-D plot of $\sigma_{J1,2}$ versus ($\phi_2 = \phi_1$) for distribution (A), (B) contour map of $\sigma_{J1,2}$ versus ($\phi_2 = \phi_1$) for distribution (A), and (C) contour map of $\sigma_{J1,2}$ for distribution (B).

amplitude V_n and phase ϕ are assumed to be the same among tiers, where $V_{n1} = V_{n2} = 90$ mV and $\phi_1 = \phi_2 = 270°$. The simulation results are illustrated in Fig. 17.25.

Similar to the effects of V_n, f_n greatly affects $\mu_{J1,2}$. For instance, $\mu_{J1,2}$ increases with f_n by up to 70% for distribution (B). The variation of skitter, however, decreases with f_n. The resulting $\Delta\sigma_{J1,2}$

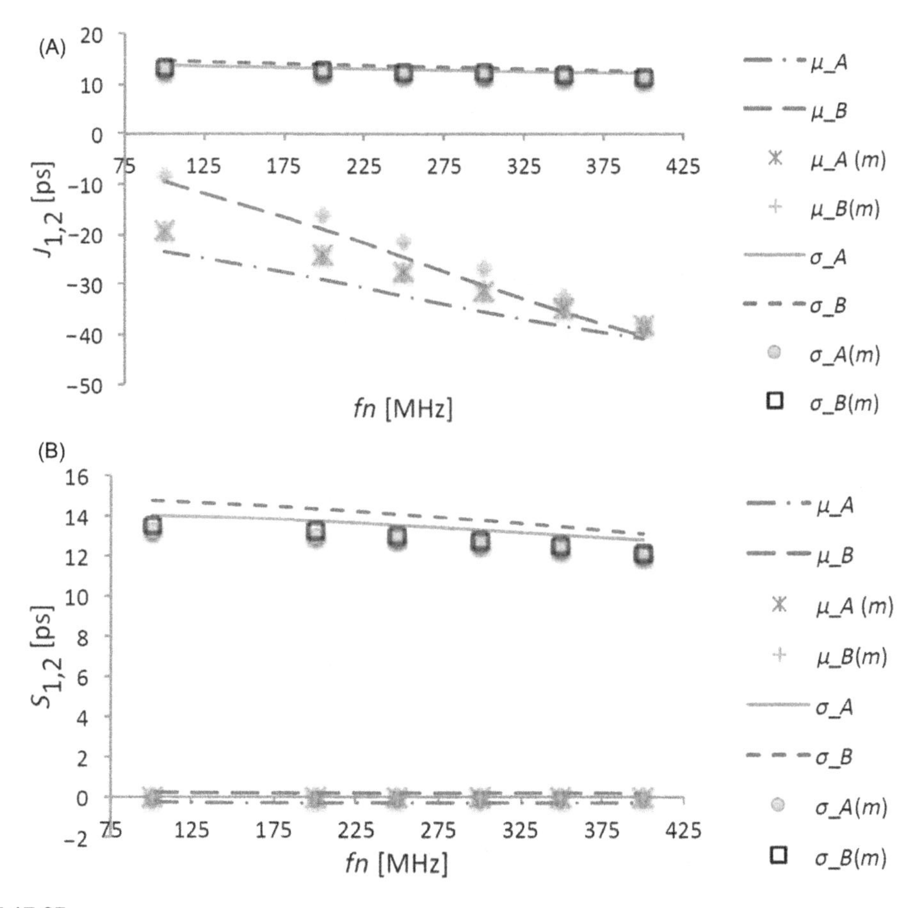

FIGURE 17.25

Skitter versus f_n. (A) Change in $J_{1,2}$, and (B) change in $S_{1,2}$.

and $\Delta\sigma_{S1,2}$ reach 15% for both distributions (A) and (B). This behavior is due to the lower voltage seen by the clock buffers when the clock propagates along the path. The change of μ_d and σ_d for the delay of two serial inverters (a clock buffer) is illustrated in Fig. 17.26A. Both μ_d and σ_d decrease with V_{dd}. As shown in Fig. 17.26A, assume that the clock edge with the worst case σ_J arrives at the input of the clock path at t_0. When f_n increases from f_{n1} to f_{n2}, the propagation time of this edge decreases from t_1 to t_2 and the supply voltage within this duration increases. This higher supply voltage introduces lower σ in the buffer delay, which lowers $\sigma_{J1,2}$ and $\sigma_{S1,2}$, see (17.65) and (17.70). Consequently, the mean setup skitter increases significantly with the frequency of the power noise, while both $\sigma_{J1,2}$ and $\sigma_{S1,2}$ decrease with this frequency.

As shown in Fig. 17.25, the statistical model for skitter is reasonably accurate as compared with SPICE-based simulations for different frequencies of the power noise. For the worst case $\mu_{J1,2}$ ($\sigma_{J1,2}$) shown in Fig. 17.25, the error is, respectively, -11% (-12%), -7% (-10%), -8% (-4%),

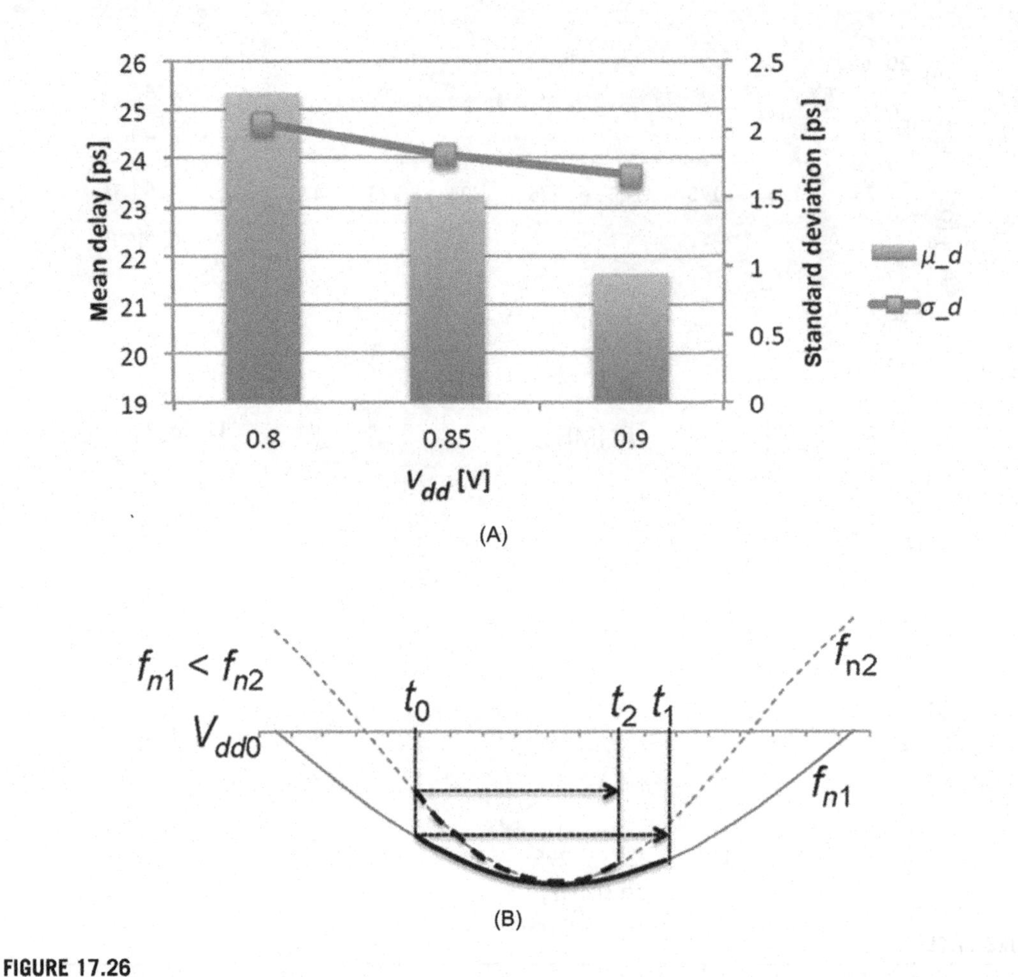

FIGURE 17.26

Change of f_n on delay variations. (A) Mean and standard deviation of buffer delay versus V_{dd}, and (B) supply voltage to the clock path during propagation of a clock edge.

and −10% (−9%). Since $\sigma_{J1,2}$ varies with the power noise, process variations and power noise need to be simultaneously modeled to correctly characterize the clock delay uncertainty. The difference in mean skitter varies by up to 60% due to the different V_n among tiers. $\sigma_{J1,2}$ can vary by up to 51% due to the different ϕ (see Figs. 17.23B and C). Decreasing the variations as well as the mean skitter therefore improves the robustness of 3-D clock distribution networks.

17.3.4 EFFECT OF SKITTER ON SYNTHESIZED CLOCK TREES

A set of guidelines is provided to support the design of robust 3-D clock distribution networks. The objective of these guidelines is to decrease skitter in 3-D circuits.

Guideline 1: Given the freedom to choose which tiers the clock paths should be placed within a 3-D circuit, the mean skitter can be decreased by placing most of the clock path in those tiers that exhibit the lowest power supply noise.
Guideline 2: For 3-D clock paths equally distributed among tiers, the worst case $\mu_{J1,2}$ and $\sigma_{J1,2}$ can be decreased by shifting ϕ among the different tiers.
Guideline 3: By decreasing the frequency of the resonant power supply noise, $\mu_{J1,2}$ can be decreased by trading off $\sigma_{J1,2}$ and $\sigma_{S1,2}$.
Guideline 4: By properly sizing the clock buffers, the tradeoff between skitter and power consumption can be exploited.

To illustrate the utility of these guidelines, several examples of synthesized 3-D clock trees are discussed here. The 3-D circuits are evaluated on IBM benchmark clock circuits [664], generated by randomly distributing the clock sinks to different tiers [656]. The 3-D clock trees are synthesized based on the means-and-medians (MMM)-TB algorithm [593] (see Chapter 15, Synchronization in Three-Dimensional ICs). The buffers are inserted assuming a constraint of 50 fF on the capacitive load. Each clock buffer is formed by an inverter ($W_n = 4.83$ μm $W_p = 2.1$ W_n). An example of the resulting three tier clock trees for the "r1" benchmark circuit (267 sinks) is illustrated in Fig. 17.27A. The clock

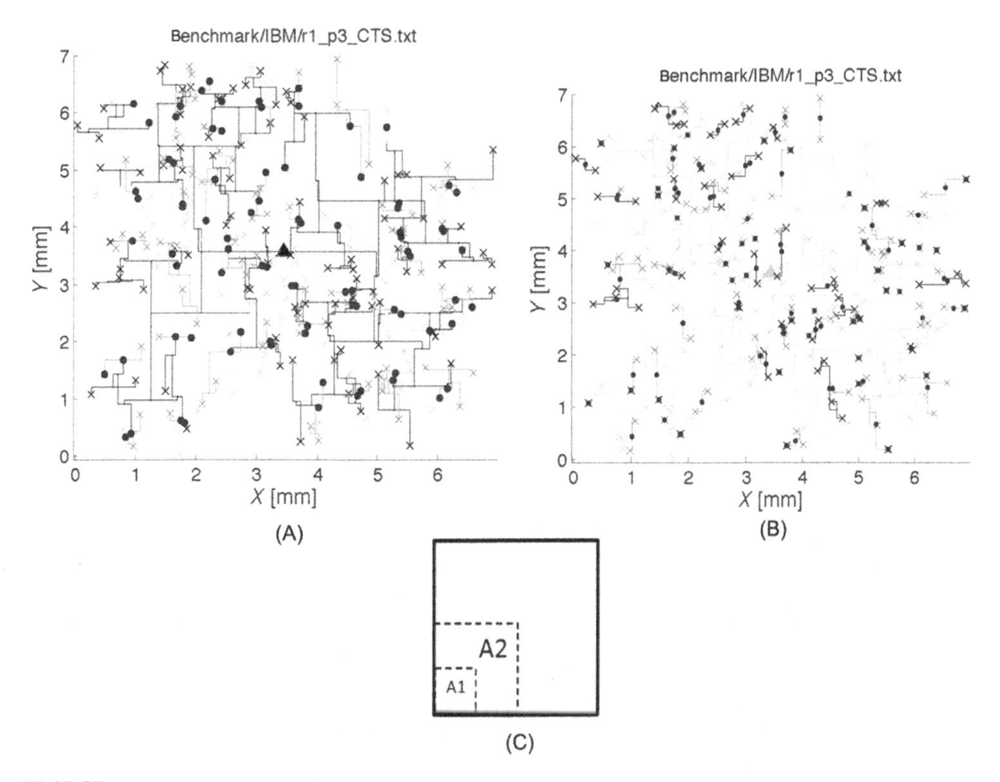

FIGURE 17.27

Synthesized 3-D clock tree. (A) Majority of clock buffers in the first tier, (B) majority of clock buffers in the third tier, and (C) regions where the skitter is measured.

Table 17.6 Simulated 3-D Clock Networks Based on IBM Benchmark Clock Network Circuits

	# of Sinks	# of Buffers	Area (mm^2)	t_S (h)	t_m (s)	Speedup
r3	862	2128	9.8 × 9.6	1.8	45	142 ×
r4	1903	4695	12.7 × 12.7	1.9	53	129 ×
r5	3101	7496	14.5 × 14.3	2.4	56	154 ×

source, clock sinks, and TSVs are denoted, respectively, by the triangle "▲," cross "×," and dot "●" markers. The clock networks in tiers 1, 2, and 3 are denoted, respectively, in blue (black in print), red (dark grey in print), and green (light grey in print).

The skitter is measured within two different regions, as illustrated in Fig. 17.27C. For both regions, A_1 and A_2, the skitter is reported between the pair of the farthest sinks. The three largest IBM benchmark circuits r3, r4, and r5, are evaluated. SPICE simulations are performed on the paths of interest with 2,000 Monte Carlo simulations. The primary features of these benchmarks are listed in Table 17.6, where the computational time is also listed. Note that the simulation time is only for the selected clock paths, not for the entire clock tree. The initial phase and frequency of the power supply noise are assumed to be the same among the three tiers ($f_{n1} = f_{n2} = f_{n3} = 400$ MHz). The amplitude V_n is assumed to differ among the tiers ($V_{n1} = 0.09$ volts, $V_{n2} = 0.08$ volts, and $V_{n3} = 0.065$ volts).

The skitter is reported in Table 17.7. The highest mean skitter occurs when $\phi_1 = \phi_2 = \phi_3 = 270°$, and the highest σ is reported for $\phi_1 = \phi_2 = \phi_3 = 200°$. Four design practices are compared with each other.

Case 1 (C1), the majority of the clock tree is located in tier 1. $\mu_{J1,2}$ is obtained by only considering power noise. $\sigma_{J1,2}$ is determined by only considering process variations.
Case 2 (C2), the majority of the clock tree is also located in tier 1, but the power noise and process variations are simultaneously modeled. $\mu_{J1,2}$ and $\sigma_{J1,2}$ are determined by simultaneously considering both types of variations.
Case 3 (C3), the majority of the clock tree is placed in the middle tier (tier 2).
Case 4 (C4), the majority of the tree is placed in tier 3. The modeling approach in C3 and C4 is the same as in C2.

In C1 and C2, most of the clock buffers are placed in tier 1, adjacent to the heat sink, to constrain the increase in temperature. In C3, the majority of the clock tree is placed in the middle tier to decrease both the number of TSVs and the power consumption, as suggested in [593]. In C4, based on Guideline 1, the majority of the clock tree is located in tier 3 (with the lowest V_n), as illustrated in Fig. 17.27B. As listed in Table 17.7, $\mu_{J1,2}$ in Case 1 is similar to Case 2. Nevertheless, $\sigma_{J1,2}$ and $\sigma_{S1,2}$ are significantly underestimated in Case 1 for both regions A_1 and A_2. As compared to Case 2, the difference in $\sigma_{J1,2}$ and $\sigma_{S1,2}$ reaches 36%. This difference demonstrates the necessity for simultaneously modeling process variations and power noise, since separately modeling process variations and power noise significantly underestimates the variations due to skitter.

The difference between the analytic model and SPICE-based Monte Carlo simulations is listed in the error column of Table 17.7. For all $\sigma_{J1,2}$ and $\sigma_{S1,2}$, the error of the proposed model is below

Table 17.7 Skitter in 3-D Clock Networks Evaluated on IBM Clock Distribution Network Benchmark Circuits

		A1							A2						
Benchmark		**C1**	**C2**	**C3**	**C4**	**Impr1[a]**	**Impr2**	**Error[b]**	**C1**	**C2**	**C3**	**C4**	**Impr1**	**Impr2**	**Error**
Setup μ (ps)	r3	−52.6	−52.1	−44.0	−35.9	31%	18%	−5%	−53.7	−53.1	−44.8	−36.4	31%	19%	−7%
	r4	−66.3	−65.0	−58.6	−48.8	25%	17%	−3%	−69.3	−68.6	−62.1	−52.0	24%	16%	−7%
	r5	−64.8	−62.9	−56.8	−47.6	24%	16%	3%	−67.3	−66.5	−59.9	−50.2	25%	16%	−1%
Setup σ (ps)	r3	8.5	11.2	9.6	10.5	7%	−9%	−10%	11.5	15.2	13.9	13.1	14%	6%	−6%
	r4	10.7	16.6	12.0	11.3	32%	6%	−8%	10.8	15.4	16.0	15.6	−2%	2%	−7%
	r5	8.5	12.9	11.6	12.5	2%	−8%	−9%	11.8	16.0	13.9	18.5	−16%	−33%	−8%
Hold σ (ps)	r3	8.5	11.4	10.1	10.3	10%	−1%	−7%	11.5	15.6	14.4	13.2	15%	9%	−7%
	r4	10.7	14.5	13.6	11.5	21%	16%	−7%	10.8	15.1	15.6	15.6	−3%	0%	−9%
	r5	8.5	11.5	11.1	11.5	0%	−4%	−5%	11.8	15.9	15.6	17.6	−10%	−13%	−6%

[a] *Impr1 and Impr2 are the improvements, respectively, of C4 over C2 and C3.*
[b] *Error is the maximum error of the model as compared with SPICE-based Monte Carlo simulations.*

10%, as compared to Monte Carlo simulations. The error in μ is below 7% for $J_{1,2}$. Considering the greater than 129× speedup in computational time, as reported in Table 17.6, this analytic model provides an efficient way to accurately estimate skitter.

In Case 2, the majority of the clock distribution network is placed in the tier adjacent to the heat sink. In Case 3, the majority of the clock distribution network is placed in the middle tier to reduce the number of TSVs and dissipate less power [593].

The number of TSVs and power consumption of the entire tree for Cases 2 to 4 are illustrated in Fig. 17.28. The results are normalized with respect to Case 4. As proposed in [593], Case 3 produces the fewest TSVs (see #TSV(C2/C4) and #TSV(C3/C4) in Fig. 17.28). The total power is similar among the three cases due to the similar number of clock buffers, as shown by power (C2/C4) and power (C3/C4) in Fig. 17.28. The power per tier, however, differs due to the different distribution of buffers among the tiers.

Case 4 mitigates the mean skitter, lowering the number of TSVs and the power consumed by the tiers. As illustrated in Figs. 17.14 and 17.15, the tier next to the package exhibits the lowest V_n. Consequently, $\mu_{J1,2}$ of Case 4 is significantly improved as compared to Cases 2 and 3, as shown by the first three rows of Impr1 and Impr2 in Table 17.7. This improvement ranges from 16% to 31%. This comparison illustrates the efficiency of Guideline 1 in decreasing the mean skitter. For several paths, however, $\sigma_{J1,2}$ and $\sigma_{S1,2}$ in Case 4 increase as compared to Cases 2 and 3. This situation is due to the change in the topology of the clock trees. For example, for the pair of paths in A2 and circuit r5, the number of buffers after the merging point of these paths increases as compared to Case 2. These buffers are located in different tiers. Consequently, $\sigma_{J1,2}$ and $\sigma_{S1,2}$ both increase.

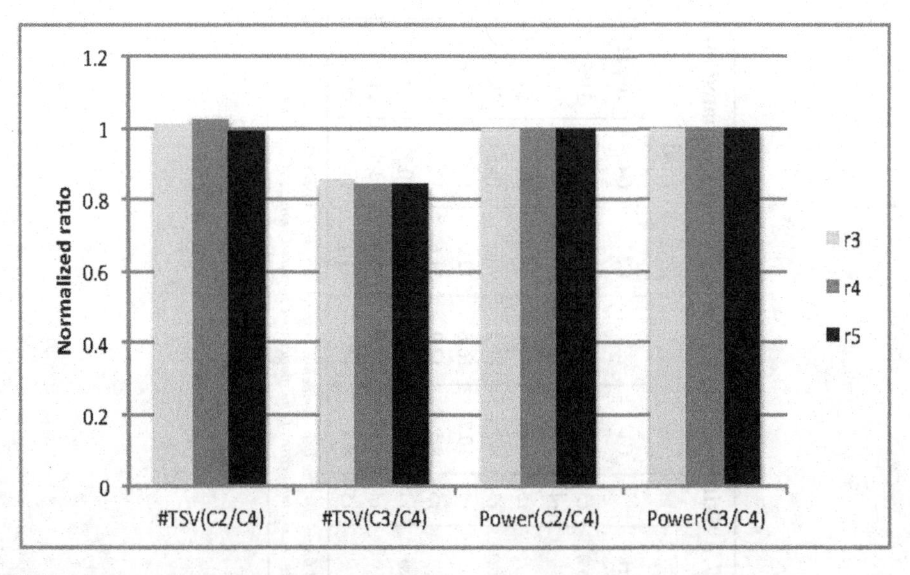

FIGURE 17.28

Normalized number of TSVs and power dissipation for Cases 2 to 4.

17.4 SUMMARY

The effect of variability due to the several D2D variations that stem from the multitier nature of 3-D circuits is discussed in this chapter. The power noise experienced by the different tiers within a 3-D circuit is also considered during the modeling process. Statistical models of the delay of the critical paths, and the skew and jitter of the clock distribution networks in 3-D circuits are described. Based on these models, the overall effects of process and power noise variations in 3-D circuits are evaluated. The primary conclusions of this chapter are:

- Modeling process variations in 3-D circuits requires the inclusion of D2D variations across all of the tiers within a stack.
- An even distribution of the critical paths among the tiers of a 3-D circuit results in the worst case delay variation as compared to an uneven distribution of the critical paths among tiers (assuming that the critical paths do not change from a planar version of the circuit).
- Assuming all of the traits of the intra and intertier paths are the same (e.g., the number of gate stages and the sensitivity of the gate delay to process variations), an intertier path always exhibits a lower likelihood of being the critical path that limits the performance of a 3-D system.
- Multi-via and single via H-trees are potentially useful global clock distribution networks for 3-D circuits.
- In multi-via 3-D H-trees, simply increasing the number of tiers does not necessarily improve the clock skew, as skew changes nonmonotonically with the number of tiers.
- In a 3-D circuit, if the data related sinks are located primarily within the same tier, the single via topology is more efficient in reducing skew variations and can therefore support higher clock frequencies.
- Alternatively, if the data related sinks are widely distributed across several tiers, the multi-via topology is more efficient in reducing skew variations and can support a higher clock frequency.
- A multi-via topology can better exploit the traits of vertical integration (i.e., shorter wires) to significantly increase operating frequencies.
- A hybrid H-tree topology (multi-group topology) combining the advantages of both the single via and multi-via topologies exhibits the lowest variability in clock skew.
- If the data related sinks are located in adjacent tiers of a 3-D circuit, the multi-group 3-D clock tree topology is more efficient in reducing skew variations than both the single and multi-via topologies.
- For 3-D circuits, as more tiers are vertically stacked, the difference in resonant power noise among tiers increases.
- Skitter is the combination of skew and jitter due to the effects of variability on clock distribution networks.
- The setup and hold skitter captures the combined effects of both process variations and power noise.
- For unequally distributed clock paths, the mean skitter is mainly determined by the tier containing the longest portion of the clock paths.
- The standard deviation of the setup and hold skitter increases with the amplitude of the resonant power noise.

- For both setup and hold skitter, σ changes considerably with the phase of the power noise. The highest σ and μ of the skitter do not occur at the same phase of the power noise.
- For equally distributed clock paths across 3-D ICs, the worst case skitter can be decreased by properly shifting ϕ among the tiers with a phase shifted clock distribution network.
- The mean setup skitter increases significantly with the frequency of the power noise, while both $\sigma_{J1,2}$ and $\sigma_{S1,2}$ decrease with this frequency.
- To decrease skitter in 3-D circuits, specific measures can be used, where each method incurs a different circuit overhead. Given the freedom to choose which tiers to place the clock paths in a 3-D circuit, the mean skitter can be decreased by placing most of the clock path on those tiers exhibiting the lowest power noise.
- For 3-D clock paths equally distributed among tiers, the worst case $\mu_{J1,2}$ and $\sigma_{J1,2}$ can be decreased by shifting ϕ among the different tiers. Alternatively, by decreasing the frequency of the resonant power noise, $\mu_{J1,2}$ is traded off for higher $\sigma_{J1,2}$ and $\sigma_{S1,2}$.
- By properly sizing the clock buffers, the tradeoff between skitter and power consumption can be exploited.

CHAPTER

POWER DELIVERY FOR THREE-DIMENSIONAL ICs

18

CHAPTER OUTLINE

Power delivery has traditionally been an important task in the design process of integrated circuits. This process is driven by two primary objectives. Abundant current should be delivered to all of the devices across a circuit, and the impedance of the power distribution network should be lower than a target value over the frequency spectrum of interest. This frequency range typically extends well above the highest operating frequency of a circuit, as the switching frequency of a device is much higher than the operating frequency of the circuit.

Maintaining the impedance of the power distribution network below a target impedance guarantees that the voltage at the terminals of the devices is sufficiently close to the nominal power supply (V_{dd}). The same requirement applies to the ground terminals (V_{gnd}), as any deviation from the nominal voltage can adversely affect the performance of a circuit.

To facilitate the delivery at a voltage (i.e., as close as possible to the nominal V_{dd} and V_{gnd}), power distribution systems have historically been hierarchically designed. This approach is also due to the vastly different physical scales and the cost of the disparate components within power distribution systems. A typical power distribution system is illustrated in Fig. 18.1, where a large number of components with different electrical characteristics are included.

A voltage regulation module (VRM) provides the proper voltage levels and supplies the current for the integrated circuit. This current flows through the metal traces and vertical plated vias of the printed circuit board (PCB) to the solder balls of the package, and through the interconnect layers of the package and solder bumps (e.g., C4) to the V_{dd} and V_{gnd} pads of the on-chip power distribution network. Several variations of this path exist as package technologies have evolved over time.

Three-Dimensional Integrated Circuit Design. DOI: http://dx.doi.org/10.1016/B978-0-12-410501-0.00018-6

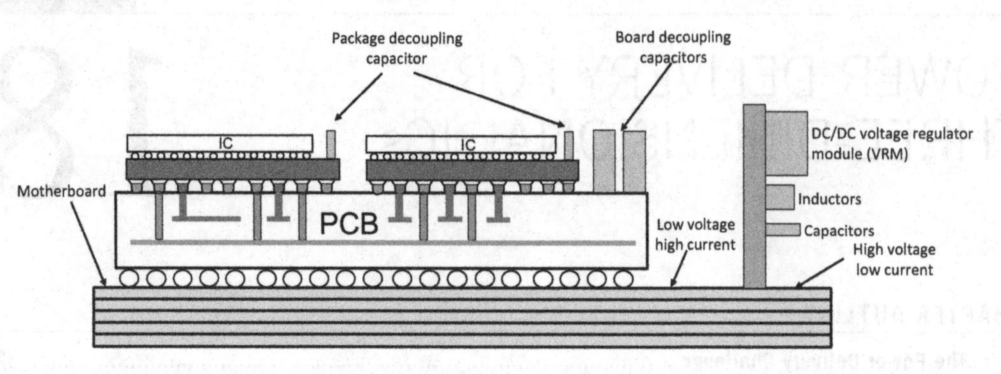

FIGURE 18.1

Cross-sectional view of power distribution system where several levels of the hierarchy, motherboard, PCB, package, and integrated circuit are shown. The VRM and the decoupling capacitors placed at all levels of the hierarchy are also illustrated.

The general hierarchy however is the same for any power distribution system. This system also includes several passive elements, primarily decoupling capacitors added to the board, package, and circuit to reduce the length of the current paths and therefore the losses on the voltage and ground rails.

Losses stem from the resistance of the wires at all levels of the power distribution system, resulting in resistive voltage drop and ground noise, thereby reducing the rail-to-rail voltage. This noise is expressed as IR and is usually referred as IR drop or resistive noise. In addition to this DC power supply noise, the inductance of the wires also contributes to losses whenever circuits switch. This dynamic component of the power supply noise is described by $L\ dI/dt$ or inductive noise, where dI/dt is the rate of the switching current of the circuit and L is the interconnect inductance. These losses degrade the speed and noise margin [665].

Different measures and practices are used to reduce these losses across the different levels of the hierarchy of a power distribution system. In 3-D circuits, these measures need to be revised for the on-chip power distribution network, as only the on-chip portion of the overall power distribution system is affected by the third dimension. Although the use of interposer technologies, which are broadly considered as another form of 3-D integration, can affect the power distribution system in a complex way, the focus of this chapter is power distribution for vertically stacked circuits rather than interposer technologies. Consequently, the on-chip power delivery problem for 3-D circuits is discussed in Section 18.1. As the analysis techniques for power distribution networks are applicable to either two-dimensional (2-D) or 3-D circuits, analysis techniques for power grids are not discussed in-depth in this chapter. Rather, models of varying complexity for 3-D power networks are reviewed in Section 18.2, as the through silicon vias (TSV) require different models as compared to the horizontal wires of the power network. The intricacies of 3-D power distribution networks are reviewed in Section 18.3, where the important role of TSVs on the behavior of the on-chip power distribution is discussed. First order tradeoffs related to the usage of TSVs are described in this section. Issues related to the insertion of decoupling capacitance in

3-D circuits are considered in Section 18.4. An optimization process for 3-D power distribution networks is described in Section 18.5. The primary concepts of the chapter are summarized in the last section.

18.1 THE POWER DELIVERY CHALLENGE

Vertically integrating multiple circuit tiers affects a power network in two ways. First, the horizontal dimensions of the power distribution network are significantly reduced as compared to the footprint of a 2-D circuit. The power distribution networks within the different tiers are connected with TSVs. These vertical interconnections increase the resistance seen by the active circuits of the stack located farther from the package. This situation is aggravated as the number of tiers increases. At the circuit package boundary, as the footprint of the stack decreases, a smaller number of power/ground (P/G) C4 connections are available to supply current for the entire stack. Consequently, a large current flows through these connections and the TSVs, producing significant resistive losses across the circuit stack.

As 3-D systems support the integration of heterogeneous technologies and functionalities within a circuit stack, an efficient scheme to deliver power across a 3-D circuit employs a single tier dedicated to power delivery [666]. This approach is similar to the strategy discussed in Chapter 15, Synchronization in Three-Dimensional ICs, where an entire tier within a 3-D circuit is utilized for deploying the synchronization circuitry. Rather than allowing a large current to flow from the VRM at a low voltage compatible with the on-chip power supply, as shown in Fig. 18.1, conversion to the voltage level of the integrated circuits takes place in one of the tiers within the 3-D stack. A lower current flowing through the power distribution system reduces the resistive losses across the board and package.

There are several advantages for this approach, where the most important advantage is the decrease in interconnect parasitic impedances between the source of power (i.e., VRM) and the logic tier. Additionally, the multiple voltage levels required in modern ICs are distributed on-chip with lower losses. This on-chip converter enables local dynamic scaling of voltages. Since all of the voltages are produced on-chip, the number of P/G pins can also be reduced.

Integration of a buck converter is demonstrated in a 180 nm SiGe bipolar CMOS (BiCMOS) process [666] where the operating frequency of the converter is 200 MHz and the control bandwidth is about 10 MHz. This converter is integrated into the tier adjacent to the logic tier (e.g., a processor), as schematically shown in Fig. 18.2. The passives required for the buck converter are placed in the third tier and can therefore be separately optimized to minimize the parasitic losses of the converter.

A first order circuit diagram of the tested converter is shown in Fig. 18.3, where the gate drivers, control switch, synchronous rectifier, output *LC* filter, and active load are illustrated. Two converters drive the logic tier. An input of 1.8 volts is converted to 0.9 volts to drive the load. An operating frequency of 200 MHz ensures that the area of the *LC* filter is not excessive. Within the power stage the PMOS control switch has an equivalent width of 16.6 mm, with an on-resistance of 152 mΩ. The NMOS synchronous rectifier has an equivalent width of 11 mm with an on-resistance of 62 mΩ. The tapered buffers driving both the control switch and synchronous rectifier have a tapering ratio of nine. The output capacitor of the *LC* filter uses a metal oxide

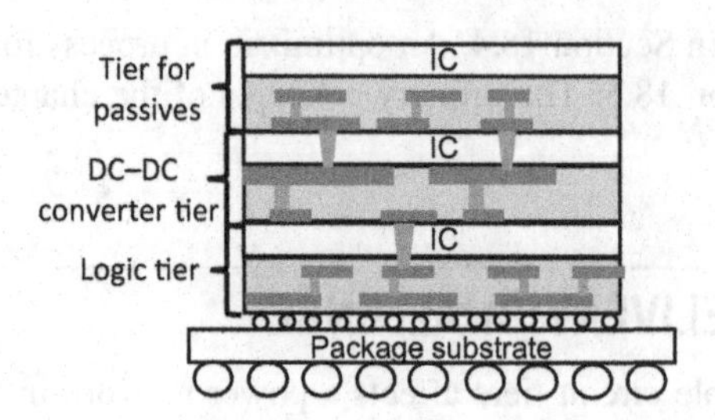

FIGURE 18.2

A three tier circuit where DC–DC conversion is integrated in the upper tiers to reduce losses within the power delivery system [666].

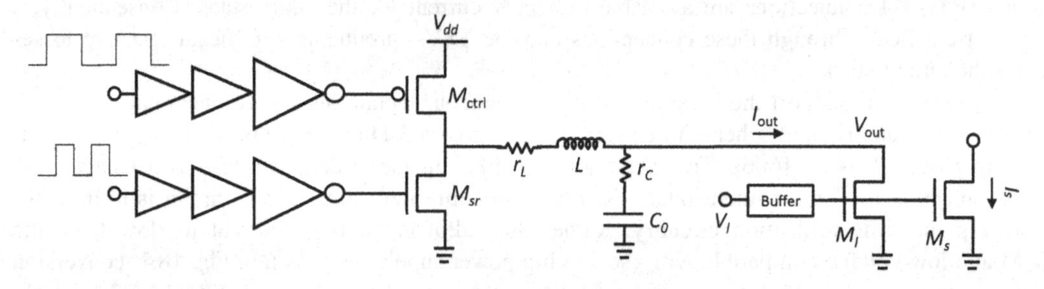

FIGURE 18.3

Buck converter integrated within a separate tier and connected to the logic tier with TSVs [666].

semiconductor (MOS) capacitance to limit the area as compared to metal–insulator–metal (MIM) capacitors. An 8.22 nF capacitance is utilized. The effective series resistance of the MOS capacitor is 1 mΩ.

For the inductor of the converter, two shunted metal layers are used. The width of the windings of the inductor is 25 μm to reduce the resistance of the inductor while maintaining a high quality factor. The DC resistance (no skin and other high frequency effects are considered) of the inductor is 201 mΩ. 3.5 windings are spaced at a distance of 5 μm. The resulting diameter of the inductor is 290 μm, yielding an inductance of 2.14 nH. Furthermore, the quality factor of the inductor is enhanced by placing a patterned metal as a ground plane to reduce eddy current losses within the substrate. The primary design parameters for this converter are listed in Table 18.1. The prototype circuit exhibits an efficiency of 64% while operating at 200 MHz, providing an output current of 500 mA.

Integration of DC–DC converters within a 3-D stack can also be performed in a more systematic approach, where these converters are placed within more than one tier of the stack. As these components, however, occupy significant area, these circuits should be appropriately modeled to better evaluate any benefits of a multitier approach. The primary benefit of on-chip buck converters is to reduce the resistance of the power distribution paths, where the contribution of the TSV resistance is not negligible. To quantify any gains from the integrated buck converters, the architecture

Table 18.1 Design Parameters of the Components of the Power Stage [666]

Control switch	Width = 16.6 mm	$R_{DS(on)} = 152\ m\Omega$
Synchronous rectifier	Width = 11 mm	$R_{DS(on)} = 62\ m\Omega$
Inductor	$L = 2.14$ nH	$R_{DC} = 201\ m\Omega$
Capacitor	$C = 8.22$ nF	ESR = 1 mΩ

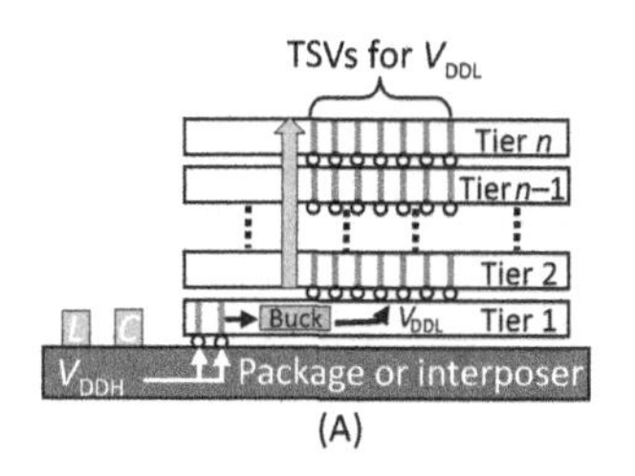

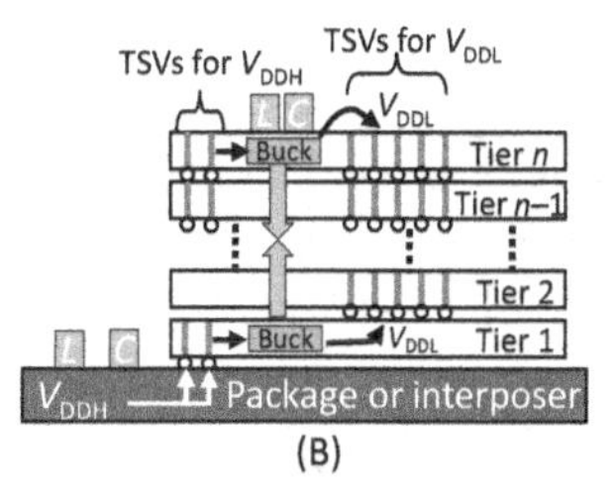

FIGURE 18.4

3-D power delivery system. (A) DC–DC buck converters are integrated within only one tier, and (B) DC–DC converters are integrated in the tiers at both ends of the stack. Two different types of TSVs are noted, those TSVs that distribute a high (off-chip) voltage (V_{DDH}) to the converters and those TSVs which distribute a low (on-chip) voltage (V_{DDL}) downstream from the output of the converters.

of a 3-D stack shown in Fig. 18.4 is assumed where the buck converters are integrated within the tiers at both ends of the 3-D system. The use of a buck converter at the uppermost tier in this example is justified by the long distance from the tier to the package.

To address the potentially high resistive voltage drop, the buck converters can be configured as follows. The buck converter within the uppermost tier supplies the upper half of the stack, while the buck converter within the bottommost tier supplies the lower half. Thus, if the 3-D circuit comprises n tiers, tiers 1 to $n/2$ are supplied by the converter in tier 1 and tiers $n/2 + 1$ to n are supplied by the converter in tier n. Consequently, the longest vertical path within the n tier stack is reduced from $n - 1$ TSVs to $n/2 - 1$ TSVs; the path is essentially halved.

To ascertain the merits of this approach, a closed-form model is provided, assuming that each tier draws the same current and the density of the P/G TSVs is the same for all of the tiers. Although these assumptions are unlikely to be precise, this assumption allows a first order model to offer insight into the efficiency of this scheme [667]. An equivalent circuit for a 3-D stack that includes only one converter in one of the tiers, as in [666], is illustrated in Fig. 18.5. The total *IR drop* across n tiers is [667]

$$V_{drop}(n, k_{TSV}) = (n-1)I\frac{r_{TSV}}{\frac{k_{TSV}}{2}} + (n-2)I\frac{r_{TSV}}{\frac{k_{TSV}}{2}} + \cdots + I\frac{r_{TSV}}{\frac{k_{TSV}}{2}} + (n-1)I\frac{r_{TSV}}{\frac{k_{TSV}}{2}}$$

$$+ (n-2)I\frac{r_{TSV}}{\frac{k_{TSV}}{2}} + \cdots + I\frac{r_{TSV}}{\frac{k_{TSV}}{2}} = \frac{2Ir_{TSV}}{k_{TSV}}n(n-1), \tag{18.1}$$

with the help of the arithmetic progression formula. In this expression, k_{TSV} is the total (i.e., both power and ground) number of TSVs in each tier, and I is the current drawn by each tier. If two

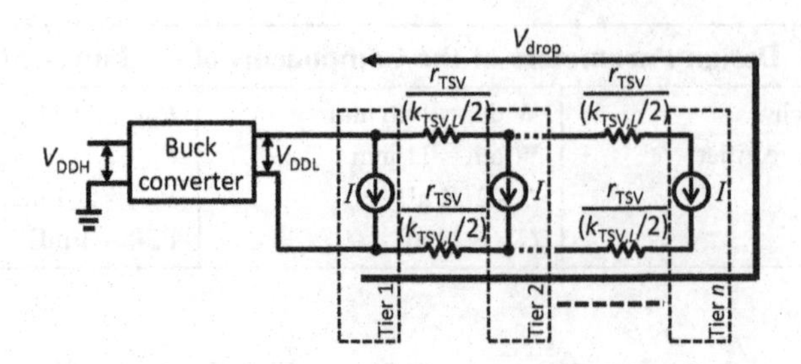

FIGURE 18.5

Equivalent circuit of the on-chip power distribution network of an n tier 3-D circuit, where the total *IR drop* across the tiers is denoted as V_{drop}. Only one buck converter is integrated in one tier and the on-chip power distribution network is modeled as a 1-D network [667].

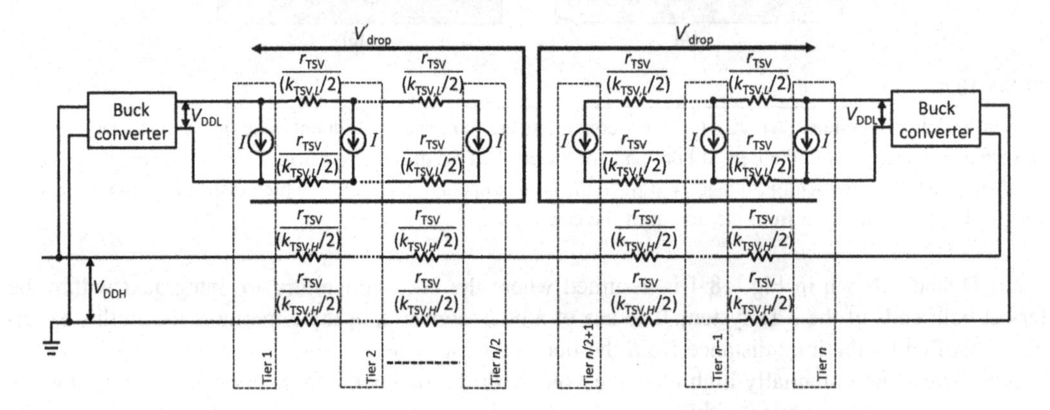

FIGURE 18.6

Equivalent circuit of the on-chip power distribution network of an n tier 3-D circuit, where the total *IR drop* across the tiers is denoted as V'_{drop}. Two buck converters are integrated within the tiers at both ends of the circuit, each supplying current to half of the tiers of the stack [667].

buck converters are employed, the equivalent 1-D model for the power distribution system is illustrated in Fig. 18.6, where each converter is responsible for providing current to only half of the tiers. Consequently, the *IR drop* decreases to $V'_{drop}(n/2, k_{TSV}) = (Ir_{TSV}/2k_{TSV})n(n-2)$. This reduction is the ratio of the voltage drop in the single converter approach [667]

$$\frac{V'_{drop}\left(\frac{n}{2}, k_{TSV}\right)}{V_{drop}(n, k_{TSV})} = \frac{1}{4} \cdot \frac{n-2}{n-1} < \frac{1}{4}. \tag{18.2}$$

This reduction, however, incurs some power and area overhead due to the TSVs connecting V_{DDH} to the uppermost tier of the 3-D circuits. To determine the power overhead of these TSVs, assume the power consumed by the resistance of these TSVs is [667],

$$P_{TSV,H} = 2\frac{r_{TSV}}{\frac{k_{TSV,H}}{2}}(n-1)I_H^2 = \frac{4r_{TSV}}{k_{TSV,H}}(n-1)I_H^2, \tag{18.3}$$

where $k_{TSV,H}$ is the number of P/G TSVs for each tier. Note that only the voltage drop on the power TSVs is computed. The number of power TSVs is $k_{TSV,H}/2$. Furthermore, as the current drawn by each tier is I, the power consumed by the load for the single buck converter approach is $P_{load} = nIV_{DDL}$. If two buck converters are used, the efficiency of the converter located within the tier farthest from the package pins, connected to $V_{DD,H}$ through the TSVs, is

$$\eta = \frac{P_{out}}{P_{in}} = \frac{P_{load}/2}{V_{DDH} I_H}. \tag{18.4}$$

Combining (18.3) and (18.4), the ratio of the power overhead of the additional TSVs for the buck converter over the total power is

$$\frac{P_{TSV,H}}{P_{load}} = \frac{r_{TSV} P_{load}(n-1)}{\eta^2 V_{DDH}^2} \left(\frac{k_{TSV,L}}{k_{TSV,H}}\right) \frac{1}{k_{TSV,L}}. \tag{18.5}$$

Similarly, the number of TSVs in the downstream power distribution network is obtained by combining $V_{drop}(n/2, k_{TSV,L})$ and P_{load} into [667],

$$\frac{1}{k_{TSV,L}} = \frac{2V_{DDL}^2}{r_{TSV} P_{load}(n-2)} \left(\frac{V_{DDL}}{V_{DDH}}\right). \tag{18.6}$$

The combination of (18.5) and (18.6) provides an analytic description of the TSV area and power overhead for different target efficiencies, which are

$$\frac{k_{TSV,H}}{k_{TSV,L}} \frac{P_{TSV,H}}{P_{load}} = \frac{2}{\eta^2} \frac{n-1}{n-2} \left(\frac{V'_{drop}}{V_{DDL}}\right) \left(\frac{V_{DDL}}{V_{DDH}}\right)^2 = \frac{1}{2\eta^2} \frac{n-2}{n-1} \left(\frac{V_{drop}}{V_{DDL}}\right) \left(\frac{V_{DDL}}{V_{DDH}}\right)^2. \tag{18.7}$$

For a large number of tiers the ratios, $(n-1)/(n-2)$ and $(n-2)/(n-1)$, can be considered equal to one. From (18.7), several tradeoffs can be explored. For example, assuming $V_{DDH} = 3.3$ V and $V_{DDL} = 1.2$ V, if an efficiency of $\eta = 0.8$ is desired, the area of the TSVs increases by 35%. These additional TSVs add a power overhead of about 3%. The *IR drop* simultaneously decreases from 10% to 2.5%, a factor of four improvement.

Using two buck converters in a large vertical stack has been evaluated through a daisy chain connection of TSVs in a single tier, where this chain connection represents a vertical path of P/G TSVs [667]. The diameter of the TSVs is 20 μm with a pitch of 50 μm. The typical resistance of a single TSV is 29 mΩ. A buck converter drives the TSV chain where several active loads are interspersed among the TSVs, mimicking the load for each tier of a 3-D stack. This structure is shown in Fig. 18.7, where the resistance of the TSV shown in the diagram corresponds to a signal path crossing eight tiers. Each of the active loads is 200 mA and, therefore, the output current of the converter is 1,600 mA.

The inductor of the buck converter exhibits a resistance of 40 mΩ and inductance of 3.9 nH. The dimensions of the inductor are 1.6 mm × 0.8 mm × 0.8 mm and the inductor is placed off-chip. The capacitor is also off-chip, exhibiting a capacitance of 1 μF with dimensions 1.0 mm × 0.5 mm × 0.5 mm. These sizes are chosen based on [668]. The fabricated converter has an efficiency of 72% and bandwidth of 62 MHz. Measurements of a prototype test circuit with the eight daisy chained TSVs demonstrate that two buck converters each supplying half of the tiers reduces the *IR* drop by 78% as compared to employing a single converter supplying all of the active loads through the entire TSV stack, consistent with the analytic model [667].

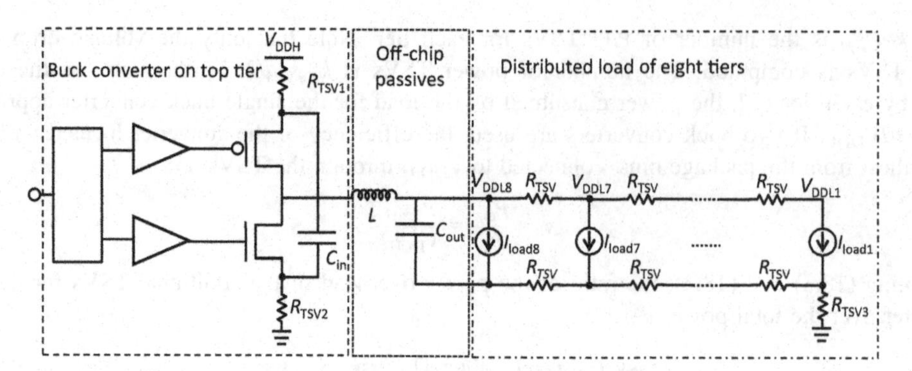

FIGURE 18.7

A converter providing current within a prototype 2-D circuit used to emulate a 3-D system comprising eight tiers, where the TSVs and active loads are connected in a daisy chain [667].

18.1.1 MULTILEVEL POWER DELIVERY FOR THREE-DIMENSIONAL ICs

The primary advantage of integrating DC–DC converters within several tiers of a 3-D circuit is lower current flow through the power distribution system, incurring lower losses. 3-D systems lend themselves to another interesting power distribution strategy. Differential rails are used between adjacent tiers [669]. Consequently, the power network of a specific tier is connected to those voltage levels that ensure a voltage difference V_{dd} appears between the power and ground lines. Assuming, for example, m power supply levels where each level is a multiple of V_{dd}, the power network of the mth and $m-1$ levels is connected, respectively, to mV_{dd} and $(m-1)\,V_{dd}$. Only one network has a ground rail connected to zero voltage. An illustration of this "multi-level" V_{dd} scheme is depicted in Fig. 18.8 for a three tier system, where in this topology each tier is connected to a pair of voltage levels. The main advantage of the multilevel approach is that charge is recycled among differential pairs of power supply rails. In this way, power supply noise is considerably reduced [669]. This charge recycle is ideal when each tier draws the same current. Although this requirement may be feasible for homogeneous 3-D circuits, for example, memory stacks, this demand cannot be satisfied in a straightforward manner for other more heterogeneous systems and requires additional measures, as discussed in this subsection.

To assess the benefits of a multilevel power delivery system employed in 3-D circuits, consider the example illustrated in Fig. 18.9 [669]. In Fig. 18.9A, the sum of the resistance of the vertical P/G paths, with $2k_{TSV}$ TSVs for a 3-D circuit composed of n tiers, is substituted by a resistance within a single tier. The load of all of the tiers is modeled as a single current source with amplitude I. The total resistance of each vertical path (i.e., power or ground) is notated as r, and is inversely proportional to the number of TSVs used for power and ground distribution, $r \propto 1/k_{TSV}$. The total number of TSVs is $2k_{TSV}$ and is assumed to be split equally among the tiers, unless mentioned otherwise. For the circuit shown in Fig. 18.9A, the resistive voltage drop and power dissipated by the vertical path of the power distribution network are, respectively, $2Ir$ and I^2r. If multilevel power delivery is applied where only one pair of voltage levels is assigned to a tier, each tier draws I/n current and the resistance of each TSV path is $0.5r(n+1)$ and consists of $2k_{TSV}/(n+1)$ TSVs, as illustrated in Fig. 18.9B.

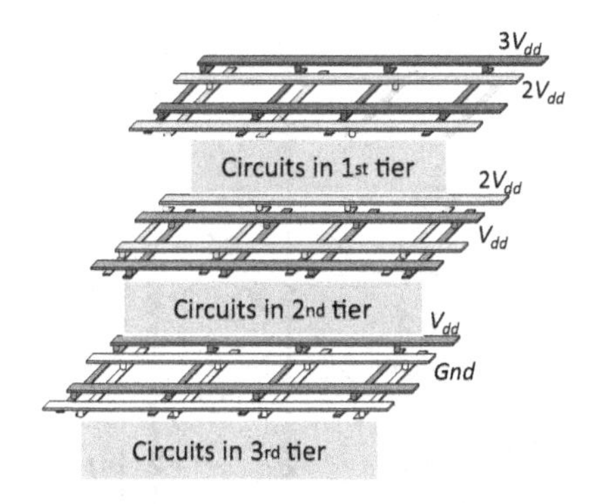

FIGURE 18.8

A multi-level power distribution network applied to a three tier circuit where each pair of power levels is mapped to a single tier.

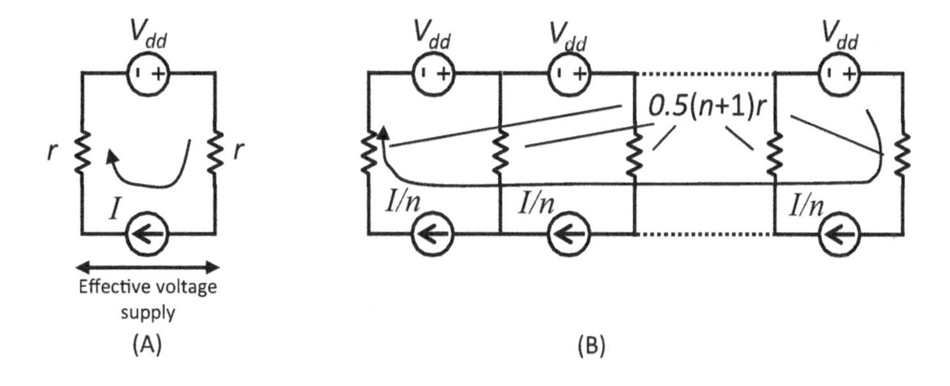

FIGURE 18.9

Equivalent circuit diagram of a power distribution network of a 3-D circuit, (A) supplied by a single V_{dd}, and (B) supplied by several pairs of V_{dd} supplies [669].

If all of the tiers are active, the current only flows through the paths at the edges of the circuit (see Fig. 18.9B), thereby incurring the smallest voltage drop. The worst case situation in terms of *IR drop* occurs when the current is not balanced among the tiers, which only happens if one tier is active. In this case, the resistive supply noise is $Ir(n+1)/n$, which is lower than a single V_{dd} for $n > 1$. Furthermore, the power consumed by the stack is maximum whenever successive tiers operate alternately, which in the case of two pairs of voltages yields the greater reduction in noise as compared to a single V_{dd}. Increasing the number of power levels returns diminishing gains.

Based on this discussion, different mappings of a multi-level power delivery scheme can be applied to a 3-D circuit. An example of a three tier system is considered where two tiers are memory tiers and another tier hosts a processor, as depicted in Fig. 18.10. The memory tiers are assumed to sink current I, while the processor tier draws twice as much current. The technology

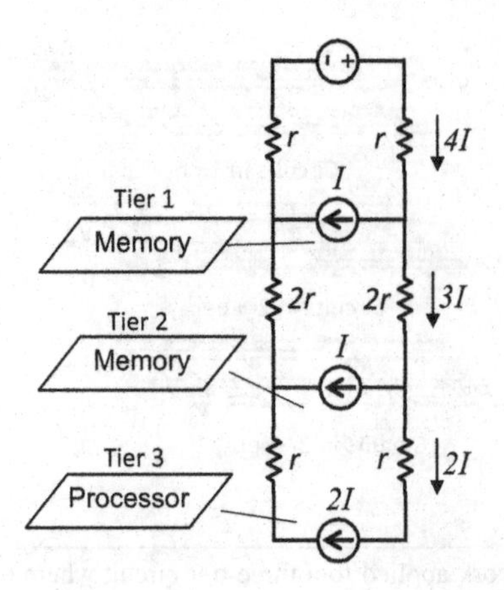

FIGURE 18.10

A 3-D circuit consisting of two memory tiers and one processor tier [669].

assumed for this system is the MIT Lincoln Laboratories (MITLL) 3-D technology [670]. In this technology, a TSV spans each of the three tiers and is comprised of two stacked single TSVs, where the extracted resistance of a single TSV is 1 Ω. The resistance of a stacked TSV, in this example, is split into three resistive segments with values of 0.5, 1, and 0.5 Ω, which are the vertical resistance of the corresponding tier.

Assuming that this circuit is manufactured with the MITLL technology and these TSV resistances, if the resistance of the vertical path in the first tier is denoted by r and since the total number of TSVs is equally distributed among the tiers, the resistance of the vertical path across the tiers is r, $2r$, and r for, respectively, the first, second, and third tier. Based on these assumptions the power supply noise in the tier farthest from the power supply tier (see Fig. 18.10) is $2(r4I + 2r3I + r2I) = 24rI$. To reduce this power supply noise, two pairs of voltage levels (i.e., $2V_{dd}$ to V_{dd} and V_{dd} to V_{gnd}) are used within each tier, as depicted in Fig. 18.11A. This two level power delivery system yields a maximum *IR* drop of $18Ir$. If some part of the circuit is power gated, as illustrated in Figs. 18.11B and C, the leakage current, assumed to be aI, of the inactive circuits opposes the current flow in the active part of the system, reducing the resulting power supply noise [669]. In both Figs. 18.11B and C, the power supply noise becomes $18Ir - 9aIr$. For large leakage currents, $a = 0.5$. The decrease in noise as compared to the fully active system of Fig. 18.11A is 44%.

The use of multiple power supply pairs within a tier, however, incurs considerable overhead, requiring, for example, many level shifters. A more straightforward power delivery topology is consequently shown in Fig. 18.12, where the two memory tiers are connected, respectively, to $2V_{dd}$ and V_{dd}, and the processor tier is connected to V_{dd} and V_{gnd}. As this tier consumes equal current in both of the memory tiers, the resulting power delivery topology is balanced, leading to a maximum decrease in *IR drop*. Similar to the previously mentioned topologies, the worst *IR drop* appears when current imbalances exist between the two power networks. Two cases are illustrated, respectively, in Figs. 18.12B and C. The power supply noise when the memory tiers are inactive is $V_{\text{drop1}} = 12Ir - 12aIr$. If the processor tier is power gated, the power supply noise is

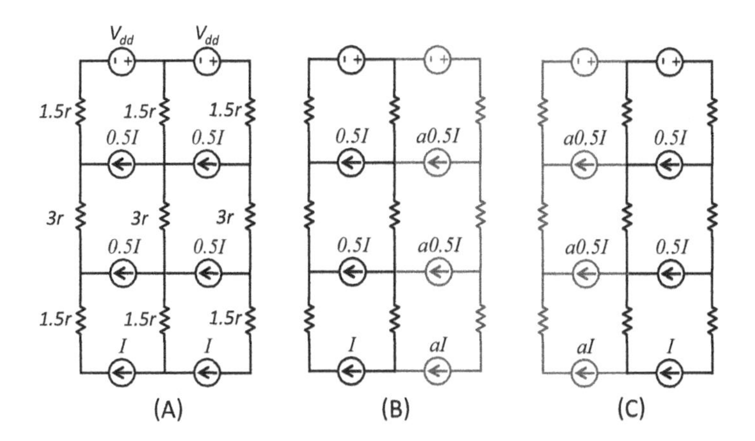

FIGURE 18.11

Multi-level power delivery system where two pairs of voltage levels are employed in each tier. (A) All of the circuits are active, (B) right half of the circuit in each tier is inactive (shown in gray), and (C) left half of the circuit in each tier is inactive (shown in gray) [669].

$V_{drop2} = 24Ir - 6aIr$. Consequently, the highest power supply noise is $V_{drop} = \max[(24Ir - 6aIr), (12Ir - 12aIr)]$.

In all of these different topologies for the multi-level power delivery system, the number of TSVs per tier is considered the same, which limits the benefits obtained from this power delivery approach. If the number of TSVs can be adjusted on a per tier basis, the problem of determining the power supply noise becomes an optimization problem, which can be roughly described as follows. Consider a three tier system similar to those discussed thus far. For each tier i, determine the number of TSVs k_{TSVi} ensuring that the total number of TSVs is equal to k_{TSV} and the power supply noise is minimum which is described by $\max(V_{drop1}, V_{drop2})$. For different leakage currents, different TSV distributions reduce the power supply noise to between 22% and 34%. The resulting TSV distribution among the tiers is typically greatest for the two highly imbalanced cases (see Figs. 18.12B and C). Alternatively, the TSVs can be optimally distributed across tiers to reduce the resistive power consumed by the power distribution network.

Although this first order analysis illustrates the potential of a multi-level power delivery system, several difficulties exist that may limit application of this approach. The predominant issue emerges from the requirement to maintain balanced currents among the tiers (assuming each tier employs a single pair of voltage levels). Furthermore, intertier signals may require voltage level conversion which adds nonnegligible latency and area overhead. Since intertier signals typically belong to the critical paths, this overhead can greatly degrade the performance of the system. Additionally, applying commonplace techniques for power savings, such as dynamic voltage scaling, is not straightforward.

One way to overcome this issue of imbalanced currents is to add converters to produce a voltage difference of V_{dd} between the power and ground rails in each tier. The power efficiency of a system that employs these converters is proportional to the imbalance in current across the system, since the larger the current imbalance, the more charge the converter needs to provide, thereby

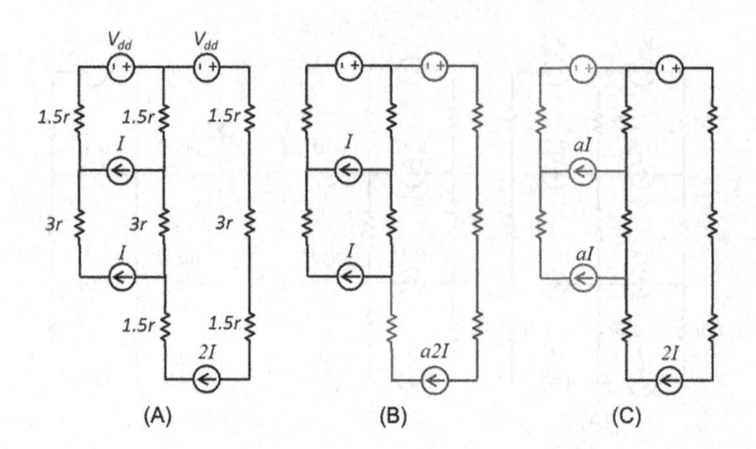

FIGURE 18.12

Multi-level power delivery system where each tier is supplied by one pair of voltage levels. (A) All of the circuits are active, (B) the processor is inactive (shown in gray), and (C) the memory tiers are inactive (shown in gray) [669].

reducing the power efficiency. Although using a large number of converters requires significant area, this overhead can be compensated by reducing the density of the TSVs used to distribute power.

Simulations of an eight tier stacked processor architecture based on the ARM Cortex A9 dual-core demonstrate that a multilevel power delivery system exhibits similar supply noise as compared to a standard single V_{dd} supply system if the workload imbalance across the tiers remains under 50% with the same area overhead [671]. The overhead in the standard supply system is the area occupied by the power TSVs while for the multilevel power delivery system, the overhead comprises the area of the converters and the fewer number of power TSVs.

Although modern EDA tools yield robust and high quality power networks in planar circuits, no provision exists for the vertical wires, which produce significant power supply noise [672,673]. The role of these wires in modeling and designing power distribution networks for 3-D circuits is discussed in the following sections.

18.2 MODELS FOR THREE-DIMENSIONAL POWER DISTRIBUTION NETWORKS

All of the 3-D power delivery techniques discussed in the previous section utilize a 1-D model of the on-chip power distribution network and emphasize resistive *IR* voltage drops. Consequently, a simple resistive network is used. Additionally, the TSV are modeled as a resistor, which adds resistance to the on-chip power distribution network. Although these models are useful in demonstrating the benefits of these power delivery approaches, these models do not provide an accurate analysis of the behavior of 3-D power distribution networks.

Several models with different accuracy and complexity have been developed to characterize the behavior of these networks. The impedance of these networks is modeled across a large range of frequencies, utilizing a combination of lumped and distributed models for the disparate components of the power distribution system. Analytic models are first discussed in this section, listing the assumptions that allow a closed-form treatment of these complex interconnection structures. More advanced models that support a more practical analysis with fewer assumptions are also presented. These models simplify the power distribution networks, enabling analysis with reasonable simulation times and accuracy over a large range of frequencies which exceed by several times the typical operating frequency of digital circuits.

As interconnect meshes are typical topologies for power distribution networks [205], most models assume that the power distribution network of each tier is composed of two or more global metal layers, which are densely connected with metal contacts. Pads are periodically inserted in one of the tiers (the tier connected to the package), forming an array of power and ground pads. The other tiers within the 3-D circuit are connected through P/G TSVs to this tier. The illustrative example shown in Fig. 18.13A includes three tiers where the uppermost tier is assumed to be connected to the package. The power and ground networks are decoupled and can be analyzed separately, as shown in Fig. 18.13B. Assuming that the same network is used in every tier and is uniformly structured, the network can be divided into multiple identical cells, which are electrically modeled as shown in the same figure. The boundary of this cell is illustrated by a dashed rectangle (see Fig. 18.13B) and is adjacent to the (V_{dd} or V_{gnd}) pads where a quarter of a pad is included at each corner of the cell to maintain symmetry across cells.

As depicted in Fig. 18.13B, each interconnect segment is modeled by a single resistor R_w, while the TSVs are modeled as a lumped *LR* element, notated as L_{TSV} and R_{TSV}. Within each cell, Δx and Δy are, respectively, the length of the x and y edges of the cell. The circuits within the area $\Delta x \Delta y$ of each cell are modeled as a current source of density $J(s)$ in the frequency domain. The capacitor connected in parallel to the current source shown in Fig. 18.13B models the capacitance

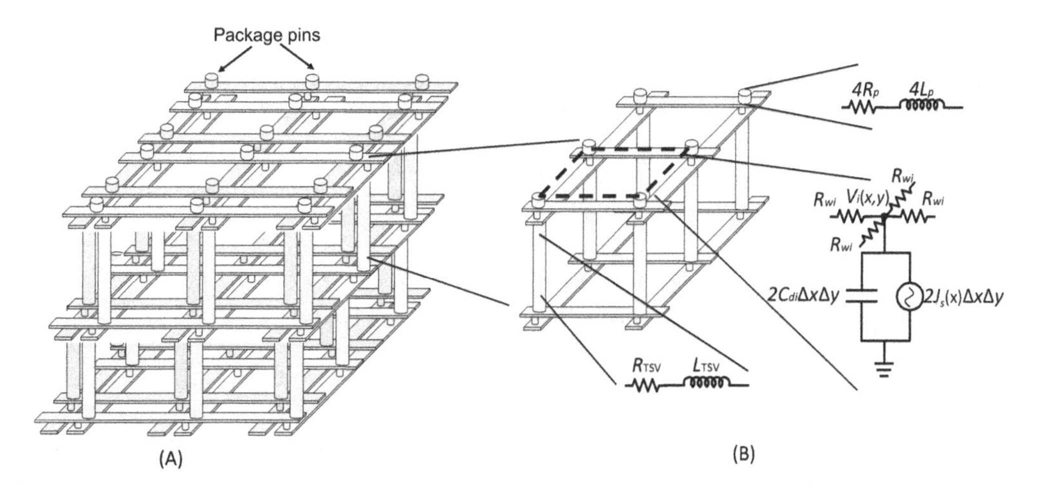

FIGURE 18.13

A 3-D power distribution netwok (not to scale), (A) the power (ground) meshes are connected by power (ground) TSVs, and (B) the equivalent circuit model of a package pin, TSV, and unit cell including the decoupling capacitance and current source [674].

of the circuit and the intentional capacitance added as a decoupling capacitance. The package pins connected to the pads of one tier (the first tier in the example of Fig. 18.13A) are modeled as a lumped resistance R_p and inductance L_p. The number of cells constituting a power mesh is typically large to model the mesh as a continuous power (or ground) plane [675].

In this case, the power supply noise $V_i(x,y,s)$ in tier i at coordinates (x, y) of the power (or ground) network and frequency s can be described by a partial differential equation [674],

$$\nabla^2 V_i(x,y,s) = R_{wi}J_i(s) + 2V_i(x,y,s)sR_{wi}C_{di} + \Phi_i(x,y,s), \tag{18.8}$$

where $\Phi_i(x,y,s)$ is the source function which is the same for all but the tier connected to the package, and is [676]

$$\Phi_i(x,y,s) = R_{wi}\sum_{j=1}^{k_{TSV}}\left(\frac{V_{(i-1)TSVj} - V_{iTSVj}}{sL_{TSVj} + R_{TSVj}} - \frac{V_{iTSVj} - V_{(i+1)TSVj}}{sL_{TSVj} + R_{TSVj}}\right)\delta(x - x_{TSVj})\delta(y - y_{TSVj}). \tag{18.9}$$

V_{iTSVj} is the voltage of the TSV j connected to tier i, k_{TSV} is the number of TSVs in each tier (which in this model is considered the same among tiers), and $\delta(x - x')$ is the step function. The source function for the tier connected to the package (here assumed to be tier 1) differs from (18.9) and is

$$\Phi_1(x,y,s) = -\frac{R_{w1}}{4sL_p}V_{pad}(s)\delta(x)\delta(y) + R_{w1}\sum_{j=1}^{k_{TSV}}\left(-\frac{V_{1TSVj} - V_{2TSVj}}{sL_{TSVj} + R_{TSVj}}\right)\delta(x - x_{TSVj})\delta(y - y_{TSVj}), \tag{18.10}$$

where V_{pad} is the voltage of the P/G pads in tier 1. To provide a closed-form solution, boundary conditions for (18.8) are defined. Assuming that a quarter of the P/G pads at the corner of the cell supply all of the current in this cell and no current flows outwards or inwards at an upright angle to the boundary of the cell, the boundary conditions are

$$\left.\frac{\vartheta V_i}{\vartheta y}\right|_{x=0} = 0, \left.\frac{\vartheta V_i}{\vartheta y}\right|_{x=\Delta x} = 0, \left.\frac{\vartheta V_i}{\vartheta x}\right|_{y=0} = 0, \left.\frac{\vartheta V_i}{\vartheta x}\right|_{y=\Delta y} = 0, \tag{18.11}$$

where the upper left corner of the cell is assumed to be located at $(x, y) = (0,0)$. To solve (18.8), the following transformation is applied to yield a Helmholtz equation [677],

$$V_i(x,y,s) = u_i(x,y,s) - \frac{J_i(s)}{2sC_{di}}. \tag{18.12}$$

The PDE in (18.8) is rewritten as

$$\nabla^2 u_i(x,y,s) = 2u_i(x,y,s)sR_{wi}C_{di} + \Phi_{ui}(x,y,s), \tag{18.13}$$

which also complies with the boundary conditions of the second kind

$$\left.\frac{\vartheta u_i}{\vartheta y}\right|_{x=0} = 0, \left.\frac{\vartheta u_i}{\vartheta y}\right|_{x=\Delta x} = 0, \left.\frac{\vartheta u_i}{\vartheta x}\right|_{y=0} = 0, \left.\frac{\vartheta u_i}{\vartheta x}\right|_{y=\Delta y} = 0. \tag{18.14}$$

The transformed PDE (18.13) can be solved in the frequency domain using the Green's function $G(x, y, u, v, s)$ [677] of the Helmholtz equation, where x, y, u, and v, correspond to the mesh coordinates and the boundary conditions in (18.14). The resulting closed-form solution for the power supply noise in tier i ($i \neq 1$) and 1 is, respectively,

$$V_i(x,y,s) = R_{wi}\sum_{j=1}^{k_{TSV}}\left(\frac{V_{(i-1)TSVj} - 2V_{iTSVj} + V_{(i+1)TSVj}}{sL_{TSVj} + R_{TSVj}}\right)G(x,y,x_{TSVj},y_{TSVj},s) - \frac{J_i(s)}{2sC_{di}}, \tag{18.15}$$

and

$$V_1(x,y,s) = \frac{R_{w1}}{4sL_p} V_{\text{pad}}(s) G(x,y,0,0,s) + R_{w1} \sum_{j=1}^{k_{\text{TSV}}} \left(-\frac{V_{1\text{TSV}j} - V_{2\text{TSV}j}}{sL_{\text{TSV}j} + R_{\text{TSV}j}} \right) G(x,y,x_{\text{TSV}j},y_{\text{TSV}j},s) - \frac{J_1(s)}{2sC_{d1}}. \quad (18.16)$$

In both of these expressions, the voltage of the pads V_{pad} and TSVs $V_{\text{TSV}j}$ is unknown and, in addition to V_i, needs to be determined. This task can be achieved by transferring these unknowns to the left hand side and solving the resulting system of equations. A unique solution exists as there is the same number of equations and unknowns. The inverse Laplace transform is applied as a final step to obtain the power supply noise in the time domain, and the overall noise is the addition of the noise component from the power and ground networks.

The accuracy of these closed-form expressions is compared with SPICE simulations of structures similar to those illustrated in Fig. 18.13A for a five tier system assuming a 45 nm CMOS technology. The electrical characteristics of each tier are based on a 65 nm Intel processor [678], where the interconnect characteristics are not scaled to the 45 nm technology node. Comparisons of both the magnitude and phase of the power supply noise with SPICE exhibit an error of less than 4% for all of the investigated scenarios [674].

Although this model provides a fast estimate of the power supply noise and is easily integrated into mathematical software packages, the assumption of uniform traits, for example, the same current demand per cell and decoupling capacitance make the model over-restrictive. More elaborate models have been developed to relax these constraints. No closed-form solution however is available for these more elaborate models. Numerical methods and/or simulations are used to estimate the power supply noise. The salient feature of these models is the order of magnitude decrease in the number of elements that comprise the power distribution network, reducing the number of nodes for which the voltage is evaluated. Consequently, these models, in terms of complexity and computational time, are between the analytic models and the numerical simulation of a fully extracted netlist characterizing a power distribution network.

These simulation based models utilize the concept of unit cells; however, the cells contain additional circuit elements. More specifically, the effect of the silicon substrate is included [679]. This effect is expected to be more significant for memory tiers in 3-D systems as memory circuits utilize fewer metal layers as compared to logic circuits. In this configuration, the distance of the metal power distribution layers in a memory tier are closer to the substrate, and are therefore more strongly coupled with the silicon substrate.

From these observations, a unit cell of a power distribution network, similar to the circuit shown in Fig. 18.13B, is modeled to ensure that the decoupling capacitance and silicon substrate are both included. In a closed-form model, the pitch of the pads determines the boundary of the unit cells, whereas the size of the unit cells in this model is based on the frequency. One guideline is to limit the size of a unit cell in the x and y directions to be shorter than one tenth of the wavelength [679], as determined by the effective dielectric constant and highest frequency. In this way, a lumped model of a unit cell does not significantly affect the accuracy. Furthermore, each unit cell is modeled as a two port electric network.

These two port networks are assembled using the segmentation method [679] to form a model of a complete power network. This method connects the two port networks in appropriate matrices to characterize the overall power distribution network. An example of the segmentation method linking the different unit cells to form a larger network is shown in Fig. 18.14. Within a unit cell, the decoupling capacitance is modeled by an *RC* lumped section while the P/G TSVs are modeled by a *RLC* lumped sections. Note that in this model the power and ground networks are not decoupled but rather are treated as a single network where each grid cell contains both power and ground segments.

An example of decomposing a unit cell into different segments along the *x* and *y* directions is shown in Fig. 18.15. Each of these segments is modeled as an *RLGC* lumped sections. Some of these segments are identical and share the same *RLGC* section; for instance, segments A_x and A_y. The corresponding *RLGC* lumped sections for segments $A_{x(y)}$, $B_{x(y)}$, and $C_{x(y)}$ are depicted in Fig. 18.16. To limit the number of elements in the *RLGC* section, certain assumptions are applied to the modeling process. The most important assumption is that the lateral electric field between

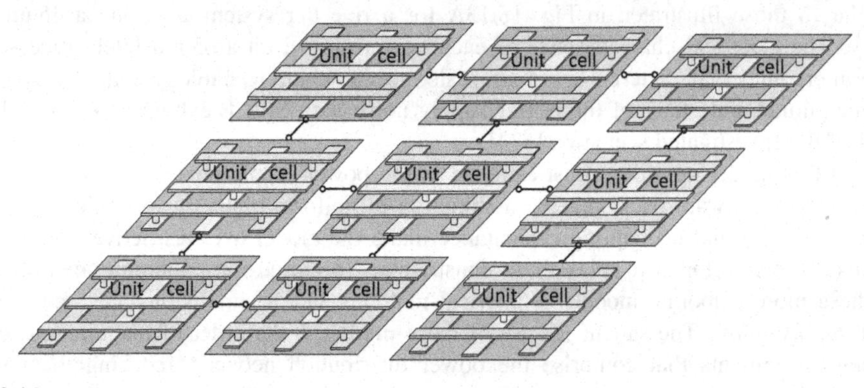

FIGURE 18.14

The segmentation method linking successive unit cells to model an entire power distribution network [679].

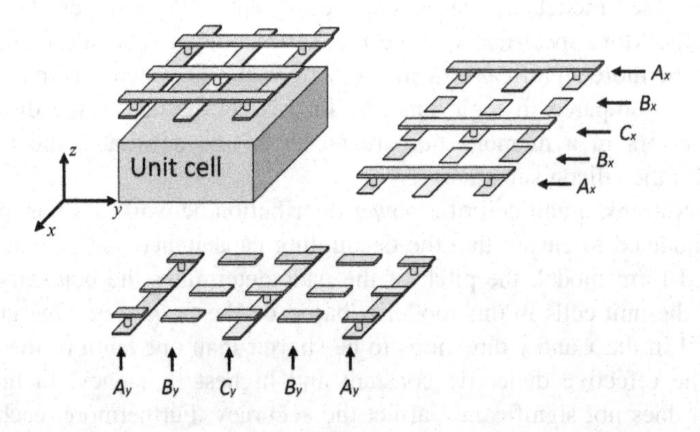

FIGURE 18.15

Decomposition of a unit cell including both power and ground lines along the *x* and *y* directions. The different structures formed by the decomposition process are also illustrated. Two metal layers are utilized for the power distribution network [679].

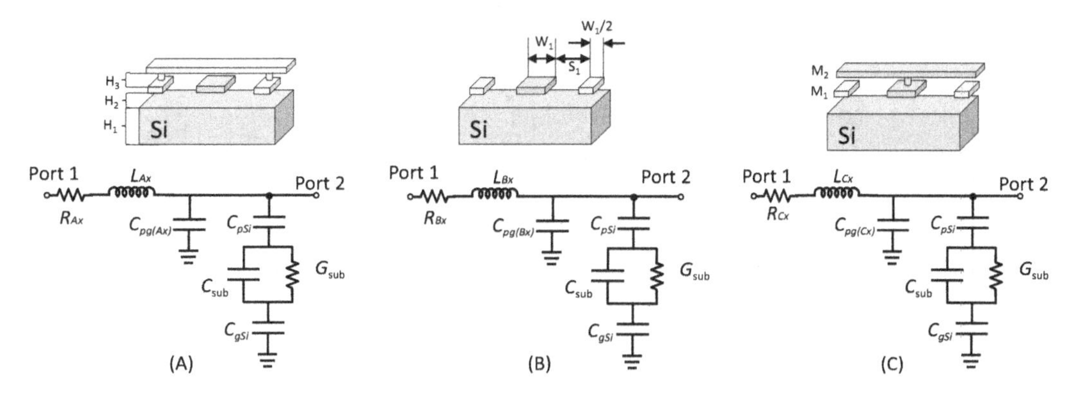

FIGURE 18.16

Decomposed structures and equivalent *RLGC* lumped sections. The notation of the physical parameters used in Table 18.2 is also defined [679].

neighboring ground and power interconnects in the first metal layer (M1) is considerably lower than the field between a conductor and the substrate. Consequently, the conductors within these segments can be modeled as coplanar waveguides with infinitely distant ground planes.

Some components of the *RLGC* section are common for all of the different segments, A, B, and C. These components include the conductance of the silicon substrate G_{sub}, the power (ground) to substrate capacitance $C_{p\text{Si}}$ ($C_{g\text{Si}}$), the substrate capacitance C_{sub}, and the coupling capacitance between power and ground lines C_{pg} which are described, respectively, by [679]

$$G_{\text{sub}} = 2\sigma_{\text{Si}} \frac{K(k_i)}{K(k_i')}, \tag{18.17}$$

$$C_{\text{psi}} = C_{\text{gsi}} = \varepsilon_{r,\text{IMD}}\varepsilon_0 \left(\frac{W_1}{H_2} + 0.77 + 1.06\left(\frac{W_1}{H_2}\right)^{0.25} + 1.06\left(\frac{T_1}{H_2}\right)^{0.5} \right), \tag{18.18}$$

$$C_{\text{sub}} = 2\varepsilon_{r,\text{Si}}\varepsilon_0 \frac{K(k_i)}{K(k_i')}, \tag{18.19}$$

and

$$C_{pg(x)} = \varepsilon_{r,\text{IMD}}\varepsilon_0 C_3'^{\text{air}}. \tag{18.20}$$

The inductance for each segment is determined from the capacitance of the structure assuming a lossless transmission line where the dielectric is air,

$$L_x = \frac{1}{c^2 C_x'^{\text{air}}}. \tag{18.21}$$

$C_x'^{\text{air}}$ is different for A, B, and C and is based on the expressions listed in Table 18.2. The resistance R_x for each segment is also listed in the same table.

To evaluate the accuracy of the model, a 3-D power distribution grid with an area of 1 mm × 1 mm is utilized as a case study. This grid is simulated with a commercial EM solver and a frequency domain model for frequencies ranging from 0.1 GHz up to 20 GHz [679].

Table 18.2 Formulae for Computing the Components of the *RLGC* Lumped Model of a Unit Cell [679]

Segment A	Segment B	Segment C
$C_x'^{\text{air}} = C_1'^{\text{air}} + C_3'^{\text{air}}$ (18.22)	$C_x'^{\text{air}} = 2C_1'^{\text{air}}$ (18.23)	$C_x'^{\text{air}} = C_1'^{\text{air}}$ (18.24)
$R_x = \frac{R_{M1}\left(\frac{W_1}{2}\right)}{W_1} + \left(\frac{0.5R_{M1}\left(\frac{W_1}{2}\right)}{\frac{W_1}{2}}\right) // \frac{R_{M2}\left(\frac{W_1}{2}\right)}{5W_1}$ (18.25)	$R_x = \frac{R_{M1}S_1}{W_1} + \left(\frac{0.5R_{M1}S_1}{\frac{W_1}{2}}\right)$ (18.26)	$R_x = \frac{R_{M1}\left(\frac{W_1}{2}\right)}{W_1} + \left(\frac{0.5R_{M1}\left(\frac{W_1}{2}\right)}{\frac{W_1}{2}}\right)$ (18.27)
$C_i'^{\text{air}} = 2\varepsilon_0 \frac{K(k_i)}{K(k_i')}$ (18.28)	$\frac{K(k_i)}{K(k_i')} = \left(\begin{array}{l} \frac{1}{\pi}\ln\left(\frac{2(1+\sqrt{k_i})}{(1-\sqrt{k_i})}\right), (0.5 \le k_i^2 \le 1) \\ \pi/\ln\left(\frac{2(1+\sqrt{k_i'})}{(1-\sqrt{k_i'})}\right), (0.0 \le k_i^2 \le 0.5) \end{array}\right)$ (18.29)	$k_i' = \sqrt{1-k_i^2}$ (18.30)
$k_1 = \frac{W_1}{W_1 + 2S_1}$ (18.31)	$k_2 = \frac{\sinh\left(\frac{\pi W_1}{4H_2}\right)}{\sinh\left(\frac{\pi(W_1+2S_1)}{4H_2}\right)}$ (18.32)	$k_3 = \frac{\tanh\left(\frac{\pi W_1}{4H_3}\right)}{\sinh\left(\frac{\pi(W_1+2S_1)}{4H_3}\right)}$ (18.33)

Any inaccuracy between the EM solver and model is low (around 10%) for frequencies far from the resonant frequency but increase significantly, exceeding 100% at the resonant frequencies. As the resonances occur at frequencies above 10 GHz, these inaccuracies are attributed to the large size of the cell. In other words, the accuracy is improved by increasing the number of cells to model the network while degrading the computational gains (as compared to a full electromagnetic (EM) solver).

18.2.1 ELECTRO-THERMAL MODEL OF POWER DISTRIBUTION NETWORKS

In Chapter 12, Thermal Modeling and Analysis, and Chapter 13, Thermal Management Strategies for Three-Dimensional ICs, the importance of thermal issues in 3-D circuits is discussed where the effects of temperature are integrated into several physical design techniques to avoid thermal hot spots. In addition to affecting the timing of signals and the leakage current, elevated temperatures also increase the resistance of the power and ground lines, thereby increasing the resistive voltage drop on these lines. Thermal issues are more important for P/G TSVs as these interconnections also act as the primary means to conduct heat towards the heat sink [673]. Consequently, in addition to thermal aware physical design, thermal issues are considered in the sign-off phase of the IC design process when the power distribution network is typically developed. Extending the models discussed in the previous section to the design of power distribution networks to consider temperature is a straightforward process due to the electro-thermal duality that enables the formulation of the heat equations as a linear system of equations.

Thermal models that utilize the electro-thermal duality are discussed in Chapter 12, Thermal Modeling and Analysis. These models utilize resistive or *RC* networks to evaluate the temperature across a circuit. The system of equations consists of known resistances and current sources, and the unknowns are the voltage at the nodes within a network. This formulation is similar to the linear system obtained for power grid analysis. By solving the thermal model to determine the temperature, and the electrical model to evaluate power supply noise, as illustrated in Fig. 18.17, the analysis converges to the temperature aware power supply noise. The thermal aware model of a power network discussed in this subsection enables a more accurate treatment of the design of power distribution grids as thermal issues affect the allocation of the P/G TSVs [673].

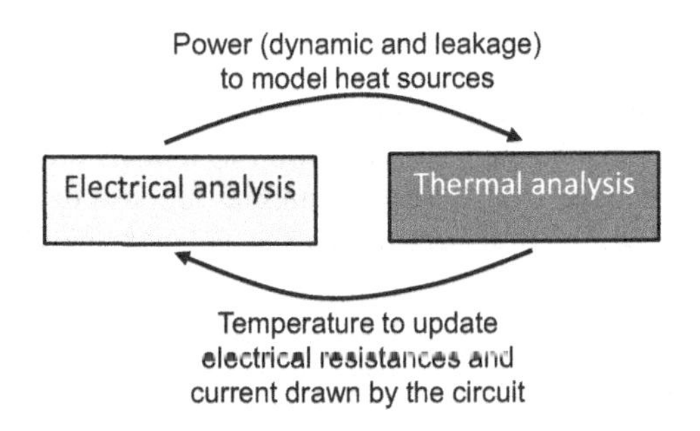

FIGURE 18.17

Iterative process for electro-thermal analysis [472].

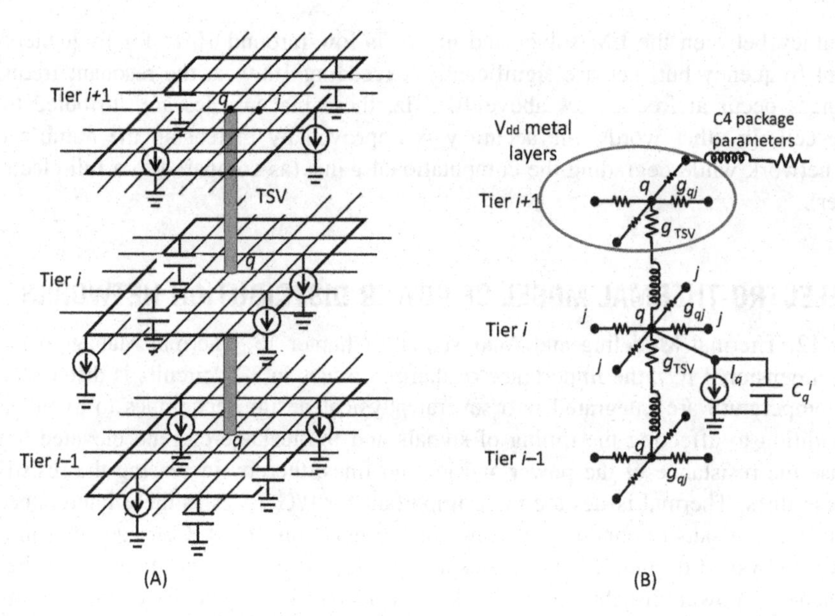

FIGURE 18.18

Overview of power grid, (A) a small segment of a power grid, and (B) corresponding electrical model including the parasitic impedance of the package [472].

A mesh structure of the power and ground networks, similar to the structure shown in Fig. 18.13, allows the electrical *RLC* model of these networks to be obtained. To determine the voltage at the nodes of the network, a modified nodal analysis can be used [680]. With this method, first order differential equations are used to formulate the problem of determining the voltage at the network nodes. A portion of the power mesh for a three tier system is illustrated in Fig. 18.18 to describe the notation used in these differential equations. With this notation, the voltage node q in tier i is [472]

$$V_q^i = \frac{\sum_{j\in s_q} V_j^i g_{qj}^i + \sum_{k\in\{i+1,i-1\}} V_q^k g_{\text{TSV}}^{i,k} + I(s)_q^i}{\sum_{j\in s_q} g_{qj}^i + \sum_{k\in\{i+1,i-1\}} g_{\text{TSV}}^{i,k} + sC_q^i}, \tag{18.34}$$

where s_q is the set of neighboring nodes of q, and $i+1$ and $i-1$ are adjacent to the ith tier. The admittance between two nodes (e.g., j and q) is notated by g_{qj}, while the admittance of a TSV connecting two tiers at node q is denoted as g_{TSV}. The term $I(s)$ models the current load and the capacitance at node q and tier i is notated as C_q^i. These quantities are also depicted in Fig. 18.18.

The voltage nodes can be written in matrix format for the entire mesh as [472]

$$\left(\begin{bmatrix} G^1 & G_{\text{TSV}}^{12} & 0 \\ G_{\text{TSV}}^{21} & G^2 & G_{\text{TSV}}^{23} \\ 0 & G_{\text{TSV}}^{32} & G^3 \end{bmatrix} + s\begin{bmatrix} C^1 & 0 & 0 \\ 0 & C^2 & 0 \\ 0 & 0 & C^3 \end{bmatrix}\right) \cdot \begin{bmatrix} V^1 \\ V^2 \\ V^3 \end{bmatrix} = \begin{bmatrix} P^1 \\ P^2 \\ P^3 \end{bmatrix}, \tag{18.35}$$

where $G_{ni\times ni}^i$, is a symmetric and sparse matrix of the conductance for the power network of tier i, while a similar system of matrices also applies to the ground network. The subscript $ni \times ni$

indicates that the network in each tier can be of different size. The terms G^{21}_{TSV} and G^{32}_{TSV} are diagonal matrices that contain the conductance of the TSVs that, respectively, connect the network of tier 1 to tier 2, and tier 2 to tier 3. The other two TSV conductance terms are the transpose of these matrices. The sparsity of the TSV conductance matrices depends upon the density of TSVs that connect the adjacent networks. The decoupling capacitance within the network of tier i is captured by the matrices $C^{i}_{ni \times ni}$, whereas the unknown voltage nodes in each tier are represented by the column vector $V^{i}_{ni \times ni}$. Finally, the current and voltage sources contained in the network of each tier are included in the vectors $P^{i}_{ni \times ni}$.

This linear system can be solved with respect to the voltage nodes with different methods, where multigrid methods as in [681] are good candidates. From the voltage nodes, the current flowing within each branch of the network in tier i (required for the thermal analysis) is

$$J^{i}_{q-j} = \frac{I^{i}_{q-j}}{h_{q-j}w_{q-j}} = \frac{(V^{i}_{q} - V^{i}_{j})g^{i}_{q-j}}{h_{q-j}w_{q-j}}, \tag{18.36}$$

where I^{i}_{q-j} is the current flowing along branch $q-j$, and h^{i}_{q-j} and w^{i}_{q-j} are, respectively, the thickness and width of the wire used for branch $q-j$.

Due to the electro-thermal duality, the heat equation describing the temperature across the power distribution network can be described by a similar system of matrices as in (18.35). This description is more computationally efficient if the same number of nodes between the two matrices is utilized [472]. Both the power distribution and thermal networks for a 3-D circuit are mapped to the same geometric mesh.

In formulating the matrices of the thermal model, each matrix element is the thermal conductance of each branch of the power network between two metal contacts. Approximating this thermal conductance with a lumped element is valid if the length of this branch is not longer than the characteristic thermal length [437]. The notion of this length is that the heat from Joule heating of the wire (from the current flowing through this wire) flows to the adjacent metal layer located closer to the heat sink through the metal contacts (or TSVs) rather than the intermetal dielectric. The temperature of the wire is therefore determined by the temperature at the contacts, which facilitates the modeling process as the thermal and electrical networks have the same structure (i.e., the same nodal voltage and temperature (thermal voltage) are determined).

The thermal conductance is analogous to the electrical conductance, as described by [682]

$$g_{th} = \rho \kappa_{\text{metal}} g_{\text{elec}}, \tag{18.37}$$

where ρ and κ_{metal} are, respectively, the electrical resistivity and thermal conductivity of the metal layer. Although both of these material properties are temperature dependent, only the dependence of the resistivity on temperature is considered in this model. The dependence of the metal resistivity on temperature is

$$\rho = \rho_0[1 + \beta(T - T_0)], \tag{18.38}$$

where ρ_0 is the thermal resistivity at a reference temperature, and T_0 and β are temperature coefficients for the resistance. Typical values for these terms are $T_0 = 27$ °C and $\beta = 0.0039$/°C. Consequently, the thermal resistance of a power (or ground) line of length L, width W, and thickness H in tier i is

$$R_i = \rho \frac{L_i}{H_i W_i}[1 + \beta(T - T_0)]. \tag{18.39}$$

The thermal capacitance models the heat capacity stored in the material as the temperature of the material changes. In ICs, heat is primarily stored within the silicon substrate. Thus, the thermal capacitance of a tier is

$$C_{th} = c_p t A, \tag{18.40}$$

where c_p is the heat capacity of silicon, and t and A are, respectively, the thickness and area of the substrate of the tier. This thermal capacitance is uniformly distributed across the nodes of the thermal network.

Similar to the parasitic impedances of the package in an electrical model, the off-chip components of the thermal network should also be considered within the thermal model. Two paths for the heat to flow are typically modeled. Transfer of heat to the heat sink is the primary heat flow path, while the package boundary is the secondary path for the propagation of heat. A thermal resistor between the heat sink and each node of the thermal network models the propagation of heat between the IC and the heat sink. For the secondary path, a thermal resistance is employed for each C4 pin [550]. Finally, the TSVs are thermally modeled as an RC section based on [683]. These individual thermal components are combined into a thermal network with the same components as the dual electrical network shown in Fig. 18.18 (except for at the boundary of the heat sink and package), enabling computation of the temperature at the node q by

$$T_q^i = \frac{\sum_{j \in s_q} T_j^i g_{th_{qj}}^i + \sum_{k \in \{i+1, i-1\}} T_q^k g_{th_{\mathrm{TSV}}}^{i,k} + Q_q^i}{\sum_{j \in s_q} g_{th_{qj}}^i + \sum_{k \in \{i+1, i-1\}} g_{th_{\mathrm{TSV}}}^{i,k} + sC_{th_q}^i}, \tag{18.41}$$

where T_j is the temperature at the neighboring nodes of q determined by the node set s_q, and $g_{th_{qj}}$ is the thermal conductance between any pair of nodes q and j. The term Q_q^i denotes the heat generated at node q, which includes the power of the circuit and the Joule heating described by I^2/g_{qj} for each wire segment between nodes q and j.

This combined electro-thermal model facilitates the exploration of several tradeoffs among the TSV density, sizing, number of horizontal wires, and decoupling capacitance, leading to the effective design of 3-D power distribution networks. The important role of TSVs in the behavior of these networks is discussed in the following section.

18.3 THROUGH SILICON VIA TECHNOLOGIES TO MITIGATE POWER SUPPLY NOISE

As discussed in Chapter 3, Manufacturing Technologies for Three-Dimensional Integrated Circuits, TSV fabrication processes can be categorized as via-last, via-middle, and via-first. In addition to the fabrication process, different bonding styles, such as face-to-back or back-to-back exist. These combinations produce a variety of vertical interconnections for 3-D systems, with the TSVs connecting: (1) local-to-global, (2) global-to-global, and (3) local-to-local interconnect layers from adjacent tiers. For power and ground TSVs, this situation can provide additional paths to distribute power to reduce the distance to the current load. Furthermore, the TSVs can be of different diameter within each tier or with TSVs that are not solid but rather ring structures [684]. Several of these rings can be manufactured, resulting in coaxial TSVs (CTSVs). The effects of these manufacturing choices on the behavior of 3-D power distribution networks are reviewed in this subsection.

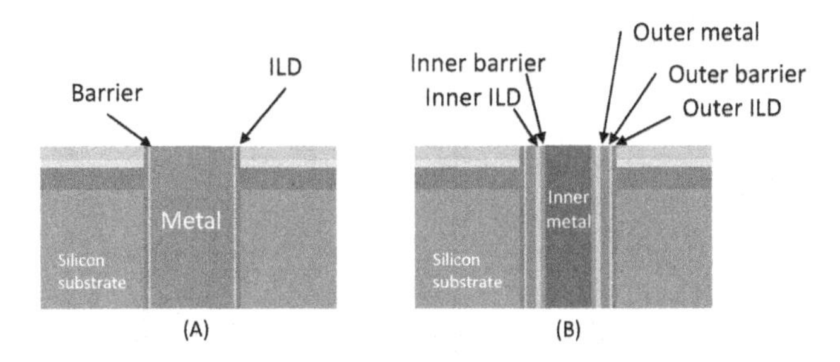

FIGURE 18.19

Cross-sectional view of a TSV. (A) A standard solid TSV, and (B) a CTSV with two layers of metal separated by a dielectric layer [684].

A CTSV consists of two or more metal rings separated by an inner layer of dielectric material. An example of one standard and one CTSV is, respectively, depicted in Figs. 18.19A and B. The choice of dielectric material and thickness of the separation layer primarily affects the electrical properties of the CTSV. The naturally higher capacitance of a CTSV does not significantly reduce the decoupling capacitance within the overall system; however, area gains exist from halving the number of required P/G TSVs. A CTSV can provide both power and ground; for example, at the outer and inner ring, requiring fewer TSVs to distribute power.

Other ways to reduce power supply noise emerge for the via-middle and via-first TSV processes. In these processes, the TSVs connect a global metal layer to an intermediate or local metal layer of an adjacent tier. For P/G TSVs, this connectivity provides an additional local path to distribute power and ground. Consider the one-dimensional model of two segments of a power distribution network within the 3-D circuit shown in Fig. 18.20, where a via-first process is assumed. A via-middle process can be treated similarly. The equivalent circuit of these two segments is illustrated in Fig. 18.21. Note that each circuit contains two paths with different impedance characteristics.

The primary difference between the two power delivery systems depicted in Figs. 18.20A and B is that the TSV in Fig. 18.20B connects only to the topmost metal layer (MT), while in Fig. 18.20A the TSV also connects to the first metal layer (M1). With the latter approach, additional paths are formed, which are shown by the thick solid curves in Fig. 18.20A. These paths are called "TSV paths" in the remainder of this section. To investigate the physical behavior of a TSV path, a current source is assumed to be connected to MT through a stack of intratier vias (i.e., metal contacts). Note that devices can only be placed at a certain distance from a TSV due to manufacturing constraints.

The current only flows through two metal layers (e.g., MT and M1). Inclusion of any other metal layer would increase the impedance of the path as MT is the least resistive layer. The on-chip inductance is omitted in this analysis due to the local (within a few micrometers) and fast decay of the effect that inductance has on the on-chip power distribution network [662]. The inductive component of the TSV impedance is, however, included in this analysis. Furthermore, the capacitance of the TSV is not considered, since for power and ground TSVs this capacitance behaves as a decoupling capacitance, improving the impedance characteristics of the power delivery system over a range of frequencies. Including the capacitance of the TSV in the analysis further demonstrates the effectiveness of the TSV path.

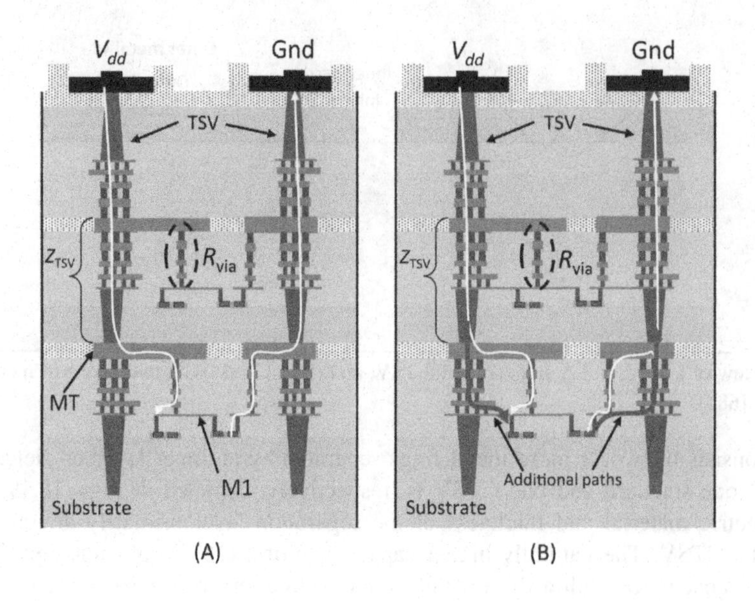

FIGURE 18.20

Current paths within a 3-D circuit. (A) Where the TSV is connected to the power lines on both the uppermost (MT) and the first (M1) metal layers, and (B) where the TSV is connected only to the topmost (MT) metal layer.

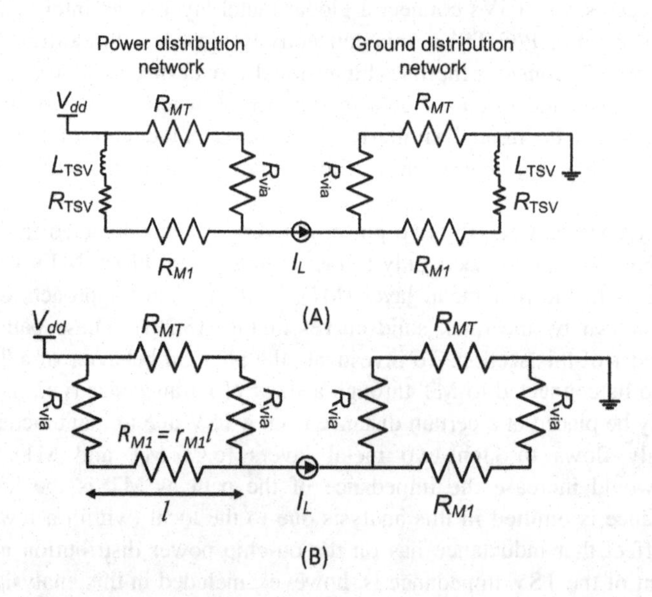

FIGURE 18.21

Equivalent circuit of the current flow paths illustrated in Fig. 18.20. (A) The TSV locally distributes current, and (B) only stacks of metal vias supply current to the load.

Based on the circuits shown in Fig. 18.21, a first order analysis is used to determine the greatest current that each of the circuits can carry while satisfying the voltage drop constraints. The maximum current that the circuits shown in Figs. 18.21A and B can carry is, respectively,

$$I_L \leq \frac{\left(1 - V_{\text{dropmax}}\right) V_{dd}}{(R_{M1} + Z_{\text{TSV}}) \| (R_{MT} + R_{\text{via}})}, \tag{18.42}$$

and

$$I_L \leq \frac{\left(1 - V_{\text{dropmax}}\right) V_{dd}}{(R_{M1} + R_{\text{via}}) \| (R_{MT} + R_{\text{via}})}. \tag{18.43}$$

Z_{TSV} and R_{via} are, respectively, the impedance of a TSV and the resistance of a stack of vias connecting the topmost and lowest metal layers (see Fig. 18.20). The resistance of the topmost and lowest metal layers are, respectively, $R_{MT} = r_{MT}l$ and $R_{Ml} = r_{M1}l$, where r_{MT} and r_1 are the resistance per length and l is the distance of the current source from the TSV. The width of the power and ground lines on the topmost layer is 20 times wider than the minimum width. The width of the lowest metal layer is twice the minimum width. Consequently, for an industrial 0.18 μm CMOS technology with six metal layers, the resistance per unit length is $r_{M1} = 29.78 r_{MT}$ and $r_{MT} = 4.62$ Ω/mm. In addition, the resistance of a stack of vias between M6 and M1 is $R_{\text{via}} = 32.5$ Ω [685].

Assuming that $V_{dd} = 1$ volt and $l = 30$ μm, the voltage drop at the current source, V_L for both circuits, is plotted in Fig. 18.22. The TSV is assumed to have a resistance of 1 Ω, including the resistance of a large number of parallel connected metal vias, and an inductance of $L_{\text{TSV}} = 20$ pH [676]. The maximum switching frequency is $f = 10$ GHz to consider the high frequency components

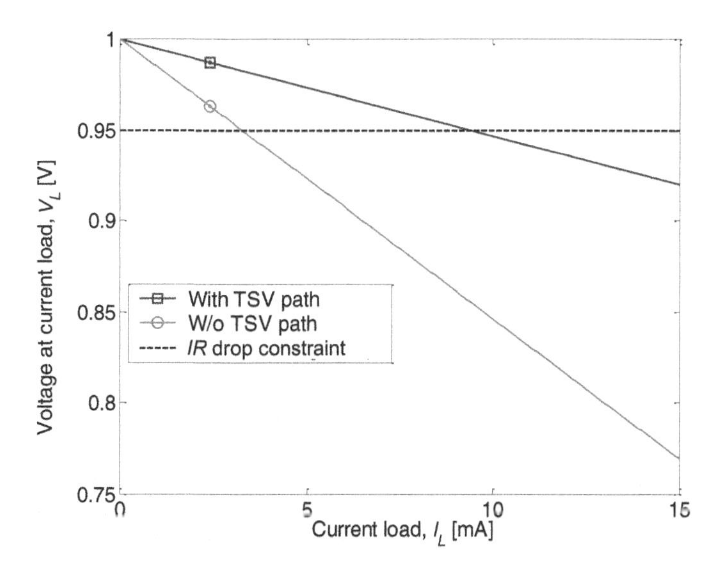

FIGURE 18.22

Voltage drop at the current source as a function of the current drawn by the power supply.

of a digital signal. Additionally, the TSV resistance is larger than typically reported, depicting a worst case resistance for the TSV [147].

From Fig. 18.22, when the TSV path is exploited, a significantly larger current is supplied to the transistors without exceeding the allowed voltage ripple. Thus, for the circuit shown in Fig. 18.21B, the maximum voltage drop ($V_{dropmax} = 5\% V_{dd}$) is reached when $I_{Lmax} = 3.3$ mA, while for the circuit that includes the TSV path, the maximum current that can be sustained is $I_{Lmax} = 9.4$ mA; a considerable 2.8 × increase.

Alternatively, the maximum distance of the current source from the TSV, where the circuit draws a fixed current, is plotted in Fig. 18.23. When the TSV path is considered, the current source can be placed farther from the power and ground pads or, equivalently, a smaller number of stacked vias can distribute the current within the circuit.

The difference in the voltage drop for a fixed current load of $I_L = 3$ mA between the two circuits shown in Fig. 18.21 is significant for small distances and gradually decreases with increasing distance. For larger distances, on the order of millimeters, both circuits produce a similar voltage drop. This behavior can be explained by considering the current flow through the different paths that exist within the circuit shown in Fig. 18.21A. In this power distribution network, the current propagates through the low resistance TSV and M1, exhibiting a higher resistance as compared to M6. Alternatively, the current flows through the less resistive M6 as compared to M1 and the stacked vias, which in turn, is at least an order of magnitude more resistive than a single TSV.

For short distances (i.e., hundreds of micrometers), the path through the TSV and M1 (e.g., the TSV path) exhibits a lower impedance as compared to the path consisting of M6 and the stacked vias. Consequently, most of the current flows through the TSV path, resulting in a substantially

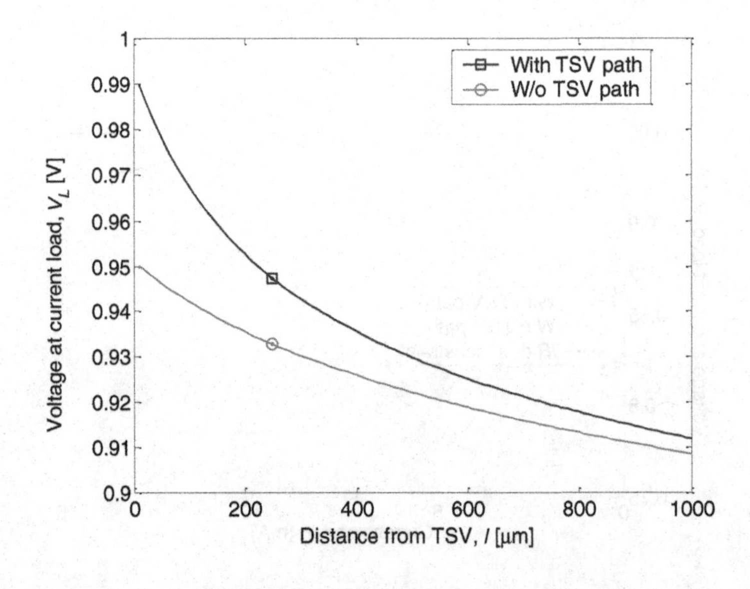

FIGURE 18.23

Voltage drop as a function of distance of the current source from the TSV.

smaller voltage drop as compared to the circuit shown in Fig. 18.21B. For the circuit shown in Fig. 18.21B, the greatest portion of the current flows through M6 and the stacked vias since the alternative path comprising M1 and stacked vias exhibits a considerably greater impedance.

As the distance between the current source and the TSV in the structure shown in Fig. 18.21A increases, the resistance along M1 also increases faster than the resistance of the path to M6. Beyond a specific distance, which depends on the impedance characteristics of the interconnects, the path that consists of M6 and a via stack exhibits a lower impedance than the path that includes M1 and the TSV. Beyond this distance, most of the current flows through M6 and, eventually, the voltage drop in both circuits is approximately the same.

This behavior, therefore, suggests that the TSV path within power and ground distribution networks has a local effect and is efficient for those transistors within a specific region around the TSV. This distance is determined by the current demand in the vicinity of the TSV and the interconnect impedance characteristics of the TSV and intratier metal layers.

These ancillary paths should not be perceived as another means to globally distribute power and ground within a 3-D circuit. These paths, however, can locally enhance the distribution of power within a circuit. Different ways exist to exploit this advantage. For example, the TSVs are a crucial element of a 3-D circuit as these interconnects provide intertier communication and power. The size of the TSVs, however, greatly increases routing congestion. To mitigate this issue, several stacked vias within the power grid in a region close to each TSV can be removed as most of the current flows through the TSV and M1. Alternatively, the intentional decoupling capacitance can be reduced since the voltage drop from the power supply to the transistors is considerably less.

18.3.1 ENHANCED POWER INTEGRITY BY EXPLOITING THROUGH SILICON VIA PATHS

The effect of the TSV paths on the behavior of a power grid is explored in this subsection. Two pairs of 10 by 10 resistive metallic grids are used to model a portion of a power distribution network. Each pair corresponds to the topmost (i.e., M6) and lowest (i.e., M1) metal layer, which is depicted, respectively, in Figs. 18.24A and B. At each grid node other than the node in the middle, a stack of vias connects the two grids. At the center of the grids, a TSV connects to a power pad. The only difference between the two pairs of grids is that in one of the grids the TSV is connected to both M6 and M1 in addition to the power pad (i.e., the TSV path). Alternatively, in the other grid, the TSV is only connected to the power pad and M6, while the node at the center of the M6 and M1 grids is connected to a stack of intratier vias. The ground distribution network is similarly modeled.

Each stack of vias is modeled as a resistor, while the impedance of the TSV includes a resistive and an inductive component. The resistance and inductance are as assumed in Section 18.3. Specific nodes of the grid that model layer M1 are connected to a current source, as shown in Fig. 18.24B. Furthermore, a decoupling capacitor notated as C_{dec} is connected to each grid node, as illustrated in Fig. 18.24B. Each current source is modeled with a triangular waveform [686]. The rise and fall times are, respectively, 30 and 70 ps and the switching period is 100 ps. No intermediate quiet time between successive switching is therefore assumed. A voltage ripple of 5% of the power supply is assumed and $V_{dd} = 1$ volt.

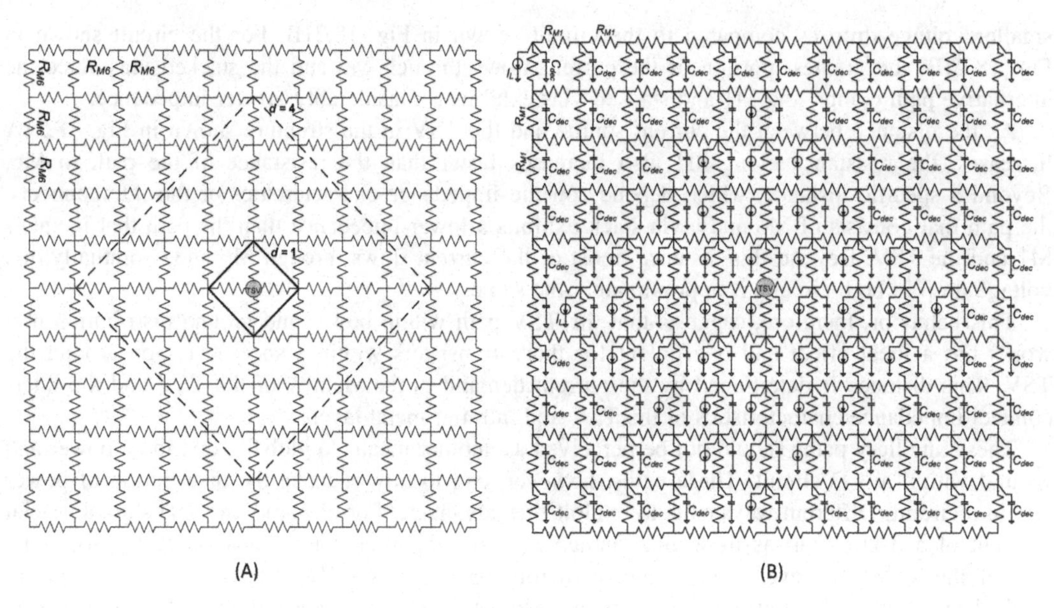

FIGURE 18.24

Resistive grid to model a segment of a power distribution system. (A) In the uppermost (M6) metal layer, and (B) in the lowest (M1) metal layer.

To demonstrate the effects of the low impedance path on the power grid formed by connecting the TSV to both M6 and M1, two different switching scenarios are considered. Initially, all of the sources shown in Fig. 18.24B draw current, while in the second scenario, only three sources switch. The length of the grid segments is notated by l and is varied to explore the resulting voltage drop.

For the first scenario, each current source draws a peak current of $I_L = 0.8$ mA. In addition, the decoupling capacitors are removed to evaluate the voltage drop across the entire grid caused by switching the current sources. Both of the grids are simulated with SPICE.

The voltage drop at specific nodes of the M1 grid (including the node where the maximum voltage drop occurs) are plotted in Fig. 18.25 for increasing length l of the grid segment. These nodes are located at the upper left corner of the grid (S1), at a $4l$ distance to the right of the TSV (S2), and at the TSV (S3). The voltage drop at these nodes is illustrated in Fig. 18.25 by the curves denoted by, respectively, the circles, squares, and triangles. For the grid where the TSV path is present (depicted in Fig. 18.25 by the group of solid curves), the voltage drop is significantly lower as compared to the grid where the TSV path is not considered. Note that the voltage drop at the current source located at the upper left corner of the grid (S1) (i.e., the pair of curves denoted by the circles) is affected less by the TSV path as compared to the other two nodes. This situation demonstrates the locality of the effect caused by the TSV path.

A negligible increase (~5 mV) of the voltage at the TSV location is noted, as shown by the solid curve with increasing l. This counterintuitive behavior can be explained by considering the current that flows through the TSV path and the neighboring paths through the stacked vias. As l increases, the impedance of each M1 segment becomes comparable to the impedance of a stack of

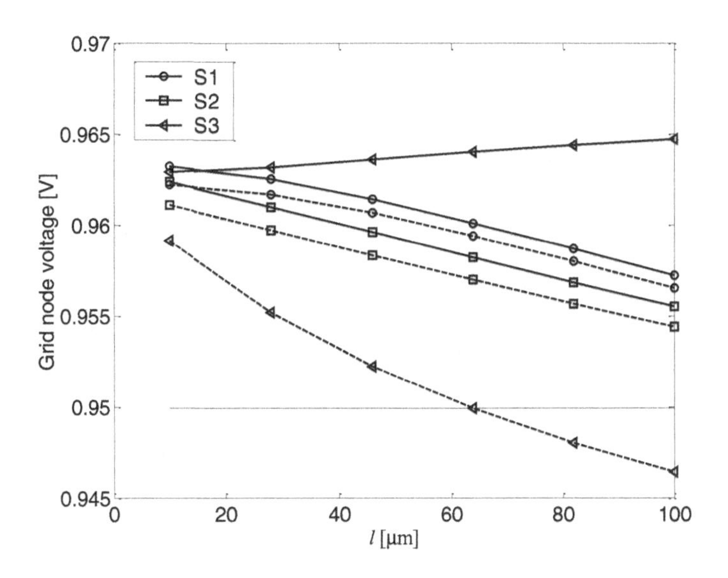

FIGURE 18.25

SPICE simulation of the voltage drop on the M1 grid for different nodes with (solid curves) and without (dashed curves) the TSV path. No stacked vias are removed ($d = 0$).

vias. Consequently, the current that flows through the TSV and M1 decreases. Alternatively, the current that flows through M6 to the TSV stacked vias increases. This change in the flow of current causes the voltage to increase at the TSV node with increasing l.

The decrease in the voltage drop due to the TSV path can be used to improve routability within a tier. This improvement is important since TSVs greatly increase routing congestion. To demonstrate that a smaller number of stacked vias within the power grid is required when the TSV path is considered, the stacked vias are removed within increasing radius from the TSV notated as d. The resulting voltage drop is depicted in Fig. 18.26 where $l = 30$ μm. Note that fewer paths provide current to the transistors. Since the stacked vias are removed from the grid, the TSV path supports greater current. The voltage drop specification is therefore maintained up to $d = 3$ or, equivalently, with 22% fewer intratier vias (see the dashed curves in Fig. 18.26).

The stacked vias are also removed from the grid when the TSV path is not present. The voltage drop on this grid is also shown in Fig. 18.26 by the solid curves. In this grid, the TSV is not connected to M1 and, consequently, the voltage drop rapidly increases as the stacked vias are removed to decrease routing congestion.

In the second scenario, only three sources switch. These three sources are located at the upper left corner of the grid (S1) and at the adjacent nodes (in the east and west direction) of the node where the TSV is connected (S2). The peak current of these sources is $I_L = 8$ mA, while the rise and fall times are the same as in the previous scenario. The voltage drop at these current sources (i.e., S1, S2) and the center of the M1 grid (S3) is depicted in Fig. 18.27. The voltage drop within the two grids with (solid curves) and without (dashed curves) the TSV path is illustrated. The simulation results indicate that the additional path again decreases the voltage drop for this current

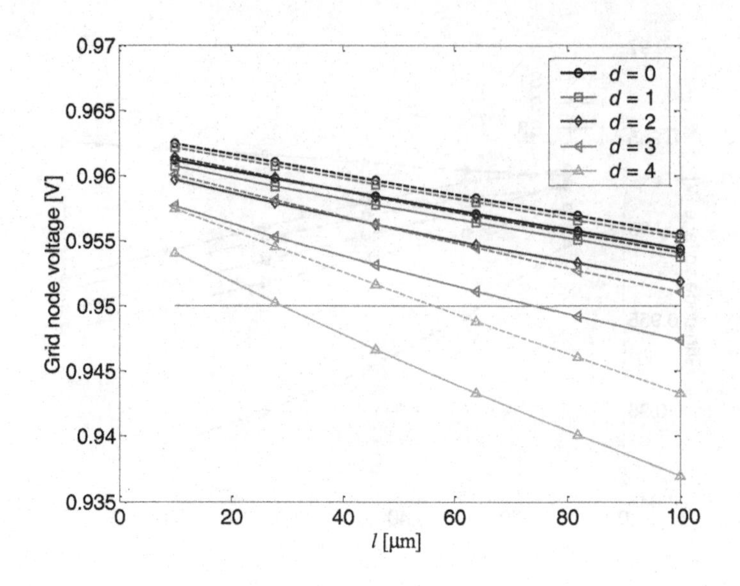

FIGURE 18.26

SPICE simulation of the maximum voltage drop on the M1 grid by successively removing the stacked vias (i.e., increasing d) with (dashed curves) and without (solid curves) the TSV path.

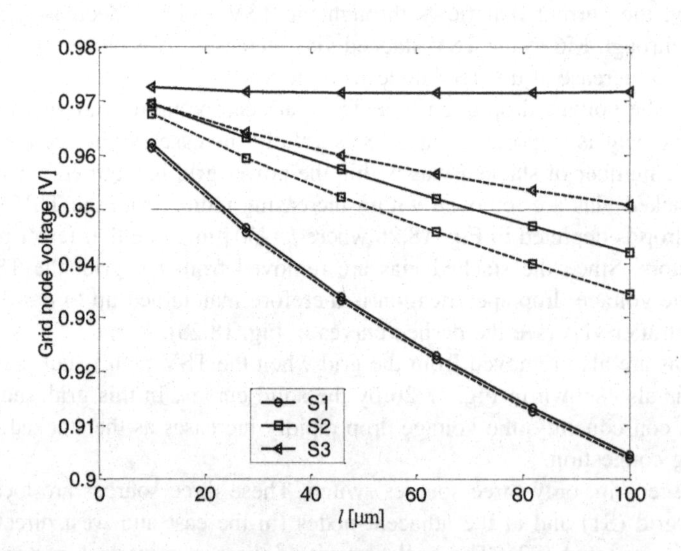

FIGURE 18.27

SPICE simulation of the voltage drop on the M1 grid for different nodes and with no stacked vias removed ($d = 0$) with (solid curves) and without (dashed curves) the TSV path. Only three current sources switch.

Table 18.3 Minimum Node Voltage Across the Power Grids With and Without Considering the TSV Path

	Minimum Node Voltage (mV)	
Total Decoupling Capacitance (nF)	**With TSV Path**	**Without TSV Path**
0.05	952	945
0.0625	956	952

source configuration. The locality of the TSV path is also demonstrated, since the voltage drop at the remote current source (S1) does not significantly change when the length of the grid segment is varied. Note the solid and dashed curves for S1 denoted by the circles in Fig. 18.27, which are practically indistinguishable.

These local power distribution paths can also decrease the extrinsic or intentional decoupling capacitance used to compensate for the voltage drop on the power grid. For the first scenario, the peak current of the sources is increased to $I_L = 1$ mA, while the decoupling capacitors satisfy the voltage drop constraint. The decoupling capacitance is listed in Table 18.3 where the TSV path is both present and not present. The grid including the TSV path requires 25% less capacitance to satisfy the voltage drop constraint, an important savings in the area of a tier within a 3-D system.

18.3.2 EFFECT OF THROUGH SILICON VIA TAPERING ON POWER DISTRIBUTION NETWORKS

In all of the power delivery schemes presented in Section 18.1 and the models of 3-D power distribution networks discussed in Section 18.2, the TSVs contribute to the power supply noise along the vertical flow of current within a 3-D circuit. To lower the parasitic impedance of the TSV, one approach is to increase the number and/or diameter of the TSVs, which increases the area occupied by these interconnects. These approaches assume that the same number and size of TSVs are employed within each tier. This constraint is however a strict and inappropriate constraint as the overall current carried by the TSVs within each tier is not the same.

Those TSVs within the tiers closer to the package typically carry current drawn from all of the other tiers. The TSVs in the upper tiers closer to the heat sink carry an increasingly lower current (assuming all of the tiers are active). This behavior supports the use of a nonuniform radius for the TSVs across the 3-D stack. For those tiers closer to the package, the TSVs exhibit a greater diameter [673], as illustrated in Fig. 18.28A. Although uniform sizing can be applied to all of the TSVs, uniform sizing wastes silicon area in those tiers where the TSVs carry less current.

The TSVs, however, constitute primary paths for the flow of heat within 3-D circuits, as discussed in Chapter 12, Thermal Modeling and Analysis. By following this same approach, to resize the TSVs to ensure the efficient flow of heat throughout a 3-D stack, increasing TSV sizing occurs in the opposite direction as compared to sizing TSVs for controlling power supply noise [673]. The electrical and thermal networks are separately evaluated to determine the optimum TSV size to

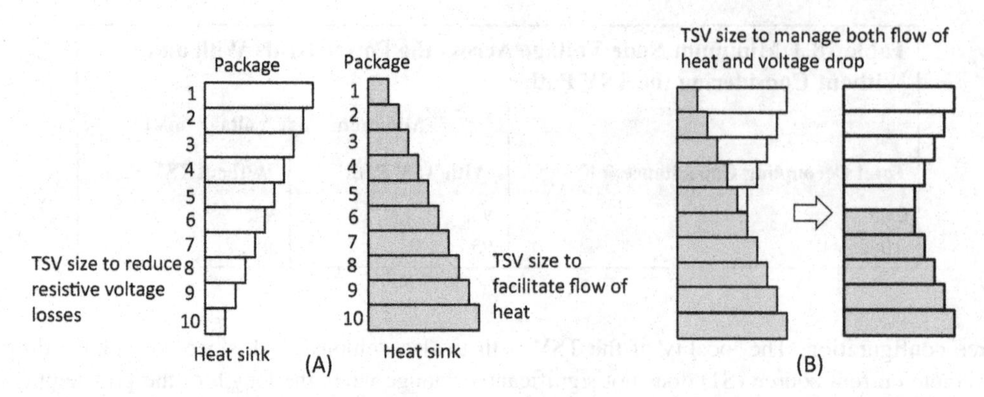

FIGURE 18.28

Nonuniform TSV tapering to address both power supply noise and temperature. (A) Opposite tapering is required to individually satisfy the power supply noise and temperature objectives, and (B) adapting the size of the TSVs across tiers to ensure that both objectives are satisfied [673].

satisfy each of these objectives. From this evaluation, those sizes that satisfy both objectives (without being optima for each individual objective) are selected, as illustrated in Fig. 18.28B.

A Thevenin network is utilized to determine the TSV size, where the same current through each tier is assumed. The maximum TSV resistance for the minimum TSV radius allowed by the target fabrication process is also known. By sweeping the TSV radius (or by applying an optimization method), the optimal TSV size to satisfy voltage constraints and minimize the area of the TSVs is determined. In a similar manner, the size of the TSVs to satisfy the temperature constraints is determined where the thermal resistance is utilized in this case.

To demonstrate the effects of nonuniform sizes of the TSVs, an exemplary 3-D circuit with ten tiers is considered [673]. Two different sizing ratios are considered. In the first case, the tapering step is 2 μm (Case 1), while in the second case, the tapering step is 0.2 μm (Case 2). In both cases, a minimum radius of a TSV of 1 μm is assumed. To satisfy both the thermal and power supply noise objectives, the resulting radii of the TSVs for the two cases are: (1) $R_{9-10} = 9$ μm, $R_{8-9} = 7$ μm, $R_{7-8} = 5$ μm, $R_{6-7} = 3$ μm, $R_{5-6} = 1$ μm, $R_{4-5} = 3$ μm, $R_{3-4} = 5$ μm, $R_{2-3} = 7$ μm, and $R_{1-2} = 9$ μm for the tapering step of 2 μm; (2) $R_{9-10} = 1.8$ μm, $R_{8-9} = 1.6$ μm, $R_{7-8} = 1.4$ μm, $R_{6-7} = 1.2$ μm, $R_{5-6} = 1$ μm, $R_{4-5} = 1.2$ μm, $R_{3-4} = 1.4$ μm, $R_{2-3} = 1.6$ μm, and $R_{1-2} = 1.8$ μm for the tapering step of 0.2 μm.

Comparing these two cases with the case of uniform sizing across all of the tiers within this stack, the uniform TSVs yield a voltage drop of 90 mV for the TSV farthest from the package tier. For Case 1, the noise in the same tier is only 65 mV [673]. Furthermore, the noise in Case 1 is lower than in Case 2, demonstrating that a relatively small difference in TSV radius produces limited gains. The temperature also drops when a nonuniform radius is applied to the TSVs. Similarly, the temperature decreases from 43°C to 30°C for the tier located farthest from the heat sink if tapering is applied. The drop in temperature due to the use of nonuniform TSVs is considerable. In addition, a greater tapering step will further lower the temperature.

Another well known method to limit power supply noise is the use of decoupling capacitance. Techniques for allocating decoupling capacitance within 3-D circuits are discussed in the following section, where the contribution of a decoupling capacitance from neighboring tiers to decrease power supply noise in some other tier is described.

18.4 DECOUPLING CAPACITANCE FOR THREE-DIMENSIONAL POWER DISTRIBUTION NETWORKS

Decoupling capacitors are an indispensable component at every level of the power distribution system. In on-chip power distribution networks, this capacitance is achieved by either a MOS capacitance or a MIM capacitance [325]. The typical difference between the two types of on-chip decoupling capacitances is a MOS capacitor provides a higher density capacitance at the expense of higher leakage current as compared to a MIM capacitor [687]. Trench decoupling capacitors can also be used as an alternative to reduce the leakage current of a MOS capacitance although this technology is fairly expensive [688]. Typical decoupling capacitors are efficient for a small distance from the switching load. This effective distance has been analyzed for 2-D systems [686]. For 3-D circuits, however, the decoupling capacitance of an adjacent tier can be located within a distance of tens of micrometers from the switching load due to the short and low resistive path of the TSVs. Allocation of the decoupling capacitance within a 3-D system should therefore consider the overall 3-D stack rather than individual tiers.

To explore the effects of the decoupling capacitance among the tiers, some simple cases are discussed based on the analytic model of power supply noise presented in Section 18.2. These systems consist of a varying number of tiers, where each tier is an Intel microprocessor manufactured in a 65 nm CMOS technology node [678]. Each tier comprises five circuit blocks with the same footprint and the same current density of 100 A/cm^2 [674]. All of the blocks switch simultaneously to produce the worst case power supply noise. The current waveform is a ramp function with a rise time of 0.7 ns.

The power network contains 43 power (ground) tracks between each pair of power (ground) pads. The resistance of a wire segment R_w in (18.8) is 0.22 Ω. The inductance of the package pins is 0.5 nH. The TSVs are 200 μm in height with a diameter of 50 μm. The effective inductance (one half of the loop inductance) for one TSV is determined from an EM solver as 0.06 nH. Finally, 20% of each tier is assumed to be available for the MOS decoupling capacitors.

Several scenarios illustrated in Fig. 18.29 are evaluated in terms of the worst case power supply noise. The 2-D system (single tier) exhibits the lowest noise, $V_{drop} = 182$ mV, as shown in Fig. 18.29A. Stacking these tiers increases the noise in the topmost tier (farthest from the package pins) to 400 mV, as depicted in Fig. 18.29B. Adding a fifth tier as a decoupling capacitance tier adjacent to the package, as illustrated in Fig. 18.29C, decreases this noise by 22% to 312 mV. The rationale for placing this tier close to the package is to offer a low recharge time for the decoupling capacitance. The efficiency of the decoupling capacitance tier however is highest if this tier is placed on the top of the 3-D stack, as shown in Fig. 18.29D. In this case, the noise is 256 mV.

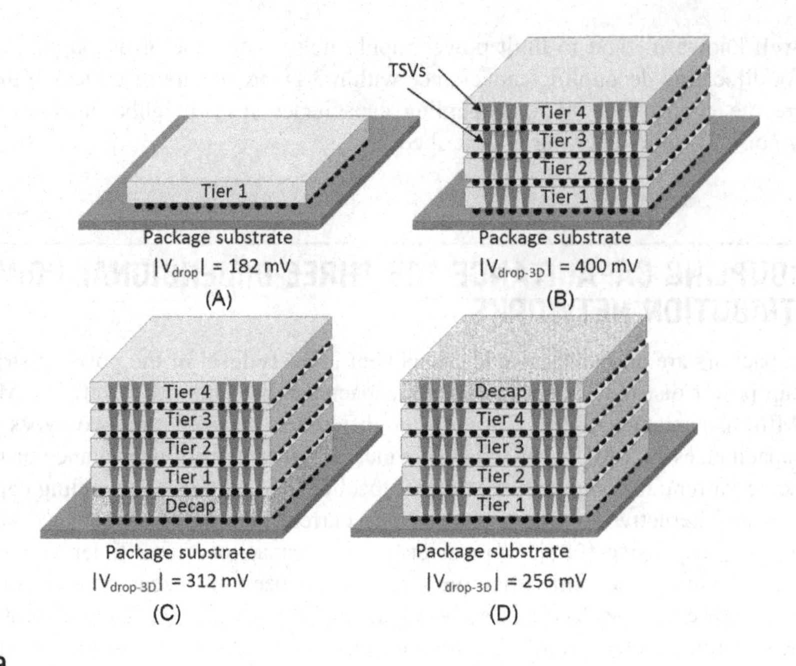

FIGURE 18.29

Power supply noise from employing one tier of decoupling capacitance. (A) A 2-D system, (B) a four tier system with no tier for the decoupling capacitance, (C) a decoupling capacitance tier close to the package, and (D) a decoupling capacitance tier on top of the 3-D system [674].

Although the noise decreases, this noise is considerably higher than in the single tier case. The use of a second tier as a decoupling capacitance is depicted in Fig. 18.30 to further lower the noise. As shown in Figs. 18.30B and C, interleaving the decoupling capacitance tiers with the processor tiers does not yield the lowest noise. Rather, placing both of these tiers on top of the stack, as illustrated in Fig. 18.30D, produces the lowest noise $V_{\text{drop-3D}} = 199$ mV, which is much closer to the noise produced by a 2-D system.

In each of these scenarios, the decoupling capacitance is the same across each tier which may not be possible for several reasons. These reasons include, for example, the limited whitespace for the decoupling capacitance in each tier, the different technology nodes among the tiers, and the operating conditions of the circuits. A particular situation arises when the circuits in the adjacent tiers are power gated to save power. In these cases, the decoupling capacitance (intentional or intrinsic) from these tiers is no longer available since those transistors act as power switches that isolate this capacitance from the power distribution network of the overall system. This situation enhances the beneficial effects of the capacitance from the neighboring tiers. Specific topologies that overcome this variation in decoupling capacitance are discussed in the following subsection.

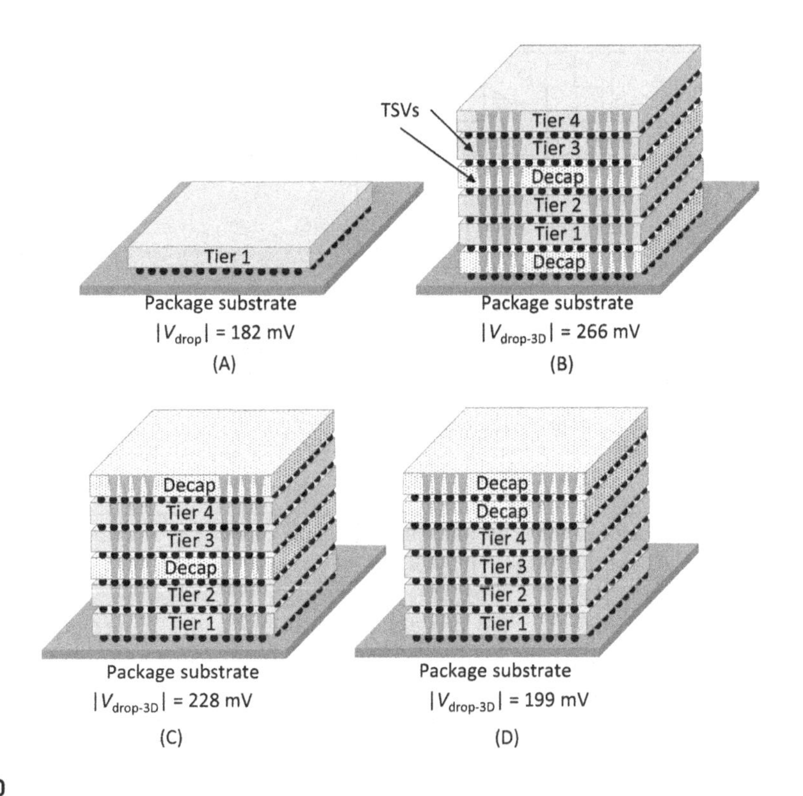

FIGURE 18.30

Power supply noise from employing two tiers of decoupling capacitance. (A) A 2-D system, (B) one decoupling capacitance tier is placed next to the package and the second tier between tiers two and three, (C) one decoupling capacitance tier is placed on top of the stack and the second tier between tiers two and three, and (D) both decoupling capacitance tiers are placed on top of the stack [674].

18.4.1 DECOUPLING CAPACITANCE TOPOLOGIES FOR POWER GATED THREE-DIMENSIONAL ICs

Power gating is a broadly used technique to greatly reduce the power of an integrated system. Those circuits that do not perform a task during a certain time are temporarily disconnected from the power supply to eliminate leakage and dynamic current. Large transistors, usually called "sleep transistors," disconnect the power supply rails from the "virtual" power supply which is connected to the circuits [689]. A side effect of power gating is the decrease in the overall capacitance of the system. Consequently, the capacitance in these circuits cannot behave as decoupling capacitors and help alleviate abrupt current surges in adjacent circuits. This situation is more subtle in 3-D circuits for two reasons. First, due to the potentially greater power densities as compared to 2-D circuits, low power methods, such as power gating, may be applied more aggressively. Second, the short and low impedance vertical paths provided by the TSVs enable the capacitance of adjacent tiers to more effectively satisfy abrupt current demands within a tier.

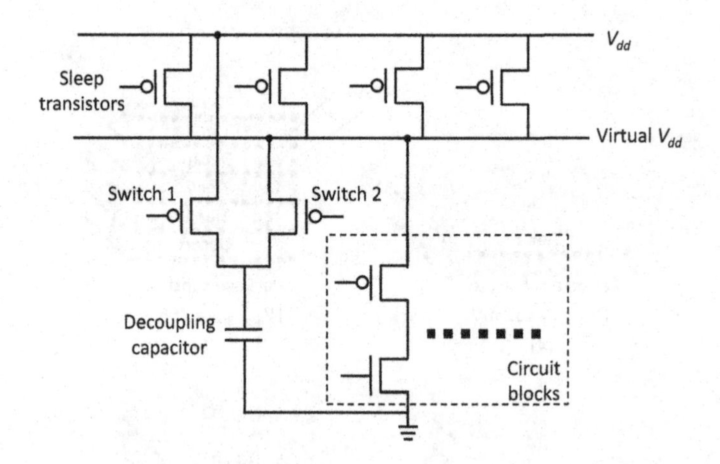

FIGURE 18.31

Reconfigurable decoupling capacitance topology where the decoupling capacitor is connected to the power rail even if the sleep transistors are switched off [691].

Another aspect of power gating is the process of transitioning the power gated circuit to an operating state. If several circuits simultaneously transition to an active state, the sudden current demand can cause a significant dip in the power supply voltage. This voltage drop appears as power supply noise on nearby active circuits [690]. Considering these two issues, methodologies to effectively utilize decoupling capacitance in 3-D circuits are discussed in this subsection.

A topology that allows the decoupling capacitance to be utilized despite the circuit block being power gated is illustrated in Fig. 18.31 [691]. The primary difference from a standard power gated circuit is that the decoupling capacitor is connected both to the global voltage supply and to the virtual power rail through two switch transistors, which select one of the two power lines.

If transistor *switch 2* is on and *switch 1* is off, the decoupling capacitor is connected to the virtual V_{dd}. If the circuit is active, the sleep transistors are on, and the decoupling capacitor provides charge to the local circuit blocks, shown in Fig. 18.31, rather than to more distant circuits as the path to the V_{dd} rail is more resistive. Alternatively, if the circuit is power gated, *switch 2* is off and *switch 1* is on, the decoupling capacitor supplies charge to the other circuits, disconnecting the V_{dd} rail from the virtual power supply.

A tradeoff between the switch transistor and decoupling capacitor exists for this topology as the overall area overhead from these components should be as small as possible while satisfying any voltage constraints. Further increasing the decoupling capacitance is not useful as the area and, consequently, the on-resistance of the switch transistors is high, limiting the efficiency of the capacitor. Alternatively, if the switch transistor is large, lowering the impedance of the discharge path of the capacitor, the leakage current is greater. Therefore, choosing an appropriate size for these two components is required to achieve a decoupling capacitor system with the highest efficiency.

Another topology for connecting the decoupling capacitance is illustrated in Fig. 18.32, where the decoupling capacitor is directly connected to the V_{dd} rail. The capacitor provides sufficient charge to the adjacent circuits irrespective of whether the local circuit blocks are power gated. The discharge path for the local circuit blocks, however, includes the sleep transistor.

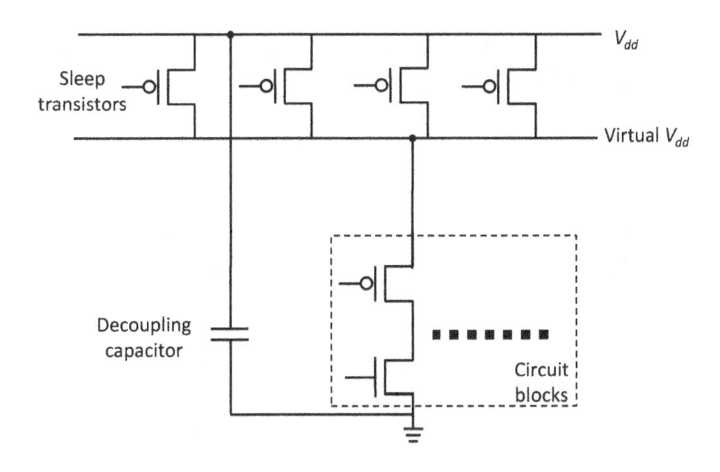

FIGURE 18.32

Always on decoupling capacitance topology. The charge provided to the local circuit blocks flows through the sleep transistors [692].

Consequently, this topology can exhibit greater power supply noise as compared to the reconfigurable topology shown in Fig. 18.31. As the sleep transistors, however, are typically large (the equivalent transistor width is on the order of millimeters), the voltage drop across these devices is extremely small.

To evaluate the efficiency and explore tradeoffs between these two topologies, a power grid for a 3-D circuit with three tiers is considered. The topmost tier is connected to a package with C4 connections. Ten metal layers are used in this topology [693]. The top two layers are the global power grid, and layers 7 and 8 are the virtual grid. In addition, a via-last manufacturing process for the TSVs is assumed. The grid is 1 mm × 1 mm. The segments of the power grid and the TSVs are modeled with multiple *RLC* π segments. The C4 connections are modeled as *RL* sections. Furthermore, the load consists of an inverter pair of different size distributed across the circuit. Finally, the voltage drop constraint is set to 5% of V_{dd}, 50 mV.

As the 3-D circuit includes three tiers, several scenarios with one or more active tiers are evaluated. These scenarios are listed in columns 1 to 3 of Table 18.4 where the maximum power supply noise for all of these scenarios is also listed. The decoupling capacitance ensures that the voltage constraint is satisfied for the standard topology with all tiers active (and, consequently, connected to global V_{dd}). As reported in Table 18.4, for the other scenarios where some tiers are power gated, the traditional topology violates the voltage constraint. This behavior is due to the lower decoupling capacitance available across the entire 3-D stack. Alternatively, both of the other topologies exhibit a considerably smaller power supply noise.

Another important parameter that affects power noise in power gated circuits is the wake-up time of the inactive circuits, which affects the operation of the active systems. This behavior is common to both 2-D and 3-D circuits. A straightforward method to decrease the current demand during the wake-up process (the transition from the power gated to the operating state) is to slightly delay the turn-on time of the sleep transistors to ensure that the drawn current increases at a slower

Table 18.4 Peak Voltage Noise for Different Scenarios of Activity Among the Three Tiers of a 3-D Circuit for the Standard, Reconfigurable, and Always On Topologies of Decoupling Capacitance [692]

Operating State of Tiers			Power Supply Noise of Decoupling Capacitance Topologies (mV)				
				Reconfigurable		Always On	
Top Tier	Middle Tier	Bottom Tier	Standard		Decrease		Decrease
On	On	On	50	50	–	50	–
On	Off	On	48.16	43.48	9.7%	43.40	9.9%
Off	Off	On	52.22	39.64	24.1%	38.07	27.1%
Off	On	On	48.55	44.50	8.3%	42.89	11.7%
Off	On	Off	52.51	38.78	26.1%	37.55	28.5%

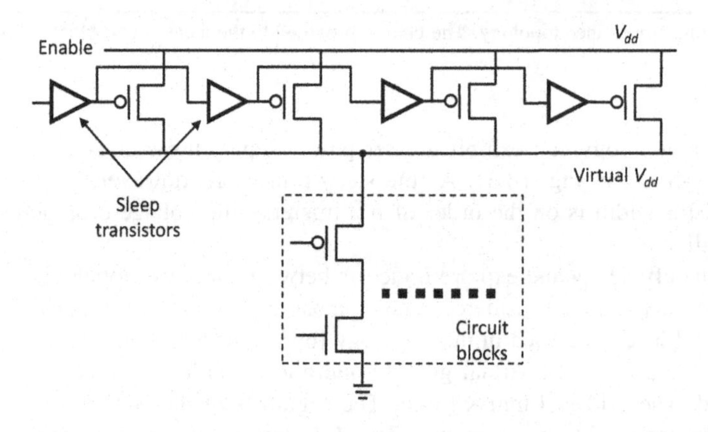

FIGURE 18.33

A daisy chain of buffers switches the sleep transistors on, subsequently ensuring that the current gradually increases, limiting the abrupt current changes within the power grid [695].

rate. A daisy chain of buffers is utilized for this approach, as illustrated in Fig. 18.33. The drawback of this approach, however, is the prolonged wake-up time of the circuit. To avoid this situation, a wake-up controller initiates the wake-up process in two steps, where a small number of sleep transistors is switched on, followed by the remaining transistors switching on [694]. The duration of each step is constant and is typically several clock cycles.

This two step method is also applicable to 3-D circuits; however, the situation is more complicated since the power supply noise exhibited on each tier strongly depends upon the location of the tier within the stack [695]. For 3-D stacks assuming that each tier is either active, inactive, or transitions to the active state, several cases exist where the power supply noise varies across all of the possible operating scenarios. Different wake-up times are therefore used between the two steps depending upon the tier that transitions to the active state and the tiers that are already active. This information is utilized to guide an adaptive wake-up controller, where the time duration of each step is carefully controlled [695]. The objective of this adaptive controller is to reduce the total wake-up time, while satisfying any power supply noise constraints.

This adaptive two step wake-up controller has been evaluated for a 3-D circuit with five tiers, where a 22 nm technology node is assumed. In the first step, 5% of the sleep transistors are switched on, whereas the remaining transistors are switched on during the second step. The duration of each step, however, is adjusted according to the operating conditions of the overall 3-D stack. A comparison of the efficiency of the wake-up controller is performed with both constant and adaptive steps. The adaptive controller exhibits an average decrease of 28% in the wake-up time of the 3-D stack, without violating the power supply noise constraint. Although not discussed here, the number of sleep transistors switched on between the two steps is another variable to control the wake-up time at the expense of a more highly complex controller.

18.5 WIRE SIZING METHODS IN THREE-DIMENSIONAL POWER DISTRIBUTION NETWORKS

A straightforward approach to design a 3-D power distribution network is to independently generate a power network for each tier of the stack and connect these networks with TSVs. Unfortunately, this approach does not consider several traits of 3-D power distribution networks, leading to suboptimal networks that either do not satisfy the power supply noise constraints throughout the stack or results in overdesign, wasting vital interconnect and power resources.

As described in the previous sections of this chapter, the behavior of 3-D power distribution networks depends on the number and size of the TSVs, the TSV technology, and the decoupling capacitance of the adjacent tiers. These features are relevant only to 3-D circuits. Moreover, most case studies are based on simple assumptions. Although these assumptions are useful to understand the behavior of 3-D power distribution networks, more formal techniques are required.

Optimizing power distribution networks for 3-D systems requires consideration of several parameters in addition to the parameters of 2-D circuits, such as the size of the horizontal wires. Moreover, specific assumptions, such as uniform power dissipation for each tier should be relaxed to consider more practical cases. An algorithm which optimizes the area of the power network within each tier while simultaneously satisfying global power noise and thermal objectives throughout a 3-D system is presented in this section. Several other constraints are also added to the optimization process.

The power supply noise objective of the algorithm is

$$V_{\text{drop}}^{i} = \frac{V_{\text{dropmax}} \text{ for all tiers}}{\text{number of tiers}} - \frac{1}{g_{\text{TSV}}^{i-1,i}} \sum_{j=1}^{\text{top tiers}} I_j - r_{\text{int}}^{i} I_i, \tag{18.44}$$

where V_{dropmax} for all tiers is the overall noise allowed across the system. The term $g_{\text{TSV}}^{i-1,i}$ is the total conductance of all of the TSVs connecting tier $i-1$ to tier i (for $i>1$). The current flowing through each of these TSVs is denoted by I_j. Furthermore, the bonding style is face-to-back, which implies that the topmost global layer of a tier connects by TSVs to the bottommost local layer in

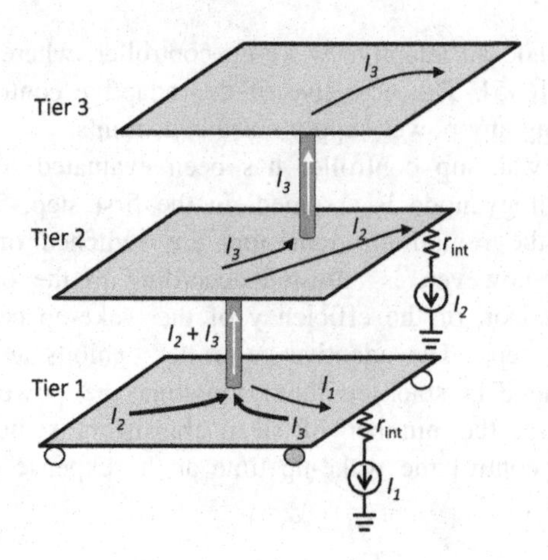

FIGURE 18.34

Current flow within a three tier stack. Note the current flowing through the TSVs of each tier [472].

the next tier. For this bonding approach, r^i_{int} notates the total impedance from all of the metal layers of the power network within each tier.

For the three tier circuit shown in Fig. 18.34, the corresponding constraints for the power supply noise based on (18.44) are [472]

$$V^1_{\text{drop}} = \frac{V_{\text{dropmax}} \text{ for all tiers}}{3} - r^1_{\text{int}} I_1, \tag{18.45}$$

$$V^2_{\text{drop}} = \frac{V_{\text{dropmax}} \text{ for all tiers}}{3} - \frac{1}{g^{12}_{\text{TSV}}}(I_2 + I_3) - r^2_{\text{int}} I_2, \tag{18.46}$$

and

$$V^3_{\text{drop}} = \frac{V_{\text{dropmax}} \text{ for all tiers}}{3} - \frac{1}{g^{23}_{\text{TSV}}} I_3 - r^3_{\text{int}} I_3, \tag{18.47}$$

where the term relating to the resistive voltage drop of the TSVs is not included in (18.45) since this tier is the bottommost tier connected to the package.

An expression for the temperature constraint is similar to (18.44) based on the electro-thermal duality, and is [472]

$$T^i_{\text{max}} = \frac{\max\Delta T_{\text{between tiers}}}{\text{number of tiers}} - \frac{1}{g^{i-1,i}_{\text{TSV}}} \sum_{j}^{\text{bottom tiers}} Q_j - r^i_{\text{thint}} Q_i, \tag{18.48}$$

where max_$\Delta T_{\text{between tiers}}$ is the maximum allowed temperature difference between the uppermost and bottom tiers. The term $g^{i-1,i}_{\text{TSV}}$ denotes the total thermal resistance of all of the TSVs connecting two adjacent tiers. The heat produced within a tier i is notated as Q_i. These constraints for the power supply noise and temperature, respectively, facilitate the analysis of the optimization problem as the size of the power grid and TSVs of each tier are treated separately. Satisfaction of these constraints (see, for example (18.44)) ensures that the noise and temperature limits are maintained across the entire 3-D stack.

With these constraints and utilizing the electro-thermal model presented in Section 18.2.1 to determine the temperature and voltage at any node within the power distribution network of a 3-D circuit, the problem of optimizing resources in these networks is stated as follows. Minimize the cost function,

$$\sum \beta w_{\text{track}} + \sum (1-\beta) D_{\text{TSV}}, \tag{18.49}$$

to ensure that the following constraints are satisfied [472]:

(C1) for the voltage of the node q in tier i (see (18.34) and (18.44)),

$$V_q^i \geq V_{dd} - V_{\text{drop}}^i, \tag{18.50}$$

(C2) for the voltage of the node q in tier i (see (18.41) and (18.48)),

$$T_q^i \leq T_{\text{amb}} + T_{\text{max}}^i, \tag{18.51}$$

(C3) for the width of the power and ground lines,

$$w_{\min} \leq w \leq w_{\max}, \tag{18.52}$$

(C4) for the diameter of the TSVs,

$$D_{\min}^* \leq D \leq D_{\max}, \tag{18.53}$$

(C5) for ensuring reliability, an electromigration constraint can be added (see (18.36)) where

$$J_{q-j}^i \leq J_{\max}, \tag{18.54}$$

and **(C6)** is a constraint that emphasizes either the voltage drop or the temperature objective. The weighted function is

$$a \cdot V_q^i + (1-a) T_q^i. \tag{18.55}$$

In the cost function (18.49), w_{track} is the width of the horizontal wire tracks within the power distribution network within each tier, and D_{TSV} is the diameter of the TSVs within each tier. Note that the TSVs at different tiers can have different diameters but the TSVs within a tier are all uniformly sized. The coefficient β is a weighting factor to determine the significance of each term in (18.49). The overall optimization process is illustrated in Fig. 18.35. Since the tiers towards the middle of the stack are more sensitive to power supply noise and thermal issues, the TSV sizing process is initially applied only to these tiers, as shown in Fig. 18.35A [472]. The resulting TSV diameter for these tiers is a lower bound for the other tiers, denoted as $D^*_{\min}$ (see C4). Note also that in step 4 of the algorithm depicted in Fig. 18.35C, the objectives $O_{Vq\text{drop}}$ and O_{Tq} are dimensionless. If any of these objectives is equal to unity, the maximum allowed voltage drop or temperature has been achieved.

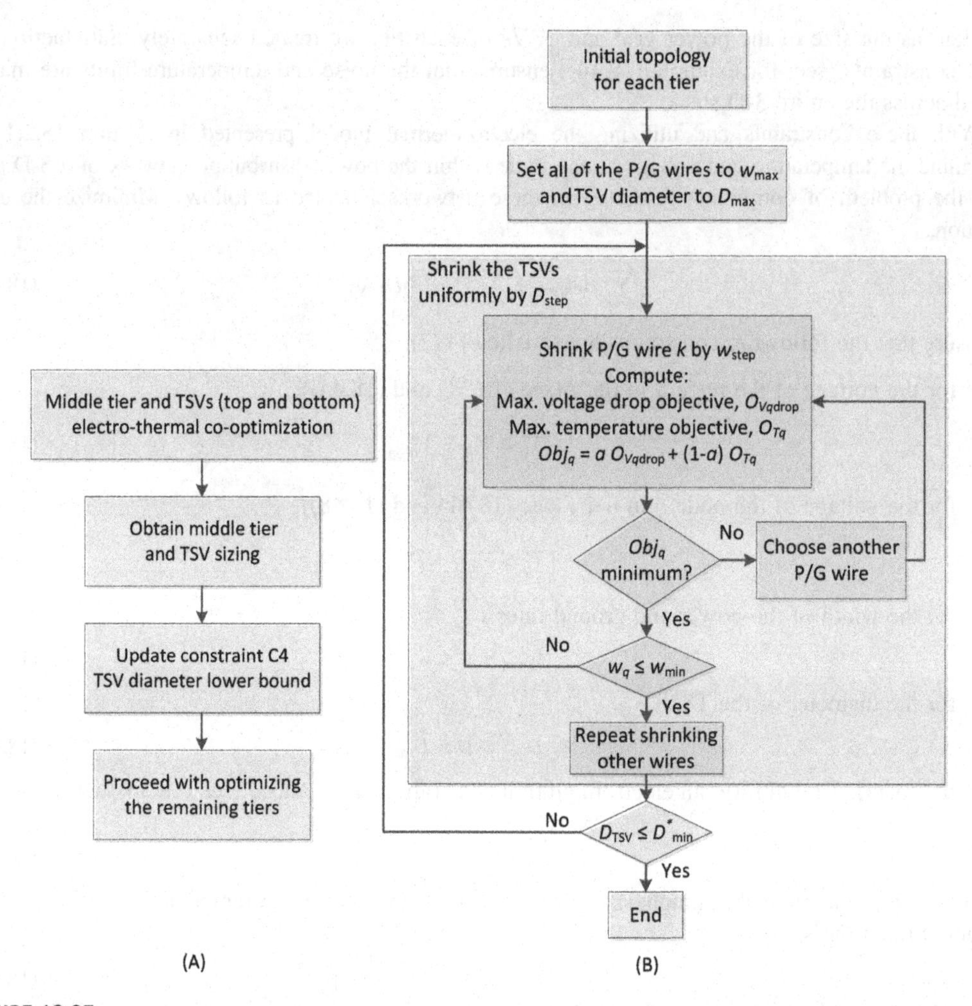

FIGURE 18.35

Optimization framework for 3-D power distribution networks where both power supply noise and temperature constraints are considered. (A) Optimal sizing process for the middle tier(s) is initially determined, (B) the flowchart of the algorithm, and (C) step by step description of the algorithm [472].

Per the cost function given by (18.49), the algorithm minimizes the area of the horizontal and vertical wires while simultaneously satisfying the objectives, O_{Vqdrop} and O_{Tq}. The diameter of the TSVs has a lower bound that is not set by the specifications of the technology but rather from the TSV diameter in the middle tiers D^*_{min}. In the flowchart depicted in Fig. 18.35B, the width of each horizontal wire within the power network of a tier is successively reduced, ensuring constraints C1 to C6 are not violated. This procedure is also applied to the TSVs and is iterated across all of the

3-D power distribution network (PDN) and TSV electro-thermal optimization

Step 1: Upsize all power network wires by w_{max} and TSVs by D_{max} to ensure that both voltage drop and temperature constraints are met

Step 2: Shrink the diameter of all of the TSVs by D_{step}

Step 3: Shrink each P/G wire q by w_{step} where k is the total number of P/G wires within the network

Step 4: Compute:

Voltage drop objective : $O_{V_{qdrop}} = \frac{V_{drop\_shrunk\_track_q}}{V^i_{drop}}$

Temperature objective: $O_{T_q} = \frac{T_{max\_shrunk\_track_q}}{T^i_{max}}$

Step 5: Compute: $Obj_q = a \cdot O_{V_{qdrop}} + (1-a) \cdot O_{T_q}$

Step 6: Compute: min(Obj_1, Obj_2,... Obj_q,.... Obj_k)
Identify wire j with the minimum electro-thermal objective Obj_j

Step 7: Does wire j have $w_j \geq w_{min}$?
- Yes, accept width reduction for wire j
  Update width of wire j to $w_j = w_j - w_{step}$
- No, go back to step 3

Step 8: Repeat shrinking the wire widths (steps 3–7) while electro-thermal constraints are met as $Obj_q \leq 1$

Step 9: Repeat shrinking TSV diameter (steps 2–8)

Step 10: End. New wire width distribution and TSV diameter are determined ensuring that the electro-thermal constraints are met

(C)

FIGURE 18.35

(Continued)

tiers. The algorithm terminates when any horizontal wire or TSVs within a tier can no longer be decreased without violating constraints C1 to C6.

This optimization algorithm is applied to a 3-D circuit with three tiers where the allowed voltage drop is 100 mV (10% of V_{dd}) and the allowed temperature difference between the two extreme tiers is 10°C. The ambient temperature is assumed to be 27°C. To demonstrate the effects of parameter *a,* the power density is set the same for all tiers. The applied power density in this case is 0.6 μW/μm^2. If temperature is emphasized ($a = 0$), the savings in area reaches 43%. If supply noise

Table 18.5 Savings in Area for Several Cases of Nonuniform Power Density Among the Three Tiers [472]

		O_{Vdrop}		O_T			
Case #	**Tier 3–Tier 2–Tier 1**	**Before**	**After**	**Before**	**After**	**Area Savings**	**Run Time (s)**
1	High–medium–low	0.87	1	0.46	0.56	25.0%	1440
2	High–low–medium	0.91	1	0.50	0.56	20.0%	2160
3	Medium–high–low	0.91	1	0.56	0.61	18.3%	1550
4	Medium–low–high	0.84	1	0.60	0.72	26.7%	1550
5	Low–medium–high	0.79	1	0.63	0.82	33.3%	1780
6	Low–high–medium	0.87	1	0.58	0.70	26.7%	1200
7	High–high–high	0.98	1	0.66	0.70	8.0%	900
8	Medium–medium–medium	0.88	1	0.45	0.90	36.0%	780
9	Low–low–low	0.86	1	0.34	0.82	41.0%	1140

is the primary focus ($a = 1$), the savings in area reaches 55%. For equally important voltage drop and temperature constraints ($a = 0.5$), the savings in area is 53%. The greatest savings in area is produced for $a = 0.9$.

The algorithm is also evaluated for several power densities to reflect different levels of power dissipation among tiers. The following power densities are considered: (1) 1.5 $\mu W/\mu m^2$ as the high power density, (2) 1 $\mu W/\mu m^2$ as the medium power density, and (3) 0.8 $\mu W/\mu m^2$ as the low power density. For $\beta = 1$ and $a = 0.5$, the area savings for the different cases of power densities are reported in Table 18.5. In columns 3 and 5, the listed objectives correspond to the use of w_{max} for all metal layers within the power distribution network. In columns 4 and 6, the same objectives are reported where the optimization algorithm is applied.

In all of the reported cases, the algorithm produces a wire size that satisfies the voltage drop target, while the temperature is considerably less than originally specified. This situation indicates that the wires can be further decreased in size without violating the maximum temperature constraint. Any further decrease in size will however violate the voltage drop constraint. The resulting gains in area, however, are considerable, reaching 41%.

18.6 SUMMARY

The behavior and related models and design issues for 3-D power distribution systems are discussed in this chapter. The primary concepts are summarized below:

- Vertical integration of the tiers affects power distribution networks in two ways. First, the horizontal dimensions of the power distribution network and the number of P/G package pins are significantly reduced as compared to the footprint of a 2-D circuit. Second, the power distribution networks within the different tiers are connected with TSVs with different electrical characteristics as compared to the horizontal wires.

- Integrating DC–DC converters into a 3-D stack greatly decreases the interconnect parasitic impedance of the power supply and logic tiers, improving the overall efficiency of the power delivery system.
- For a large number of tiers, integrating converters within the tiers at the ends of the stack lowers the voltage drop by more than 75% as compared to using converters in only one tier.
- An alternative power delivery method called "multi-level" utilizes differential rails of supply voltages between adjacent tiers. The primary advantage of the multilevel approach is that charge is recycled among differential pairs of power supply rails, greatly reducing the power supply noise.
- Several challenges relating to multi-level power delivery exist. The primary challenge is to maintain balanced currents among the tiers.
- Several models for 3-D power distribution networks have been developed to support design space exploration. These models range from closed-form analytic models based on the continuous plane approximation for the power and ground grids to compact models that replace segments of the power grids and TSVs with passive impedance sections.
- Models of the power distribution network within the memory tiers are more sensitive to substrate losses as the metal layers in these circuits are close to the substrate. These losses should therefore be included within the models.
- Electro-thermal models are based on the duality of voltage to temperature and current to heat flux to efficiently analyze power distribution networks.
- TSVs efficiently distribute power and ground in 3-D systems. Coaxial TSVs can half the number of TSVs used for P/G distribution in addition to reducing the required decoupling capacitance.
- Via-first and via-middle TSV technologies support ancillary paths that can improve local power integrity. These paths decrease routing congestion, save area, or lower the added decoupling capacitance.
- Nonuniform TSV diameters among tiers facilitates both power and thermal integrity.
- The low impedance of the TSV paths allows the decoupling capacitance of a tier to provide charge to the switching circuits in other tiers. The tiers contain decoupling capacitance for the neighboring circuits, where this capacitance is more efficient if placed farthest from the package.
- Power gating disconnects the neighboring decoupling capacitance, degrading power integrity. Specific circuits with low overhead can avoid this situation.
- A controller adapts the switching pattern and wake-up time of the sleep transistors. This technique avoids high current demands when the circuits transition from a power gated state to an operating state.
- Power and thermal integrity in 3-D systems is achieved by iteratively optimizing the power network of each tier to satisfy global constraints across the 3-D stack.
- An iterative algorithm, where the width of the horizontal P/G wires and the diameter of the TSVs are reduced, provides a means to satisfy noise and temperature constraints while reducing wire area.

CHAPTER

CASE STUDY: 3-D POWER DISTRIBUTION TOPOLOGIES AND MODELS

19

CHAPTER OUTLINE

An important issue for 3-D integrated circuits (ICs) is the design of a robust power distribution network that can provide sufficient current to every load within a system. In planar ICs, where flip-chip packaging is adopted, an array of power and ground pads is placed throughout an integrated circuit. Increasing current densities and faster current transients, however, complicate the power distribution design process. Three-dimensional integration provides additional metal layers for the power distribution networks through topologies unavailable in 2-D circuits. With 3-D technologies, individual planes can potentially be dedicated to delivering power.

The challenges of efficiently delivering power across a 2-D circuit while satisfying local current requirements have been explored for decades [205, 244, 696]. Two-dimensional power distribution networks are designed to achieve specific noise requirements. A variety of techniques have been developed to minimize both *IR* drops and *L di/dt* noise [697, 698], such as multi-tier decoupling placement schemes [399, 699, 700] and power gating [701]. These techniques have supported the increasing current demands of each progressive technology node; however, 3-D integrated systems are in a state of infancy, and much work is required to design efficient power distribution topologies for these vertically integrated systems.

Power delivery in 3-D integrated systems presents difficult new challenges for delivering sufficient current to each device plane. Specialized techniques are required to ensure that each device plane is operational while not exceeding a target output impedance [205, 244]. The focus of this chapter is on a primary issue in 3-D power delivery, the power distribution network, and it provides a quantitative experimental analysis of the noise measured on each tier within a three tier 3-D integrated stack.

Three-Dimensional Integrated Circuit Design. DOI: http://dx.doi.org/10.1016/B978-0-12-410501-0.00019-8

The effects of the through silicon vias (TSVs) on *IR* voltage drops and *L di/dt* noise are significant, as the impedance of a 3-D power distribution network is greatly affected by the TSVs [702]. In addition, the electrical characteristics of a TSV can vary greatly based on the 3-D via diameter, length, and dielectric thickness, as discussed in Chapter 4, Electrical Properties of Through Silicon Vias. A comparison of two different via densities for identical power distribution networks is also provided in this chapter, and implications of the 3-D via density on the power network design process are discussed.

The proper placement of decoupling capacitors can potentially reduce noise within the power network while enhancing performance. The effects of the board level decoupling capacitors on *IR* and *L di/dt* noise in 3-D circuits are also described here. Methods for placing decoupling capacitors at the interface between tiers to minimize the effects of intertier noise coupling are also suggested.

The 3-D test circuit is described in the following section. A brief update on the MITLL 3-D process is also provided in this section as specific features of this 3-D process have been enhanced as compared with the process described in Chapter 16, Case Study: Clock Distribution Networks for Three-Dimensional ICs. Experimental results of the noise characteristics of the power distribution networks are presented in Section 19.2. A discussion of the experimental results and the effects of the choice of power distribution topology on the noise characteristics of the power network are provided in Section 19.3. A short summary is offered in Section 19.4.

19.1 3-D POWER DISTRIBUTION NETWORK TEST CIRCUIT

The fabricated test circuit is 2 mm × 2 mm and composed of four equal area quadrants. Three quadrants are used to evaluate the effects of the topology of the power distribution network on the noise propagation characteristics and one quadrant is dedicated to DC-to-DC conversion. Each stacked power network (530 μm × 500 μm) includes three discrete 2-D power networks, one network within each of the three device planes. The total area occupied by each block is less than 0.3 mm^2, representing a portion of a power delivery network. Each block includes the same logic circuit but utilizes a different power distribution architecture. The power supply voltage is 1.5 volts for all of the blocks. The different power distribution architectures are reviewed in Section 19.1.1, an overview of the various circuit layouts and schematics is provided in Section 19.1.2, and the logic circuitry common to each power module is described in Section 19.1.3.

19.1.1 3-D POWER TOPOLOGIES

Interdigitated power/ground lines are used in each of the power network topologies. There are five main objectives for the test circuit: (i) determine the peak and average noise within the power and ground distribution networks, (ii) determine the effects of the board level decoupling capacitors on reducing power noise, (iii) explore the effects of a dedicated power/ground plane on power noise, (iv) investigate the effects of the TSV density on the noise characteristics of the power network, and (v) evaluate the resonant characteristic frequency of each power network topology. The three topologies are illustrated in Fig. 19.1. The difference between the left (Block 1) and central (Block 2) topologies is the number of TSVs, where the latter topology contains two-thirds the number of TSVs. The third

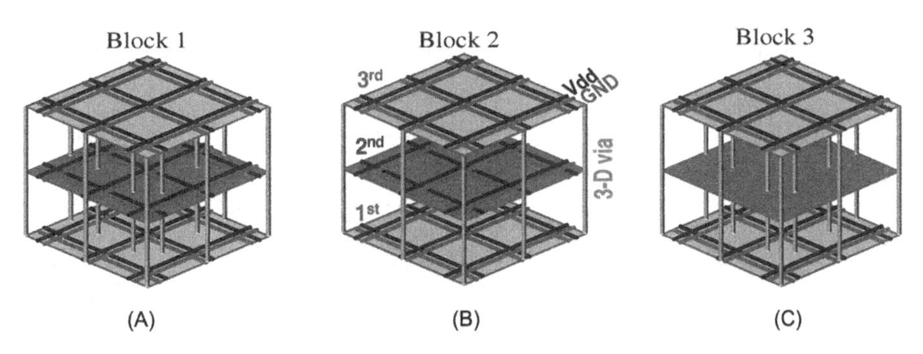

FIGURE 19.1

Power distribution network topologies. (A) interdigitated power network on all tiers with the 3-D vias distributing current on the periphery and through the middle of the circuit, (B) interdigitated power network on all tiers with the 3-D vias distributing current on the periphery, and (C) interdigitated power network on tiers 1 and 3 and power/ground planes on tier 2 with the 3-D vias distributing current on the periphery and through the middle of the circuit.

topology (Block 3) replaces the interdigitated power and ground lines on the second device plane with two metal planes to assess the benefit of allocating dedicated power and ground planes to deliver current to the loads within a 3-D system. The interdigitated power and ground lines are both 15 μm wide and separated by a 1 μm space for all three power distribution networks. The power and ground planes of Block 3 are separated by a 1 μm space of inter-layer dielectric composed of plasma etched, chemical vapor deposited tetraethyl orthosilicate (TEOS). Based on the experimental results, greater insight into the noise propagation properties of these 3-D power network topologies is provided, as discussed below.

19.1.2 LAYOUTS AND SCHEMATICS OF THE 3-D TEST CIRCUIT

The components of both the noise generation and noise detection circuits are described in this section. Layout and schematic views of the various sections of the test circuit are provided. The transistor dimensions of the various circuits are also included in the circuit schematics. The minimum sized transistor length is 150 nm. Most transistors are designed, however, with a 200 nm channel length (matching the dimensions of the standard library). A few circuits, such as the ring oscillators (RO), are designed with larger channel lengths (see Fig. 19.5 and Table 19.1). A layout view of the overall test circuit is shown in Fig. 19.2. Also shown in this figure are the three blocks discussed in this chapter and a fourth block for evaluating a 3-D DC-to-DC buck converter [703].

As previously mentioned, each of the three blocks includes two sets of circuits: noise generation and noise detection circuits. In addition to these circuits, calibration circuits are included to evaluate the gain and bandwidth of the source follower noise sensing devices. Each block contains the calibration circuits, but one set of calibration circuitry is sufficient to adequately characterize the gain and bandwidth of the sense circuits. The calibration circuits are identical to the detection circuits other than the input to these circuits being directly supplied from an external source. The layout of Block 1 is depicted in Fig. 19.3. As previously noted, Block 1 includes interdigitated power and ground lines on all three device planes. The four layouts depict the three stacked device planes

Table 19.1 Length and Width of the Transistors within the Ring Oscillator shown in Fig. 19.5A

Ring Oscillator	Device	Type	L (nm)	W (nm)
250 MHz	M1, M2	NMOS	300	1500
	M3, M5, M9, M11	PMOS	1225	1000
	M4, M6, M10, M12	NMOS	1225	600
	M7	PMOS	300	3000
	M8	NMOS	300	1500
500 MHz	M1, M2	NMOS	300	1500
	M3, M5, M9, M11	PMOS	800	1000
	M4, M6, M10, M12	NMOS	800	600
	M7	PMOS	300	3000
	M8	NMOS	300	1500
1 GHz	M1, M2	NMOS	300	1500
	M3, M5, M9, M11	PMOS	500	1000
	M4, M6, M10, M12	NMOS	500	600
	M7	PMOS	300	3000
	M8	NMOS	300	1500

(Fig. 19.3A), the bottom device plane (Fig. 19.3B), the middle device plane (Fig. 19.3C), and the top device plane (Fig. 19.3D). The three different subcircuits are labeled in Fig. 19.3A. The RF pads shown in Fig. 19.3D are associated with the calibration circuits.

The layout of the logic block used to generate sequence patterns for the current mirrors is shown in Fig. 19.4. The components of the sequence generator are enclosed in boxes and labeled in Fig. 19.4B. The components included in the test circuit are the three ring oscillators (RO), buffers for the ring oscillators, four pseudorandom number generators (PRNG), and buffers for the number generators. The RO and PRNG buffers are identically sized except that the enable signal for the RO is connected to the external reset signal pin, whereas the PRNG enable signal is tied to the 1.5 volt supply. A schematic of each component and the corresponding transistor sizes are included in Figs. 19.5 and 19.6.

The layout and schematic of the current mirror and transmission gate switch used to modulate the current mirror are shown, respectively, in Figs. 19.7 and 19.8. The four figures depict the three stacked device planes (Fig. 19.7A), bottom device plane (Fig. 19.7B), middle device plane (Fig. 19.7C), and top device plane (Fig. 19.7D). The output from the PRNGs is the input to the switches of the current mirror, as shown in Fig. 19.8.

The layout of the power and ground noise detection circuits with the corresponding control logic is shown in Fig. 19.9. The four figures depict the circuits on each of the three device planes (Fig. 19.9A), bottom device plane (Fig. 19.9B), middle device plane (Fig. 19.9C), and top device plane (Fig. 19.9D). As shown in the figure, the control logic is only present on the second device plane. The power and ground detection circuits and the control logic are both shown in Fig. 19.9A.

A schematic of the logic that controls the RF output pads among the three device planes is shown in Fig. 19.10. The signal that controls the RF pads is provided for both the power and ground detection signals on each device plane and is rotated every 218 cycles. The control of the

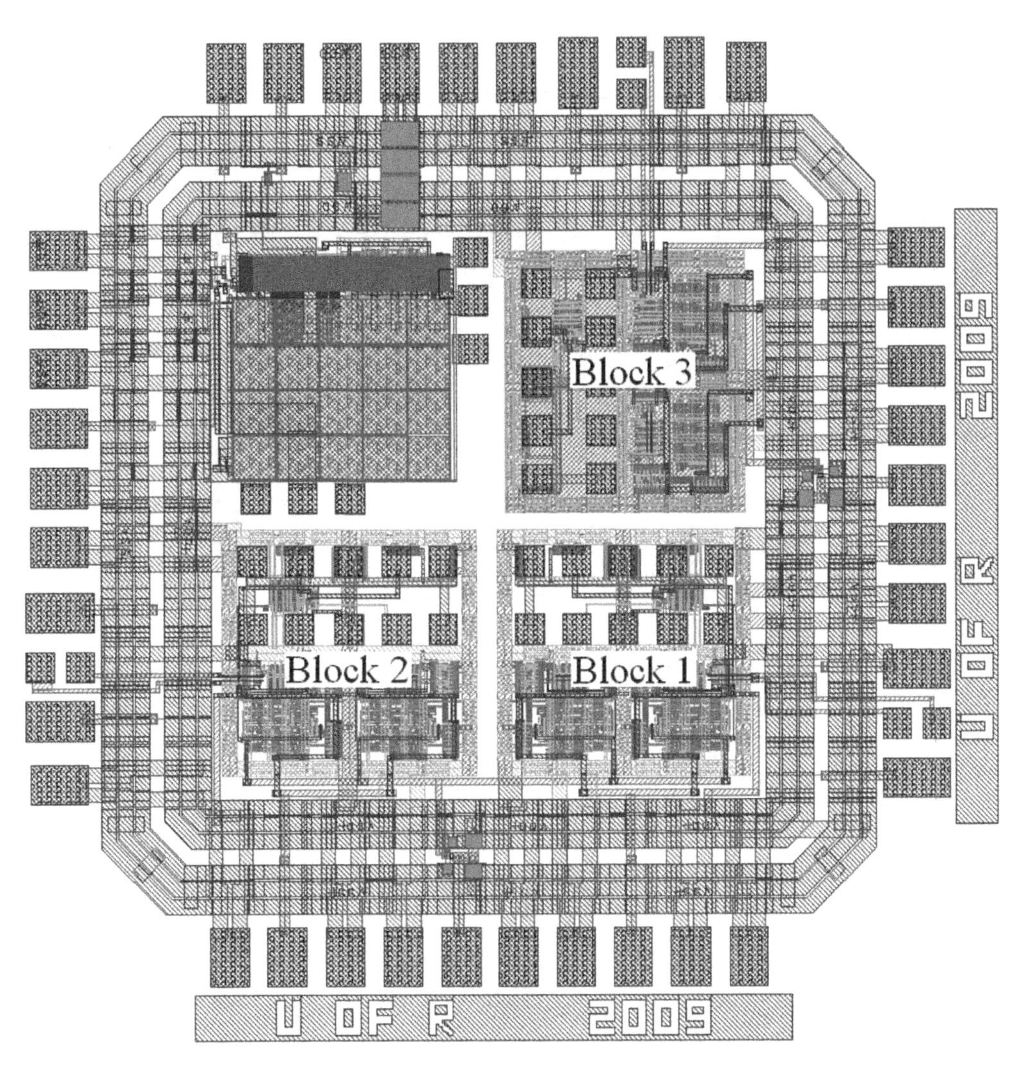

FIGURE 19.2

Layout of the power distribution network test circuit.

output pads has four phases; three phases rotate control amongst the three device planes, and a fourth phase connects the RF output pads for both the power and ground detection circuits to ground. The fourth phase provides a delineating time during measurement to distinguish data from each of the three device planes as the four phases are cycled.

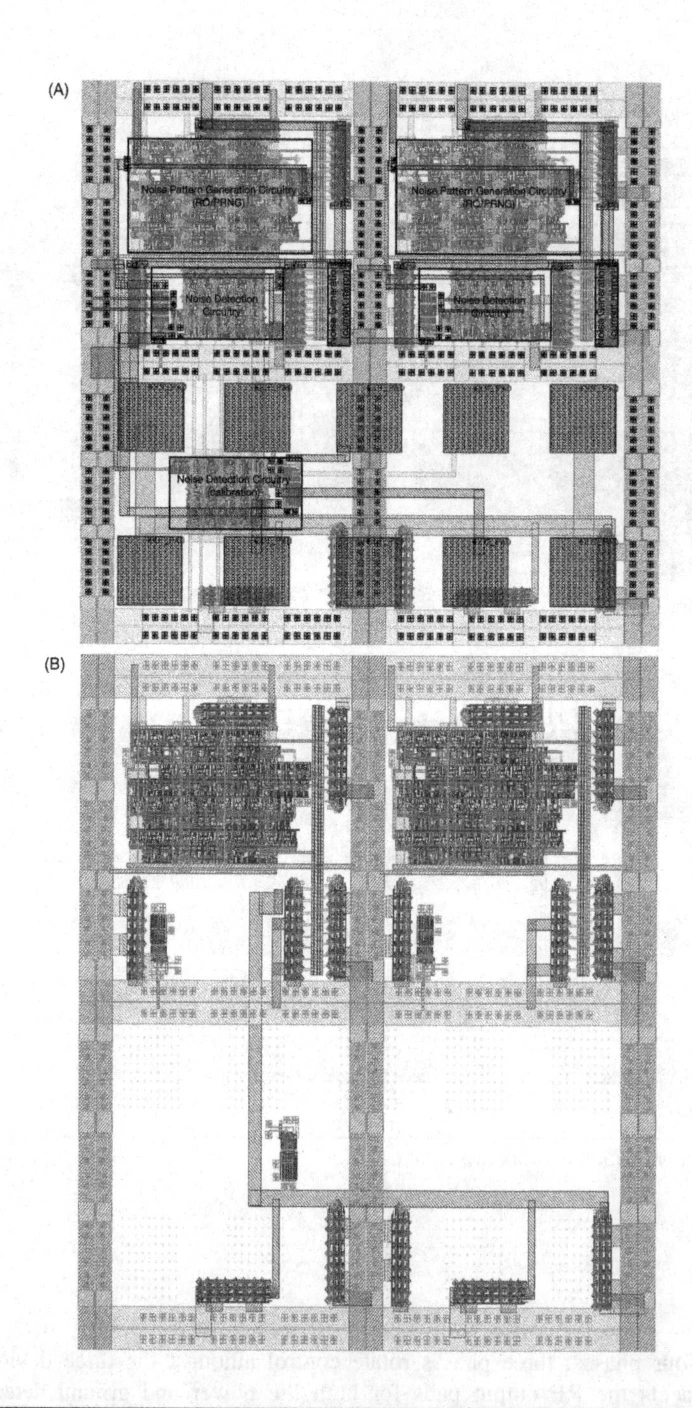

FIGURE 19.3

Layout of the test circuit containing three interdigitated power and ground networks and test circuits for generating and measuring noise. (A) Overlay of all three device planes, (B) power and ground networks of the bottom tier (tier 1), (C) power and ground networks of the middle tier (tier 2), and (D) power and ground networks of the top tier (tier 3).

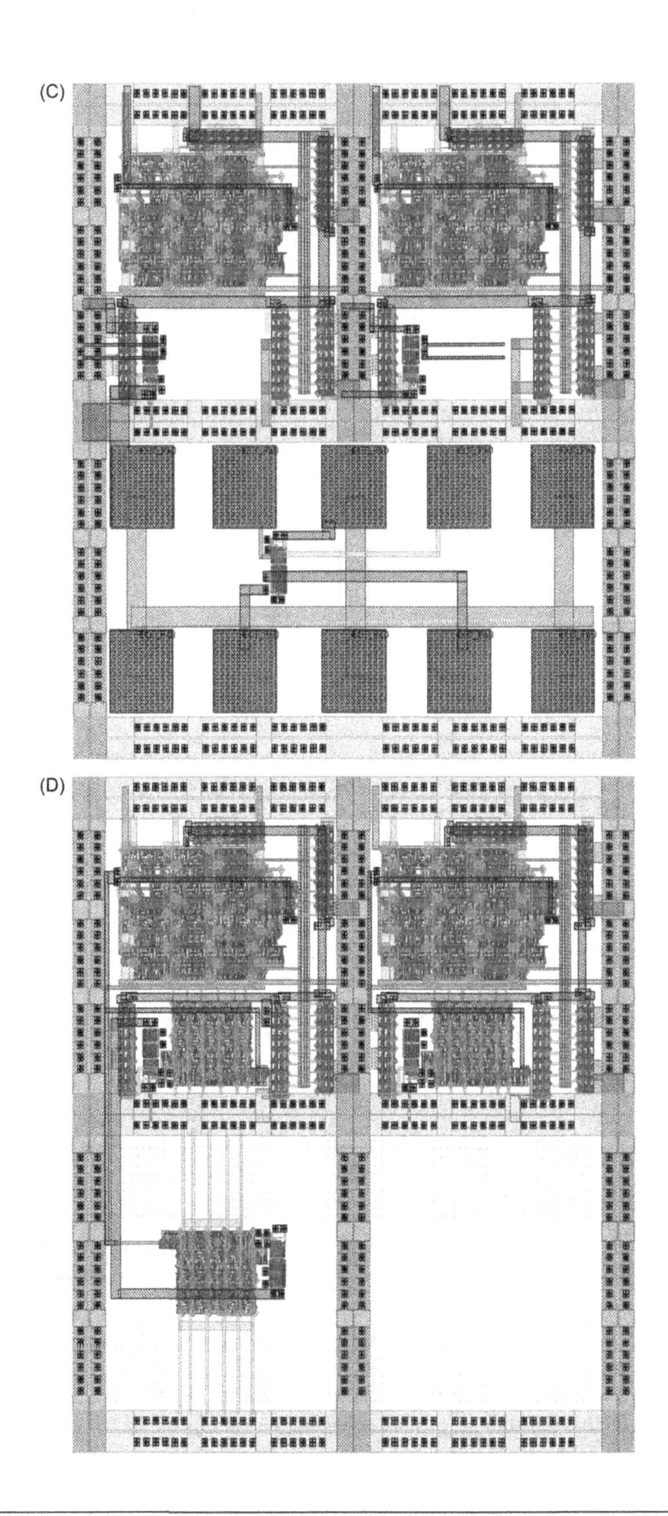

FIGURE 19.3

(Continued)

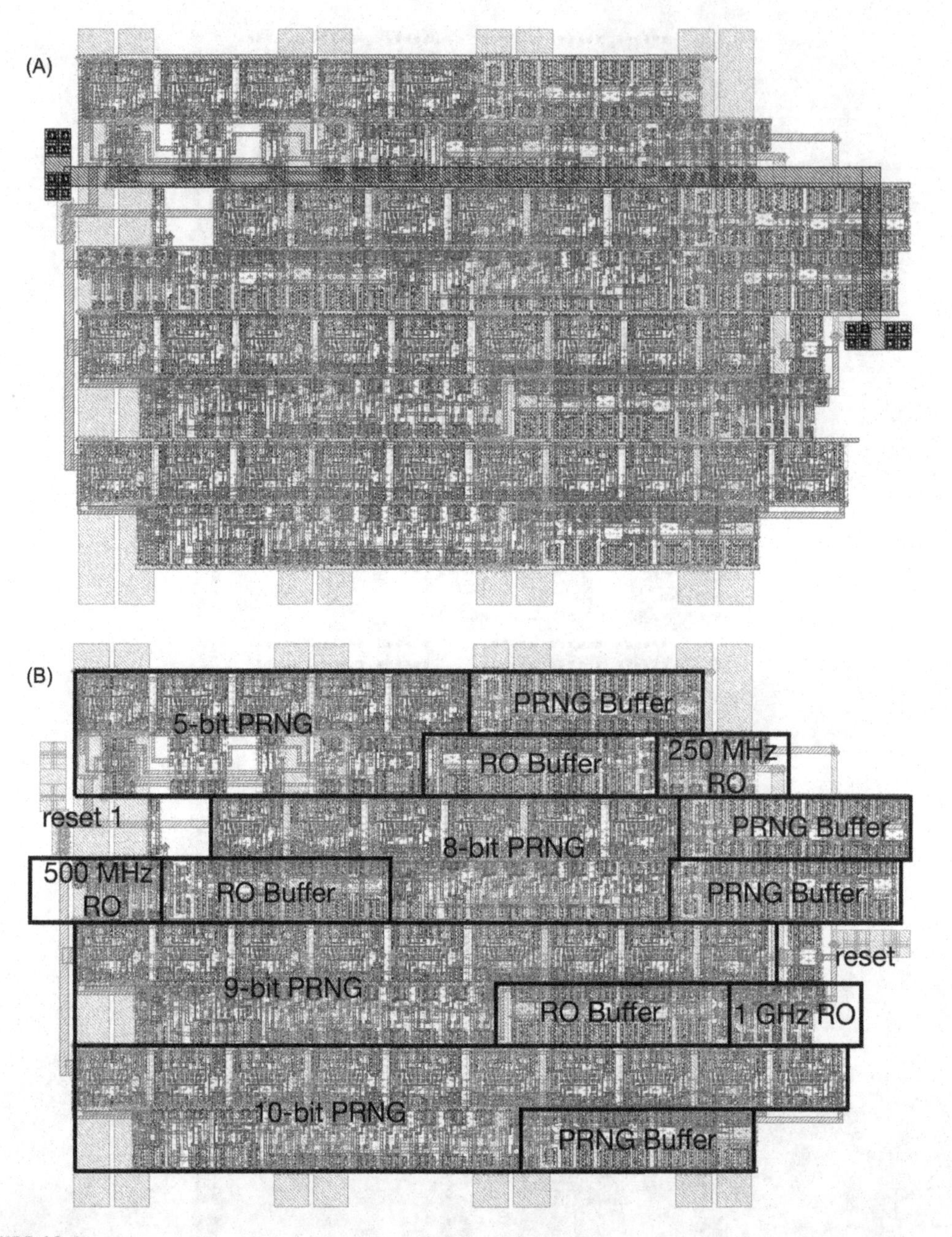

FIGURE 19.4

Layout of the pattern sequence source for the noise generation circuits. (A) All three device planes, (B) noise generation circuits on the bottom tier (tier 1), (C) noise generation circuits on the middle tier (tier 2), and (D) noise generation circuits on the top tier (tier 3).

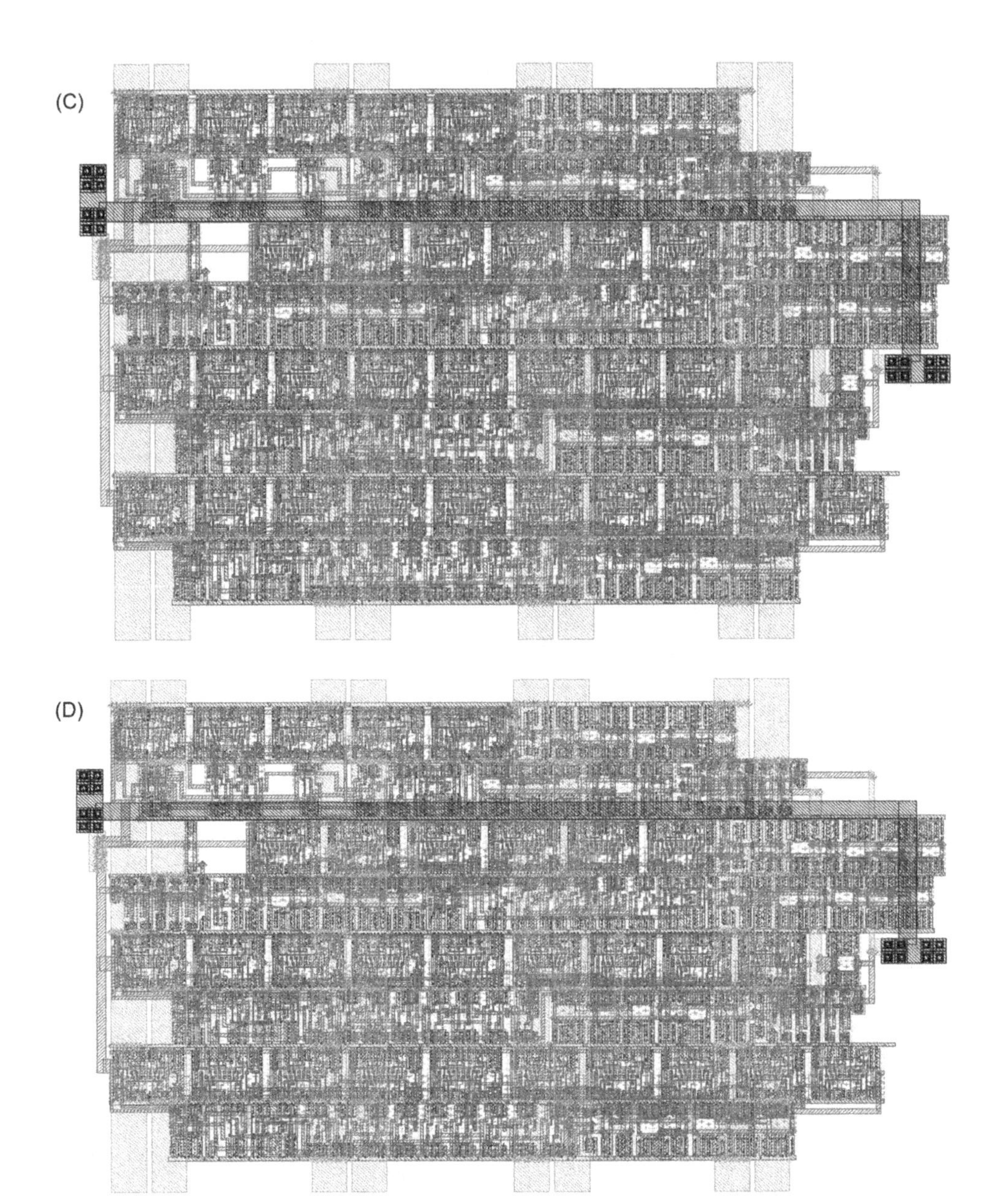

FIGURE 19.4

(Continued)

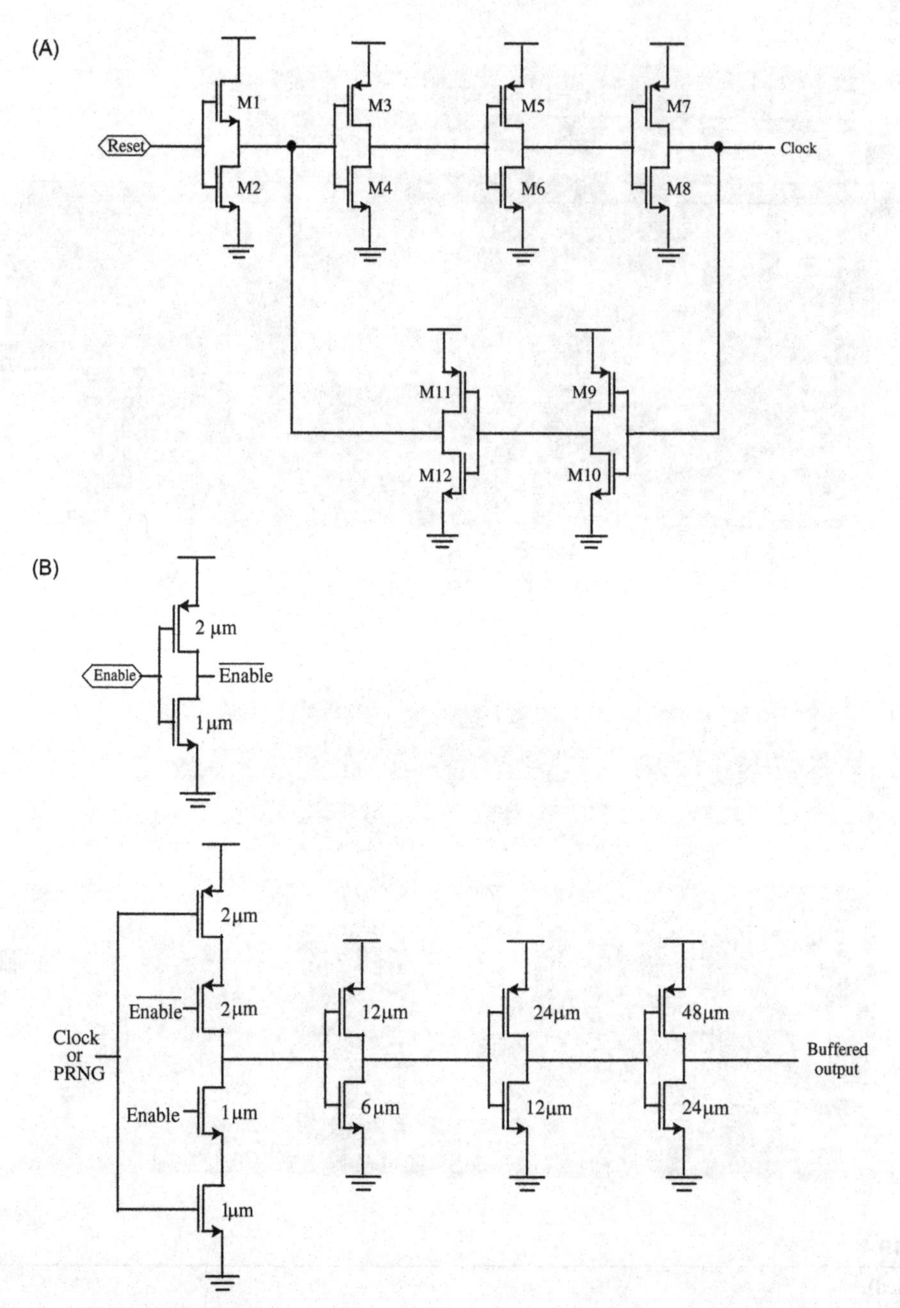

FIGURE 19.5

Pattern sequence source for the noise generation circuits. (A) Ring oscillator, (B) buffer used for the RO and PRNG, (C) 5-bit PRNG, (D) 6-bit PRNG, (E) 9-bit PRNG, and (F) 10-bit PRNG.

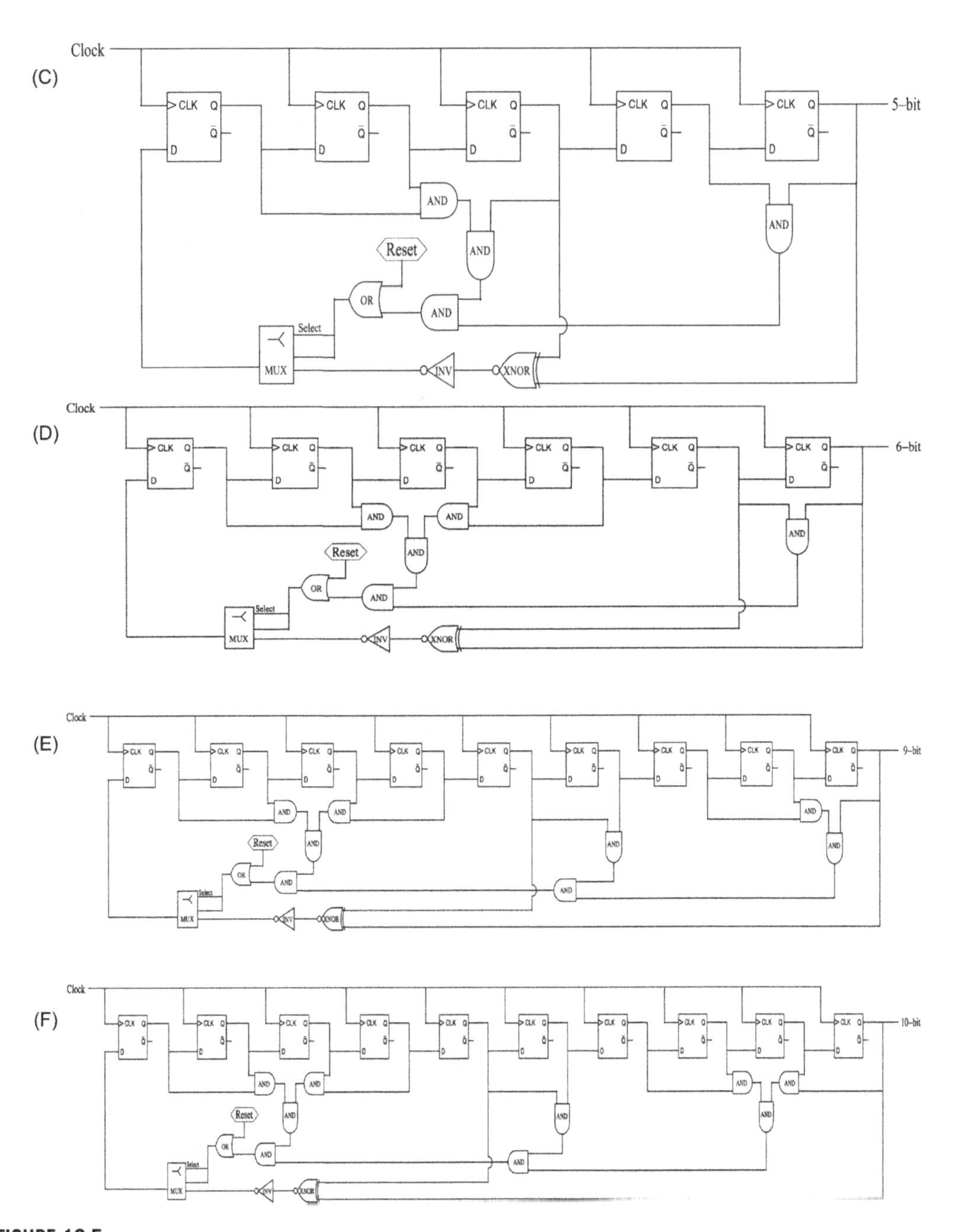

FIGURE 19.5

(Continued)

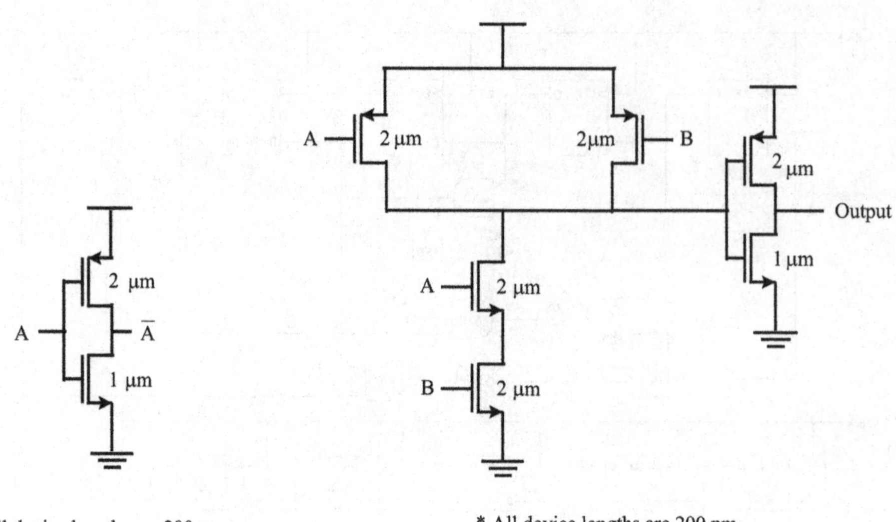

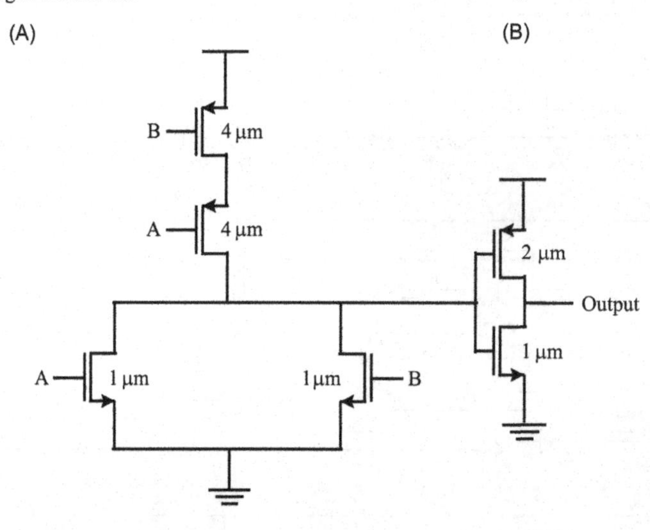

FIGURE 19.6

Individual components in Figs. 19.5C–F with the corresponding transistor sizes. (A) Inverter, (B) AND gate, (C) OR gate, (D) XNOR gate, (E) 2-to-1 MUX, and (F) D flip-flop.

The test circuit occupies an area of 2 mm × 2 mm. Each power distribution network is 530 μm × 500 μm. DC and RF pads surround the four test blocks. A block diagram of the DC and RF pads is shown in Fig. 19.11. The DC pads corresponding to the numbered pads shown in the figure are listed in Table 19.2. These DC pads provide the various bias voltages, reset

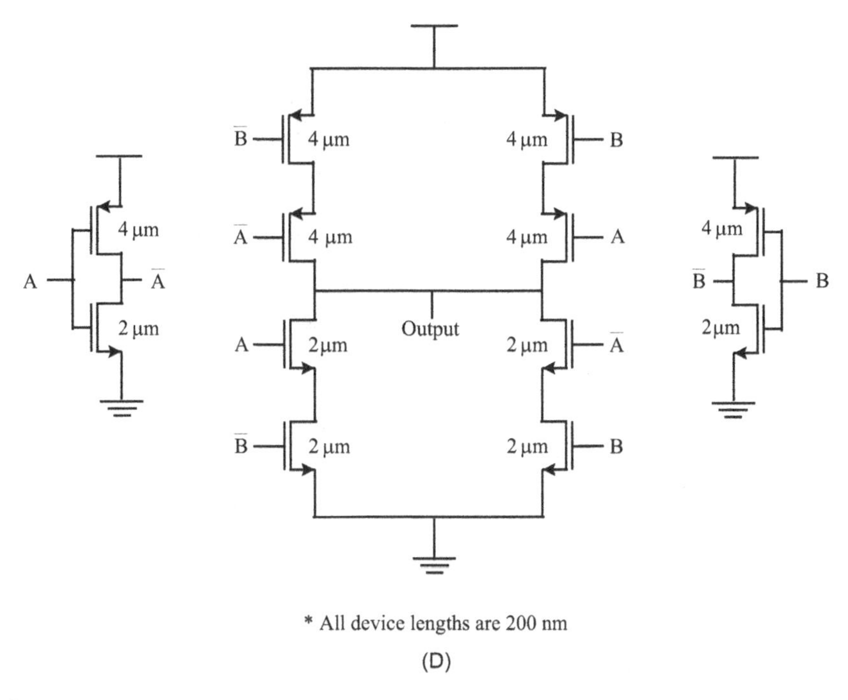

* All device lengths are 200 nm

(D)

FIGURE 19.6

(Continued)

signals, and power and ground signals. The RF pads surrounding the circuit blocks provide a pathway to probe and detect noise on the power and ground distribution networks. As shown in Fig. 19.11, there are three sets of two RF pads surrounding the blocks, one set for each of the three different 3-D power distribution networks. The RF pads internal to each block are also shown in Fig. 19.11. These RF pads are used to calibrate the source follower power and ground noise detection circuits.

There are two sets of five RF pads to calibrate the noise detection circuits, as shown in Fig. 19.11. Each set of five pads is allocated as a ground-signal-ground-signal-ground (GSGSG) pattern. Two sets of GSG RF pads are within the set of five pads with the middle ground pad shared between the two sets of pads. One set of GSG pads is dedicated to the sense circuit detecting noise on the power lines, and the other GSG is dedicated to the sense circuit detecting noise on the ground line. The set of five RF pads closest to the center of the block is used to pass a sinuisodial signal from an Agilent E8364A general purpose network analyzer (PNA) to the noise sensing circuits. The other set of five RF pads is used as an output from the sense circuits to the PNA. The input and output RF pads for calibrating both the power and ground line noise detection circuits are enclosed in boxes to emphasize that the middle pad is a shared ground pad between the two circuits, as shown in Fig. 19.11.

A microphotograph of the post wire bonded test circuit is shown in Fig. 19.12. Wirebonding is performed on a West Bond 7400 bonder with ultrasonic bonding. The aluminum wirebonds range in length from approximately 0.7 cm to 1 cm.

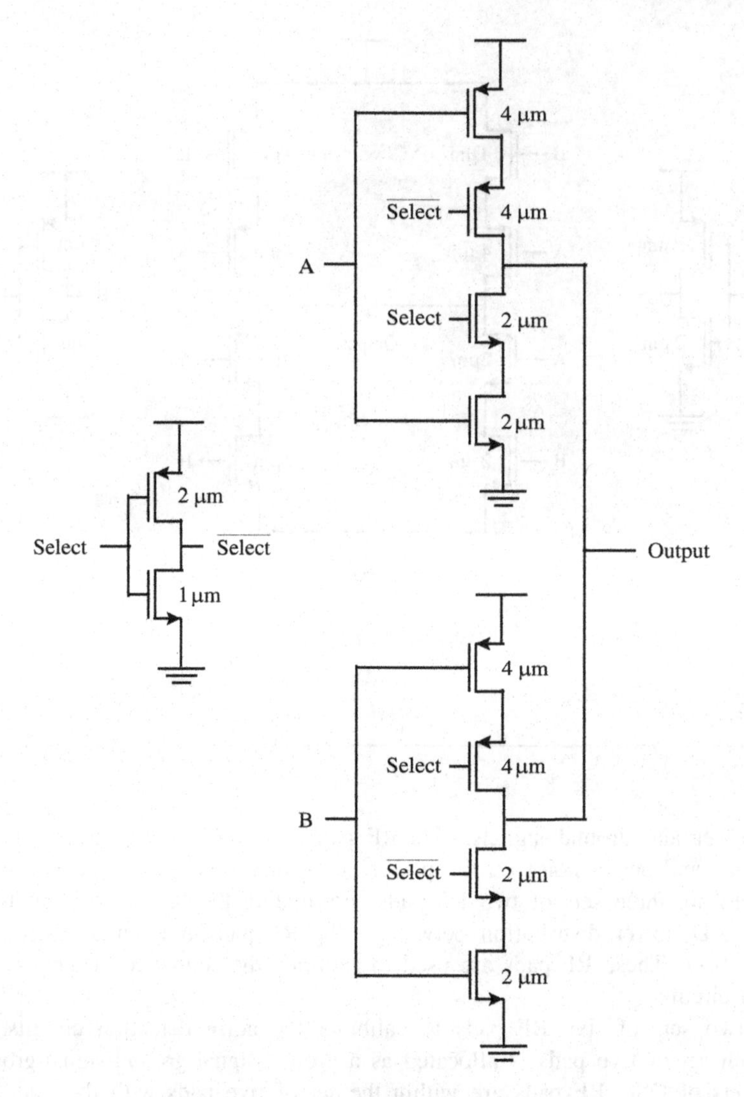

FIGURE 19.6

(Continued)

19.1.3 3-D CIRCUIT ARCHITECTURE

The three power networks utilize identical on-chip circuitry, with two of the topologies only differing by the total number of TSVs required to distribute power to the lower two device planes.

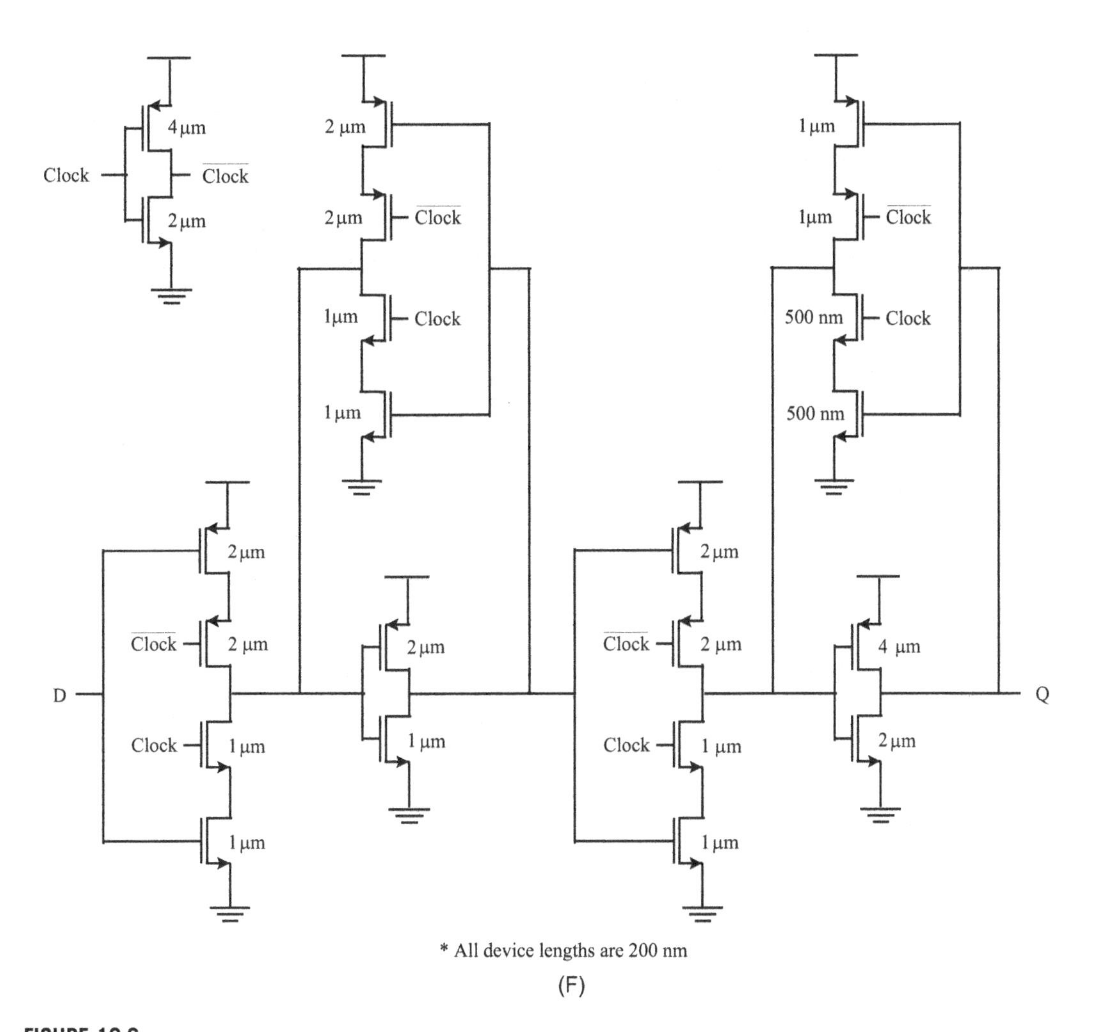

FIGURE 19.6

(Continued)

With one network, 1,728 TSVs are placed along both the inner and outer interdigitated power/ground lines, as shown in Fig. 19.1A, while the second network includes 1,152 TSVs located along the outer peripheral interdigitated power/ground lines, as depicted in Fig. 19.1B. The third power network topology also includes 1,728 TSVs, but the interdigitated power network on the second stacked die is replaced with two metal planes, one plane for ground and one plane for power.

The basic logic blocks are illustrated in Fig. 19.13. The power supply noise generators sink different current from the power lines. These noise generators are placed on each tier of each power network topology. Voltage sense circuitry is included on each of the tiers of each test block to measure the noise on both the power and ground lines. The second tier of the power delivery network, illustrated in Fig. 19.1C, does not include noise sense circuitry or noise generation circuitry as this

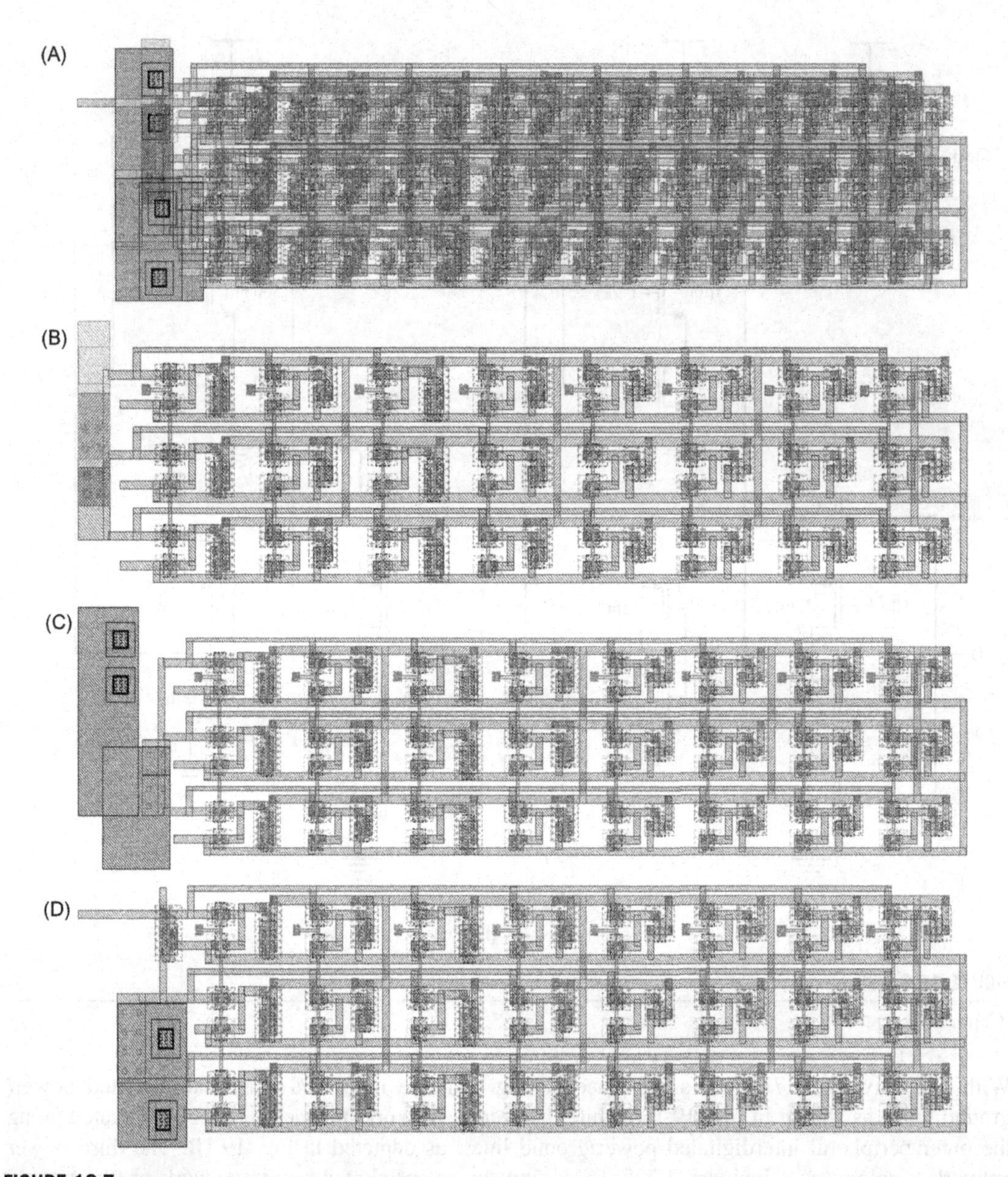

FIGURE 19.7

Layout of the noise generation circuits. (A) All three device planes, (B) noise generation circuits on the bottom tier (tier 1), (C) noise generation circuits on the middle tier (tier 2), and (D) noise generation circuits on the top tier (tier 3).

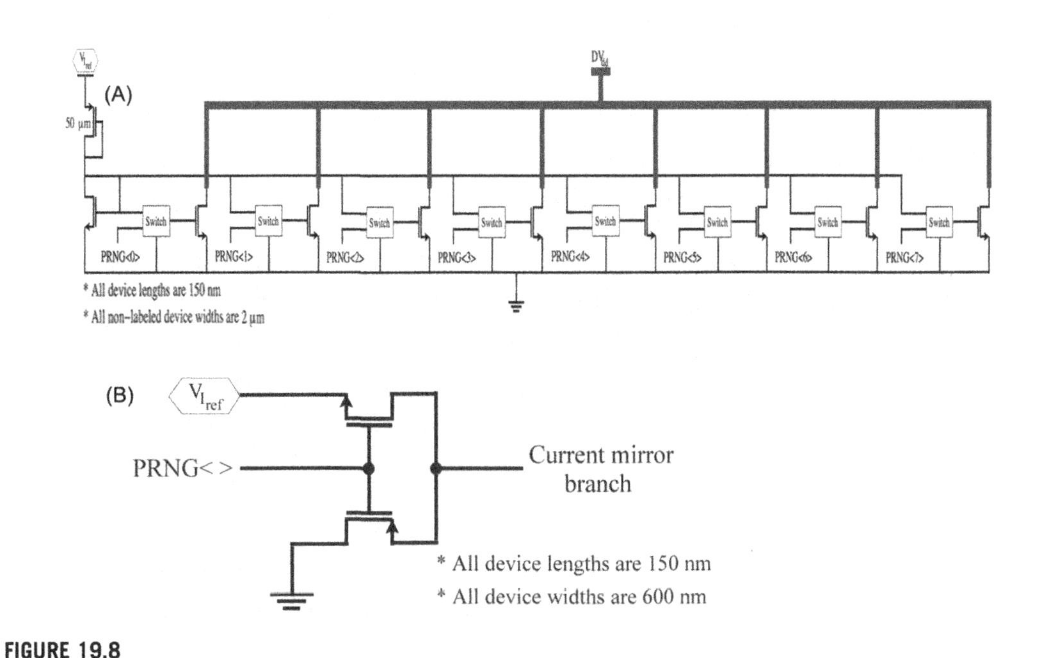

FIGURE 19.8

Schematic view of the (A) current mirror, and (B) switches that vary the total current through the current mirror.

tier is dedicated to power and ground. The voltage range and average voltage of the sense circuitry on each tier for each test block are compared for the three topologies shown in Fig. 19.1.

A schematic of the on-chip circuitry on each tier of each power network is shown in Fig. 19.13. The circuit is designed to emulate a power load drawing current, generating noise within the power and ground networks. Two sets of circuits are present: noise generating circuits and noise detection circuits. The noise generating circuits include ten current mirrors per device plane (15 for the second device plane); 250 MHz, 500 MHz, and 1 GHz ring oscillators; and five-, six-, nine-, and ten-bit PRNGs. The clock pulses generated by the ring oscillators are buffered and fed into the PRNGs, which are again buffered before randomly driving the current mirrors as enable bits. Each current mirror requires eight enable bits, with each enable bit turning on one branch of the current mirror, and draws a maximum current of 4 mA (0.5 mA per branch). With ten current mirrors on each of the first and third device planes and 15 current mirrors on the second tier, the maximum current drawn by each power network is 140 mA at 1.5 volts, producing a maximum power density of 79 W/cm^2 in each 530 μm × 500 μm topology.

Separate noise detection circuits are included for evaluating power and ground noise. A schematic of the power and ground detection circuits is shown in Fig. 19.14. These single stage amplifier circuits utilize an array structured noise detection circuit, as suggested in [704]. Each device plane contains both power and ground noise detection circuits. An off-chip resistor, labeled R_{bias} in Fig. 19.14, biases the output node of the PMOS current mirror (with a PMOS threshold voltage of −0.56 volts) and ensures that the current mirror operates in the saturation region. The resistor is chosen to ensure that the maximum voltage at the output node of the current mirror does not exceed 400 mV, the voltage at which the current mirror enters the triode region. The measured current at

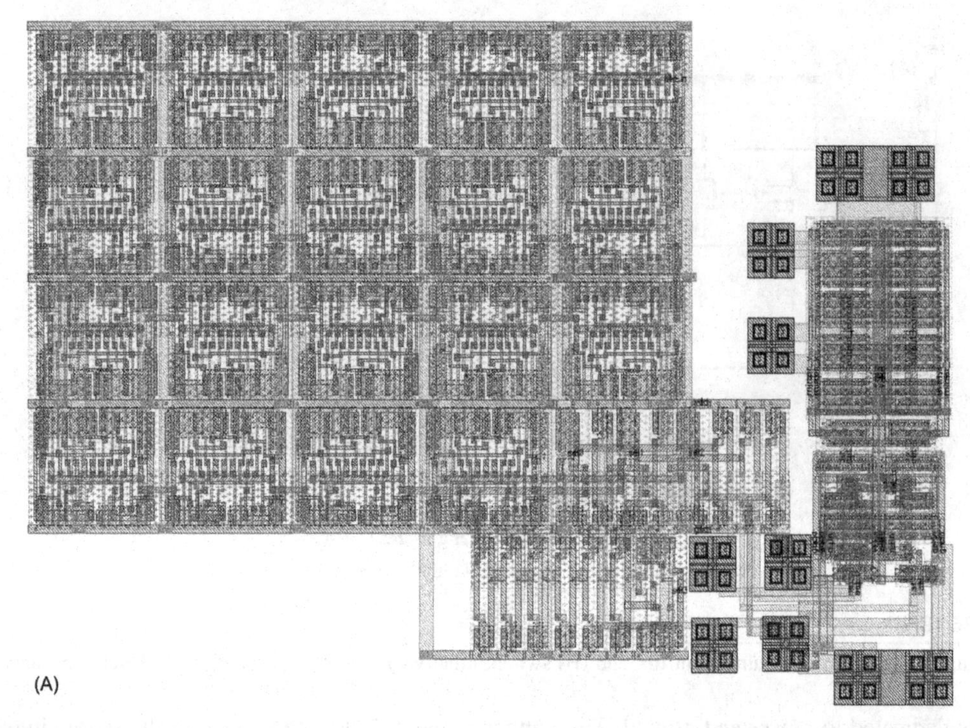

FIGURE 19.9

Layout of the power and ground noise detection circuits including the control circuit. (A) All three device planes, (B) power and ground sense circuits for the bottom tier (tier 1), (C) power and ground sense circuits for the middle tier (tier 2) and control circuit for all three tiers, and (D) power and ground sense circuits for the top tier (tier 3).

the output of the current mirror ranges between 200 milliamperes and 300 milliamperes. Since R_{bias} is chosen as 100 Ω, the maximum voltage at the output node is 300 mV. In addition, the intrinsic impedance Z_o, shown in Fig. 19.14, is 50 Ω.

Two sets of ground-signal-ground (GSG) output pads source current from the final stage of the noise detection circuits of all three tiers, as shown in Fig. 19.13. One set of GSG output pads is dedicated to the three noise detection circuits that monitor the power network, while the other set of GSG pads supports the three noise detection circuits that monitor the ground network. Counter logic rotates the control of the output pads among the three device planes every 218 cycles (at a 250 MHz clock) for both the power and ground noise detection circuits. A second isolated power and ground network with a 2 μF board level decoupling capacitor ensures that the generated noise is not injected into the detection circuits.

19.1.4 3-D IC FABRICATION TECHNOLOGY

The manufacturing process developed by MITLL for fully depleted silicon-on-insulator (FDSOI) 3-D circuits is described extensively in Section 16.1. This process has evolved into a more

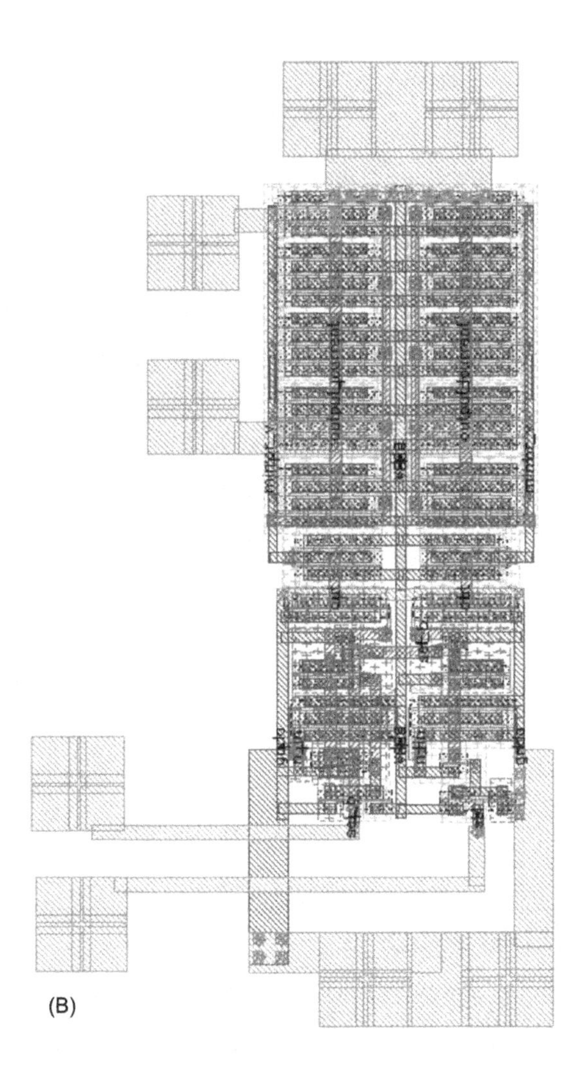

FIGURE 19.9

(Continued)

advanced technology node of 150 nm [670]. The operating voltage of the SOI transistors is 1.5 volts. The back end of the line remains the same, including one polysilicon layer and three metal layers interconnecting the devices on each wafer. A backside metal layer also exists on the upper two tiers, providing the starting and landing pads for the TSVs, and the I/O, power supply, and ground pads for the overall 3-D circuit. An enhanced feature of this process is the high density TSVs. The dimensions of the TSVs are 1.25 μm × 1.25 μm and 10 μm deep, much smaller than many existing 3-D technologies [150,155]. The step-by-step fabrication of the three tier stack is shown in Figs. 16.1 to 16.6. A microphotograph of the fabricated die is shown in Fig. 19.15.

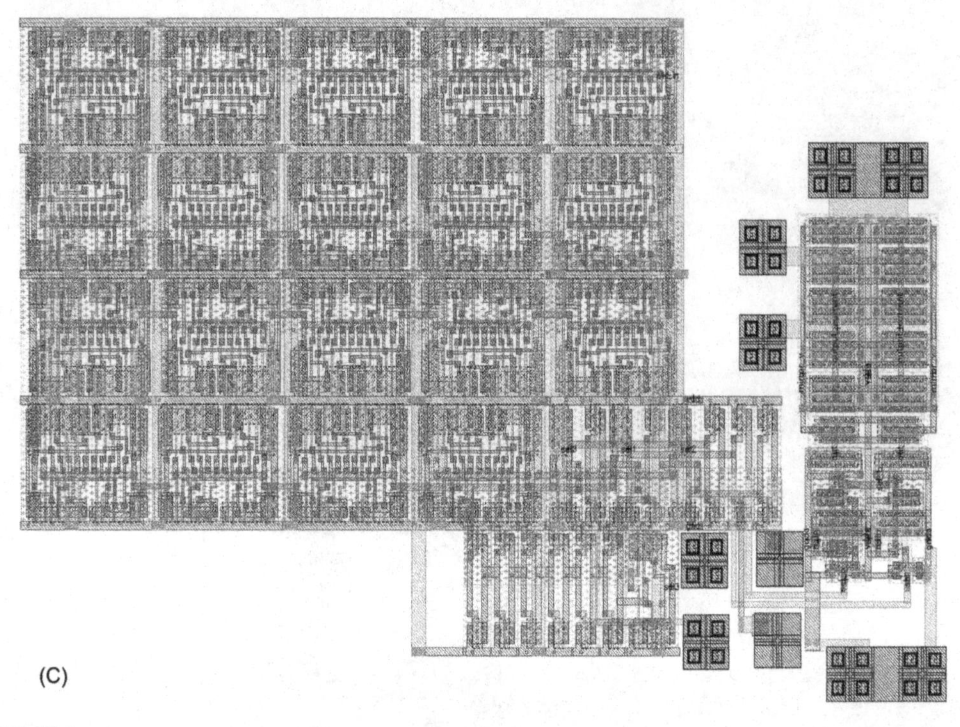

FIGURE 19.9

(Continued)

19.2 EXPERIMENTAL RESULTS

The noise generated within the power distribution network is detected by a source follower amplifier. A schematic of the amplifier circuit is depicted in Fig. 19.14. The sense circuits can detect a minimum voltage of 165 μV (from simulation). Noise from the digital circuit blocks is coupled into the sense circuit through the node labeled DV_{dd}. The gain of the circuit is controlled by adjusting the analog voltage, labeled AV_{dd} in Fig. 19.14.

The gain and bandwidth of both the power and ground network noise detection circuits are calibrated by S-parameter extraction. The measured results are shown in Fig. 19.16. The simulated DC gain and 3 dB bandwidth of the power network detection circuit are, respectively, −3.8 dB and 1.4 GHz. The measured DC gain and 3 dB bandwidth are, respectively, −4.1 dB and 1.3 GHz. Similarly, the simulated DC gain and 3 dB bandwidth of the ground network detection circuit are, respectively, −4.0 dB and 1.35 GHz, and the measured DC gain and 3 dB bandwidth are, respectively, −4.25 dB and 1.15 GHz. For both the power and ground detection circuits, the measured gain is within 3.4% of simulations. The models include the impedance of the on-chip interconnect (15 Ω), bias-T inductance (340 μH) and capacitance (3 μF), resistance to bias the output node (100 Ω), and intrinsic impedance Z_o of the network analyzer (50 Ω), as shown in Fig. 19.14.

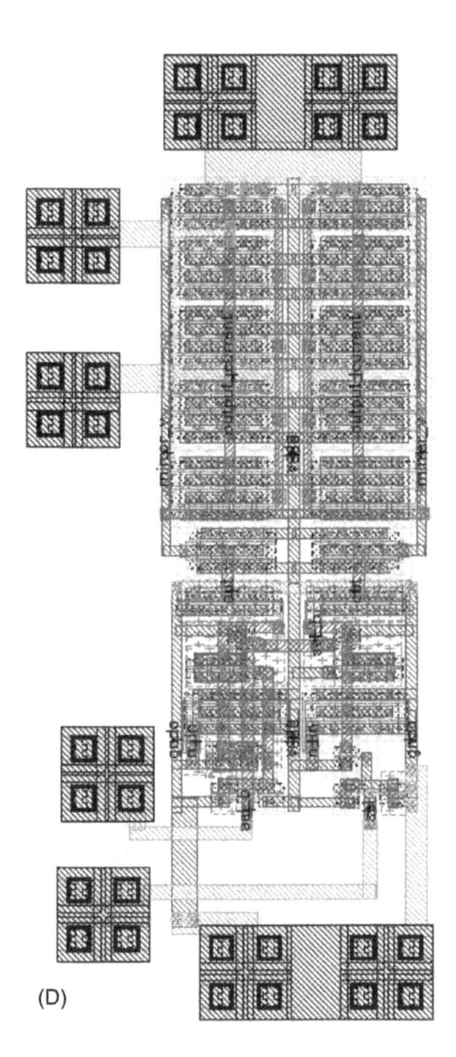

FIGURE 19.9

(Continued)

The power spectral density of the generated noise within the power network with the voltage bias on the current mirrors set to 0.75 volts is shown in Fig. 19.17. The noise data shown on the top half of Fig. 19.17 include the effect of a 4 μF board level decoupling capacitance, whereas the noise data on the bottom half of Fig. 19.17 are without any decoupling capacitance between the power and ground networks. Both plots illustrate the three noise components produced by the 250 MHz, 500 MHz, and 1 GHz ring oscillators. No on-chip decoupling capacitance is added to the three power distribution topologies other than the intrinsic capacitance of the power and ground networks. The peak noise power does not precisely match the ring oscillator frequencies as the ring oscillators are not tuned to the target 250, 500, and 1,000 MHz frequencies. The peak noise

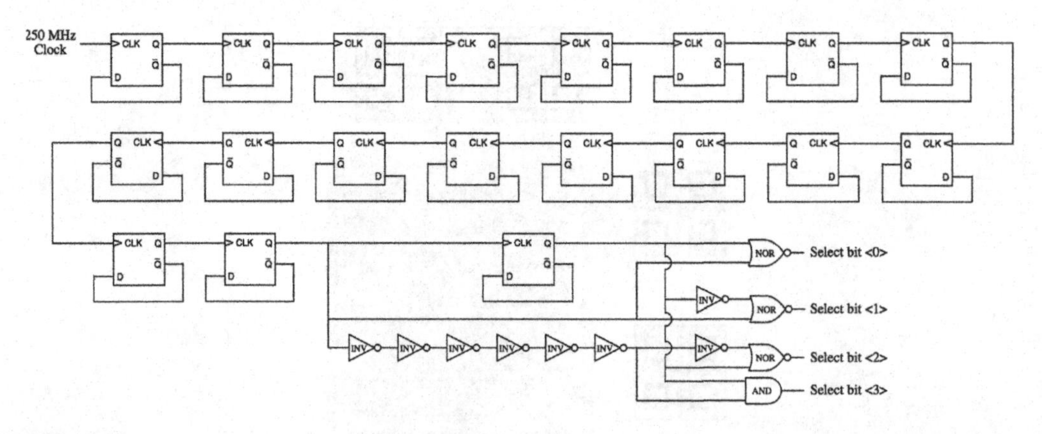

FIGURE 19.10

Rotating control logic to manage the RF output pads among the three device planes. The control signals to the RF pads are provided for both the power and ground detection signals for each device plane.

therefore occurs at 97 MHz (−49 dB), 480 MHz (−47 dB), and 960 MHz (−50 dB) with a board level decoupling capacitor, and at 96 MHz (−27 dB), 520 MHz (−29 dB), and 955 MHz (−33 dB) without a decoupling capacitor. The inclusion of a board level decoupling capacitor reduces the peak noise within the power networks by approximately 20 dB.

A time domain analysis of the generated noise is used to compare the three different power networks. The detected noise voltage is measured after the intrinsic capacitance of the bias-T junction that couples the noise into the oscilloscope and spectrum analyzer (see node labeled port0 in Fig. 19.14). The noise amplitude with increasing current mirror bias voltage (from 0 to 1 volt) is depicted in Fig. 19.18. These results indicate that the current mirrors function properly. The 4,096 data points used to generate each subfigure shown in Fig. 19.18 are centered around 0 volts for both the power and ground networks, as only the RF component is passed to the oscilloscope.

The average noise for each topology, with or without a board level decoupling capacitance, and for both the power and ground networks as a function of the applied bias voltage to the current mirrors, is shown in Fig. 19.19. A total of 4,096 data points are used to generate each average noise value for each topology at each of the six bias voltages. Since the current mirrors are activated using a random bit sequence generated from the random number generators, the 4,096 data points include voltages as low as the nominal off state (all zeroes bit sequence) to a peak voltage when all current mirrors are in the on state. Therefore, the lowest voltage (and the highest voltage as shown in Fig. 19.20) in most cases is similar, however, the average voltage amplitude of the data set increases with increasing bias voltage applied to the current mirrors. In all cases, with or without decoupling capacitors, the network topology that includes the metal planes exhibits lower average noise as compared with the other two topologies, as the metal planes behave as an additional decoupling capacitor. Results from a statistical analysis of the noise generated on the power network including the mean, median, and 25th and 75th quartiles are listed in Table 19.3. The number of TSVs also affects the magnitude of the noise, as shown in Fig. 19.19. Reducing the number of TSVs from 1,728 to 1,152 increases the average amplitude of the noise by 2% to 14.2%, as the parasitic impedance of the 3-D power network is larger. The amplitude of the noise is less than twice

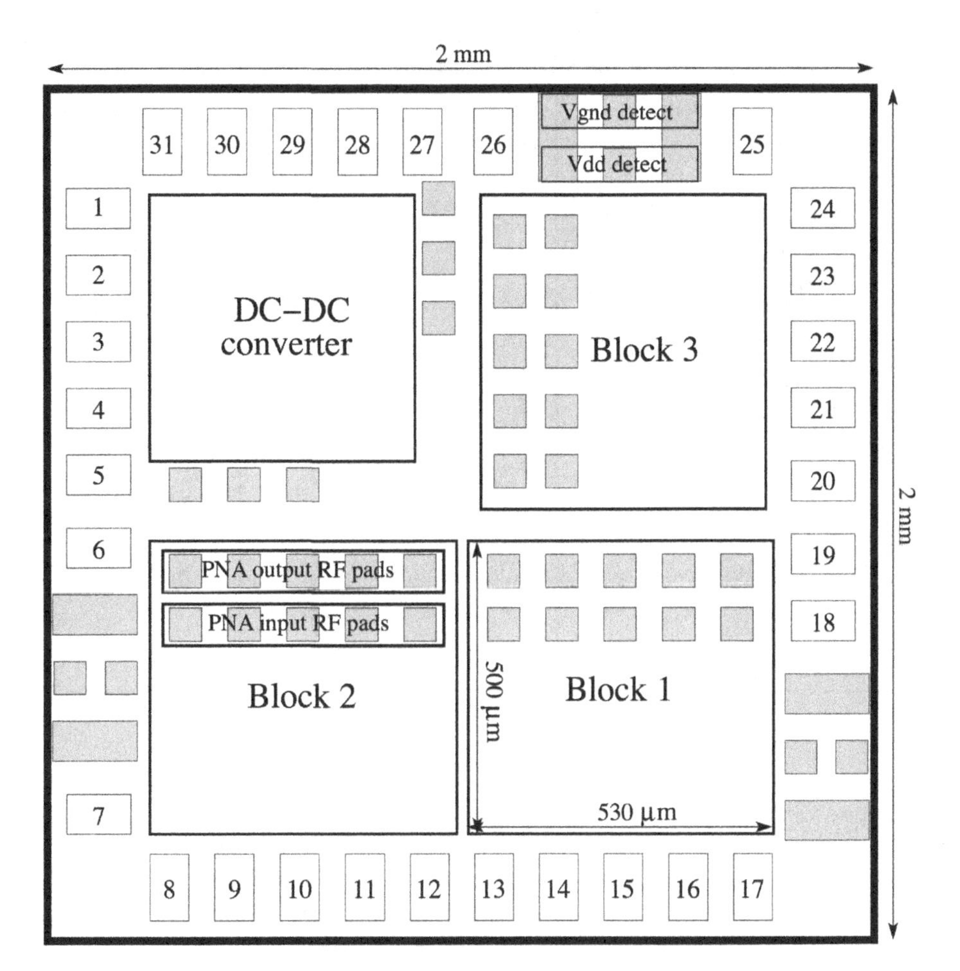

FIGURE 19.11

Block and I/O pin diagram of the DC and RF pad layout. The numbered rectangles are DC pads providing power and ground, and DC bias points for the current mirrors, reset signals, and electrostatic discharge protection. The light colored squares and rectangles are RF pads used to calibrate the sense circuits (internal to the labeled blocks) and measure noise on the power/ground networks (external to the labeled blocks).

as great since the portion of the impedance contributed by the TSVs along the path from the power supply, through the cables and wirebonds, into the 3-D power network, and back is a small fraction of the total impedance.

The peak noise for each topology, with or without a board level decoupling capacitance and for both the power and ground networks as a function of the applied bias voltage to the current mirrors, is shown in Fig. 19.20. Unlike the average noise shown in Fig. 19.19, the peak voltage detected from each topology does not follow a distinct pattern. No single topology contributes the largest noise voltage at any specific current mirror bias voltage. These voltages represent single data points

Table 19.2 DC Pad Assignment of the 3-D Test Circuit

Index	Pad Connectivity	Index	Pad Connectivity
1	3.3V V_{dd}	17	Analog V_{gnd}
2	1.5V V_{dd}	18	1.5V V_{dd}
3	V_{gnd}	19	V_{gnd}
4	V_{Iref}	20	V_{reset}
5	V_{gnd}	21	1.5V V_{dd}
6	1.5V V_{dd}	22	V_{gnd}
7	Analog V_{gnd}	23	V_{Iref}
8	Analog V_{dd}	24	Analog V_{dd}
9	V_{Iref}	25	Analog V_{gnd}
10	V_{gnd}	26	1.5V V_{dd}
11	1.5V V_{dd}	27	V_{gnd}
12	V_{reset}	28	ESD V_{gnd}
13	V_{gnd}	29	ESD V_{dd}
14	1.5V V_{dd}	30	V_{reset}
15	V_{Iref}	31	V_{gp}
16	Analog V_{dd}		

The pad index is shown in Fig. 19.11.

indicative of the maximum peak-to-peak noise voltage for each topology at each bias voltage. For the average noise voltage, Block 3 produces a lower average noise than Block 1, which produces a lower average noise than Block 2. Interestingly, the average noise for each topology is approximately 75% to 90% lower than the peak-to-peak noise voltage, indicating that a majority of the noise data are located within close proximity of the nominal power and ground voltages. In addition, the saturation voltage of the detection circuitry at the output node (port0) is approximately 230 mV when the gain is −4.2 dB. The noise detection range is approximately 600 mV centered around 1.5 volts and 0 volts, respectively, for the power and ground lines. The detection circuits for the power network, therefore, detect noise that ranges between 1.2 volts and 1.8 volts, and for the ground networks, between −0.3 volts and 0.3 volts.

In addition to the peak and average noise voltage measurements for each power network topology, an analysis of the current drawn by the digital power network is provided here. The measured current for each block within the test circuit as a function of the bias voltage applied to the current mirrors is listed in Table 19.4. Each topology draws a similar amount of current since the total number of current mirrors for each power network is identical. The data listed in Table 19.4 reveal that Blocks 1 and 2 sink similar currents. Block 3 draws about 30% to 45% more current as compared to the other two topologies. The Block 3 topology includes power and ground planes on the second device plane; therefore, in this topology, leakage current between these two large planes as well as leakage current from the TSVs distributing power to the bottom tier contribute additional current. Based on the currents listed in Table 19.4 and a 1.5 volt power supply, a peak power density of 59.8 W/cm^2 is achieved for Block 3 when the current mirrors are biased with a voltage of 1.25 volts.

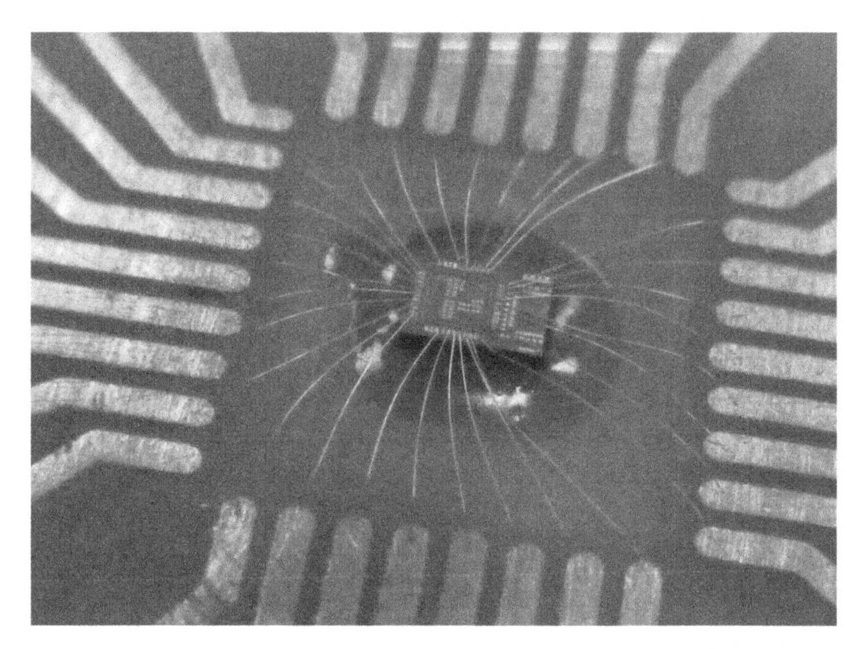

FIGURE 19.12

Microphotograph of the wire bonded test circuit.

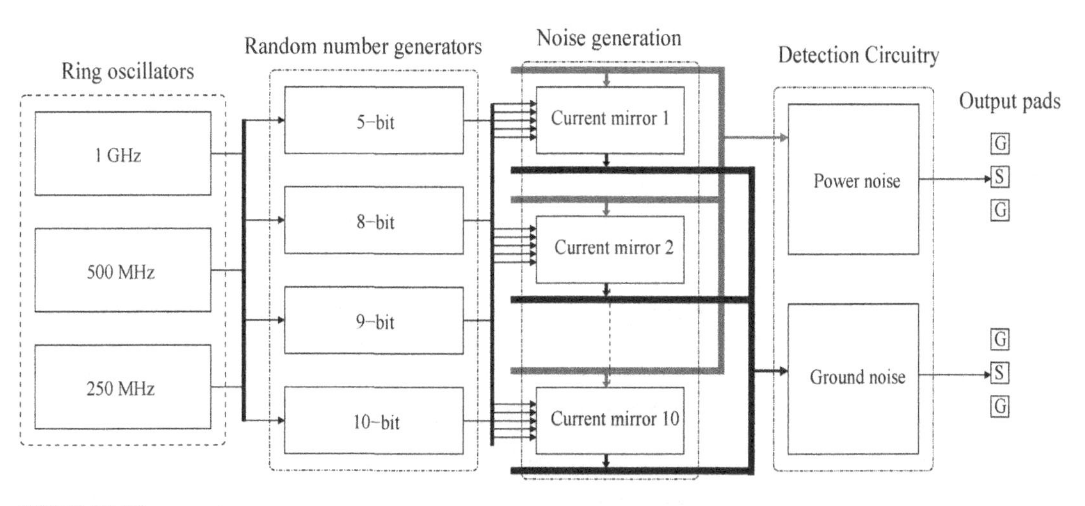

FIGURE 19.13

Block level schematic of noise generation and detection circuits.

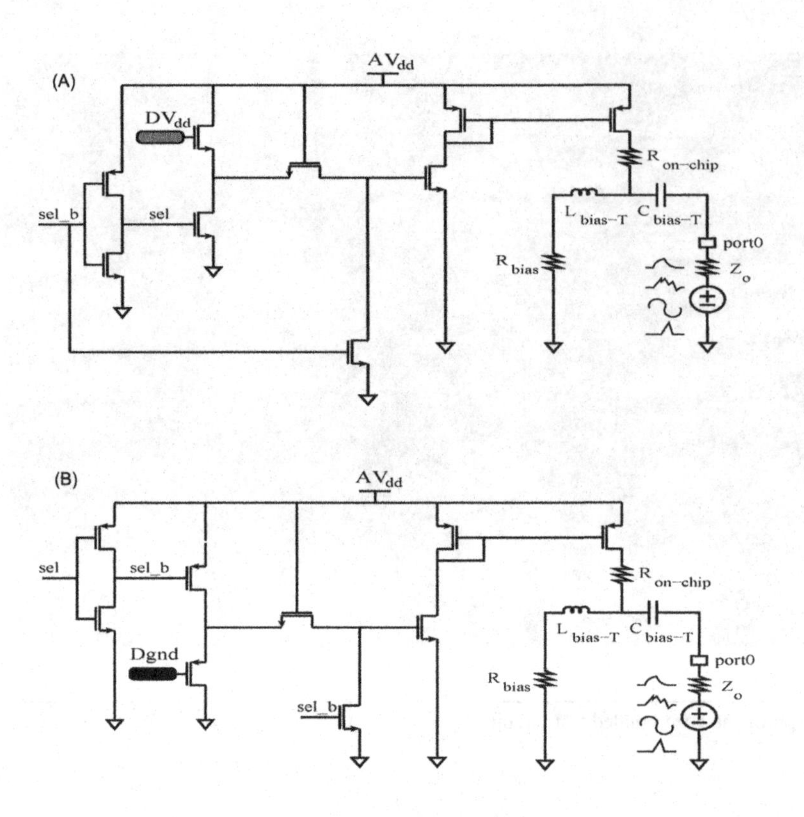

FIGURE 19.14

Source follower noise detection circuits detect noise on both the digital (A) power lines, and (B) ground lines.

19.3 CHARACTERISTICS OF 3-D POWER DISTRIBUTION TOPOLOGIES

The choice of power distribution topology affects the noise propagation characteristics of the power network, as indicated by the results described in Section 19.2. Additional insight into the design of the power networks (see Section 19.3.1), and a discussion of the characteristics of the 3-D power distribution network based on experimental results (see Section 19.3.2) are provided in this section.

19.3.1 PRE-LAYOUT DESIGN CONSIDERATIONS

Several considerations are accounted for in the design of the power distribution topologies. One issue is the placement of a sufficient number of TSVs to satisfy electromigration constraints between planes. The total number of TSVs required to satisfy a target current density of 1×10^6 A/cm^2, a current load of 0.14 amperes, a TSV diameter of 1.25 μm, and a footprint of 530 μm by 500 μm (the footprint area of each power distribution network) verifies that the 576 and 864 TSVs per tier for, respectively, Block 2 and Blocks 1 and 3 far exceed the necessary twelve

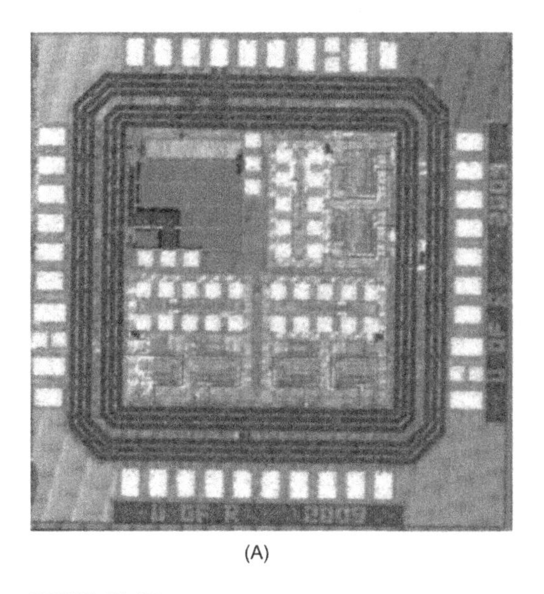

(A)

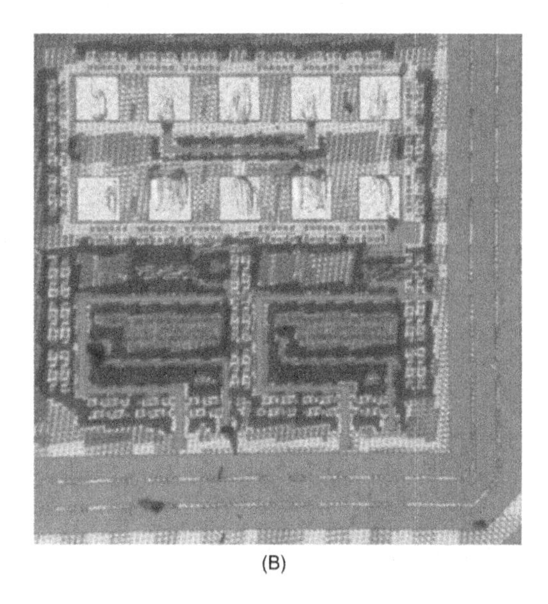

(B)

FIGURE 19.15

Fabricated test circuit examining noise propagation within three different power distribution networks, and a distributed DC-to-DC rectifier. (A) Microphotograph of the 3-D test circuit, and (B) an enlarged image of Block 1.

TSVs to satisfy any electromigration constraints. The number of TSVs to satisfy electromigration constraints is

$$\text{Area}_{\text{target current density}} = \frac{I_{\text{load}}}{J_{\text{electomigration}}}, \tag{19.1}$$

$$\text{Number of TSVs} = \frac{\text{Area}_{\text{target current density}}}{\text{Area}_{\text{TSV}}}. \tag{19.2}$$

The diameter of the TSV determines the "keep out zone" surrounding the TSV, an area where no devices can be fabricated. For the MIT Lincoln Laboratories 3-D technology, the keep out zone is approximately two times the diameter D. A square area with a width of $2 \times D$ and depth of $2 \times D$ is therefore used to determine the area penalty per TSV. An area of 1.36% and 2.04% of the 530 μm × 500 μm area is occupied by each power distribution topology on each device plane for, respectively, 576 and 864 TSVs. The area is the same for tiers 2 and 3, where all I/Os originate from tier 3. No area penalty on device plane 1 exists as no TSVs are necessary in this front-to-back bonded tier (the TSVs land on metal 3 on tier 1). For a keep out zone equivalent to $3 \times D$, the area penalty is 3.06% and 4.58% for, respectively, 576 and 864 TSVs.

To determine the resonant frequency of each topology, a model of the 3-D power distribution networks includes the impedance of the cables, board, wirebonds, on-chip DC pads, TSVs, and the power distribution network on each device plane. The equivalent electrical parameters are listed in Table 19.5, and the capacitance of the 3-D power distribution topologies is listed in Table 19.6. The impedances listed in Table 19.5 for the cables, board, and wirebonds are divided into three equivalent π model sections to characterize the distributed nature of the lines, as shown in Fig. 19.21.

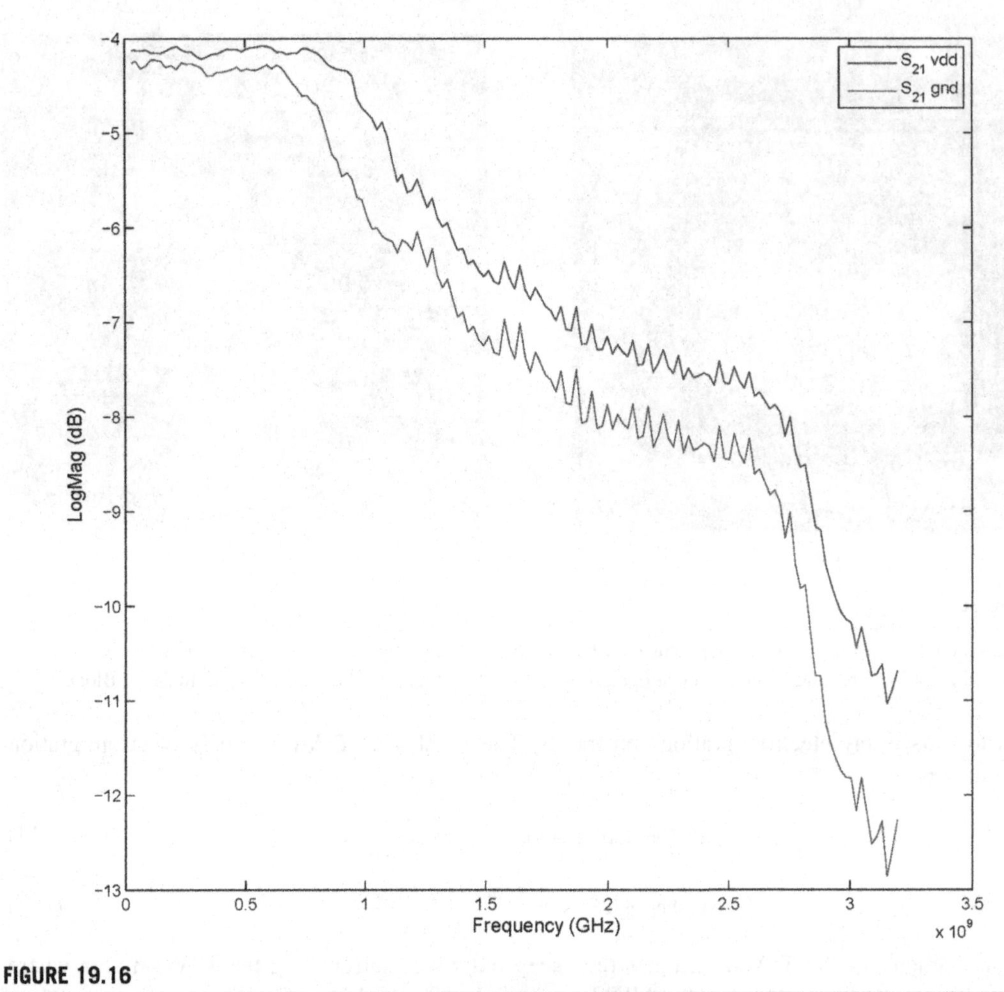

FIGURE 19.16

S-parameter characterization of the power and ground noise detection circuits.

The interdigitated power and ground lines in metal 3 are 530 μm long, 15 μm wide, and exhibit a sheet resistance of 0.08 Ω per sheet, producing a 2.83 Ω resistance. Similarly, the metal 2 lines are 500 μm long and 15 μm wide with a sheet resistance of 0.12 Ω per sheet, producing a 4 Ω resistance. Each power and ground line is divided into eight equal *RLC* π model sections to represent the distributed nature of the on-chip metal lines. An inductance of 1 pH is included with each π model section. Three horizontal pairs (metal 3) of power and ground lines as well as three vertical pairs (metal 2) of power and ground lines are considered for each interdigitated power distribution network on each device plane. The *RLC* parameters for a single π section of the interdigitated topology also model the dedicated plane topology, but the connections between sections are modified to better characterize the power and ground planes, as illustrated in Fig. 19.21.

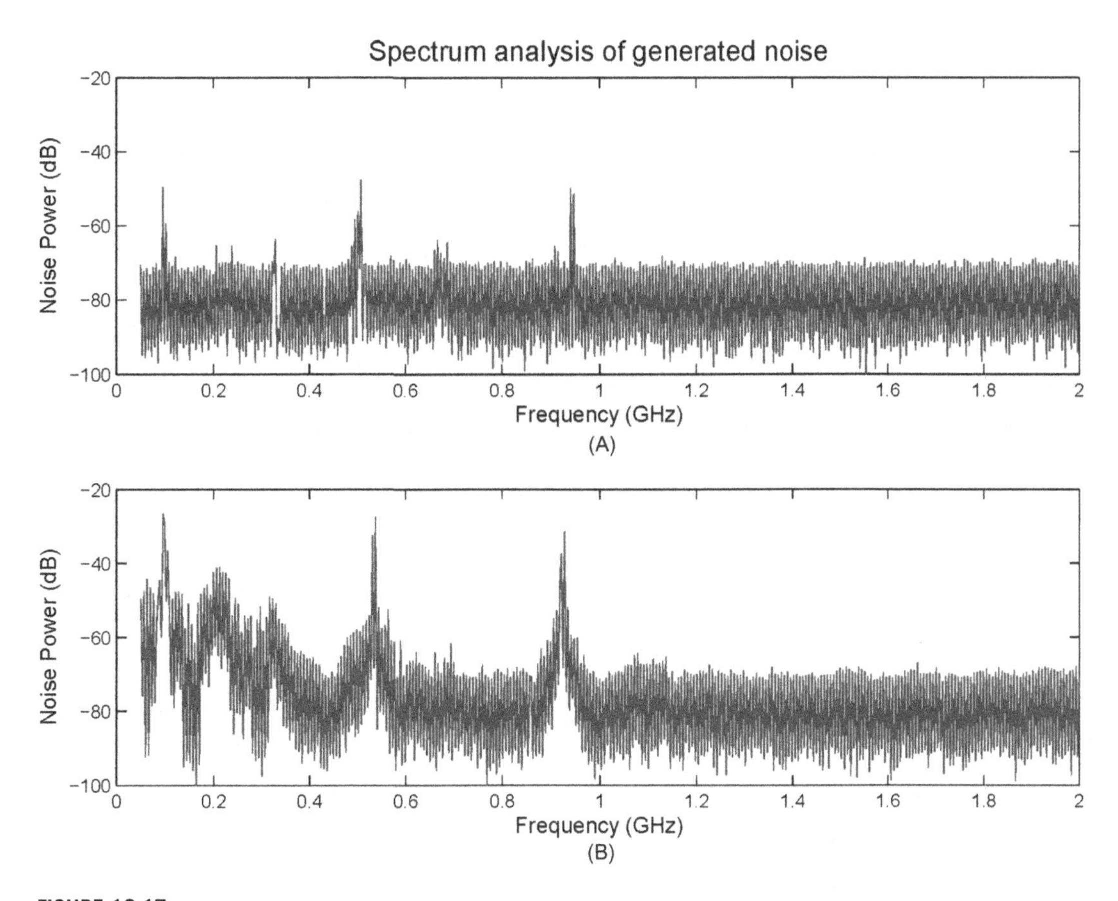

FIGURE 19.17

Spectral analysis of the noise generated on the power line of Block 2, (A) board level decoupling capacitance, and (B) without board level decoupling capacitance.

The resonant frequency for the three different power distribution networks is determined with and without board level decoupling capacitors and with and without an intentional on-chip load capacitance (separate from the intrinsic capacitance of the power distribution network). The resulting resonant frequency for the three power distribution topologies for the different capacitive configurations is listed in Table 19.7. The resonant frequencies reported in Table 19.7 consider the distributed inductance of the cables, board, wirebonds, and on-chip power distribution network. The on-chip inductance produces the highest resonant frequency, as reported in Table 19.7 for each block. The other three resonant frequencies, from the smallest to the largest, are dependent, respectively, on the cable, board, and wirebond inductances. The presence of a board level decoupling capacitor eliminates all but the highest resonant frequency, the resonance caused by the on-chip inductance.

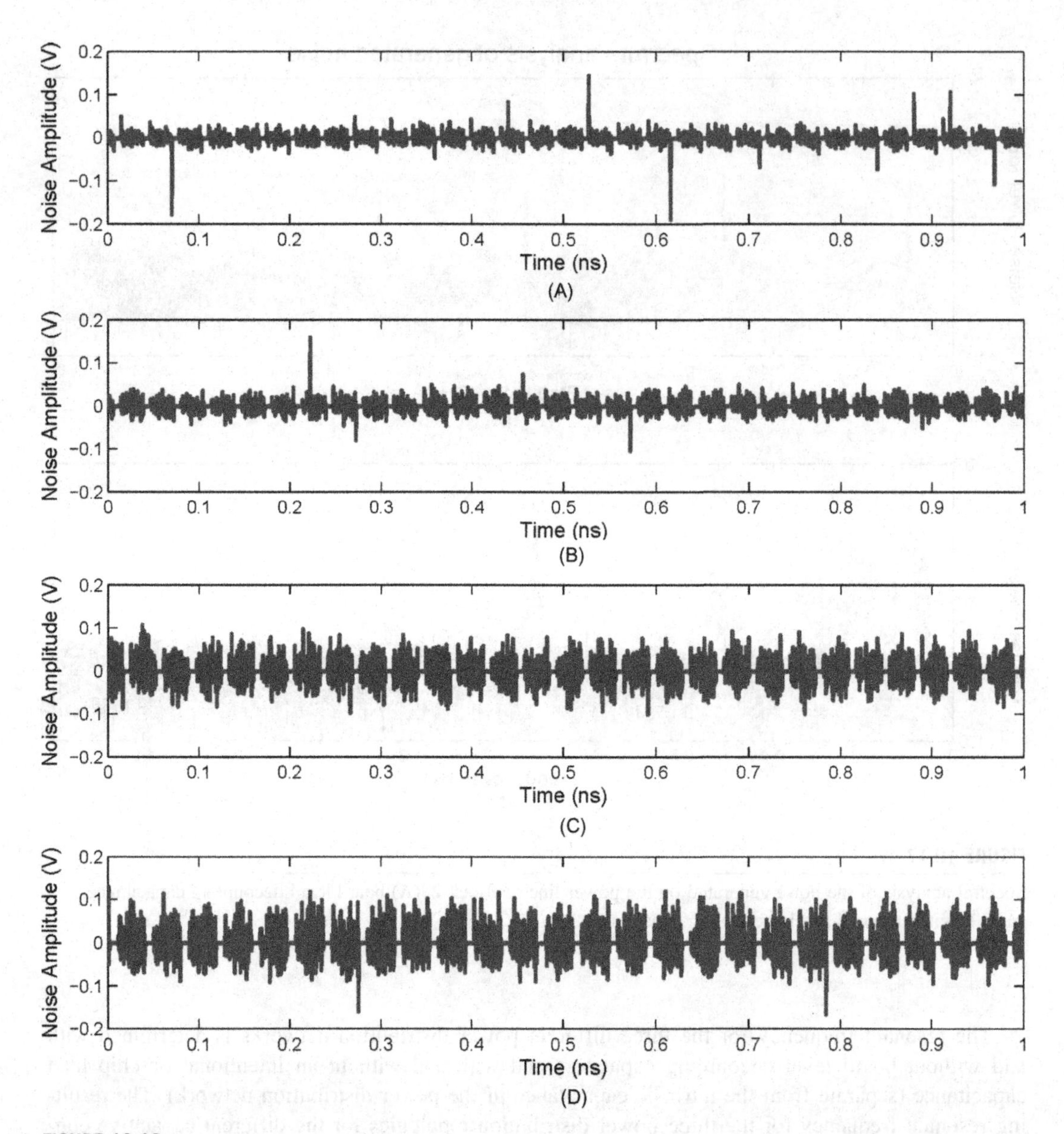

FIGURE 19.18

Time domain measurement of the generated noise on the power line of Block 2 without board level decoupling capacitance for a voltage bias on the current mirrors of (A) 0 volts, (B) 0.5 volt, (C) 0.75 volts, and (D) 1 volt.

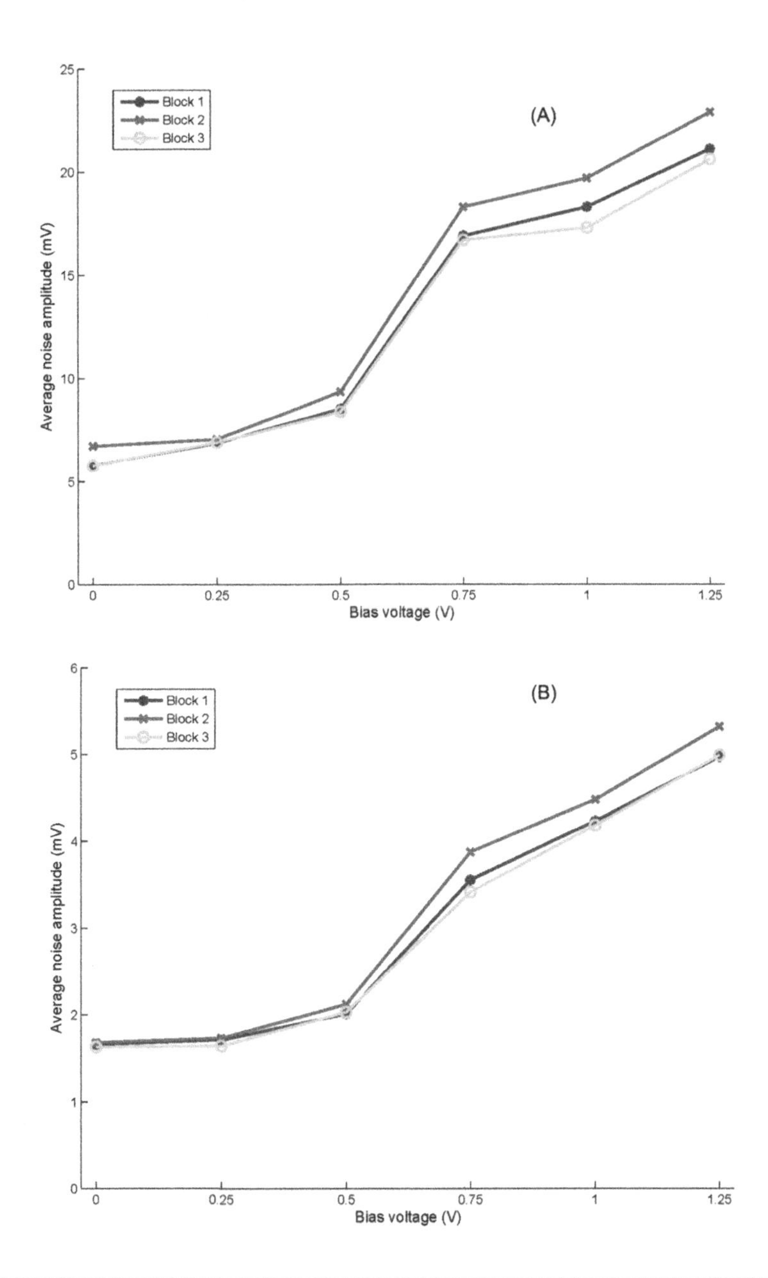

FIGURE 19.19

Average noise voltage on the power and ground distribution networks with and without board level decoupling capacitance. (A) Average noise of power network without decoupling capacitance, (B) average noise of power network with decoupling capacitance, (C) average noise of ground network without decoupling capacitance, and (D) average noise of ground network with decoupling capacitance. A total of 4,096 data points are used to calculate the average noise for each topology at each current mirror bias voltage.

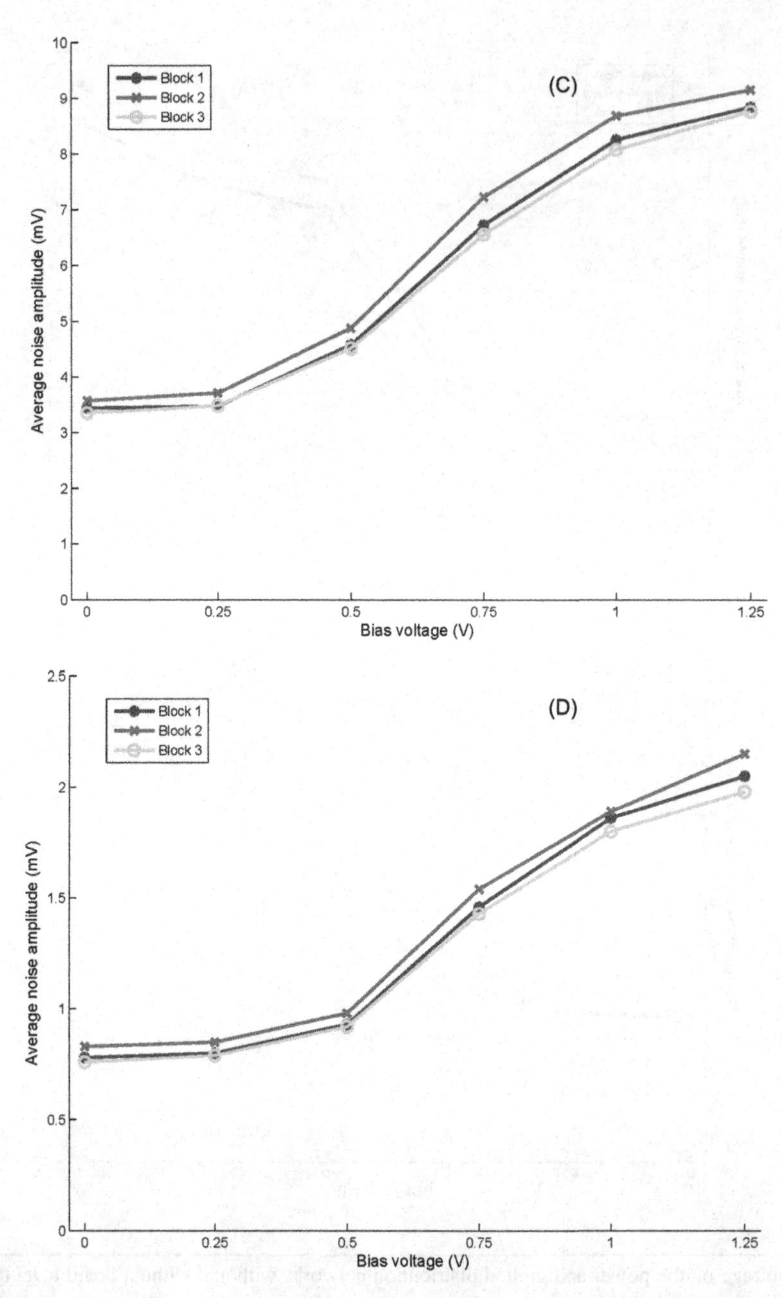

FIGURE 19.19

(Continued)

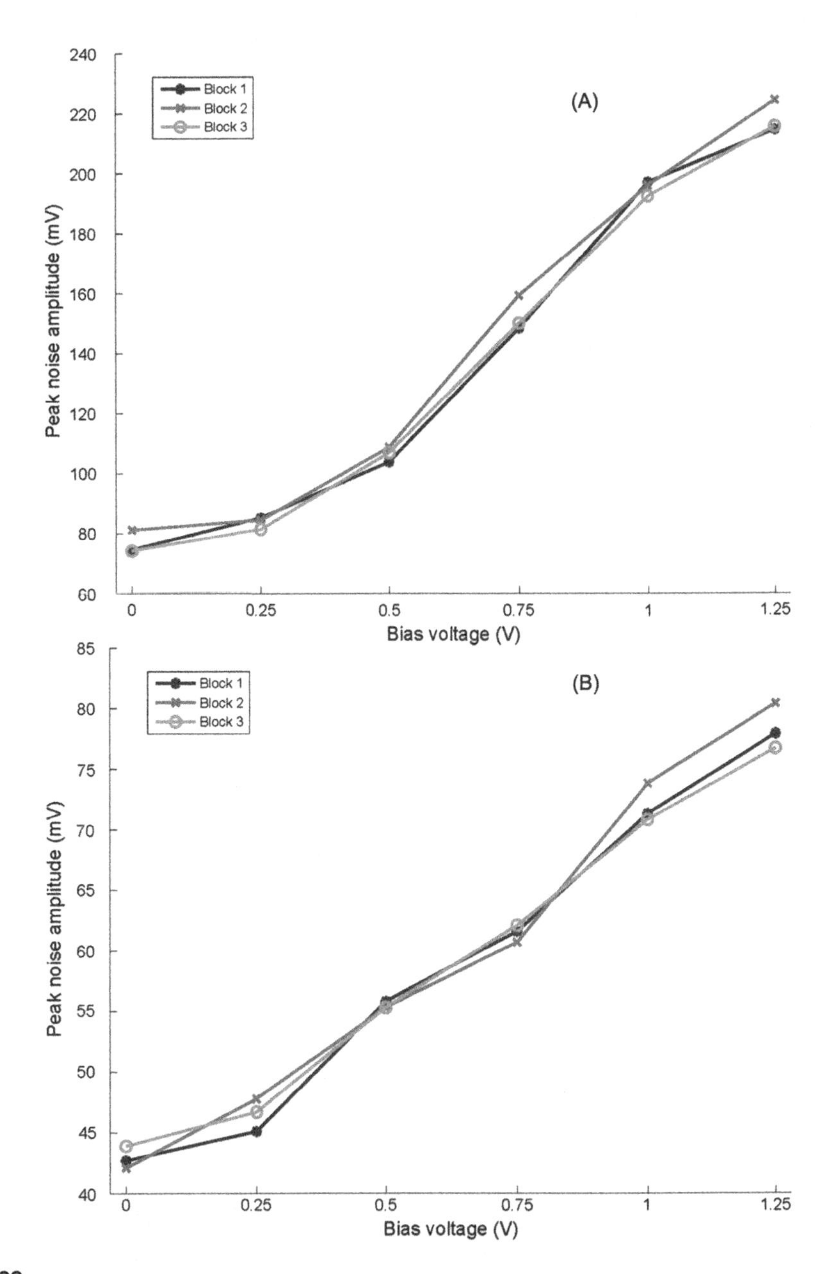

FIGURE 19.20

Peak noise voltage on the power and ground distribution networks with and without board level decoupling capacitance. (A) Peak noise of power network without decoupling capacitance, (B) peak noise of power network with decoupling capacitance, (C) peak noise of ground network without decoupling capacitance, and (D) peak noise of ground network with decoupling capacitance. A single peak data point (from 4,096 points) is determined for each topology at each current mirror bias voltage.

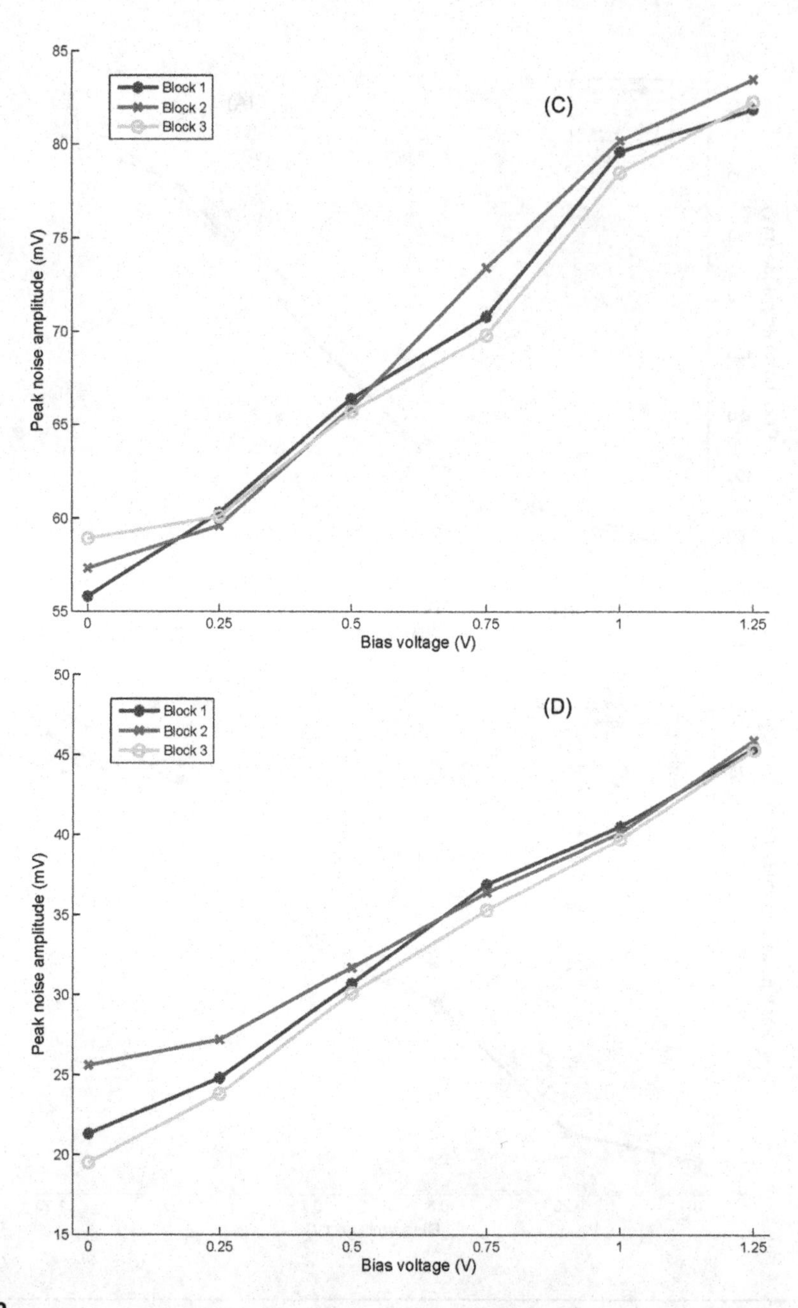

FIGURE 19.20

(Continued)

Table 19.3 Statistical Analysis of the Noise Generated on the Power Network

Power Network Block	On-Chip Capacitor	Current Mirror Bias Voltage (V)	Mean	25th Quartile	Median	75th Quartile
Block 1	No	0	5.77	1.65	4.50	8.84
		0.25	6.85	1.67	4.64	9.89
		0.5	8.52	1.93	6.64	13.17
		0.75	16.90	2.44	12.30	24.78
		1.0	18.30	2.18	12.95	26.50
		1.25	21.10	2.57	15.89	30.28
	Yes	0	1.66	0.80	1.10	2.93
		0.25	1.71	0.82	1.20	2.56
		0.5	2.01	0.99	1.74	3.18
		0.75	3.55	1.34	2.51	5.06
		1.0	4.23	1.59	3.09	6.74
		1.25	4.97	1.82	3.77	7.98
Block 2	No	0	6.66	1.74	4.56	9.54
		0.25	7.05	1.74	4.76	10.21
		0.5	9.31	2.03	6.72	13.87
		0.75	18.30	2.45	12.77	28.38
		1.0	19.01	2.22	13.03	28.94
		1.25	22.86	2.64	16.55	35.56
	Yes	0	1.68	0.87	1.24	2.99
		0.25	1.73	0.88	1.31	2.85
		0.5	2.12	1.14	1.88	3.55
		0.75	3.87	1.56	2.58	5.37
		1.0	4.48	1.77	3.23	7.05
		1.25	5.32	1.85	3.89	8.37
Block 3	No	0	5.76	1.65	4.48	8.80
		0.25	6.88	1.63	4.59	9.87
		0.5	8.37	1.94	6.60	12.89
		0.75	16.70	2.44	12.09	24.18
		1.0	17.30	2.06	12.89	26.22
		1.25	20.60	2.50	15.58	29.94
	Yes	0	1.63	0.77	1.10	2.88
		0.25	1.64	0.81	1.14	2.52
		0.5	2.03	0.99	1.70	3.11
		0.75	3.41	1.31	2.50	5.04
		1.0	4.18	1.60	3.02	6.69
		1.25	4.99	1.81	3.77	8.01

Sample size of 4,096 points at each bias voltage.

Table 19.4 Power Supply Current within the Different Power Distribution Topologies as a Function of Bias Voltage on the Current Mirrors

Power Network Block	Current as Function of Bias Voltage (mA)					
	0 V	0.25 V	0.5 V	0.75 V	1 V	1.25 V
Block 1	29.2 to 34.6	31.4 to 38.7	40.1 to 45.3	56.2 to 59.8	73.6 to 79.5	86.4 to 91.3
Block 2	32.6 to 38.5	33.6 to 40.1	43.7 to 47.8	59.2 to 66.3	75.5 to 82.5	88.6 to 93.8
Block 3	43.6 to 55.3	44.1 to 56.3	52.9 to 65.1	68.7 to 78.7	83.4 to 91.9	97.5 to 105.7

Table 19.5 Physical and Electrical Parameters of the Cables, Board, Wirebonds, On-Chip DC Pads, Power Distribution Networks, and TSVs

Component	Width (μm)	Diameter (μm)	Length (mm)	Thickness (μm)	Resistance (Ω)	Capacitance (pF)	Inductance (nH)
Cables	—	1020	965.2	—	0.072	10.59	1445.75
Board	760	—	40	43.2	0.021	1.88[a]	26.48
Wirebonds	—	25.4	4 to 5	—	0.278	0.0233	5.92
DC pads	80	—	0.12	2	—	0.0717	—
Pad to power grid	60	—	0.260	0.63	0.52	—	—
TSV	—	1.25	0.009	—	0.4	0.124×10^{-3}[b]	5.55×10^{-3}

[a] *An additional 4 μF when board level decoupling capacitors are added.*
[b] *Coupling capacitance between two TSVs [703].*

Table 19.6 Capacitance of the Three Different Power Distribution Blocks, Interdigitated, and Power/Ground Planes

Block or Topology	Power or Ground	Capacitance (fF)	Inductance (nH)
Interdigitated	V_{dd}	330.09	—
	V_{gnd}	330.08	—
Planes	V_{dd}	1000.35	—
	V_{gnd}	1158.24	—
Block 1	V_{dd}	993.38	1631.68
	V_{gnd}	993.39	1596.38
Block 2	V_{dd}	964.37	1610.37
	V_{gnd}	964.38	1574.18
Block 3	V_{dd}	1862.74	2487.10
	V_{gnd}	1949.60	2590.36

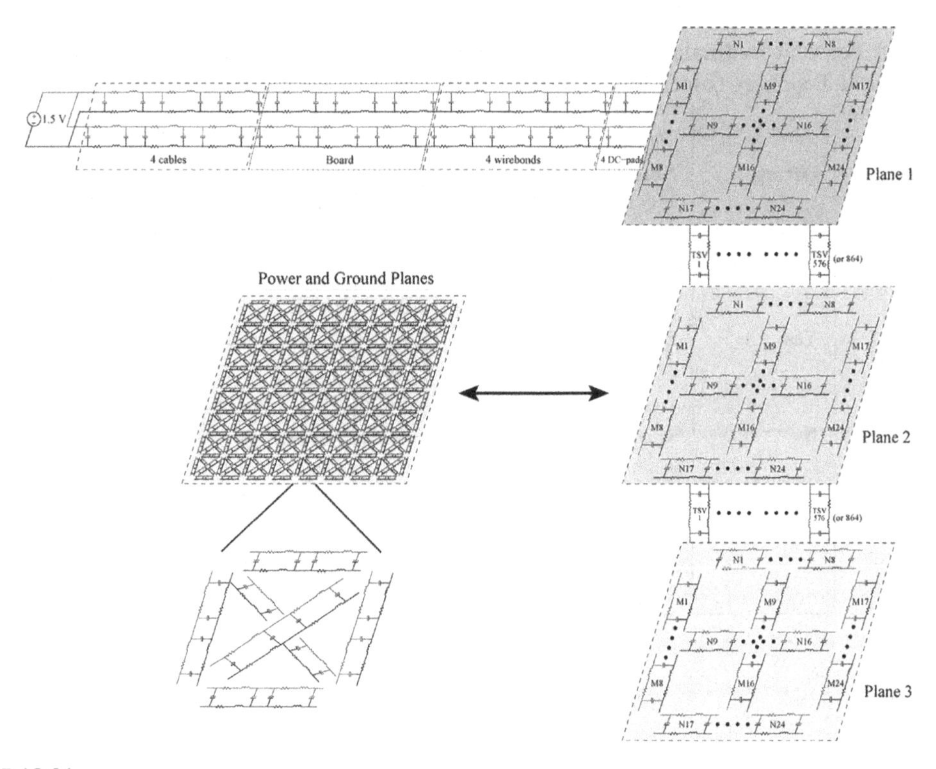

FIGURE 19.21

Equivalent electrical model of the cables, board, wirebonds, on-chip DC pads, power distribution networks, and TSVs.

Table 19.7 Resonant Frequency of the Three Different Power Distribution Networks with and without Board Level Decoupling Capacitors and with and without an On-Chip Load Capacitance

Block	Decoupling Capacitance (μF)	Extracted Load (fF)	Resonant Frequencies (GHz)
Block 1	none	none	0.224, 0.447, 0.794, 1.25
	4	none	1.25
	none	638	0.199, 0.447, 0.708, 1.25
	4	638	1.25
Block 2	none	none	0.223, 0.447, 0.794, 1.78
	4	none	1.78
	none	646	0.199, 0.447, 0.708, 1.78
	4	646	1.78
Block 3	none	none	0.224, 0.447, 0.794, 1.67
	4	none	1.67
	none	624	0.199, 0.447, 0.708, 1.67
	4	624	1.67

Table 19.8 Per cent Reduction in Average Noise from the Noisiest Topology (Block 2) to the Least Noisiest Topology (Block 3) as a Function of Bias Voltage on the Current Mirrors

Power or Ground	DeCap Present	Noisier Block	Quieter Block	Per cent Noise Reduction as Function of Bias Voltage (volts)					
				0	0.25	0.5	0.75	1	1.25
Power	No	2	1	14.0	2.7	8.9	7.7	7.1	7.9
		2	3	14.2	2.3	10.5	8.7	12.2	10.0
		1	3	0.2	−0.4	1.8	1.2	5.5	2.4
Power	Yes	2	1	1.2	1.2	5.2	8.3	5.6	6.6
		2	3	3.0	5.2	4.2	11.9	6.7	6.2
		1	3	1.8	4.1	−1.0	3.9	1.2	−0.4
Ground	No	2	1	3.6	6.5	6.1	6.9	4.9	3.4
		2	3	6.2	6.2	7.4	9.3	7.0	4.3
		1	3	2.6	−0.3	1.3	2.5	2.2	0.9
Ground	Yes	2	1	6.0	5.9	5.1	5.2	1.6	4.7
		2	3	8.4	7.1	6.1	7.1	4.8	7.9
		1	3	2.6	1.3	1.1	2.1	3.3	3.4

19.3.2 DESIGN CONSIDERATIONS BASED ON EXPERIMENTAL RESULTS

Based on the experimental results, both increasing the number of TSVs by 50% and utilizing a dedicated power and ground plane lower the power noise. Although the noise is lower with both a greater number of TSVs and dedicated power planes (2% to 14.2% lower noise, as listed in Table 19.8), the power noise is limited by the small series resistance of the 3-D power network as compared with the larger series resistance of the cables and wirebonds, as described in Section 19.3.1.

Two issues in 3-D power distribution networks are considered based on the experimental results: (1) the benefits of the power and ground planes to justify the use of two metallization levels, and (2) the benefits and drawbacks of increased TSV density on area and noise. Addressing (1), the power planes provide an additional reduction of 0.2% to 5.5% in average noise as compared with the fully interdigitated topology. The reduction in average noise is primarily due to the increased capacitance of the power and ground networks, as indicated in Table 19.6. The average noise characteristics of the planes topology can be further improved by increasing the size of the power/ground planes or using multiple metal layers for both power and ground. The use of full planes to deliver current to the load does; however, require significant metal resources and therefore this issue must be considered. In addition, the power planes complicate the design of the signal interconnects, where holes are required in the power/ground planes to pass signals between device planes. For those applications where greater noise reduction is required, additional metallization and design complexity may be justified.

The choice of TSV density provides the greatest reduction in average noise, ranging from 2.7% to 14.0%, as listed in Table 19.8. The additional area penalty when increasing the number of TSVs by 50% from 576 to 864 is 0.68% (1.36% as compared with 2.04%), noting that the area penalty is also dependent on the keep out region, as described in Section 19.3.1. The increased area due to the higher TSV density is small as compared to the large reduction in average noise.

19.4 SUMMARY

The design of a power distribution network for application to 3-D circuits is considerably more complex than the design of a 2-D power network. The primary issues in the design process of a 3-D test circuit and related measurements are summarized as follows:

- The preferable topology of a power distribution network is not dictated by a single design objective but, rather, by the overall 3-D system level requirements, including the availability of I/O pins and number of bonded tiers. This test circuit provides enhanced understanding of topology dependent noise generation and propagation within 3-D power delivery systems.
- A test circuit examining power grid noise in a 3-D integrated stack has been designed, fabricated, and tested. Three topologies to distribute power within a 3-D circuit have been evaluated, and an analysis of the peak noise voltage, voltage range, average noise voltage, and resonant frequency characteristics for both power and ground is discussed.
- Fabrication and vertical bonding were performed by MIT Lincoln Laboratory for a 150 nm, three metal layer SOI process. Three wafers are vertically bonded to form a 3-D stack.
- A noise analysis of three power delivery topologies is described. Calibration circuits for a source follower sense circuit compare the different power delivery topologies as well as the individual 3-D circuits. The effects of the TSV density on the noise profile of a 3-D power delivery network are experimentally characterized.
- A comparison of the peak noise and resonant behavior for each topology with and without board level decoupling capacitors is provided, and suggestions for enhancing the design of the power delivery network are offered.

CHAPTER

3-D CIRCUIT ARCHITECTURES

20

CHAPTER OUTLINE

Technological, physical, and thermal design methodologies have been presented in the previous chapters. The architectural implications of adding the third dimension in the integrated circuit (IC) design process are discussed in this chapter. Various wire limited integrated systems are explored where the third dimension can mitigate many interconnect issues. Primary examples of this circuit category are the microprocessor memory system, on-chip networks, and field programmable gate arrays (FPGAs). Although networks-on-chip (NoC) and FPGAs are generic communication fabrics as compared to microprocessors, the effects of the third dimension on the performance of a microprocessor are discussed here due to the importance of this circuit type.

The performance enhancements that originate from the 3-D implementation of wire dominated circuits are presented in the following sections. These results are based on analytic models and academic design tools under development with exploratory capabilities. Existing performance limitations of these circuits are summarized in the following section, where a categorization of these circuits is also provided. 3-D architectural choices and corresponding tradeoffs for microprocessors, memories, and microprocessor memory systems are discussed in Section 20.2. 3-D topologies for on-chip networks are presented and evaluated in Section 20.3. Both analytic expressions and simulation tools are utilized to explore these topologies. Finally, the extension to the third dimension of an important design solution, namely FPGAs, is analyzed in Section 20.4. A brief summary of the analysis of these 3-D architectures is offered in Section 20.4.

Three-Dimensional Integrated Circuit Design. DOI: http://dx.doi.org/10.1016/B978-0-12-410501-0.00020-4

20.1 CLASSIFICATION OF WIRE LIMITED 3-D CIRCUITS

Any 2-D IC can be vertically fabricated with one or more processes developed for 3-D circuits. The benefits, however, vary for different circuits [291]. Wire dominated circuits are good candidates for vertical integration since these circuits greatly benefit from the significant decrease in wirelength. Performance projections as previously discussed in Chapter 7, Interconnect Prediction Models, highlight this situation. Consequently, only communication centric circuits are emphasized in this discussion. The different circuit categories considered in this chapter are illustrated in Fig. 20.1.

The first category includes application specific ICs (ASIC) that are typically part of a larger computing system. A criterion for a circuit to belong to this category is whether the third dimension can considerably improve the primary performance characteristics of a circuit, such as speed, power, and area. A fast Fourier transform circuit is an example of a 3-D ASIC [705]. Memory arrays and microprocessors are other circuit examples that belong to this category. 3-D integration is amenable to the interconnect structures connecting these components comprising more complex integrated systems.

Communication fabrics, such as NoC and FPGAs, are particularly appropriate for vertical integration. For instance, several low latency and high throughput multi-dimensional topologies have been presented in the past for traditional interconnection networks, such as 3-D meshes and tori [706]. These topologies, depicted in Fig. 20.2, are not usually considered for on-chip networks due to the long interconnects that hamper the overall performance of the network. These limitations are naturally circumvented when a third degree of design freedom is added.

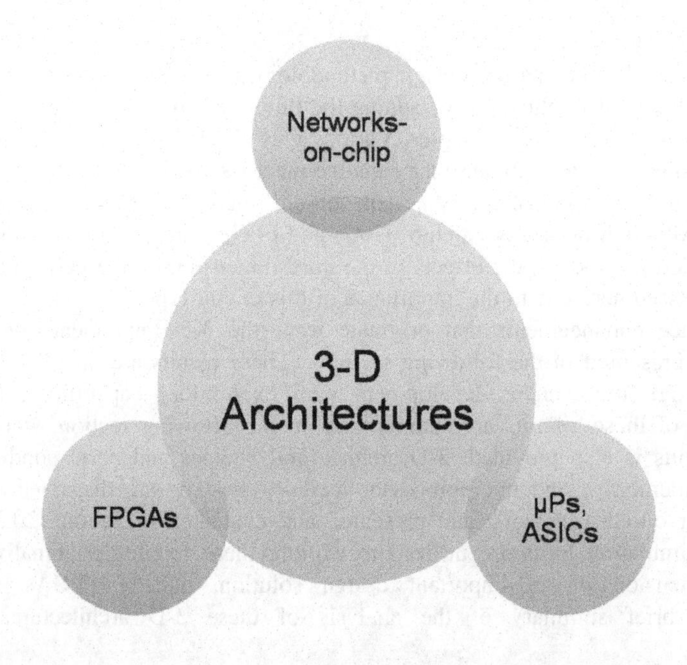

FIGURE 20.1

Taxonomy of 3-D architectures for wire limited circuits.

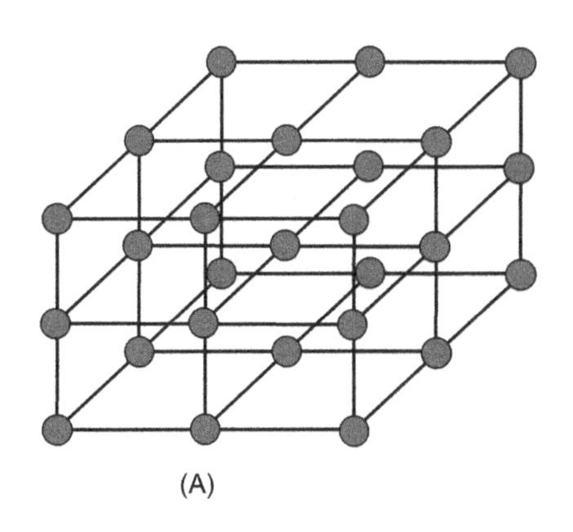

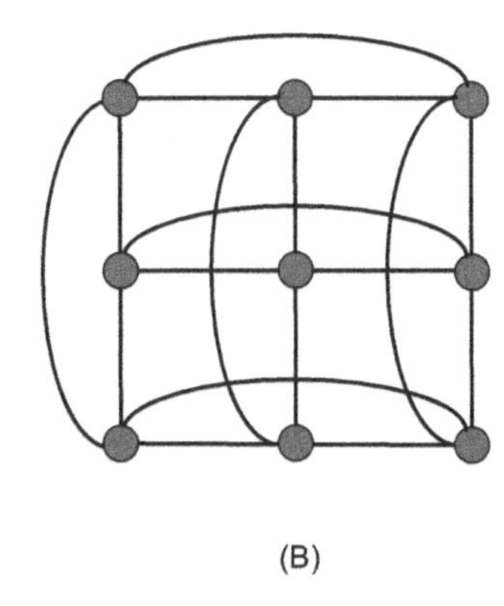

FIGURE 20.2

Popular interconnection network topologies, (A) 3-D mesh, and (B) 2-D torus.

In the case of FPGAs, a design style for an increasing number of applications, vertical integration offers a twofold opportunity; increased interconnectivity among the logic blocks (LBs) as well as shorter distances among these blocks. These advantages will enhance the performance of FPGAs which is the major impediment of this design style. Consequently, 3-D architectures are indispensable for contemporary FPGAs. Each of these circuit categories is successively reviewed in the remainder of the chapter. Related design tools are also discussed and further issues in the design of these circuits are highlighted.

20.2 3-D MICROPROCESSORS AND MEMORIES

Microprocessor and memory circuits constitute a fundamental component of every computing system. Due to the use of these circuits in myriads of applications, the effects of 3-D integration on this system are of significant interest. Both of these types of circuits are amenable to a variety of 3-D architectural alternatives. The partitioning scheme used on standard 2-D circuits drastically affects the characteristics of the resulting 3-D architectures. The different partitioning levels and related building elements based on these partitions are illustrated, respectively in Figs. 20.3 and 20.4. A finer partitioning level typically requires a larger design effort and higher vertical interconnect densities. Consequently, each partitioning level is only compatible with a specific 3-D technology. Note that the intention here is not to determine an effective partitioning methodology specific to 3-D microprocessors and memories but rather to determine the architectural granularity that improves the performance of these circuits. The physical design techniques described in Chapter 9, Physical Design Techniques for Three-Dimensional ICs, can be used to achieve the optimum design objectives for a particular architecture.

Different architectures for several blocks not including the on-chip cache memory of a microprocessor are discussed in Section 20.2.1. The 3-D organization of the cache memory is described in Section 20.2.2. Finally, the 3-D integration of a combined microprocessor and memory is discussed in Section 20.2.3.

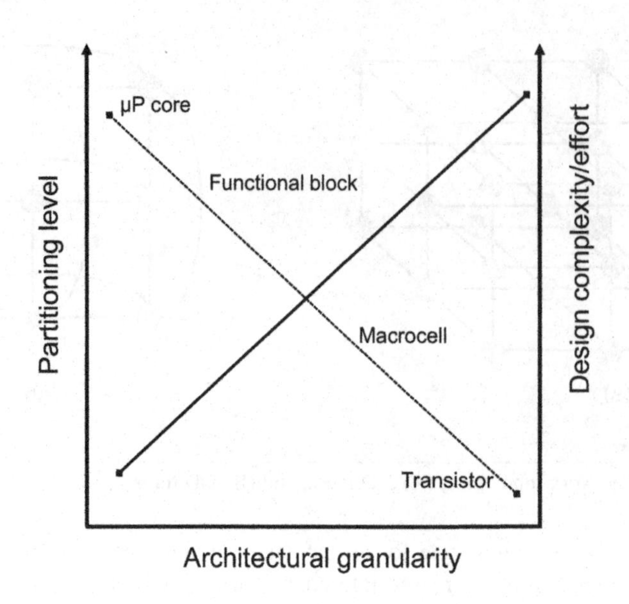

FIGURE 20.3

Different partitioning levels and related design complexity vs the architectural granularity for 3-D microprocessors.

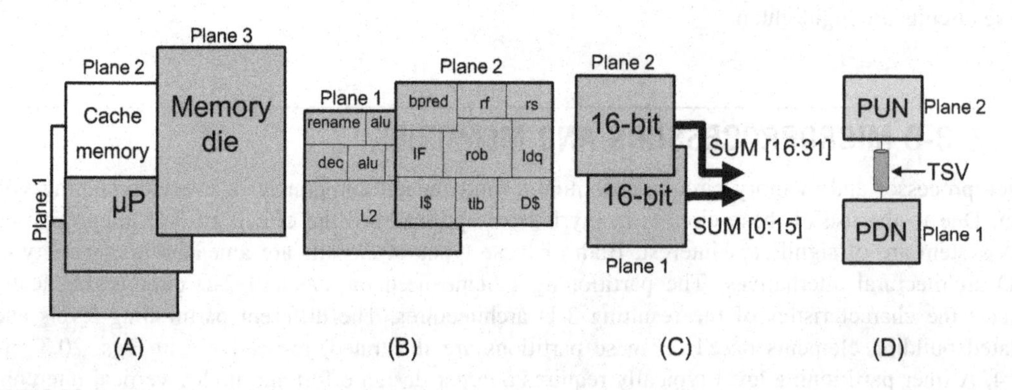

FIGURE 20.4

An example of different partitions levels for a 3-D microprocessor system at the (A) core, (B) functional unit block (FUB), (C) macrocell, and (D) transistor levels.

20.2.1 3-D MICROPROCESSOR LOGIC BLOCKS

The microprocessor circuit analyzed herein consists of one logic core and an on-chip cache memory. The architectures can be extended to multicore microprocessor systems. For a microprocessor circuit, partitioning the functional blocks or macrocells (i.e., circuits within the functional blocks) to several tiers (see Fig. 20.4) is meaningful and can improve the performance of the individual functional blocks and the microprocessor. For a microprocessor system, partitioning at the core level is also possible. The resulting effects on a microprocessor are higher speed, additional instructions per cycle (IPC), and a decrease in the number of pipeline stages.

In general, the 3-D implementation of a wire limited functional block (i.e., partitioning at the macrocell level) decreases the dissipated power and delay of the blocks. The benefits, however, are minimal for blocks that do not include relatively long wires. In this case, some performance improvement exists due to the decrease in the area of the block and, consequently, in the length of the wires traversing the block. In addition to the cache memory, many other blocks within the processor core can be usefully placed on more than one physical tier. These blocks include, for example, the instruction scheduler, arithmetic circuits, content addressable memories, and register files.

The instruction scheduler is a critical component constraining the maximum clock frequency and dissipates considerable power [707]. Placing this block in two tiers results in a lower delay and power by, respectively 44% and 16%, [708]. Additional savings in both delay and power are achieved by using three or four tiers; however, the savings saturate rapidly. This situation is more pronounced in the case of arithmetic units, such as adders and logarithmic shifters. Delay and power improvements for Brent Kung [709] and Kogge Stone [710] adders are listed in Table 20.1. As indicated from these results, the benefits of utilizing more than two tiers are negligible. Further reductions in delay are not possible for more than two tiers since the delay of the logic dominates the delay of the interconnect.

Consequently, partitioning at the macrocell level is not necessarily helpful for every microprocessor component and should be carefully applied to ensure that only the wire dominated blocks are designed as 3-D blocks. Alternatively, another architectural approach does not split but simply stacks the functional blocks on adjacent physical tiers, decreasing the length of the wires shared by these blocks. As an example of this approach, consider the two tier 3-D design of the Intel Pentium 4 processor where 25% of the pipeline stages in the 2-D architecture are eliminated, improving performance by almost 15% [711]. The power consumption is also decreased by almost 15%. A similar architectural approach for the Alpha 21364 [712] processor resulted in a 7.3% and 10.3% increase in the IPC for, respectively two and four tiers [713].

A potential hurdle in the performance improvement is the introduction of new hotspots or an increase in the peak temperature as compared to a 2-D version of the microprocessor. Thermal analysis of the 3-D Intel Pentium 4 has shown that the maximum temperature is only 2 °C greater as compared to the 2-D counterpart, reaching a temperature of 101 °C [713]. If maintaining the

Table 20.1 Performance and Power Improvements of 3-D over 2-D Architectures [708]

	Kogge Stone Adder			Brent Kung Adder
	16 bits		32 bits	32 bits
# of Input Bits	Delay	Power	Delay	Delay
Two tiers	20.2%	8%	9.6%	13.3%
Three tiers	23.6%	15%	20.0%	18.1%
Four tiers	32.7%	22%	20.0%	21.7%

same thermal profile is a primary objective, the power supply can be scaled, partially limiting the improved performance. For this specific example, voltage scaling pacified any temperature increase while providing an 8% performance improvement and 34% decrease in power (for two tiers) [711]. In these case studies, a 2-D architecture has been redesigned into a 3-D microprocessor architecture. Greater enhancements can be realized if the microprocessor is designed from scratch, initially targeting a 3-D technology as compared to simply migrating a 2-D circuit into a multi-tier stack. A considerable portion of the overall performance improvement in a microprocessor can be achieved by distributing the cache memory onto several physical tiers, as described in the following section.

20.2.2 3-D DESIGN OF CACHE MEMORIES

Data exchange between the processor logic core and the memory has traditionally been a fundamental performance bottleneck. Therefore, a small amount of memory, specifically cache memory, is integrated with the logic circuitry offering very fast data transfer, while the majority of the main memory is placed off-chip. The size and organization of the cache memory greatly depend upon the architecture of the microprocessor and has steadily increased over the past several microprocessor generations [714,715]. Due to the small size of the cache memory, a common problem is that data are often fetched from the main memory. This situation, widely known as a cache miss, is a high latency task. Increasing the cache memory size can partly lower the cache miss rate. 3-D integration supports both larger and faster cache memories. The former characteristic is achieved by adding more memory on the upper tiers of a 3-D stack, and the latter objective can be enhanced by constructing novel cache architectures with shorter interconnects.

A schematic view of a 2-D 32 KB cache is illustrated in Fig. 20.5, where only the data array is depicted. The memory is arranged into smaller arrays (subarrays) to decrease both the access time and power dissipation. Each memory subarray i is denoted as Block i. The size of the subarrays is determined by two parameters, N_{dwl} and N_{dbl}, which correspond, respectively, to the divisions of the initial number of word and bit lines. In this example, each subarray contains 128×256 bits. In addition to the SRAM arrays, other circuits are shown in Fig. 20.5. The local word line decoders and drivers are placed on the left side of each subarray. The multiplexers and sense amplifiers are located on the bottom and top side of each row. The word line predecoder is placed at the center of the entire memory.

Several ways exist to partition this structure into multiple tiers. An example of a 2-D and 3-D organization of a 32 Kb cache memory is schematically illustrated in Fig. 20.6. The memory can be stacked at the functional block, macrocell, and transistor levels, where a functional block is considered in this case to be equivalent to a memory subarray. Halving the memory and placing each half on a separate tier decreases the length of the global wires, such as the address input to the word line predecoder nets, the data output lines, and the wires used for synchronization. Partitioning at the functional block level does not improve the delay and power of the subarrays which adds to the total access time and power dissipation.

Dividing each subarray can therefore result in lower access time and power consumption. Partitioning occurs along the x and y directions, halving, respectively, the length of the bit and word lines at each division. An example of a subarray partition is depicted in Fig. 20.7. The number of partitions along each direction is characterized by the parameters N_x and N_y. For word line partitions ($N_x > 1$), the word lines are replicated on the upper tiers, as shown in Fig. 20.7. The

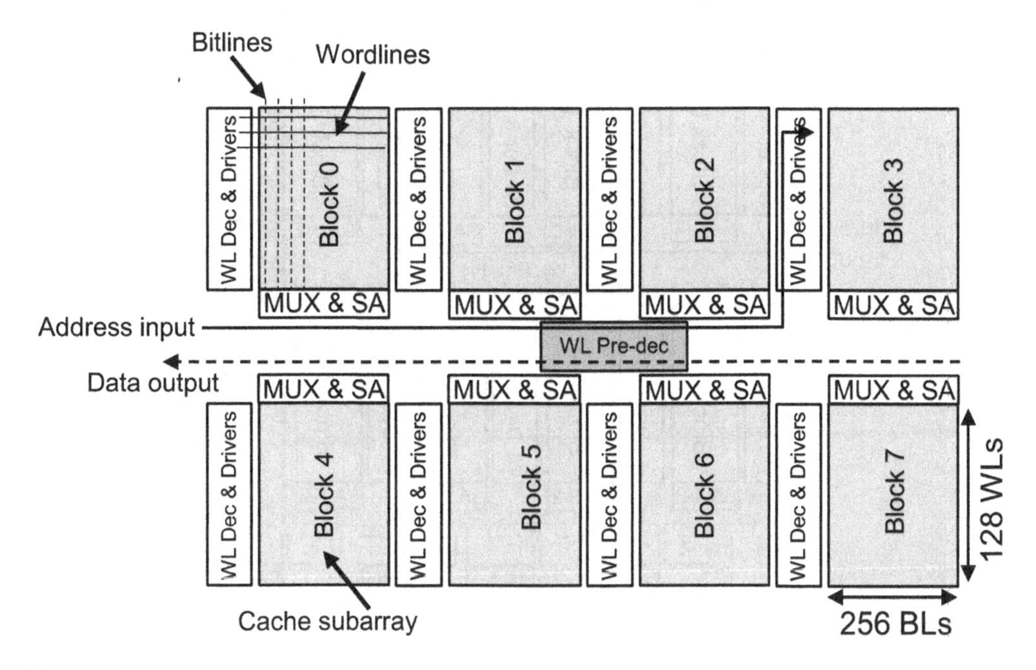

FIGURE 20.5

2-D organization of a cache memory with additional circuitry [716].

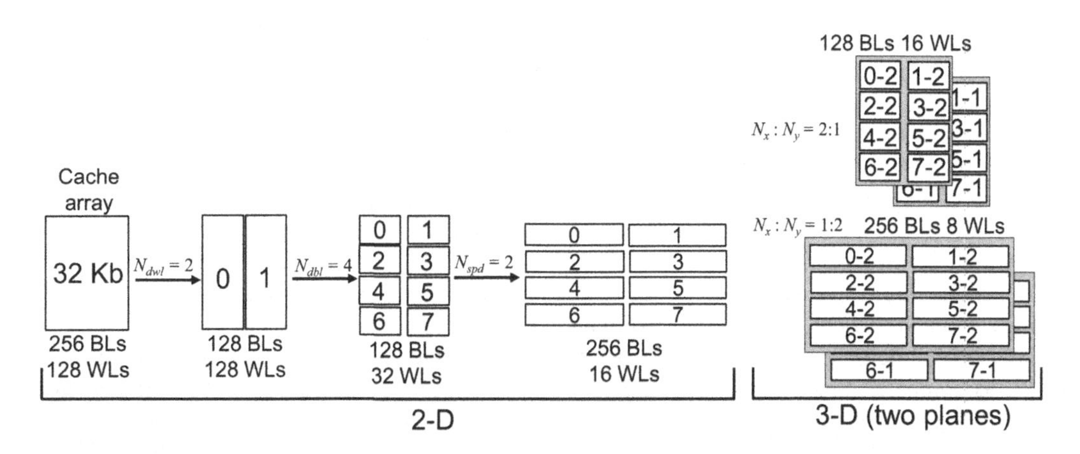

FIGURE 20.6

2-D and 3-D organization of a 32 Kb cache memory array. N_{spd} is the number of sets connected to a word line [713].

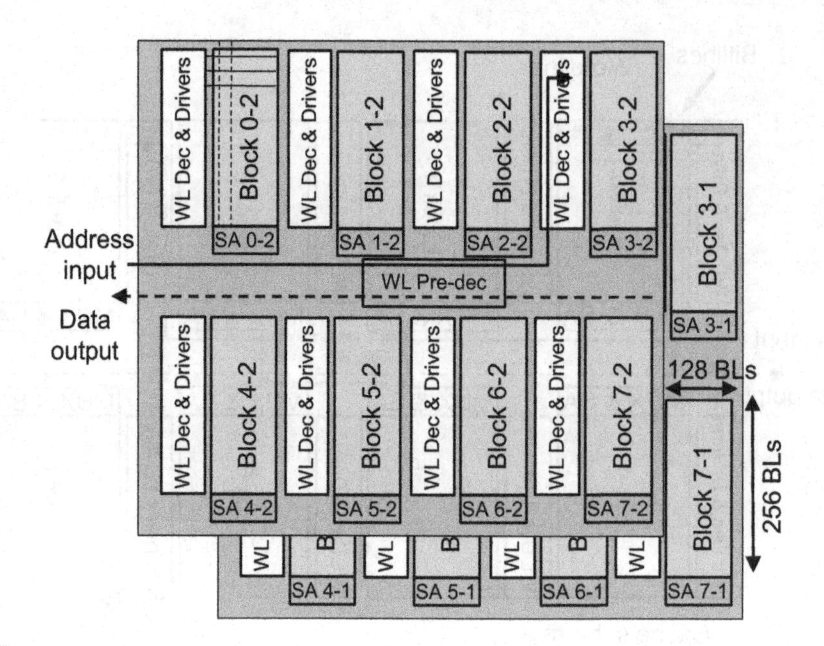

FIGURE 20.7

Word line partitioning onto two tiers of the 2-D cache memory shown in Fig. 20.5 [713].

length of the word lines, however, decreases, resulting in smaller device sizes within the drivers and local decoders. Furthermore, the area of the overall array decreases, leading to shorter global wires, such as the input address from the predecoder to the local decoders and data output lines.

In the case of bit line partitioning ($N_y > 1$), the length of these lines and the number of pass transistors tied to each bit line are reduced, as illustrated in Fig. 20.8. The sense amplifiers can either be replicated on the upper tiers or shared among the bit lines from more than one tier. In the former case, the leakage current increases but the access time is improved, while in the latter case, the power savings is greater, however, the speed enhancement is not significant due to the requirement for bit multiplexing.

In general, word line partitioning results in a smaller delay but not necessarily a larger power savings. More specifically, for high performance caches, partitioning the word lines offers greater savings in both delay and energy. For instance, a 1 MB cache where $N_x=4$ and $N_y=1$ is faster by 16.3% as compared to a partition where $N_x=1$ and $N_y=4$ [716]. Alternatively, bit line partitioning is more efficient for low power memories. When the memory is designed for low power, bit line partitioning decreases the power by approximately 14% as compared to the reduction achieved by word line partitioning [716]. This behavior can be explained by considering the original 2-D version of the cache memory. High performance memories favor wide arrays, which implies longer word lines, while low power memories exhibit a greater height resulting in longer bit lines.

Finally, partitioning is also possible at the transistor level where the basic six transistor SRAM cell can be split among the tiers of a 3-D stack. This extra fine granularity, however, has a negative

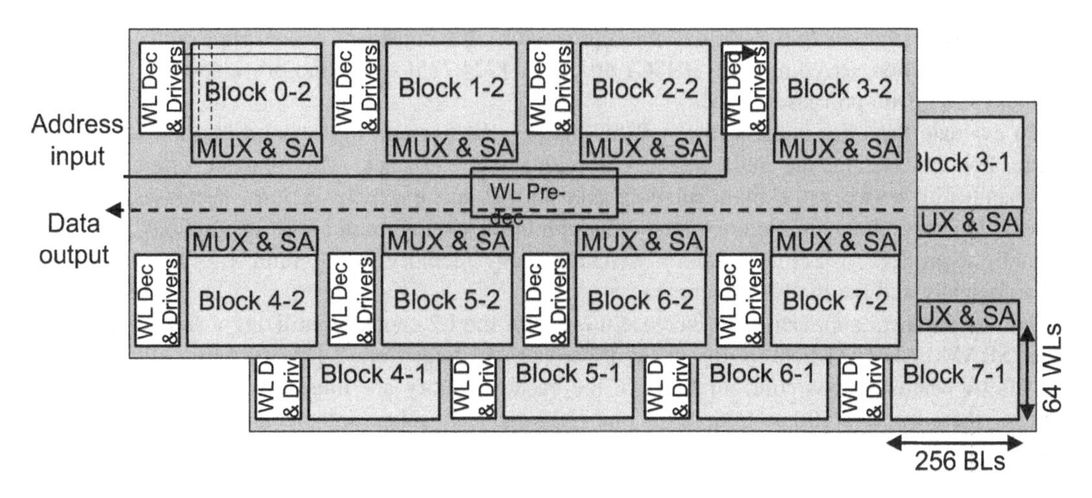

FIGURE 20.8

Bit line partitioning onto two tiers of the 2-D cache memory shown in Fig. 20.5 [713].

effect on the total area of the memory, since the size of a TSV is typically larger than the area of an SRAM cell [717]. Consequently, partitioning at the macrocell level offers the greatest advantages for 3-D cache memory architectures when TSVs are utilized. However, monolithic 3-D circuits can support the effective integration of memories at the SRAM cell level [115].

Analyzing the performance of these architectures is a multivariable task and design aids to support this analysis are needed. PRACTICS [718] and 3-D CACTI [716] offer exploratory capabilities for cache memories. The cache cycle time models in both of these tools are based on CACTI, an exploratory tool for 2-D memories [719]. These tools utilize delay [252], energy [720], and thermal models [721] as well as extracted delay and power profiles from SPICE simulations to characterize a cache memory. Several parameters are treated as variables. The optimum value of these parameters is determined based on the primary design objective of the architecture, such as high speed or low power. Although these tools are not highly accurate, early architectural decisions can be explored. Choosing an appropriate 3-D architecture for the cache memory further increases the performance of the microprocessor in addition to the performance improvements offered by the 3-D design of specific blocks within the processor. In this case, the primary limitation of the system is the off-chip main memory discussed in the following section.

20.2.3 ARCHITECTING A 3-D MICROPROCESSOR MEMORY SYSTEM

Although partitioning the cache memory on multiple tiers enhances the performance of the microprocessor, data transfer to and from the main memory remains a significant hindrance. The ultimate 3-D solution is to stack the main memory on the upper tiers of a 3-D microprocessor system. This option may be feasible for low performance processors with low memory requirements [722]. For modern computing systems with considerable memory demands, increasing the size of the on-chip cache, mainly the second level (L2) cache, is an efficient approach to improve performance [723].

Two systems based on two different microprocessors are considered; one approach contains a Reduced instruction set computing (RISC) processor [724,725] while the other approach includes an Intel Core 2 Duo processor [726].

To evaluate the effectiveness of the RISC system, the average time per instruction is a useful metric. For this system, the main memory is within the 3-D stack. This practice offers a higher buss bandwidth between the main memory and the L2 cache which, in turn, decreases the time required to access the main memory. The reduction in the average number of instructions, however, is small, about 6.1% [722]. In addition, stacking many memory tiers within a 3-D stack can be technologically and thermally challenging.

A more practical approach is to increase the size of the L2 cache by utilizing a small number of either SRAM or DRAM tiers on top of the processor. An Intel Core 2 Duo system is illustrated in Fig. 20.9A where various configurations of the cache memory are illustrated [711]. Note that in some of these configurations, both level one (L1) and L2 caches are included on the same tier. Increasing the size of the L2 cache increases the time to access this memory. The advantage of the reduced cache miss rates due to more data and instructions available on-chip considerably outweighs, however, the increase in access time. Each of the architectures illustrated in Fig. 20.9 decreases the number of cycles per memory access for a large number of benchmarks [711]. The only exceptions are those benchmarks where the 4 MB memory in the baseline system is sufficient. Additionally, the power consumed by the microprocessor memory system decreases since a smaller number of transactions takes place over the off-chip high capacitive buss. Furthermore, the required bandwidth of the off-chip buss drops as much as three times as compared to a 2-D system [711].

The presence of a second memory tier naturally increases the on-chip power consumption as compared to a 2-D microprocessor. The estimated power consumed by each configuration, shown in Fig. 20.9, is listed in Table 20.2. Despite the increase in power dissipation, the highest increase in temperature as compared to the baseline system does not exceed 5 °C with a maximum (and manageable) temperature of 92.9 °C [711,727].

Larger on-chip memories can decrease cache miss rates for single or double core microprocessors [728]. For large scale systems with tens of cores, however, the memory access time and the related buss bandwidth can be a performance bottleneck. An on-chip network is an effective way to overcome these issues. 3-D architectures for NoC are therefore the subject of the following section.

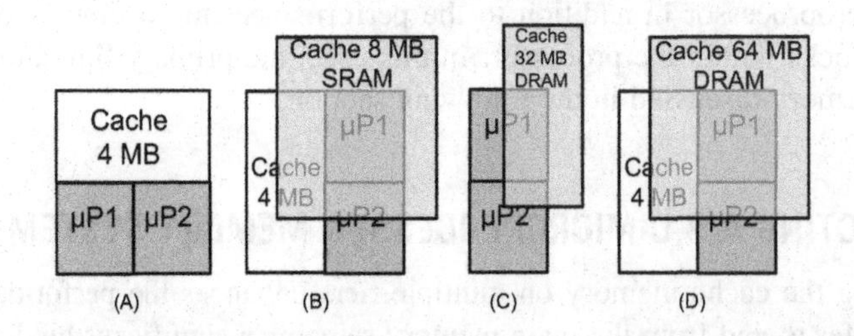

FIGURE 20.9

Different organizations of a microprocessor system, (A) 2-D baseline system, (B) a second tier with 8 MB SRAM cache memory, (C) a second tier with 32 MB SRAM cache memory, and (D) a second tier with 64 MB DRAM cache memory [711].

Table 20.2 Power Dissipation of the 3-D Microprocessor Architectures [711]

Architecture	Power Consumption [W]		
	Tier 1	Tier 2	Total
Fig. 20.9A (2-D)	92	–	92
Fig. 20.9B	92	14	106
Fig. 20.9C	85	3.1	88.1
Fig. 20.9D	92	6.2	98.2

20.3 3-D NETWORKS-ON-CHIP

NoC is a design paradigm to enhance interconnections within complex integrated systems. These networks have an interconnect structure which provide Internet like communication among various elements of the network; however, on-chip networks differ from traditional interconnection networks in that communication among the network elements is achieved through the on-chip routing layers rather than the metal tracks of the package or printed circuit board (PCB).

NoC offer high flexibility and regularity, supporting simpler interconnect models and greater fault tolerance. The canonical interconnect backbone of the network combined with appropriate communication protocols enhance the flexibility of these systems [37]. NoC provide communication among a variety of functional intellectual property (IP) blocks or processing elements (PEs), such as processor and Digital signal processing (DSP) cores, memory blocks, FPGA blocks, and dedicated hardware, serving a plethora of applications that include image processing, personal devices, and mobile handsets, [729–731] (the terms IP block and PEs are interchangeably used in this chapter to describe functional structures connected by a NoC). The intra-PE delay, however, cannot be reduced by the network. Furthermore, the length of the communication channel is primarily determined by the area of the PE, which is typically unaffected by the network structure. By merging vertical integration with NoC, many of the individual limitations of 3-D ICs and NoC are circumvented, yielding a robust design paradigm with unprecedented capabilities.

Research in 3-D NoC has progressed considerably in the past few years where several 3-D topologies have been explored [732], and design methods and synthesis tools [733] have been published [503,734,735]. Several methodologies to manage vertical links with TSVs have also been reported [736].

Addo-Quaye [503] presented an algorithm for the thermal aware mapping and placement of 3-D NoC including regular mesh topologies. Li et al. [735] proposed a similar 3-D NoC topology employing a buss structure for communicating among PEs located on different physical tiers. Targeting multiprocessor systems, the proposed scheme in [735] considerably reduces cache latencies by utilizing the third dimension. Multidimensional interconnection networks have been studied under various constraints, such as constant bisection-width and pin-out constraints [706]. NoC differ from generic interconnection networks, however, in that NoC are not limited by the channel width or pin-out. Alternatively, physical constraints specific to 3-D NoC, such as the number of nodes that can be placed in the third dimension and the asymmetry in the length of the channels of the network, have to be considered.

In this chapter, various possible topologies for 3-D NoC are presented. Additionally, analytic models for the zero-load latency and power consumption with delay constraints of these networks that capture the effects of the topology on the performance of 3-D NoC are described [737]. Although these models are applied to mesh topologies, the interconnect models remain valid for other topologies as long as the parameters, such as the number of hops for a topology, are properly adapted to the features of the target topology. Optimum topologies are shown to exist that minimize the zero-load latency and power consumption of a network. These optimum topologies depend upon a number of parameters characterizing both the router and the communication channel, such as the number of ports of the network, the length of the communication channel, and the impedance characteristics of the interconnect. Several tradeoffs among these parameters that determine the minimum latency and power consumption topology of a network are described for different network sizes. A cycle accurate simulator for 3-D topologies is also discussed. This tool is used to evaluate the behavior of several 3-D topologies under broad traffic scenarios.

Several interesting topologies, which are the topic of this chapter, emerge by incorporating the third dimension in NoC. In the following section, several topological choices for 3-D NoC are reviewed. In Section 20.3.2, an analytic model of the zero-load latency of traditional interconnection networks is adapted for each of the proposed 3-D NoC topologies, while the power consumption model of these network topologies is described in Section 20.3.3. In Section 20.3.4, the 3-D NoC topologies are compared in terms of the zero-load network latency and power consumption with delay constraints, and guidelines for the optimum design of speed driven or power driven NoC structures are provided. An advanced NoC simulator, which is used to evaluate the performance of a broad variety of 3-D network topologies, is presented in Section 20.3.5.

20.3.1 3-D NOC TOPOLOGIES

Several topologies for 3-D networks are presented and related terminology is introduced in this section. Mesh structures have been a popular network topology for conventional 2-D NoC [738,739]. A fundamental element of a mesh network is illustrated in Fig. 20.10A, where each PE is connected to the network through a router. A PE can be integrated either on a single physical tier (2-D IC) or on several physical tiers (3-D IC). Each router in a 2-D NoC is connected to a neighboring router in one of four directions. Consequently, each router has five ports. Alternatively, in a 3-D NoC, the router typically connects to two additional neighboring routers located on the adjacent physical tiers. The architecture of the router is considered here to be a canonical router with input and output buffering [740]. The combination of a PE and router is called a network node. For a 2-D mesh network, the total number of nodes N is $N=n_1 \times n_2$, where n_i is the number of nodes included in the ith physical dimension.

Integration in the third dimension introduces a variety of topological choices for NoCs. For a 3-D NoC as shown in Fig. 20.10B, the total number of nodes is $N=n_1 \times n_2 \times n_3$, where n_3 is the number of nodes in the third dimension. In this topology, each PE is on a single yet possibly different physical tier (2-D IC–3-D NoC). Alternatively, a PE can be implemented on only one of the n_3 physical tiers of the system and, therefore, the 3-D system contains $n_1 \times n_2$ PEs on each

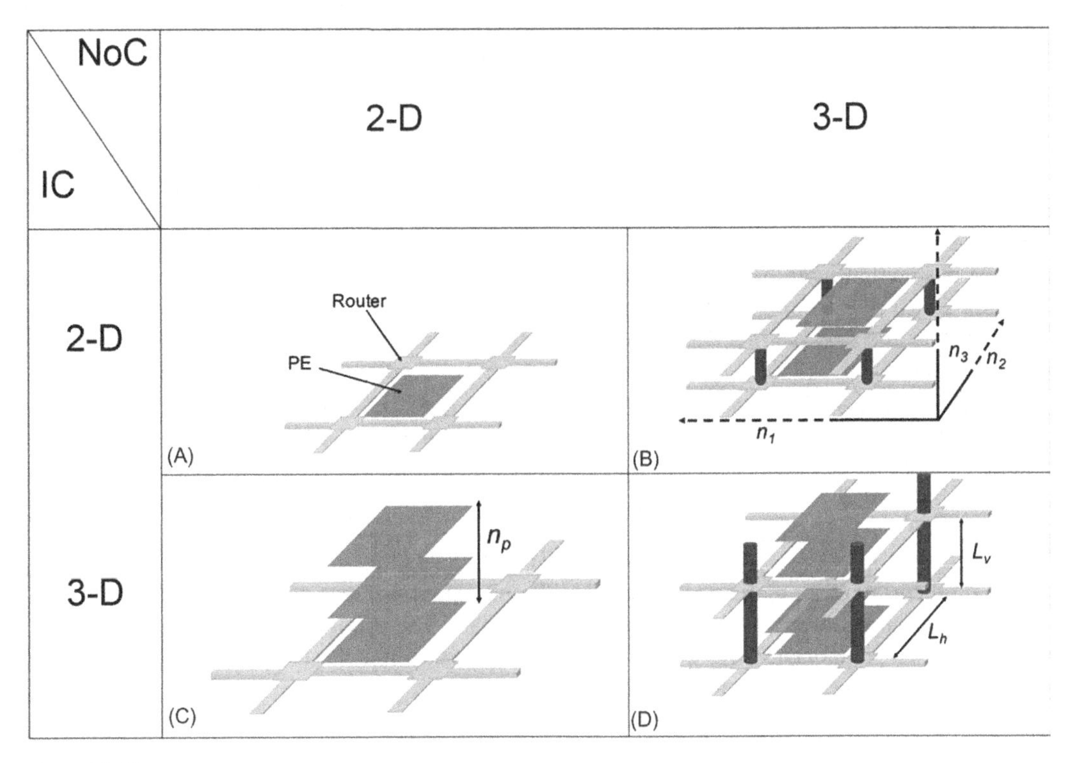

FIGURE 20.10

Several NoC topologies (not to scale), (A) 2-D IC–2-D NoC, (B) 2-D IC–3-D NoC, (C) 3-D IC–2-D NoC, and (D) 3-D IC–3-D NoC.

one of the n_3 physical tiers such that the total number of nodes is N. This topology is discussed in [503] and [735]. A 3-D NoC topology is illustrated in Fig. 20.10C, where the interconnect network is contained within one physical tier (i.e., $n_3 = 1$), while each PE is integrated on multiple tiers, notated as n_p (3-D IC–2-D NoC). Finally, a hybrid 3-D NoC based on the two previous topologies is depicted in Fig. 20.10D. In this NoC topology, both the interconnect network and the PEs can span more than one physical tier within the stack (3-D IC–3-D NoC). In the following section, latency expressions for each of the NoC topologies are described, assuming a zero-load model.

20.3.2 ZERO-LOAD LATENCY FOR 3-D NOC

In this section, analytic models of the zero-load latency of each of the 3-D NoC topologies are described. The zero-load network latency is widely used as a performance metric in traditional interconnection networks [741]. The zero-load latency of a network is the latency where only one packet traverses the network. Although such a model does not consider contention among packets, the zero-load latency model can be used to describe the effect of a topology on the performance of a network. The zero-load latency of an NoC with wormhole switching is [741]

$$T_{network} = hops \cdot t_r + t_c + \frac{L_p}{b}, \tag{20.1}$$

where the first term is the routing delay, t_c is the propagation delay along the wires of the communication channel, which is also called a buss here for simplicity, and the third term is the serialization delay of the packet. *hops* is the average number of routers that a packet traverses to reach the destination node, t_r is the router delay, L_p is the length of the packet in bits, and b is the bandwidth of the communication channel defined as $b \equiv w_c f_c$, where w_c is the width of the channel in bits and f_c is the inverse of the propagation delay of a bit along the longest communication channel.

Since the number of tiers that can be stacked in a 3-D NoC is constrained by the target technology, n_3 is also constrained. Furthermore, n_1, n_2, and n_3 are not necessarily equal. The average number of hops in a 3-D NoC is

$$hops = \frac{n_1 n_2 n_3 (n_1 + n_2 + n_3) - n_3(n_1 + n_2) - n_1 n_2}{3(n_1 n_2 n_3 - 1)}, \tag{20.2}$$

assuming dimension order routing to ensure that the minimum distance paths are used for the routing of packets between any source destination node pair. The number of hops in (20.2) can be divided into two components, the average number of hops within the two dimensions n_1 and n_2, and the average number of hops within the third dimension n_3,

$$hops_{2-D} = \frac{n_3(n_1 + n_2)(n_1 n_2 - 1)}{3(n_1 n_2 n_3 - 1)}, \tag{20.3}$$

$$hops_{3-D} = \frac{(n_1 n_2)(n_3^2 - 1)}{3(n_1 n_2 n_3 - 1)}. \tag{20.4}$$

The delay of the router t_r is the sum of the delay of the arbitration logic t_a and the delay of the switch t_s, which in this chapter is considered to be a classic crossbar switch [741],

$$t_r = t_a + t_s. \tag{20.5}$$

The delay of the arbiter can be described from [742],

$$t_\alpha = \left(21\left(\frac{1}{4}\right)\log_2 p + 14\left(\frac{1}{12}\right) + 9\right)\tau, \tag{20.6}$$

where p is the number of ports of the router and τ is the delay of a minimum sized inverter for the target technology. Note that (20.6) exhibits a logarithmic dependence on the number of router ports. The length of the crossbar switch also depends upon the number of router ports and the width of the buss,

$$l_s = 2(w_t + s_t)w_c p, \tag{20.7}$$

where w_t and s_t are, respectively, the width and spacing or, alternatively, the pitch of the interconnect and w_c is the width of the communication channel in bits. Consequently, the worst case delay of the crossbar switch is determined by the longest path within the switch, which is equal to (20.7).

The delay of the communication channel t_c is

$$t_c = t_v hops_{3-D} + t_h hops_{2-D}, \tag{20.8}$$

where t_v and t_h are, respectively, the delay of the vertical and horizontal channels (see Fig. 20.10B). Note that if $n_3=1$, (20.8) describes the propagation delay of a 2-D NoC. Substituting (20.8) and (20.5) into (20.1), the overall zero-load network latency for a 3-D NoC is

$$T_{network} = hops(t_a + t_s) + hops_{2-D} t_h + hops_{3-D} t_v + \frac{L_p}{w_c} t_h. \tag{20.9}$$

To characterize t_s, t_h, and t_v, the models described in [743] are adopted, where repeaters implemented as simple inverters are inserted along the interconnect. According to these models, the propagation delay and rise time of a single interconnect stage for a step input, respectively, are

$$t_{di} = 0.377 \frac{r_1 c_1 l_i^2}{k_i^2} + \left(R_{d0} C_0 + \frac{R_{d0} c_i l_i}{h_i k_i} + \frac{r_i l_i C_{g0} h_i}{k_i} \right), \tag{20.10}$$

$$t_{di} = 1.1 \frac{r_1 c_1 l_i^2}{k_i^2} + 2.75 \left(R_{r0} C_0 + \frac{R_{r0} c_i l_i}{h_i k_i} + \frac{r_i l_i C_{g0} h_i}{k_i} \right), \tag{20.11}$$

where r_i (c_i) is the per unit length resistance (capacitance) of the interconnect and l_i is the total length of the interconnect. The index i is used to notate the different interconnect delays included in the network (i.e., $i \in \{s,v,h\}$). h_i and k_i denote the number and size of the repeaters, respectively, and C_{g0} and C_0 represent, respectively, the gate and total input capacitance of a minimum sized device. C_0 is the summation of the gate and drain capacitance of the device. R_{d0} and R_{r0} describe, respectively, the equivalent output resistance of a minimum sized device for the propagation delay and transition time of a minimum sized inverter where the output resistance is approximated as

$$R_{r(d)0} = K_{r(d)} \frac{V_{dd}}{I_{dn0}}. \tag{20.12}$$

K denotes a fitting coefficient and I_{dn0} is the drain current of an NMOS device at both V_{ds} and V_{gs} equal to V_{dd}. The value of these device parameters are listed in Table 20.3. A 45 nm technology node is assumed and SPICE simulations of the predictive technology library are used to determine the individual parameters [252,432].

To include the effect of the input slew rate on the total delay of an interconnect stage, (20.10) and (20.11) are further refined by including an additional coefficient γ as in [744],

$$\gamma_r = \frac{1}{2} - \frac{1 - \frac{V_{tn}}{V_{dd}}}{1 - a_n}. \tag{20.13}$$

By substituting the subscript n with p, the corresponding value for a falling transition is obtained. The average value γ of γ_r and γ_f is used to describe the effect of the transition time on the interconnect delay. The overall interconnect delay can therefore be described as

$$t_i = k(t_{di} + \gamma t_{r,i}) = a_1 \frac{r_i}{c_i} + a_2 \left(R_0 C_0 k + \frac{R_0 c_i l_i}{h} + R_i C_{g0} h \right), \tag{20.14}$$

where R_0, a_1, and a_2 are described in [655] and the index i denotes the different interconnect structures such as the crossbar switch ($i \equiv s$), horizontal buss ($i \equiv h$), and vertical buss ($i \equiv v$).

Table 20.3 Interconnect and Design Parameters, 45 nm Technology

Parameter	Value	
	NMOS	PMOS
W_{min}	100 nm	250 nm
I_{dsat}/W	1115 μA/μm	349 μA/μm
V_{dsat}	478 mV	−731 mV
V_t	257 mV	−192 mV
A	1.04	1.33
I_{sub0}	48.8 nA	
I_{g0}	0.6 nA	
V_{dd}	1.1 Volts	
Temp.	110°C	
K_d	0.98	
K_r	0.63	
C_{g0}	512 fF	
C_{d0}	487 fF	
τ	17 ps	

For minimum delay, the size h and number k of repeaters are determined, respectively, by setting the partial derivative of t_i with respect to h_i and k_i equal to zero and solving for h_i and k_i,

$$k_i^* = \sqrt{\frac{a_1 r_i c_i l_i^2}{a_2 R_0 C_0}}, \tag{20.15}$$

$$h_i^* = \sqrt{\frac{R_0 c_i l_i}{r_i C_{g0}}}. \tag{20.16}$$

The expression in (20.14) only considers RC interconnects. An RC model is sufficiently accurate to characterize the delay of a crossbar switch since the length of the longest wire within the crossbar switch and the signal frequencies ensures that inductive behavior is not prominent. For the buss lines, however, inductive behavior can appear. For this case, suitable expressions for the delay and repeater insertion characteristics can be adopted from [655]. For the target operating frequencies (1 to 2 GHz) and buss length (<2 mm) considered in this chapter, an RC interconnect model provides sufficient accuracy [588]. Additionally, for the vertical buss, $k_v = 1$ and $h_v = 1$, meaning that no repeaters are inserted and minimum sized drivers are utilized. Repeaters are not necessary due to the short length of the vertical buss. Driving a buss with minimum sized inverters can affect the resulting minimum latency and power dissipation topology, as discussed in the following sections. Note that the latency expression includes the effects of the input slew rate. Additionally, since a repeater insertion methodology for minimum latency is applied, any further reduction in latency is due to the network topology.

The length of the vertical communication channel for the 3-D NoC shown in Fig. 20.10 is

$$l_v = \begin{cases} L_v, & \text{for 2D IC–3D NoC} \\ n_p L_v, & \text{for 3D IC–3D NoC} \\ 0, & \text{for 2D IC–2D NoC and 3D IC–2D NoC,} \end{cases} \tag{20.17a – c}$$

where L_v is the length of a through silicon (intertier) via connecting two routers on adjacent physical tiers. n_p is the number of physical tiers used to integrate each PE. The length of the horizontal communication channel is assumed to be

$$l_h = \begin{cases} \sqrt{A_{PE}}, & \text{for 2D IC–2D NoC and 2D IC–3D NoC} \\ 1.12\sqrt{A_{PE}/n_p}, & \text{for 3D IC–2D NoC and 3D IC–3D NoC}(n_p > 1), \end{cases} \tag{20.18a,b}$$

where A_{PE} is the area of the PE. The area of all of the PEs and, consequently, the length of each horizontal channel are assumed to be equal. For those cases where the PE is placed in multiple physical tiers, a coefficient is included to consider the effect of the intertier vias on the reduction in the ideal wirelength due to utilization of the third dimension. The value of this coefficient (= 1.12) is based on the layout of a crossbar switch manufactured in the fully depleted silicon-on-insulator (FD-SOI) 3-D technology from MIT Lincoln Laboratory (MITLL) [307]. The same coefficient is also assumed for the PEs placed on more than one physical tier. In the following section, expressions for the power consumption of a network with delay constraints are presented.

20.3.3 POWER CONSUMPTION IN 3-D NOC

Power dissipation is a critical issue in 3-D circuits. Although the total power consumption of 3-D systems is expected to be lower than that of mainstream 2-D circuits (since the global interconnects are shorter [308]), the increased power density is a challenging issue for this novel design paradigm. Therefore, those 3-D NoC topologies that offer low power characteristics should be of significant interest.

The different power consumption components for interconnects with repeaters are briefly discussed in this section. Due to specified performance characteristics, a low power design methodology with delay constraints for the interconnect in an NoC is adopted from [655]. An expression for the total power consumption per bit of a packet transferred between a source destination node pair is used as the basis for characterizing the power consumption of an NoC for the 3-D topologies.

The power consumption components of an interconnect line with repeaters are:

a. *Dynamic power consumption* is the dissipated power due to the charge and discharge of the interconnect and input gate capacitance during a signal transition, and can be described by

$$P_{di} = a_s f(c_i l_i + h_i k_i C_0) V_{dd}^2, \tag{20.19}$$

where f is the clock frequency and a_s is the switching factor [745]. A value of 0.15 is assumed here; however, for NoC, the switching factor can vary considerably. This variation, however, does not affect the power comparison for the various topologies as the same switching factor is incorporated in each term for the total power consumed per bit of the network (the absolute value of the power consumption, however, changes).

b. *Short-circuit power* is due to the DC current path that exists in a CMOS circuit during a signal transition when the input signal voltage changes between V_{tn} and $V_{dd} + V_{tp}$. The power consumption due to this current is described as short-circuit power and is modeled in [746] by

$$P_{si} = \frac{4a_s f I_{d0}^2 t_{ri}^2 V_{dd} k_i h_i^2}{V_{dsat} G C_{effi} + 2H I_{d0} t_{ri} h_i}, \tag{20.20}$$

where I_{d0} is the average drain current of the NMOS and PMOS devices operating in the saturation region and the value of the coefficients G and H are described in [747]. Due to resistive shielding of the interconnect capacitance, an effective capacitance is used in (20.20) rather than the total interconnect capacitance. This effective capacitance is determined from the methodology described in [748] and [749].

c. *Leakage current power* comprises two power components, the subthreshold and gate leakage currents. The subthreshold power consumption is due to current flowing during the cut-off region (below threshold), causing I_{sub} current to flow. The gate leakage component is due to current flowing through the gate oxide, denoted as I_g. The total leakage current power can be described as

$$P_{li} = h_i k_i V_{dd}(I_{sub0} + I_{g0}), \tag{20.21}$$

where the average subthreshold I_{sub0} and gate I_{g0} leakage current of the NMOS and PMOS transistors is used in (20.21).

The total power consumption with delay constraint T_0 for a single line of a crossbar switch P_{stotal}, horizontal buss P_{htotal}, and vertical buss P_{vtotal} is, respectively,

$$P_{stotal}(T_0 - t_a) = P_{di} + P_{si} + P_{li}, \tag{20.22}$$

$$P_{htotal}(T_0) = P_{di} + P_{si} + P_{li}, \tag{20.23}$$

$$P_{htotal}(T_0) = P_{di} + P_{si} + P_{li}. \tag{20.24}$$

The power consumption of the arbitration logic is not included in (20.22), since most of the power is consumed by the crossbar switch and the buss interconnect, as discussed in [750]. Note that for a crossbar switch, the additional delay t_a of the arbitration logic poses a stricter delay constraint on the power consumption of the switch, as shown in (20.22). The minimum power consumption with delay constraints is determined by the methodology described in [655], for which the optimum size h^*_{powi} and number k^*_{powi} of the repeaters for a single interconnect line is determined. Consequently, the minimum power consumption per bit between a source destination node pair in a NoC with a delay constraint is

$$P_{bit} = hopsP_{stotal} + hops_{2-D}P_{htotal} + hops_{3-D}P_{vtotal}. \tag{20.25}$$

The effect of resistive shielding is also considered in determining the effective interconnect capacitance. Furthermore, since the repeater insertion methodology in [655] minimizes the power consumed by the repeater system, any additional decrease in power consumption is due only to the network topology. In the following section, those 3-D NoC topologies that exhibit the maximum performance and minimum power consumption with delay constraints are presented. Tradeoffs in determining these topologies are discussed and the impact of the network parameters on the resulting optimum topologies are demonstrated for different network sizes.

20.3.4 PERFORMANCE AND POWER ANALYSIS FOR 3-D NOC

Several network parameters characterizing the topology of a network can significantly affect the speed and power of a system. The evaluation of these network parameters is discussed in subsection 20.3.4.1. The improvement in network performance achieved by the 3-D NoC topologies is explored in subsection 20.3.4.2. The distribution of nodes that produces the maximum performance is also discussed. The power consumption with delay constraints of a 3-D NoC and the topologies that yield the minimum power consumption of a 3-D NoC are presented in subsection 20.3.4.3.

20.3.4.1 Parameters of 3-D networks-on-chip

The physical layer of a 3-D NoC consists of different interconnect structures, such as a crossbar switch, the horizontal buss connecting neighboring nodes on the same physical tier and the vertical buss connecting nodes on different, not necessarily adjacent, physical tiers. The device parameters characterizing the receiver, driver, and repeaters are listed in Table 20.3. The interconnect parameters reported in Table 20.4 are different for each type of interconnect within a network.

A typical interconnect structure is shown in Fig. 20.11, where three parallel metal lines are sandwiched between two ground planes. This interconnect structure is considered for the crossbar switch (at the network nodes) where the intermediate metal layers are assumed to be utilized. The horizontal buss is implemented on the global metal layers and, therefore, only the lower ground plane is present in this structure for a 2-D NoC. For a 3-D NoC, however, the substrate (back-to-face tier bonding) or a global metal layer of an upper tier (face-to-face tier bonding) behaves as a second ground plane. To incorporate this additional ground plane, the horizontal bus capacitance is changed by the coefficient $a_{3\text{-}D}$. A second ground plane decreases the coupling capacitance to an adjacent line, while the line-to-ground capacitance increases. The vertical buss is different from the other structures in that this buss uses through silicon vias. These intertier vias can exhibit significantly different impedance characteristics as compared to traditional horizontal interconnect structures, as discussed in [436] and also verified by extracted impedance parameters. The electrical interconnect parameters are extracted using a commercial impedance extraction tool [423], while the physical parameters are extrapolated from the predictive technology library [252], [432] and the 3-D integration technology developed by MITLL for a 45 nm technology node [307]. The physical and electrical interconnect parameters are listed in Table 20.4. For each of the interconnect structures, a buss width of 64 bits is assumed. In addition, n_3 and n_p are constrained by the maximum number of physical tiers n_{max} that can be vertically stacked. A maximum of eight tiers is assumed. The constraints that apply for each of the 3-D NoC topologies shown in Fig. 20.10 are

$$n_3 \leq n_{max}, \quad \text{for 2D IC} - \text{3D NoC}, \tag{20.26a}$$

$$n_p \leq n_{max}, \quad \text{for 3D IC} - \text{2D NoC}, \tag{20.26b}$$

$$n_3 n_p \leq n_{max}, \quad \text{for 3D IC} - \text{3D NoC}. \tag{20.26c}$$

A small set of parameters is used as variables to explore the performance and power consumption of the 3-D NoC topologies. This set includes the network size or, equivalently, the number of nodes

Table 20.4 Interconnect Parameters

Interconnect Structure	Parameter: Electrical	Parameter: Physical
Crossbar switch	$\rho = 3.07\ \mu\Omega$-cm $k_{ILD} = 2.7$ $r_s = 614\ \Omega$/mm $c_s = 157.6$ fF/mm	$w = 200$ nm $s = 200$ nm $t = 250$ nm $h = 500$ nm
Horizontal bus	$\rho = 2.53\ \mu\Omega$-cm $k_{ILD} = 2.7$ $r_h = 46\ \Omega$/mm ch = 332.6 (192.5) fF/mm $a_{3\text{-}D} = 1.02$ (1.06)	$w = 500$ nm $s = 250$ (500) nm $t = 1100$ nm $h = 800$ nm –
Vertical bus	$\rho = 5.65\ \mu\Omega$-cm $r_v = 51.2\ \Omega$/mm $c_v = 600$ fF/mm	$w = 1050$ nm $L_v = 10\ \mu$m –

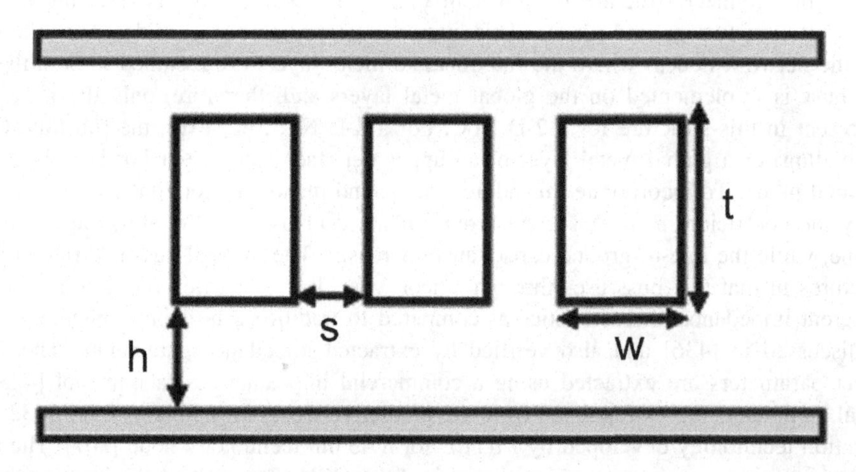

FIGURE 20.11

Typical interconnect structure for intermediate metal layers.

within the network N, the area of each PE A_{PE}, which is directly related to the buss length as described in (20.18), and the maximum allowed interconnect delay when evaluating the minimum power consumption with delay constraints. The range of values for these variables is listed in Table 20.5. Depending upon the network size, the NoC are roughly divided as small ($N = 16$ to 64 nodes), medium ($N = 128$ to 256 nodes), and large ($N = 512$ to 2048 nodes) networks. For multiprocessor SoC networks, sizes of up to $N = 256$ is expected to be feasible in the near future [735,751], whereas

Table 20.5 Network Parameters

Parameter	Values
N	16, 32, 64, 128, 256, 512, 1,024, 2,048
A_{PE} [mm^2]	0.5, 0.64, 0.81, 1.00, 1.56, 2.25, 4.00
T_O [ps]	1,000, 500

for NoC with a finer granularity, where the PEs each corresponds to hardware blocks of approximately 100,000 gates, network sizes over a few thousands nodes are predicted at the 45 nm technology node [752]. Note that this classification of the networks is not strict and is only intended to facilitate the discussion in the following sections.

20.3.4.2 Performance tradeoffs for 3-D NoC

The performance enhancements that can be achieved in NoC by utilizing the third dimension are discussed in this subsection. Each of the 3-D topologies decreases the zero-latency of the network by reducing different delay components, as described in (20.9). In addition, the distribution of network nodes in each physical dimension that yields the minimum zero-load latency is shown to significantly change with the network and interconnect parameters.

2-D IC–3-D NoC

Utilizing the third dimension to implement a NoC directly results in a decrease in the average number of hops for packet switching. The average number of hops on the same tier $hops_{2\text{-}D}$ (the intratier hops) and the average number of hops in the third dimension $hops_{3\text{-}D}$ (the intertier hops) are also reduced. Interestingly, the distribution of nodes n_1, n_2, and n_3 that yields the minimum total number of hops is not always the same as the distribution that minimizes the number of intratier hops. This situation occurs particularly for small and medium networks, while for large networks, the distribution of n_1, n_2, and n_3 which minimizes the $hops$ also minimizes $hops_{2\text{-}D}$.

In a 3-D NoC, the number of router ports increases from five to seven, increasing, in turn, both the switch and arbiter delay. Furthermore, a short vertical buss generally exhibits a lower delay than a relatively long horizontal buss. In Fig. 20.12, the zero-load latency of the 2-D IC–3-D NoC is compared to that of the 2-D IC–2-D NoC for different network sizes. A decrease in latency of 15.7% and 20.1% can be observed for, respectively, $N = 128$ and $N = 256$ nodes with $A_{PE} = 0.81$ mm^2.

The node distribution that produces the lowest latency varies with network size. For example, $n_{3max} = 8$ is not necessarily the optimum for small and medium networks, although by increasing n_3, more hops occur through the short, low latency vertical channel. This result can be explained by considering the reduction in the number of hops that originate from utilizing the third dimension for packet switching. For small and medium networks, the decrease in the number of hops is small and cannot compensate the increased routing delay due to the greater number of router ports in a 3-D NoC. As the horizontal buss length becomes longer, however, (e.g., approaching 2 mm), $n_3 > 1$, and a slight decrease in the number of hops significantly decreases the overall delay, despite the increase in the routing delay for a 3-D NoC. As an example, consider a network with $log_2N = 4$ and A_{PE}=0.81 mm^2. The minimum latency node distribution is $n_1 = n_2 = 4$ and $n_3 = 1$

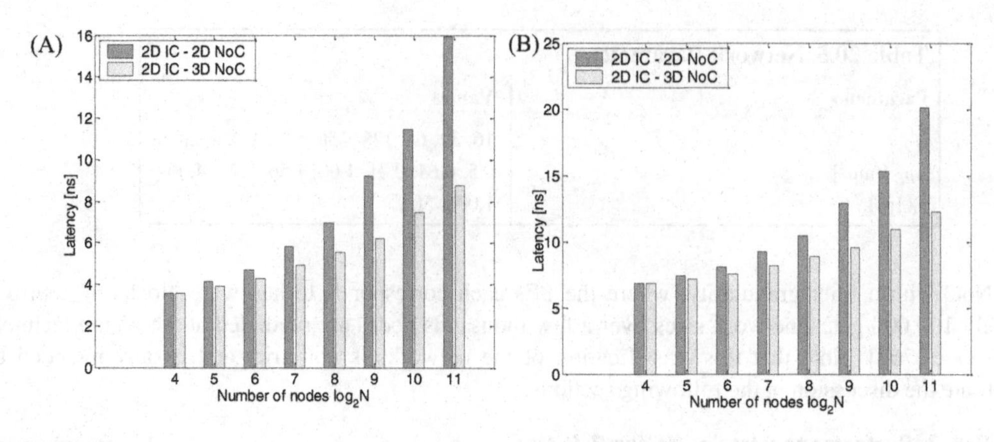

FIGURE 20.12

Zero-load latency for several network sizes. (A) $A_{PE} = 0.81\ \text{mm}^2$ and $c_h = 332.6$ fF/mm, and (B) $A_{PE} = 4\ \text{mm}^2$ and $c_h = 332.6$ fF/mm.

(identical to a 2-D IC–2-D NoC, as shown in Fig. 20.12), while for $A_{PE}=4\ \text{mm}^2$, $n_1 = n_2 = 2$ and $n_3 = 4$.

The optimum node distribution can also be affected by the delay of the vertical channel. The repeater insertion methodology for minimum delay as described in Section 20.3.2 can significantly reduce the delay of the horizontal buss by inserting large sized repeaters (i.e., $h > 300$). In this case, the delay of the vertical buss becomes comparable to that of the horizontal buss with repeaters. Consider a network with $N = 128$ nodes. Two different node distributions yield the minimum average number of hops, specifically, $n_1 = 4$, $n_2 = 4$, and $n_3 = 8$ and $n_1 = 8$, $n_2 = 4$, and $n_3 = 4$. The first of the two distributions also results in the minimum number of intratier $hops_{2\text{-}D}$, thereby reducing the latency of the horizontal buss, as described by (20.9). Simulation results, however, indicate that this distribution is not the minimum latency node distribution, as the delay due to the vertical channel is nonnegligible. For this reason, the latter distribution with $n_3 = 4$ is preferable since a smaller number of $hops_{3\text{-}D}$ occurs, resulting in the minimum network latency.

3-D IC–2-D NoC

For this type of 3-D network, the PEs are allowed to span multiple physical tiers while the network effectively remains 2-D (i.e., $n_3=1$). Consequently, the network latency is only reduced by decreasing the length of the horizontal buss, as described in (20.18). The routing delay component remains constant with this 3-D topology. Decreasing the horizontal buss length lowers both the communication channel delay and the serialization delay. In Fig. 20.13, the decrease in latency that can be achieved by a 3-D IC–3-D NoC is illustrated. A latency decrease of 30.2% and 26.4% can be observed for, respectively, $N = 128$ and $N = 256$ nodes with $A_{PE} = 2.25\ \text{mm}^2$. The use of multiple physical tiers reduces the latency; therefore, the optimum value for $n_p = n_{max}$, regardless of the network size and buss length.

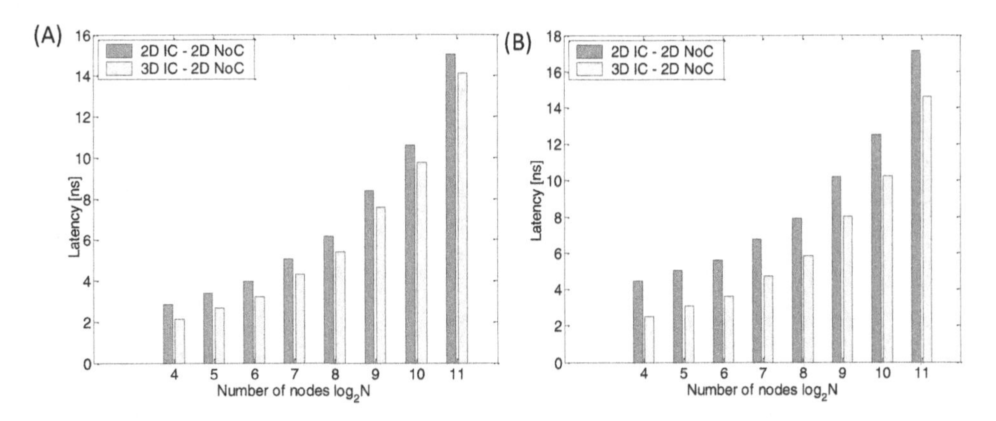

FIGURE 20.13

Zero-load latency for various network sizes. (A) $A_{PE} = 0.64\ \text{mm}^2$ and $c_h = 192.5$ fF/mm, (B) $A_{PE} = 2.25\ \text{mm}^2$ and $c_h = 192.5$ fF/mm.

In Figs. 20.13A and 20.13B, the improvement in the network latency over a 2-D IC–2-D NoC for several network sizes and for different PE areas (i.e., different horizontal buss length) is illustrated for, respectively, the 2-D IC–3-D NoC and 3-D IC–2-D NoC topologies. Note that for the 2-D IC–3-D NoC topology, the improvement in delay is smaller for PEs with a larger area or, equivalently, with longer buss lengths independent of the network size. For longer buss lengths, the buss latency comprises a larger portion of the total network latency. Since for a 2-D IC–3-D NoC only the hop count is reduced, the improvement in latency is lower for longer buss lengths. Alternatively, the improvement in latency is greater for PEs with a larger area independent of the network size for 3-D IC–2-D NoC. This situation is due to the significant reduction in the PE area (or buss length) achieved with this topology. Consequently, there is a tradeoff in the latency of a NoC that depends both on the network size and the area of the PEs. In Fig. 20.14A, the improvement is not significant for small networks (all of the curves converge to approximately zero) in 2-D IC–3-D NoC while this situation does not occur for 3-D IC–2-D NoC. This behavior is due to the increase in the delay of the network router as the number of ports increases from five to seven for 2-D IC–3-D NoC, which is a considerable portion of the network latency for small networks. Note that for 3-D IC–2-D NoC, the network essentially remains 2-D and therefore the delay of the router for this topology does not increase. To achieve the minimum delay, a 3-D NoC topology that exploits these tradeoffs is described in the following subsection.

3-D IC–3-D NoC

This topology offers the greatest decrease in latency over the aforementioned 3-D topologies. The 2-D IC–3-D NoC topology decreases the number of hops while the buss and serialization delays remain constant. With the 3-D IC–2-D NoC, the buss and serialization delay is smaller but the number of hops remains unchanged. With the 3-D IC–3-D NoC, all of the latency components can

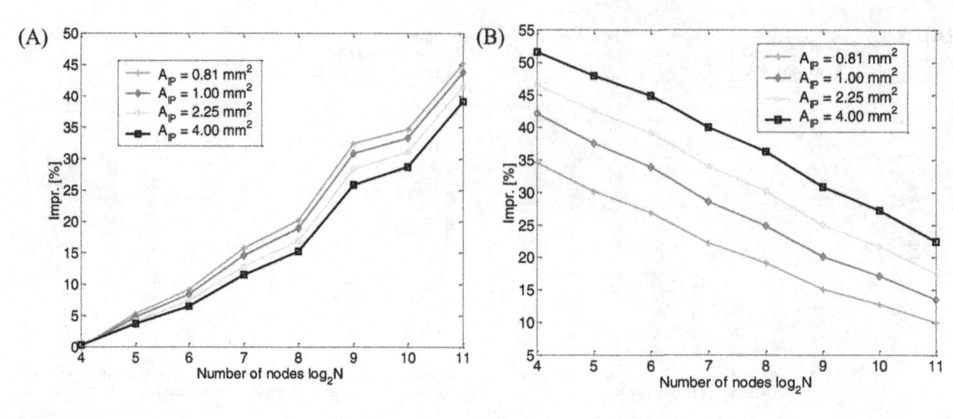

FIGURE 20.14

Improvement in zero-load latency for different network sizes and PE areas (i.e., buss lengths). (A) 2-D IC–3-D NoC, and (B) 3-D IC–2-D NoC.

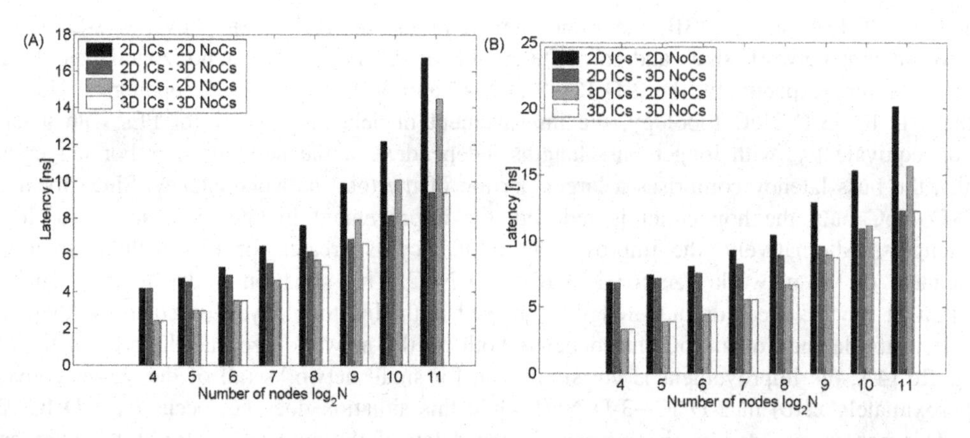

FIGURE 20.15

Zero-load latency for various network sizes. (A) $A_{PE} = 1\ \text{mm}^2$ and $c_h = 332.6$ fF/mm, and (B) $A_{PE} = 4\ \text{mm}^2$ and $c_h = 332.6$ fF/mm.

be decreased by assigning a portion of the available physical tiers for the network while the remaining tiers of the stack are used for the PE. The resulting decrease in network latency as compared to a standard 2-D IC–2-D NoC and the other two 3-D topologies is illustrated in Fig. 20.15. A decrease in latency of 40% and 36% can be observed, respectively, for $N = 128$ and $N = 256$ nodes with $A_{PE} = 4\ \text{mm}^2$. Note that the 3-D IC–3-D NoC topology achieves the greatest savings in latency by optimally balancing n_3 with n_p.

For certain network sizes, the performance of the 3-D IC–2-D NoC is identical to either the 2-D IC–3-D NoC or 3-D IC–2-D NoC. This behavior occurs because for large network sizes,

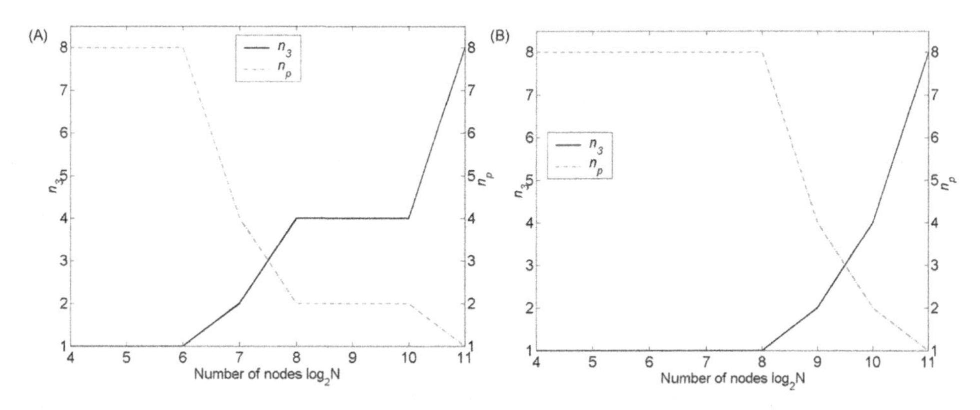

FIGURE 20.16

n_3 and n_p values for minimum zero-load latency for various network sizes. (A) $A_{PE} = 1$ mm^2 and $c_h = 332.6$ fF/mm, and (B) $A_{PE} = 4$ mm^2 and $c_h = 332.6$ fF/mm.

the delay due to the large number of hops dominates the total delay and, therefore, the latency can be primarily reduced by decreasing the average number of hops ($n_3 = n_{max}$). For small networks, the buss delay is large and the latency savings is typically achieved by reducing the buss length ($n_p = n_{max}$). For medium networks, though, the optimum topology is obtained by dividing n_{max} between n_3 and n_p to ensure that (20.26c) is satisfied. This distribution of n_3 and n_p as a function of the network size and buss length is illustrated in Fig. 20.16.

Note the shift in the value of n_3 and n_p as the PE area A_{PE} or, equivalently, the buss length increases. For long busses, the delay of the communication channel becomes dominant and therefore the smaller number of hops for medium-sized networks cannot significantly decrease the total delay. Alternatively, further decreasing the buss length by placing the PEs within a greater number of physical tiers leads to a larger savings in delay.

The suggested optimum topologies for different network sizes (namely, small, medium, and large networks) also depend upon the interconnect parameters of the network. Consequently, a change in the optimum topology for different network sizes can occur when different interconnect parameters are considered. Despite the sensitivity of the topologies on the interconnect parameters, the tradeoff between the number of hops and the buss length for different 3-D topologies (see Figs. 20.14 and 20.16) can be exploited to improve the performance of an NoC. In the following subsection, the topology that yields the minimum power consumption while satisfying the delay constraints is described. The distribution of nodes for that topology is also discussed.

20.3.4.3 Power consumption in 3-D NoC

The different power consumption components for the interconnect within an NoC are described in Section 20.3.3. The methodology presented in [655] is applied here to minimize the power consumption of these interconnects while satisfying the specified operating frequency of the network. Since a power minimization methodology is applied to the buss lines, the power consumed by the network can only be further reduced by the choice of network topology. Additionally, the power consumption also depends upon the target operating frequency, as discussed later in this section.

As with the zero-load latency, each topology affects the power consumption of the network in a different way. From (20.25), the power consumption can be reduced by either decreasing the number of hops for the packet or by decreasing the buss length. Note that by reducing the buss length, the interconnect capacitance is not only reduced but also the number and size of the repeaters required to drive the lines are decreased, resulting in a greater savings in power. The effect of each of the 3-D topologies on the power consumption of an NoC is investigated in this section.

2-D IC—3-D NoC

Similar to the network latency, the power consumption is decreased in this topology by reducing the number of hops for packet switching. Again, the increase in the number of ports is significant; however, the effect of this increase is not as important as in the latency of the network. A 3-D network, therefore, can reduce power even in small networks. The power savings achieved with this topology is depicted in Fig. 20.17 for several network sizes, where the savings is greater in larger networks. This situation occurs because the reduction in the average number of hops for a 3-D network increases for larger network sizes. A power savings of 26.1% and 37.9% is achieved, respectively, for $N = 128$ and $N = 512$ with $A_{PE} = 1\ \text{mm}^2$.

3-D IC—2-D NoC

With this topology, the number of hops in the network is the same as for a 2-D network. The horizontal buss length, however, is shorter since the PEs are placed in more than one physical tier. The greater the number of physical tiers that can be integrated into a 3-D system, the larger the power savings, meaning that the optimum value for n_p with this topology is always n_{max} regardless of the

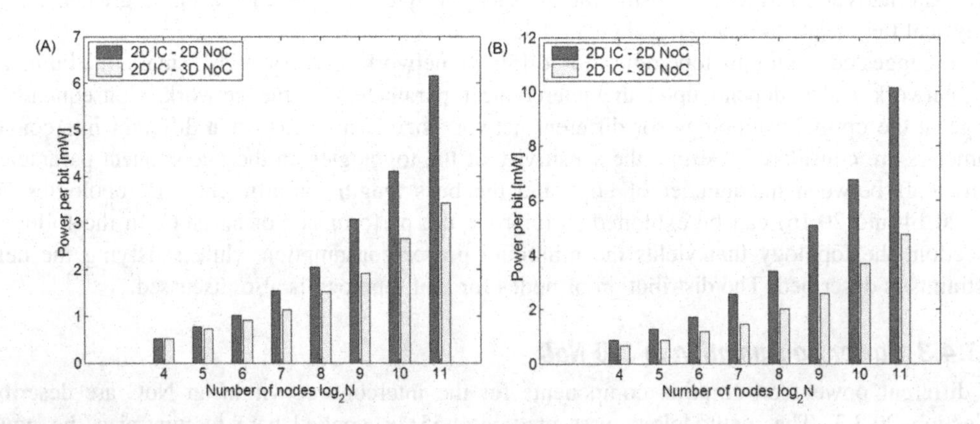

FIGURE 20.17

Power consumption with delay constraints for several network sizes. (A) $A_{PE} = 1\ \text{mm}^2$, $c_h = 332.6$ fF/mm, and $T_O = 500$ ps, and (B) $A_{PE} = 4\ \text{mm}^2$, $c_h = 332.6$ fF/mm, and $T_O = 500$ ps.

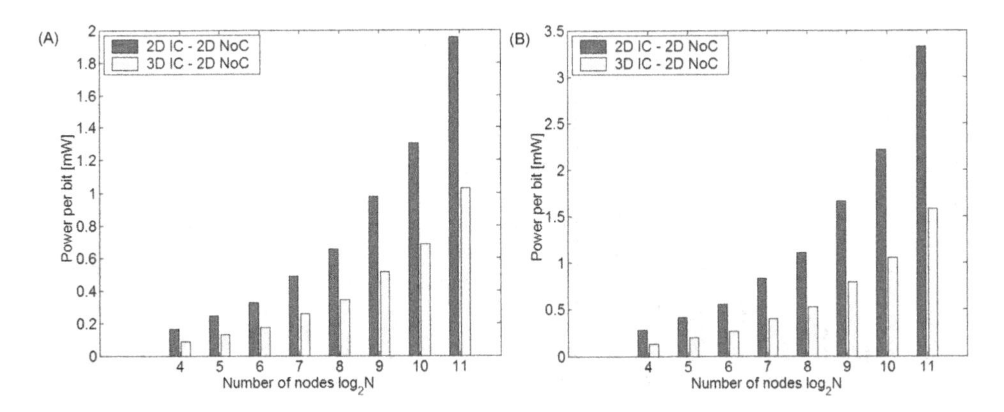

FIGURE 20.18

Power consumption with delay constraints for several network sizes. (A) $A_{PE} = 0.64$ mm^2, $c_h = 192.5$ fF/mm, and $T_O = 1000$ ps, and (B) $A_{PE} = 2.25$ mm^2, $c_h = 192.5$ fF/mm, and $T_O = 1000$ ps.

network size and operating frequency. The savings is limited by the number of physical tiers that can be integrated in a 3-D technology. The power savings for various network sizes are shown in Fig. 20.18. Note that for this type of NoC, the maximum performance topology is identical to the minimum power consumption topology, as the key element of both objectives originates solely from the shorter buss length. The savings in power is approximately 35% when $A_{PE}=0.64$ mm^2 for every network size as the per cent reduction in the buss length is the same for each network size.

3-D IC–3-D NoC

Allowing the available physical tiers to be utilized either for the third dimension of the network or for the PEs, the 3-D IC–3-D NoC scheme achieves the greatest savings in power in addition to the minimum delay, as discussed in the previous subsection. The distribution of nodes along the physical dimensions, however, that produces either the minimum latency or the minimum power consumption for every network size is not necessarily the same. This nonequivalence is due to the different degree of importance of the average number of hops and the buss length in determining the latency and power consumption of a network. In Fig. 20.19, the power consumption of the 3-D IC–3-D NoC topology is compared to the previously discussed 3-D topologies. A power savings of 38.4% is achieved for $N = 128$ with $A_{PE}=1$ mm^2. For certain network sizes, the power consumption of the 3-D IC–3-D NoC topology is the same as the 2-D IC–3-D NoC and 3-D IC–2-D NoC topologies. For the 2-D IC–3-D NoC, the power consumption is less by reducing the number of hops for packet switching, while for the 3-D IC–2-D NoC, the NoC power dissipation is decreased by shortening the buss length. The former approach typically benefits small networks, while the latter approach dissipates less power in large networks. For medium sized networks and depending upon the network and interconnect parameters, nonextreme values for the n_3 and n_p parameters (e.g., $1 < n_3 < n_{max}$ and $1 < n_p < n_{max}$) are required to produce the minimum power consumption topology.

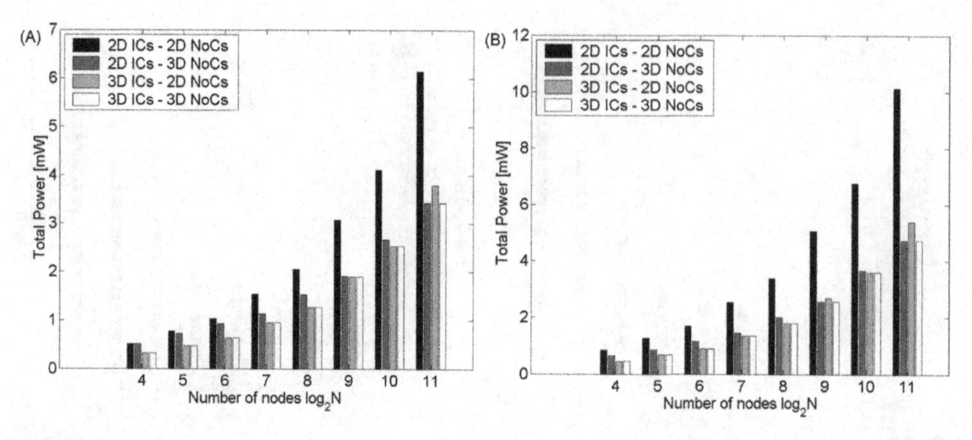

FIGURE 20.19

Power consumption with delay constraints for various network sizes. (A) $A_{PE} = 1\ \text{mm}^2$, $c_h = 332.6$ fF/mm, and $T_O = 500$ ps, and (B) $A_{PE} = 4\ \text{mm}^2$, $c_h = 332.6$ fF/mm, and $T_O = 500$ ps.

Note that this work emphasizes the latency and power consumption of a network, neglecting the performance requirements of the individual PEs. If the performance of the individual PEs is important, only one 3-D topology may be available; however, even with this constraint, a significant savings in latency and power can be achieved since in almost every case the network latency and power consumption can be decreased as compared to a 2-D IC–2-D NoC topology. Furthermore, as previously mentioned, if the available topology is the 2-D IC–3-D NoC, setting n_3 equal to n_{max} is not necessarily the optimum choice.

The proposed zero-load network latency and power consumption expressions capture the effect of the topology; yet these models do not incorporate the effects of the routing scheme and traffic load. Alternatively, these models can be treated as lower bounds for both the latency and the power consumption of the network. Since minimum distance paths and no contention are implicitly assumed in these expressions, nonminimal path routing schemes and heavy traffic loads will increase both the latency and power consumption of the network. The routing algorithm is managed by the upper layers, other than the physical layer, comprising the communication protocol of the network. In addition, the traffic patterns depend upon the executed application of the network. The effect of each of the parameters on the performance of 3-D NoCs is explored in the following section by utilizing a network simulator.

20.3.4.4 Design aids for 3-D NoCs[1]

To evaluate the performance of emerging 3-D topologies for different applications, effective design aids are required. A network simulator that evaluates the effectiveness of different 3-D topologies is described in this section. Simulations of a broad variety of 3-D mesh- and torus-based topologies as well as traffic patterns are also discussed.

[1]This section, 20.3.4.4, was contributed by Professor Dimitrios Soudris of the National Technical University of Athens.

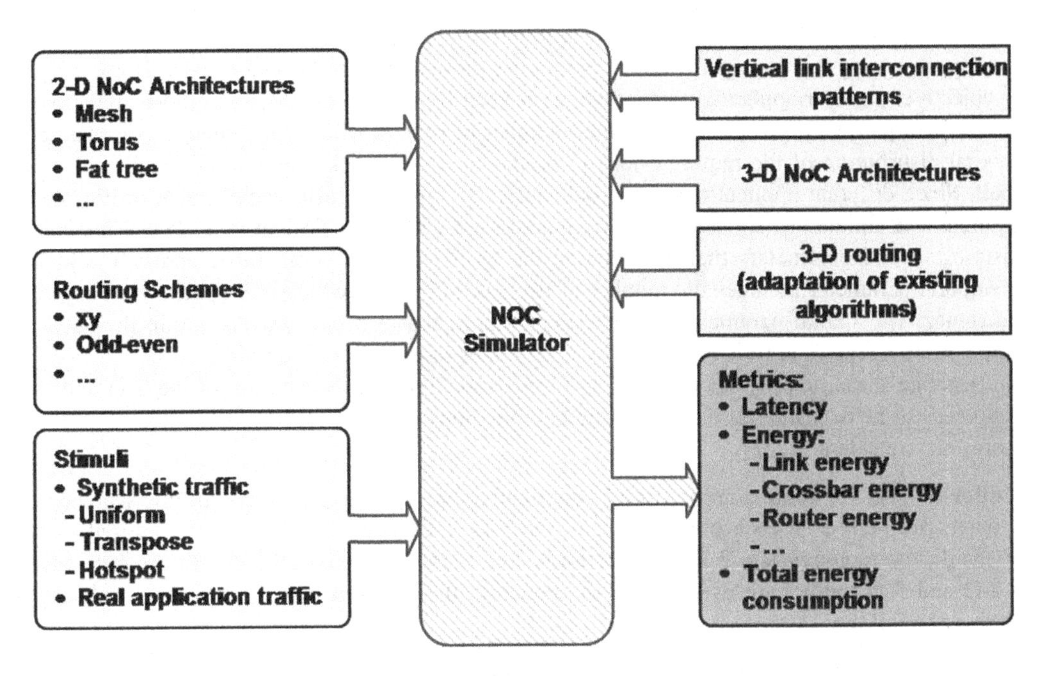

FIGURE 20.20

An overview of the 3-D NoC simulator.

3-D NoC simulator

The core of the 3-D NoC simulator is based on the *Worm_Sim* NoC simulator [753], which has been extended to support 3-D topologies. In addition to 3-D meshes and tori, variants of these topologies are supported. Related routing schemes have also been adapted to provide routing in the vertical direction. An overview of the characteristics and capabilities of the simulator is depicted in Fig. 20.20.

Several fundamental characteristics of a network topology are reported, including the energy consumption, average packet latency, and router area. The reported energy consumption includes the energy consumed by each component of the network, such as the crossbar switch and other circuitry within the network router, and the interconnect busses (i.e., link energy). Consequently, this simulator is a useful tool for exploring early decisions related to the system architecture.

An important capability of this tool is that variations of a basic 3-D mesh and torus topology can be efficiently explored. These topologies are characterized by heterogeneous interconnectivity, combining 2-D and 3-D network routers within the same network. The primary difference between these routers is that since a 3-D network router is connected to network routers on adjacent tiers, a 3-D router has two more ports than a 2-D network router. Consequently, a 3-D network router consumes a larger area and dissipates more power, yet provides greater interconnectivity.

Although these topologies typically have a greater delay than a straightforward 3-D topology, the savings in energy can be significant; in particular, those applications where speed is not the primary objective. Another application area that can benefit from these topologies is those applications where the data packets propagate over a small number of routers within the network. Alternatively, the spatial distribution of the required hops to propagate a data packet within the on-chip network is small. Since different applications can produce diverse types of traffic, an efficient traffic model is required. The simulator includes the traffic model used within the Trident tool [754]. This model consists of several parameters that characterize the spatial and temporal traffic across a network. The temporal parameter includes the number of packets and rate at which these packets are injected into a router. The spatial parameters include the distance that a packet travels within the network and the portion of the total traffic injected by each router into the network [754].

To generate these heterogeneous mesh and torus topologies, a distribution of the 2-D and 3-D routers needs to be determined. The following combinations of 2-D and 3-D routers for a 3-D mesh and torus are considered [755]:

Uniform: The 3-D routers are uniformly distributed over the different tiers. In this scheme, the 3-D routers are placed on each physical tier of the 3-D NoC. If no 2-D routers are inserted within the network, the topology is a 3-D mesh or torus (see Figs. 20.10 and 20.21B). In the case where both 2-D and 3-D routers are used within the network, the location of the routers on each tier is determined as follows:

- Place the first 3-D router on the (X, Y, Z) position of each tier.
- The four neighboring 2-D routers are placed in positions $(X + r + 1, Y, Z)$, $(X - r - 1, Y, Z)$, $(X, Y + r + 1, Z)$ and $(X, Y - r - 1, Z)$. The parameter r represents the periodicity (or frequency) of the 2-D routers within each tier. Consequently, a 2-D router is inserted every r 3-D routers in each direction within a tier. This scheme is exemplified in Fig. 20.21B, where one tier of a 3-D NoC with $r = 3$ is shown.

Center: The 3-D routers are located at the center of each tier, as shown in Fig. 20.21C. Since 3-D routers only exist at the center of a tier, only 2-D routers are placed along the outer region of each tier to connect the neighboring network nodes within the same tier.

Periphery: The 3-D routers are located at the periphery of each tier (see Fig. 20.21D). This combination is complementary to the **Center** placement.

Full Custom: The position of the 3-D routers is fully customized based on the requirements of the application, while minimizing the area occupied by the routers, since fewer 3-D routers are used. This configuration, however, produces an irregular structure that does not adapt well to changes in network functionality and application.

As illustrated in Fig. 20.20, different routing schemes are supported by the simulator. These algorithms are extended from the *Worm_Sim* simulator to support routing in the third dimension:

i. **XYZ-OLD,** which is an extended version of the shortest path XY routing algorithm.
ii. **XYZ,** which is based on XY routing where this algorithm determines which direction produces a lower delay, forwarding the packet in this direction.
iii. **ODD–EVEN** is the odd–even routing scheme as presented in [756]. In this scheme, the packets can take turns to avoid deadlock situations.

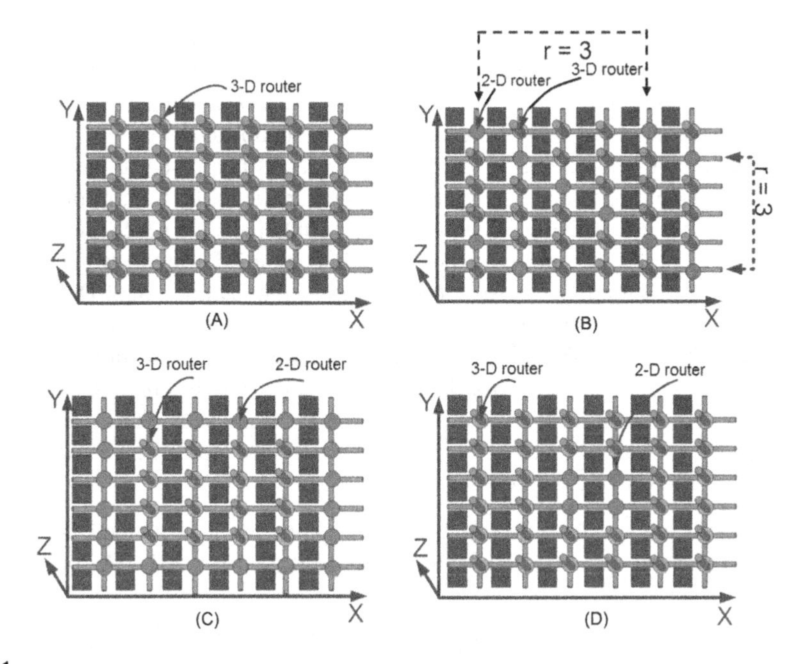

FIGURE 20.21

Position of the vertical interconnection links for each tier within a 3-D NoC (each tier is a 6 × 6 mesh), (A) fully connected 3-D NoC, (B) uniform distribution of vertical links, (C) vertical links at the center of the NoC, and (D) vertical links at the periphery of the NoC.

Consequently, the performance of a 3-D NoC can be evaluated under a broad and diverse set of parameters which reinforce the exploratory capabilities of the tool. In general, the following parameters of the simulator are configured:

- The NoC architecture, which can be a 2- or 3-D mesh or torus, where the dimensions of the network in the x, y, and z directions are, respectively, n_1, n_2, and n_3.
- The type of input traffic and traffic load.
- The routing algorithm.
- The vertical link configuration file which defines whether a vertical link (required in 3-D routers) is present.
- The router model and related energy and delay models.

In the following section, several 3-D NoC topologies are explored under different traffic patterns and loads.

Evaluation of 3-D NoCs under different traffic scenarios

To compare the performance of 3-D topologies with conventional 2-D meshes and tori, two different network sizes with 64 and 144 nodes are considered. Each of these case studies is evaluated both in two and three dimensions. In two dimensions, the network nodes are connected to form 8 × 8 and 12 × 12 2-D, respectively, meshes and tori. Alternatively, the 3-D topologies of the target networks are placed on four physical tiers (i.e., $n_3 = 4$). The dimensions of the two networks ($n_1 \times n_2 \times n_3$), consequently, are, respectively, 4 × 4 × 4 and 6 × 6 × 4, respectively. Metrics for comparing the 2-D and 3-D topologies are the average packet latency, dissipated energy, and physical area. The area of the PEs is excluded since this area is the same in both 2-D and 3-D topologies. Topologies that contain PEs on multiple tiers (see Fig. 20.10C) are not considered in this analysis.

Although the 3-D topologies exhibit superior performance as compared to the 2-D meshes and tori, a topology that combines 2-D and 3-D routers can be beneficial for certain traffic patterns and loads. Using a fewer number of 3-D routers in a network results in smaller area and possibly power. This situation is due to the fewer number of ports required by a 2-D router, which has two fewer ports as compared to a 3-D router. Consequently, several combinations of 2-D and 3-D routers within a 3-D topology have been evaluated, some of which are illustrated in Fig. 20.21. Note that for each combination of 2-D and 3-D routers, the location of these routers within a 3-D on-chip network is the same on each tier of the network. In the case studies, ten different combinations of 2-D and 3-D routers are compared in terms of energy, delay, and area. These combinations are described below, where for the sake of clarity the number (and per cent in parenthesis) of 2-D and 3-D routers within a 4 × 4 × 4 NoC is provided:

- **Full:** All of the PEs are connected to 3-D routers [number of 3-D routers: 64 (100%)]. This combination corresponds to a fully connected mesh (see Fig. 20.10A) or torus 3-D network.
- **Uniform-based:** 2-D routers are connected to specific PEs within each tier of the 3-D network. The distribution of the 2-D routers is controlled by the parameter r, as discussed in the previous section. The chosen values are three (*by_three*), four (*by_four*), and five (*by_five*). The corresponding number of 3-D routers is: 44 (68.75%), 48 (75%), and 52 (81.25%). These combinations have a decreasing number of 2-D routers, approaching a fully connected 3-D mesh or torus.
- **Odd:** In this combination, all of the routers within a row are of the same type (i.e., either 2-D or 3-D). The type of router alternates among rows [number of 3-D routers: 32 (50%)].
- **Edges:** A portion of the PEs is located at the center of each tier and connected to 2-D routers with dimensions $n_{x(2\text{-}D)} \times n_{x(2\text{-}D)}$ while the remaining network nodes are connected to 3-D routers. For the example network, $n_{x(2\text{-}D)} = 2$ and, consequently, the number of 3-D routers is 48 (75%).
- **Center**: A segment of the PEs located at the center of each tier is connected to 3-D routers with dimensions $n_{x(3\text{-}D)} \times n_{x(3\text{-}D)}$ while the remaining PEs are connected to 2-D routers. For the example network, the number of 3-D routers is 16 (25%).
- **Side-based**: The PEs along a side (e.g., an outer row) of each tier are connected to 2-D routers. The combinations have the PEs along one (*one_side*), two (*two_side*), or three (*three_side*) sides connected to 2-D routers. The number of 3-D routers for each pattern, consequently, is, respectively, 48 (75%), 36 (56.25%), and 24 (37.5%). These combinations have an increasing number of 2-D routers, approaching the same number of routers as in a 2-D mesh or torus.

The interconnectivity for the two network sizes is evaluated under different traffic patterns and loads. The parameters of the traffic model used in the 3-D NoC simulator have been adjusted to produce the following common types of traffic patterns [757]:

- **Uniform:** The traffic is uniformly distributed across the network with the network nodes receiving approximately the same number of packets.

- **Transpose:** In this traffic scheme, packets originating from a node at location (a, b, c) reach the node at the destination (n_1-a, n_2-b, n_3-c), where n_1, n_2, n_3 are the dimensions of the network.
- **Hot spot:** A small number of network nodes (i.e., hot spot nodes) receive an increasing number of packets as compared to the majority of the nodes, which is modeled as uniformly receiving packets. The hot spot nodes within a 2-D NoC are positioned in the middle of each quadrant of the network. Alternatively, in a 3-D NoC, a hot spot is located in the middle of each tier.

The traffic loads are low, normal, and high. The heavy load has a 50% increase in traffic, whereas the low load has a 90% decrease in traffic as compared to a normal load.

The energy consumption in Joules and the average packet latency in cycles are compared for each of these patterns. The energy model of the NoC simulator is an architectural level model that determines the energy consumed by propagating a single bit across the network [758]. This model includes the energy of a bit through the various components of the network, such as the buffer and switch within a router and the interconnect buss among two neighboring routers. Note that only the dynamic component of the consumed energy is considered in this model [758]. Additionally, for each topology and combination of 2-D and 3-D routers, the total area of the routers is determined based on the gate equivalent of the switching fabric [759]. The objective is to determine which of these combinations results in higher network performance as compared to the 2-D and fully connected 3-D NoC. All of the simulations are performed for 200,000 cycles.

The average packet latency of the differently interconnected NoCs for a torus architecture is depicted in Fig. 20.22. The latency is normalized to the average packet latency of a fully connected 3-D NoC under normal load condition and for each traffic scheme. As expected, the network latency increases proportionally with traffic load.

Mesh topologies exhibit similar behavior, though the latency is higher due to the decreased connectivity as compared to the torus topologies. This behavior is depicted in Fig. 20.23, where the latency of a 64-node mesh and torus NoC are compared (the basis for the latency normalization is the average packet latency of a fully connected 3-D torus). The mesh topologies exhibit an increased packet latency of 34% as compared to the torus topology for the same traffic pattern, traffic load, and routing algorithm.

The results of employing partial vertical connectivity (i.e., a combination of 2-D and 3-D routers) within a 3-D mesh network with uniform traffic, medium traffic load, and XYZ-OLD routing are illustrated in Fig. 20.24. The energy consumption, average packet latency, router area, and percent of 2-D routers for the $4 \times 4 \times 4$ and $6 \times 6 \times 4$ mesh architectures are illustrated, respectively, in Figs. 20.24A and 20.24B. All of these metrics are normalized to a fully connected 3-D NoC.

The advantages of a 3-D NoC as compared to a 2-D NoC are depicted in Fig. 20.24A. In this case, the 8×8 mesh dissipates 39% more energy and exhibits a 29% higher packet delivery latency as compared to a fully connected 3-D NoC. The overall area of the routers, however, is 71% of the area of a fully connected 3-D NoC, since all of the routers are 2-D. Employing the *by_five* combination results in a 3% reduction in energy and 5% increase in latency. In this combination, only 81% of the routers are 3-D, resulting in 5% smaller area for the switching logic. For a larger network (see Fig. 20.24B), several router combinations are superior to a fully connected 3-D NoC.

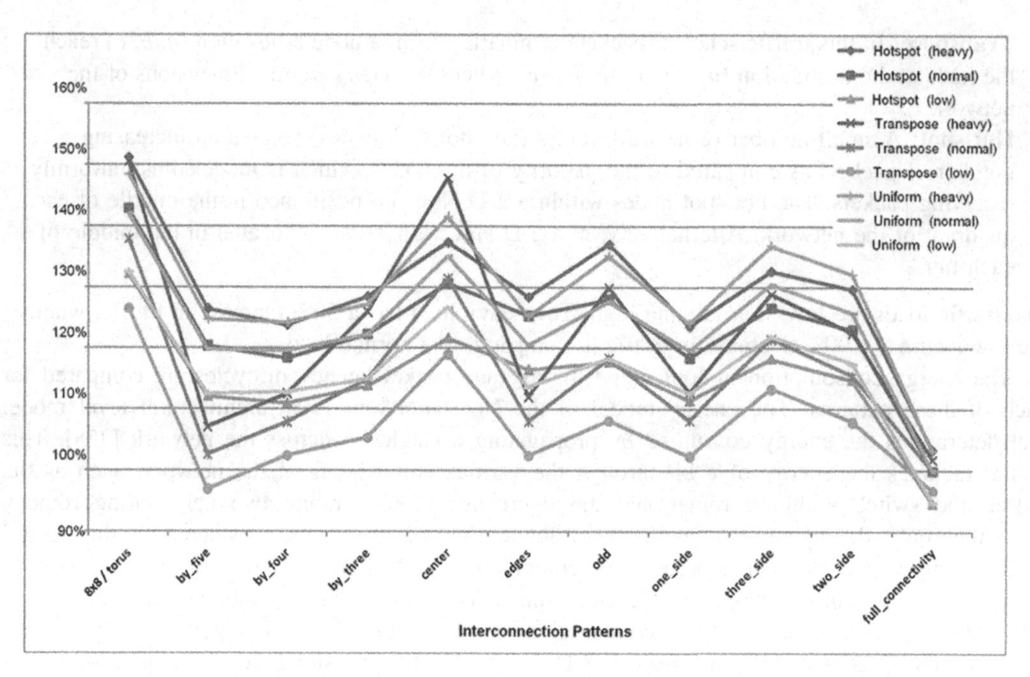

FIGURE 20.22

Effect of traffic load on the latency of a 2-D and 3-D torus NoC for each type of traffic and XYZ routing.

Summarizing, the overall performance of a 2-D NoC is significantly lower, exhibiting an increased energy and latency of approximately 50%.

When the traffic load is increased by 50%, the performance of all of the router combinations degrades as compared to the fully connected 3-D NoC. This behavior occurs since in a 3-D NoC containing both 2-D and 3-D routers, a fewer number of 3-D routers is used to save energy, reducing, in turn, the interconnectivity within the network. This lower interconnectivity increases the number of hops required to propagate data packets, increasing the overall network latency.

In the case of low traffic loads, alternative combinations can be beneficial since the requirements for communication resources are low. The simulation results for a 64 and 144 node 2-D and 3-D NoC under low uniform traffic and XYZ routing are illustrated in Fig. 20.25. An exception is the "edges" combination in the 64 node 3-D NoC (see Fig. 20.25A), where all of the 3-D routers reside along the edges of each tier within the 3-D NoC. This arrangement produces a 7% increase in packet latency. The performance of the 2-D NoC again decreases with increasing NoC dimensions. This behavior is depicted in Fig. 20.25B, where the 2-D NoC dissipates 38% more energy while the latency increases by 37%.

Finally, the energy and latency of the various interconnection patterns are compared to that of a fully connected 3-D NoC in Table 20.6. The three types of traffic are shown in the first column. The minimum and maximum value of the energy dissipation is listed in the next two columns. The minimum and maximum value of the average packet latency is, respectively, reported in the

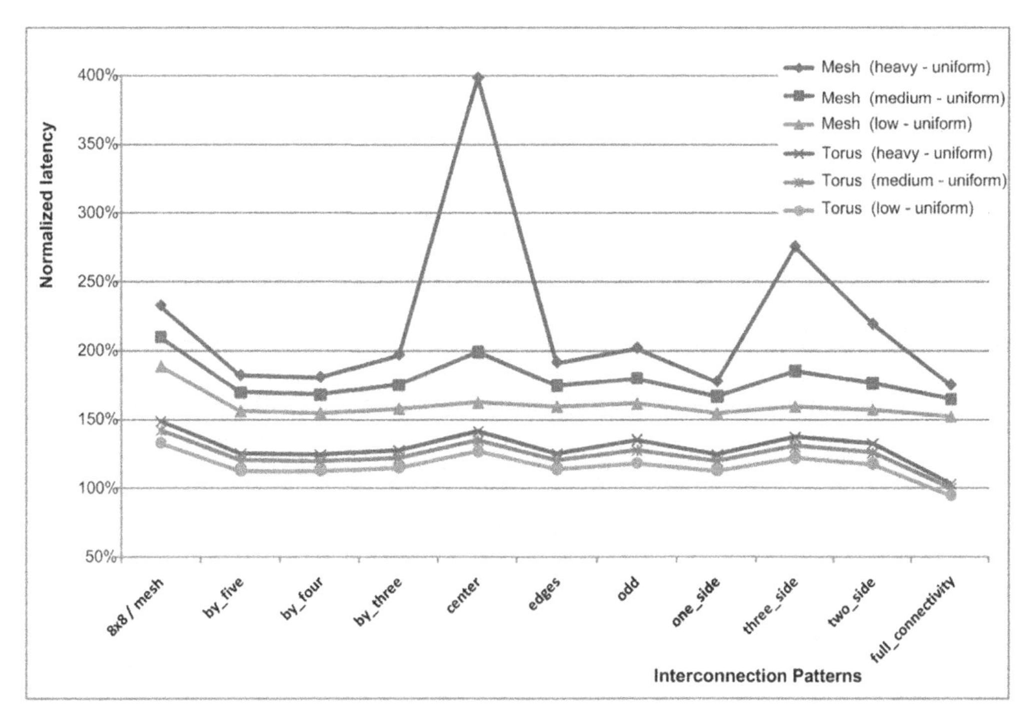

FIGURE 20.23

Latency of 64 node 2-D and 3-D meshes and tori NoCs under uniform traffic, XYZ routing, and several traffic loads.

fourth and fifth column. Although only the latency increases, which is expected as the alternative interconnection patterns decrease the interconnectivity of the network, certain traffic patterns produce a considerable savings in energy. This savings in energy is important due to the significance of thermal effects in 3-D circuits. In addition to on-chip networks, another design style that greatly benefits from vertical integration is FPGAs, which is the topic of the following section.

20.4 3-D FPGAs

FPGAs are programmable ICs that implement abstract logic functions with a considerably smaller design turnaround time as compared to other design styles, such as application-specific (ASIC) or full custom ICs. Due to this flexibility, the share of the IC market for FPGAs has steadily increased. The tradeoff for the reduced time to market and versatility of the FPGAs is lower speed and increased power consumption as compared to ASICs. A traditional physical structure of an FPGA is depicted in Fig. 20.26, where the logic blocks (LB) can implement any digital logic function with some sequential elements and arithmetic units [760]. The switch boxes (SBs) provide the

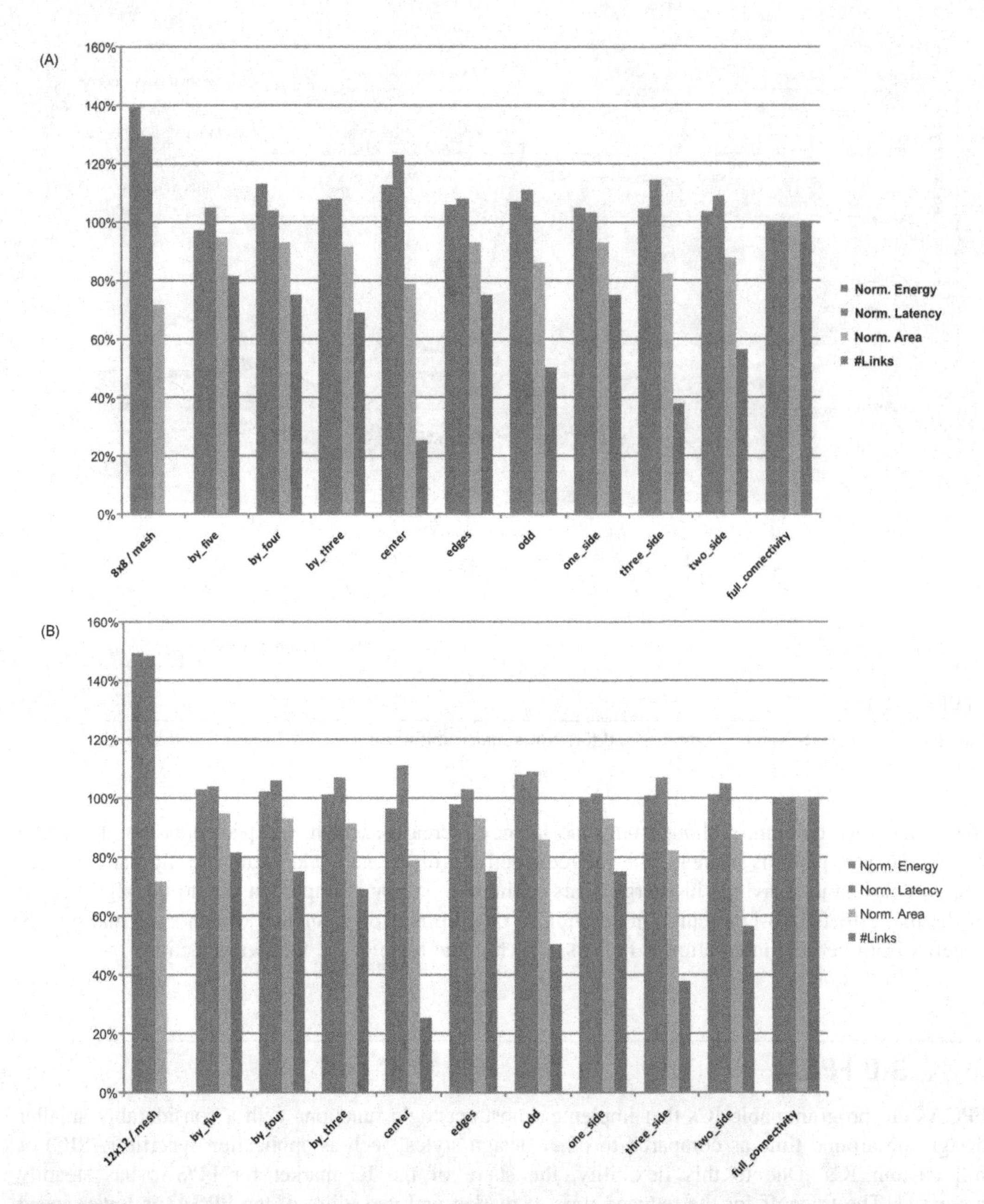

FIGURE 20.24

Different performance metrics under uniform traffic and a normal traffic load of a 3-D NoC for alternative interconnection topologies with XYZ-OLD routing, (A) 64 network nodes, and (B) 144 network nodes.

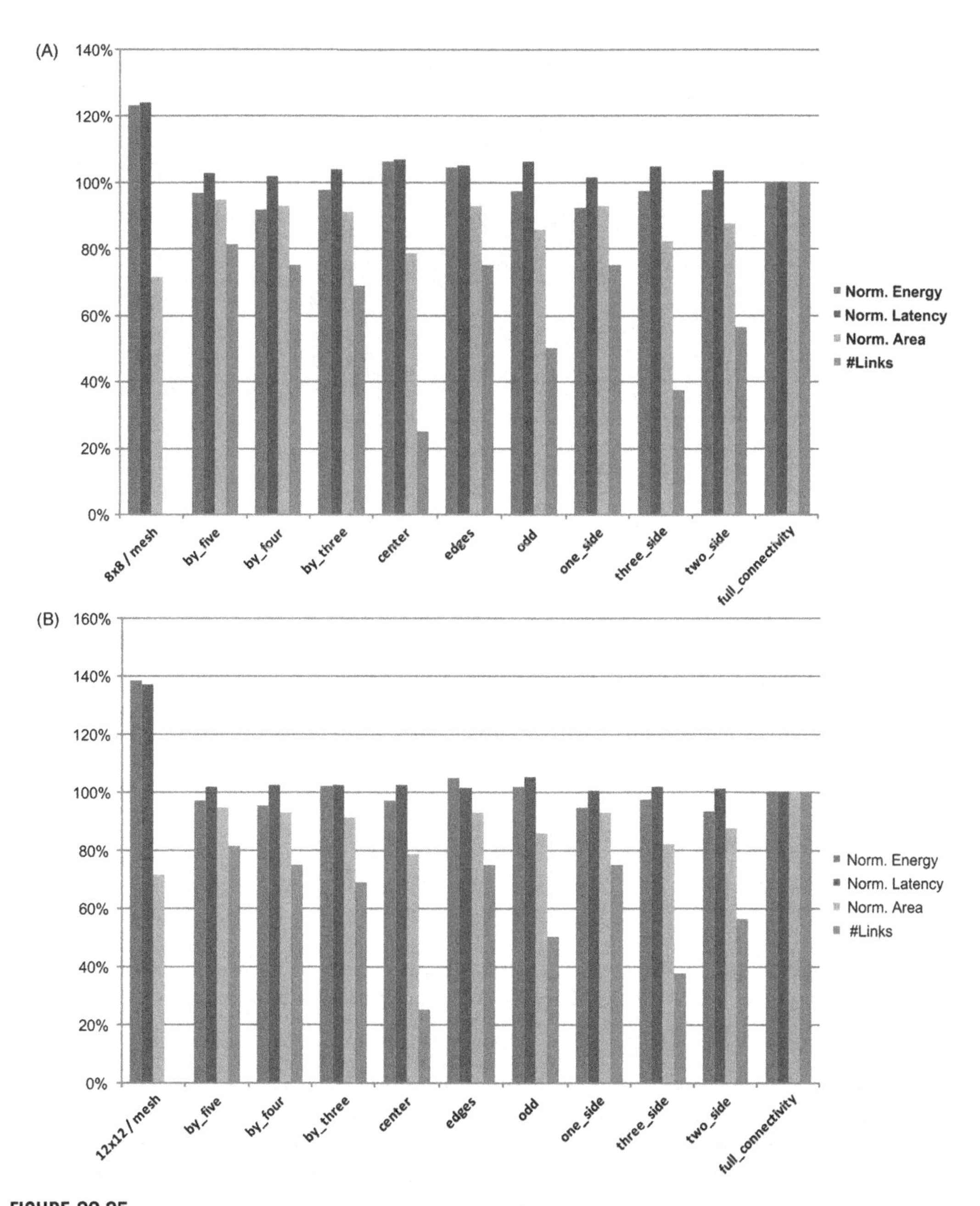

FIGURE 20.25

Several performance metrics under uniform traffic and a low traffic load of a 3-D NoC for alternative interconnection topologies with XYZ routing, (A) a 4 × 4 × 4 3-D mesh, and (B) a 6 × 6 × 4 3-D mesh.

Table 20.6 Min–Max Variation in 3-D NoC Latency and Power Dissipation as Compared to 2-D NoC under Different Traffic Patterns and a Normal Traffic Load

Min–Max Energy and Latency				
	Normalized			
	Energy		Latency	
Traffic Pattern	Min	Max	Min	Max
Uniform	92%	108%	98%	113%
Transpose	88%	116%	100%	354%
Hotspot	71%	116%	100%	134%

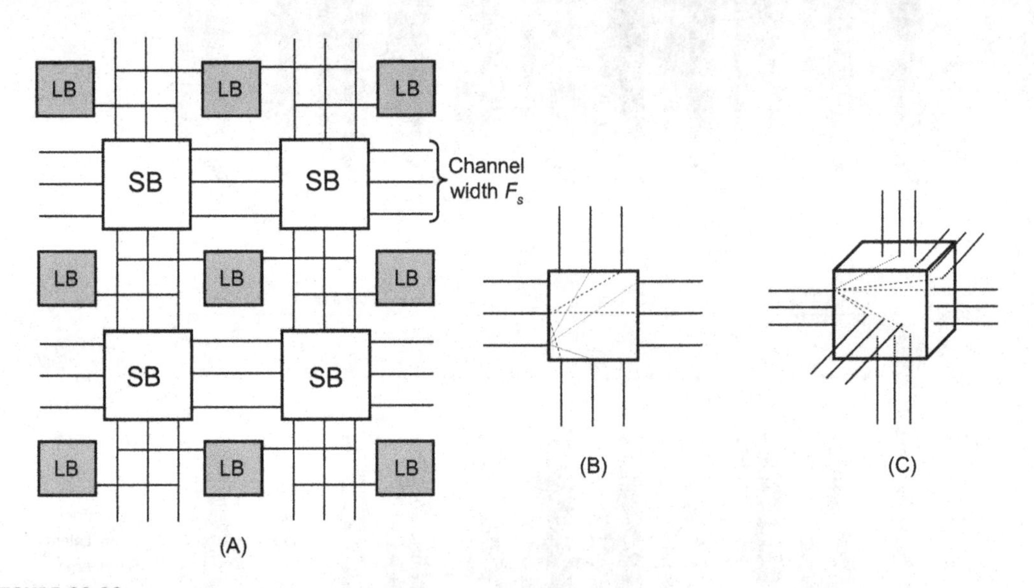

FIGURE 20.26

Typical FPGA architecture, (A) 2-D FPGA, (B) 2-D switch box, and (C) 3-D switch box. A routing track can connect three outgoing tracks in a 2-D SB, while in a 3-D SB, a routing track can connect five outgoing routing tracks.

interconnections among the LBs. The SBs include pass transistors, which connect (or disconnect) the incoming routing tracks with the outgoing routing tracks. Memory circuits control these pass transistors and program the LBs for a specific application. In FPGAs, the SBs constitute the primary delay component of the interconnect delay between the LBs and can consume a great amount of power.

Extending FPGAs to the third dimension can improve performance while decreasing power consumption as compared to conventional planar FPGAs. A generalization of FPGAs to the third dimension would include multiple planar FPGAs, wafer or die bonded to form a 3-D system.

The crucial difference between a 2-D and 3-D FPGA is that the SB provides communication to five LBs in a 3-D system rather than three neighboring LBs as in a 2-D FPGA (see Figs. 20.26B and 20.26C). Consequently, each incoming interconnect segment connects to five outgoing segments rather than three outgoing segments. The situation is somewhat different for the bottom and top-most tier of a 3-D FPGA, but in the following discussion, for simplicity this difference is neglected. Since the connectivity of a 3-D SB is greater, additional pass transistors are required in each SB, increasing the power consumption, memory requirements to configure the SB, and, possibly, the interconnect delay. The decreased interconnect length and greater connectivity can compensate, however, for the added complexity and power of the 3-D SBs.

To estimate the size of the array beyond which the third dimension is beneficial, the shorter average interconnect length offered by the third dimension and the increased complexity of the SBs should be simultaneously considered [761]. Incorporating the hardware resources (e.g., the number of transistors) required for each SB and the average interconnect length for a 2-D and 3-D FPGA, the minimum number of LBs for a 3-D FPGA to outperform a 2-D FPGA is determined from the solution of the following equation,

$$F_{s,2-D}\frac{2}{3}N^{1/2} = F_{s,3-D}N^{1/3}, \tag{20.27}$$

where $F_{s,\text{2-D}}$ and $F_{s,\text{3-D}}$ are, respectively, the channel width of a 2-D and 3-D FPGA, respectively, and N is the number of LBs. Solving (20.27) yields $N = 244$, a number that is well exceeded in modern FPGAs.

Since the pass transistors, employed both in 2-D and 3-D SBs, contribute significantly to the interconnect delay, degrading the performance of an FPGA, those interconnects that span more than one LB can be utilized. These interconnect segments are named after the number of LBs that is traversed by these segments, as shown in Fig. 20.27. Wires that span two, four, or even six LBs are quite common in contemporary FPGAs. Interconnects that span one fourth to a half of an IC edge are also possible [762].

The opportunities that the third dimension offers in SRAM based 2-D FPGAs have also been investigated [763]. Analytic models that estimate the channel width in 2-D FPGAs have been

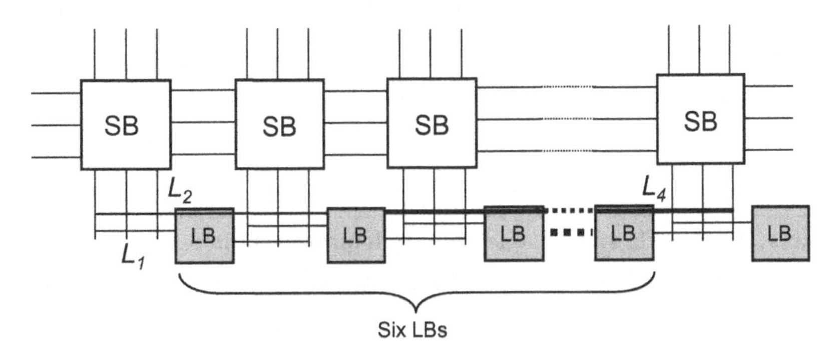

FIGURE 20.27

Interconnects that span more than one logic block. L_i denotes the length of these interconnects and i is the number of LBs traversed by these wires.

extended to 3-D FPGAs. Hence, the channel width W for an FPGA with N LBs, exclusively consisting of unit length interconnect segments and implemented in n physical tiers, can be described by

$$W = \frac{\sum_{l=1}^{2\sqrt{N/n}-2+(n-1)d_v} \chi_{\text{fpga}}}{\left(2N + \frac{(n-1)N}{n}\right)e_t}, \tag{20.28}$$

where $f_{3\text{-}D}(l)$ is a stochastic interconnect length distribution similar to those discussed in Chapter 7, Interconnect Prediction Models. χ_{fpga} converts a point-to-point distance into an interconnect length and e_t is the utilization parameter of the wiring tracks. These two factors can be determined from statistical data characterizing the placement and routing of benchmarks circuits on FPGAs. Note that these factors depend both on the architecture of the FPGA and the automated layout algorithm used to route the FPGA.

Several characteristics of FPGAs have been estimated from benchmarks circuits and randomized netlists placed and routed with the SEGment Allocator (SEGA) [764], the versatile place and route (VPR) [765] tools, and analytic expressions, such as (20.28). In these benchmark circuits, each FPGA is assumed to contain 20,000 four input LBs and is manufactured in a 0.25 μm CMOS technology. The area, channel density, and average wirelength are measured in LB pitches, which is the distance between two adjacent LBs, and are listed in Table 20.7 for different number of physical tiers.

The improvement in the interconnect delay with a length equal to the die edge is depicted in Fig. 20.28 for different number of tiers. Those wires that span multiple LBs use unit length segments (i.e., no SBs are interspersed along these wires) whereas the die edge long wires use interconnect segments with a length equal to a quarter of the die edge. A significant decrease in delay is projected; however, these gains diminish for more than four tiers, as indicated by the saturated portion of the delay curves depicted in Fig. 20.28. The components of the power dissipated in a 3-D FPGA assuming a 2.5 volt power supply are shown in Fig. 20.29. The power consumed by the LBs remains constant since the structure of the LBs does not vary with the third dimension. However, due to the shorter interconnect length, the power dissipated by the interconnects is less. This improvement, however, is smaller than the improvement in the interconnect delay, as indicated by the slope of the curves illustrated in Figs. 20.28A and 20.28B. This behavior is attributed to the extra pass transistors in a 3-D SB, which increase the power consumption, compromising the benefit of the shorter interconnect length. Due to the reduced interconnect length, the power dissipated by the clock distribution network is also less.

Table 20.7 Area, Wirelength, and Channel Density Improvement in 3-D FPGAs

Number of Tiers	Area (cm^2)	Channel Density	Avg. Wirelength (LB Pitch)
1 (2-D)	7.84	41	8
2 (3-D)	3.1	24	6
3 (3-D)	1.77	20	5
4 (3-D)	1.21	18	5

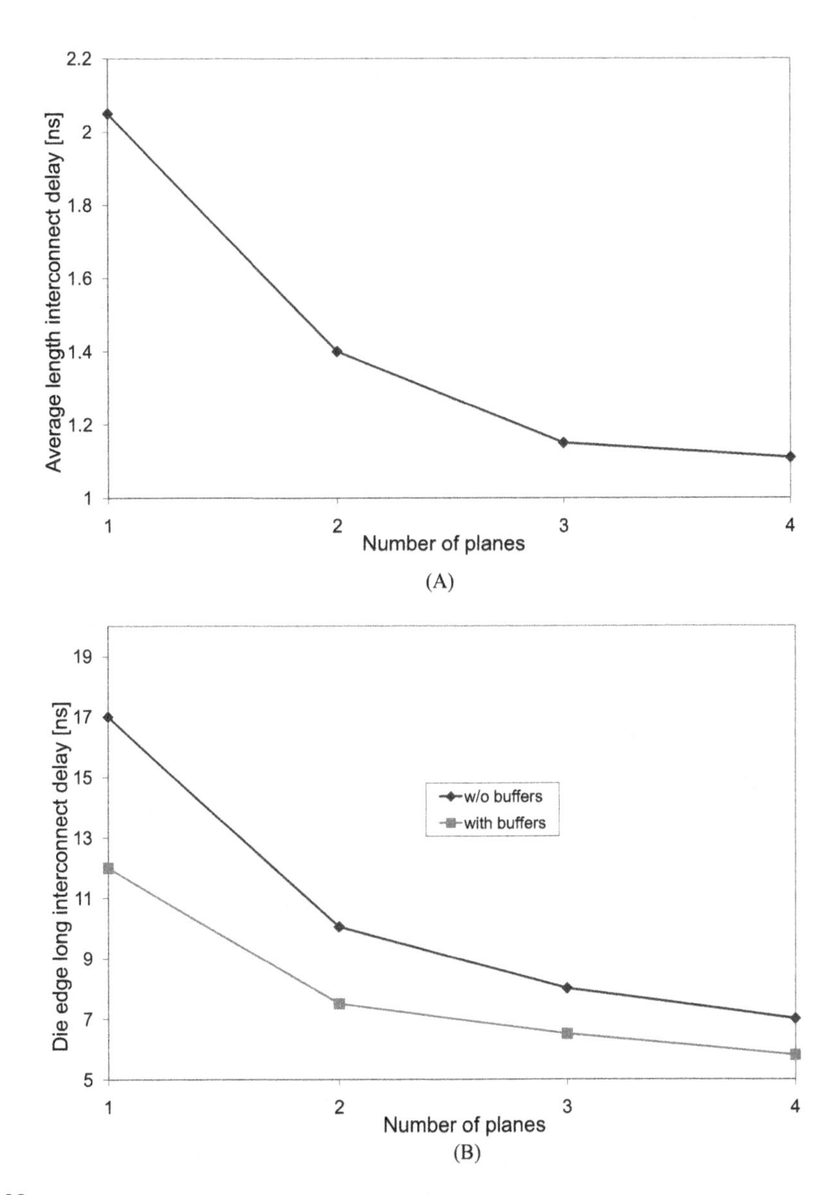

FIGURE 20.28

Interconnect delay for several number of physical tiers, (A) average length wires, and (B) die edge length interconnects.

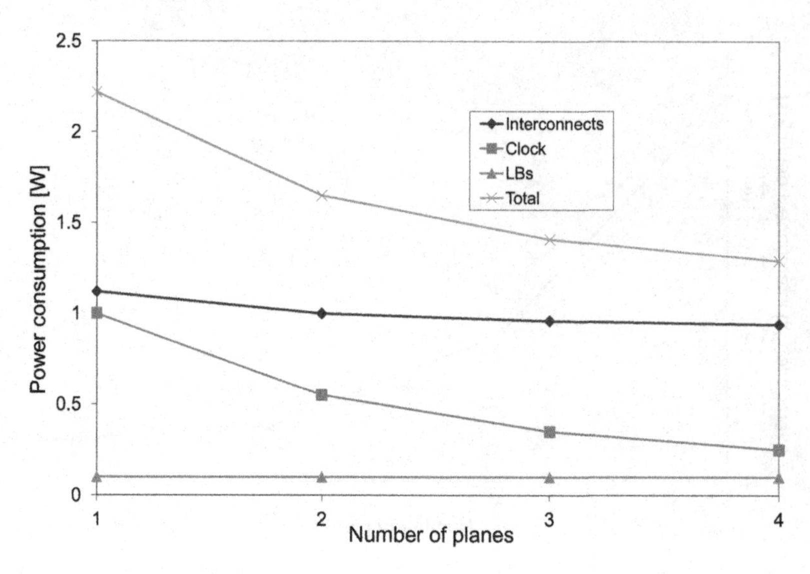

FIGURE 20.29

Power dissipated by 2-D and 3-D FPGAs.

20.5 SUMMARY

Different classes of circuit structures that can greatly benefit from 3-D integration are described in this chapter. Several architectures and related designs aids for these circuits are discussed. The primary points of this discussion can be summarized as follows:

- Vertical integration can improve both the latency and power consumption of wire limited and communication centric circuits. These circuits include the microprocessor memory system, on-chip networks, and FPGAs.
- Implementing several components of a microprocessor across multiple physical tiers decreases the power consumption and improves the speed by utilizing fewer redundant pipeline stages.
- Partitioning a cache memory into a 3-D structure can reduce the time required to access the memory.
- Stacking additional memory tiers on a microprocessor can improve the overall performance of a microprocessor memory system without exceeding the thermal budget of the system.
- 3-D NoC are a natural evolution of 2-D NoC and exhibit superior performance.
- The minimum latency and power consumption can be achieved in 3-D NoC by reducing both the number of hops per packet and the length of the communications channels.
- Expressions for the zero-load network latency and power consumption are described for 3-D NoC.
- NoC implemented by a 3-D mesh reduce the number of hops required to propagate a data packet between two nodes within a network.

- 2-D IC–3-D NoC decrease the latency and power consumption by reducing the number of hops.
- 3-D IC–2-D NoC decrease the latency and power consumption by reducing the buss length or, equivalently, the distance between adjacent network nodes.
- Large networks are primarily benefited by the 2-D IC–3-D NoC topology, while small networks are enhanced by the 3-D IC–2-D NoC topology.
- The distribution of nodes corresponding to either the minimum latency or power consumption of a network depends upon both the interconnect parameters and network size.
- A tradeoff exists between the number of tiers utilized for a network and the number of PEs. Consequently, and not surprisingly, the 3-D IC–3-D NoC topology achieves the greatest improvement in latency and power consumption by most effectively exploiting the third dimension.
- A simulator that supports the exploration of 3-D NoC topologies with different routing schemes, traffic patterns, and traffic loads has been developed.
- A multi-tier on-chip network that includes both 2-D and 3-D routers can produce a significant savings in energy and area with a tolerable increase in latency as compared to a full 3-D topology.
- 3-D FPGAs have been proposed to improve the performance of 2-D FPGAs, where multiple planar FPGAs are bonded to form a 3-D stack of FPGAs.
- A critical issue in 3-D FPGA is the greater complexity of the 3-D SB that can negate the benefits from the shorter interconnect length and enhanced connectivity among the LBs.

CHAPTER

CONCLUSIONS

21

With FET channel lengths below 10 nm, modern silicon integrated systems support gigascale densities composed of billions of transistors. Modern on-chip systems often include several heterogeneous processing cores within a microprocessor system or a mixture of different silicon technologies, such as analog, digital, and RF, in a mixed-signal SoC. A fundamental requirement for this system-on-chip (SoC) paradigm is efficient and reliable communication among the many different system components.

High speed, low power, and low noise intercomponent communication is fundamentally limited by the increasing impedance characteristics and length of the interconnect. Three-dimensional (3-D) integration is an effective solution to the deleterious effects of these long horizontal interconnects. In vertically integrated systems, the long interconnects spanning several millimeters are replaced by orders of magnitude shorter vertical wires. This inherent reduction in wirelength provides opportunities for increased speed, enhanced noise margins, and lower power.

3-D integrated systems can also comprise a variety of different silicon technologies, such as digital, analog, RF CMOS, and SOI circuits, non-silicon semiconductor technologies, such as InP, SiGe, HgCadTd, and GaAs, nonvolatile technologies, such as Spin transfer torque magnetic tunnel junction (STT-MTJ), magnetic and resistive RAMs, and photonic integrated circuits, making 3-D systems highly suitable for a broad spectrum of applications. This salient characteristic of heterogeneous integration has greatly increased interest in 3-D systems as compared to other, more incremental solutions that simply scale CMOS technologies.

Within the context of an electronic product, circuit level enhancements can profoundly affect fundamental system requirements, such as form factor, portability, and computing power. Since the application space for these systems is broad, several different forms of 3-D technologies have been developed while other 3-D technologies are currently under development. These approaches range from monolithic 3-D circuits to polylithic and heterogeneous multi-tier systems. Interposer based approaches complete the spectrum of non-planar integrated systems with galvanic connections between dies. The fundamental element of a 3-D silicon system can be a single transistor, a functional block, a bare die, or a packaged circuit. The structural granularity of the system, in turn, determines the form of vertical interconnection which links the elements of a 3-D circuit in the vertical direction. The cost of manufacturing is directly related to the processing capabilities offered by these technologies. Fabrication processes that utilize TSVs for vertical interconnects are an economic form of vertical integration, while exhibiting considerable improvement in system performance.

An alternative to TSV-based 3-D systems is the contactless integration of physical tiers utilizing capacitive or inductive coupling. Between these two schemes, inductive coupling can support higher communication distances and, therefore, greater integration densities. Although the silicon area required by the on-chip inductors within the links is significant, interest exists for this type of 3-D

Three-Dimensional Integrated Circuit Design. DOI: http://dx.doi.org/10.1016/B978-0-12-410501-0.00021-6

integration in niche applications where the manufacturing complexity of the TSV cannot be justified. As discussed in this book, the performance of the inductive links can be comparable to the TSVs, particularly at large TSV diameters. The capabilities of contactless 3-D circuits have not yet been sufficiently explored and design methodologies are far from complete and systematic.

A major focus of this book is that effective physical design methodologies and algorithms for 3-D circuits have been developed, emphasizing the significant role of the TSVs. In the 3-D design process, the through silicon vias are treated as a means to improve signaling among the physical tiers within a 3-D stack. The TSVs in a 3-D circuit provide synchronization, power delivery, and signaling among the tiers and—in contrast to mainstream two-dimensional (2-D) circuits—thermal cooling. The primary challenges and novel solutions for satisfying these objectives are discussed throughout the book.

The electrical characteristics of the TSVs affect all of these objectives. Consequently, the impedance characteristics of the TSV structures are considered in design algorithms and techniques to enhance the speed, power, and area of 3-D systems. These methods increase the benefits of the shorter interconnect length. The effectiveness of physical design techniques is considerably improved by exploiting the electrical behavior of these vias. Interestingly, not all TSV technologies yield improvements in speed, power, and area. For those circuits where the speed is primarily determined by the logic gates, the gains from 3-D integration are constrained. Furthermore, an electronic storm is produced within this multi-tier system. The TSVs can produce significant noise coupling with adjacent devices in specific types of substrates. As 3-D integration targets highly heterogeneous systems, coupling between a TSV and different types of substrates is modeled and explored. Low resistivity substrates, for example, can produce high coupling, which needs to be considered during the design process.

To assess the gains and limitations of TSVs in integrated systems, systematic approaches are provided for inserting and distributing the vertical interconnects to improve the overall thermal conductivity of a 3-D stack. By including the thermal objective in 3-D physical design algorithms and techniques, significant thermal gradients and high temperatures, which can degrade the reliability and performance of a 3-D circuit, are significantly reduced. Due to the high power densities, the thermal objective is an integral element of the physical design process for 3-D integrated systems.

Thus, thermal issues are tackled during every step of the design process, including floorplanning, placement, and routing, as described in several chapters of this book. Thermal management strategies applied during runtime are also beneficial. These methods exploit different techniques for lowering the power dissipated by a system to better manage the local temperatures and distribute the heat within a 3-D stack. Another method to address thermal issues is to cool between the tiers of a 3-D stack. This approach is efficient but is currently unavailable in production systems as the cooling channels in the thinned substrate within the tiers significantly complicate the manufacturing process.

The lack of advanced packaging technologies has increased the role of on-chip solutions in removing heat from multi-tier systems. Design methodologies for circuit/package thermal codesign offer superior solutions caused by the high temperatures within 3-D circuits. Consequently, although difficulties exist in applying some of these approaches, the material presented in this book offers broad coverage of different thermal management techniques to reinforce research into this paramount issue for 3-D systems.

The role of TSVs in enhancing power delivery and distribution is also described. An interesting outcome of several studies is that those 3-D technologies that support more than one TSV diameter have a distinct advantage in addressing power integrity issues in multi-tier systems. Each use of TSVs, including for signaling, synchronization, power delivery, and heat transfer, is evaluated in several case studies of three fabricated 3-D circuits. Different methods to efficiently distribute the clock signal in the gigahertz regime are explored, both temporally and within the three spatial dimensions. In addition, several topologies to distribute power and ground are evaluated. In the third test circuit, the important issue of intra and intertier thermal coupling is experimentally evaluated and validated. In a nutshell, enhanced intertier signaling is a prerequisite for high performance, high bandwidth 3-D computing.

A topic in 3-D circuits that has, somewhat surprisingly, received limited interest to date is process variability, which has been extensively explored in 2-D circuits. The diverse technologies comprising a 3-D system and the disparate behavior of the circuits within these tiers require statistical and probabilistic models to characterize process, environmental, and power noise variations, while methods that exploit the broader sources of variability to mitigate variations across a 3-D system are currently nascent and require attention. These topics are discussed in great detail in this book.

High performance applications typically include communications limited architectures, such as a processor-memory system. Communication fabrics, such as on-chip networks, and FPGAs also greatly benefit from the short vertical interconnects. Different architectural configurations are explored in this book, demonstrating the many possible power and latency tradeoffs that can result from exploiting the third dimension. Appropriate latency and power models supporting this discussion are also presented.

The primary achievement of this book is the thorough and integrated exploration of the multiple aspects of 3-D integration, ranging from manufacturing to physical design and algorithms, to thermal analysis and management, to system level architectures. This comprehensive approach targets the design and analysis of 3-D integration, primarily from a circuits perspective. With this point of view, the primary objective of this book is to provide physical and electrical intuition into the most sensitive issues regarding the vertical interconnect, which is intertwined within each step of the 3-D design process.

As a final concluding remark, consider that although early 3-D systems have already appeared in the marketplace, the commercial growth of 3-D integration will be greatly enhanced once an appropriate business model and supply chain are in place. Unfortunately, the major stakeholders have not yet managed to present a successful paradigm for the design and manufacture of 3-D systems; rather, an *ad hoc* mix-and-match approach is applied. This wait for an effective business model has greatly slowed the evolution of this fascinating and promising systems integration platform.

Looking beyond these unresolved non-technical issues, the material described in this book is intended to shed light on those areas related to the design of 3-D integrated systems. The primary objective of the entire 3-D ecosystem is to develop large scale multi-functional, multi-tier heterogeneous systems while continuing the microelectronics revolution.

Appendix A: Enumeration of Gate Pairs in a 3-D IC

The starting gates N_{start} used in the derivation of the interconnect length distribution for three-dimensional (3-D) circuits in [291] are described in this appendix. The starting gates are those gates that form manhattan hemispheres of radius l, where l is the interconnect length connecting two gates. For a 3-D circuit consisting of n tiers and including $N_n = N/n$ gates per tier and those gates located close to the periphery of each tier, only partial manhattan hemispheres are formed. Consequently, the number of gates encircled by these partial manhattan hemispheres varies with the interconnect length:

Region I: $l = 0$

$$N_{start} = N_n n \tag{A-1}$$

Region II: $0 < l \leq (n-1)d_v$

$$N_{start} = n(N_n - l) \tag{A-2}$$

Region III: $(n-1)d_v < l \leq \frac{\sqrt{N_n}}{2} + 1$

$$N_{start} = nN - l \tag{A-3}$$

Region IV: $\frac{\sqrt{N_n}}{2} + 1 < l \leq \sqrt{N_n}$

$$N_{start} = nN_n - l - \left(l - \frac{\sqrt{N_n}}{2} - 1\right)\left(l - \frac{\sqrt{N_n}}{2}\right) \tag{A-4}$$

Region V: $\sqrt{N_n} < l \leq \sqrt{N_n} + (n-1)d_v$

$$N_{start} = \frac{7N_n}{4} - \frac{\sqrt{N_n}}{2} - \sqrt{N_n}l - f\left[l, g\left[l - \sqrt{N_n} - 1\right], 1\right] \tag{A-5}$$

Region VI: $\sqrt{N_n} + (n-1)d_v < l \leq \frac{3\sqrt{N_n}}{2}$

$$N_{start} = \frac{7N_n}{4} - \frac{\sqrt{N_n}}{2} - \sqrt{N_n}l + f[l, n-1, 2] - 2f\left[l, n-1, \frac{3}{2}\right] \tag{A-6}$$

Region VII: $\frac{3\sqrt{N_n}}{2} < l \leq \frac{3\sqrt{N_n}}{2} + (n-1)d_v$

$$N_{start} = \left\{ \begin{array}{l} \left(2\sqrt{N_n} - l\right)\left(2\sqrt{N_n} - l - 1\right) + f[l, n-1, 2] - 2f\left[l, n-1, \frac{3}{2}\right] \\ + 2f\left[l, g\left[l - \frac{3}{2}\sqrt{N_n} - 1, d_v\right], \frac{3}{2}\right] \end{array} \right\} \tag{A-7}$$

Region VIII: $\frac{3\sqrt{N_n}}{2} + (n-1)d_v < l \leq 2\sqrt{N_n}$

$$N_{\text{start}} = \left(2\sqrt{N_n} - l\right)\left(2\sqrt{N_n} - l - 1\right) + f[l, n-1, 2] \tag{A-8}$$

Region IX: $2\sqrt{N_n} < l \leq 2\sqrt{N_n} + (n-1)d_v$

$$N_{\text{start}} = f[l, n-1, 2] - f\left[l, g\left[l - 2\sqrt{N_n} - 1, d_v\right], 2\right] \tag{A-9}$$

The function $g[x,y]$ is the discrete quotient function while function $f[x,y,z]$ is

$$f[l,y,z] = 2\left\{\begin{array}{l} \frac{d_v^2 y}{6}\left(2y^2 + 3y + 1\right) + \frac{y}{2}\left(2z\sqrt{N_n}d_v - 2d_v l - d_v\right)(y+1) \\ + 4y\left(9N_n - 3\sqrt{N_n}l - \frac{3}{2}\sqrt{N_n} + l + l^2\right) \end{array}\right\} \tag{A-10}$$

d_v denotes the intertier distance between two adjacent tiers within a 3-D system.

Appendix B: Formal Proof of Optimum Single Via Placement

In this appendix, the Lemma used to determine the optimum via location for an intertier interconnect that includes only one TSV for different values of r_{21}, c_{12}, and l_1 is stated and proved.

Lemma 1: If $f(x) = Ax^2 + Bx + C$ and $\dfrac{d^2f(x)}{dx^2} < 0$

a. for $x_{max} \in [x_0, x_1], \quad x_1 > x_0 > 0$

  i. if $x_{max} > \dfrac{x_1 + x_0}{2}, \quad f(x_0) < f(x_1)$,

  ii. if $x_{max} < \dfrac{x_1 + x_0}{2}, \quad f(x_1) < f(x_0)$,

b. for $x_{max} < x_0, f(x_0) > f(x_1)$,

c. for $x_{max} > x_1, f(x_0) < f(x_1)$.

Proof:

a.

  i. $f(x)$ is a parabola with a symmetry axis at $x = x_{max}$. Thus, $f(x_{max} - x) = f(x_{max} + x)$. For $x = x_{max} - x_0, f(x_0) = f(2x_{max} - x_0)$. Since $\left(d^2f(x)/dl_1^2\right) < 0, f(x)$ is decreasing for $x > x_{max}$. From the hypothesis, $x_{max} > (x_1 + x_0)/2 \Leftrightarrow 2x_{max} - x_0 > x_1 > x_{max} \Leftrightarrow f(2x_{max} - x_0) < f(x_1)$, and since $f(x_0) = f(2x_{max} - x_0), f(x_0) < f(x_1)$.

  ii. $f(x_0) = f(2x_{max} - x_0)$, and $2x_{max} - x_0 > x_{max}$ since from the hypothesis $x_{max} \in [x_0, x_1]$. By similar reasoning as in (i), $x_{max} < (x_1 + x_0)/2 \Leftrightarrow x_1 > 2x_{max} - x_0 \Leftrightarrow f(x_1) < f(2x_{max} - x_0) \Leftrightarrow f(x_1) < f(x_0)$.

b. $f(x)$ decreases for $x > x_{max}$. Since $x_{max} < x_0$, for $x_{max} < x_0 < x_1, f(x_{max}) > f(x_0) > f(x_1)$.

c. $f(x)$ increases for $x < x_{max}$. Since $x_{max} > x_0$, for $x_0 < x_1 < x_{max}, f(x_0) < f(x_1) < f(x_{max})$.

Appendix C: Proof of the Two-Terminal Via Placement Heuristic

A formal proof of the two terminal heuristic for placing intertier vias (e.g., TSVs) is described in this appendix. Consider the following expression that describes the critical point (i.e., the derivative of the delay is set equal to zero) for placing a via v_j, as illustrated in Fig. C.1,

$$x_j^* = -\left[\frac{l_{vj}\left(r_j c_{vj} - r_{vj}c_{j+1} + r_{j+1}c_{j+1} - r_j c_{j+1}\right) + R_{uj}\left(c_j - c_{j+1}\right) + \Delta x_j\left(r_j - r_{j+1}\right)c_{j+1} + C_{dj}\left(r_j - r_{j+1}\right)}{r_j c_j - 2r_j c_{j+1} + r_{j+1}c_{j+1}}\right]. \quad \text{(C.1)}$$

From this expression, the critical point x_j is a monotonic function of the upstream resistance and downstream capacitance of the allowed interval for via v_j, denoted respectively, as R_{uj} and C_{dj},

$$x_j^* = f\left(R_{uj}^*, C_{dj}^*\right). \quad \text{(C.2)}$$

These quantities (R_{uj}^* and C_{dj}^*) depend upon the location of the other vias along the net and are unknown. However, as the allowed intervals for the vias and the impedance characteristics of the line are known, the minimum and maximum value of these impedances, $R_{uj\ \min}$, $R_{uj\ \max}$, $C_{dj\ \min}$, and $C_{dj\ \max}$, can be determined. Without loss of generality, assume that $r_j > r_{j+1}$ and $c_j > c_{j+1}$ (the other cases are similarly treated). For this case, the critical point (i.e., $\partial T/\partial x_j = 0$) is a strictly increasing function of R_{uj} and C_{dj}. Consequently, the minimum and maximum value for the critical point $x_{j\ \min}^*$ and $x_{j\ \max}^*$ are determined from, respectively,

$$x_{j\ \min}^* = f\left(R_{uj\ \min}, C_{dj\ \min}\right), \quad \text{(C.3)}$$

$$x_{j\ \max}^* = f\left(R_{uj\ \max}, C_{dj\ \max}\right). \quad \text{(C.4)}$$

The final value of the upstream (downstream) capacitance for via v_j, which is determined after placing all of the remaining vias of the net denoted as R_{uj}^* $\left(C_{dj}^*\right)$ within the range, is

$$R_{uj\ \min} < R_{uj}^* < R_{uj\ \max}, \quad \left(C_{dj\ \min} < C_{dj}^* < C_{dj\ \max}\right). \quad \text{(C.5)}$$

Due to the monotonic relationship of the critical point x_j on R_{uj} and C_{dj},

$$x_{j\ \min}^* = f\left(R_{uj\ \min}, C_{dj\ \min}\right) < x_j^* = f\left(R_{uj}^*, C_{dj}^*\right) < x_{j\ \max}^* = f\left(R_{uj\ \max}, C_{dj\ \max}\right). \quad \text{(C.6)}$$

Consequently, by iteratively decreasing the range of the x_j^* according to (C.6), the location for v_j can be determined.

To better explain this iterative procedure, an example is offered in Chapter 10, Timing Optimization for Two-Terminal Interconnects, where the vias, v_i, v_j, and v_k, shown in Fig. C.1, have not yet been placed. In this example, vias v_i and v_k are assumed to belong to case (iii) of the heuristic. Since the allowed intervals for vias v_i, v_j, and v_k and the impedance characteristics of the respective horizontal segments are known, the minimum $x_{\min}^{*0}$ and maximum $x_{\max}^{*0}$ critical point for all of the segments i, j, and k are obtained. The minimum and maximum values of R_{ui}^0, R_{uj}^0, R_{uk}^0, C_{di}^0, C_{dj}^0, and C_{dk}^0 are determined, where the superscript represents the number of iterations.

From (C.6), the via location of segments i and k is contained within the limits determined by (C.1). As the interval for placing the vias v_i and v_k decreases, the minimum (maximum) value of the upstream resistance and downstream capacitance of segment j increases (decreases), that is, $R^0_{uj\ \min} < R^1_{uj\ \min}$, $C^0_{dj\ \min} < C^1_{dj\ \min}$, $R^1_{uj\ \max} < R^0_{uj\ \max}$, and $C^1_{dj\ \max} < C^0_{dj\ \max}$. Due to the monotonicity of x_j^* [see (C.2)–(C.4) and (C.6)] on R_{uj} and C_{dj}, $x^{*0}_{j\ \min} < x^{*1}_{j\ \min}$ and $x^{*1}_{j\ \max} < x^{*0}_{j\ \max}$, the range of values for x_j^* therefore also decreases and, typically, after two or three iterations, the optimum location for the corresponding via is determined.

The above example is extended to each of the other possible cases that can occur for segments i and k. Specifically,

a. i and k belong to either case (i) or (ii). Both R_{uj} and C_{dj} are precisely determined or, equivalently, $R_{uj\ \min} = R_{uj\ \max}$ and $C_{dj\ \min} = C_{dj\ \max}$. Consequently, the placement of both vias v_i and v_k is known and $x^{*0}_{j\ \min} = x^{*0}_{j\ \max} = x_j^*$. The placement of v_j is also determined within the first iteration.

b. i belongs to case (i) or (ii) and k belongs to case (iii). R_{uj} is precisely determined or, equivalently, $R_{uj\ \min} = R_{uj\ \max}$ and the placement of via v_i is known. Since v_i is placed and k belongs to case (iii), $C^0_{dj\ \min} < C^1_{dj\ \min}$ and $C^1_{dj\ \max} < C^0_{dj\ \max}$. The placement of via v_j converges faster, as only the placement of segment k remains unknown after the first iteration.

c. k belongs to case (i) or (ii) and i belongs to case (iii). C_{dj} is precisely determined or, equivalently, $C_{dj\ \min} = C_{dj\ \max}$ and the placement of via v_k is known. Since v_k is placed and i belongs to case (iii), $R^0_{uj\ \min} < R^1_{uj\ \min}$ and $R^1_{uj\ \max} < R^0_{uj\ \max}$. The placement of via v_j converges faster as only the placement of segment i remains unknown after the first iteration.

d. i belongs to any of the cases (i) to (iii) and k belongs to case (iv). R_{uj} is readily determined [cases (i) and (ii)] or converges, as described in the previous example, $R^0_{uj\ \min} < R^1_{uj\ \min}$ and $R^1_{uj\ \max} < R^0_{uj\ \max}$. As k belongs to case (iv), however, C_{dj} does not change as in the cases above. If the decrease in the upstream resistance is sufficient to determine x_j^* according to (C.1), v_j is marked as processed, otherwise v_j is marked as unprocessed and the algorithm continues to the next via. In the latter case, the placement approach is described by case (iv) of the heuristic.

e. k belongs to any of the cases (i) to (iii) and i belongs to case (iv). C_{dj} is readily determined [cases (i) and (ii)] or converges, as described in the previous example, implying $C^0_{dj\ \min} < C^1_{dj\ \min}$ and $C^1_{dj\ \max} < C^0_{dj\ \max}$. As i belongs to case (iv), however, R_{uj} does not change as in the aforementioned cases. Overall, if the decrease in the downstream capacitance is sufficient to determine x_j^* according to (C.1), v_j is marked as processed, otherwise v_j is marked as unprocessed and the algorithm continues to the next via. In the latter case, the placement approach is described by case (iv) of the heuristic.

f. Both i and k belong to case (iv). Therefore, both R_{uj} and C_{dj} cannot be bounded. Consequently, v_j is marked as unprocessed and the next via is processed. Alternatively, this sub-case degenerates to case (iv) of the heuristic presented in Chapter 10, Timing Optimization for Two-Terminal Interconnects.

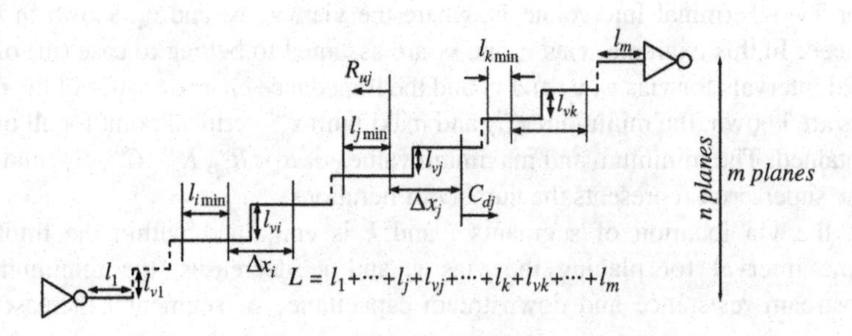

FIGURE C.1

Intertier interconnect consisting of m segments connecting two circuits located n tiers apart.

Appendix D: Proof of Condition for Via Placement of Multi-terminal Nets

In this appendix, a proof for necessary *condition 1* is provided (Fig. D.1).
Condition 1: If $r_j > r_{j+1}$, only a *type-1* move for v_j can reduce the delay of a tree.

Proof: Consider Fig. D.1, where the intertier via v_j (the solid square) can be placed in any direction d_e, d_s, and d_n within, respectively the interval l_{d_e}, l_{d_s}, and l_{d_n}. For the tree shown in Fig. D.1 and removing the terms that are independent of v_j, (11.1) is

$$T_w = \sum_{v_i \in U_{0j}} \sum_{s_p \in \overline{P_{spU_{ij}}}} w_{s_p} R_{uij}\left(c_{v_j} l_{v_j} + C_{d_j}\right) + \sum_{s_p \in P_{spv_j}} w_{s_p}\left(R_{u_j}\left(c_{v_j} l_{v_j} + C_{d_j}\right) + r_{v_j} l_{v_j} C_{d_j} + \frac{r_{v_j} c_{v_j} l_{v_j}^2}{2}\right), \tag{D.1}$$

where

$$C_{d_j} = \sum_{\forall k} C_{dv_j d_k} + c_{j+1}\left(l_{d_e} + l_{d_s} + l_{d_n}\right). \tag{D.2}$$

Suppose that a *type-2* move is required, shifting v_j by x toward the d_e direction (the dashed square). Expression (11.1) becomes

$$T'_w = \left[\begin{array}{l}\left(\sum_{v_i \in U_{ij}} \sum_{s_p \in \overline{P_{spU_{ij}}}} w_{s_p} R_{uij} + \sum_{s_p \in P_{spv_j}} w_{s_p} R_{u_j}\right)\left(c_{v_j} l_{v_j} + c_j x + C_{d_j}\right) + \sum_{s_p \in P_{spv_j}} w_{s_p} \\ \times\left[\left(r_j - r_{j+1}\right) x C_{d_j} + r_{j+1} l_{d_e}\left(C_{d_j} - \frac{1}{2} c_{j+1} l_{d_e}\right) + r_j x\left(c_{v_j} l_{v_j} + C_{d_j}\right) + \frac{1}{2}\left(r_{v_j} c_{v_j} l_{v_j}^2 + r_j c_j x^2\right)\right]\end{array}\right]. \tag{D.3}$$

For a *type-2* move to reduce the weighted delay of the tree, shifting v_j should decrease T_w, or, equivalently, $\Delta T = T'_w - T_w < 0$. Subtracting (D.1) from (D.3) yields

$$\Delta T = \left\{\begin{array}{l}\sum_{s_p \in P_{spv_j}} w_{s_p}\left[r_j x\left(c_{v_j} l_{v_j} + C_{d_j}\right) + R_{u_j} c_j x + r_{j+1} l_{d_e}\left(C_{d_j} - \frac{c_{j+1} l_{d_e}}{2}\right)\right. \\ \left. + \left(r_j - r_{j+1}\right) x C_{d_j} + \frac{r_j c_j x_j^2}{2}\right] + \sum_{v_i \in U_{ij}} \sum_{s_p \in \overline{P_{spU_{ij}}}} w_{s_p} R_{uij} c_j x\end{array}\right\}. \tag{D.4}$$

Since $r_j > r_{j+1}$ and $C_{dj} > \left(c_{j+1} l_{d_e}/2\right)$ from (D.2), (D.3) is always positive and a *type-2* move cannot reduce the delay of a tree.

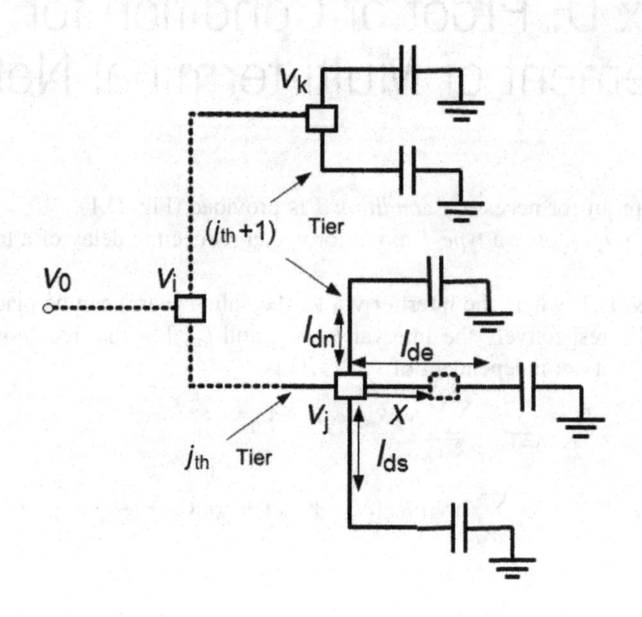

FIGURE D.1

Portion of an interconnect tree.

Appendix E: Correlation of WID Variations for Intratier Buffers

In this appendix, the uncorrelated WID variations and multilevel correlation model are used to characterize the skew distribution for several 3-D clock trees. The correlation between WID variations of buffers i and j within one tier (intratier buffers) is characterized by corr(i, j). Two types of correlations are considered, uncorrelated and multi-level spatial correlation. In the former type, WID variations of the buffers within one tier are considered, independent of the other tiers. For any pair of buffers i and j, corr(i,j) = 0. Consequently, from (17−37), the standard deviation of the delay of sink u due to WID variations is

$$\sigma^2_{D_u^{\mathrm{WID}}} = \sum_{j=1}^{n_u} \sigma^2_{d_i^{\mathrm{WID}}}. \tag{E.1}$$

Alternatively, the standard deviation is described differently if spatial correlations are considered. Based on the multilevel correlation model described in [640], a multi-level quad-tree partition is used, and the intra-die variations of a device are divided into l levels, as illustrated in Fig. E.1. At the lth level, there are 4^{l-1} regions. An independent variable is assigned to each region to represent a component of the WID variations of a device. The overall WID variations of buffer k are composed of the sum of these independent components at different levels,

$$\Delta L_{\mathrm{WID},k} = \sum_{\substack{1 \le i \le l \\ \text{region } r \text{ intersects } k}} \Delta L_{i,r}, \tag{E.2}$$

where $\Delta L_{i,r}$ is the random variable associated with the quad-tree at level i, region (i,r). The distribution of $\Delta L_{\mathrm{WID},k}$ is captured by the elementary circuit shown in Fig. 17.6. This distribution is obtained by assigning the same probability distribution to all of the random variables associated with a particular level, and by dividing the total intra-die variability among the different levels. Consequently, the spatial correlation among devices in the same tier is modeled to ensure that devices located close to each other are highly correlated, while those devices located at a large horizontal distance from each other exhibit low correlation (Fig. E.1).

The correlation between the WID variations of buffers i and j is described by the sum of the correlations at all of the levels,

$$\mathrm{corr}(i,j) = \frac{1}{l}\sum_{k=1}^{l} \mathrm{corr}_k(i,j), \tag{E.3}$$

where $\mathrm{corr}_k(i,j)$ is the correlation between buffers i and j at the kth level. As illustrated in Fig. E.1, assuming buffers i and j are located, respectively, in zones (k, region$_i$) and (k, region$_j$),

$$\mathrm{corr}_k(i,j) = \begin{cases} 1, & \text{if } (k, \text{region}_i) = (k, \text{region}_j) \\ 0, & \text{if } (k, \text{region}_i) \neq (k, \text{region}_j). \end{cases} \tag{E.4a and b}$$

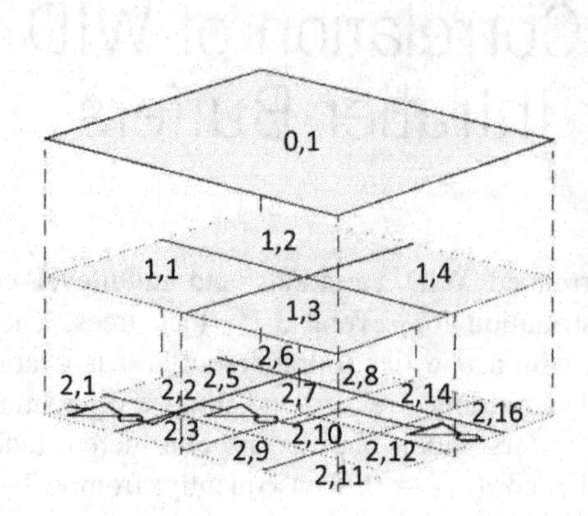

FIGURE E.1

Modeling spatial correlations using quad-tree partitioning [640].

Appendix F: Extension of the Proposed Model to Include Variations of Wires

The model described in Section 17.2 can be extended to include variations in the horizontal wires. This extended model is presented in this appendix. Consider the 3-D clock tree shown in Fig. 17.8, where the delay variations of a buffer stage $\Delta d_{\text{stage}(i)}$ include the variations due to the capacitance ΔC_{int} and resistance ΔR_{int} of the wires,

$$\begin{aligned}\Delta d_{\text{stage }(i)} = \Delta d_i + 0.69\left(R'_{b(i)} + \Delta R_{b(i)}\right)\Delta C_{\text{int}} + 0.38\left(R'_{\text{int}}\Delta C_{\text{int}} + \Delta R_{\text{int}}C'_{\text{int}} + \Delta R_{\text{int}}\Delta C_{\text{int}}\right) \\ + 0.69\left(R'_{\text{int}}\Delta C_{b(i+1)} + \Delta R_{\text{int}}C'_{b(i+1)} + \Delta R_{\text{int}}\Delta C_{b(i+1)}\right).\end{aligned} \tag{F.1}$$

According to the definition of Δd_i in (17.24), the term $0.69R'_{\text{int}}\left(\Delta C_{b(i+1)}\right)$ is included in Δd_{i+1}. Consequently, $\Delta d_{\text{stage}(i)}$ is rewritten as

$$\begin{aligned}\Delta d_{\text{stage}(i)} = \Delta d_i + 0.69\left(R'_{b(i)} + \Delta R_{b(i)}\right)\Delta C_{\text{int}} + 0.38\left(R'_{\text{int}}\Delta C_{\text{int}} + \Delta R_{\text{int}}C'_{\text{int}} + \Delta R_{\text{int}}\Delta C_{\text{int}}\right) \\ + 0.69\left(\Delta R_{\text{int}}C'_{b(i+1)} + \Delta R_{\text{int}}\Delta C_{b(i+1)}\right) = \Delta d_i + \Delta d_{\text{int}(i)},\end{aligned} \tag{F.2}$$

where delay variations due to the wires are denoted by $\Delta d_{\text{int}(i)}$.

As discussed in [637], since the variations of the characteristics of the metal wires are relatively low as compared with nominal values, the variations of the wire delay can be approximated by a first order Taylor series expansion without significant loss of accuracy. Similar to (17.26), $\Delta d_{\text{int}(i)}$ can be approximated as

$$d_{\text{int}(i)} \approx \sum_{p_j \in \vec{P}} \left(\left[\frac{\partial \Delta d_{\text{int}(i)}}{\partial R_{\text{int}}}\frac{\partial \Delta R_{\text{int}}}{\partial p_j}\right]_0 \Delta p_j + \left[\frac{\partial \Delta d_{\text{int}(i)}}{\partial C_{\text{int}}}\frac{\partial \Delta C_{\text{int}}}{\partial p_j}\right]_0 \Delta p_j\right), \tag{F.3}$$

where p_j is the jth parameter of the wire, and P is the vector of the parameters of those wires affected by process variations. For example, consider the variations in the width and thickness of the metal and the thickness of the ILD [637,638], $P(W_{\text{int}}, h_{\text{int}}, t_{\text{ILD}})$. Assuming these parameters are modeled by a Gaussian distribution and are independent of each other [637], the distribution of $\Delta d_{\text{int}(i)}$ can be approximated by a Gaussian distribution,

$$\Delta d_{\text{int}(i)} \sim \mathcal{N}\left(0, \sigma^2_{\text{dint}(i)}\right), \tag{F.4}$$

$$\sigma^2_{\text{dint}(i)} = \sum_{p_j \in \vec{P}} \left(\left[\frac{\partial \Delta d_{\text{int}(i)}}{\partial R_{int}}\frac{\partial \Delta R_{\text{int}}}{\partial p_j}\right]_0 + \left[\frac{\partial \Delta d_{\text{int}(i)}}{\partial C_{\text{int}}}\frac{\partial \Delta C_{\text{int}}}{\partial p_j}\right]_0\right)^2 \sigma^2_{p_j}, \tag{F.5}$$

$$R_{\text{int}} = \frac{\rho l}{h_{\text{int}} W_{\text{int}}}, \tag{F.6}$$

$$C_{\text{int}} = 2\left(C_g + C_c\right)l, \tag{F.7}$$

where ρ, l, h_{int}, and W_{int} are, respectively, the resistivity, length, thickness, and width of the interconnect, C_g includes both the ground and fringe capacitance, and C_c is the coupling capacitance. Expressions of C_g and C_c are available, respectively, in [252] and [766].

Table F.1 Parameters of the Horizontal Interconnects

Parameters	W_m [nm]	t_m [nm]	t_{ILD} [nm]
Nominal	430	1000	160
$3\sigma_{D2D}$	43	50	12
$3\sigma_{WID}$	21.5	25	6

Table F.2 Skew Variation in 3-D Circuits Considering Wire Variations

Topology Skew Variation	Multi-via Topology				Single-via Topology			
	$\sigma_{s1,2}$	$\sigma_{s1,3}$	$\sigma_{s1,4}$	$\sigma_{s1,5}$	$\sigma_{s1,2}$	$\sigma_{s1,3}$	$\sigma_{s1,4}$	$\sigma_{s1,5}$
Model [ps]	7.01	15.09	7.45	15.3	3.99	13.94	56.46	56.46
Specter [ps]	7.19	16.44	7.64	16.55	4.00	13.77	56.38	56.30
Error [%]	−3	−8	−2	−8	<1	1	<1	<1

Considering the delay variations caused by both the clock buffers and wires in (F.2), the skew variations $\Delta s_{u,v}$ includes two terms,

$$\Delta s_{u,v} = \Delta s_{b(u,v)} + \Delta s_{\text{int}(u,v)}. \tag{F.8}$$

The distribution of $\Delta s_{b(u,v)}$ is obtained from (17.24)–(17.45). The distribution of $\Delta s_{\text{int}(u,v)}$ is obtained from (17.30)–(17.45) by substituting $\Delta d_{\text{int}(i)}$ for Δd_i. Consequently, $\Delta s_{u,v}$ can be described by a Gaussian distribution,

$$\Delta s_{u,v} \sim \mathcal{N}\left(0, \sigma^2_{s_{b(u,v)}} + \sigma^2_{s_{\text{int}(u,v)}}\right). \tag{F.9}$$

The extended model is compared with Monte Carlo simulations including the variations of r_{int} and c_{int} in a π interconnect model. Based on the parameters in [637], the nominal value and standard deviation of the parameters of the wires are listed in Table F.1. The multi-via and single via trees described in Section 17.2 are used to verify the accuracy of the extended model. The results for the independent WID variations are reported in Table F.2, where the accuracy of the model including variations in the wires is reasonably high ($<8\%$ for a multi-via clock network topology).

Glossary of Terms

2-D	Two-dimensional
3-D	Three-dimensional
ADVP	Alternating direction TSV planning algorithm
ALD	Atomic layer deposition
a-Si	Amorphous-Si
ASIC	Application-specific integrated circuit
BCB	Benzocyclobutene
BEOL	Back-end-of-line
BOX	Buried oxide
C4	Controlled collapse chip connect
CAD	Computer-aided design
CBA	Combined bucket and 2-D array
CMOS	Complementary metal oxide semiconductor
CMP	Chemical mechanical planarization
CPD	Coarse pin distribution
CTE	Coefficient of thermal expansion
CVD	Chemical vapor deposition
D2D	Die-to-die
DC	Direct current
DME	Deffered-merge embedding
DPD	Detailed pin distribution
DRIE	Deep reactive ion etching
DVFS	Dynamic votlage and frequency scaling
EDP	Energy delay product
eDRAM	Embedded dynamic random access memory
ELO	Epitaxial lateral overgrowth
ESD	Electrostatic discharge
FDSOI	Fully depleted silicon-on-insulator
FEM	Finite element method
FEOL	Front-end-of-line
FPGA	Field programmable gate array
GSG	Ground-signal-ground
GSGSG	Ground-signal-ground-signal-ground
HPWL	Half perimeter wirelength
ICP	Inductively plasma enhanced
ILD	Interlayer dielectric
IoT	Internet of things
IP	Intellectual property
IPC	Instructions per cycle
ITRS	International technology roadmap for semiconductors
ITVPA	Interconnect tree via placement algorithm
JMOS	Junction MOS
KGD	Known good die
KOZ	Keep out zone

LB	Logic block
LDPC	Low density parity check
LPCVD	Low pressure chemical vapor deposition
LTO	Low temperature oxide
MEMS	Micro-electro-mechanical systems
MIM	Metal insulator metal
MITLL	MIT Lincoln Laboratory
MMM	Method of means and medians
MOCVD	Metalorganic chemical vapor deposition
MOM	Method of moments
MOSFET	Metal oxide semiconductor field effect transistor
MST	Minimum spanning tree
NAPC	Normalized average power consumption
NCP	Non-conductive particle paste
NLP	Nonlinear programming
NoC	Network-on-chip
OPRSOC	Open RISC platform system-on-chip
OSAT	Outsourced semiconductor assembly and test
PCB	Printed circuit board
PE	Processing element
PEALD	Plasma enhanced atomic layer deposition
PECVD	Plasma enhanced chemical vapor deposition
PI	Performance improvement
PL	Performance loss
PLL	Phase locked loop
PPA	Performance, power, area
PRNG	Pseudorandom number generator
PSG	Phosphosilicate glass
PTM	Predictive technology model
PVD	Physical vapor deposition
PWB	Printed wire board
RDL	Redistribution layer
RF	Radio frequency
RIE	Reactive ion etching
RISC	Reduced instruction set computer
RMST	Rectilinear minimum spanning tree
SA	Simulated annealing
SB	Switch box
SCSVPA	Single critical sink interconnect tree via placement algorithm
SEG	Selective epitaxial growth
SiP	System-in-package
SoC	System-on-chip
SOI	Silicon-on-insulator
SoP	System-on-package
SOS	Silicon-on-sapphire
SPICE	Simulation program with integrated circuit emphasis
SST	Steady-state temperature
SW/HW	Software / Hardware

TAB	Tape adhesive bonding
TCG	Transition closure graph
TEOS	Tetraethylorthosilicate
TFC	TSV fault-tolerant component
TFT	Thin film transistors
TGV	Through glass via
TPR	Three-dimensional place and route
TSV	Through silicon via
TTSV	Thermal through silicon via
TTVPA	Two terminal via placement algorithm
VPR	Versatile place and route
VRM	Voltage regulation module
WDM	Wavelength division multiplexing
WID	Within die

References

[1] R. N. Noyce, "Microelectronics," *Scientific American*, Vol. 237, No. 3, pp. 62–69, September 1977.

[2] C. Weiner, "How the Transistor Emerged," *IEEE Spectrum*, Vol. 10, No. 1, pp. 24–33, January 1973.

[3] C. Mack, "The Multiple Lives of Moore's Law," *IEEE Spectrum*, Vol. 52, No. 4, pp. 31–37, April 2015.

[4] L. Su, "Architecting the Future Through Heterogeneous Computing," *IEEE Solid-State Circuits Magazine*, Vol. 5, No. 2, pp. 63–72, Spring 2013.

[5] X. Wu *et al.*, "Stacked 3-D Fin-CMOS Technology," *IEEE Electron Device Letters*, Vol. 26, No. 6, pp. 416–418, June 2005.

[6] N. Planes *et al.*, "28nm FDSOI Technology Platform for High-Speed Low-Voltage Digital Applications," *Proceedings of the IEEE Symposium on VLSI Technology*, pp. 133–134, June 2012.

[7] R. H. Dennard, "Past Progress and Future Challenges in LSI Technology: From DRAM and Scaling to Ultra-Low-Power CMOS," *IEEE Solid-State Circuits Magazine*, Vol. 7, No. 2, pp. 29–38, Spring 2015.

[8] S. Borkar and A. A. Chien, "The Future of Microprocessors," *Communications of the ACM*, Vol. 54, No. 5, pp. 67–77, May 2011.

[9] [Online] Gartner, Available: http://www.gartner.com/newsroom/id/2954317.

[10] [Online] *International Technology Roadmap for Semiconductors ITRS*, Available: http://www.itrs2.net/uploads/4/9/7/7/49775221/irc-itrs-mtm-v2_3.pdf.

[11] P. Kogge and J. Shalf, "Exascale Computing Trends: Adjusting to the "New Normal" for Computer Architecture," *Computing in Science & Engineering*, Vol. 15, No. 6, pp. 16–26, November/December 2013.

[12] M. D. Assunçãoa *et al.*, "Big Data Computing and Clouds: Trends and Future Directions," *Journal of Parallel and Distributed Computing*, Vol. 79–80, pp. 3–15, May 2015.

[13] K. C. Chun *et al.*, "A Scaling Roadmap and Performance Evaluation of In-Plane and Perpendicular MTJ Based STT-MRAMs for High-Density Cache Memory," *IEEE Journal of Solid-State Circuits*, Vol. 48, No. 22, pp. 598–610, February 2013.

[14] G. Lapidus, "Transistor Family History," *IEEE Spectrum*, pp. 34–35, January 1977.

[15] K. A. Chen, "Computer Aided Memory Design Using Transmission Line Models," *IEEE Transactions on Computers*, Vol. C-17, No. 7, pp. 640–648, July 1968.

[16] R. N. Noyce, "Large Scale Integration: What is Yet to Come," *Science Magazine*, Vol. 195, No. 4283, pp. 1102–1106, March 1977.

[17] K. C. Saraswat and F. Mohammadi, "Effect of Scaling of Interconnections on the Time Delay of VLSI Circuits," *IEEE Transactions on Electron Devices*, Vol. ED-29, No. 4, April 1982.

[18] [Online] International Technology Roadmap for Semiconductors ITRS, 2005 Edition. Available: http://www.itrs2.net.

[19] C. Akrout *et al.*, "A 480-MHz RISC Microprocessor in a 0.12-μm L_{eff} CMOS Technology with Copper Interconnects," *IEEE Journal of Solid-State Circuits*, Vol. 33, No. 11, pp. 1609–1616, November 1998.

[20] D. H. Allen *et al.*, "A 0.2 μm 1.8 V SOI 550 MHz 64 b Power PC Microprocessor with Copper Interconnects," *Proceedings of the IEEE International Solid-State Circuits Conference*, pp. 438–439, February 1999.

[21] M. Naik *et al.*, "Process Integration of Double Level Copper-Low k (k = 2.8) Interconnect," *Proceedings of the IEEE International Interconnect Technology Conference*, pp. 181–183, May 1999.

[22] P. Zarkesh-Ha *et al.*, "The Impact of Cu/Low k on Chip Performance," *Proceedings of the IEEE International ASIC/SoC Conference*, pp. 257–261, September 1999.

[23] Y. Takao *et al.*, "A 0.11 μm Technology with Copper and Very-Low-k Interconnects for High Performance System-on-Chip Cores," *Proceedings of the IEEE International Electron Device Meeting*, pp. 559–562, December 2000.

[24] J. D. Meindl, "Interconnect Opportunities for Gigascale Integration," *IEEE Micro*, Vol. 23, No. 3, pp. 28–35, May/June 2003.

[25] R. Venkatesan, J. A. Davis, K. A. Bowman, and J. D. Meindl, "Optimal *n*-Tier Interconnect Architectures for Gigascale Integration (GSI)," *IEEE Transactions on Very Large Integration (VLSI) Systems*, Vol. 9, No. 6, pp. 899–912, December 2001.

[26] K. M. Lepak, I. Luwandi, and L. He, "Simultaneous Shield Insertion and Net Ordering Under Explicit *RLC* Noise Constraint," *Proceedings of the IEEE/ACM Design Automation Conference*, pp. 199–202, June 2001.

[27] P. Fishburn, "Shaping a VLSI Wire to Minimize Elmore Delay," *Proceedings of the IEEE European Design and Test Conference*, pp. 244–251, March 1997.

[28] M. A. El-Moursy and E. G. Friedman, "Exponentially Tapered H-Tree Clock Distribution Networks," *IEEE Transactions on Very Large Scale Integration (VLSI) Systems*, Vol. 13, No. 8, pp. 971–975, August 2005.

[29] H. B. Bakoglu and J. D. Meindl, "Optimal Interconnection Circuits for VLSI," *IEEE Transactions on Electron Devices*, Vol. ED-32, No. 5, pp. 903–909, May 1985.

[30] Y. I. Ismail, E. G. Friedman, and J. L. Neves, "Exploiting On-Chip Inductance in High Speed Clock Distribution Networks," *IEEE Transactions on Very Large Scale Integration (VLSI) Systems*, Vol. 9, No. 6, pp. 963–973, December 2001.

[31] V. Adler and E. G. Friedman, "Uniform Repeater Insertion in *RC* Trees," *IEEE Transactions on Circuits and Systems I: Fundamental Theory and Applications*, Vol. 47, No. 10, pp. 1515–1523, October 2000.

[32] M. Ghoneima *et al.*, "Reducing the Effective Coupling Capacitance in Buses Using Threshold Voltage Adjustment Techniques," *IEEE Transactions on Circuits and Systems I: Fundamental Theory and Applications*, Vol. 53, No. 9, pp. 1928–1933, September 2006.

[33] M. R. Stan and W. P. Burleson, "Bus-Invert Coding for Low-Power I/O," *IEEE Transactions on Very Large Scale Integration (VLSI) Systems*, Vol. 3, No. 1, pp. 49–58, March 1998.

[34] R. Bashirullah, L. Wentai, and R. K. Cavin, III, "Current-Mode Signaling in Deep Submicrometer Global Interconnects," *IEEE Transactions on Very Large Scale Integration (VLSI) Systems*, Vol. 11, No. 3, pp. 406–417, June 2003.

[35] V. V. Deodhar and J. A. Davis, "Optimization of Throughput Performance for Low-Power VLSI Interconnects," *IEEE Transactions on Very Large Scale Integration (VLSI) Systems*, Vol. 13, No. 3, pp. 308–318, March 2005.

[36] H. Zhang, V. George, and J. M. Rabaey, "Low-Swing On-Chip Signaling Techniques: Effectiveness and Robustness," *IEEE Transactions on Very Large Scale Integration (VLSI) Systems*, Vol. 8, No. 3, pp. 264–272, June 2000.

[37] L. Benini and G. De Micheli, "Networks on Chip: A New SoC Paradigm," *IEEE Computer*, Vol. 31, No. 1, pp. 70–78, January 2002.

[38] M. Haurylau *et al.*, "On-Chip Optical Interconnect Roadmap: Challenges and Critical Directions," *IEEE Journal of Selected Topics on Quantum Electronics*, Vol. 12, No. 6, pp. 1699–1705, November/December 2006.

[39] G. Chen *et al.*, "On-Chip Copper-Based vs. Optical Interconnects: Delay Uncertainty, Latency, Power, and Bandwidth Density Comparative Predictions," *Proceedings of the IEEE International Interconnect Technology Conference*, pp. 39–41, June 2006.

[40] G. Chen *et al.*, "Predictions of CMOS Compatible On-Chip Optical Interconnect," *Integration, The VLSI Journal*, Vol. 40, No. 4, pp. 434–446, July 2007.

[41] W. Shockley, "Semiconductive Wafer and Method of Making the Same," U.S. Patent No. 3,044,909, July 1962.

[42] M. G. Smith and S. Emanuel, "Methods of Making Thru-Connections in Semiconductor Wafers," U.S. Patent No. 3,343,256, September 1967.

[43] [Online], National Academy of Sciences. Available: http://www.nasonline.org/publications/biographical-memoirs/memoir-pdfs/shockley-w.pdf (accessed in September 2016).
[44] J. H. Lau, "Evolution, Challenges, and Outlook of TSV, 3D Integration and 3D Silicon Integration," *Proceedings of the IEEE International Symposium on Advanced Packaging Materials*, pp. 462–488, October 2011.
[45] G. T. Goele *et al.,* "Vertical Single Gate CMOS Inverters on Laser-Processed Multilayer Substrates," *Proceedings of the IEEE International Electron Device Meetings*, Vol. 27, pp. 554–556, December 1981.
[46] J. F. Gibbons and K. F. Lee, "One-Gate-Wide CMOS Inverter on Laser-Recrystallized Polysilicon," *IEEE Electron Device Letters*, Vol. EDL-1, No. 6, pp. 117–118, June 1980.
[47] Y. Akasaka, "Three-Dimensional IC Trends," *Proceedings of the IEEE*, Vol. 74, No. 12, pp. 1703–1714, December 1986.
[48] R. Thom, "High Density Infrared Detector Arrays," U.S. Patent No. 4,039,833, February 1977.
[49] W. R. Davis *et al.,* "Demystifying 3D ICs: The Pros and Cons of Going Vertical," *IEEE Design and Test of Computers*, Vol. 22, No. 6, pp. 498–510, November/December 2005.
[50] J. W. Joyner, P. Zarkesh-Ha, J. A. Davis, and J. D. Meindl, "A Three-Dimensional Stochastic Wire-Length Distribution for Variable Separation of Strata," *Proceedings of the IEEE International Interconnect Technology Conference*, pp. 126–128, June 2000.
[51] M. Koyanagi *et al.,* "Future System-on-Silicon LSI Chips," *IEEE Micro*, Vol. 18, No. 4, pp. 17–22, July/August 1998.
[52] V. K. Jain, S. Bhanja, G. H. Chapman, and L. Doddannagari, "A Highly Reconfigurable Computing Array: DSP Plane of a 3D Heterogeneous SoC," *Proceedings of the IEEE International SoC Conference*, pp. 243–246, September 2005.
[53] J. Jeddeloh and B. Keeth, "Hybrid Memory Cube New DRAM Architecture Increases Density and Performance," *Proceedings of the IEEE Symposium on VLSI Technology*, pp. 87–88, June 2012.
[54] J. Kim and Y. Kim, "HBM: Memory Solution for Bandwidth-Hungry Processors," *Proceedings of the IEEE Hot Chips Symposium*, pp. 1–24, August 2014.
[55] T.-Y. Oh *et al.,* "A 7 Gb/s/pin 1 Gbit GDDR5 SDRAM With 2.5 ns Bank to Bank Active Time and No Bank Group Restriction," *IEEE Journal of Solid-State Circuits*, Vol. 46, No. 1, pp. 107–118, January 2011.
[56] W.-S. Kwon *et al.,* "Enabling a Manufacturable 3D Technologies and Ecosystem Using 28nm FPGA with Stack Silicon Interconnect Technology," *Proceedings of the International Symposium on Microelectronics*, pp. 217–222, September/October 2013.
[57] V. H. Nguyen and P. Christie, "The Impact of Interstratal Interconnect Density on the Performance of Three-Dimensional Integrated Circuits," *Proceedings of the IEEE/ACM International Workshop on System Level Interconnect Prediction*, pp. 73–77, April 2005.
[58] I. Savidis and E. G. Friedman, "Electrical Modeling and Characterization of 3-D Vias," *Proceedings of the IEEE International Symposium on Circuits and Systems*, pp. 784–787, May 2008.
[59] R. R. Tummala *et al.,* "The SOP for Miniaturized, Mixed-Signal Computing, Communication, and Consumer Systems of the Next Decade," *IEEE Transactions on Advanced Packaging*, Vol. 27, No. 2, pp. 250–267, May 2004.
[60] R. R. Tummala, "SOP: What is it and Why? A New Microsystem-Integration Technology Paradigm-Moore's Law for System Integration of Miniaturized Convergent Systems of the Next Decade," *IEEE Transactions on Advanced Packaging*, Vol. 27, No. 2, pp. 241–249, May 2004.
[61] V. Sundaram *et al.,* "Next-Generation Microvia and Global Wiring Technologies for SOP," *IEEE Transactions on Advanced Packaging*, Vol. 27, No. 2, pp. 315–325, May 2004.
[62] [Online] Hybrid Memory Cube. Available: http://www.hybridmemorycube.org/.
[63] [Online] High Bandwidth Memory. Available: http://www.amd.com/en-us/innovations/software-technologies/hbm.
[64] [Online] Xilinx. Available: http://www.xilinx.com/products/silicon-devices/3dic.html.

[65] H. P. Hofstee, "Future Microprocessors and Off-Chip SOP Interconnect," *IEEE Transactions on Advanced Packaging*, Vol. 27, No. 2, pp. 301–303, May 2004.
[66] S. F. Al-Sarawi, D. Abbott, and P. D. Franzon, "A Review of 3-D Packaging Technology," *IEEE Transactions on Components, Packaging, and Manufacturing Technology–Part B*, Vol. 21, No. 1, pp. 2–14, February 1998.
[67] P. Garrou, "Future ICs Go Vertical," *Semiconductor International*, [online], February 2005
[68] M. Karnezos, "3-D Packaging: Where All Technologies Come Together," *Proceedings of the IEEE/SEMI International Electronics Manufacturing Technology Symposium*, pp. 64–67, July 2004.
[69] E. Beyne, "The Rise of the 3rd Dimension for System Integration," *Proceedings of the IEEE International Interconnect Technology Conference*, pp. 1–5, June 2006.
[70] C. Fox III and M. Warren, "High-Density Electronic Package Comprising Stacked Sub-Modules Which are Electrically Interconnected by Solder-Filled Vias," U.S. Patent No. 5,128,831, July 1992.
[71] I. Miyano *et al.*, "Fabrication and Thermal Analysis of 3-D Located LSI Packages," *Proceedings of the European Hybrid Microelectronics Conference*, pp. 184–191, June 1993.
[72] J. Miettinen, M. Mantysalo, K. Kaija, and E. O. Ristolainen, "System Design Issues for 3D System-in-Package (SiP)," *Proceedings of the IEEE Electronic Components and Technology Conference*, pp. 610–615, June 2004.
[73] S. Stoukach *et al.*, "3D-SiP Integration for Autonomous Sensor Nodes," *Proceedings of the IEEE Electronic Components and Technology Conference*, pp. 404–408, June 2006.
[74] N. Tamaka *et al.*, "Low-Cost Through-Hole Electrode Interconnection for 3D-SiP Using Room-Temperature Bonding," *Proceedings of the IEEE Electronic Components and Technology Conference*, pp. 814–818, June 2006.
[75] W. J. Howell *et al.*, "Area Array Solder Interconnection Technology for the Three-Dimensional Silicon Cube," *Proceedings of the IEEE Electronic Components and Technology Conference*, pp. 1174–1178, May 1995.
[76] K. Hatada, H. Fujimoto, T. Kawakita, and T. Ochi, "A New LSI Bonding Technology 'Micron Bump Bonding Assembly Technology'," *Proceedings of the IEEE International Electronic Manufacturing Technology Symposium*, pp. 23–27, October 1988.
[77] J.-C. Souriau, O. Lignier, M. Charrier, and G. Poupon, "Wafer Level of 3D System in Package for RF and Data Applications," *Proceedings of the IEEE Electronic Components and Technology Conference*, pp. 356–361, June 2005.
[78] K. Tanida *et al.*, "Ultra-High-Density 3D Chip Stacking Technology," *Proceedings of the IEEE Electronic Components and Technology Conference*, pp. 1084–1089, May 2003.
[79] J. A. Minahan, A. Pepe, R. Some, and M. Suer, "The 3D Stack in Short Form," *Proceedings of the IEEE Electronic Components and Technology Conference*, pp. 340–344, May 1992.
[80] S. P. Larcombe, J. M. Stern, P. A. Ivey, and L. Seed, "Utilizing a Low Cost 3D Packaging Technology for Consumer Applications," *IEEE Transactions on Consumer Electronics*, Vol. 41, No. 4, pp. 1095–1102, November 1995.
[81] J. M. Stem *et al.*, "An Ultra Compact, Low-Cost, Complete Image-Processing System," *Proceedings of the IEEE International Solid-State Circuits Conference*, pp. 230–231, February 1995.
[82] J. U. Knickerbocker *et al.*, "Development of Next-Generation System-on-Package (SOP) Technology Based on Silicon Carriers with Fine Pitch Interconnects," *IBM Journal of Research and Development*, Vol. 49, No. 4/5, pp. 725–753, July/September 2005.
[83] J. H. Lau *et al.*, "Low-Cost TSH (Through-Silicon Hole) Interposers for 3D IC Integration," *Proceedings of the Electronic Components and Technology Conference*, pp. 290–296, May 2014.
[84] K. Ruhmer, "Lithography Challenges for 2.5D Interposer Manufacturing," *Proceedings of the Electronic Components and Technology Conference*, pp. 523–527, May 2014.

[85] A. Yu *et al.*, "Development of Fine Pitch Solder Microbumps for 3D Chip Stacking," *Proceedings of the IEEE Electronics Packaging Technology Conference*, pp. 350–354, December 2008.
[86] J.-S. Kim *et al.*, "A 1.2 V 12.8 GB/s 2 Gb Mobile Wide-I/O DRAM With 4 128 I/Os Using TSV Based Stacking," *IEEE Journal of Solid-State Circuits*, Vol. 47, No. 1, pp. 107–116, January 2012.
[87] B. Sawyer *et al.*, "Modeling, Design, Fabrication and Characterization of First Large 2.5D Glass Interposer as a Superior Alternative to Silicon and Organic Interposers at 50 Micron Bump Pitch," *Proceedings of the Electronic Components and Technology Conference*, pp. 742–747, May 2014.
[88] J. Keech *et al.*, "Fabrication of 3D-IC Interposers," *Proceedings of the Electronic Components and Technology Conference*, pp. 1829-1833, May 2013.
[89] V. Sundaran *et al.*, "Low Cost, High Performance, and High Reliability 2.5D Silicon Interposer," *Proceedings of the Electronic Components and Technology Conference*, pp. 342–347, May 2013.
[90] S. Kuramochi, S. Koiwa, K. Suzuki, and Y. Fukuoka, "Cost Effective Interposer for Advanced Electronic Packages," *Proceedings of the Electronic Components and Technology Conference*, pp. 1673–1678, May 2014.
[91] D.-C. Hu *et al.*, "Embedded Glass Interposer for Heterogeneous Multi-Chip Integration," *Proceedings of the Electronic Components and Technology Conference*, pp. 314–317, May 2015.
[92] S. Goodwin *et al.*, "Process Integration, Improvements, and Testing of Si Interposers for Embedded Computing Applications," *Proceedings of the Electronic Components and Technology Conference*, pp. 8–12, May 2014.
[93] W. Flack, G. Kenyon, and M. Ranjan, "Large Area Interposer Lithography," *Proceedings of the Electronic Components and Technology Conference*, pp. 26–32, May 2014.
[94] J. M. Yook, D. Kim, and J. C. Kim, "High Performance IPDs (Integrated Passive Devices) and TGV (Through Glass Via) Interposer Technology Using the Photosensitive Glass," *Proceedings of the Electronic Components and Technology Conference*, pp. 41–46, May 2014.
[95] A. Shorey *et al.*, "Advancements in Fabrication of Glass Interposers," *Proceedings of the Electronic Components and Technology Conference*, pp. 20–25, May 2014.
[96] P. Batude *et al.*, "3D Sequential Integration Opportunities and Technology Optimization," *Proceedings of the IEEE International Interconnect Technology Conference*, pp. 373–376, May 2014.
[97] M. W. Geis, D. C. Flanders, D. A. Antoniadis, and H. I. Smith, "Crystalline Silicon on Insulators by Graphoepitaxy," *Proceedings of the IEEE International Electron Devices Meeting*, pp. 210–212, December 1979.
[98] S. Akiyama *et al.*, "Multilayer CMOS Device Fabricated on Laser Recrystallized Silicon Islands," *Proceedings of the IEEE International Electron Devices Meeting*, pp. 352–355, December 1983.
[99] S. Kawamura *et al.*, "Three-Dimensional CMOS IC's Fabricated by Using Beam Recrystallization," *IEEE Electron Device Letters*, Vol. EDL-4, No. 10, pp. 366–368, October 1983.
[100] K. Sugahara *et al.*, "SOI/SOI/Bulk-Si Triple-Level Structure for Three-Dimensional Devices," *IEEE Electron Device Letters*, Vol. EDL-7, No. 3, pp. 193–194, March 1986.
[101] K. F. Lee, J. F. Gibbons, and K. C. Saraswat, "Thin Film MOSFET's Fabricated in Laser-Annealed Polycrystalline Silicon," *Applied Physics Letters*, Vol. 35, No. 2, pp. 173–175, July 1979.
[102] H. Hazama *et al.*, "Application of E-beam Recrystallization to Three-Layer Image Processor Fabrication," *IEEE Transactions on Electron Devices*, Vol. 38, No. 1, pp. 47–54, January 1991.
[103] V. Subramanian and K. C. Saraswat, "High-Performance Germanium-Seeded Laterally Crystallized TFT's for Vertical Device Integration," *IEEE Transactions on Electron Devices*, Vol. 45, No. 9, pp. 1934–1939, September 1998.
[104] V. W. Chan, P. C. H. Chan, and M. Chan, "Three-Dimensional CMOS Integrated Circuits on Large Grain Polysilicon Films," *Proceedings of the IEEE International Electron Devices Meeting*, pp. 161–164, December 2000.

[105] G. W. Neudeck, S. Pae, J. P. Denton, and T. C. Su, "Multiple Layers of Silicon-on-Insulator for Nanostructure Devices," *Journal of Vacuum Science Technology B*, Vol. 17, No. 3, pp. 994–998, May/June 1999.

[106] N. Hirashita, T. Katoh, and H. Onoda, "Si-Gate CMOS Devices on a Si Lateral Solid-Phase Epitaxial Layer," *IEEE Transactions on Electron Devices*, Vol. 36, No. 3, pp. 548–552, March 1989.

[107] X. Lin, S. Zhang, X. Wu, and M. Chan, "Local Clustering 3-D Stacked CMOS Technology for Interconnect Loading Reduction," *IEEE Transactions on Electron Devices*, Vol. 53, No. 6, pp. 1405–1410, June 2006.

[108] H.-S. P. Wong, K. K. Chan, and Y. Tuar, "Self-Aligned (Top and Bottom) Double-Gate MOSFET with a 25 nm Thick Silicon Channel," *Proceedings of the IEEE International Electron Devices Meeting*, pp. 427–430, December 1997.

[109] R. S. Shenoy and K. C. Saraswat, "Novel Process for Fully Self-Aligned Planar Ultrathin Body Double-Gate FET," *Proceedings of the IEEE International Silicon on Insulator Conference*, pp. 190–191, October 2004.

[110] P. Batude *et al.*, "Advances, Challenges, and Opportunities in 3D CMOS Sequential Integration," *Proceedings of the IEEE International Electron Devices Meeting*, pp. 151–154, December 2011.

[111] P. Batude *et al.*, "Advances in 3D CMOS Sequential Integration," *Proceedings of the IEEE International Electron Devices Meeting*, pp. 345–348, December 2009.

[112] P. Batude *et al.*, "GeOI and SOI 3D Monolithic Cells Integrations for High Density Applications," *Proceedings of the Symposium on VLSI Technology*, pp. 166–167, June 2009.

[113] B. Yu *et al.*, "FinFet Scaling to 10 nm Gate Length," *Proceedings of the IEEE International Electron Devices Meeting*, pp. 251–254, December 2002.

[114] X. Wu *et al.*, "A Three-Dimensional Stacked Fin-CMOS Technology for High-Density ULSI Circuits," *IEEE Transactions on Electron Devices*, Vol. 52, No. 9, pp. 1998–2003, September 2005.

[115] P. Batude *et al.*, "3D CMOS Integration: Introduction of Dynamic Coupling and Application to Compact and Robust 4T SRAM," *Proceedings of the IEEE International Conference on Integrated Circuit Design and Technology*, pp. 281–284, June 2008.

[116] A. Fan, A. Rahman, and R. Reif, "Copper Wafer Bonding," *Electrochemical and Solid-State Letters*, Vol. 2, No. 10, pp. 534–536, October 1999.

[117] R. Reif, A. Fan, K. N. Chen, and S. Das, "Fabrication Technologies for Three-Dimensional Integrated Circuits," *Proceedings of the IEEE International Symposium on Quality Electronic Design*, pp. 33–37, March 2002.

[118] R. J. Gutmann *et al.*, "Three-Dimensional (3D) ICs: A Technology Platform for Integrated Systems and Opportunities for New Polymeric Adhesives," *Proceedings of the IEEE International Conference on Polymers and Adhesives in Microelectronics and Photonics*, pp. 173–180, October 2001.

[119] J.-Q. Lu *et al.*, "Stacked Chip-to-Chip Interconnections Using Wafer Bonding Technology with Dielectric Bonding Glues," *Proceedings of the IEEE International Interconnect Technology Conference*, pp. 219–221, June 2001.

[120] A. Klumpp, R. Merkel, R. Wieland, and P. Ramm, "Chip-to-Wafer Stacking Technology for 3D System Integration," *Proceedings of the IEEE Electronic Components and Technology Conference*, pp. 1080–1083, May 2003.

[121] C. A. Bower *et al.*, "High Density Vertical Interconnect for 3-D Integration of Silicon Integrated Circuits," *Proceedings of the IEEE Electronic Components and Technology Conference*, pp. 399–403, June 2006.

[122] T. Fukushima, Y. Yamada, H. Kikuchi, and M. Koyanagi, "New Three-Dimensional Integration Using Self-Assembly Technique," *Proceedings of the IEEE International Electron Devices Meeting*, pp. 348–351, December 2005.

[123] A. W. Topol *et al.*, "Enabling SOI-Based Assembly Technology for Three-Dimensional (3D) Integrated Circuits (ICs)," *Proceedings of the IEEE International Electron Devices Meeting*, pp. 352–355, December 2005.

[124] S. Tiwari *et al.*, "Three-Dimensional Integration for Silicon Electronics," *Proceedings of the IEEE Lester Eastman Conference on High Performance Devices*, pp. 24–33, August 2002.

[125] A. C. Fischer *et al.*, "Very High Aspect Ratio Through-Silicon Vias (TSVs) Fabricated Using Automated Magnetic Assembly of Nickel Wires," *Journal of Micromechanics and Microengineering*, Vol. 22, No. 10, August 2012.

[126] D. Malta *et al.*, "Optimization of Chemistry and Process Parameters for Void-Free Copper Electroplating of High Aspect Ratio Through-Silicon-Vias for 3D Integration," *Proceedings of the IEEE Electronic Components and Technology Conference*, pp. 1301–1306, May 2009.

[127] R. Beica, C. Sharbono, and T. Ritzdorf, "Through Silicon Via Copper Electrodeposition for 3D Integration," *Proceedings of the IEEE Electronic Components and Technology Conference*, pp. 577–583, May 2008.

[128] S. Skordas *et al.*, "Wafer-Scale Oxide Fusion Bonding and Wafer Thinning Development for 3D Systems Integration," *Proceedings of the IEEE International Workshop on Low Temperature Bonding for 3D Integration*, pp. 203–208, May 2012.

[129] A. R. Mirza, "One Micron Precision, Wafer-Level Aligned Bonding for Interconnect, MEMS and Packaging Applications," *Proceedings of the IEEE Electronic Components and Technology Conference*, pp. 676–680, May 2000.

[130] K. Sakuma *et al.*, "Bonding Technologies for Chip Level and Wafer Level 3D Integration," *Proceedings of the IEEE Electronic Components and Technology Conference*, pp. 647–654, May 2014.

[131] Y. H. Yu *et al.*, "Process Development to Enable 3D IC Multi-Tier Die Bond for 20 μm Pitch and Beyond," *Proceedings of the IEEE Electronic Components and Technology Conference*, pp. 572–575, May 2014.

[132] Z. Li, Y. Li, and J. Xie, "Design and Package Technology Development of Face-to-Face Die Stacking as a Low Cost Alternative for 3D IC Integration," *Proceedings of the IEEE Electronic Components and Technology Conference*, pp. 338–341, May 2014.

[133] N. Watanabe, T. Kojima, and T. Asano, "Wafer-Level Compliant Bump for Three-Dimensional LSI with High-Density Area Bump Connections," *Proceedings of the IEEE International Electron Devices Meeting*, pp. 671–674, December 2005.

[134] P. Batra *et al.*, "Three-Dimensional Wafer Stacking Using Cu TSV Integrated with 45 nm High Performance SOI-CMOS Embedded DRAM Technology," *Proceedings of the IEEE SOI-3D-Subthreshold Microelectronics Technology Unified Conference*, October 2013.

[135] R. N. Vrtis, K. A. Heap, W. F. Burgoyne, and L. M. Roberson, "Poly (Arylene Ethers) as Low Dielectric Constant Materials for ULSI Interconnect Applications," *Proceedings of the Materials Research Society Symposium*, Vol. 443, p. 171, December 1997.

[136] N. H. Hendricks, "The Status of Low-k Materials Development," *Proceedings of the IEEE International VLSI Multilevel Interconnect Conference*, p. 17, June 2000.

[137] S. F. Hahn, S. J. Martin, and M. L. McKelvy, "Thermally Induced Polymerization of an Arylvinylbenzocyclobenzene Monomer," *Macromolecules*, Vol. 25, No. 5, pp. 1539–1545, September 1992.

[138] D. Oben, P. Weigand, M. J. Shapiro, and S. A. Cohen, "Influence of the Cure Process on the Properties of Hydrogen Silsesquioxane Spin on Glass," *Proceedings of the Materials Research Society Symposium*, Vol. 443, p. 195, December 1997.

[139] T.-M. Lu and J. A. Moor, "Vapor Deposition of Low-Dielectric-Constant Polymeric Thin Films," *Materials Research Bulletin*, Vol. 22, No. 10, pp. 28–32, October 1997.

[140] [Online] JEDEC Solid State Technology Association, Electronic Industry Association. Available: http://www.jedec.org/home.

[141] D. J. Na *et al.*, "TSV MEOL (Mid End of Line) and Packaging Technology of Mobile 3D-IC Stacking," *Proceedings of the IEEE Electronic Components and Technology Conference*, pp. 596–600, May 2014.

[142] E. Culurciello and A. G. Andreou, "Capacitive Inter-Chip Data and Power Transfer for 3-D VLSI," *IEEE Transactions on Circuits and Systems II: Express Briefs*, Vol. 53, No. 12, pp. 1348–1352, December 2006.

[143] S. A. Kühn, M. B. Kleiner, R. Thewes, and W. Weber, "Vertical Signal Transmission in Three-Dimensional Integrated Circuits by Capacitive Coupling," *Proceedings of the IEEE International Symposium on Circuits and Systems*, Vol. 1, pp. 37–40, May 1995.

[144] A. Fazzi *et al.*, "3-D Capacitive Interconnections for Wafer-Level and Die-Level Assembly," *IEEE Journal of Solid-State Circuits*, Vol. 42, No. 10, pp. 2270–2282, October 2007.

[145] J. Xu *et al.*, "AC Coupled Interconnect for Dense 3-D ICs," *IEEE Transactions on Nuclear Science*, Vol. 51, No. 5, pp. 2156–2160, October 2004.

[146] R. S. Patti, "Three-Dimensional Integrated Circuits and the Future of System-on-Chip Designs," *Proceedings of the IEEE*, Vol. 94, No. 6, pp. 1214–1224, June 2006.

[147] D. Henry *et al.*, "Low Electrical Resistance Silicon Through Vias: Technology and Characterization," *Proceedings of the IEEE Electronic Components and Technology Conference*, pp. 1360–1365, June 2006.

[148] D. M. Jang *et al.*, "Development and Evaluation of 3-D SiP with Vertically Interconnected Through Silicon Vias (TSV)," *Proceedings of the IEEE Electronic Components and Technology Conference*, pp. 847–850, June 2007.

[149] B. Kim *et al.*, "Factors Affecting Copper Filling Process within High Aspect Ratio Deep Vias for 3D Chip Stacking," *Proceedings of the IEEE Electronic Components and Technology Conference*, pp. 838–843, June 2006.

[150] M. W. Newman *et al.*, "Fabrication and Electrical Characterization of 3D Vertical Interconnects," *Proceedings of the IEEE Electronic Components and Technology Conference*, pp. 394–398, June 2006.

[151] N. T. Nguyen *et al.*, "Through-Wafer Copper Electroplating for Three-Dimensional Interconnects," *Journal of Micromechanics and Microengineering*, Vol. 12, No. 4, pp. 395–399, July 2002.

[152] C. S. Premachandran *et al.*, "A Vertical Wafer Level Packaging Using Through Hole Filled Via Interconnect by Lift-Off Polymer Method for MEMS and 3D Stacking Applications," *Proceedings of the IEEE Electronic Components and Technology Conference*, pp. 1094–1098, June 2005.

[153] F. Laermer, P. Schilp, and R. Bosch Gmbh, "Method of Anisotropically Etching Silicon," U.S. Patent No. 5,501,893, 1996; German Patent No. 4,241,045C1, 1994.

[154] R. Nagarajan *et al.*, "Development of a Novel Deep Silicon Tapered Via Etch Process for Through-Silicon Interconnection in 3-D Integrated Systems," *Proceedings of the IEEE Electronic Components and Technology Conference*, pp. 383–387, June 2006.

[155] P. Dixit and J. Miao, "Fabrication of High Aspect Ratio 35 μm Pitch Interconnects for Next Generation 3-D Wafer Level Packaging by Through-Wafer Copper Electroplating," *Proceedings of the IEEE Electronic Components and Technology Conference*, pp. 388–393, June 2006.

[156] N. Ranganathan *et al.*, "High Aspect Ratio Through-Wafer Interconnect for Three-Dimensional Integrated Circuits," *Proceedings of the IEEE Electronic Components and Technology Conference*, pp. 343–348, June 2005.

[157] S. X. Zhang, S.-W. R. Lee, L. T. Weng, and S. So, "Characterization of Copper-to-Silicon for the Application of 3D Packaging with Through Silicon Vias," *Proceedings of the IEEE International Conference on Electronic Packaging Technology*, pp. 51–56, September 2005.

[158] C. Odoro *et al.*, "Analysis of the Induced Stresses in Silicon During Thermocompression Cu-Cu Bonding of Cu-Through-Vias in 3D-SIC Architecture," *Proceedings of the IEEE Electronic Components and Technology Conference*, pp. 249–255, June 2007.
[159] D. Sabuncuoglu-Tezcan *et al.*, "Sloped Through Wafer Vias for 3D Wafer Level Packaging," *Proceedings of the IEEE Electronic Components and Technology Conference*, pp. 643–647, June 2007.
[160] L. W. Nagel and D. O. Pederson, "Simulation Program with Integrated Circuit Emphasis (SPICE)," *Proceedings of the IEEE Midwest Symposium on Circuit Theory*, pp. 1–64, April 1973.
[161] P. S. Andry *et al.*, "Fabrication and Characterization of Robust Through-Silicon Vias for Silicon-Carrier Applications," *IBM Journal of Research and Development*, Vol. 52, No. 6, pp. 571–581, November 2008.
[162] P. Leduc *et al.*, "Enabling Technologies for 3D Chip Stacking," *Proceedings of the IEEE International Symposium on VLSI Technology, Systems and Applications*, pp. 76–78, April 2008.
[163] MITLL Low-Power FDSOI CMOS Process Design Guide, MIT Lincoln Laboratories, June 2006.
[164] Z. Xu and J.-Q. Lu, "Through-Silicon-Via Fabrication Technologies, Passives Extraction, and Electrical Modeling for 3-D Integration/Packaging," *IEEE Transactions on Semiconductor Manufacturing*, Vol. 26, No. 1, pp. 23–34, February 2013.
[165] M. Koyanagi *et al.*, "Three-Dimensional Integration Technology Based on Wafer Bonding with Vertical Buried Interconnections," *IEEE Transactions on Electron Devices*, Vol. 53, No. 11, pp. 2799–2808, November 2006.
[166] M. Koyanagi, T. Fukushima, and T. Tanaka, "High-Density Through Silicon Vias for 3-D LSIs," *Proceedings of the IEEE*, Vol. 97, No. 1, pp. 49–59, January 2009.
[167] M. Kawano *et al.*, "Three-Dimensional Packaging Technology for Stacked DRAM With 3-Gb/s Data Transfer," *IEEE Transactions on Electron Devices*, Vol. 55, No. 7, pp. 1614–1620, July 2008.
[168] J. Bea, T. Fukushima, T. Tanaka, and M. Koyanagi, "Evaluation of Cu Diffusion From Cu Through-Silicon Via (TSV) in Three-Dimensional LSI by Transient Capacitance Measurement," *IEEE Transactions on Electron Devices Letters*, Vol. 32, No. 7, pp. 940–942, July 2011.
[169] K. A. Jenkins and C. S. Patel, "Copper-Filled Through Wafer Vias with Very Low Inductance," *Proceedings of the IEEE International Interconnect Technology Conference*, pp. 144–146, June 2005.
[170] C. Okoro *et al.*, "Insertion Bonding: A Novel Cu-Cu Bonding Approach for 3D Integration," *Proceedings of the IEEE Electronic Components and Technology Conference*, pp. 1370–1375, June 2010.
[171] Y. Liang and Y. Li, "Closed-Form Expressions for the Resistance and the Inductance of Different Profiles of Through-Silicon Vias," *IEEE Transactions on Electron Device Letters*, Vol. 32, No. 3, pp. 393–395, March 2011.
[172] J. Knickerbocker *et al.*, "Three-Dimensional Silicon Integration," *IBM Journal of Research and Development*, Vol. 52, No. 6, pp. 553–569, November 2008.
[173] J. Knickerbocker *et al.*, "3D Silicon Integration," *Proceedings of the IEEE Electronic Components and Technology Conference*, pp. 538–543, May 2008.
[174] Y. Civale *et al.*, "Spin-on Dielectric Liner TSV for 3D Wafer Level Packaging Applications," *Proceeding of the IEEE International Interconnect Technology Conference*, pp. 3–5, June 2010.
[175] K. Y. Au *et al.*, "3D Chip Stacking & Reliability Using TSV-Micro C4 Solder Interconnection," *Proceedings of the IEEE Electronic Components and Technology Conference*, pp. 1376–1384, June 2010.
[176] N. Sillon *et al.*, "Enabling Technologies for 3D Integration: From Packaging Miniaturization to Advanced Stacked ICs," *Proceedings of the IEEE International Electron Devices Meeting*, No. 1, pp. 1–4, December 2008.
[177] J. Burns *et al.*, "A Wafer-Scale 3-D Circuit Integration Technology," *IEEE Transactions on Electron Devices*, Vol. 53, No. 10, pp. 2507–2516, October 2006.

[178] H. Kitada *et al.*, "Development of Low Temperature Dielectrics Down to 150°C for Multiple TSVs Structure with Wafer-on-Wafer (WOW) Technology," *Proceedings of the IEEE International Interconnect Technology Conference*, pp. 1–3, June 2009.

[179] Y. P. R. Lamy *et al.*, "RF Characterization and Analytical Modelling of Through Silicon Vias and Coplanar Waveguides for 3D Integration," *IEEE Transactions on Advanced Packaging*, Vol. 33, No. 4, pp. 1072–1079, November 2010.

[180] S. Arkalgud, "Stacking the Chips," *SEMATECH/ISMI Symposium*, 2009.

[181] G. Katti, M. Stucchi, K. De Meyer, and W. Dehaene, "Electrical Modeling and Characterization of Through Silicon Via for Three-Dimensional ICs," *IEEE Transactions on Electron Devices*, Vol. 57, No. 1, pp. 256–262, January 2010.

[182] T. Sakurai and K. Tamaru, "Simple Formulas for Two- and Three-Dimensional Capacitances," *IEEE Transactions on Electron Devices*, Vol. ED-30, No. 2, pp. 183–185, February 1983.

[183] M. I. Elmasry, "Capacitance Calculations in MOSFET VLSI," *IEEE Electron Device Letters*, Vol. EDL-3, No. 1, pp. 6–7, January 1982.

[184] A. Ruehli, "Inductance Calculations in a Complex Integrated Circuit Environment," *IBM Journal of Research and Development*, Vol. 16, No. 5, pp. 470–481, September 1972.

[185] A. Mezhiba and E. Friedman, "Inductive Properties of High-Performance Power Distribution Grids," *IEEE Transactions on Very Large Scale Integration (VLSI) Systems*, Vol. 10, No. 6, pp. 762–776, December 2002.

[186] A. Rahman, J. Trezza, B. New, and S. Trimberger, "Die Stacking Technology for Terabit Chip-to-Chip Communications," *Proceedings of the IEEE Custom Integrated Circuits Conference*, pp. 587–590, September 2006.

[187] L. L. W. Leung and K. J. Chen, "Microwave Characterization and Modeling of High Aspect Ratio Through-Wafer Interconnect Vias in Silicon Substrates," *IEEE Transactions on Microwave Theory and Techniques*, Vol. 53, No. 8, pp. 2472–2480, August 2005.

[188] F. M. Finkbeiner *et al.*, "Development of Ultra-Low Impedance Through-Wafer Micro-Vias," *Nuclear Instruments and Methods in Physics Research Section A*, Vol. 520, No. 1–3, pp. 463–465, March 2004.

[189] E. M. Chow *et al.*, "Process Compatible Polysilicon-Based Electrical Through-Wafer Interconnects in Silicon Substrates," *Journal of Microelectromechanical Systems*, Vol. 11, No. 6, pp. 631–640, December 2002.

[190] I. Luusua *et al.*, "Through-Wafer Polysilicon Interconnect Fabrication with In-Situ Boron Doping," *Micro- and Nanosystems – Materials and Devices*, Vol. 872, pp. 77–81, March 2005.

[191] J. H. Wu, *Through-Substrate Interconnects for 3-D Integration and RF Systems*, Ph.D. Dissertation, Massachusetts Institute of Technology, October 2006.

[192] S. M. Alam, R. E. Jones, S. Rauf, and R. Chatterjee, "Inter-Strata Connection Characteristics and Signal Transmission in Three-Dimensional (3D) Integration Technology," *Proceedings of the IEEE International Symposium on Quality Electronic Design*, pp. 580–585, March 2007.

[193] R. Weerasekera *et al.*, "Closed-Form Equations for Through-Silicon Via (TSV) Parasitics in 3-D Integrated Circuits (ICs)," *Proceedings of the IEEE Design, Automation & Test in Europe Conference*, pp. 1–3, April 2009.

[194] A. Y. Weldezion *et al.*, "Bandwidth Optimization for Through Silicon Via (TSV) Bundles in 3D Integrated Circuits," *Proceedings of the IEEE Design, Automation & Test in Europe Conference*, pp. 1–4, April 2009.

[195] R. Weerasekera, *System Interconnection Design Trade-offs in Three-Dimensional Integrated Circuits*, Ph.D. Dissertation, KTH School of Information and Communication Technologies, Stockholm, Sweden, December 2008.

[196] I. Savidis *et al.*, "Electrical Modeling and Characterization of Through-Silicon Vias (TSVs) for 3-D Integrated Circuits," *Microelectronics Journal*, Vol. 41, No. 1, pp. 9–16, January 2010.
[197] R. Weerasekera *et al.*, "Compact Modelling of Through-Silicon Vias (TSVs) in Three-Dimensional (3-D) Integrated Circuits," *Proceedings of the IEEE International Conference on 3D System Integration*, pp. 1–8, September 2009.
[198] K. Salah, H. Ragai, Y. Ismail, and A. El Rouby, "Equivalent Lumped Element Models for Various n-Port Through Silicon Vias Networks," *Proceedings of the IEEE International Asia and South Pacific Design Automation Conference*, pp. 176–183, January 2011.
[199] Z. Guo and G. G. Pan, "On Simplified Fast Modal Analysis for Through Silicon Vias in Layered Media Based Upon Full-Wave Solutions," *IEEE Transactions on Advanced Packaging*, Vol. 33, No. 2, pp. 517–523, May 2010.
[200] T. H. Lee, *Planar Microwave Engineering*, Cambridge University Press, 2004.
[201] I. Savidis and E. G. Friedman, "Closed-Form Expressions of 3-D Via Resistance, Inductance, and Capacitance," *IEEE Transactions on Electron Devices*, Vol. 56, No. 9, pp. 1873–1881, September 2009.
[202] K. Banerjee, S. K. Souri, P. Kapour, and K. C. Saraswat, "3-D ICs: A Novel Chip Design Paradigm for Improving Deep-Submicrometer Interconnect Performance and Systems-on-Chip Integration," *Proceedings of the IEEE*, Vol. 89, No. 5, pp. 602–633, May 2001.
[203] J. E. Sergent and A. Krum, Eds., *Thermal Management Handbook for Electronic Assemblies*, McGraw-Hill, 1998.
[204] M. R. Ward, *Electrical Engineering Science*, McGraw-Hill, 1971.
[205] R. Jakushokas, M. Popovich, A. V. Mezhiba, S. Kose, and E. G. Friedman, *Power Distribution Networks with On-Chip Decoupling Capacitors, Second Edition,* Springer Science + Businsess Media, 2011.
[206] E. B. Rosa, "The Self and Mutual Inductances of Linear Conductors," *Bulletin of the Bureau of Standards*, Vol. 4, No. 2, pp. 301–344, 1908.
[207] J. Kim *et al.*, "High-Frequency Scalable Electrical Model and Analysis of a Through Silicon Via (TSV)," *IEEE Transactions on Components, Packaging, and Manufacturing Technology*, Vol. 1, No. 2, pp. 181–195, February 2011.
[208] M. W. Beattie and L. T. Pileggi, "On-Chip Induction Modeling: Basics and Advanced Methods," *IEEE Transactions on Very Large Scale Integration (VLSI) Systems*, Vol. 10, No. 6, pp. 712–729, December 2002.
[209] R. A. Pucel, *Gallium Arsenide Technology*, D. Feny, Ed., Chapter 6, p. 216, 1985.
[210] M. E. Goldfarb and R. A. Pucel, "Modeling Via Hole Grounds in Microstrip," *IEEE Microwave and Guided Wave Letters*, Vol. 1, No. 6, pp. 135–137, June 1991.
[211] C. Xu, V. Kourkoulos, R. Suaya, and K. Banerjee, "A Fully Analytical Model for the Series Impedance of Through-Silicon Vias With Consideration of Substrate Effects and Coupling With Horizontal Interconnects," *IEEE Transactions on Electron Devices*, Vol. 58, No. 10, pp. 3529–3540, October 2011.
[212] [Online] *ANSYS Quick 3-D (Q3D) Extractor*, ANSYS. Available: http://www.ansys.com/Products/Simulation + Technology/Electromagnetics/High-Performance + Electronic + Design/ANSYS + Q3D + Extractor.
[213] K. Nabors and J. White, "FastCap: A Multipole-Accelerated 3-D Capacitance Extraction Program," *IEEE Transactions on Computer-Aided Design of Integrated Circuits and Systems*, Vol. 10, No. 11, pp. 1447–1459, November 1991.
[214] C. Xu, R. Suaya, and K. Banerjee, "Compact Modeling and Analysis of Through-Si-Via-Induced Electrical Noise Coupling in Three-Dimensional ICs," *IEEE Transactions on Electron Devices*, Vol. 58, No. 11, pp. 4024–4034, November 2011.
[215] G. Katti *et al.*, "Through-Silicon-Via Capacitance Reduction Technique to Benefit 3-D IC Performance," *IEEE Electron Device Letters*, Vol. 31, No. 6, pp. 549–551, June 2010.

[216] S. M. Sze and K. K. Ng, *Physics of Semiconductor Devices, Third Edition,* John Wiley & Sons Press, 2006.
[217] Y. Tsividis and C. McAndrew, *Operation and Modeling of the MOS Transistor, Third Edition,* Oxford University Press, 2010.
[218] C. Jueping *et al.,* "Through-Silicon Via (TSV) Capacitance Modeling for 3D NoC Energy Consumption Estimation," *Proceedings of the IEEE International Conference on Solid-State and Integrated Circuit Technology*, pp. 815–817, November 2010.
[219] F. T. Ulaby, *Fundamentals of Applied Electromagnetics*, Prentice Hall, 2004.
[220] B. Curran, I. Ndip, S. Guttovski, and H. Reichl, "The Impacts of Dimensions and Return Current Path Geometry on Coupling in Single Ended Through Silicon Vias," *Proceedings of the IEEE Electronic Components and Technology Conference*, pp. 1092–1097, May 2009.
[221] N. N. Rao, *Elements of Engineering Electromagnetics, Fifth Edition*, Prentice Hall, 2000.
[222] S. W. Ho *et al.,* "High RF Performance TSV Silicon Carrier for High Frequency Application," *Proceedings of the IEEE Electronic Components and Technology Conference*, pp. 1946–1952, May 2008.
[223] Z. Xu and J.-Q. Lu, "High-Speed Design and Broadband Modeling of Through-Strata-Vias (TSVs) in 3D Integration," *IEEE Transactions on Components, Packaging, and Manufacturing Technology*, Vol. 1, No. 2, pp. 154–162, February 2011.
[224] J. Kim *et al.,* "Modeling and Analysis of Differential Signal Through Silicon Via (TSV) in 3D IC," *Proceedings of the IEEE Components, Packaging, and Manufacturing Technology Symposium Japan*, pp. 1–4, August 2010.
[225] T. H. Lee, *The Design of CMOS Radio-Frequency Integrated Circuits, Second Edition*, Cambridge University Press, 2004.
[226] Y. I. Ismail and E. G. Friedman, "On the Extraction of On-Chip Inductance," *Journal of Circuits, Systems and Computers*, Vol. 12, No. 1, pp. 31–40, February 2003.
[227] H.-Y. Liao and H.-K. Chiou, "RF Model and Verification of Through-Silicon Vias in Fully Integrated SiGe Power Amplifier," *IEEE Electron Device Letters*, Vol. 32, No. 6, pp. 809–811, June 2011.
[228] K. Salah *et al.,* "Compact Lumped Element Model for TSV in 3D-ICs," *Proceedings of the IEEE International Symposium on Circuits and Systems*, pp. 2321–2324, May 2011.
[229] C. Bermond *et al.,* "High Frequency Characterization and Modeling of High Density TSV in 3D Integrated Circuits," *Proceedings of the IEEE Workshop on Signal Propagation on Interconnects*, pp. 1–4, May 2009.
[230] [Online] *ANSYS HFSS*, ANSYS. Available: http://www.ansys.com/Products/Simulation + Technology/Electromagnetics/High-Performance + Electronic + Design/ANSYS + HFSS.
[231] M. Ravenstahl and M. Kopp, Application Brief: ANSYS HFSS for ECAD, ANSYS, 2013.
[232] [Online] EIP Electromagnetic Field Solver Suite of Tools, IBM, March 2010. Available: http://www.alphaworks.ibm.com/tech/eip.
[233] M. Kamon, M. J. Tsuk, and J. K. White, "FastHenry: A Multipole-Accelerated 3-D Inductance Extraction Program," *IEEE Transactions on Microwave Theory and Techniques*, Vol. 42, No. 9, pp. 1750–1758, September 1994.
[234] [Online] *EMPro 3D EM Simulation Software*, Agilent Technologies. Available: http://www.home.agilent.com/en/pc-1297143/empro-3d-em-simulation-software.
[235] *Agilent EEs of EDA EMPro*, Agilent Technologies, February 2012.
[236] [Online] *Sonnet Suites: High Frequency Electromagnetic Software*, Sonnet. Available: http://www.sonnetsoftware.com/products/sonnet-suites.
[237] Going With the Flow: Sonnet Professional Suite Release 14, Sonnet, 2012.
[238] [Online] *3D Electromagnetic Simulation Software*, CST. Available: http://www.cst.com.
[239] A. R. Hall, *Generalized Method of Moments*, Oxford University Press, 2005.
[240] E. K. Miller, L. Medgyesi-Mitschang, and E. H. Newman, Eds., *Computational Electromagnetics: Frequency-Domain Method of Moments*, IEEE Press, 1992.

[241] D. L. Logan, *A First Course in the Finite Element Method, Fourth Edition*, Nelson, A Division of Thomson Canada Limited, 2007.

[242] P. Schneider *et al.,* "Towards a Methodology for Analysis of Interconnect Structures for 3D-Integration of Micro Systems," *Analog Integrated Circuits and Signal Processing*, Vol. 57, No. 3, pp. 205–211, February 2008.

[243] MITLL Low-Power FDSOI CMOS Process Application Notes, MIT Lincoln Laboratories, June 2006.

[244] E. Salman and E. G. Friedman, *High Performance Integrated Circuit Design*, McGraw-Hill, 2012.

[245] S. M. Satheesh and E. Salman, "Effect of TSV Fabrication Technology on Power Distribution in 3D ICs," *Proceedings of the ACM/IEEE Great Lakes Symposium on VLSI*, pp. 287–292, May 2013.

[246] E. Salman, "Noise Coupling Due To Through Silicon Vias (TSVs) in 3-D Integrated Circuits," *Proceedings of the IEEE International Symposium on Circuits and Systems*, pp. 1411–1414, May 2011.

[247] L. Shifren *et al.,* "Predictive Simulation and Benchmarking of Si and Ge pMOS FinFETs for Future CMOS Technology," *IEEE Transactions on Electron Devices*, Vol. 61, No. 7, pp. 2271–2277, July 2014.

[248] G. K. Dalapati *et al.,* "Impact of Buffer Layer on Atomic Layer Deposited TiAlO Alloy Dielectric Quality for Epitaxial-GaAs/Ge Device Application," *IEEE Transactions on Electron Devices*, Vol. 60, No. 1, pp. 192–199, January 2013.

[249] A. Rogalski, "New Trends in Infrared and Terahertz Detectors," *Proceedings of the IEEE Optoelectronic and Microelectronic Materials Devices Conference*, pp. 218–220, December 2014.

[250] A. L. Betz and R. T. Boreiko, "Space Applications for HgCdTe at FIR Wavelengths between 50 and 150 um," *Proceedings of the SPIE Materials for Infrared Detectors*, pp. 1–9, November 2001.

[251] M. G. Farooq *et al.,* "3-D Copper TSV Integration, Testing and Reliability," *Proceedings of the IEEE International Electron Devices Meeting*, pp. 7.1.1–7.1.4, December 2011.

[252] [Online] NIMO Group, "Predictive Technology Model (PTM)," Available: http://ptm.asu.edu/.

[253] T. Sakurai, "Approximation of Wiring Delay in MOSFET LSI," *IEEE Journal of Solid-State Circuits*, Vol. SC-18, No. 4, pp. 418–426, August 1983.

[254] K. Agarwal, D. Sylvester, and D. Blaauw, "Modeling and Analysis of Crosstalk Noise in Coupled *RLC* Interconnects," *IEEE Transactions on Computer-Aided Design of Integrated Circuits and Systems*, Vol. 25, No. 5, pp. 892–901, May 2006.

[255] R. Gharpurey and R. G. Meyer, "Modeling and Analysis of Substrate Coupling in Integrated Circuits," *IEEE Journal of Solid-State Circuits*, Vol. 31, No. 3, pp. 344–353, March 1996.

[256] S. Spiesshoefer *et al.,* "Process Integration for Through-Silicon Vias," *Journal of Vacuum Science & Technology A*, Vol. 23, No. 4, pp. 824–829, June 2005.

[257] I. A. Papistas and V. F. Pavlidis, "Bandwidth-to-Area Comparison of Through Silicon Vias and Inductive Links for 3-D ICs," *Proceedings of the IEEE European Conference on Circuit Theory and Design*, pp. 1–4, August 2015.

[258] H. Ishikuro, N. Miura, and T. Kuroda, "Wideband Inductive-Coupling Interface for High-Performance Portable System," *Proceedings of the IEEE Custom Integrated Circuits Conference*, pp. 13–20, September 2007.

[259] N. Miura *et al.,* "Analysis and Design of Transceiver Circuit and Inductor Layout for Inductive Inter-Chip Wireless Superconnect," *Proceedings of the IEEE Symposium on VLSI Circuits*, pp. 246–249, June 2004.

[260] N. Miura *et al.,* "A 1 Tb/s 3 W Inductive-Coupling Transceiver for 3D-Stacked Inter-Chip Clock and Data Link," *IEEE Journal of Solid-State Circuits*, Vol. 42, No. 1, pp. 111–122, January 2007.

[261] N. Miura, T. Sakurai, and T. Kuroda, *"Inductive Coupled Communications," Coupled Data Communication Techniques for High Performance and Low Power Computing*, Springer, 2010.

[262] N. Miura *et al.,* "A 195-Gb/s 1.2-W Inductive Inter-Chip Wireless Superconnect with Transmit Power Control Scheme for 3-D-Stacked System in a Package," *IEEE Journal of Solid-State Circuits*, Vol. 41, No. 1, pp. 23–34, January 2006.

[262a] X. Chen and S. Kiaei, "Monocycle Shapes for Ultra Wideband System," *Proceedings of the IEEE International Symposium on Circuits and Systems*, pp. 597–600, May 2002.

[263] N. Miura *et al.*, "Analysis and Design of Inductive Coupling and Transceiver Circuit for Inductive Inter-Chip Wireless Superconnect," *IEEE Journal of Solid-State Circuits*, Vol. 40, No. 4, pp. 829–837, April 2005.
[264] N. Miura *et al.*, "An 11 Gb/s Inductive-Coupling Link with Burst Transmission," *Proceedings of the IEEE International Solid-State Circuits Conference*, pp. 298–614, February 2008.
[265] [Online] Wide I/O 2 DRAM Standard. Available https://www.jedec.org.
[266] [Online] Hybrid Memory Cube Specification. Available http://www.hybridmemorycube.org.
[267] T. Sekiguchi *et al.*, "Inductorless 8.9 mW 25 Gb/s 1:4 DEMUX and 4 mW 13 Gb/s 4:1 MUX in 90 nm CMOS," *Journal of Semiconductor Technology and Science*, Vol. 10, No. 3, pp. 176–184, September 2010.
[268] N. Miura, T. Sakurai, and T. Kuroda, "Crosstalk Countermeasures for High-Density Inductive-Coupling Channel Array," *IEEE Journal of Solid-State Circuits*, Vol. 42, No. 2, pp. 410–421, February 2007.
[269] K. Niitsu *et al.*, "Interference from Power/Signal Lines and to SRAM Circuits in 65nm CMOS Inductive-Coupling Link," *Proceedings of the IEEE Asian Solid-State Circuits Conference*, pp. 131–134, November 2007.
[270] I. A. Papistas and V. F. Pavlidis, "Crosstalk Noise Effects of On-Chip Inductive Links on Power Delivery Networks," *Proceedings of the IEEE International Symposium on Circuits and Systems*, pp. 1938–1941, May 2016.
[271] [Online] Ansys RedHawk. Available https://www.apache-da.com/products/redhawk (accessed in May 2016).
[272] [Online] Ansys Electronics Desktop. Available http://www.ansys.com/products/electronics/ansys-electronics-desktop.
[273] W. Koh *et al.*, "Copper Pillar Bump Technology Progress Overview," *Proceedings of the IEEE International Conference on Electronic Packaging Technology and High Density Packaging*, pp. 1–5, August 2011.
[274] I. A. Papistas and V. F. Pavlidis, "Inter-Tier Crosstalk Noise on Power Delivery Networks for 3-D ICs with Inductively-Coupled Interconnects," *Proceedings of the ACM Great Lakes Symposium of VLSI*, pp. 257–262, May 2016.
[275] M. Popovich, A. V. Mezhiba, and E. G. Friedman, *Power Distribution Networks with On-Chip Decoupling Capacitors*, Springer Verlag, 2008.
[276] A. V. Mezhiba and E. G. Friedman, "Impedance Characteristics of Power Distribution Grids in Nanoscale Integrated Circuits," *IEEE Transactions on Very Large Scale Integration (VLSI) Systems*, Vol. 12, No. 11, pp. 1148–1155, November 2004.
[277] S. Han and D. Wentzloff, "Performance Improvement of Resonant Inductive Coupling for Wireless 3D IC Interconnect," *Proceedings of the IEEE International Symposium on Antennas and Propagation*, pp. 1–4, July 2010.
[278] K. Onizuka *et al.*, "Chip-to-Chip Inductive Wireless Power Transmission System for SiP Applications," *Proceedings of the IEEE Custom Integrated Circuits Conference*, pp. 575–578, September 2006.
[279] Y. Yuxiang, Y. Yoshida, and T. Kuroda, "Non-Contact 10% Efficient 36 mW Power Delivery Using On-Chip Inductor in 0.18-μm CMOS," *Proceedings of the IEEE Asian Solid-State Circuits Conference*, pp. 115–118, November 2007.
[280] S. Han and D. Wentzloff, "Wireless Power Transfer Using Resonant Inductive Coupling for 3D Integrated ICs," *Proceedings of the IEEE International 3D Systems Integration Conference*, pp. 1–5, November 2010.
[281] A. Kurs *et al.*, "Wireless Power Transfer via Strongly Coupled Magnetic Resonances," *Science*, Vol. 317, No. 5834, pp. 83–86, July 2007.
[282] S. Han and D. Wentzloff, "0.61 W/mm^2 Resonant Inductively Coupled Power Transfer for 3D ICs," *Proceedings of the IEEE Custom Integrated Circuits Conference*, pp. 1–4, September 2012.

[283] B. S. Landman and R. L. Russo, "On a Pin Versus Block Relationship for Partitions of Logic Graphs," *IEEE Transactions on Computers*, Vol. C-20, No. 12, pp. 1469–1479, December 1971.

[284] W. E. Donath, "Placement and Average Interconnection Lengths of Computer Logic," *IEEE Transactions on Circuits and Systems*, Vol. 26, No. 4, pp. 272–277, April 1979.

[285] P. Christie and D. Stroobandt, "The Interpretation and Application of Rent's Rule," *IEEE Transactions on Very Large Scale Integration (VLSI) Systems*, Vol. 8, No. 6, pp. 639–648, December 2000.

[286] P. Verplaetse, D. Stroobandt, and J. Van Campenhout, "A Stochastic Model for the Interconnection Topology of Digital Circuits," *IEEE Transactions on Very Large Scale Integration (VLSI) Systems*, Vol. 9, No. 6, pp. 938–942, December 2001.

[287] A. B. Kahng, S. Mantik, and D. Stroobandt, "Toward Accurate Models of Achievable Routing," *IEEE Transactions on Computer-Aided Design of Integrated Circuits and Systems*, Vol. 20, No. 5, pp. 648–659, May 2001.

[288] D. Stroobandt, *A Priori Wire Length Estimates for Digital Design*, Kluwer Academic Publishers, Netherlands, 2001.

[289] J. A. Davis, V. K. De, and J. D. Meindl, "A Stochastic Wire-Length Distribution for Gigascale Integration (GSI) – Part I: Derivation and Validation," *IEEE Transactions on Electron Devices*, Vol. 45, No. 3, pp. 580–589, March 1998.

[290] J. A. Davis, V. K. De, and J. D. Meindl, "A Stochastic Wire-Length Distribution for Gigascale Integration (GSI) – Part II: Applications to Clock Frequency, Power Dissipation, and Chip Size Estimation," *IEEE Transactions on Electron Devices*, Vol. 45, No. 3, pp. 590–597, March 1998.

[291] J. W. Joyner *et al.*, "Impact of Three-Dimensional Architectures on Interconnects in Gigascale Integration," *IEEE Transactions on Very Large Scale Integration (VLSI) Systems*, Vol. 9, No. 6, pp. 922–928, December 2001.

[292] J. W. Joyner, P. Zarkesh-Ha, J. A. Davis, and J. D. Meindl, "Vertical Pitch Limitations on Performance Enhancement in Bonded Three-Dimensional Interconnect Architectures," *Proceedings of the ACM International System Level Interconnect Prediction Conference*, pp. 123–127, April 2000.

[293] J. W. Joyner, P. Zarkesh-Ha, and J. D. Meindl, "A Stochastic Global Net-Length Distribution for a Three-Dimensional System-on-a-Chip (3D-SoC)," *Proceedings of the IEEE International ASIC/SOC Conference*, pp. 147–151, September 2001.

[294] J. W. Joyner, *Opportunities and Limitations of Three-Dimensional Integration for Interconnect Design*, Ph.D. Dissertation, Georgia Institute of Technology, Atlanta, Georgia, July 2003.

[295] A. Rahman, A. Fan, J. Chung, and R. Reif, "Wire-Length Distribution of Three-Dimensional Integrated Circuits," *Proceedings of the IEEE International Interconnect Technology Conference*, pp. 233–235, May 1999.

[296] A. Rahman and R. Reif, "System Level Performance Evaluation of Three-Dimensional Integrated Circuits," *IEEE Transactions on Very Large Scale Integration (VLSI) Systems*, Vol. 8, No. 6, pp. 671–678, December 2000.

[297] A. Rahman, A. Fan, and R. Reif, "Comparison of Key Performance Metrics in Two- and Three-Dimensional Integrated Circuits," *Proceedings of the IEEE International Interconnect Technology Conference*, pp. 18–20, June 2000.

[298] R. Zhang, K. Roy, C.-K. Koh, and D. B. Janes, "Stochastic Interconnect Modeling, Power Trends, and Performance Characterization of 3-D Circuits," *IEEE Transactions on Electron Devices*, Vol. 48, No. 4, pp. 638–652, April 2001.

[299] R. Zhang, K. Roy, C.-K. Koh, and D. B. Janes, "Power Trends and Performance Characterization of 3-Dimensional Integration," *Proceedings of the IEEE International Symposium on Circuits and Systems*, Vol. IV, pp. 414–417, May 2001.

[300] D. Stroobandt and J. Van Campenhout, "Accurate Interconnection Lengths in Three-Dimensional Computer Systems," *IEICE Transactions on Information and System, Special Issue on Physical Design in Deep Submicron*, Vol. 10, No. 1, pp. 99–105, April 2000.

[301] D. Stroobandt, H. Van Marck, and J. Van Campenhout, "On the Use of Generating Polynomials for the Representation of Interconnection Length Distributions," *Proceedings of the International Workshop on Symbolic Methods and Applications in Circuit Design*, pp. 74–78, October 1996.

[302] D. Stroobandt, *"Improving Donath's Technique for Estimating the Average Interconnection Length in Computer Logic,"* ELIS Department *Technical Report DG 96-01*, Ghent University, Belgium, 1996

[303] W. E. Donath, "Wire Length Distribution for Placements of Computer Logic," *IBM Journal of Research and Development*, Vol. 25, No. 2/3, pp. 152–155, May 1981.

[304] K. C. Saraswat, S. K. Souri, K. Banerjee, and P. Kapour, "Performance Analysis and Technology of 3-D ICs," *Proceedings of the ACM International System Level Interconnect Prediction Conference*, pp. 85–90, April 2000.

[305] J. W. Joyner, P. Zarkesh-Ha, and J. D. Meindl, "A Global Interconnect Design Window for a Three-Dimensional System-on-a-Chip," *Proceedings of the IEEE International Interconnect Technology Conference*, pp. 154–156, June 2001.

[306] J. W. Joyner and J. D. Meindl, "Opportunities for Reduced Power Distribution Using Three-Dimensional Integration," *Proceedings of the IEEE International Interconnect Technology Conference*, pp. 148–150, June 2002.

[307] *"FDSOI Design Guide,"* MIT Lincoln Laboratory, Cambridge, 2006.

[308] H. Hua *et al.*, "Performance Trend in Three-Dimensional Integrated Circuits," *Proceedings of the IEEE International Interconnect Technology Conference*, pp. 45–47, June 2006.

[309] OpenRISC Reference Platform System-on-a-Chip and OpenRISC I200 IP Core Specification, online [http://www.opencores.org/projects.cgi/web/orlk/orpso].

[310] K. Bernstein *et al.*, "Interconnects in the Third Dimension: Design Challenges for 3-D ICs," *Proceedings of the IEEE/ACM Design Automation Conference*, pp. 562–567, June 2007.

[311] S. A. Kühn, M. B. Kleiner, P. Ramm, and W. Weber, "Performance Modeling of the Interconnect Structure of a Three-Dimensional Integrated RISC Processor/Cache System," *IEEE Transactions on Components, Packaging, and Manufacturing Technology – Part B*, Vol. 19, No. 4, pp. 719–727, November 1996.

[312] E. Beyne *et al.*, "Through-Silicon Via and Die Stacking Technologies for Microsystems Integration," *Proceedings of the IEEE International Electron Devices Meeting*, pp. 1–4, December 2008.

[313] A. Redolfi *et al.*, "Implementation of an Industry Compliant, 5×50 μm, Via-Middle TSV Technology on 300 mm Wafers," *Proceedings of the IEEE Electronic Components and Technology Conference*, pp. 1384–1388, May/June 2011.

[314] S. Van Huylenbroeck *et al.*, "Small Pitch, High Aspect Ratio Via-last TSV Module," *Proceedings of the IEEE Electronic Components and Technology Conference*, pp. 43–49, May/June 2016.

[315] D. Velenis, E. J. Marinissen, and E. Beyne, "Cost Effectiveness of 3D Integration Options," *Proceedings of the 3D Systems Integration Conference*, pp. 1–6, November 2010.

[316] Y. Civale *et al.*, "Enhanced Barrier Seed Metallization for Integration of High-Density High Aspect-Ratio Copper-Filled 3D Through-Silicon Via Interconnects," *Proceedings of the IEEE Electronic Components and Technology Conference*, pp. 822–826, May/June 2012.

[317] S. Van Huylenbroeck *et al.*, "Advanced Metallization Scheme for 3×50 μm Via Middle TSV and Beyond," *Proceedings of the IEEE Electronic Components and Technology Conference*, pp. 66–72, May 2015.

[318] A. Radisi *et al.*, "Copper Plating of Through-Si Vias for 3D-Stacked Integrated Circuits," *Symposium on Materials and Technologies for 3-D Integration held at the 2008 MRS Fall Meeting*, Vol. 1112, pp. 159–164, December 2008.

[319] A. Jourdain *et al.*, "Integration of TSVs, Wafer Thinning and Backside Passivation on Full 300 mm CMOS Wafers for 3D Applications," *Proceedings of the IEEE Electronic Components and Technology Conference*, pp. 1122–1125, May 2011.

[320] K. Vandersmissen *et al.*, "Demonstration of a Cost Effective Cu Electroless TSV Metallization Scheme," *Proceedings of the International Interconnect Technology Conference & Materials for Advanced Metallization Conference*, pp. 197–200, May 2015.

[321] M. Detalle *et al.*, "Interposer Technology for High Bandwidth Interconnect Applications," *Proceedings of the IEEE Electronic Components And Technology Conference*, pp. 323–328, May 2012.

[322] E. Beyne, "Electrical, Thermal and Mechanical Impact of 3D TSV and 3D Stacking Technology on Advanced CMOS Devices-Technology Directions," *Proceedings of the IEEE International Conference on 3D System Integration*, pp. 1–6, January-February 2012.

[323] R. Chaware, K. Nagarajan, K. Ng and S. Y. Pai, "Assembly Process Integration Challenges and Reliability Assessment of Multiple 28 nm FPGAs Assembled on a Large 65 nm Passive Interposer," *IEEE International Reliability Physics Symposium*, pp. 2B.2.1-2B.2.5, April 2012.

[324] M. Detalle *et al.*, "Fat Damascene Wires for High Bandwidth Routing in Silicon Interposer," *International Conference on Solid State Devices and Materials*, September 2012.

[325] I. P. Vaisband, R. Jakushokas, M. Popovich, A. V. Mezhiba, S. Köse, and E. G. Friedman, *On-Chip Power Delivery and Management, Fourth Edition,* Springer, 2016.

[326] J. De Vos *et al.*, "Key Elements for Sub-50 μm Pitch Micro Bump Processes," *Proceedings of the IEEE Electronic Components and Technology Conference*, pp. 1122–1126, May 2013.

[327] C. N. Berglund, "A Unified Yield Model Incorporating both Defect and Parametric Effects," *IEEE Transactions on Semiconductor Manufacturing*, Vol. 9, No. 3, pp. 447–454, August 1996.

[328] A. B. Kahng, J. Lienig, I. L. Markov, and J. Hu, *VLSI Physical Design: From Graph Partitioning to Timing Closure*, Springer, 2011.

[329] R. H. J. M. Otten, "Automatic Floorplan Design," *Proceedings of the IEEE/ACM Design Automation Conference*, pp. 261–267, June 1982.

[330] X. Hong *et al.*, "Corner Block List: An Effective and Efficient Topological Representation of Non-Slicing Floorplan," *Proceedings of the IEEE/ACM International Conference on Computer-Aided Design*, pp. 8–11, November 2000.

[331] E. F. Y. Young, C. C. N. Chu, and C. S. Zion, "Twin Binary Sequences: A Non-Redundant Representation for General Non-Slicing Floorplan," *IEEE Transactions on Computer-Aided Design of Integrated Circuits and Systems*, Vol. 22, No. 4, pp. 457–469, April 2003.

[332] S. N. Adya and I. L. Markov, "Fixed-Outline Floorplanning: Enabling Hierarchical Design," *IEEE Transactions on Very Large Scale Integration (VLSI) Systems*, Vol. 11, No. 6, pp. 1120–1135, December 2003.

[333] S. Chen and T. Yoshimura, "Fixed-Outline Floorplanning: Block-Position Enumeration and a New Method for Calculating Area Costs," *IEEE Transactions on Computer-Aided Design of Integrated Circuits and Systems*, Vol. 27, No. 5, pp. 858–871, May 2008.

[334] K.-C. Chan, C.-J. Hsu, and J.-M. Lin, "A Flexible Fixed-Outline Floorplanning Methodology for Mixed-Size Modules," *Proceedings of the IEEE Asia and South Pacific Design Automation Conference*, pp. 435–440, January 2013.

[335] J. M. Lin and Y. W. Chang, "TCG: A Transitive Closure Graph Based Representation for Non-Slicing Floorplans," *Proceedings of the IEEE/ACM Design Automation Conference*, pp. 764–769, June 2001.

[336] Y. Pang, F. Balasa, K. Lampaert, and C. K. Cheng, "Block Placement with Asymmetry Constraint Based on the *O*-tree Nonslicing Representation," *Proceedings of the IEEE/ACM Design Automation Conference*, pp. 464–468, June 2000.

[337] X. Hong *et al.*, "Non-Slicing Floorplan and Placement Using Corner Block List Topological Representation," *IEEE Transactions on Circuits and Systems II: Express Briefs*, Vol. 51, No. 5, pp. 228–233, May 2004.

[338] H. Murata, K. Fujiyoshi, S. Nakatake, and Y. Kajitani, "VLSI Module Placement Based on Rectangle-Packing by the Sequence-Pair," *IEEE Transactions on Computer-Aided Design of Integrated Circuits and Systems*, Vol. 15, No. 12, pp. 1518–1524, December 1996.

[339] J. Knechtel, I. L. Markov, and J. Lienig, "Assembling 2-D Blocks into 3-D Chips," *IEEE Transactions on Computer-Aided Design of Integrated Circuits and Systems*, Vol. 31, No. 2, pp. 228–241, February 2012.

[340] X. Tang, R. Tian, and D. F. Wong, "Fast Evaluation of Sequence Pair in Block Placement by Longest Common Subsequence Computation," *Proceedings of the IEEE Conference on Design, Automation and Test in Europe*, pp. 106–111, March 2000.

[341] X. Tang and D. F. Wong, "FAST-SP: A Fast Algorithm for Block Placement Based on Sequence Pair," *Proceedings of the IEEE Asia and South Pacific Design Automation Conference*, pp. 521–524, February 2001.

[342] A. B. Kahng, "Classical Floorplanning Harmful?" *Proceedings of the ACM International Symposium on Physical Design*, pp. 207–213, May 2000.

[343] S. Kirkpatrick, C. D. Gelatt, and M. P. Vecchi, "Optimization by Simulated Annealing," *Science*, Vol. 220, No. 4598, pp. 671–680, May 1983.

[344] J. Bhasker and S. Sun, *Static Timing Analysis for Nanometer Designs: A Practical Approach*, Springer, 2009.

[345] J. H. Y. Law, E. F. Y. Young, and R. L. S. Ching, "Block Alignment in 3D Floorplan Using Layered TCG," *Proceedings of the ACM Great Lakes Symposium on VLSI*, pp. 376–380, April/May 2006.

[346] H. Yamazaki, K. Sakanushi, S. Nakatake, and Y. Kajitani, "The 3D-Packing by Meta Data Structure and Packing Heuristics," *IEICE Transactions on Fundamentals of Electronics, Communications and Computer Sciences*, Vol. E83-A, No. 4, pp. 639–645, April 2000.

[347] L. Cheng, L. Deng, and D. F. Wong, "Floorplanning for 3-D VLSI Design," *Proceedings of the IEEE Asia and South Pacific Design Automation Conference*, pp. 405–411, January 2005.

[348] S. Salewski and E. Barke, "An Upper Bound for 3D Slicing Floorplans," *Proceedings of the IEEE Asia and South Pacific Design Automation Conference*, pp. 567–572, January 2002.

[349] Y. Deng and W. P. Maly, "Interconnect Characteristics of 2.5-D System Integration Scheme," *Proceedings of the IEEE International Symposium on Physical Design*, pp. 341–345, April 2001.

[350] P. H. Shiu, R. Ravichandran, S. Easwar, and S. K. Lim, "Multi-Layer Floorplanning for Reliable System-on-Package," *Proceedings of the IEEE International Symposium on Circuits and Systems*, Vol. V, pp. 69–72, May 2004.

[351] J. Cong, J. Wei, and Y. Zhang, "A Thermal-Driven Floorplanning Algorithm for 3-D ICs," *Proceedings of the IEEE/ACM International Conference on Computer-Aided Design*, pp. 306–313, November 2004.

[352] Z. Li *et al.*, "Hierarchical 3-D Floorplanning Algorithm for Wirelength Optimization," *IEEE Transactions on Circuits and Systems I: Regular Papers*, Vol. 53, No. 12, pp. 2637–2646, December 2006.

[353] G. Karypis and V. Kumar, "Multilevel k-Way Hypergraph Partitioning," *Proceedings of the IEEE/ACM Design Automation Conference*, pp. 343–348, June 1999.

[354] T. Yan, Q. Dong, Y. Takashima, and Y. Kajitani, "How Does Partitioning Matter for 3D Floorplanning?" *Proceedings of the ACM International Great Lakes Symposium on VLSI*, pp. 73–76, April/May 2006.

[355] [Online]. Available: http://www.cse.ucsc.edu/research/surf/GSRC/progress.html.

[356] C. M. Fiduccia and R. M. Mattheyses, "A Linear-Time Heuristic for Improving Network Partitions," *Proceeding of the IEEE/ACM Design Automation Conference*, pp. 175–181, June 1982.

[357] M.-C. Tsai, T.-C. Wang, and T. T. Hwang, "Through-Silicon Via Planning in 3-D Floorplanning," *IEEE Transactions on Very Large Scale Integration (VLSI) Systems*, Vol. 19, No. 8, pp. 1448–1457, August 2011.

[358] A. V. Goldberg, "An Efficient Implementation of a Scaling Minimum-Cost Flow Algorithm," *Journal of Algorithms*, Vol. 22, No. 1, pp. 1–29, January 1997.
[359] R. Fowler, M. S. Paterson, and S. L. Tanimoto, "Optimal Packing and Covering in the Plane are NP-Complete," *Information Processing Letters*, Vol. 12, No. 3, pp. 133–137, June 1981.
[360] X. He, S. Dong, X. Hong, and S. Goto, "Integrated Interlayer Via Planning and Pin Assignment for 3D ICs," *Proceedings of the IEEE/ACM System Level Interconnect Prediction Workshop*, pp. 99–104, July 2009.
[361] J. Cong, T. Kong, and D. Z. Pan, "Buffer Block Planning for Interconnect Driven Floorplanning," *Proceedings of the IEEE/ACM International Conference on Computer-Aided Design*, pp. 358–363, November 1999.
[362] P. Sarkar and C.-K. Koh, "Routability-Driven Repeater Block Planning for Interconnect-Centric Floorplanning," *IEEE Transactions on Computer-Aided Design of Integrated Circuits and Systems*, Vol. 20, No. 5, pp. 660–671, May 2001.
[363] X. He, S. Dong, Y. Ma, and X. Hong, "Simultaneous Buffer and Interlayer Via Planning for 3D Floorplanning," *Proceedings of the IEEE International Symposium on Quality Electronic Design*, pp. 740–745, March 2009.
[364] M. Healy *et al.*, "Multiobjective Microarchitectural Floorplanning for 2-D and 3-D ICs," *IEEE Transactions on Computer-Aided Design of Integrated Circuits and Systems*, Vol. 26, No. 1, pp. 38–52, January 2007.
[365] P. Shivakumar and N. P. Jouppi, "CACTI 3.0: An Integrated Cache Timing, Power, and Area Model," HP Western Research Labs, Palo Alto, CA, Technical Report 2001.2, 2001.
[366] J. C. Eble, V. K. De, D. S. Wills, and J. D. Meindl, "A Generic System Simulator (GENESYS) for ASIC Technology and Architecture Beyond 2001," *Proceedings of the IEEE International ASIC Conference*, pp. 193–196, September 1996.
[367] [Online]. T. M. Austin, *Simplescalar Tool Suite*. Available: http://www.simplescalar.com.
[368] D. Brooks, V. Tiwari, and M. Martonosi, "Wattch: A Framework for Architectural-Level Power Analysis and Optimizations," *Proceedings of the ACM/IEEE International Symposium on Computer Architecture*, pp. 83–94, June 2000.
[369] N. A. Sherwani, *Algorithms for VLSI Physical Design Automation, Third Edition*, Kluwer Academic Publishers, 2002.
[370] W.-J. Sun and C. Sechen, "Efficient and Effective Placement for Very Large Circuits," *IEEE Transactions on Computer-Aided Design of Integrated Circuits and Systems*, Vol. 14, No. 3, pp. 349–359, March 1995.
[371] H. Eisenmann and F. M. Johannes, "Generic Global Placement and Floorplanning," *Proceedings of the IEEE/ACM Design Automation Conference*, pp. 269–274, June 1998.
[372] J. A. Roy *et al.*, "Capo: Robust and Scalable Open-Source Min-Cut Floorplacer," *Proceedings of the ACM/SIGDA International Symposium on Physical Design*, pp. 224–226, April 2005.
[373] A. R. Agnihorti *et al.*, "Mixed Block Placement via Fractional Cut Recursive Bisection," *IEEE Transactions on Computer-Aided Design of Integrated Circuits and Systems*, Vol. 25, No. 5, pp. 748–761, May 2005.
[374] T.-C. Chen *et al.*, "NTUplace3: An Analytical Placer for Large-Scale Mixed-Size Designs with Preplaced Blocks and Density Constraints," *IEEE Transactions on Computer-Aided Design of Integrated Circuits and Systems*, Vol. 27, No. 7, pp. 187–192, July 2008.
[375] T. Chan, J. Cong, and K. Sze, "Multilevel Generalized Force-Directed Method for Circuit Placement," *Proceedings of the ACM/SIGDA International Symposium on Physical Design*, pp. 185–192, April 2005.
[376] U. Brenner and M. Struzyna, "Faster and Better Global Placement by a New Transportation Algorithm," *Proceedings of the IEEE/ACM Design Automation Conference*, pp. 591–596, June 2005.

[377] N. Viswanathan and C. C.-N. Chu, "Fastplace: Efficient Analytical Placement Using Cell Shifting, Iterative, Local Refinement and a Hybrid Net Model," *IEEE Transactions on Computer-Aided Design of Integrated Circuits and Systems*, Vol. 24, No. 5, pp. 722–733, May 2005.

[378] B. Hu and M. Marek-Sadowska, "Multilevel Fixed-Point-Addition-Based VLSI Placement," *IEEE Transactions on Computer-Aided Design of Integrated Circuits and Systems*, Vol. 24, No. 8, pp. 1188–1203, August 2005.

[379] Z. Xiu, J. D. Ma, S. M. Fowler, and R. A. Rutenbar, "Large-Scale Placement by Grid-Warping," *Proceedings of the IEEE/ACM Design Automation Conference*, pp. 351–356, July 2004.

[380] N. Quinn and M. Breuer, "A Force Directed Component Placement Procedure for Printed Circuit Boards," *IEEE Transactions on Circuits and Systems*, Vol. 26, No. 6, pp. 377–388, June 1979.

[381] K. M. Hall, "An *r*-Dimensional Quadratic Placement Algorithm," *Management Science*, Vol. 17, No. 3, pp. 219–229, November 1970.

[382] A. Kennings and K. P. Vorwerk, "Force-Directed Methods for Generic Placement," *IEEE Transactions on Computer-Aided Design of Integrated Circuits and Systems*, Vol. 25, No. 10, pp. 2076–2087, October 2006.

[383] P. Spindler, U. Schlichtmann, and F. M. Johannes, "Kraftwerk2 – A Fast Force-Directed Quadratic Placement Approach Using an Accurate Net Model," *IEEE Transactions on Computer-Aided Design of Integrated Circuits and Systems*, Vol. 27, No. 8, pp. 1398–1411, August 2008.

[384] M. Ohmura, "An Initial Placement Algorithm for 3-D VLSI," *Proceedings of the IEEE International Symposium on Circuits and Systems*, Vol. IV, pp. 195–198, May 1998.

[385] T. Tanprasert, "An Analytical 3-D Placement that Preserves Routing Space," *Proceedings of the IEEE International Symposium on Circuits and Systems*, Vol. III, pp. 69–72, May 2000.

[386] I. Kaya, M. Olbrich, and E. Barke, "3-D Placement Considering Vertical Interconnects," *Proceedings of the IEEE International SOC Conference*, pp. 257–258, September 2003.

[387] R. Hentschke and R. Reis, "A 3D-Via Legalization Algorithm for 3D VLSI Circuits and its Impact on Wire Length," *Proceedings of the IEEE International Symposium on Circuits and Systems*, pp. 2036–2039, May 2007.

[388] D. H. Kim, K. Athikulwongse, and S. K. Lim, "A Study of Through-Silicon-Via Impact on the 3D Stacked Layout," *Proceedings of the IEEE/ACM International Conference on Computer-Aided Design*, pp. 674–680, November 2009.

[389] [Online]. Available: http://www.iwls.org/iwls2005.

[390] C. Serafy and A. Srivastava, "TSV Replacement and Shield Insertion for TSV-TSV Coupling Reduction in 3-D Global Placement," *IEEE Transactions on Computer-Aided Design of Integrated Circuits and Systems*, Vol. 34, No. 4, pp. 554–562, April 2015.

[391] C. Serafy, B. Shi, and A. Srivastava, "A Geometric Approach to Chip-Scale TSV Shield Placement for the Reduction of TSV Coupling in 3D-ICs," *Integration, The VLSI Journal*, Vol. 47, No. 3, pp. 307–317, June 2014.

[392] W. C. Naylor, R. Donelly and L. Sha, "Non-Linear Optimization System and Method for Wire Length and Delay Optimization for an Automatic Electric Circuit Placer," U.S. Patent No. 6,301,693, October 2001.

[393] J. Cong and G. Luo, "A Multilevel Analytical Placement for 3D ICs," *Proceedings of the IEEE Asia and South Pacific Design Automation Conference*, pp. 361–366, January 2009.

[394] M.-K. Hsu, V. Balabanov, and Y.-W. Chang, "TSV-Aware Analytical Placement for 3-D IC Designs Based on a Novel Weighted-Average Wirelength Model," *IEEE Transactions on Computer-Aided Design of Integrated Circuits and Systems*, Vol. 32, No. 4, pp. 497–509, April 2013.

[395] A. B. Kahng and Q. Wang, "Implementation and Extensibility of an Analytic Placer," *IEEE Transactions on Computer-Aided Design of Integrated Circuits and Systems*, Vol. 24, No. 5, pp. 734–747, May 2005.

[396] [Online]. Available: http://er.cs.ucla.edu/benchmarks/ibm-place.

[397] E. Wong, J. Minz, and S. K. Lim, "Multi-Objective Module Placement for 3-D System-On-Package," *IEEE Transactions on Very Large Scale Integration (VLSI) Systems*, Vol. 14, No. 5, pp. 553–557, May 2006.

[398] L. Zhou, C. Wakayama, and C.-J. R. Shi, "CASCADE: A Standard Supercell Design Methodology With Congestion-Driven Placement for Three-Dimensional Interconnect-Heavy Very Large-Scale Integrated Circuits," *IEEE Transactions on Computer-Aided Design of Integrated Circuits and Systems*, Vol. 26, No. 7, pp. 1270–1282, July 2007.

[399] M. Popovich and E. G. Friedman, "Decoupling Capacitors for Multi-Voltage Power Distribution Systems," *IEEE Transactions on Very Large Scale Integration (VLSI) Systems*, Vol. 14, No. 3, pp. 217–228, March 2006.

[400] A. Mezhiba and E. G. Friedman, *Power Distribution Networks in High Speed Integrated Circuits*, Kluwer Academic Publishers, 2004.

[401] [Online]. Available: http://www.gtcad.gatech.edu.

[402] E. Wong, J. Minz, and S. K. Lim, "Power Supply Noise-Aware 3D Floorplanning for System-on-Package," *Proceedings of the IEEE Topical Meeting on Electrical Performance on Electronic Packaging*, pp. 259–262, October 2005.

[403] M. Hanan, "On Steiner's Problem with Rectilinear Distance," *SIAM Journal of Applied Mathematics*, Vol. 14, No. 2, pp. 255–265, Mach 1966.

[404] J. Lou, S. Thakur, S. Krishnamoorthy, and H. S. Sheng, "Estimating Routing Congestion Using Probabilistic Analysis," *IEEE Transactions on Computer-Aided Design of Integrated Circuits and Systems*, Vol. 21, No. 1, pp. 32–41, January 2002.

[405] L. Cheng, W. N. N. Hung, G. Yang, and X. Song, "Congestion Estimation for 3-D Circuit Architectures," *IEEE Transactions on Circuits and Systems II, Express Briefs*, Vol. 51, No. 12, pp. 655–659, December 2004.

[406] R. J. Enbody, G. Lynn, and K. H. Tan, "Routing the 3-D Chip," *Proceedings of the IEEE/ACM Design Automation Conference*, pp. 132–137, June 1991.

[407] S. Tayu and S. Ueno, "On the Complexity of Three-Dimensional Channel Routing," *Proceedings of the IEEE International Symposium on Circuits and Systems*, pp. 3399–3402, May 2007.

[408] C. C. Tong and C.-L. Wu, "Routing in a Three-Dimensional Chip," *IEEE Transactions on Computers*, Vol. 44, No. 1, pp. 106–117, January 1995.

[409] J. Minz and S. K. Lim, "Block-Level 3-D Global Routing With an Application to 3-D Packaging," *IEEE Transactions on Computer-Aided Design of Integrated Circuits and Systems*, Vol. 25, No. 10, pp. 2248–2257, October 2006.

[410] A. Hashimoto and J. Stevens, "Wire Routing by Optimizing Channel Assignment within Large Apertures," *Proceedings of the IEEE/ACM Design Automation Conference*, pp. 155–169, June 1971.

[411] T. Ohtsuki, *Advances in CAD for VLSI: Vol. 4, Layout Design and Verification*, Elsevier, 1986.

[412] M. Pathak and S. K. Lim, "Performance and Thermal-Aware Steiner Routing for 3-D Stacked ICs," *IEEE Transactions on Computer-Aided Design of Integrated Circuits and Systems*, Vol. 28, No. 9, pp. 1373–1386, September 2009.

[413] K. Boese, A. Kahng, B. McCoy, and G. Robins, "Near-Optimal Critical Sink Routing Tree Constructions," *IEEE Transactions on Computer-Aided Design of Integrated Circuits and Systems*, Vol. 14, No. 12, pp. 1417–1436, December 1995.

[414] W. C. Elmore, "The Transient Response of Damped Linear Network with Particular Regard to Wideband Amplifiers," *Journal of Applied Physics*, Vol. 19, No. 1, pp. 55–63, January 1948.

[415] J. Cong and Y. Zhang, "Thermal-Driven Multilevel Routing for 3-D ICs," *Proceedings of the IEEE Asia and South Pacific Design Automation Conference*, pp. 121–126, January 2005.

[416] J. Cong, K.-S. Leung, and D. Zhou, "Performance Driven Interconnect Design Based on Distributed *RC* delay model," *Proceedings of the ACM Design Automation Conference*, pp. 606–611, June 1993.

[417] A. Harter, *Three-Dimensional Integrated Circuit Layout*, Cambridge University Press, 1991.

[418] B. Hoefflinger, S. T. Liu, and B. Vajdic, "A Three-Dimensional CMOS Design Methodology," *IEEE Transactions on Electron Devices*, Vol. ED-31, No. 2, pp. 171–173, February 1984.

[419] S. M. Alam, D. E. Troxel, and C. V. Thompson, "A Comprehesive Layout Methodology and Layout-Specific Circuit Analyses for Three-Dimensional Integrated Circuits," *Proceedings of the IEEE International Symposium on Quality Electronic Design*, pp. 246–251, March 2002.

[420] S. Das, A. Chandrakasan, and R. Reif, "Design Tools for 3-D Integrated Circuits," *Proceedings of the IEEE Asia and South Pacific Design Automation Conference*, pp. 53–56, January 2003.

[421] [Online]. Available: http://www.ece.ncsu.edu/erl/3DIC/pub.

[422] G. Chen and E. G. Friedman, "An *RLC* Interconnect Model Based on Fourier Analysis," *IEEE Transactions on Computer-Aided Design of Integrated Circuits and Systems*, Vol. 24, No. 2, pp. 170–183, February 2005.

[423] Metal User's Guide, www.oea.com.

[424] P. Ramn *et al.*, "InterChip Via Technology for Vertical System Integration," *Proceedings of the IEEE International Interconnect Technology Conference*, pp. 160–162, June 2001.

[425] K. D. Boese *et al.*, "Fidelity and Near-Optimality of Elmore-Based Routing Constructions," *Proceedings of the IEEE International Conference on Computer Design*, pp. 81–84, October 1993.

[426] A. I. Abou-Seido, B. Nowak, and C. Chu, "Fitted Elmore Delay: A Simple and Accurate Interconnect Delay Model," *IEEE Transactions on Very Large Scale Integration (VLSI) Systems*, Vol. 12, No. 7, pp. 691–696, July 2004.

[427] J. P. Fishburn and C. A. Svenvon, "Shaping a Distributed-*RC* Line to Minimize Elmore Delay," *IEEE Transactions on Circuits and Systems I: Fundamental Theory and Applications*, Vol. 42, No. 12, pp. 1020–1022, December 1995.

[428] J. Cong and K.-S. Leung, "Optimal Wiresizing under Elmore Delay Model," *IEEE Transactions on Computer-Aided Design of Integrated Circuits and Systems*, Vol. 14, No. 3, pp. 321–336, March 1995.

[429] J. D. Cho *et al.*, "Crosstalk-Minimum Layer Assignment," *Proceedings of the IEEE Conference on Custom Integrated Circuits*, pp. 29.7.1–29.7.4, May 1993.

[430] J. G. Ecker, "Geometric Programming: Methods, Computations and Applications," *SIAM Review*, Vol. 22, No. 3, pp. 338–362, July 1980.

[431] S. Boyd, S. J. Kim, L. Vandenberghe, and A. Hassibi, "A Tutorial on Geometric Programming," *Optimization and Engineering*, Vol. 8, No. 1, pp. 67–127, March 2007.

[432] W. Zhao and Y. Cao, "New Generation of Predictive Technology Model for Sub-45nm Design Exploration," *Proceedings of the IEEE International Symposium on Quality Electronic Design*, pp. 585–590, March 2006.

[433] J. Löfberg, "YALMIP: A Toolbox for Modeling and Optimization in MATLAB," *Proceedings of the IEEE International Symposium on Computer-Aided Control Systems Design*, pp. 284–289, September 2004.

[434] D. Henrion and J. B. Lasserre, "GloptiPoly: Global Optimization over Polynomials with Matlab and SeDuMi," *ACM Transactions on Mathematical Software*, Vol. 29, No. 2, pp. 165–194, June 2003.

[435] C.-P. Che, H. Zhou, and D. F. Wong, "Optimal Non-Uniform Wire-Sizing under the Elmore Delay Model," *Proceedings of the IEEE/ACM International Conference on Computer-Aided Design*, pp. 38–43, November 1996.

[436] C. Ryu *et al.*, "High Frequency Electrical Circuit Model of Chip-to-Chip Vertical Via Interconnection for 3-D Chip Stacking Package," *Proceedings of the IEEE Topical Meeting on Electrical Performance of Electronic Packaging*, pp. 151–154, October 2005.

[437] M. Pedram and S. Nazarian, "Thermal Analysis of Heterogeneous 3-D ICs with Various Integration Scenarios," *Proceedings of the IEEE*, Vol. 94, No. 8, pp. 1487–1501, August 2006.

[438] K. Banerjee, A. Mehrotra, A. Sangiovanni-Vincentelli, and C. Hu, "On Thermal Effects in Deep Sub-Micron VLSI Interconnects," *Proceedings of the IEEE/ACM Design Automation Conference*, pp. 885–890, June 1999.

[439] C. H. Tsai and S.-M. Kang, *Cell-Level Placement for Improving Substrate Thermal Distribution*, Vol. 19, No. 2, pp. 253–266, February 2000.

[440] V. Szekely, M. Rencz, and B. Courtois, "Tracing the Thermal Behavior of ICs," *IEEE Design and Test of Computers*, Vol. 15, No. 2, pp. 14–21, April/June 1998.

[441] M. B. Kleiner, S. A. Kühn, P. Ramn, and W. Weber, "Thermal Analysis of Vertically Integrated Circuits," *Proceedings of the IEEE International Electron Devices Meeting*, pp. 487–490, December 1995.

[442] Y. Zhang, H. Oh, and M. S. Bakir, "Within-Tier Cooling and Thermal Isolation Technologies for Heterogeneous 3D ICs," *Proceedings of the IEEE International 3D Systems Integration Conference*, pp. 1–6, October 2013.

[443] M. K. Tiwari *et al.*, "Waste Heat Recovery in Supercomputers and 3D Integrated Liquid Cooled Electronics," *Proceedings of the IEEE Intersociety Conference on Thermal and Thermomechanical Phenomena in Electronic Systems*, pp. 545–551, May/June 2012.

[444] F. P. Incropera, D. P. Dewitt, T. L. Bergman, and A. S. Lavine, *Introduction to Heat Transfer, Fifth Edition*, John Wiley & Sons, 2007.

[445] S.-T. Wu *et al.*, "Thermal and Mechanical Design and Analysis of 3D IC Interposer with Double-Sided Active Chips," *Proceedings of the IEEE Electronic Components and Technology Conference*, pp. 1471–1479, May 2013.

[446] R. Bazaz, J. Xie, and M. Swamanathan, "Electrical and Thermal Analysis for Design Exchange Formats in Three-Dimensional Integrated Circuits," *Proceedings of the IEEE International Symposium on Quality Electronic Design*, pp. 308–315, March 2013.

[447] H. C. Chien *et al.*, "Thermal Evaluation and Analyses of 3D IC Integration SiP with TSVs for Network System Applications," *Proceedings of the IEEE Electronic Components and Technology Conference*, pp. 1866–1873, May/June 2012.

[448] H. Qian *et al.*, "Thermal Simulator of 3D-IC with Modeling of Anisotropic TSV Conductance and Microchannel Entrance Effects," *Proceedings of the IEEE Asia and South Pacific Design Automation Conference*, pp. 485–490, January 2013.

[449] C. Torregiani *et al.*, "Compact Thermal Model of Hot Spots in Advanced 3D-Stacked ICs," *Proceedings of the IEEE Electronics Packaging Technology Conference*, pp. 131–136, December 2009.

[450] J. Xie and M. Swaminathan, "3D Transient Solver using Non-Conformal Domain Decomposition Approach," *Proceedings of the IEEE/ACM International Conference on Computer-Aided Design*, pp. 333–340, November 2012.

[451] M. S. Pittler, D. M. Powers, and D. L. Schnabel, "System Development and Technology Aspects of the IBM 3081 Processor Complex," *IBM Journal of Research and Development*, Vol. 26, No. 1, pp. 2–11, January 1982.

[452] D. B. Tuckerman and R. F. W. Pease, "High-Performance Heat Sinking for VLSI," *IEEE Electron Device Letters*, Vol. EDL-2, No. 5, pp. 126–129, May 1981.

[453] B. Dang, M. S. Bakir, and J. D. Meindl, "Integrated Thermal-Fluidic I/O Interconnects for an On-Chip Microchannel Heat Sink," *IEEE Electron Device Letters*, Vol. 27, No. 2, pp. 117–119, February 2006.

[454] J. G. Koomey, "Worldwide Electricity Used in Date Centers," *Environmental Research Letters*, Vol. 3, pp. 1–8, September 2008.

[455] Y. J. Kim *et al.*, "Thermal Characterization of Interlayer Microfluidic Cooling of Three-Dimensional Integrated Circuits With Nonuniform Heat Flux," *ASME Journal of Heat Transfer*, Vol. 132, pp. 041009-1–041009-9, April 2010.

[456] R. W. Knight, D. J. Hall, J. S. Goodling, and R. C. Jaeger, "Heat Sink Optimization with Application to Microchannels," *IEEE Transactions on Components and Hybrids, and Manufacturing Technologies*, Vol. 15, No. 5, pp. 832–842, October 1992.

[457] J. Li and G. P. Peterson, "Geometric Optimization of a Micro Heat Sink With Liquid Flow," *IEEE Transactions on Components and Packaging Technologies*, Vol. 29, No. 1, pp. 145–154, March 2006.

[458] S. Im and K. Banerjee, "Full Chip Thermal Analysis of Planar (2-D) and Vertically Integrated (3-D) High Performance ICs," *Proceedings of the IEEE International Electron Devices Meeting*, pp. 727–730, December 2000.

[459] A. Jain, R. E. Jones, R. Chatterjee, and S. Pozder, "Analytical and Numerical Modeling of the Thermal Performance of Three-Dimensional Integrated Circuits," *IEEE Transactions on Components and Packaging Technologies*, Vol. 33, No. 1, pp. 56–63, March 2010.

[460] T.-Y. Chiang, S. J. Souri, C. O. Chui, and K. C. Saraswat, "Thermal Analysis of Heterogeneous 3-D ICs with Various Integration Scenarios," *Proceedings of the IEEE International Electron Devices Meeting*, pp. 681–684, December 2001.

[461] C. C. Liu, J. Zhang, A. K. Datta, and S. Tiwari, "Heating Effects of Clock Drivers in Bulk, SOI, and 3-D CMOS," *IEEE Transactions on Electron Device Letters*, Vol. 23, No. 12, pp. 716–728, December 2002.

[462] [Online]. Available: http://www.uk.comsol.com/heat-transfer-module.

[463] [Online]. Available: http://www.ansys.com/Products/Simulation + Technology/Fluid + Dynamics/Specialized + Products/ANSYS + Icepak.

[464] H. Oprins *et al.*, "Thermal Test Vehicle for the Validation of Thermal Modeling of Hot-Spot Dissipation in 3D Stacked ICs," *Proceedings of the Electronic System-Integration Technology Conference*, pp. 1–6, September 2010.

[465] Z. Tan, M. Furmanczyk, M. Turowski, and A. Przekwas, "CFD-Micromesh: A Fast Geometrical Modeling and Mesh Generation Tool for 3D Microsystem Simulations," *Proceedings of the International Conference on Modeling and Simulation of Microsystems*, pp. 712–715, March 2000.

[466] P. Wilkerson, M. Furmanczyk, and M. Turowski, "Compact Thermal Model Analysis for 3-D Integrated Circuits," *Proceedings of the International Conference on Mixed Design of Integrated Circuits and Systems*, pp. 277–282, June 2004.

[467] P. Wilkerson, M. Furmanczyk, and M. Turowski, "Fast, Automated Thermal Simulation of Three-Dimensional Integrated Circuits," *Proceedings of the Intersociety Conference on Thermal and Thermomechanical Phenomena in Electronic Systems*, pp. 706–713, June 2004.

[468] H.-C. Chien *et al.*, "Estimation for Equivalent Thermal Conductivity of Silicon-Through Vias (TSVs) Used for 3D IC Integration," *Proceedings of the IEEE Microsystems, Packaging, Assembly and Circuits Technology Conference*, pp. 153–156, October 2011.

[469] H. Xu, V. F. Pavlidis, and G. De Micheli, "Analytical Heat Transfer Model for Thermal Through-Silicon Vias," *Proceedings of the Conference on Design, Automation, and Test, in Europe*, pp. 395–400, March 2011.

[470] Z. Liu, S. Swarup, and S. X.-D. Tan, "Compact Lateral Thermal Resistance Modeling and Characterization for TSV and TSV Array," *Proceedings of the IEEE/ACM International Conference on Computer-Aided Design*, pp. 275–280, November 2013.

[471] X. P. Wang, W.-Y. Yin, and S. He, "Multiphysics Characterization of Transient Electromechanical Responses of Through-Silicon Vias Applied With a Periodic Voltage Pulse," *IEEE Transactions on Electron Devices*, Vol. 57, No. 6, pp. 1382–1389, June 2010.

[472] A. Todri-Sanial *et al.*, "Globally Constrained Locally Optimized 3-D Power Delivery Network," *IEEE Transactions on Very Large Scale Integration (VLSI) Systems*, Vol. 22, No. 10, pp. 2131–2144, October 2014.

[473] M. E. Van Valkenburg, *Network Analysis, Third Edition*, Prentice-Hall, 1974.

[474] D. Liu and S. V. Garimella, "Analysis and Optimization of the Thermal Performance of Microchannel Heat Sinks," *International Journal for Numerical Methods in Heat & Fluid Flow*, Vol. 15, No. 1, pp. 7–26, 2005.

[475] B. Shi, A. Srivastava, and P. Wang, "Non-Uniform Micro-Channel Design for Stacked 3D-ICs," *Proceedings of the ACM/IEEE Design Automation Conference*, pp. 658–663, June 2011.

[476] National Research Council, Research Analysis Group, *Physics of Sound in the Sea: Part IV Acoustic Properties of Wakes*, Technical Report, Division 6, Vol. 8, The Murray Printing Company, 1949, reprinted in 1969.

[477] H. Mizunuma, Y.-C. Lu, and C. L. Yang, "Thermal Modeling and Analysis for 3-D ICs with Integrated Microchannel Cooling," *IEEE Transactions on Computer-Aided Design of Integrated Circuits and Systems*, Vol. 30, No. 9, pp. 1293–1306, September 2011.

[478] Y. Cheng, C. Tsai, C. Teng, and S. Kang, *Electrothermal Analysis of VLSI Systems*, Kluwer Academic Publishers, 2000.

[479] S. Wunsche, C. Claub, and P. Schwarz, "Electro-Thermal Circuit Simulation Using Simulator Coupling," *IEEE Transactions on Very Large Scale Integration (VLSI) Systems*, Vol. 5, No. 3, pp. 277–282, September 1997.

[480] M. N. Ozisik, *Finite Difference Methods in Heat Transfer*, CRC, 1994.

[481] G. Digele, S. Lindenkreuz, and E. Kasper, "Fully Coupled Dynamic Electro-Thermal Simulation," *IEEE Transactions on Very Large Scale Integration (VLSI) Systems*, Vol. 5, No. 3, pp. 250–257, September 1997.

[482] J. Xie and M. Swaminathan, "Electrical-Thermal Co-Simulation of 3D Integrated Systems With Micro-Fluidic Cooling and Joule Heating Effects," *IEEE Transactions on Components, Packaging, and Manufacturing Technologies*, Vol. 1, No. 2, pp. 234–246, February 2011.

[483] K. Stuben and U. Trottenberg, *Multigrid Methods: Fundamental Algorithms, Model Problem Analysis and Applications, Lecture Notes in Mathematics*, Springer-Verlag, 1982.

[484] H. C. Elman, M. D. Mihajlović, and D. J. Silvester, "Fast Iterative Solvers for Buoyancy Driven Flow Problems," *Journal of Computational Physics*, Vol. 230, pp. 3900–3914, May 2011.

[485] W. Briggs, *A Multigrid Tutorial*, SIAM, 1987.

[486] J. Ruge and K. Stuben, *Algebraic Multigrid (AMG) Methods, Frontiers in Applied Mathematics*, SIAM, 1987.

[487] P. Li, L. T. Pileggi, and M. Asheghi, "IC Thermal Simulation and Modeling via Efficient Multigrid-Based Approaches," *IEEE Transactions on Computer-Aided Design of Integrated Circuits and Systems*, Vol. 25, No. 9, pp. 1763–1776, September 2006.

[488] Z. Feng and P. Li, "Fast Thermal Analysis on GPU for 3D ICs With Integrated Microchannel Cooling," *IEEE Transactions on Computer-Aided Design of Integrated Circuits and Systems*, Vol. 21, No. 8, pp. 1526–1539, August 2013.

[489] D. Oh, C. C. P. Chen, and Y. H. Hu, "Efficient Thermal Simulation for 3-D IC With Thermal Through-Silicon Vias," *IEEE Transactions on Computer-Aided Design of Integrated Circuits and Systems*, Vol. 31, No. 11, pp. 1767–1771, November 2012.

[490] D. Oh, C. C. P. Chen, and Y. H. Hu, "3DFFT: Thermal Analysis of Non-Homogeneous IC Using 3D FFT Green Function Method," *Proceedings of the IEEE International Symposium on Quality Electronic Design*, pp. 567–572, March 2007.

[491] S. Melamed *et al.*, "Junction-Level Thermal Analysis of 3-D Integrated Circuits Using High Definition Power Blurring," *IEEE Transactions on Computer-Aided Design of Integrated Circuits and Systems*, Vol. 31, No. 5, pp. 676–689, May 2012.

[492] T. Kemper, Y. Zhang, Z. Bian, and A. Shakouri, "Ultrafast Temperature Profile Calculation in IC Chips," *Proceedings of the International Workshop on Thermal Investigations of ICs and Systems*, pp. 1–5, September 2006.

[493] R. C. Goncalez and R. E. Woods, *Digital Image Processing*, Prentice-Hall Inc., 2002.

[494] J.-H. Park, A. Shakouri, and S.-M. Kang, "Fast Thermal Analysis of Vertically Integrated Circuits (3-D ICs) Using Power Blurring Method," *Proceedings of the ASME InterPACK Conference*, pp. 701–707, July 2009.

[495] K. Puttaswamy and G. H. Loh, "Thermal Analysis of a 3-D Die Stacking High-Performance Microprocessor," *Proceedings of the ACM International Great Lakes Symposium on VLSI*, pp. 19–24, April/May 2006.

[496] W.-L. Hung *et al.*, "Interconnect and Thermal-Aware Floorplanning for 3-D Microprocessors," *Proceedings of the IEEE International Symposium on Quality Electronic Design*, pp. 98–103, March 2006.

[497] C. Zhu *et al.*, "Three-Dimensional Chip-Multiprocessor Run-Time Thermal Management," *IEEE Transactions on Computer-Aided Design of Integrated Circuits and Systems*, Vol. 27, No. 8, pp. 1479–1492, August 2008.

[498] X. Zhou *et al.*, "Thermal Management for 3D Processors via Task Scheduling," *Proceedings of the IEEE International Conference on Parallel Processing*, pp. 115–122, September 2008.

[499] [Online]. Available: https://en.wikipedia.org/wiki/DEC_Alpha (accessed in November 2016).

[500] P. Zhou *et al.*, "3D-STAF: Scalable Temperature and Leakage Aware Floorplanning for Three-Dimensional Integrated Circuits," *Proceedings of the IEEE/ACM International Conference on Computer-Aided Design*, pp. 590–597, November 2007.

[501] Y. Zang *et al.*, "ISAC: Integrated Space and Time Adaptive Chip-Package Thermal Analysis," *IEEE Transactions on Computer-Aided Design of Integrated Circuits and Systems*, Vol. 26, No. 1, pp. 86–99, January 2007.

[502] M. D. Moffitt, A. N. Ng, I. L. Markov, and M. E. Pollack, "Constraint-Driven Floorplan Repair," *Proceedings of the IEEE/ACM Design Automation Conference*, pp. 1103–1108, June 2006.

[503] C. Addo-Quaye, "Thermal-Aware Mapping and Placement for 3-D NoC Designs," *Proceedings of the IEEE International SOC Conference*, pp. 25–28, September 2005.

[504] D. E. Goldberg, *Genetic Algorithms in Search, Optimization, and Machine Learning*, Addison-Wesley, 1989.

[505] B. Goplen and S. Sapatnekar, "Efficient Thermal Placement of Standard Cells in 3-D ICs using a Force Directed Approach," *Proceedings of the IEEE/ACM International Conference on Computer-Aided Design*, pp. 86–89, November 2003.

[506] [Online]. Available: http://vlsicad.eecs.umich.edu/BK/Slots/cache/er.cs.ucla.edu/benchmarks/ibm-place2/.

[507] L. Benini and G. De Micheli, *Dynamic Power Management: Design Techniques and CAD Tools*, Springer, 1998.

[508] J. Donald and M. Martonosi, "Techniques for Multicore Thermal Management: Classification and New Exploration," *Proceedings of the IEEE International Symposium on Computer Architecture*, pp. 78–88, June 2006.

[509] C. Isci and M. Martonosi, "Runtime Power Monitoring in Highend Processors: Methodology and Empirical Data," *Proceedings of the IEEE/ACM International Symposium on Microarchitecture*, pp. 98–104, December 2003.

[510] J. Choi, "Thermal-Aware Task Scheduling at the System Software Level," *Proceedings of the IEEE/ACM International Symposium on Low Power Electronic Design*, pp. 213–218, August 2007.

[511] A. Kumar, L. Shang, L.-S. Peh, and N. Jha, "HybDTM: A Coordinated Hardware-Software Approach for Dynamic Thermal Management," *Proceedings of the IEEE/ACM Design Automation Conference*, pp. 548–553, July 2006.

[512] M. Gomaa, M. D. Powell, and T. N. Vijaykumar, "Heat-and-Run: Leveraging SMT and CMP to Manage Power Density through the Operating System," *Proceedings of the International Conference on Architectural Support for Programming Languages and Operating Systems*, pp. 260–270, October 2004.

[513] S. Sharifi, A. K. Coskun, and T. Simunic-Rosing, "Hybrid Dynamic Energy and Thermal Management in Heterogeneous Embedded Multiprocessor SoCs," *Proceedings of the ACM Asia and South Pacific Design Automation Conference*, pp. 873–878, January 2010.

[514] S. O. Memik, R. Mukherjee, M. Ni, and J. Long, "Optimizing Thermal Sensor Allocation for Microprocessors," *IEEE Transactions on Computer-Aided Design of Integrated Circuits and Systems*, Vol. 27, No. 3, pp. 516–527, March 2008.

[515] M. Ghosh and H.-H. S. Lee, "Smart Refresh: An Enhanced Memory Controller Design for Reducing Energy in Conventional and 3D Die-Stacked DRAMs," *Proceedings of the IEEE/ACM International Symposium on Microarchitecture*, pp. 134–145, December 2007.

[516] [Online]. Micron DRAM Power Calculator. Available: https://www.micron.com/support/power-calc (accessed in January 2016).

[517] [Online] Standard Performance Evaluation Corporation. Available: www.specbench.org.

[518] N. L. Binkert *et al.*, "The M5 Simulator: Modeling Networked Systems," *IEEE Micro*, Vol. 26, No. 4, pp. 52–60, July/August 2006.

[519] D. Brooks, V. Tiwari, and M. Martonosi, "Wattch: A Framework for Architectural-Level Power Analysis and Optimizations," *Proceedings of the ACM/IEEE International Symposium on Computer Architecture*, pp. 83–94, June 2000.

[520] D. Tarjan, S. Thoziyoor and N. P. Jouppi, "CACTI 4.0," HP Laboratories, Palo Alto, CA, Technical Report, HPL-2006-86, June 2006.

[521] J. Srinivasan, S. V. Adve, P. Bose, and J. A. Rivers, "Exploiting Structural Duplication for Lifetime Reliability Enhancement," *Proceedings of the ACM/IEEE International Symposium on Computer Architecture*, pp. 520–531, June 2005.

[522] Y. Yang *et al.*, "Adaptive Multi-Domain Thermal Modeling and Analysis for Integrated Circuit Synthesis and Design," *Proceedings of the IEEE International Conference on Computer-Aided Design*, pp. 575–582, November 2006.

[523] K. Kang, J. Kim, S. Yoo, and C.-M. Kyung, "Runtime Power Management of 3-D Multi-Core Architectures Under Peak Power and Temperature Constraints," *IEEE Transactions on Computer-Aided Design of Integrated Circuits and Systems*, Vol. 30, No. 6, pp. 905–918, June 2011.

[524] K. Choi, R. Soma, and M. Pedram, "Fine-Grained Dynamic Voltage and Frequency Scaling for Precise Energy and Performance Tradeoff Based on the Ratio of Off-Chip Access to On-Chip Computation Times," *IEEE Transactions on Computer-Aided Design of Integrated Circuits and Systems*, Vol. 24, No. 1, pp. 18–28, January 2005.

[525] [Online] *Performance Application Programming Interface*. Available: http://icl.cs.utk.edu/papi/.

[526] M.-L. Li *et al.*, "The ALPBench Benchmark Suite for Complex Multimedia Applications," *Proceedings of the IEEE International Symposium on Workload Characterization*, pp. 34–35, October 2005.

[527] S. Lee, K. Kang, and C.-M. Kyung, "Runtime Thermal Management for 3-D Chip-Multiprocessors With Hybrid SRAM/MRAM L2 Cache," *IEEE Transactions on Very Large Scale Integration (VLSI) Systems*, Vol. 23, No. 3, pp. 520–533, March 2015.

[528] G. De Micheli and L. Benini, *Networks on Chips: Technology and Tools*, Morgan Kaufmann Publishers, 2006.

[529] G. Sun *et al.*, "A Novel Architecture of the 3D Stacked MRAM L2 Cache for CMPs," *Proceedings of the IEEE International Symposium on High Performance Computer Architecture*, pp. 239–249, February 2009.

[530] D. H. Albonesi, "Selective Cache Ways: On-Demand Cache Resource Allocation," *Proceedings of the IEEE/ACM International Symposium on Microarchitecture*, pp. 248–259, November 1999.

[531] S. Pinel *et al.*, "Thermal Modeling and Management in Ultrathin Chip Stack Technology," *IEEE Transactions on Components and Packaging Technologies*, Vol. 25, No. 2, pp. 244–253, June 2002.

[532] S. Hu *et al.*, "A Thermal Isolation Technique Using Through-Silicon Vias for Three-Dimensional ICs," *IEEE Transactions on Electron Devices*, Vol. 60, No. 3, pp. 1282–1287, March 2013.

[533] E. Wong and S. K. Lim, "3D Floorplanning with Thermal Vias," *Proceedings of the IEEE Design, Automation, and Test Conference in Europe*, March 2006.

[534] L. Xiao, S. Sinha, J. Xu, and E. F. Y. Young, "Fixed-Outline Thermal-Aware 3D Floorplanning," *Proceedings of the IEEE Asia and South Pacific Design Automation Conference*, pp. 561−567, January 2010.

[535] P. Doyle and J. Snell, *Random Walks and Electric Networks*, Mathematical Association of America, 1984.

[536] B. Goplen and S. Sapatnekar, "Placement of Thermal Vias in 3-D ICs Using Various Thermal Objectives," *IEEE Transactions on Computer-Aided Design of Integrated Circuits and Systems*, Vol. 25, No. 4, pp. 692−709, April 2006.

[537] J. Cong and Y. Zhang, "Thermal Driven Multilevel Routing for 3-D ICs," *Proceedings of the IEEE Asia and South Pacific Design Automation Conference*, pp. 121−126, June 2005.

[538] J. Cong and Y. Zhang, "Thermal Via Planning for 3-D ICs," *Proceedings of the IEEE/ACM International Conference on Computer-Aided Design*, pp. 744−751, November 2005.

[539] J. Cong, M. Xie, and Y. Zhang, "An Enhanced Multilevel Routing System," *Proceedings of the IEEE/ACM International Conference on Computer-Aided Design*, pp. 51−58, November 2002.

[540] J. Cong, J. Fang, and Y. Zhang, "Multilevel Approach to Full-Chip Gridless Routing," *Proceedings of the IEEE/ACM International Conference on Computer-Aided Design*, pp. 234−241, November 2001.

[541] Z. Li *et al.*, "Efficient Thermal Via Planning Approach and its Application in 3-D Floorplanning," *IEEE Transactions on Computer-Aided Design of Integrated Circuits and Systems*, Vol. 26, No. 4, pp. 645−658, April 2007.

[542] T. H. Cormen, C. E. Leiserson, and R. L. Rivest, *Introduction to Algorithms*, The MIT Press, 1990.

[543] K. Wang *et al.*, "Rethinking Thermal Via Planning with Timing-Power-Temperature Dependence for 3D ICs," *Proceedings of the IEEE Asia and South Pacific Design Automation Conference*, pp. 261−266, January 2011.

[544] T. Zhang, Y. Zhang, and S. Sapatnekar, "Temperature-Aware Routing in 3-D ICs," *Proceedings of the IEEE Asia and South Pacific Design Automation Conference*, pp. 309−314, January 2006.

[545] A. Leon *et al.*, "A Power-Efficient High-Throughput 32-Thread SPARC Processor," *Proceedings of the IEEE International Solid-State Circuits Conference*, pp. 295−304, February 2006.

[546] M. M. Sabry *et al.*, "Energy-Efficient Multiobjective Thermal Control for Liquid-Cooled 3-D Stacked Architectures," *IEEE Transactions on Computer-Aided Design of Integrated Circuits and Systems*, Vol. 30, No. 12, pp. 1883−1896, December 2011.

[547] H. Qian, X. Huang, H. Yu, and C. Chang, "Cyber-Physical Thermal Management of 3D Multi-Core Cache-Processor System with Microfluidic Cooling," *Journal of Low Power Electronics*, Vol. 7, No. 1, pp. 110−121, February 2011.

[548] M.-C. Hsieh and C.-K. Yu, "Thermo-Mechanical Simulations for 4-Layer Stacked IC Packages," *Proceedings of the EuroSimE-International Conference on Thermal, Mechanical and Multi-Physics Simulation and Experiments in Microelectronics and Micro-Systems*, pp. 1−7, April 2008.

[549] L. Jiang *et al.*, "Thermal Modeling of On-Chip Interconnects and 3D Packaging Using EM Tools," *Proceedings of the IEEE Electrical Performance of Electronic Packaging*, pp. 279−282, October 2008.

[550] A. Jain, R. E. Jones, and S. Pozder, "Thermal Modeling and Design of 3D Integrated Circuits," *Proceedings of the Intersociety Conference on Thermal and Thermomechanical Phenomena in Electronic Systems*, pp. 1139−1145, May 2008.

[551] A. Sridhar *et al.*, "3D-ICE: Fast Compact Transient Thermal Modeling for 3D ICs with Inter-Tier Liquid Cooling," *Proceedings of the IEEE International Conference on Computer-Aided Design*, pp. 463−470, November 2010.

[552] W. Huang *et al.*, "Compact Thermal Modeling for Temperature-Aware Design," *Proceedings of the IEEE/ACM Design Automation Conference*, pp. 878−883, June 2004.

[553] W. Huang *et al.*, "HotSpot: A Compact Thermal Modeling Methodology for Early-Stage VLSI Design," *IEEE Transactions on Very Large Scale Integration (VLSI) Systems*, Vol. 14, No. 5, pp. 501−513, May 2006.

[554] J. Cong, G. Luo, J. Wei, and Y. Zhang, "Thermal-Aware 3D IC Placement Via Transformation," *Proceedings of the IEEE Asia and South Pacific Design Automation Conference*, pp. 780–785, January 2007.

[555] K. Balakrishnan, V. Nanda, and S. Easwar, "Wire Congestion and Thermal Aware 3D Global Placement," *Proceedings of the IEEE Asia and South Pacific Design Automation Conference*, pp. 1131–1134, January 2005.

[556] K. Puttaswamy and G. H. Loh, "Thermal Herding: Microarchitecture Techniques for Controlling Hotspots in High-Performance 3D-Integrated Processors," *Proceedings of the IEEE International Symposium on High Performance Computer Architecture*, pp. 193–204, February 2007.

[557] M. S. Bakir *et al.*, "3D Heterogeneous Integrated Systems: Liquid Cooling, Power Delivery, and Implementation," *Proceedings of the IEEE Custom Integrated Circuits Conference*, pp. 663–670, September 2008.

[558] C. R. King *et al.*, "3D Stacking of Chips with Electrical and Microfluidic I/O Interconnects," *Proceedings of the IEEE Electronic Components and Technology Conference*, pp. 1–7, May 2008.

[559] B. Dang *et al.*, "Integrated Microfluidic Cooling and Interconnects for 2D and 3D Chips," *IEEE Transactions on Advanced Packaging*, Vol. 33, No. 1, pp. 79–87, February 2010.

[560] M. M. Sabry, D. Atienza, and A. K. Coskun, "Thermal Analysis and Active Cooling Management for 3D MPSoCs," *Proceedings of the IEEE International Symposium on Circuits and Systems*, pp. 2237–2240, May 2011.

[561] H. Oprins *et al.*, "Numerical and Experimental Characterization of the Thermal Behavior of a Packaged DRAM-on-Logic Stack," *Proceedings of the IEEE Electronic Components and Technology Conference*, pp. 1081–1088, June 2012.

[562] Tezzaron Semiconductor, http://www.tezzaron.com/.

[563] C. Yibo *et al.*, "Through Silicon Via Aware Design Planning for Thermally Efficient 3-D Integrated Circuits," *IEEE Transactions on Computer-Aided Design of Integrated Circuits and Systems*, Vol. 32, No. 9, pp. 1335–1346, September 2013.

[564] C. Santos *et al.*, "System-Level Thermal Modeling for 3D Circuits: Characterization with a 65nm Memory-on-Logic Circuit," *Proceedings of the IEEE International 3-D Systems Integration Conference*, pp. 1–6, October 2013.

[565] I. Vaisband and E. G. Friedman, "Heterogeneous Methodology for Energy Efficient Distribution of On-Chip Power Supplies," *IEEE Transactions on Power Electronics*, Vol. 28, No. 9, pp. 4267–4280, September 2013.

[566] T. M. Andersen *et al.*, "A 4.6 W/mm^2 Power Density 86% Efficiency On-Chip Switched Capacitor DC-DC Converter in 32 nm SOI CMOS," *Proceedings of the IEEE Applied Power Electronics Conference and Exposition*, pp. 692–699, March 2013.

[567] I. Vaisband, A. Mahmood, E. G. Friedman, and S. Kose, "Digitally Controlled Pulse Width Modulator for On-Chip Power Management," *IEEE Transactions on Very Large Scale Integration (VLSI) Systems*, Vol. 22, No. 12, pp. 2527–2534, December 2014.

[568] B. Ciftcioglu *et al.*, "3-D Integrated Heterogeneous Intra-Chip Free-Space Optical Interconnect," *Optics Express*, Vol. 20, No. 4, pp. 4331–4345, February 2012.

[569] J. Xue *et al.*, "An Intra-Chip Free-Space Optical Interconnect," *Proceedings of the ACM/IEEE International Symposium on Computer Architecture*, pp. 94–105, June 2010.

[570] B. Ciftcioglu *et al.*, "A 3-D Integrated Intrachip Free-Space Optical Interconnect for Many-Core Chips," *IEEE Photonics Technology Letters*, Vol. 23, No. 3, pp. 164–166, February 2011.

[571] J. Wang, I. Savidis, and E. G. Friedman, "Thermal Analysis of Oxide-Confined VCSEL Arrays," *Microelectronics Journal*, Vol. 42, No. 5, pp. 820–825, May 2011.

[572] K. Skadron *et al.*, "Temperature-Aware Microarchitecture," *Proceedings of the IEEE International Symposium on Computer Architecture*, pp. 2–13, May 2003.

[573] J. Meng, K. Kawakami, and A. K. Coskun, "Optimizing Energy Eciency of 3-D Multicore Systems with Stacked DRAM under Power and Thermal Constraints," *Proceedings of the ACM/IEEE Design Automation Conference*, pp. 648–655, March 2012.
[574] E. G. Friedman, Ed., *Clock Distribution Networks in VLSI Circuits and Systems*, IEEE Press, 1995.
[575] A. Deutsch and P. J. Restle, "Designing the Best Clock Distribution Network," *Proceedings of the IEEE Symposium on VLSI Circuits*, pp. 2–5, June 1998.
[576] E. G. Friedman, "Clock Distribution Design in VLSI Circuits-an Overview," *Proceedings of the IEEE International Symposium on Circuits and Systems*, pp. 1475–1478, May 1993.
[577] J. L. Neves and E. G. Friedman, "Design Methodology for Synthesizing Clock Distribution Networks Exploiting Non-Zero Clock Skew," *IEEE Transactions on Very Large Scale Integration (VLSI) Systems*, Vol. VLSI-4, No. 2, pp. 286–291, June 1996.
[578] M. A. B. Jackson, A. Srinivasan, and E. S. Kuh, "Clock Routing for High-Performance ICs," *Proceedings of the ACM/IEEE Design Automation Conference*, pp. 573–579, June 1990.
[579] R.-S. Tsay, "An Exact Zero-Skew Clock Routing Algorithm," *IEEE Transactions on Computer-Aided Design of Integrated Circuits and Systems*, Vol. 12, No. 2, pp. 242–249, February 1993.
[580] M. Edahiro, "An Efficient Zero-Skew Routing Algorithm," *Proceedings of the ACM/IEEE Design Automation Conference*, pp. 375–380, June 1994.
[581] K. D. Boese and A. B. Kahng, "Zero-Skew Routing Trees With Minimum Wirelength," *Proceedings of the International ASIC Conference*, pp. 17–21, September 1992.
[582] T.-H. Chao *et al.*, "Zero Skew Clock Routing with Minimum Wirelength," *IEEE Transactions on Circuits and Systems II: Analog and Digital Signal Processing*, Vol. 39, No. 11, pp. 799–814, November 1992.
[583] E. G. Friedman, "Clock Distribution Networks in Synchronous Digital Integrated Circuits," *Proceedings of the IEEE*, Vol. 89, No. 5, pp. 665–692, May 2001.
[584] C. J. Alpert *et al.*, "Minimum Buffered Routing with Bounded Capacitive Load for Slew Rate and Reliability Control," *IEEE Transactions on Computer-Aided Design of Integrated Circuits and Systems*, Vol. 22, No. 3, pp. 241–253, March 2003.
[585] A. Rajaram and D. Z. Pan, "Robust Chip-Level Clock Tree Synthesis," *IEEE Transactions on Computer-Aided Design of Integrated Circuits and Systems*, Vol. 30, No. 6, pp. 877–890, June 2011.
[586] J. Cong, A. B. Kahng, C.-K. Koh, and C.-W. A. Tsao, "Bounded-Skew Clock and Steiner Routing," *ACM Transactions on Design Automation of Electronic Systems*, Vol. 3, No. 3, pp. 341–388, July 1998.
[587] P. J. Restle *et al.*, "A Clock Distribution Network for Microprocessors," *IEEE Journal of Solid-State Circuits*, Vol. 36, No. 5, pp. 792–799, May 2001.
[588] Y. I. Ismail, E. G. Friedman, and J. L. Neves, "Figures of Merit to Characterize the Importance of On-Chip Inductance," *IEEE Transactions on Very Large Scale Integration (VLSI) Systems*, Vol. 7, No. 4, pp. 442–449, December 1999.
[589] V. Arunachalam and W. Burleson, "Low-Power Clock Distribution in a Multilayer Core 3D Microprocessor," *Proceedings of the ACM Great Lakes Symposium on VLSI*, pp. 429–434, May 2008.
[590] R. E. Kessler, "The Alpha 21264 Microprocessor," *IEEE Micro*, Vol. 19, No. 2, pp. 24–36, March/April 1999.
[591] L.-T. Pang *et al.*, "A Shorted Global Clock Design for Multi-GHz 3D Stacked Chips," *Proceedings of the Symposium on VLSI Circuits*, pp. 170–171, June 2012.
[592] M. Wordeman, J. Silberman, G. Maier, and M. Scheuermann, "A 3D System Prototype of an eDRAM Cache Stacked Over Processor-Like Logic Using Through-Silicon Vias," *Proceedings of the IEEE International Solid-State Circuits Conference*, pp. 186–187, February 2012.
[593] X. Zhao, J. Minz, and S. K. Lim, "Low-Power and Reliable Clock Network Design for Through-Silicon Vias (TSV) Based 3D ICs," *IEEE Transactions on Components, Packaging, and Manufacturing Technology*, Vol. 1, No. 2, pp. 247–259, February 2011.

[594] T.-Y. Kim and T. Kim, "Clock Tree Embedding for 3D ICs," *Proceedings of the IEEE Asia and South Pacific Design Automation Conference*, pp. 486–491, January 2010.

[595] T.-Y. Kim and T. Kim, "Clock Tree Synthesis with Pre-bond Testability for 3D Stacked IC Designs," *Proceedings of the ACM/IEEE Design Automation Conference*, pp. 723–728, June 2010.

[596] Z. Zhao, D. L. Lewis, H.-H. S. Lee, and S. K. Lim, "Low-Power Clock Tree Design for Pre-Bond Testing of 3-D Stacked ICs," *IEEE Transactions on Computer-Aided Design of Integrated Circuits and Systems*, Vol. 30, No. 5, pp. 732–745, May 2011.

[597] W. Liu *et al.*, "Whitespace-Aware TSV Arrangement in 3-D Clock Tree Synthesis," *IEEE Transactions on Very Large Scale Integration (VLSI) Systems*, Vol. 23, No. 9, pp. 1842–1853, September 2015.

[598] C.-L. Lung *et al.*, "Through-Silicon Via Fault-Tolerant Clock Networks for 3-D ICs," *IEEE Transactions on Computer-Aided Design of Integrated Circuits and Systems*, Vol. 32, No. 7, pp. 1100–1109, July 2013.

[599] W. R. Bottoms, "Test Challenges for 3D Integration," *Proceedings of the IEEE International Custom Integrated Circuits Conference*, pp. 1–8, September 2011.

[600] X. Zhao and S. K. Lim, "Through-Silicon-Via-Induced Obstacle-Aware Clock Tree Synthesis for 3D ICs," *Proceedings of the IEEE Asia and South Pacific Design Automation Conference*, pp. 347–352, January/February 2012.

[601] G. Di Natale, M.-L. Flottes, B. Rouzeyre, and H. Zimouche, "Built-in Self-Test for Manufacturing TSV Defects Before Bonding," *Proceedings of the IEEE VLSI Test Symposium*, pp. 1–6, April 2014.

[602] G. E. Tellez and M. Sarrafzadeh, "Minimal Buffer Insertion in Clock Trees with Skew and Slew Rate Constraints," *IEEE Transactions on Computer-Aided Design of Integrated Circuits and Systems*, Vol. 16, No. 4, pp. 333–342, April 1997.

[603] [Online] *GSRC Benchmark Circuits*. Available: http://vlsicad.ucsd.edu/GSRC/bookshelf/Slots/BST.

[604] K. Chakrabarty, S. Deutsch, H. Thapliyal, and F. Ye, "TSV Defects and TSV-Induced Circuit Failures: The Third Dimension in Test and Design for Test," *Proceedings of the IEEE International Reliability Physics Symposium*, pp. 5F.1.1-5F.1.12, April 2012.

[605] E. J. Marinissen and Y. Zorian, "Testing 3-D Chips Containing Through-Silicon-Vias," *Proceedings of the IEEE International Test Conference*, pp. 1–11, November 2009.

[606] D. L. Lewis and H.-H. S. Lee, "A Scan-Island Based Design Enabling Pre-Bond Testability in Die-Stacked Microprocessors," *Proceedings of the IEEE International Test Conference*, October 2007.

[607] [Online] *RMST-Pack*. Available: http://vlsicad.ucsd.edu/GSRC/bookshelf/Slots/RSMT/ RMST.

[608] [Online] *ISPD Contest 2009*. Available: http://ispd.cc/contests/09/ispd09cts.html.

[609] M. Laisne, K. Arabi, and T. Petrov, "System and Methods Utilizing Redundancy in Semiconductor Chip Interconnects," U.S. Patent No. 8,384,417 B2, February 2013.

[610] J. Kim, F. Wang, and M. Nowak, "Method and Apparatus for Providing Through Silicon Via (TSV) Redundancy," U.S. Patent No. 8,988,130 B2, March 2015.

[611] A. C. Hsieh *et al.*, "TSV Redundancy: Architecture and Design Issues in 3-D IC," *Proceedings of the IEEE Conference on Design, Automation and Test in Europe*, pp. 1206–1211, March 2010.

[612] U. Kang *et al.*, "8 Gb 3-D DDR3 DRAM Using Through-Silicon-Via Technology," *Proceedings of the IEEE International Solid-State Circuits Conference*, pp. 130–132, February 2009.

[613] W. Cui, H. Chen, and Y. Han, "VLSI Implementation of Universal Random Number Generator," *Proceedings of the IEEE Asia-Pacific Conference on Circuits and Systems*, Vol. 1, pp. 465–470, October 2002.

[614] N. Hedenstierna and K. O. Jeppson, "CMOS Circuit Speed and Buffer Optimization," *IEEE Transactions on Computer-Aided Design*, Vol. CAD-6, No. 2, pp. 270–281, March 1987.

[615] N. C. Li, G. L. Haviland, and A. A. Tuszynski, "CMOS Tapered Buffer," *IEEE Journal of Solid-State Circuits*, Vol. 25, No. 4, pp. 1005–1008, 1990.

[616] C. Punty and L. Gal, "Optimum Tapered Buffer," *IEEE Journal of Solid-State Circuits*, Vol. 27, No. 1, pp. 1005–1008, January 1992.

[617] B. S. Cherkauer and E. G. Friedman, "A Unified Design Methodology for CMOS Tapered Buffers," *IEEE Transactions on Very Large Scale Integration (VLSI) Systems*, Vol. 3, No. 1, pp. 99–111, March 1995.
[618] I. Savidis, V. F. Pavlidis, and E. G. Friedman, "Clock Distribution Models of 3-D Integrated Systems," *Proceedings of the IEEE International Symposium on Circuits and Systems*, pp. 2225–2228, May 2011.
[619] S. G. Duvall, "Statistical Circuit Modeling and Optimization," *Proceedings of the International Workshop on Statistical Metrology*, pp. 56–63, June 2000.
[620] S. Asai and Y. Wada, "Technology Challenges for Integration Near and Below 0.1 μm," *Proceedings of the IEEE*, Vol. 85, No. 4, pp. 505–520, April 1997.
[621] D. Sylvester and C. Wu, "Analytical Modeling and Characterization of Deep-Submicrometer Interconnect," *Proceedings of the IEEE*, Vol. 89, No. 5, pp. 634–664, May 2001.
[622] S. Nassif, "Delay Variability: Sources, Impact and Trends," *Proceedings of the IEEE International Solid-State Circuits Conference*, pp. 368–369, February 2000.
[623] M. Dietrich, In J. Haase, Ed., *Process Variations and Probabilistic Integrated Circuit Design*, Springer, 2012.
[624] T. A. Bruner, "Impact of Lens Aberrations on Optical Lithography," *Journal of IBM Research and Development*, Vol. 41, No. 1, pp. 57–67, January 1997.
[625] A. K. K. Wong, *Resolution Enhancement Techniques in Optical Lithography*, SPIE Press, 2001.
[626] I. Matthew *et al.*, "Design Restrictions for Patterning with Off Axis Illumination," *Proceedings of SPIE*, Vol. 5754, pp. 1574–1585, May 2004.
[627] R. Chang, Y. Cao, and C. Spanos, "Modeling the Electrical Effects of Metal Dishing due to CMP for On-Chip Interconnect Optimization," *IEEE Transactions on Electron Devices*, Vol. 51, No. 10, pp. 1577–1583, October 2004.
[628] J. Cain and C. Spanos, "Electrical Linewidth Metrology for Systematic CD Variation Characterization and Causal Analysis," *Proceedings of SPIE*, Vol. 5038, pp. 350–361, June 2003.
[629] M. Orshansky, S. R. Nassif, and D. Boning, *Design for Manufacturability and Statistical Design*, Springer Science + Business Media, LLC, 2008.
[630] T. McConaghy, K. Breen, J. Dyck, and A. Gupta, *Variation-Aware Design of Custom Integrated Circuits: A Hands-on Field Guide*, Springer Science + Business Media, LLC, 2013.
[631] A. Shreider, *The Monte Carlo Method*, Pergamon Press, 1966.
[632] S. Garg and D. Marculescu, "3D-GCP: An Analytical Model for the Impact of Process Variations on the Critical Path Delay Distribution of 3D ICs," *Proceedings of the IEEE International Symposium on Quality of Electronic Design*, pp. 147–155, March 2009.
[633] K. A. Bowman, S. G. Duvall, and J. D. Meindl, "Impact of Die-to-Die and Within-Die Parameter Fluctuations on the Maximum Clock Frequency Distribution for Gigascale Integration," *IEEE Journal of Solid-State Circuits*, Vol. 37, No. 2, pp. 183–190, February 2002.
[634] M. Eisele, J. Berthold, D. Schmitt-Landsiedel, and R. Mahnkopf, "The Impact of Intra-Die Device Parameter Variations on Path Delays and on the Design for Yield of Low Voltage Digital Circuits," *IEEE Transactions on Very Large Scale Integration (VLSI) Systems*, Vol. 5, No. 4, pp. 360–368, April 1997.
[635] X. Jiang and S. Horiguchi, "Statistical Skew Modeling for General Clock Distribution Networks in Presence of Process Variations," *IEEE Transactions on Very Large Scale Integration (VLSI) Systems*, Vol. 9, No. 5, pp. 704–717, May 2001.
[636] H. Xu, V. F. Pavlidis, and G. De Micheli, "Process-Induced Skew Variation for Scaled 2-D and 3-D ICs," *Proceedings of the ACM/IEEE System Level Interconnect Prediction Workshop*, pp. 17–24, June 2010.

[637] H. Chang and S. Sapatnekar, "Statistical Timing Analysis Under Spatial Correlations," *IEEE Transactions on Computer-Aided Design of Integrated Circuits and Systems*, Vol. 24, No. 9, pp. 1467–1482, September 2005.

[638] K. A. Bowman *et al.*, "Impact of Die-to-Die and Within-Die Parameter Variations on the Clock Frequency and Throughput of Multi-Core Processors," *IEEE Transactions on Very Large Scale Integration (VLSI) Systems*, Vol. 17, No. 12, pp. 1679–1690, December 2009.

[639] S. Garg and D. Marculescu, "System-Level Process Variability Analysis and Mitigation for 3D MPSoCs," *Proceedings of the Design, Automation and Test in Europe Conference*, pp. 604–609, April 2009.

[640] A. Agarwal, D. Blaauw, and V. Zolotov, "Statistical Timing Analysis for Intra-Die Process Variations with Spatial Correlations," *Proceedings of the ACM/IEEE International Conference on Computer-Aided Design*, pp. 900–907, November 2003.

[641] M. Orshansky *et al.*, "Impact of Systematic Spatial Intra-Chip Gate Length Variability on Performance of High-Speed Digital Circuits," *Proceedings of the ACM/IEEE International Conference on Computer-Aided Design*, pp. 62–67, November 2000.

[642] M. Hashimoto, T. Yamamoto, and H. Onodera, "Statistical Analysis of Clock Skew Variation in H-tree Structure," *Proceedings of the IEEE International Symposium on Quality of Electronic Design*, Vol. 88, pp. 402–407, March 2005.

[643] [Online] International Technology Roadmap for Semiconductors ITRS, 2009 Edition. Available: http://www.itrs2.net.

[644] H. Xu, V. F. Pavlidis, and G. De Micheli, "Effects of Process Variations on 3-D Global Clock Distribution Networks," *ACM Journal on Emerging Technologies in Computing Systems*, Vol. 8, No. 3, p. Article 20, August 2012.

[645] T. Xanthopoulos, Ed., *Clocking in Modern VLSI Systems*, Springer, 2009.

[646] [Online] JEDEC Standard, Definition of Skew Specifications for Standard Logic Devices. Available: http://www.jedec.org/sites/default/files/docs/jesd65b.pdf.

[647] B. Razavi, *Phase-Locking in High-Performance Systems: From Devices to Architectures*, John Wiley & Sons, 2003.

[648] M. Saint-Laurent and M. Swaminathan, "Impact of Power-Supply Noise on Timing in High-Frequency Microprocessors," *IEEE Transactions on Advanced Packaging*, Vol. 27, No. 1, pp. 135–144, February 2004.

[649] J. Jang, O. Franza, and W. Burleson, "Compact Expressions for Supply Noise Induced Period Jitter of Global Binary Clock Trees," *IEEE Transactions on Very Large Scale Integration (VLSI) Systems*, Vol. 20, No. 1, pp. 66–79, December 2010.

[650] K. L. Wong, T. Rahal-Arabi, M. Ma, and G. Taylor, "Enhancing Microprocessor Immunity to Power Supply Noise With Clock-Data Compensation," *IEEE Journal of Solid-State Circuits*, Vol. 41, No. 4, pp. 749–758, April 2006.

[651] H. Xu, V. F. Pavlidis, X. Tang, W. Burleson, and G. De Micheli, "Timing Uncertainty in 3-D Clock Trees Due to Process Variations and Power Supply Noise," *IEEE Transactions on Very Large Scale Integration (VLSI) Systems*, Vol. 21, No. 12, pp. 2226–2239, December 2013.

[652] R. Franch *et al.*, "On-Chip Timing Uncertainty Measurements on IBM Microprocessors," *Proceedings of the IEEE International Test Conference*, pp. 1–7, October 2007.

[653] K. Shinkai, M. Hashimoto, A. Kurokawa, and T. Onoye, "A Gate Delay Model Focusing on Current Fluctuation over Wide-Range of Process and Environmental Variability," *Proceedings of the ACM/IEEE International Conference on Computer-Aided Design*, pp. 47 53, November 2006.

[654] Y. Ismail, E. G. Friedman, and J. Neves, "Equivalent Elmore Delay for *RLC* Trees," *IEEE Transactions on Computer-Aided Design of Integrated Circuits and Systems*, Vol. 19, No. 1, pp. 83–97, January 2000.

[655] G. Chen and E. G. Friedman, "Low-Power Repeaters Driving *RC* and *RLC* Interconnects with Delay and Bandwidth Constraints," *IEEE Transactions on Very Large Scale Integration (VLSI) Systems*, Vol. 14, No. 2, pp. 161–172, February 2006.

[656] X. Zhao, S. Mukhopadhyay, and S. K. Lim, "Variation-Tolerant and Low-Power Clock Network Design for 3D ICs," *Proceedings of the IEEE Electronic Components and Technology Conference*, pp. 2007–2014, June 2011.

[657] J. Yang *et al.*, "Robust Clock Tree Synthesis with Timing Yield Optimization for 3D-ICs," *Proceedings of the Asia and South Pacific Design Automation Conference*, pp. 621–626, January 2011.

[658] L. Yu *et al.*, "Methodology for Analysis of TSV Stress Induced Transistor Variation and Circuit Performance," *Proceedings of the IEEE International Symposium on Quality Electronic Design*, pp. 216–222, March 2012.

[659] P. Friedberg, J. Cain, and C. Spanos, "Modeling Within-Die Spatial Correlation Effects for Process-Design Co-Optimization," *Proceedings of the IEEE International Symposium on Quality of Electronic Design*, pp. 516–521, March 2005.

[660] A. Agarwal, V. Zolotov, and D. Blaauw, "Statistical Clock Skew Analysis Considering Intradie-Process Variations," *IEEE Transactions on Computer-Aided Design of Integrated Circuits and Systems*, Vol. 23, No. 8, pp. 1231–1242, August 2004.

[661] [Online] International Technology Roadmap for Semiconductors ITRS, 2010 Edition. Available: http://www.itrs2.net.

[662] S. Pant and E. Chiprout, "Power Grid Physics and Implications for CAD," *Proceedings of the ACM/IEEE Design Automation Conference*, pp. 199–204, July 2006.

[663] D. Jiao, J. Gu, and C. Kim, "Circuit Design and Modeling Techniques for Enhancing the Clock-Data Compensation Effect Under Resonant Supply Noise," *IEEE Journal of Solid-State Circuits*, Vol. 45, No. 10, pp. 2130–2141, October 2010.

[664] [Online] R. S. Tsay, IBM Clock Benchmarks. Available: http://vlsicad.ucsd.edu/GSRC/bookshelf/Slots/BST/#III.

[665] G. Bai, S. Bobba, and I. N. Hajj, "Static Timing Analysis Including Power Supply Noise Effect on Propagation Delay in VLSI Circuits," *Proceedings of the ACM/IEEE Design Automation Conference*, pp. 295–300, June 2011.

[666] J. Sun *et al.*, "3D Power Delivery for Microprocessors and High-Performance ASICs," *Proceedings of the IEEE Applied Power Electronics Conference*, pp. 127–133, March 2007.

[667] Y. Shinozuka *et al.*, "Reducing *IR* Drop in 3D Integration to Less than ¼ Using Buck Converter on Top Die (BCT) Scheme," *Proceedings of the IEEE International Symposium on Quality Electronic Design*, pp. 210–215, March 2013.

[668] G. Schrom *et al.*, "Optimal Design of Monolithic Integrated DC-DC Converters," *Proceedings of the IEEE International Conference on IC Design and Technology*, pp. 65–67, May 2006.

[669] P. Jain, T.-H. Kim, J. Keane, and C. H. Kim, "A Multi-Story Power Delivery Technique for 3D Integrated Circuits," *Proceedings of the ACM/IEEE International Symposium on Low Power Electronic Design*, pp. 57–62, August 2008.

[670] MITLL Low-Power FDSOI CMOS Process Design Guide, MIT Lincoln Laboratories, September 2008.

[671] R. Zhang *et al.*, "A Cross-Layer Design Exploration of Charge-Recycled Power-Delivery in Many-Layer 3D-IC," *Proceedings of the ACM/IEEE Design Automation Conference*, pp. 1–6, June 2015.

[672] V. F. Pavlidis and G. De Micheli, "Power Distribution Paths for 3-D ICs," *Proceedings of the International ACM Great Lakes Symposium on Very Large Scale Integration*, pp. 263–268, May 2009.

[673] A. Todri-Sanial *et al.*, "A Study of Tapered 3-D TSVs for Power and Thermal Integrity," *IEEE Transactions on Very Large Scale Integration (VLSI) Systems*, Vol. 21, No. 2, pp. 306–319, February 2013.

[674] G. Huang, M. S. Bakir, A. Naeemi, and J. D. Meindl, "Power Delivery for 3-D Chip Stacks: Physical Modeling and Design Implication," *IEEE Transactions on Components, Packaging, and Manufacturing Technology*, Vol. 2, No. 5, pp. 852–859, May 2012.

[675] G. Huang *et al.*, "Compact Physical Models for Power Supply Noise and Chip/Package Co-Design of Gigascale Integration," *Proceedings of the Electronic Components and Technology Conference*, pp. 1659–1666, May/June 2007.

[676] G. Huang *et al.*, "Power Delivery for 3D Chip Stacks: Physical Modeling and Design Implications," *Proceedings of the IEEE Electrical Performance of Electronic Packaging*, pp. 205–208, October 2007.

[677] A. D. Polyamin, *Handbook of Linear Partial Differential Equations for Engineers and Scientists*, Chapman & Hall, 2002.

[678] P. Bai *et al.*, "A 65 nm Logic Technology Featuring 35 nm Gate Lengths, Enhanced Channel Strain, 8 Cu Interconnect Layers, Low-k ILD and 0.57 cm^2 SRAM Cell," *Proceedings of the IEEE International Electron Device Meeting*, pp. 657–660, December 2004.

[679] K. Kim *et al.*, "Modeling and Analysis of a Power Distribution Network in TSV-Based 3-D Memory IC Including P/G TSVs, On-Chip Decoupling Capacitors, and Silicon Substrate Effects," *IEEE Transactions on Components, Packaging, and Manufacturing Technology*, Vol. 2, No. 12, pp. 2057–2070, December 2012.

[680] H. Chen and D. Ling, "Power Supply Noise Analysis Methodology for Deep-Submicron VLSI Chip Design," *Proceedings of the ACM/IEEE Design Automation Conference*, pp. 638–643, June 1997.

[681] J. N. Kozhaya, S. R. Nassif, and F. N. Najm, "A Multigrid Like Technique for Power Grid Analysis," *IEEE Transactions on Computer-Aided Design of Integrated Circuits and Systems*, Vol. 21, No. 10, pp. 1148–1160, October 2002.

[682] Y. Zhong and M. D. F. Wong, "Thermal-Area *IR* Drop Analysis in Large Power Grid," *Proceedings of the IEEE International Symposium in Quality Electronic Design*, pp. 194–199, March 2008.

[683] G. Katti *et al.*, "Temperature-Dependent Modeling and Characterization of Through-Silicon Via Capacitance," *IEEE Electron Device Letters*, Vol. 32, No. 4, pp. 563–565, April 2011.

[684] S. Adamshick, D. Coolbaugh, and M. Liehr, "Feasibility of Coaxial Through Silicon Via 3D Integration," *Electronics Letters*, Vol. 49, No. 16, pp. 1028–1030, August 2013.

[685] R. Doering and Y. Nishi, Eds., *Handbook of Semiconductor Manufacturing Technology*, CRC Press, 2008.

[686] M. Popovich, M. Sotman, A. Kolodny, and E. G. Friedman, "Effective Radii of On-Chip Decoupling Capacitors," *IEEE Transactions on Very Large Scale Integration (VLSI) Systems*, Vol. 16, No. 7, pp. 894–907, July 2008.

[687] S.-U. Park *et al.*, "Analysis of Reliability Characteristics of High Capacitance Density MIM Capacitors with SiO_2–HfO_2–SiO_2 Dielectrics," *Microelectronic Engineering*, Vol. 88, No. 12, pp. 3389–3392, December 2011.

[688] C. Pei *et al.*, "A Novel, Low-Cost Deep Trench Decoupling Capacitor for High-Performance, Low-Power Bulk CMOS Applications," *Proceedings of the International Solid-State and Integrated Circuit Technology*, pp. 146–1149, October 2008.

[689] Y. Shin, J. Seomun, K.-M. Choi, and T. Sakurai, "Power Gating: Circuits, Design Methodologies, and Best Practice for Standard-Cell VLSI Designs," *ACM Transactions on Design Automation of Electronic Systems*, Vol. 15, No. 4, pp. 1–37, September 2010.

[690] Z. Zhang, X. Kavousianos, K. Chakrabarty, and Y. Tsiatouhas, "A Robust and Reconfigurable Multi-Mode Power Gating Architecture," *Proceedings of the International Conference on VLSI Design*, pp. 280–285, January 2011.

[691] T. Xu, P. Li, and B. Yan, "Decoupling for Power Gating: Sources of Power Noise and Design Strategies," *Proceedings of the ACM/IEEE Design Automation Conference*, pp. 1002–1007, June 2011.

[692] H. Wang and E. Salman, "Decoupling Capacitor Topologies for TSV-Based 3-D ICs With Power Gating," *IEEE Transactions on Very Large Scale Integration (VLSI) Systems*, Vol. 23, No. 13, pp. 2983–2991, December 2012.

[693] [Online] *FreePDK45.* Available: http://www.eda.ncsu.edu/wiki/NCSU_EDA_Wiki (accessed in May 2016).

[694] K. Jeong *et al.*, "MAPG: Memory Access Power Gating," *Proceedings of IEEE Conference on Design, Automation and Test in Europe*, pp. 1054–1059, March 2012.

[695] S. Kim, S. Kang, K. J. Han, and Y. Kim, "Novel Adaptive Power Gating Strategy of TSV-Based Multi-Layer 3D IC," *Proceedings of the IEEE International Symposium on Quality Electronic Design*, pp. 537–541, March 2015.

[696] S. Song and A. Glasser, "Power Distribution Techniques for VLSI Circuits," *IEEE Journal of Solid-State Circuits*, Vol. SC-21, No. 1, pp. 150–156, February 1986.

[697] K. T. Tang and E. G. Friedman, "Simultaneous Switching Noise in On-Chip CMOS Power Distribution Networks," *IEEE Transactions on Very Large Scale Integration (VLSI) Systems*, Vol. 10, No. 4, pp. 487–493, August 2002.

[698] K. T. Tang and E. G. Friedman, "Incorporating Voltage Fluctuations of the Power Distribution Network into the Transient Analysis of CMOS Logic Gates," *Analog Integrated Circuits and Signal Processing*, Vol. 31, No. 3, pp. 249–259, June 2002.

[699] S. Zhao, C. Koh, and K. Roy, "Decoupling Capacitance Allocation and its Application to Power Supply Noise Aware Floorplanning," *IEEE Transactions on Computer-Aided Design of Integrated Circuits and Systems*, Vol. 21, No. 1, pp. 81–92, January 2002.

[700] M. Popovich, E. G. Friedman, M. Sotman, and A. Kolodny, "On-Chip Power Distribution Grids with Multiple Supply Voltages for High Performance Integrated Circuits," *IEEE Transactions on Very Large Scale Integration (VLSI) Systems*, Vol. 16, No. 7, pp. 908–921, July 2008.

[701] A. Mukheijee and M. Marek-Sadowska, "Clock and Power Gating with Timing Closure," *IEEE Transactions on Design and Test of Computers*, Vol. 20, No. 3, pp. 32–39, May 2003.

[702] I. Savidis, S. Kose, and E. G. Friedman, "Power Noise in TSV-Based 3-D Integrated Circuits," *IEEE Journal of Solid-State Circuits*, Vol. 48, No. 2, pp. 587–597, February 2013.

[703] J. Rosenfeld and E. G. Friedman, "A Distributed Filter Within a Switching Converter for Application to 3-D Integrated Circuits," *IEEE Transactions on Very Large Scale Integration (VLSI) Systems*, Vol. 19, No. 6, pp. 1075–1085, June 2011.

[704] M. Nagata, T. Okumoto, and K. Taki, "A Built-in Technique for Probing Power Supply and Ground Noise Distribution within Large-Scale Digital Integrated Circuits," *IEEE Journal of Solid-State Circuits*, Vol. 40, No. 4, pp. 813–819, April 2005.

[705] M. Sule, *Design of Pipeline Fast Fourier Transform Processors Using 3 Dimensional Integrated Circuit Technology*, Ph.D. Dissertation, North Carolina State University, December 2007.

[706] W. J. Dally, "Performance Analysis of *k*-ary *n*-cube Interconnection Networks," *IEEE Transaction on Computers*, Vol. 39, No. 6, pp. 775–785, June 1990.

[707] S. Palacharla, N. D. Jouppi, and J. E. Smith, "Complexity-Effective Superscalar Processors," *Proceedings of the IEEE International Conference on Computer Architecture*, pp. 206–218, June 1997.

[708] B. Vaidyanathan *et al.*, "Architecting Microprocessor Components in 3-D Design Space," *Proceedings of the IEEE International Conference on VLSI Design*, pp. 103–108, January 2007.

[709] R. P. Brent and H. T. Kung, "A Regular Layout for Parallel Adders," *IEEE Transactions on Computers*, Vol. C-31, No. 3, pp. 260–264, March 1982.

[710] P. M. Kogge and H. S. Stone, "A Parallel Algorithm for the Efficient Solution of a General Class of Recurrence Equations," *IEEE Transactions on Computers*, Vol. C-22, No. 8, pp. 786–793, August 1973.

[711] B. Black *et al.*, "Die Stacking (3D) Microarchitecture," *Proceedings of the IEEE/ACM International Symposium on Microarchitecture*, pp. 469–479, December 2006.

[712] S. S. Mukherjee *et al.*, "The Alpha 21364 Network Architecture," *IEEE Micro*, Vol. 22, No. 2, pp. 26–35, January/February 2002.

[713] Y. Xie, G. H. Loh, B. Black, and K. Bernstein, "Design Space Exploration for 3D Architectures," *ACM Journal on Emerging Technologies in Computing Systems*, Vol. 2, No. 2, pp. 65–103, April 2006.

[714] A. J. Smith, "Cache Memories," *ACM Computing Surveys*, Vol. 14, No. 3, pp. 473–530, September 1982.

[715] J. Sahuquillo and A. Pont, "Splitting the Data Cache: A Survey," *IEEE Concurrency*, Vol. 8, No. 3, pp. 30–35, July/September 2000.

[716] Y.-F. Tsai *et al.*, "Design Space Exploration for 3-D Cache," *IEEE Transactions on Very Large Scale Integration (VLSI) Systems*, Vol. 16, No. 4, pp. 444–455, April 2008.

[717] K. Zhang *et al.*, "A SRAM Design on 65 nm CMOS Technology with Integrated Leakage Reduction Scheme," *Proceedings of the IEEE International Symposium on VLSI Circuits*, pp. 294–295, June 2004.

[718] A. Zeng, J. Lü, K. Rose, and R. J. Gutmann, "First-Order Performance Prediction of Cache Memory with Wafer-Level 3D Integration," *IEEE Design and Test of Computers*, Vol. 22, No. 6, pp. 548–555, November/December 2005.

[719] S. J. E. Wilton and N. D. Jouppi, "CACTI: An Enhanced Cache Access and Cycle Time Model," *IEEE Journal of Solid-State Circuits*, Vol. 31, No. 5, pp. 677–688, May 1996.

[720] M. Mamidipaka, K. Khouri, N. Dutt, and M. Abadir, "Analytical Models for Leakage Power Estimation of Memory Array Structures," *Proceedings of the IEEE/ACM International Conference on Hardware/Software Codesign and System Synthesis*, pp. 146–151, September 2004.

[721] G. M. Link and N. Vijaykrishnan, "Thermal Trends in Emergent Technologies," *Proceedings of the IEEE International Symposium on Quality Electronic Design*, pp. 625–632, March 2006.

[722] M. B. Kleiner, S. A. Kühn, P. Ramn, and W. Weber, "Performance Improvement of the Memory Hierarchy of RISC-Systems by Application of 3-D Technology," *IEEE Transactions on Components, Packaging, and Manufacturing Technology – Part B*, Vol. 19, No. 4, pp. 709–718, November 1996.

[723] D. H. Albonesi and I. Koren, "Improving the Memory Bandwidth of Highly-Integrated, Wide-Issue, Microprocessor-Based Systems," *Proceedings of the IEEE International Conference on Paraller Architectures and Compilation Techniques*, pp. 126–135, November 1997.

[724] K. Suzuki *et al.*, "A 500 MHz, 32 bit, 0.4 μm CMOS RISC Processor," *IEEE Journal of Solid-State Circuits*, Vol. 29, No. 12, pp. 1464–1473, December 1994.

[725] C. E. Gimarc and V. M. Milutinovic, "A Survey of RISC Processors and Computers of the Mid-1980s," *IEEE Computer*, Vol. 20, No. 9, pp. 59–69, September 1987.

[726] [Online] Intel. Available: http://www.intel.com/products/processor/core2/index.htm.

[727] G. H. Loh, Y. Xie, and B. Black, "Processor Design in 3D Die-Stacking Technologies," *IEEE Micro*, Vol. 27, No. 3, pp. 31–48, May/June 2007.

[728] Z. Guz, I. Keidar, A. Kolodny, and U. C. Weiser, "Nahalal: Cache Organization for Chip Multiprocessors," *Computer Architecture Letters*, Vol. 6, No. 1, pp. 21–24, January 2007.

[729] D. Bertozzi *et al.*, "NoC Synthesis Flow for Customized Domain Specific Multiprocessor Systems-on-Chip," *IEEE Transactions on Parallel and Distributed Systems*, Vol. 16, No. 2, pp. 113–129, February 2005.

[730] J. C. Koob *et al.*, "Design of a 3-D Fully Depleted SOI Computational RAM," *IEEE Transactions on Very Large Scale Integration (VLSI) Systems*, Vol. 13, No. 3, pp. 358–368, March 2005.

[731] S. Kumar *et al.*, "A Network on Chip Architecture and Design Methodology," *Proceedings of the IEEE International Annual Symposium on VLSI*, pp. 105–112, April 2002.

[732] B. S. Feero and P. P. Pande, "Networks-on-Chip in a Three-Dimensional Environment: A Performance Evaluation," *IEEE Transactions on Computers*, Vol. 58, No. 1, pp. 32–45, January 2009.

[733] C. Seiculescu, S. Murali, L. Benini, and G. De Micheli, "SunFloor 3D: A Tool for Networks On Chip Topology Synthesis for 3D Systems on Chips," *Proceedings of the ACM/IEEE Design, Automation and Test in Europe Conference*, pp. 9–14, April 2009.

[734] V. F. Pavlidis and E. G. Friedman, "3-D Topologies for Networks-on-Chip," *Proceedings of the IEEE International SOC Conference*, pp. 285–288, September 2006.

[735] F. Li *et al.*, "Design and Management of 3D Chip Multiprocessors Using Network-in-Memory," *Proceedings of the IEEE International Symposium on Computer Architecture*, pp. 130–142, June 2006.

[736] Y. Wang *et al.*, "Economizing TSV Resources in 3-D Network-on-Chip Design," *IEEE Transactions on Very Large Scale Integration (VLSI) Systems*, Vol. 23, No. 3, pp. 493–506, March 2015.

[737] V. F. Pavlidis and E. G. Friedman, "3-D Topologies for Networks-on-Chip," *IEEE Transactions on Very Large Scale Integration (VLSI) Systems*, Vol. 15, No. 10, pp. 1081–1090, October 2007.

[738] A. Jantsch and H. Tenhunen, *Networks on Chip*, Kluwer Academic Publishers, 2003.

[739] M. Millberg *et al.*, "The Nostrum Backbone-A Communication Protocol Stack for Networks on Chip," *Proceedings of the IEEE International Conference on VLSI Design*, pp. 693–696, January 2004.

[740] J. M. Duato, S. Yalamanchili, and L. Ni, *Interconnection Networks: An Engineering Approach*, Morgan Kaufmann, 2003.

[741] W. J. Dally and B. Towles, *Principles and Practices of Interconnection Networks*, Morgan Kaufmann, 2004.

[742] L.-S. Peh and W. J. Dally, "A Delay Model for Router Microarchitectures," *IEEE Micro*, Vol. 21, No. 1, pp. 26–34, January/February 2001.

[743] T. Sakurai, "Closed-Form Expressions for Interconnection Delay, Coupling, and Crosstalk in VLSIs," *IEEE Transactions on Electron Devices*, Vol. 40, No. 1, pp. 118–124, January 1993.

[744] T. Sakurai and A. R. Newton, "Alpha-Power Law MOSFET Model and its Applications to CMOS Inverter Delay and other Formulas," *IEEE Journal of Solid-State Circuits*, Vol. 25, No. 2, pp. 584–594, April 1990.

[745] K. Banerjee and A. Mehrotra, "A Power-Optimal Repeater Insertion Methodology for Global Interconnects in Nanometer Design," *IEEE Transactions on Electron Devices*, Vol. 49, No. 11, pp. 2001–2007, November 2002.

[746] H. J. M. Veendrick, "Short-Circuit Dissipation of Static CMOS Circuitry and its Impact on the Design of Buffer Circuits," *IEEE Journal of Solid-State Circuits*, Vol. SC-19, No. 4, pp. 468–473, August 1984.

[747] K. Nose and T. Sakurai, "Analysis and Future Trend of Short-Circuit Power," *IEEE Transactions on Computer-Aided Design of Integrated Circuits and Systems*, Vol. 19, No. 9, pp. 1023–1030, September 2000.

[748] G. Chen and E. G. Friedman, "Effective Capacitance of *RLC* Loads for Estimating Short-Circuit Power," *Proceedings of the IEEE International Symposium on Circuits and Systems*, pp. 2065–2068, May 2006.

[749] P. R. O'Brien and T. L. Savarino, "Modeling the Driving-Point Characteristic of Resistive Interconnect for Accurate Delay Estimation," *Proceedings of the IEEE/ACM International Conference on Computer-Aided Design*, pp. 512–515, April 1989.

[750] H. Wang, L.-S. Peh, and S. Malik, "Power-Driven Design of Router Microarchitectures in On-Chip Networks," *Proceedings of the IEEE International Symposium on Microarchitecture*, pp. 105–116, December 2003.

[751] C. Marcon *et al.*, "Exploring NoC Mapping Strategies: An Energy and Timing Aware Technique," *Proceedings of the ACM/IEEE Design, Automation and Test in Europe Conference*, Vol. 1, pp. 502–507, March 2005.

[752] P. P. Pande *et al.*, "Performance Evaluation and Design Trade-Offs for Network-on-Chip Interconnect Architectures," *IEEE Transactions on Computers*, Vol. 54, No. 8, pp. 1025–1039, August 2005.

[753] R. Marculescu, U. Y. Ogras, and N. H. Zamora, "Computation and Communication Refinement for Multiprocessor SoC Design: A System-Level Perspective," *ACM Transactions on Design Automation of Electronic Systems*, Vol. 11, No. 3, pp. 564–592, July 2006.

[754] V. Soteriou, H. Wang, and L.-S. Peh, "A Statistical Trace Model for On-Chip Interconnection Networks," *Proceedings of the IEEE International Symposium on Modeling, Analysis, and Simulation of Computer and Telecommunication Systems*, pp. 104–116, September 2006.

[755] K. Siozios, K. Sotiriadis, V. F. Pavlidis, and D. Soudris, "Exploring Alternative 3D FPGA Architectures: Design Methodology and CAD Tool Support," *Proceedings of the IEEE International Conference on Field Programmable Logic and Applications*, pp. 652–655, August 2007.

[756] G.-M. Chiu, "The Odd-Even Turn Model for Adaptive Routing," *IEEE Transactions on Parallel and Distributed Systems*, Vol. 11, No. 7, pp. 729–738, July 2000.

[757] K. Lahiri *et al.*, "Evaluation of the Traffic-Performance Characteristics of System-on-Chip Communication Architectures," *Proceedings of the Conference on VLSI Design*, pp. 29–35, October 2000.

[758] T. T. Ye, L. Benini, and G. De Micheli, "Analysis of Power Consumption on Switch Fabrics in Network Routers," *Proceedings of the IEEE/ACM Design Automation Conference*, pp. 524–529, June 2002.

[759] B. Feero and P. P. Pande, "Performance Evaluation for Three-Dimensional Networks-on-Chip," *Proceedings of the IEEE International Symposium on VLSI*, pp. 305–310, March 2007.

[760] [Online]. Available: http://www.xilinx.com.

[761] M. J. Alexander *et al.*, "Placement and Routing for Three-Dimensional FPGAs," *Proceedings of the Canadian Workshop on Field-Programmable Devices*, pp. 11–18, May 1996.

[762] [Online]. Available: http://www.xilinx.com/products/silicon_solutions/fpgas/spartan_series/spartan3_fpgas/index.htm.

[763] A. Rahman, S. Das, A. P. Chandrakasan, and R. Reif, "Wiring Requirement and Three-Dimensional Integration Technology for Field Programmable Gate Arrays," *IEEE Transactions on Very Large Scale Integration (VLSI) Systems*, Vol. 11, No. 1, pp. 44–53, February 2003.

[764] G. G. Lemieux and S. D. Brown, "A Detailed Routing Algorithm for Allocating Wire Segments in Field-Programmable Gate Arrays," *Proceedings of the IEEE Physical Design Workshop*, pp. 215–226, April 1993.

[765] V. Betz and J. Rose, "VPR: A New Packing, Placement, and Routing Tool for FPGA Research," *Proceedings of the International Workshop on Field Programmable Logic Applications*, pp. 213–222, September 1997.

[766] S. C. Wong, G.-Y. Lee, and D.-J. Ma, "Modeling of Interconnect Capacitance, Delay, and Crosstalk in VLSI," *IEEE Transactions on Semiconductor Manufacturing*, Vol. 13, No. 1, pp. 108–111, February 2000.

Index

Note: Page numbers followed by "*f*" and "*t*" refer, respectively, to figures and tables.

A

B

C

D

M

N

R

S

V

W

X

Y

Printed in the United States
By Bookmasters